LIFE: THE SCIENCE OF BIOLOGY

FOURTH EDITION
LIFE
The Science of Biology

William K. Purves
Harvey Mudd College
Claremont, California

Gordon H. Orians
The University of Washington
Seattle, Washington

H. Craig Heller
Stanford University
Stanford, California

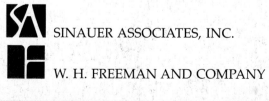 SINAUER ASSOCIATES, INC.

W. H. FREEMAN AND COMPANY

THE COVER

Elephants at a water hole in northern Botswana, Africa.
Photograph by Frans Lanting/Minden Pictures.

THE FRONTISPIECE

Scarlet ibis and cattle egrets at Hato el Frio, Venezuela.
Photograph by Art Wolfe.

LIFE: THE SCIENCE OF BIOLOGY, Fourth Edition
Copyright © 1995 by Sinauer Associates, Inc.
All rights reserved. This book may not be reproduced
in whole or in part without permission.

Address editorial correspondence to Sinauer Associates, Inc.,
Sunderland, Massachusetts 01375 U.S.A.

Address orders to W. H. Freeman and Co. Distribution Center,
4419 West 1980 South, Salt Lake City, Utah 84104 U.S.A.

Library of Congress Cataloging-in-Publication Data

Purves, William. K. (William Kirkwood), 1934-
        Life, the science of biology / William K. Purves,
    Gordon H. Orians, H. Craig Heller. -- 4th ed.
                p.   cm.
        Includes bibliographical references and index.
        ISBN 0-7167-2629-7
            1. Biology.   I. Orians, Gordon H.   II. Heller, H. Craig.
III. Title.
QH305.2.P87   1995
574--dc20   94-24802
                                                        CIP

ABOUT THE BOOK

Editor: Andrew D. Sinauer

Project Editor: Carol J. Wigg

Developmental Editor: Elmarie Hutchinson

Copy Editor: Stephanie Hiebert

Production Manager: Christopher Small

Book Layout and Production: Janice Holabird

Art Editing and Illustration Program: J/B Woolsey Associates

Photo Research: Jane Potter

Book and Cover Design: Rodelinde Graphic Design

Composition: DEKR Corporation

Color Separations: Vision Graphics, Inc.

Prepress: Lanman Lithotech

Cover Manufacture: John P. Pow Company

Book Manufacture: R. R. Donnelley & Sons Company, Willard, OH

Printed in U.S.A.

4 3 2 1

To Jean, Betty, and Renu

ABOUT THE AUTHORS

William K. Purves

Bill Purves is Stuart Mudd Professor of Biology as well as founder and chair of the Department of Biology at Harvey Mudd College in Claremont, California. He received his Ph.D. from Yale University in 1959 under Arthur Galston. A Fellow of the American Association for the Advancement of Science, Professor Purves has served as head of the Life Sciences Group at the University of Connecticut, Storrs, and as chair of the Department of Biological Sciences, University of California, Santa Barbara, where he won the Harold J. Plous Award for teaching excellence. His research interests focus on the chemical and physical regulation of plant growth and flowering.

Professor Purves has taught introductory biology each year for over thirty years and considers teaching the course the most interesting and important of his professional activities. "I can't imagine a year without teaching it," he says. In describing his teaching philosophy, Purves states, "Students learn biological concepts much more rapidly and effectively if they understand where the concepts come from—what the experimental and conceptual background is. 'Facts' by themselves can be boring or incomprehensible, but they become exciting if given a context."

Gordon H. Orians

Gordon Orians is Professor of Zoology at the University of Washington. He received his Ph.D. from the University of California, Berkeley, in 1960 under Frank Pitelka. Professor Orians has been elected to the National Academy of Sciences and the American Academy of Arts and Sciences. He was President of the Organization for Tropical Studies from 1982 to 1994, and is currently President-elect of the Ecological Society of America. He is a recipient of the Brewster Medal from the American Ornithologists' Union, and in 1994 he received the Distinguished Service Award of the American Institute of Biological Sciences.

Professor Orians is a leading authority in ecology and evolution, with research interests in behavioral ecology, plant–herbivore interactions, community structure, and the biology of rare species. Like the other authors, he draws from his research to bring an added dimension to his teaching and writing. "Teachers who understand research because they are engaged in it can more easily communicate the excitement researchers feel as they discover new things," Orians says. "All three authors of *Life* have spent considerable time doing research and we have tried throughout the book to show the sources of our current understanding of biology."

H. Craig Heller

Craig Heller is Lorry Lokey/Business Wire Professor of Biological Sciences and Human Biology and Associate Dean of Research at Stanford University, and is a popular lecturer on animal and human physiology. He received his Ph.D. from Yale University in 1970 and did postdoctoral research at the Scripps Institution of Oceanography on brain regulation of body temperature in mammals. He has continued this research since coming to Stanford in 1972, studying a variety of phenomena ranging from hibernating squirrels to sleeping college students to diving seals to meditating yogis. Professor Heller is a Fellow of the American Association for the Advancement of Science and a recipient of the Walter J. Gores Award for Excellence in Teaching.

"A first course in biology requires the student to learn more new words than a first course in a foreign language," Heller says. "The secret to teaching and learning biology is to focus on central and overarching concepts. Once you grasp the concept of how something works, you have a framework on which the facts and vocabulary fall into place. Conceptual understanding also helps you relate what you learn to the real world."

In revising this book, we have once again examined our goals and our hopes for it. Above all, we want to help students understand biological concepts and see where the concepts originate. For this reason we display biology as an experimental and observational science. Frequently we offer the student a chance to think—to figure out the next step rather than wait passively to learn it from us. For example, we ask the student to interpret experimental data used in elucidating the genetic code, in understanding the role of homeotic genes in flower development, and in working out the Calvin–Benson cycle. In Chapter 21, students are presented with real species with identified character traits and work through how to construct a phylogeny using cladistic methods. We use the cladistic material again in Chapter 27, where we explore human relationships with chimpanzees and gorillas. As yet another example, we lead the student to discover the physiological differences between reptiles and mammals by taking the reader through a series of experiments on the thermal biology of a mouse and a lizard.

Even when we present topics directly, we prefer to explain new material rather than serving it up as a cut-and-dried collection of "facts." Still, the study of biology requires exposure to a stunning number of new facts, and students are easily deterred by a bewildering excess of information. How can we deal with this problem? Our approach is to emphasize fascinating examples wherever possible, using these to engage the student's interest so that she or he wants to learn other related material.

The creation of a new edition provides opportunities to rethink how best to present existing material, as well as what from the exploding array of new information to include. Current advances in biomedical sciences present special challenges to the organizers of a course or a textbook. Diseases such as cancer and AIDS are of deep interest in relation to a wide variety of biological topics, including immunology, genetics, evolution, membrane biology, and virology; instructors often want to use the diseases as examples in discussing these topics. Yet students and instructors also want to see a disease and its biology considered in a single place in the book. For this edition, we have adopted an approach that we hope is effective. Recent advances in gene therapy and in the cloning of genes for particular diseases lend themselves to consideration in an all-new Chapter 15, "Genetic Disease and Modern Medicine." We consider some general principles and some widely applicable techniques in this chapter, which also includes much of our coverage of cancer. We build from the close of this chapter to begin the next, "Defenses against Disease," with an overview of AIDS as a worldwide problem. Before the end of Chapter 16, the student knows enough about the immune system to understand AIDS in greater biological detail.

Developmental biology continues to be one of the fastest-moving areas of biology; the newer work is reflected in Chapter 17. We pay particular attention to recent developments in the *Drosophila* larva and *Caenorhabditis elegans* systems. After giving genetic "instructions" for building a fly, we conclude the molecular section of the book with a transition from *Drosophila* larval genetics to evolutionary biology.

In the section on evolution we have expanded the coverage of methods of reconstructing phylogenies. We provide new treatments of cladistic methods and show how phylogenies are used to shed light on a wide variety of evolutionary questions. We use phylogenetic trees a number of times in subsequent chapters in the book to show how traits of organisms evolve under the influence of evolutionary agents.

Our restructuring of the chapters on plant anatomy and physiology resulted in the creation of a new Chapter 31, "Environmental Challenges to Plants," which deals with some of the ways plants cope with harsh envi-

UP-TO-DATE COVERAGE REFLECTS NEW DEVELOPMENTS IN BIOLOGY

ronments, predators, and pathogens. We have updated the plant chapters to include material on patch clamping, homeotic mutations, and other important phenomena and techniques.

Three aspects of the animal biology chapters that were significant in the third edition have been strengthened. First, we frequently use a comparative approach to help students understand basic principles and mechanisms as well as their evolutionary variations. Second, we emphasize experimental approaches so that the student learns *how* we know as well as *what* we know. Third, a capstone to the treatment of each physiological system is a discussion of how its contributions to homeostasis are controlled and regulated. Chapter 45, "Animal Behavior," has been revised to focus more strongly on the physiological mechanisms underlying behavior.

Because ecology is an increasingly experimental science, we have added descriptions of well-designed experiments that have been performed to demonstrate the causes of the evolution of traits used in courtship among animals and to assess the importance of predation and competition in structuring ecological communities. To keep pace with the increasing importance of environmental problems, we have expanded our treatment of lake eutrophication and of overexploitation of commercially important species.

PEDAGOGICAL INNOVATIONS ENRICH THE LEARNING PROCESS

Both as textbook authors and as teachers we want to help students in every way possible. In the next section ("To the Student") we offer some helpful advice we'd like students to consider as they begin their study of biology. This advice has helped many of our own students. Following "To the Student" is "*Life* at a Glance," in which we illustrate some of the pedagogic improvements that we discuss in the next few paragraphs.

A major source of the success of the third edition of this textbook was the exciting new art program developed by J/B Woolsey Associates. We have upgraded that already fine art for this edition. We have added many entirely new drawings and graphs, and virtually all the drawings in the book have been improved in one way or another by the development of a new artistic vocabulary for this edition.

Because learning is a visual as well as a verbal process, we have given much attention to creating illustrations that explain biological concepts clearly. To facilitate learning we use color consistently from illustration to illustration; for example, the outside of the cell is always represented by light red and the cytoplasm by pale blue. We have developed a set of icons (see "*Life* at a Glance") to represent biologically important molecules (such as water or ATP), active forms of enzymes, or activation and inhibition of pathways. We have used blocks of color to distinguish major pieces of information from details or to separate an illustration into parts representing discrete ideas; we think of these as "visual paragraphs." Finally, we have flagged via marginal arrowheads (also shown in "*Life* at a Glance") illustrations that are particularly significant. These figures illustrate and synthesize important concepts—protein synthesis, for example, or the life cycle of a flowering plant.

We have developed a new type of chapter-end summary, one that should help students in at least three ways. First, the student can skim the summary for orientation before reading the chapter. Second, after reading the chapter, the student can use the summary to review the material. Finally, each summary identifies the illustrations that give the best overview of the chapter and its most important concepts.

Each chapter starts with a brief introduction intended to catch the reader's interest by discussing a fascinating bit of biology. Each of these new introductions is supported by a striking photograph; an example is shown in "*Life* at a Glance."

Before beginning this edition, we asked 36 biologists, most of whom were using the third edition, to maintain "diaries" and record their ideas for improving the book. These diarists were enormously helpful in getting us started on the right foot for the new edition. Their guidance influenced the decisions to create the two new chapters mentioned earlier and to give priority to simplifying our prose. We are indebted to them.

As with the first three editions, many of our colleagues reviewed chapters or entire sections of this edition in manuscript. They and the diarists are listed below. The reviewers were helpful, thoughtful, and clearly dedicated to the success of this book, and we thank them all. We particularly thank Bob Cleland, Richard Cyr, Pat DeCoursey, Rob Dorit, Art Dunham, Margaret Fusari, Harry Green, Ray Huey, Bob Jeffries, Jim Manser, William Milsom, Ron O'Dor, Dianna Padilla, Ronald Patterson, Zoe Roizen, Seri Rudolph, Michael Ryan, Iain Taylor, and David Woodruff. They gave us explicit recommendations for extensive improvements, helped simplify our writing style, and did so in ways that encouraged us to do our very best to live up to their expectations.

The third edition profited greatly from the prodigious efforts of its outstanding developmental editor, Elmarie Hutchinson. We were delighted when Elmarie agreed to take an even stronger role in the development of the fourth edition. Among her innovations are the new type of chapter summary and its suggested use as a chapter preview. Elmarie also helped us respond to diarists' requests for simpler language, better topic sentences, and more restrained use of boldface terms. Her suggestions and guidelines were implemented by our copyeditor, Stephanie Hiebert, whose sharp and prescriptive line editing has helped streamline the book's prose.

In a book like this the illustrations are as important as the prose, and here again Elmarie gave us outstanding input, scrutinizing every figure with an eye for its internal consistency and its agreement with the text. Her suggestions were incorporated by artists John Woolsey and Patrick Lane as they met with the authors to reconceptualize artwork. The task of coordinating and checking the changes made by editors, artists, and authors fell to Carol Wigg, who got the job once again of putting all the pieces together. In addition, previous users of the book will see a significant improvement in the photography program. Jane Potter has tapped important new sources, and we have gone to great lengths to seek out new photographs to illustrate important concepts and enliven the book's appearance.

We wish to thank W. H. Freeman's entire marketing and sales group. Their enthusiasm for *Life* helped bring the book to a wider audience and the efforts of several of the sales representatives put us in touch with a number of our colleagues who had specific questions or criticisms of the book. This contact has been fruitful, and we look forward to more of this "firing line" interaction with the fourth edition.

Finally, the opportunity to work with a publishing company whose president provides frequent personal contacts and feedback that is scientifically useful as well as production-wise, is a great privilege for us. Andy Sinauer is the ideal person for authors to deal with—firm but kind, involved but not overbearing, and friendly—hence, motivating. He also has a superb eye for good associates—the Sinauer team is first-rate!

CAREFULLY REVIEWED WITH STUDENTS' NEEDS AND TEACHERS' CONCERNS IN MIND

THE EFFORTS OF MANY PEOPLE HELPED THE REVISION

William K. Purves   **Gordon H. Orians**   **H. Craig Heller**
September 1994

TO THE STUDENT

Welcome to the study of life! In our student days—and ever since—we have enjoyed studying the fascinating and fast-changing field of biology, and we hope that you will, too.

There are a few things you can do to help you get the most from this book and from your course. For openers, read the book actively—don't just read passively, but do things that force you to think as you read. If we pose questions, stop and think about them. If a passage reminds you of something that has gone before, think about that, or even check back to refresh your memory. Ask questions of the text as you go. Do you understand what is being said? Does it relate to something you already know? Is it supported by experimental or other evidence? Does that evidence convince you? How does this passage fit into the chapter as a whole? Annotate the book—write down comments in the margins about things you don't understand, or about how one part relates to another, or even when you find an idea particularly interesting. The point of doing these things is that they will help you learn. People remember things they think about much better than they remember things they have read passively. Highlighting is passive; copying is drudge work; questioning and commenting are active and well worthwhile.

For this edition we have developed new ways to help you read the book actively. The chapter-end summaries have been redesigned so that they may be used as both summaries and previews. To find out what a chapter covers, try reading the summary at the end of the chapter before you begin reading the chapter itself. Don't worry about unfamiliar terms in the summary, but notice them as terms you will need to learn. Just read all the statements as an overview and preview without studying the cited illustrations. Then, after reading the chapter, use the summary as a framework for your review. It is essential that you do study the cited illustrations and their captions as you review because important information that is covered in illustrations has been left out of the summary statements. Add concepts and details to the framework by reviewing the text.

Take advantage of our use of color and symbols in the illustrations. We generally use colors to mean the same thing from illustration to illustration, and we have developed a set of symbols (see the section "*Life* at a Glance") to represent biologically important molecules and phenomena. In many of the illustrations you will see blocks of color used to help you separate the illustration into parts representing discrete ideas. Also, some figures are identified by an arrow in the margin. These figures are particular significant; they illustrate and synthesize important concepts and retell visually the story you read in the text. Studying them will help you learn important biological concepts and systems. Going back to review them will help you to remember these concepts.

The chapter summaries will help you quickly review the high points of what you have read. A summary identifies particular illustrations that you should study to help organize the material in your mind. A way to review the material in slightly more detail after reading the chapter is to go back and look at the boldfaced terms. You can use the boldfaced terms to pose questions—and see if you can answer those questions. The boldfacing will probably be more useful on a second reading than on the first.

Use the self-quizzes and study questions at the end of each chapter. The self-quizzes are meant to help you remember some of the more detailed material and to help you sort out the information we have laid before you. Answers to all self-quizzes are in the Appendix. The study questions, on the other hand, are often fairly open-ended and are intended to cause you to reflect on the material.

Two parts of a textbook that are, unfortunately, often underused or even ignored are the glossary and the index. Both can help you a great deal. When you are uncertain of the meaning of a term, check the glossary first—there are more than 1,500 definitions in it. If you don't find a term in the glossary, or if you want a more thorough discussion of the term, use the index to find where it's discussed.

What if you'd like to pursue some of the topics in greater detail? At the end of each chapter there is a short, annotated list of supplemental readings. We have tried to choose readings from books and magazines, especially *Scientific American*, that should be available in your college library.

Most students occasionally have difficulty in courses, including biology courses. If you find that you are slipping behind in the course, or if a particular topic is giving you an unreasonable amount of trouble, here are some useful steps you might take. First, the basics: attend class, take careful lecture notes, and read the textbook assignments. Second, note that one of the most important roles of studying is to discover what you don't know, so that you can do something about it. Use the index, the glossary, the chapter summaries, and the text itself to try to answer any questions you have and to help you organize the material. Make a habit of looking over your lecture notes within 24 hours of when you take them—find out right away what points are unclear, and get them straightened out in your mind. We also call your attention to the Study Guide that accompanies *Life*. It is by Jon Glase at Cornell and Jerry Waldvogel at Clemson. It parallels this textbook and each chapter contains learning objectives, key concepts, activities, and questions with full answers and explanations.

If none of these self-help remedies does the trick, get help! Other students are often a good source of help, because they are dealing with the material at the same level as you are. Study groups can be very useful, as long as the participants are all committed to learning the material. Tutors are almost always helpful and useful, as are faculty members. The main thing is to get help when you need it. It is not a good idea to be strong and silent and drift into a low grade.

But don't make the grade the point of this or any other course. You are in college to learn, to pursue interesting subjects, and to enjoy the subjects you are pursuing. We hope you'll enjoy the pursuit of biology.

Bill Purves   **Gordon Orians**   **Craig Heller**

Life, Fourth Edition, is accompanied by a comprehensive set of supplements:

Study Guide . . . reviewed by students and extensively revised by Jon Glase of Cornell University and Jerry Waldvogel of Clemson University to help students master the textbook material.

Instructor's Manual . . . by Roberta Meehan of the University of Northern Colorado, featuring chapter objectives, chapter outlines, teaching hints and strategies, references, resources, and key terms.

Overhead transparencies and **slides** . . . a package of 300 full-color images from the book.

Transparency masters . . . of all text art figures not included in the overhead transparency/slide set.

Test bank . . . revised and updated, with at least 10 new questions per chapter and over 4,000 questions in total. Available in printed form, and in IBM and Macintosh formats.

Videodisc . . . for the first time the magnificent art program in *Life* comes to your classroom via laserdisc technology. The disc includes all the line art from the new edition, over 1,500 carefully selected still images, and more than 25 outstanding motion and animation sequences ranging from traffic through the membrane and electrophoresis to ecological succession.

Laboratory options . . . chosen from the following:

• The complete Abramoff and Thomson's *Laboratory Outlines in Biology VI*. The popular, critically acclaimed lab manual from W. H. Freeman, now in its new (1995) sixth edition.

• *Laboratory separates.* Select only those experiments you need from Abramoff and Thomson.

• *Customized laboratory package.* The separates of your choice, combined with your own laboratory exercises, notes, and other materials.

For information regarding policy on the educational use of these supplements, please contact your local W. H. Freeman representative.

SUPPLEMENTS

Videodisc Focus Group Participants
Tad Day, University of West Virginia
Guy Cameron, University of Houston
Valerie Flechtner, John Carroll University
Arnold Karpoff, University of Louisville
William Eickmeir, Vanderbilt University
Paul Ramp, University of Tennessee

Videodisc Reviewers
Stephen C. Adolph, Harvey Mudd College
Sally S. De Groot, St. Petersburg Junior College (retired)
Rachel Fink, Mt. Holyoke College
Nancy V. Hamlett, Harvey Mudd College
Brian A. Hazlett, University of Michigan
Martinez J. Hewlett, University of Arizona
Dan Lajoie, University of Western Ontario
Alfred R. Loeblich III, University of Houston
James R. Manser, Harvey Mudd College
Catherine S. McFadden, Harvey Mudd College
T. J. Mueller, Harvey Mudd College

Study Guide Reviewer
Wayne Hughes, University of Georgia

Instructor's Manual Reviewers
Erica Bergquist, Holyoke Community College
Nels H. Granholm, South Dakota State University

Transparency Reviewers
William S. Cohen, University of Kentucky
Anne M. Cusic, University of Alabama
Bruce Felgenhauer, University of Southwestern Louisiana
Alice Jacklett, State University of New York at Albany
Susan Koptur, Florida International University
Charles H. Mallery, University of Miami
Stephen P. Vives, Georgia Southern University

In addition, Tad Day, William Eickmeier, and Paul Ramp conducted student reviews of the videodisc, and Jon Glase had his students review the Study Guide. We greatly appreciate their efforts.

LIFE

They Are Not All the Same
When observed closely, the individuals in a population of red and green macaws vary a great deal.

We are aware that no two people (unless they are identical twins) look exactly alike. We also recognize our pets as distinct individuals. But we have great difficulty in seeing differences among individuals of most other species of organisms. The brilliant red and green macaws feeding on a clay cliff in the Peruvian jungle may all appear identical to the untrained eye; scientists who study them closely, however, realize that each is unique. The colored feathers display many slight variations in pattern, and the black-and-white feathers that surround the birds' eyes form patterns that, like human fingerprints, are unique to the individual. Members of many groups, particularly among behaviorally sophisticated animals such as vertebrates, readily recognize one another and adjust their behavior accordingly.

Differences among individuals in local populations, even if they are subtle, are the raw material upon which evolutionary mechanisms act to produce the striking variability revealed by the multitude of organisms living on Earth today. A good fossil record can reveal much about when and how the forms of organisms changed. Fossils may also provide clues about the reasons for those changes, but they provide only indirect evidence of the causes of evolutionary change. To obtain direct evidence we must study evolutionary changes happening today. The study of variability is at the heart of investigations into the mechanisms of evolution.

In this chapter we discuss the agents of evolution and the short-term studies designed to investigate them. By testing hypotheses observationally and experimentally we can answer key questions about the processes guiding evolutionary changes. In later chapters we consider how we use this information to explain longer-term features of the evolutionary record.

Although ideas about evolution have been put forth for centuries, until the last one hundred years none of the h_____ _____ ab___ the _____ __ ev_____ tionary change_ _____ ____ ____ ___ _____ his hypotheses ____ ____ ___ ____ __ ____ 1), but he did ____ ____ _____ ____ ____ basis of evolut____ ____ ____ _____ ____ vided by Greg__ ____ ____ ____ ____ and Darwin's ____ ____ ____ ____ twentieth cent___ ____ ____ ____ ____ evolutionary h____ ____ ____ ____ test them.

WHAT IS EVOL___

The fossil recor_ ___ ___ ___ ___ over time. The ___ ___ ___ ___ anisms shared ___ ___ ___ ___

19

The Mechanisms of Evolution

426

148 CHAPTER SEVEN

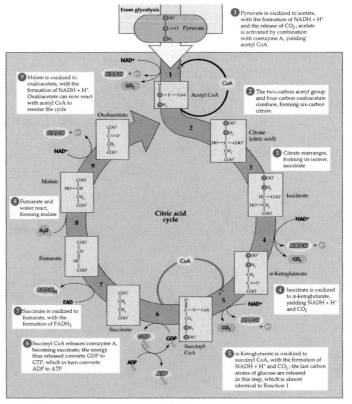

1 Pyruvate is oxidized to acetate, with the formation of NADH + H+ and the release of CO_2; acetate is activated by combination with coenzyme A, yielding acetyl CoA

2 The two-carbon acetyl group and four-carbon oxaloacetate combine, forming six-carbon citrate

3 Citrate rearranges, forming its isomer, isocitrate

4 Isocitrate is oxidized to α-ketoglutarate, yielding NADH + H+ and CO_2

5 α-Ketoglutarate is oxidized to succinyl CoA, with the formation of NADH + H+ and CO_2; the last carbon atoms of glucose are released in this step, which is almost identical to Reaction 1

6 Succinyl CoA releases coenzyme A, becoming succinate; the energy thus released converts GDP to GTP, which in turn converts ADP to ATP

7 Succinate is oxidized to fumarate, with the formation of $FADH_2$

8 Fumarate and water react, forming malate

9 Malate is oxidized to oxaloacetate, with the formation of NADH + H+. Oxaloacetate can now react with acetyl CoA to reenter the cycle

Citric acid cycle

7.13 The Citric Acid Cycle
The first reaction produces the two-carbon acetyl CoA. Notice that the two carbons from acetyl CoA are traced with color through reaction 5, after which they may be at either end of the molecule (note the symmetry of succinate and fumarate). Reactions 1, 4, 5, 7, and 9 accomplish the major overall effect of the cycle—the storing of energy—by passing electrons to the carrier molecule NAD. Reaction 6 also stores energy.

AT A GLANCE

Dynamic visual paragraphs dramatize important concepts in a way that the text alone cannot. These are flagged with marginal arrows so the student is able to refer to them readily.

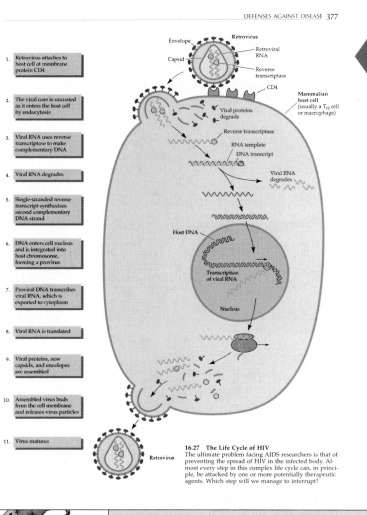

1. Retrovirus attaches to host cell at membrane protein CD4

2. The viral core is uncoated as it enters the host cell by endocytosis

3. Viral RNA uses reverse transcriptase to make complementary DNA

4. Viral RNA degrades

5. Single-stranded reverse transcript synthesizes second complementary DNA strand

6. DNA enters cell nucleus and is integrated into host chromosome, forming a provirus

7. Proviral DNA transcribes viral RNA, which is exported to cytoplasm

8. Viral RNA is translated

9. Viral proteins, new capsids, and envelopes are assembled

10. Assembled virus buds from the cell membrane and releases virus particles

11. Virus matures

16.27 The Life Cycle of HIV
The ultimate problem facing AIDS researchers is that of preventing the spread of HIV in the infected body. Almost every step in this complex life cycle can, in principle, be attacked by one or more potentially therapeutic agents. Which step will we manage to interrupt?

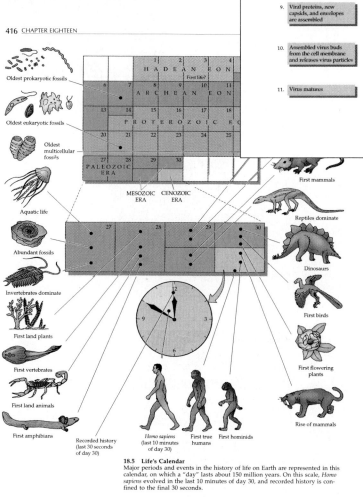

18.5 Life's Calendar
Major periods and events in the history of life on Earth are represented in this calendar, on which a "day" lasts about 150 million years. On this scale, *Homo sapiens* evolved in the last 10 minutes of day 30, and recorded history is confined to the final 30 seconds.

Stunning photos capture the excitement of *Life*, illustrating important concepts and enlivening the book's appearance.

25.31 Dicots
(a) The cactus family is a large grou... about 1,500 species in the Americas. ... takes its name from its scarlet flowe... *Anterrhinum majus*. (c) These wood ... stone National Park are members of ... as are the familiar roses from your ... 25.25 and 25.26a show other dicots.

Origin and Evolution of the Angio...

How did the angiosperms aris... analyses (see Chapter 21) have s... ing question. It is widely agreed t... and two groups of gymnosperm... and the long-extinct cycadeoids... cycads, arose from a single ances... rise to no other groups. A close ... the angiosperms and the Gneto...

pected, primarily ...
phyta have vess...
angiosperms. In ...
by the light-micr...
tilization in *Eph*...
The cycadeoids, ...
same time as dic...
portant character...
angiosperms. Th...
cycadeoids, altho...
with naked seed...
flower of *Magnol*...
The next grea...
the question, Wh...

(a)

Hydrogen bond
between water
molecules

(c)

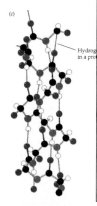

Hydrog...
in a pro...

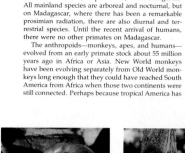

lineage gave rise to the prosimians—lemurs, tarsiers, pottos, and lorises (Figure 27.29). Prosimians were formerly found on all continents, but today they are restricted to Africa, tropical Asia, and Madagascar. All mainland species are arboreal and nocturnal, but on Madagascar, where there has been a remarkable prosimian radiation, there are also diurnal and terrestrial species. Until the recent arrival of humans, there were no other primates on Madagascar.

The anthropoids—monkeys, apes, and humans—evolved from an early primate stock about 55 million years ago in Africa or Asia. New World monkeys have been evolving separately from Old World monkeys long enough that they could have reached South America from Africa when those two continents were still connected. Perhaps because tropical America has

27.29 Prosimians
(a) The sifaka lemur, *Propithecus verreauxi*, is one of many lemur species of Madagascar, where they are part of a unique assemblage of plants and animals. (b) *Loris tardigradis*, the slender loris, of southern India. (c) In the rainforests of Borneo, this tarsier (*Tarsius bancanus*) seems otherworldly to our eyes.

27.30 Monkeys
(a) Golden lion tamarins (*Leontopithecus rosalia*) are New World monkeys, living in the trees of the coastal Brazilian rainforest. (b) Some Old World species, such as these Japanese macaques (*Macaca fuscata*) live and travel in groups.

634

2.18 Surface Tension
(a) A water strider "skates" along, supported by the surface tension of the water that is its home. (b) Surface tension demonstrated by a soap bubble on a teacup.

Boxes describe fascinating biological phenomena, highlighting special interest topics and expanding the discussion of many issues.

BOX 27.A

The Four-Minute Mile

Many mammals can run much faster, yet we humans are proud to have achieved a four-minute mile. Terrestrial vertebrates did not achieve such speeds easily. Amphibians and reptiles fill and empty their lungs using some of the same muscles they use for walking. In addition, because the limbs protrude laterally, their movement generates a strong lateral force that bends the body from side to side. Recent studies have shown that these animals cannot breathe while they walk or run. Therefore, they can operate aerobically only briefly. Because they depend upon anaerobic glycolysis while running, they tire rapidly.

In the lineage leading to dinosaurs and birds and in the lineage leading to mammals, the legs assumed more vertical positions, which reduced the lateral forces on the body during locomotion. Special ventilatory muscles that can operate independently of locomotory muscles also evolved. These muscles are visible in living birds and mammals. We can infer their existence in dinosaurs from the structure of the vertebral column and the capability of many dinosaurs for bounding, bipedal (using two legs) locomotion. The ability to breathe and run simultaneously, a capability we take for granted, was a major innovation in the evolution of terrestrial vertebrates.

A future four-minute miler?

Figure 28.12). The ability to move actively on land was not achieved easily. The first terrestrial vertebrates probably moved only very slowly, much more slowly than their aquatic relatives. The reason is that they apparently could not walk and breathe at the same time. Not until evolution of the lineages leading to the mammals, dinosaurs, and birds did special muscles evolve enabling the lungs to be filled and emptied while the limbs moved (Box 27.A). This ability enabled its bearers to maintain steady, high levels of activity, which generated enough heat to result in

(subclass Aves) e embodies an scendants of a evolved in the (Figure 27.22), s and modern *pteryx* was cov- eloped wings, much reduced may have been

vs features re- he modern

TABLE 49.4
Areas, Biomass of Plants, and Net Primary Production of Earth's Major Vegetation Zones

VEGETATION ZONE	AREA		MASS OF PLANTS		NET PRIMARY PRODUCTION	
	10^6 Km^2	PERCENT	10^9 TONS	PERCENT	10^9 TONS	PERCENT
Polar	8.05	1.6	13.77	0.6	1.33	0.6
Conifer forest	23.20	4.5	439.06	18.3	15.17	6.5
Temperate	22.53	4.5	278.67	11.5	17.97	7.7
Subtropical	24.26	4.8	323.90	13.5	34.55	14.8
Tropical	55.85	10.8	1,347.10	56.1	102.53	44.2
Total land	133.89	26.2	2,402.5	100	171.55	73.8
Glaciers	13.9	2.7	0	0	0	0
Lakes and rivers	2.0	0.4	0.04	<0.01	1.0	0.4
All continents	149.79	29.3	2402.54	100	172.55	74.2
Ocean	361.0	70.7	0.17	<0.001	60.0	25.8
Earth total	510.79	100	2,402.71	100	232.55	100

from the vents. Most of the other organisms of these ecosystems live directly or indirectly on the sulfur-oxidizing bacteria (see Figure 22.2).

This overview of the global pattern of biological production on Earth is sufficient to identify which processes limit primary production and nutrient cycling in different climatic zones and how they operate, but it does not give you a picture of what these ecosystems look like and how they function. Describing ecosystems is one of the goals of the next chapter.

SUMMARY of Main Ideas about Ecosystems

Ecosystems are powered by solar energy that first enters living organisms via photosynthesis at rates controlled by temperature and precipitation.
Review Figure 49.1

Food webs summarize who eats whom in ecological communities.
Review Figures 49.2 and 49.3

Because much of the energy taken in by an organism is used for maintenance and is eventually dissipated as heat, the efficiency of energy transfer to higher trophic levels is usually very low.
Review Figures 49.4 and 49.5

The main elements of living organisms—carbon, nitrogen, phosphorus, sulfur, hydrogen, and oxygen—cycle between organisms and other compartments of the global ecosystem.
Review Figures 49.10, 49.11, 49.12, 49.13, and Table 49.3

Human activity greatly modifies cycles of basic minerals on local, regional, and global scales.
Review Figures 49.14, 49.15, and 49.16

Earth's climate is determined primarily by the pattern of solar energy input at different latitudes and by Earth's rotation on its axis.

The directions of prevailing winds differ over the surface of Earth.
Review Figure 49.19

Surface winds drive global oceanic circulation.
Review Figure 49.20

The distribution of primary production on Earth is determined primarily by Earth's climate.
Review Figure 49.21

New chapter summaries synthesize information in the text and the art. Important concepts are encapsulated in clearly written summary sentences, with references back to key illustrations.

CONTENTS

CHAPTER 1
The Science of Biology 1

I. THE CELL

CHAPTER 2
Small Molecules 19

CHAPTER 3
Large Molecules 39

CHAPTER 4
Organization of the Cell 61

CHAPTER 5
Membranes 92

CHAPTER 6
Energy, Enzymes, and Catalysis 114

CHAPTER 7
Pathways That Release Energy in Cells 136

CHAPTER 8
Photosynthesis 162

II. INFORMATION AND HEREDITY

CHAPTER 9
Chromosomes and Cell Division 191

CHAPTER 10
Mendelian Genetics and Beyond 214

CHAPTER 11
Nucleic Acids as the Genetic Material 241

CHAPTER 12
Molecular Genetics of Prokaryotes 273

CHAPTER 13
Molecular Genetics of Eukaryotes 293

CHAPTER 14
Recombinant DNA Technology 311

CHAPTER 15
Genetic Disease and Modern Medicine 334

CHAPTER 16
Defenses against Disease 353

CHAPTER 17
Animal Development 381

III. EVOLUTIONARY PROCESSES

CHAPTER 18
The Origin of Life on Earth 411

CHAPTER 19
The Mechanisms of Evolution 426

CHAPTER 20
Species and Their Formation 446

CHAPTER 21
Systematics and Reconstructing
Phylogenies 462

IV. THE EVOLUTION OF DIVERSITY

CHAPTER 22
Bacteria and Viruses 483

CHAPTER 23
Protists 503

CHAPTER 24
Fungi 528

CHAPTER 25
Plants 545

IN BRIEF

CHAPTER 26
Sponges and Protostomate Animals 574

CHAPTER 27
Deuterostomate Animals 610

CHAPTER 28
Patterns in the Evolution of Life 644

V. THE BIOLOGY OF VASCULAR PLANTS

CHAPTER 29
The Flowering Plant Body 665

CHAPTER 30
Transport in Plants 689

CHAPTER 31
Environmental Challenges to Plants 704

CHAPTER 32
Plant Nutrition 717

CHAPTER 33
Regulation of Plant Development 733

CHAPTER 34
Plant Reproduction 757

VI. THE BIOLOGY OF ANIMALS

CHAPTER 35
Physiology, Homeostasis, and Temperature Regulation 777

CHAPTER 36
Animal Hormones 800

CHAPTER 37
Animal Reproduction 824

CHAPTER 38
Neurons and the Nervous System 849

CHAPTER 39
Sensory Systems 881

CHAPTER 40
Effectors 909

CHAPTER 41
Gas Exchange in Animals 933

CHAPTER 42
Internal Transport and Circulatory Systems 956

CHAPTER 43
Animal Nutrition 979

CHAPTER 44
Salt and Water Balance and Nitrogen Excretion 1008

CHAPTER 45
Animal Behavior 1029

VII. ECOLOGY AND BIOGEOGRAPHY

CHAPTER 46
Behavioral Ecology 1057

CHAPTER 47
Structure and Dynamics of Populations 1080

CHAPTER 48
Interactions within Ecological Communities 1100

CHAPTER 49
Ecosystems 1127

CHAPTER 50
Biogeography 1152

CHAPTER 51
Conservation Biology 1176

CONTENTS

1. THE SCIENCE OF BIOLOGY 1

Characteristics of Living
Organisms 1
The Hierarchical Organization
of Biology 3
From Molecules to Tissues 4
From Organisms to Biomes 4
Life's Emergent Properties 4
The Methods of Science 6
The Hypothetico-Deductive
Approach 7
Experimentation and Animals 9

Why People Do Science 9
Size Scales 9
Time Scales 10
Major Organizing Concepts
in Biology 11
Evolutionary Concepts 12
Darwin's Scientific Study
of Evolution 13
The Importance of a World
View 14
Life's Six Kingdoms 14
Science and Religion 15

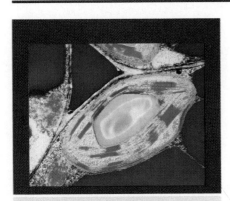

PART ONE
The Cell

2. SMALL MOLECULES 19

Atoms 19
Elements 20
How Elements Differ 20
Isotopes 21
The Behavior of Electrons 22
Box 2.A Probing an Embryo
with a Radioisotope 22
Chemical Bonds 23
Covalent Bonds 23
The Covalent Bonds of Different
Elements 25
Ions and Ionic Bonds 25
Molecules 26
Box 2.B The Gas that Says NO 27
Chemical Fractions 27
Water 28
Acids, Bases, and pH 29
Buffers 30
Polarity 30
Hydrogen Bonds 31
Interactions between Nonpolar
Molecules 33
Some Simple Organic Compounds
and Groups 33

3. LARGE MOLECULES 39

Lipids 39
Triglycerides 40
Phospholipids 42
Other Lipids 42
Box 3.A Making a Leukemia
Grow Up 44
From Monomers to Polymers 44
Carbohydrates 45
Monosaccharides 45
Disaccharides 46
Polysaccharides 46
Derivative Carbohydrates 47
Proteins 47
Amino Acids 49
Peptide Linkages 49
Levels of Protein Structure 51
Primary Structure 51
Secondary Structure 52
Tertiary Structure 52
Quaternary Structure 55
Nucleic Acids 55
Glycolipids, Glycoproteins,
Lipoproteins,
and Nucleoproteins 59

4. ORGANIZATION OF THE CELL 61

Cells and the Cell Theory 61
Common Characteristics of Cells 61
Box 4.A The Sizes of Things 63
Prokaryotic Cells 63
Features Shared by All Prokaryotic
Cells 63
Other Features of Prokaryotic
Cells 64
Kinds of Prokaryotes 66
Probing the Subcellular World:
Microscopy 66
Box 4.B The Best-Known Prokaryote:
Escherichia coli 67
The Eukaryotic Cell 68
Roles of Membranes in Eukaryotic
Cells 68

Information-Processing
Organelles 69
The Nucleus 69
Nucleus and Cytoplasm 69
Ribosomes 73
Energy-Processing Organelles 74
Mitochondria 74
Plastids 75
The Origins of Plastids,
Mitochondria, and
Eukaryotes 76
The Endomembrane System 78
Endoplasmic Reticulum 78
The Golgi Apparatus 78
Exocytosis and Endocytosis 80
Lysosomes 81
Other Organelles 82
The Cytoskeleton 83
Box 4.C Molecular "Motors" Carry
Vesicles along Microtubules 85
The Outer "Skeleton"—The Cell
Wall 86
Eukaryotes, Prokaryotes, and
Viruses 87
Fractionating the Eukaryotic Cell:
Isolating Organelles 87
Rupturing the Cell 87
Separating the Organelles 87

5. MEMBRANES 92

Membrane Structure
and Composition 92
Membrane Lipids 93
Membrane Proteins 94
Membrane Carbohydrates 97
Microscopic Views of Biological
Membranes 97
Where Animal Cells Meet 98
Tight Junctions 99
Desmosomes 100
Gap Junctions 100
Diffusion 100
Crossing the Membrane Barrier 101

Simple Diffusion 102
Membrane Transport Proteins 103
Facilitated Diffusion 104
Active Transport 104
Osmosis 105
More Activities of Membranes 107
Energy Transformations 107
Recognition and Binding 108
Cell Adhesion 109
Membrane Integrity under Stress 109
Membrane Formation and
Continuity 109

6. ENERGY, ENZYMES, AND CATALYSIS 114

Energy and the Laws of
Thermodynamics 114
Chemical Equilibrium 115
The Equilibrium Constant 116
Free Energy and Equilibria 117
Free Energy, Heat,
and Entropy 118
Reaction Rates 119
Getting over the Energy
Barrier 119
Rate Constants 121
The Highly Specific Catalysts
of Enzymes 121
Substrate Concentration
and Reaction Rate 122
Coupling of Reactions 123
Molecular Structure of Enzymes 124
Structures and Actions of Protein-
Digesting Enzymes 125
Prosthetic Groups and
Coenzymes 125
Regulation of Enzyme Activity 126
Irreversible Inhibition 127
Reversible Inhibition 127
Allosteric Enzymes 129

Mechanism of Allosteric
Effects 129
Control of Metabolism Through
Allosteric Effects 130
Sensitivity of Enzymes
to the Environment 132
Enzymes, Ribozymes,
and Abzymes 133

7. PATHWAYS THAT RELEASE ENERGY IN CELLS 136

Glycolysis, Cellular Respiration,
and Fermentation 136
ATP 137
Spending, Making, and Saving
ATP 137
Box 7.A Some Organisms Use ATP
to Make Light 138
The Energy Content of ATP 139
The Transfer of Hydrogen Atoms
and Electrons 139
How Do Cells Produce ATP? 141
The Release of Energy from
Glucose 142
Glycolysis 144
The Beginning of Cellular Respiration:
The Citric Acid Cycle 147
Continuation of Cellular Respiration:
The Respiratory Chain 149
Oxidative Phosphorylation and
Mitochondrial Structure 152
Box 7.B Dissecting the
Mitochondrion 153
The Chemiosmotic Mechanism of
Mitochondria Summarized 153
Fermentation 155
Comparative Energy Yields 156
Connections with Other Pathways 157
Feedback Regulation 157

8. PHOTOSYNTHESIS 162

Sunlight and Life on Earth 162
Early Studies of Photosynthesis 163
The Pathways of Photosynthesis 164
Light and Pigments 164
Basic Physics of Light 165
Pigments 166
Absorption Spectra and Action
Spectra 167
The Photosynthetic Pigments 168
Photophosphorylation 169
Using the Excited Chlorophyll
Molecule 169
Noncyclic Photophosphorylation:
Formation of ATP and
NADPH 169
Cyclic Photophosphorylation:
Formation of ATP but Not
NADPH 171
The Mechanism and Location
of ATP Formation 172
Noncyclic and Cyclic
Photophosphorylation
Revisited 173
Box 8.A Photosynthesis
in the Halobacteria 174
The Calvin–Benson Cycle 175
The First Stable Product of Carbon
Dioxide Fixation 175
Box 8.B Tools that Cracked
the Calvin–Benson Cycle 176
What Is the Carbon Dioxide
Acceptor? 176
Filling the Gaps in
the Calvin–Benson Cycle 180
Photorespiration 181
Alternate Modes of Carbon Dioxide
Fixation 182
Photosynthesis and Cellular
Respiration 185

9. CHROMOSOMES AND CELL DIVISION 191

The Divisions of Eukaryotic Cells 191
Eukaryotic Cells and
Chromosomes 192
The Cell Cycle 194
Mitosis 196
Development of the Chromosomes
and Spindle 196
Dancing Chromosomes 198
The End of Mitosis 201
Cytokinesis 201
Sex and Reproduction 202
The Karyotype 203
Meiosis 204
The First Meiotic Division 204

The Second Meiotic Division 208
Synapsis, Reduction, and
Diversity 208
Meiotic Errors and Their
Consequences 209
Ploidy, Mitosis, and Meiosis 210
Cell Division in Prokaryotes 211

10. MENDELIAN GENETICS AND BEYOND 214

Mendel's Discoveries 214
Mendel's Strategy 215
Experiment 1 216
Terminology for Mendelian
Genetics 217
Segregation of Alleles 217
The Test Cross 218

**PART TWO
Information
and Heredity**

Box 10.A Elements
    of Probability 219
    Independent Assortment
    of Alleles 219
Box 10.B Probabilities
    in the Dihybrid Cross 221
Genetics after Mendel: Alleles
    and Their Interactions 222
    Incomplete Dominance
    and Codominance 222
    Pleiotropy 224
    The Origin of Alleles:
    Mutation 224
    Multiple Alleles 224
Focus on Chromosomes 225
    Linkage 225
    Sex Determination 226
    Sex Linkage 228
    Mendelian Ratios Are Averages,
    Not Absolutes 230
    Special Organisms for Special
    Studies 230
    Meiosis in Neurospora 230
    Recombination in Eukaryotes 231
Box 10.C Gene Mapping
    in Eukaryotes 232
    Cytogenetics 232
Interactions of Genes
    with Other Genes
    and with the Environment 233
    Epistasis 234
    Quantitative Inheritance
    and Environmental Effects 235
Non-Mendelian Inheritance 236

**11. NUCLEIC ACIDS AS THE
    GENETIC MATERIAL 241**

What Is the Gene? 241
    The Transforming Principle 241
    The Transforming Principle
    is DNA 242
    The Genetic Material
    of a Virus 243
Nucleic Acid Structure 243
    Evidence from X-Ray
    Crystallography
    and Biochemistry 243
    Watson, Crick, and the Double
    Helix 245
    Key Elements of DNA
    Structure 245
    Alternative Structures
    for DNA 245

Structure of RNA 246
    Implications of the Double-Helical
    Structure of DNA 247
Replication of the DNA Molecule 248
    Demonstration
    of Semiconservative
    Replication 250
    Replicating an Antiparallel Double
    Helix 251
Proofreading and DNA Repair 252
Box 11.A Collaboration of Proteins
    at the Replication Fork 253
What Do Genes Control? 255
From DNA to Protein 256
    The Central Dogma of Molecular
    Biology 258
    RNA Viruses and the Central
    Dogma 258
    Transcription 259
    Transfer RNA 260
    The Ribosome 261
    Translation 263
    The Role of the Endoplasmic
    Reticulum 264
Box 11.B Making the Most of Your
    DNA 265
    The Genetic Code 266
Mutations 268
    Point Mutations 268
    Chromosomal Mutations 269
    The Frequency of Mutations 270
The Origin of New Genes 270

**12. MOLECULAR GENETICS
    OF PROKARYOTES 273**

Mutations on Bacteria and
    Bacteriophages 274
Bacterial Conjugation 274
    Isolating Specific Bacterial
    Mutants 276
    The Bacterial Fertility Factor 278
    Transfer of Male Genetic
    Elements 278
    Sexduction 280
Bacteriophages 281
    Recombination in Phages 281
    Lysogeny and the Disappearing
    Phages 281
    Transduction 282
Episomes and Plasmids 283
Transposable Elements 284
Control of Transcription in
    Prokaryotes 285

Processing a Polycistronic
    Messenger 286
Promoters 286
Operons 286
Operator-Repressor Control
    That Induces Transcription:
    The lac Operon 287
Operator-Repressor Control
    That Represses Transcription:
    The Tryptophan Operon 288
Control by Increasing Promoter
    Efficiency 289
Preventing versus Enhancing
    Transcription 289

**13. MOLECULAR GENETICS
    OF EUKARYOTES 293**

Eukaryotes and Eukaryotic Cells 293
Hybridization of Nucleic Acids 294
Eukaryotic Gene Structure 294
Repetitive DNA in Eukaryotes 295
    Disadvantages and Advantages
    of Ends: Telomeres 296
    Transposable Elements in
    Eukaryotes 297
Gene Duplication and Gene
    Families 298
RNA Processing in Eukaryotes 299
    Capping and Tailing RNA 299
    Splicing RNA 299
    The Stability of mRNAs 299
Control of Gene Expression in
    Eukaryotes 300
    Transcriptional Control:
    Gene Inactivation 300
    Transcriptional Control:
    Gene Amplification 302
    Transcriptional Control:
    Selective Gene
    Transcription 303
    Translational Control 305
    Posttranslational Control 306
Box 13.A Cassettes and the Mating
    Type of Yeast 308

**14. RECOMBINANT DNA
    TECHNOLOGY 311**

The Pillars of Recombinant DNA
    Technology 312
Cleaving and Splicing DNA 312
    Restriction Endonucleases 312
    Sticky Ends and DNA
    Splicing 313

Cloning Genes 314
Host Cells 314
Vectors 314
Inserting Foreign DNA into
Host Cells 315
Selecting Transgenic Cells 316
Controlling a Cloned Gene 318
Sources of Genes for Cloning 318
Gene Libraries and
Shotgunning 318
Complementary DNA 318
Synthetic DNA 320
Exploring DNA Organization 320
Separation of Intact
Chromosomes 320
Separation and Purification
of DNA Fragments 320
Detection of Specific DNA
Fragments 321
Autoradiography 323
Localization of Genes on
Chromosomes 323
Restriction Mapping 323
Chromosome Walking 323
Sequencing DNA 324
Box 14.A Determining the Base
Sequence of DNA 326
Gene Copies by the Billion 326
Prospects 328
Plant Agricultural
Biotechnology 329
Recombinant DNA Technology
and the Environment 330
Genome Projects and Modern
Medicine 331

**15. GENETIC DISEASE AND
MODERN MEDICINE 334**

Some Inherited Diseases 334
Triplet Repeats and the Fragility
of Some Human Genes 336
Dealing with Genetic Disease:
An Overview 337
Finding a Defective Gene:
Mapping Human
Chromosomes 337
Pedigree Analysis 338
Somatic-Cell Genetics 338
Restriction Fragment Length
Polymorphisms 340
Narrowing the Search 340
Dealing with a Defective Gene 342
Knockout Mice 342

Gene Therapy 342
Screening for Harmful Alleles 344
Cancer 345
Genes, Viruses, and Cancer 345
Growth Factors and Cancer-
Related Genes 345
Origins of Oncogenic Viruses
and Transposable Elements 346
Treating Cancer 348
Preventing Cancer 349
The Ames Test 349
Can We Cheat Death? 350

**16. DEFENSES
AGAINST DISEASE 353**

Nonspecific Defenses against
Pathogens 353
If Pathogens Evade the Blocks
to Entry 354
Viral Diseases and Interferon 355
Nonspecific Defenses
of Plants 356
Specific Defenses: The Immune
System 356
Responses of the Immune
System 356
Immunological Memory and
Immunization 359
Clonal Selection and Its
Consequences 359
Self, Nonself, and Tolerance 360
Development of Plasma Cells 361
Immunoglobulins: Agents of the
Humoral Response 361
Box 16.A Monoclonal
Antibodies 364
Antibodies and Nonspecific Defenses
Working Together 366
The Origin of Antibody Diversity 367
How a B Cell Produces a
Particular Heavy Chain 368
The Constant Region and Class
Switching 369
T Cells: Agents of Both
Responses 370
The Major Histocompatibility
Complex 371
T Cells and the Humoral Immune
Response 371
T Cells and the Cellular Immune
Response 372
The MHC and Tolerance
of Self 373

Transplants 373
Interleukins 374
Disorders of the Immune System 374
Immune Deficiency Disorders 375
Prospects for an AIDS Cure—
or an AIDS Vaccine 376

17. ANIMAL DEVELOPMENT 381

The Study of Animal
Development 381
Cleavage 382
Becoming Multicellular 382
Formation of the Blastula 385
Gastrulation 385
Gastrulation in Embryos
with Much Yolk 387
Gastrulation in Birds
and Mammals 387
Neurulation 388
Later Stages of Development 389
Growth 389
Larval Development
and Metamorphosis 391
Looking Closer at Development 391
Differentiation 393
Differentiation Irreversible? 393
Transdetermination of Imaginal
Discs 394
Determination by Cytoplasmic
Segregation 394
Polarity in the Egg
and Zygote 394
Cytoplasmic Factors in Polarity
in *Drosophila* 395
Germ-Line Granules
in *Caenorhabditis* 396
Determination by Embryonic
Induction 397
Induction and the Organizer 397
Induction at the Cellular
Level 399
Box 17.A Genes Interact
in Induction 400
Pattern Formation 400
Positional Information in
Developing Limbs 400
Establishing Body
Segmentation 401
Homeotic Mutations 402
The Homeobox 402
You've Seen the Pieces;
Now Let's Build a Fly 403
Developmental Biology
and Evolution 404

PART THREE
Evolutionary Processes

18. THE ORIGIN OF LIFE ON EARTH 411

Is Life Evolving from Nonlife
  Today? 411
When Did Life on Earth Arise? 412
How Did Life Evolve
  from Nonlife? 414
  Conditions on Earth at the Time
    of Life's Origin 414
  Box 18.A A Concise Scenario
    for the Origin and Early
    Evolution of Life 415
  An Overview of Earth's
    History 415
  Beginning the Sequence of Events:
    The Synthesis of Organic
    Compounds 415
  Continuing the Sequence:
    More Polymerization 419
  RNA: The First Biological
    Catalyst 419
  From Ribozymes to Cells 421
  The Evolution of DNA 421

Metabolism of Early Organisms 421
  Anaerobic Photosynthesis 422
  Aerobic Photosynthesis:
    Source of the Atmosphere's
    Free Oxygen 422
How Probable Was the Evolution
  of Life? 422

19. THE MECHANISMS OF EVOLUTION 426

What Is Evolution? 426
  Environmentally Induced
    Variation 427
  Genetically Based Variation 428
The Structure of Populations 428
  Measures of Genetic Variation 428
  Polymorphic Populations 430
The Hardy-Weinberg Rule 430
Changing the Genetic Structure
    of Populations 432
  Mutation 432
  Gene Flow 432
  Genetic Drift 432
  Nonrandom Mating 434
  Natural Selection 434
  Fitness 437
Genetic Variation and Evolution 437
  Distribution of Genetic
    Variation 437
  Maintenance of Genetic
    Variation 439
  Adaptive and Adaptively Neutral
    Genetic Variation 440
Interpreting Long-Term
    Evolution 442

20. SPECIES AND THEIR FORMATION 446

What Are Species? 446
The Cohesiveness of Species 447

How Do New Species Arise? 448
  Allopatric Speciation 448
  Parapatric Speciation 449
  Sympatric Speciation 449
  How Important Is Each Speciation
    Mode? 452
Reproductive Isolating
    Mechanisms 452
Hybrid Zones 453
How Much Do Species Differ
    Genetically? 455
How Long Does Speciation
    Take? 455
  Behavior and Speciation Rates 455
  Ecology and Speciation Rates 456
  Evolutionary Radiation and
    Speciation Rates 456
The Significance of Speciation 458

21. SYSTEMATICS AND RECONSTRUCTING PHYLOGENIES 462

The Importance of
    Classifications 462
The Hierarchy of the Linnaean
    System 464
Classification Systems
    and
    Evolution 465
  Reclassifying Organisms 465
  Cladistic Classification 465
Traits Used in Restructuring
    Phylogenies 473
  Structure and Behavior 473
  Molecular Traits 474
  Combinations of Traits 475
Evaluating Different Methods 476
The Future of Systematics 477
Other Uses of Phylogenies 477

PART FOUR
The Evolution of Diversity

22. BACTERIA AND VIRUSES 483

General Biology of the Bacteria 483
  Metabolic Diversity
    in the Bacteria 483
  Dividing the Bacteria into
    Kingdoms and Phyla 485
  Prokaryotes versus
    Eukaryotes 485
  Structural Characteristics
    of Bacteria and Bacterial
    Associations 486
Box 22.A The Gram Stain and
    Bacterial Cell Walls 487
  Reproduction and Resting 488
  Bacteria and Disease 488
Phyla of the Archaebacteria 489
  Thermoacidophiles 489

Methanogens 489
  Strict Halophiles 490
Phyla of the Eubacteria 490
  Gram-Negative Bacteria 490
Box 22.B Gliding through
    the Soil 491
  Gram-Positive Bacteria 494
  Mycoplasmas 496
Viruses 497
  Discovery of the Viruses 497
  Viral Structure 497
  Reproduction of Viruses 497
  Classification of Viruses 498
  Viroids: RNA without
    a Capsid 499
  Scrapie-associated Fibrils:
    Infectious Proteins? 499

23. PROTISTS 503

Protista and the Other Eukaryotic
    Kingdoms 503
General Biology of the Protists 504
    Vesicles 504
    The Cell Surface 505
    Sensitivity to the Environment 505
    Endosymbiosis 505
    Reproduction 506
    Alternation of Generations 506
Protozoans 507
    Phylum Zoomastigophora 507
    Phylum Rhizopoda 508
    Phylum Actinopoda 509
    Phylum Foraminifera 509
    Phylum Apicomplexa 509
    Phylum Ciliophora 511
Funguslike Protists 513
    Phylum Myxomycota 514
    Phylum Acrasiomycota 515
    Phylum Protomycota 516
    Phylum Oomycota 516
Algae 517
    Phylum Pyrrophyta 517
    Phylum Chrysophyta 518
    Phylum Euglenophyta 520
    Phylum Phaeophyta 520
    Phylum Rhodophyta 521
    Box 23.A Algae in a Turbulent
        Environment 522
    Phylum Chlorophyta 523
    Evolutionary Trends
        in the Algae 526

24. FUNGI 528

General Biology of the Fungi 528
    The Fungal Body 529
    Fungi and Their Environment 530
    Nutrition of Fungi 530
    Box 24.A Plant Roots and Fungi 532
    Reproduction in Fungi 532
    Dikaryon Formation in Fungi 533
    Multiple Hosts in a Fungal Life
        Cycle 533
Phylum Eumycota 536
    Class Zygomycetes 536
    Class Ascomycetes 537
    Class Basidiomycetes 539
    Class Deuteromycetes 540
Lichens 540

25. PLANTS 545

The Plant Kingdom 545
    Alternation of Generations
        in Plants 546
    Classification of Plants 546
    Plant Colonizers of the Land 547
    Origins of the Plant Kingdom 547

Two Groups within the Plant
    Kingdom 547
Nonvascular Plants 548
    The Nonvascular Plant
        Life Cycle 548
    Phylum Hepaticophyta:
        Liverworts 549
    Phylum Anthocerophyta:
        Hornworts 550
    Phylum Bryophyta: Mosses 551
Vascular Plants 552
    Evolution of the Vascular
        Plants 553
    Ancient Vascular Plants 554
    Further Developments in the
        Ancient Vascular Plants 554
Surviving Seedless Vascular
    Plants 556
    Phyla Lycophyta
        and Sphenophyta:
        Club Mosses and
        Horsetails 556
    Phylum Psilophyta:
        Two Simple Plants 557
    Plants with Complex Leaves 558
    Phylum Pterophyta: Ferns 558
    The Fern Life Cycle 559
The Seed Plants 559
The Gymnosperms 561
    Fossil Gymnosperms 561
    Phylum Coniferophyta:
        Conifers 562
    The Gymnosperm Life Cycle 563
The Angiosperms:
    Flowering Plants 563
    The Flower 563
    Evolution of the Flower 565
    Pollen and the Coevolution of
        Angiosperms and Animals 566
    The Fruit 568
    The Angiosperm Life Cycle 569
    Classes of Angiosperms 569
    Origin and Evolution of the
        Angiosperms 570
    Where Herbaceous Plants
        Predominate 571

26. SPONGES AND PROTOSTOMATE ANIMALS 574

How Are Animals Classified? 575
    Body Symmetry 575
    Developmental Pattern 575
    Body Cavities 576
The Origins of Animals 578
    Oxygen and the Evolution
        of Animals 578
    Early Animal Evolution 580
Simple Aggregations 580
    Phylum Porifera 580

Phylum Placozoa 582
The Evolution of Diploblastic
    Animals 582
    Phylum Cnidaria 583
    Phylum Ctenophora 587
The Evolution of Bilateral
    Symmetry 588
    Phylum Platyhelminthes 588
    Phylum Nemertea 589
The Development of Body
    Cavities 589
Pseudocoelomate Animals 591
    Phylum Nematoda 591
    Phylum Rotifera 592
Coelomate Animals 592
    Phylum Pogonophora:
        Losing the Gut 593
    Phylum Annelida:
        Many Subdivisions
        of the Coelom 594
The Evolution of External
    Skeletons 595
    Diversification of
        the Arthropods 596
    Ancient Arthropod Lineages 596
    Phylum Chelicerata 597
    Phylum Crustacea 598
    Phylum Uniramia 600
Calcified Protection 602
    Phylum Mollusca 602
Themes in Protostome Evolution 606

27. DEUTEROSTOMATE ANIMALS 610

Tripartite Deuterostomes 610
    Phylum Phoronida 611
    Phylum Bryozoa (Ectoprocta) 611
    Phylum Brachiopoda 612
Innovations in Feeding 613
    Phylum Hemichordata 613
Active Food Seekers 614
    Phylum Chaetognatha 614
Calcifying the Skeleton 614
    Phylum Echinodermata 614
Evolution of the Chordate
    Pharynx 617
    Phylum Chordata 617
Sucking Mud:
    The Rise of the Vertebrates 618
    Jaws: A Key Evolutionary
        Novelty 619
    Fins and Mobility 620
    Mobility and Buoyancy 621
Breathing Air and Exploring
    the Land 621
    In and Out of the Water:
        The Amphibians 622
    Colonization of the Land 625
    Modern Reptiles 625
    Dinosaurs and Their Relatives 625

Box 27.A The Four-Minute Mile 627
The Origin of Mammals 629
Human Evolution 632
   The Rise of *Homo* 638
   Homo sapiens Evolves 638
   The Evolution of Language
      and Culture 639
   The Human Lineage 640
Themes in Deuterostome
      Evolution 640

28. PATTERNS IN THE EVOLUTION OF LIFE 644

How Earth Has Changed 645
   Major Changes from Internal
      Processes 646
   Changes from External Events 647
The Fossil Record 648
   The Completeness of the Fossil
      Record 648
   Patterns in the Fossil Record 649

Life in the Remote Past 649
   Species-Rich Precambrian Life 649
   Even Richer Paleozoic Life 652
   Provincialization during the
      Mesozoic Era 654
   The Cenozoic Era 655
The Timing of Evolutionary
      Change 655
   Rates of Evolutionary Change 656
   Extinction Rates 658
   Origins of Evolutionary
      Novelties 659

PART FIVE
The Biology of Vascular Plants

29. THE FLOWERING PLANT BODY 665

Flowering Plants 665
Four Examples 666
   Coconut Palm 666
   Red Maple 666
   Rice 667
   Soybean 668
Classes of Flowering Plants 668
An Overview of the Plant Body 669
Organs of the Plant Body 669
   Roots 670
   Stems 671
   Leaves 672
Levels of Organization
      in the Plant Body 672
Plant Cells 673
   Cell Walls 673
   Parenchyma Cells 674
   Sclerenchyma Cells 674
   Collenchyma Cells 674
   Water-Conducting Cells
      of the Xylem 674
   Sieve Tube Elements 676

Plant Tissues and Tissue Systems 677
Growth and Meristems 677
The Meristems and Their
      Products 677
   The Young Root 677
   Tissues of the Stem 680
   Secondary Growth of Stems
      and Roots 681
   Leaf Anatomy 683
Support in a Terrestrial
      Environment 684
   Reaction Wood 684
   Quonset Huts and Marble
      Palaces 686

30. TRANSPORT IN PLANTS 689

Uptake and Transport of Water
      and Minerals 689
   Osmosis 689
   Uptake of Minerals 690
   Water and Ion Movement
      from Soil to Xylem 692
   Ascent of Sap in Xylem:
      The Magnitude
      of the Problem 693
   Early Models of Transport
      in the Xylem 693
   Root Pressure 693
Box 30.A There Are No Pressure
      Pumps in the Xylem 694
   The Evaporation–Cohesion–
      Tension Mechanism 694
   Measuring Tension in the Xylem
      Sap 696
Transpiration and the Stomata 696
Crassulacean Acid Metabolism
      and the Stomatal Cycle 697
Translocation of the Substances
      in the Phloem 698
   The Pressure Flow Model 699
   Are the Sieve Plates Clogged
      or Free? 700
   Loading and Unloading
      of the Phloem 700
   Sucrose in the Xylem 701

31. ENVIRONMENTAL CHALLENGES TO PLANTS 704

Threats to Plant Life 704
Dry Environments 705
   Special Adaptations of Leaves
      to Dry Environments 705
   Other Adaptations to a Limited
      Water Supply 706
Where Water Is Plentiful and Oxygen
      Is Scarce 706
Saline Environments 707
   Salt Accumulation and Salt
      Glands 708
   Adaptations Common
      to Halophytes and
      Xerophytes 709
   A Versatile Halophyte 709
Habitats Laden with Heavy
      Metals 709
Serpentine Soils 710
Plants and Herbivores 711
   Grazing and Plant
      Productivity 711
   Chemical Defenses against
      Herbivores 712
   A Versatile Secondary Product 712
Box 31.A A Protein for Defense
      against Insects 713
Protection against Fungi, Bacteria,
      and Viruses 713
Box 31.B Take Two Aspirin and Call
      Me in the Morning 714
Why Don't Plants Poison
      Themselves? 714
A Tradeoff: Protection or Taste? 715

32. PLANT NUTRITION 717

Acquiring Nutrients 717
   Autotrophs and Heterotrophs 718
   How Does a Sessile Organism
      Find Nutrients? 718
   Bulk Ingestion versus Selective
      Uptake of Nutrients 718
Which Nutrients Are Essential? 719

Soils 721
  Soils and Plant Nutrition 722
  Fertilizers and Lime 722
  Soil Formation 723
  Effects of Plants on Soils 723
Nitrogen Fixation 724
  Chemistry of Nitrogen
    Fixation 725
  Symbiotic Nitrogen Fixation 726
  The Need to Augment Biological
    Nitrogen Fixation 726
Box 32.A Biotechnology
  in a Plant 728
Denitrification 728
Nitrification 728
Nitrate Reduction 729
Sulfur Metabolism 729
Heterotrophic Seed Plants 730

33. REGULATION OF PLANT DEVELOPMENT 733

What Regulates Plant
  Development? 733
An Overview of Development 734
  From Seed to Seedling 734
  Reproductive Development 734
  Dormancy, Senescence,
    and Death 735
Seed Dormancy and
  Germination 735

Adaptive Advantages of Seed
  Dormancy 737
Seed Germination 738
Mobilizing Food Reserves 738
Gibberellins 738
  Discovery of the Gibberellins 739
  Why So Many Gibberellins? 741
  Other Activities
    of
    Gibberellins 741
Auxin 742
  Discovery of Auxin 742
  Auxin Transport 743
  Phototropism
    and Gravitropism 743
  Auxin and Vegetative
    Development 743
Box 33.A Strategies for Killing
  Weeds 745
  Auxin and Fruit Development 746
  Are There Master Reactions? 746
  Cell Walls and Growth 746
  Auxin and the Cell Wall 747
  Auxin Receptors 748
  Differentiation and Organ
    Formation 748
Cytokinins 748
  Discovery of the Cytokinins 749
  Other Activities of the
    Cytokinins 749
Ethylene 749
Abscisic Acid 750
Environmental Cues 751

Light and Phytochrome 751
Studying the Plant Genome 754

34. PLANT REPRODUCTION 757

Many Ways to Reproduce 757
Sexual Reproduction 758
  Gametophytes 758
  Pollination 758
  Double Fertilization 760
  Embryo Formation in Flowering
    Plants 760
Box 34.A In Vitro Fertilization:
  Test Tube Plants 762
Fruit 762
Box 34.B Maternal "Care"
  in Plants 763
Transition to the Flowering State 764
Photoperiodic Control
  of Flowering 764
  Other Patterns of
    Photoperiodism 766
  Importance of Night Length 766
  Circadian Rhythms and the
    Biological Clock 767
  A Flowering Hormone? 768
  The Search for Florigen 769
Vernalization and Flowering 769
Asexual Reproduction 770
  Asexual Reproduction
    in Nature 770
  Asexual Reproduction
    in Agriculture 771

35. PHYSIOLOGY, HOMEOSTASIS, AND TEMPERATURE REGULATION 777

Homeostasis 778
Organs and Organ Systems 778
  The Structure of Organs 778
  Organs for Information and
    Control 779
  Organs for Protection, Support,
    and Movement 780
  Organs of Reproduction 780
  Organs of Nutrition 781
  Organs of Transport 781
  Organs of Excretion 781
General Principles of
  Homeostasis 782
  Set Points and Feedback 782
  Regulatory Systems 782
The Effects of Temperature of Living
  Systems 784
  The Q_{10} Concept 784
  Metabolic Compensation 784
Thermoregulatory Adaptations 785

Laboratory Studies of Ectotherms
  and Endotherms 785
Field Study of an Ectotherm 786
Behavioral Thermoregulation 787
Control of Blood Flow
  to the Skin 787
Metabolic Heat Production 788
Biological Heat Exchangers 789
Thermoregulation in
  Endotherms 790
  Active Heat Production
    and Heat Loss 791
  Living in the Cold 792
  Changing Thermal Insulation 792
  Evaporative Water Loss 793
The Vertebrate Thermostat 793
  Set Points and Feedback 793
Box 35.A Fevers and "Feeling
  Crummy" 795
Turning Down the Thermostat 795
  Shallow Torpor 795
  Hibernation 796

**PART SIX
The Biology
of Animals**

36. ANIMAL HORMONES 800

Chemical Messages 800
  Local Hormones 801
  Circulating Hormones 801
  Glands and Hormones 802
Hormonal Control in
  Invertebrates 802

Molting in Insects 802
Development in Insects 803
Vertebrate Hormones 805
　Posterior Pituitary Hormones 806
　Anterior Pituitary Hormones 808
　Hypothalamic
　　Neurohormones 809
　Thyroid Hormones 810
　Parathormone 811
　Pancreatic Hormones 812
　Adrenal Hormones 813
　The Sex Hormones 814
Box 36.A The Rat Race 815
Box 36.B Muscle-Building Anabolic
　　Steroids 817
　Other Hormones 817
Receiving and Responding
　　to Hormones 817
　Receptors for Water-Soluble
　　Hormones and Second
　　Messengers 817
　The Role of G-Proteins
　　in the Production
　　of Second Messengers 818
　cAMP Targets and Response
　　Cascades 818
　Other Second Messengers 820
　Calcium Ions 821
　Lipid-Soluble Hormones 821
　Hormones in Control
　　and Regulation 821

37. ANIMAL REPRODUCTION 824

Asexual Reproduction 824
　Budding 824
　Regeneration 825
　Parthenogenesis 825
Sexual Reproductive Systems
　　of Animals 826
　Gametogenesis 826
　Sex Types 828
　Getting Eggs and Sperm
　　Together 828
Reproductive Systems in Humans
　　and Other Mammals 830
　The Male 830
　The Female 830
　The Ovarian Cycle 830
　The Menstrual Cycle 833
　Hormonal Control of the
　　Menstrual Cycle 833
　Human Sexual Responses 835
Fertilization 836
　Activating the Sperm 836
　Activating the Egg and Blocking
　　Polyspermy 836
Care and Nurture of the Embryo 837
　Oviparity 837
Box 37.A The Technology of Birth
　　Control 838

Viviparity 840
The Extraembryonic Membranes
　　and the Beginning
　　of Development 841
Pregnancy 842
Birth 844
　Labor 844
　Delivery 845
　Lactation 845

38. NEURONS AND
　　THE NERVOUS SYSTEM 849

Communication and Complexity 849
Cells of the Nervous System 851
　Neurons 851
　Glia 852
　Cells in Circuits 852
From Stimulus to Response:
　　Information Flow
　　in the Nervous System 852
　Development of the Vertebrate
　　Nervous System 853
　The Spinal Cord 854
　The Reticular System 855
　The Limbic System 856
　The Cerebrum 856
The Electrical Properties
　　of Neurons 858
　Pumps and Channels 858
　The Resting Potential 859
　Changes in Membrane
　　Potentials 859
Box 38.A Calculating Membrane
　　Potentials 860
　Action Potentials 861
　Propagation of the Action
　　Potential 861
Box 38.B Studying the Electrical
　　Properties of Membranes
　　with Voltage Clamps and Patch
　　Clamps 864
Synaptic Transmission 866
　The Neuromuscular Junction 866
　Events in the Postsynaptic
　　Membrane 867
　Excitatory and Inhibitory
　　Synapses 867
　Summation 868
　Other Synapses 868
　Neurotransmitters and
　　Receptors 869
　Clearing the Synapse of
　　Neurotransmitter 870
Neurons in Circuits 870
　The Autonomic Nervous
　　System 870
　Monosynaptic Reflexes 872
　Polysynaptic Reflexes 873
Higher Brain Functions 874

Sleeping and Dreaming 874
Learning and Memory 875
Language, Lateralization,
　　and Human Intellect 877

39. SENSORY SYSTEMS 881

What Is a Sensor? 881
　Sensation 882
　Sensory Organs 882
　Sensory Transduction 883
　Changing Sensitivity 884
Chemosensors 884
　Chemosensation
　　in Arthropods 884
　Olfaction 885
　Gustation 887
Mechanosensors 888
　Touch and Pressure 888
　Stretch Sensors 889
　Hair Cells 889
　Auditory Systems 890
Photosensors and Visual Systems 894
　Rhodopsin 894
　Visual Systems of
　　Invertebrates 897
　Vertebrate and Cephalopod
　　Eyes 897
　The Vertebrate Retina 899
　Binocular Vision 903
Other Sensory Worlds 904
　Infrared and Ultraviolet
　　Detection 904
　Echolocation 904
　Detection of Electric Fields 905
　Magnetic Sense 906

40. EFFECTORS 909

How Animals Respond to
　　Information from Sensors 909
Cilia, Flagella, and Microtubules 910
　Ciliated Cells 910
　Flagellated Cells 911
　How Cilia and Flagella Move 911
　Microtubules as Intracellular
　　Effectors 912
Microfilaments and Cell
　　Movement 912
Muscles 914
　Smooth Muscle 914
　Skeletal Muscle 915
　Controlling the Actin–Myosin
　　Interaction 917
　Twitches, Graded Contractions,
　　and Tonus 917
　Fast- and Slow-Twitch Fibers 919
　Cardiac Muscle 920
Skeletal Systems 920
　Hydrostatic Skeletons 921
　Exoskeletons 921

Vertebrate Endoskeletons 922
Types of Bone 924
Joints and Levers 926
Other Effectors 927
Chromatophores 928
Glands 929
Sound and Light Producers 929
Electric Organs 929

41. GAS EXCHANGE IN ANIMALS 933

Breathing and Life 933
Limits to Gas Exchange 934
Breathing Air or Water 934
Temperature 935
Altitude 935
Carbon Dioxide Exchange
with the Environment 935
Respiratory Adaptations 936
Surface Area 936
Ventilation and Perfusion 937
Insect Respiration 937
Fish Gills 938
Bird Lungs 939
Box 41.A Countercurrent
Exchangers 941
Breathing in Mammals 942
Surfactant and Mucus 944
Mechanics of Ventilation 946
Transport of Respiratory Gases
by the Blood 946
Hemoglobin 947
Myoglobin 948
Regulation of Hemoglobin
Function 948
Transport of Carbon Dioxide
from the Tissues 950
Regulation of Ventilation 951
The Ventilatory Rhythm 951
Matching Ventilation to Metabolic
Needs 951

42. INTERNAL TRANSPORT AND CIRCULATORY SYSTEMS 956

Transport Systems 956
Gastrovascular Cavities 956
Open Circulatory Systems 957
Closed Circulatory Systems 957
Circulatory Systems of Fishes 958
Circulatory Systems of
Amphibians 960
Circulatory Systems of Reptiles
and Crocodilians 960
Circulatory Systems of Birds
and Mammals 960
The Human Heart 961
Structure and Function 961
Cardiac Muscle and the
Heartbeat 963

Box 42.A The Electrocardiogram 964
Control of the Heartbeat 965
The Vascular System 966
Arteries and Arterioles 966
Box 42.B Cardiovascular
Disease 967
Capillaries 967
Exchange in Capillary Beds 969
The Lymphatic System 969
Venous Return 970
The Blood 971
Red Blood Cells 971
White Blood Cells 972
Platelets and Blood Clotting 972
Plasma 973
Control and Regulation of
Circulation 973
Autoregulation 974
Control and Regulation
by the Endocrine
and Nervous Systems 974

43. ANIMAL NUTRITION 979

Defining Organisms
by What They Eat 979
Nutrient Requirements 981
Nutrients as Fuel 981
Nutrients as Building Blocks 983
Mineral Nutrients 984
Vitamins 984
Box 43.A Beriberi and the Vitamin
Concept 987
Box 43.B Scurvy and Vitamin C 987
Nutritional Deficiency Diseases
in Humans 988
Adaptations for Feeding 988
Food Acquisition by
Carnivores 988
Food Acquisition by
Herbivores 988
Food Acquisition by Filter
Feeders 989
Vertebrate Teeth 990
Digestion 990
Tubular Guts 991
Digestive Enzymes 992
Structure and Function
of the Vertebrate Gut 993
The Tissue Layers 993
Movement of Food in the Gut 995
Digestion in the Mouth and
Stomach 995
The Small Intestine, Gall Bladder,
and Pancreas 996
Absorption in the Small
Intestine 997
The Large Intestine 999
Digestion by Herbivores 1000

Control and Regulation of
Digestion 1001
Neural Reflexes 1001
Hormonal Controls 1001
Control and Regulation of Fuel
Metabolism 1002
The Role of the Liver 1002
Hormonal Control of Fuel
Metabolism 1002
Box 43.C Lipoproteins: The Good,
the Bad, and the Ugly 1003

44. SALT AND WATER BALANCE AND NITROGEN EXCRETION 1008

Internal Environment 1008
Water, Salts, and the
Environment 1009
The Excretion of Nitrogenous
Wastes 1010
Invertebrate Excretory Systems 1012
Protonephridia 1012
Metanephridia 1013
Malpighian Tubules 1013
Green Glands 1014
Vertebrate Excretory Systems 1015
The Structure and Functions
of the Nephron 1015
The Evolution of the
Nephron 1017
Water Conservation
in Vertebrates 1018
Structure and Function of the
Mammalian Kidney 1019
Anatomy 1019
Glomerular Filtration Rate 1021
Tubular Reabsorption 1021
The Countercurrent
Multiplier 1021
Regulation of Kidney Functions 1023
Box 44.A Artificial Kidneys 1023
Autoregulation of the Glomerular
Filtration Rate 1024
Regulation of Blood Volume 1024
Regulation of Blood
Osmolarity 1025
Box 44.B Water Balance
in the Vampire Bat 1025
Regulatory Flexibility 1026

45. ANIMAL BEHAVIOR 1029

Genetically Determined
Behavior 1030
Deprivation Experiments 1030
Triggering Fixed Action
Patterns 1030
Motivation 1032
Fixed Action Patterns versus
Learned Behavior 1033
Hormones and Behavior 1034

Sexual Behavior in Rats 1034
Bird Brains and Bird Song 1036
The Genetics of Behavior 1036
Hybridization Experiments 1037
Selection and Crossing of Selected
Strains 1038
Molecular Genetics
of Behavior 1039
Communication 1040

Chemical Communication 1040
Visual Communication 1040
Auditory Communication 1042
Tactile Communication 1042
Electrocommunication 1043
Origins of Communication
Signals 1044
The Timing of Behavior:
Biological Rhythms 1045

Circadian Rhythms 1045
Circannual Rhythms 1047
How Do They Find Their Way? 1048
Piloting 1048
Homing 1048
Migration 1049
Navigation 1049
Human Behavior 1052

PART SEVEN
Ecology and
Biogeography

46. BEHAVIORAL ECOLOGY 1057

Costs and Benefits of Behavior 1058
Dealing with the Nonsocial
Environment 1059
Dealing with Individuals of the Same
Species 1063
Benefits of Social Life 1063
Costs of Group Living 1064
Types of Social Acts 1064
Box 46.A Calculating the Coefficient
of Relatedness 1066
Choosing Mating Partners 1066
Male Mating Tactics 1067
Female Mating Tactics 1069
Mate Choice by
Hermaphrodites 1070
Roles of the Sexes 1070
Sexual Selection 1071
The Evolution of Animal
Societies 1072
Eusocial Behavior 1074
Ecology and Social
Organization 1076

47. STRUCTURE AND DYNAMICS
OF POPULATIONS 1080

Life Histories 1081
Growth 1081
Change of Form 1081

Dispersal 1081
Life History Trade-Offs 1081
Timing of Reproduction 1083
Phylogenetic Constraints on Life
History Evolution 1084
Population Structure 1084
Population Density 1084
Spacing 1084
Age Distribution 1085
Population Dynamics 1086
Births and Deaths 1086
Life Tables 1087
Modular Organisms 1088
Population Growth When Resources
Are Abundant 1089
Population Growth When Resources
Are Scarce 1090
Dynamics of Species Ranges 1090
Population Regulation 1091
Density Dependence 1091
Disturbance 1092
Coping with Habitat
Changes 1093
Population Regulation under Human
Management 1094
Managing the Human
Population 1097

48. INTERACTIONS
WITHIN ECOLOGICAL
COMMUNITIES 1100

Types of Ecological Interactions 1101
Resources and Consumers 1101
Predator–Prey Interactions 1103
Types of Predators 1103
Short-Term Predator–Prey
Dynamics 1104
Long-Term Predator–Prey
Interactions 1107
Competition 1110
Other Interspecific Interactions 1113
Amensalism and
Commensalism 1113
Mutualism 1113
Interaction and Coevolution 1116
How Species Affect Ecological
Communities 1118
The Role of Plants 1118

Succession 1118
The Role of Animals 1120
The Role of Microorganisms 1121
Patterns of Species Richness 1122
Local Species Richness 1122
Influence of Regional
Species Richness
on Local Species Richness 1123

49. ECOSYSTEMS 1127

Energy Flow through
Ecosystems 1127
Gross Primary Production 1128
Net Primary Production 1128
Trophic Levels 1128
The Maintenance of Organisms
and Energy Flow 1129
Chemical Cycling in
Ecosystems 1133
Oceans 1133
Fresh Waters 1134
Atmosphere 1134
Land 1136
Biogeochemical Cycles 1136
The Hydrological Cycle 1137
The Carbon Cycle 1137
The Nitrogen Cycle 1138
The Phosphorus Cycle 1138
The Sulfur Cycle 1139
Human Alterations of
Biogeochemical Cycles 1140
Lake Eutrophication:
A Local Effect 1140
Acid Precipitation:
A Regional Effect 1142
Alterations of the Carbon Cycle:
A Global Effect 1143
Alteration of the Chlorine
Cycle 1143
Agriculture and Ecosystem
Productivity 1144
Climates on Earth 1145
Solar Energy Inputs 1145
Global Atmospheric
Circulation 1145
Global Oceanic Circulation 1147
Global Ecosystem
Production 1148

50. BIOGEOGRAPHY 1152

The Goals of Biogeography 1152
Historical Biogeography 1154
    Parsimony 1154
    Vicariance and Dispersal 1155
    Reconstructing Biogeographic
       Histories 1156
    Major Terrestrial Biogeographic
       Regions 1156
Ecological Biogeography 1157
    An Equilibrium Model of Species
       Richness 1157
    Tests of the Species Richness
       Equilibrium Model 1159
Terrestrial Biomes 1160
    Tundra 1163
    Boreal Forest 1163
    Temperate Deciduous Forest 1164
    Grassland 1165
    Cold Desert 1166
    Hot Desert 1166
    Chaparral 1167
    Thorn Forest and Tropical
       Savanna 1167
    Tropical Deciduous Forest 1167
    Tropical Evergreen Forest 1167
    Tropical Montane Forest 1169
Aquatic Ecosystems 1169
    Freshwater Biomes 1170
    Marine Biomes 1170

51. CONSERVATION BIOLOGY 1176

Causes of Extinctions 1177
    Overexploitation 1177
    Introduced Pests, Predators,
       and Competitors 1177
    Loss of Mutualists 1178
    Habitat Destruction 1178
Box 51.A Forest Analogs 1179
Studies of Individual Species 1180
    The Probability of Survival 1180
    Captive Propagation 1182
Biology of Rare Species 1184
    Conservation and Climate
       Change 1185
Community-Level
    Conservation 1187
    Endemism 1187
    Keystone Species 1188
    Habitat Fragmentation 1188
Habitat and Ecosystem
    Management 1190
    Megareserves 1190
    Forest Reserves 1191
    Restoration Ecology 1192
Ecosystem Services 1193

GLOSSARY G-1

ANSWERS TO SELF-QUIZZES A-1

ILLUSTRATION CREDITS C-1

INDEX I-1

A monarch butterfly feeds on nectar from flowers. She uses some of the energy from the nectar to power her flight and some of it to produce eggs. She lays her eggs on a milkweed plant, the prime food source for the caterpillars that will hatch from the eggs. After feeding, growing, and shedding its skin (under which a new, larger skin has developed), each caterpillar eventually changes into a pupa. The pupa is a nonfeeding stage encased within a protective cocoon held fast to a plant. Within the cocoon drastic reorganization changes the animal, and it emerges as an adult butterfly. At the end of the northern hemisphere's summer, surviving adult butterflies migrate south to traditional wintering areas in California and the mountains of Mexico. A suitable wintering site must have cool temperatures because the butterflies, which do not feed during the winter, must survive on stored food reserves that are used up more slowly at cool temperatures. If temperatures are too cold, however, the insects freeze. Very few places provide winter temperatures that are within the necessary narrow range most of the time. Butterflies that survive the winter migrate north in spring to initiate another annual reproductive cycle. We have no difficulty recognizing these colorful butterflies, and their pupae and caterpillars, as organisms—that is, as things that are alive. But how do we make this judgment?

The traits that lead us to say that certain things are alive are the processes that these things carry out. Monarch butterflies illustrate these processes, which we will give closer attention in order to understand more completely what it means to be alive.

CHARACTERISTICS OF LIVING ORGANISMS

Three processes characterize all organisms: metabolism, regulated growth, and reproduction. **Metabolism** is the sum of the chemical reactions taking place in an organism. All organisms depend on external sources of energy to fuel their chemical reactions. Some organisms—called **autotrophs**, which means "self-feeders"—synthesize their own organic molecules from simple raw materials. They do not need to eat other organisms to sustain themselves. Autotrophs obtain the energy to synthesize complex molecules from sunlight or, in a few cases, from the conversion of some very simple mineral substances (Figure 1.1). The remaining organisms, called **heterotrophs** ("other-feeders"), obtain energy from foods: complex chemical substances that were synthesized by other organisms (Figure 1.2). Heterotrophs break down these substances to release energy and to make the chemical building blocks for synthesizing other substances.

Monarch Butterflies Winter in the Highlands of Mexico

1

The Science of Biology

1.1 Exposed to the Sun
Autotrophs, such as these ferns, horsetails, and grasses growing in Alaska, usually present large surfaces to the sun, the source of their energy.

The metabolism of organisms proceeds well only within narrow ranges of internal physical and chemical conditions. Many of the mechanisms that organisms possess serve to maintain this relative internal constancy even when there are large changes in the surrounding environment. This maintenance of conditions at a steady state is called **homeostasis**. Homeostasis depends on an organism's ability to respond to the environment by changing the rates of its internal reactions or processes—in short, to regulate its metabolism. As we saw at the beginning of this chapter, monarch butterflies regulate their metabolism in winter by seeking places where temperatures remain within a relatively narrow range.

As a result of their metabolic activities, organisms may increase the number of molecules of which they are composed—that is, they grow. This is the second vital characteristic of living organisms: **regulated growth**. Organisms cannot achieve their adult shapes or function effectively unless their growth is carefully regulated. Uncontrolled growth, one example of which is cancer, ultimately destroys life.

Reproduction is the third major process characteristic of organisms. All organisms replicate themselves, but in many cases with variation—there are differences between parents and offspring (Figure 1.3). There are many modes of reproduction. Some organisms reproduce simply by dividing into two daughter cells. Other organisms reproduce by budding off small portions of their bodies to form new individuals. Most large organisms reproduce by means of special cells produced specifically for that purpose. All the information necessary to form a new individual is transmitted via these cells. Although it is a key characteristic of life, you should remember that reproduction is not essential for the survival of

(a)

(b)

1.2 Food from Many Sources
Heterotrophs feed on food substances synthesized by other organisms. *(a)* The African lion eats other animals. *(b)* The black-tailed deer feeds directly on plants.

1.3 Offspring Differ from Their Parents
These nursing puppies are members of a single litter, produced by the white female and fathered by a single male. Genetic variability among the offspring of two parents is the norm.

an organism. In fact, reproduction usually reduces survival rates.

Whatever its effects on the survival of individuals, reproduction with change makes possible the evolution of life and has produced one of the most distinctive features of life: **adaptation**. When we say that an organism is adapted to its environment, we mean that it possesses characteristics that enhance its survival and reproductive success in that particular environment. The caterpillars of monarch butterflies are adapted for feeding on milkweed leaves; the adults are adapted to extract nectar from flowers, to find mates, to find milkweed plants upon which to lay their eggs, and to migrate to suitable wintering areas.

Many adaptations fit organisms to their environments. The wings of birds, for example, are adapted for various types of flight, depending on the specific needs of the bird (Figure 1.4). Camouflaged animals blend into their backgrounds and are difficult for visually hunting predators to locate. Early efforts to explain adaptation implied that some purpose or foresight was involved, and these explanations did not provide testable hypotheses. Today biology proceeds from hypotheses that can be tested by scientific methods. Nearly a century and a half ago, Charles Darwin and Alfred Russel Wallace proposed the first scientifically testable theory about adaptation: evolution by natural selection (see Chapter 19).

THE HIERARCHICAL ORGANIZATION OF BIOLOGY

Biologists study processes ranging from the structure and function of simple molecules to the interactions among the hundreds or thousands of different types of organisms that live together in a particular region, as well as how these organisms evolve. Biology can be visualized as ordered into a hierarchy in which the units, from smallest to largest, are molecules, cells, tissues, organs, organisms, populations, communities, and biomes. These units interact with and adjust to one another as they seek and exchange matter, energy, and information. The organism is the central unit in biological investigations, even when a

(a)

1.4 Wings Adapted for Flight
Most birds use their wings for flight, but different birds fly in different ways. *(a)* The long, broad wings of the red-tailed hawk allow it to sustain a gliding flight above open country while it searches for prey with its keen eyes. *(b)* The action of hummingbird wings allows them to hover in front of flowers while they extract nectar.

(b)

study focuses on a unit distant from the organism in the hierarchy. Today many biologists investigate interactions among molecules and are concerned with processes that are closely related to physics and chemistry. However, when biologists study chemical structures and reactions, they ask different types of questions than chemists do. Biologists study chemical structures and reactions to discover the mechanisms upon which the life of an organism depends. At the higher levels in the hierarchy, biologists study how organisms interact with and adjust to one another to form social systems, populations, ecological communities, and biomes.

From Molecules to Tissues

The smallest entities biologists study are **molecules**, the basic structural units of chemical compounds (see Chapter 2). The principal chemical compounds constituting all living organisms are proteins, nucleic acids, lipids, and carbohydrates (see Chapter 3). Within living organisms many molecules join to form complex aggregates, such as membranes (see Chapter 5). These aggregates have properties that are essential to their functioning within the organism. These properties appear only when the aggregates form; that is, the isolated molecules do not exhibit them.

The **cell** is the fundamental unit of life because it is the simplest unit capable of independent existence and reproduction and because all organisms are composed of cells. Some organisms are single cells. Cells have many features in common (see Chapter 4), but they come in a wide variety of types, many of which are adapted for specific functions, such as secreting substances, storing and transmitting information, or capturing the energy of sunlight. In organisms consisting of many cells, cells of specific types are organized to form **tissues**. Familiar tissues are the muscles of animals and the wood of trees. Tissues may be organized into **organs,** such as skin, hearts, livers, roots, and leaves—structures composed of more than one tissue type.

From Organisms to Biomes

Each creature—each tree, each bacterium, each frog, each person—is an **organism**. Organisms are highly variable and they come in an enormous diversity of forms, to which we may apply names such as wine yeast, robins, garter snakes, howler monkeys, and sugar maple trees. Biologists call the different forms of organisms **species**; but, as we will see in Chapter 20, species are not easy to define. For our present purposes, a species can be defined as a total group of organisms, possibly living in many separate **pop-**

1.5 From Molecules to the Biosphere
The fish—an organism—belongs to a species that is a member of a coral reef community. The fish's molecules are organized into organelles (structures found in or on cells), cells, tissues, and organs, which work together in such a way that the fish can extract food from its environment, avoid its predators, and reproduce. The coral reef community exchanges energy and materials with other communities. Such exchanges unite communities in our biosphere.

ulations, capable of breeding with each other but not with other organisms. Individuals of many different species typically live together and interact to form **ecological communities** (all the different species living in a particular area) and **biomes** (the major types of ecological communities) (Figure 1.5). Together, Earth's biomes form the **biosphere**.

Life's Emergent Properties

Each level of biological organization has properties, called **emergent properties**, that are not found at lower levels. For example, cells and individual organisms exhibit properties that regulate their functioning within strict limits, but the molecules that make up the cells and organisms lack these properties. Individuals are born and they die, but an individual does not have a birth rate and a death rate. A population does. Other emergent properties of populations include age distributions, densities, and distribution patterns. Ecological communities may be described in terms of the number of species in them and the relative abundances of those species.

Emergent properties such as birth rates and species richness that appear at higher levels of organization do not violate principles that operate at lower levels of organization. Usually, however, emergent properties cannot be detected or even suspected from a study at lower levels. Biologists could never discover the existence of such emotions as hate, fear, and love by studying single nerve cells, even though they may eventually be able to explain those emotions in terms of interactions among nerve cells. The properties that emerge at the level of organization at which a biologist works often determine the types of research questions that are most appropriate.

Suppose, for example, a biologist walks along the edge of a pond and startles a frog that jumps into the water (Figure 1.6). Two obvious questions to ask are, *Why* did the frog jump into the water? and *How* did the frog jump? A biochemist might focus on the second question and answer it in terms of the molecular mechanisms underlying muscular contraction. A physiologist might also choose the second question, answering that the muscles in the frog's

Molecule (ATP)

Organelle (mitochondrion)

Cell (neuron)

Tissue (ganglion)

Organ (brain)

Organism (fish)

Biosphere

Population

Community

1.6 Why and How Did the Frog Jump?
Scientists from different disciplines are likely to answer only one of these questions and their answers are likely to be very different.

legs contracted because they were stimulated by motor nerves that synapse with the muscles, and that those nerves, in turn, fired as a result of stimuli initially received by the frog's eyes. A developmental biologist might answer the same question in terms of the embryological development of the neuromuscular wiring that underlies the physiological responses. An ecologist would be more likely to answer the first question and suggest that the frog jumped to escape from a potential predator. Finally, an evolutionist, also answering the first question, might offer some suggestions as to why frogs hop instead of walking and why they hide in water rather than in some protected site on land.

You might think that an act as familiar as a frog's jump would be completely understood, but you would be wrong. The answers biologists can give to either of these two questions at any level in the hierarchy are still incomplete. There is still much to learn about the living world.

Is either of the two questions about the frog's jumping more basic or important than the other? Is any one of the answers more fundamental or more important than the others? Not really. Both questions and all the different answers are essential parts of the full explanation of the jumping of frogs. This richness of answers to apparently simple questions makes biology a complex science, but also an exciting field. In this book, we begin with events and processes at molecular and cellular levels and end with processes at the levels of ecological communities and the biosphere. At all levels we pose both *how* and *why* questions.

THE METHODS OF SCIENCE

Science is a uniquely human activity (on this planet at least). As far as we are aware, no other animal practices science. Science, contrary to much popular opinion, is not merely a collection of facts about the world. Science is a process, a set of ways of discovering things about the universe.

Scientists employ a variety of methods in attempting to understand the structure and functioning of the universe and its components. Diverse methods are necessary because different scientific disciplines study such different subjects. Studying molecules, glaciers, climate, trees, or human social systems—to name only a few objects of scientific study—requires methods appropriate to each subject. Also, within disciplines many different questions can be asked about the same subject, as we saw with the jumping frog. And, not surprisingly, because they are people, scientists differ. They differ in the types of questions they are comfortable asking, in their skills with particular methods, and in how they interpret their results. Scientists agree, however, that the methods they use must be honest. Therefore, scientists worldwide will often accept each other's measurements and data, even though they may disagree on the conclusions to be drawn from those facts. Nonetheless, experiments yielding surprising or especially important data are often repeated by other scientists; the results may become widely accepted only after they have been obtained repeatedly.

Although science employs many methods, one general approach underlies most scientific work. This approach developed slowly over the centuries as scientists realized how it helped them make discoveries. One of its essential elements is that the methods used allow us to modify and correct our beliefs as new evidence becomes available. What is this approach?

The Hypothetico-Deductive Approach

The basic approach underlying most of science, despite the great variety of subjects studied and the methods by which they are investigated, is the **hypothetico-deductive approach**. This procedure has four stages: (1) making observations, (2) forming **hypotheses**, which are tentative answers to questions, (3) making predictions from the hypotheses, and (4) testing the predictions. Testing predictions, in turn, generates new observations, so the cycle continues indefinitely.

The following example shows how biologists apply the hypothetico-deductive approach. Biologists have long known that some caterpillars, such as those of monarch butterflies, are conspicuously colored. Others, such as the caterpillars of peppered moths, are cryptically colored—that is, they blend in with their backgrounds (Figure 1.7). Observations also revealed that conspicuously colored caterpillars often live in groups but are seldom attacked by birds. These initial observations were used to develop hypotheses, make predictions, and devise tests of those predictions. Let's examine how this was done.

GENERATING A HYPOTHESIS. The conspicuous appearance of some caterpillars, together with the observation that potential predators usually avoid them, suggested to some biologists that the bright color patterns of these caterpillars signal to potential predators that the caterpillars are distasteful or toxic. Another hypothesis is that inconspicuous caterpillars are good to eat (palatable) and their coloration reduces the chance that predators will discover and eat them. For each hypothesis of an effect there is a corresponding **null hypothesis** that asserts that the proposed effect is absent. The null hypothesis for the hypotheses we have just stated is that there is no difference in palatability between colorful and cryptic caterpillars.

Notice that these hypotheses depend on certain assumptions or on previous knowledge. We assume, for example, that birds have color vision and can learn about the qualities of their prey by encountering and tasting them. If such assumptions are uncertain, they should be tested before predictions are made from the hypotheses.

MAKING AND TESTING PREDICTIONS. The hypotheses about colorful and cryptic caterpillars led to obvious predictions. Try to formulate them for yourself before you read further. The predictions were tested by presenting captive blue jays with both brightly colored monarch butterfly caterpillars and cryptically colored caterpillars. The blue jays were first deprived of food long enough to make them hungry, so they readily attacked the caterpillars. Ingesting even part of one monarch caterpillar caused a blue jay to vomit. Because the birds were housed individually, the experimenters knew which ones had previously tasted monarchs and which ones had not. They found that a single experience with a monarch caterpillar was enough to cause a blue jay to reject all other monarch caterpillars presented to it. In nature, monarch caterpillars live in groups, so a predator readily learns to avoid all group members after having tasted one. Cryptically colored caterpillars, on the other hand, were readily attacked and eaten, and the jays continued to eat these caterpillars without showing any

(a)

1.7 Caterpillars Can Be Easy or Hard to See
(a) Many caterpillars—like this peppered moth larva, which resembles a small green twig—blend into their surroundings because they are cryptically colored. (b) This larva will become a large tropical moth. Its defensive spines make it particularly conspicuous.

(b)

(a)

1.8   Research Is Essential in Biology
Research and experimentation in biology are carried out in the field and in laboratories. *(a)* Biologists who study the canopies of rainforest trees use special climbing equipment that allows them to collect data and carry out vital studies in the field. *(b)* Some scientists conduct laboratory studies of the properties of potentially dangerous chemicals or other substances that must be kept isolated and protected from contamination before using the substances in field experiments.

signs of sickness or discontent. These results supported the palatability hypothesis. The null hypothesis was thus rejected.

Hypothetico-deductive science uses a variety of methods to test predictions. Among these are laboratory and field experiments and carefully focused observations. Each method has its strengths and weaknesses. The key feature of **experimentation** is the control of most factors that might affect a result so that the influence of the factors that do vary can be seen more clearly. The advantage of working in a laboratory is that control of the environment is easier. Field experiments are more difficult because it is usually impossible to control more than a small part of the total environment. The conditions under which the laboratory experiments with blue jays were run allowed the investigators to reject alternative explanations. Their results, for example, could not have been due to lack of hunger on the part of the birds or their failure to see the caterpillars.

Nonetheless, field experiments have one important advantage. Their results are more readily applicable to what happens where the organisms actually live and evolve. Just because an organism does something in the laboratory does not mean that it behaves the same way in nature. A laboratory experiment demonstrates the *potential* for the organism to act in a certain way, but it does not demonstrate that it will or will not act that way. Because we usually wish to explain nature, not the behavior of organisms in the laboratory, combinations of laboratory and field experiments are needed to test most hypotheses about organismic and higher-level attributes of organisms (Figure 1.8).

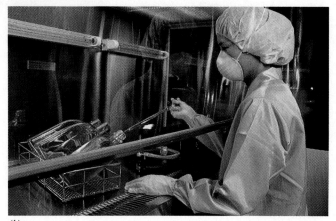

(b)

ACCEPTING A HYPOTHESIS. Scientists may disagree about the adequacy of the evidence in support of a particular hypothesis. Also, different scientists may interpret the same set of observations in different ways. Whereas most new factual discoveries in science are accepted quickly, scientists are slow to accept new *theories*, particularly those that are based on concepts that differ strikingly from ones that are commonly accepted. For example, Isaac Newton's theory that the motions of planets are determined by gravitational attraction was not universally accepted until about 80 years after he first proposed it. About 50 years elapsed between Alfred Wegener's first proposal of the theory of continental drift in 1912 and its general acceptance (see Chapter 28).

At any given moment, there are many hypotheses that are generally accepted as true or false by most scientists in the field. Others are regarded as not yet convincingly confirmed or rejected. The history of science shows us that generally accepted hypotheses are often overturned by newer discoveries. Sometimes hypotheses that have been "convincingly" rejected are resurrected by new discoveries. We can neither prove nor reject any hypothesis with absolute certainty. Nonetheless, the features of the hypothetico-deductive method allow us to develop a high degree of confidence in our interpretations of how the world works.

A single piece of evidence supporting a hypothesis rarely leads to its widespread acceptance, just as a single contrary result rarely leads to the abandonment of a hypothesis. Negative results can be obtained for many reasons, only one of which is that the hypothesis is wrong. For example, the error may be that incorrect predictions were made from a correct hypothesis. A negative result can also occur because of poor experimental design or because an inappropriate organism—one that does not fit the assumptions of the hypothesis—is chosen for the test. For example, a predator lacking color vision, or one that uses primarily its sense of smell, would not be appropriate for testing hypotheses about the colors of caterpillars.

A general textbook like this one is based on hypotheses that have been extensively tested and that are generally accepted. An extensively tested hypothesis is sometimes called a theory, but there is no rule that states how much evidence must be gathered before a hypothesis earns the status of a theory. When possible in this text, we illustrate hypotheses and theories with observations and experiments that support them, but we cannot, because of space constraints, detail all the evidence. Remember, as you read, that statements of biological "fact" are mixtures of observations, predictions, and interpretations.

Experimentation and Animals

Obtaining answers to many of the questions posed by biologists requires experimenting with plants, animals, fungi, and microorganisms. To study the antipredator adaptations of caterpillars, investigators had to keep blue jays in cages, make them hungry by depriving them of food, and feed them caterpillars. This procedure resulted in the deaths of some caterpillars and temporary stress for the jays. Determining which chemicals make the caterpillars toxic required the deaths of more caterpillars.

No amount of observation without intervention could possibly substitute for experimental manipulation. This does not mean, however, that scientists are insensitive to the welfare of the organisms with which they work. Most scientists who work with animals are continually alert to finding ways of getting answers that use the smallest number of experimental subjects and that cause the subjects the least pain and suffering.

WHY PEOPLE DO SCIENCE

Describing *how* people do science does not explain *why* they do it. The most important motivator of most biologists is curiosity. People are fascinated by the richness and diversity of life and want to learn more about organisms and how they function. Curiosity is an adaptive trait. Humans who were not curious about their surroundings and, therefore, were not motivated to learn about them, probably survived and reproduced less well, on average, than their more curious relatives. We hope this book will help you share the excitement biologists feel as they learn more about living organisms, develop hypotheses, and test them. Who knows, perhaps your curiosity will lead to an important new idea.

SIZE SCALES

Multicellular organisms are composed of many types of cells that perform different functions. Among the benefits of multicellularity are improved protection, the possibility of evolving into any of a wide variety of shapes, and the possibility of evolving to be large. Why can't evolution produce large unicellular (one-celled) organisms? A major reason relates to the changes in the surface-to-volume ratio of an object as it increases in size. A cubic cell measuring 100 micrometers (μm) along each edge has a volume (and mass) 1,000 times that of a cell measuring 10 μm along each edge, but its surface area is only 100 times greater than that of the smaller cell. Thus the surface-to-volume ratio of the smaller cell is greater than that of the larger cell (Figure 1.9).

How does the surface-to-volume ratio of a cell influence its functioning? As a cell metabolizes, it must exchange materials and heat with its environment. Although the number of chemical reactions a cell can carry out per unit time is proportional to its volume, its rate of exchange of nutrients and waste products with its environment is limited by its *surface area*. As a cell grows larger, its rate of production of wastes and its need for resources increase faster than the surface through which the cell must obtain resources and void its wastes. Substantial increases in overall sizes of organisms require the development of structures that transport food, oxygen, and waste materials to and from parts that are distant from their external surfaces. Multicellularity, with specialized cells to provide these functions, is what enabled large organisms to evolve.

Because the heat generated by the chemical reactions within an organism must be dissipated for the organism to survive, large animals have lower metabolic rates than small animals have. In total quantity, an elephant requires more food per day than a mouse

Surface area = 60,000 μm²
Volume = 1,000,000 μm³
Surface-to-volume ratio = 0.06

Surface area = 600 μm²
Volume = 1,000 μm³
Surface-to-volume ratio = 0.6

100 μm

10 μm

1.9 Small Objects Have Large Surface Areas

A cube 10 μm along each side has a much greater surface area in relation to its internal volume than a cube 100 μm along a side. Although the single large cube has the same volume as 1,000 of the small cubes, 1,000 small cubes have 10 times more surface area.

does; but gram for gram, the mouse requires more. Thus the metabolic rate of the mouse must be faster than that of the elephant. If the elephant had the same metabolic rate as the mouse, it would not be able to dissipate the heat produced by metabolism fast enough to avoid cooking itself! Conversely, a large animal does not cool off as rapidly as a small one. Thus a large lizard can heat its body in a warm area and then forage for a longer time in a cool place than a small lizard can.

Other properties of organisms also change with size. For example, the weight of an animal is related to its volume and is proportional to the cube of its linear dimensions. The legs of an organism, however, must be able to support its weight. The strength of bones is proportional to their cross-sectional area, which in turn is proportional to the square of their lengths. Therefore, an increase in size must be accompanied by a proportionally greater increase in leg diameter. Delicate, slender legs are characteristic of

lightweight impalas, not of heavy elephants (Figure 1.10).

Many ecological interactions among organisms are strongly affected by size. Size determines the food an organism can use, where it can hide, what it can mimic, the area it requires to obtain its food, and its abundance. Not surprisingly, then, biologists pay a great deal of attention to the distributions of sizes among organisms.

TIME SCALES

When Galileo studied the motion of a ball rolling down an inclined plane 375 years ago, he used his pulse to mark off equal intervals of time because no better device existed for measuring time. The rise of modern science has depended upon the invention of instruments capable of measuring time accurately. Scientists need to measure time spans both longer

1.10 Proportions Change with Size

Elephants have proportionally much more massive legs than do the slender impalas. An animal the size of an elephant with legs the shape of an impala's would collapse under its own weight.

and shorter than those we can perceive accurately with our unaided senses. In biology, the longer time spans have caused the greatest conceptual difficulties. Earth is a very old planet; the difficulty in perceiving its age delayed the recognition of evolutionary change for a long time.

Today, scientists measure long time spans by using naturally radioactive materials (see Chapter 18). Certain radioactive materials are incorporated into rocks and other materials when they are formed. As time passes, the radioactivity in these materials decreases at a regular rate. Scientists compare the proportions of radioactive and corresponding nonradioactive materials in particular rocks. From the observed ratio, they can estimate how long ago the rocks formed. This method gives us *absolute* ages. Indirect observations, such as the vertical positions of rock layers in relation to one another, help us assess the *relative* ages of materials. Young rocks lie on top of older ones (unless the rocks have been subjected to dramatic deformations, which are usually evident). By studying the remains of living things found in different layers of rock and correlating their distributions across many sites, we can determine the relative ages of different deposits even if absolute ages are not known.

Biologists working at different levels of organization may think about and study problems in different time scales. Biochemists and physiologists are concerned primarily with the time span relevant to the intervals required for chemical reactions and physiological changes within an organism. These intervals range from fractions of seconds to a day or a year. Studies of physiological processes such as aging may require observations extending over the lifetime of organisms, which may range up to centuries for some long-lived plants.

Studies of populations may extend over many generations of the organisms. The time span of generations ranges from less than an hour for some bacteria to centuries for some plants, but for any particular type of organism, the time required for changes in the sizes and distributions of populations is much longer than the time required for physiological changes in single individuals.

The study of changes in the genetic constitution of populations requires us to think in terms of microevolutionary time. Significant genetic changes usually take many generations, but they can happen quite rapidly when environmental conditions change abruptly. Organisms often change more substantially over spans of macroevolutionary time, covering thousands of generations or more. We can often observe microevolutionary changes directly, but we must measure macroevolutionary changes indirectly.

MAJOR ORGANIZING CONCEPTS IN BIOLOGY

Knowledge about organisms can be organized in many ways, and we use several methods in this book. In some chapters we look at the molecules of which organisms are composed. In other chapters we look at different groups of organisms and see how their structures, activities, and adaptations are related to their particular lifestyles. Other chapters focus on major processes carried out by organisms, such as digestion, movement, and reproduction. In still other chapters, we focus on the mechanisms of evolutionary change. Through all of the discussions, however, several major organizing concepts guide us.

The first of these concepts is that *all properties of organisms can be explained in physical and chemical terms.* Organisms appear to be triumphs of organic chemistry. The most complex biological activities, including the mysterious richness of our human emotions, are probably manifestations of underlying physicochemical systems. Because we use this concept, however, does not imply that we can understand all complex biological phenomena simply through the study of chemistry. Rather, it implies that biological phenomena must conform to the laws of physics and chemistry. Not too many years ago the idea that the properties of organisms resulted from physical and chemical interactions was vigorously debated, and lengthy books were written attacking and defending it. Today, however, most biologists accept the chemical basis of life and are attempting to identify the physical and chemical processes underlying specific biological phenomena. Some basic chemistry and biochemistry is presented in Chapters 2 and 3.

The second key organizing concept in biology came from physics and is a refinement of the understanding of energy. *Organisms can be viewed as systems that take in energy from their environments and convert it to biologically useful forms* (see Chapters 7 and 8). The notion of energy is highly abstract. Energy is weightless and occupies no space, yet it exists in many forms. Using experimentally derived formulas, we can calculate the equivalence of these forms. Energy can never be created or destroyed, but it can be converted from one form to another. If energy is weightless and occupies no space, how can we measure it and use it in meaningful ways? We measure energy by its *effects upon matter*, which can be weighed and measured.

The third organizing concept in biology is that *genetic information encodes and transmits information between generations.* You are probably familiar with this concept, having already heard many times that char-

1.11 Hooke's Microscope and What He Saw
The microscope used by Robert Hooke. In the circle are Hooke's drawings of the empty plant cells found in the bark of a cork oak.

acteristics are transmitted from parents to their offspring by a genetic code, but widespread acceptance of the concept came only in this century. The discovery of the molecules that contain the hereditary information is even more recent (see Chapter 11).

The fourth organizing concept in biology, called the **cell theory**, states that *organisms are composed of cells* and that *the cell is therefore the basic building block of life*. Cells differ greatly in size and in the complexity of their internal structures, but all cells have certain characteristics that enable us to discuss many of the properties of organisms in terms of the way cells perform work and organize themselves. Viruses, which may or may not be considered alive, are not cells; they are too simple to survive without using the machinery of cells.

Magnifying devices existed for hundreds of years before anyone thought to look carefully at organisms with them. It was not until 1665 that the Englishman Robert Hooke noticed that cork, wood, and other plant tissues are made up of small, regularly shaped cavities surrounded by walls (Figure 1.11). Hooke called these cavities cells. Living, single-celled organisms were first observed a few years later by the Dutch naturalist Anton van Leeuwenhoek, who used a simple microscope of his own design. What these two men had observed was not fully appreciated for a long time. The first strong statement that *all organisms consist of cells* was made by the German physiologist Theodor Schwann in 1839. In 1858 the German physician Rudolf Virchow suggested that *all cells*

come from preexisting cells. Experiments by the French chemist and microbiologist Louis Pasteur between 1859 and 1861 provided generally accepted proof of this assertion. Since then the cell theory, as summarized by the two statements in the previous paragraph, has been a basic principle of biology.

The fifth organizing concept in biology is that *evolution by natural selection results in adaptation*. We will consider the mechanisms of evolution in detail in Part Three, but you need some knowledge of the mechanisms to understand material in chapters that precede Part Three. Fortunately, even though the details are complex, the basic principles are simple.

EVOLUTIONARY CONCEPTS

For more than 200 years biologists have suspected that the organisms they observed evolved from types unlike those currently living. In the 1760s, the French naturalist Count George-Louis Leclerc de Buffon wrote his *Natural History of Animals*, which contained a clear statement of the possibility of evolution. Buffon originally believed that all organisms had been specially created for different ways of life, but as he studied animals he observed that the limb bones of all mammals, no matter what their way of life, are remarkably similar in many details (Figure 1.12). If these limbs were specifically created for different ways of locomotion, Buffon reasoned, they should have been built upon different plans rather than being modifications of a single plan. He also noticed that the legs of certain animals, such as pigs, have

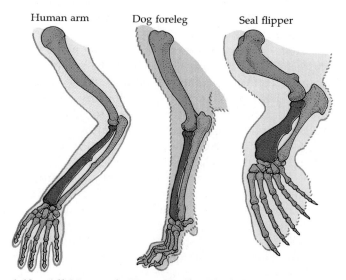

1.12 All Mammals Have Similar Limb Bones
Mammalian forelimbs have different purposes—humans use theirs for manipulating objects, dog forelimbs are used for walking on, and seals for swimming—but the number and type of their bones are similar. This shows bones of the same type in the same color.

toes that never touch the ground and appear to be of no use. Buffon found it difficult to explain the presence of these seemingly useless small toes by special creation. Both of these troubling facts could be explained if mammals had not been specially created in their present forms but had been modified from a common ancestor. Buffon therefore suggested that pigs have two functionless toes because they inherited them from ancestors in which the toes were fully formed and functional.

Buffon's student Jean Baptiste de Lamarck wrote extensively about evolution. Lamarck was the first person to support the idea of evolution with logical arguments. He was also the first person to propose a hypothesis concerning the mechanisms of evolutionary change. He suggested that lineages of organisms may change gradually over many generations as offspring inherit structures that have become larger and more highly developed as a result of continued use or, conversely, have become smaller and less developed as a result of disuse. For example, Lamarck suggested that aquatic birds extend their toes while swimming, stretching the skin between them. This stretched condition, he thought, could be inherited by the offspring, who would further stretch their skin during their lifetimes and would also pass this condition along to their offspring. According to Lamarck, birds with webbed feet would thereby evolve over a number of generations (Figure 1.13). He explained many other examples of adaptations in a similar way and showed how many domestic plants and animals have departed from the forms of their wild ancestors. We do not now believe that evolutionary changes are produced by the mechanisms proposed by Lamarck. However, Lamarck's ideas deserved more attention than they received from his contemporaries, most of whom believed in a young and unchanging universe.

1.14 Charles Darwin as a Young Man

By 1858 the climate of opinion, among biologists at least, was receptive to the theory of evolutionary processes proposed independently by Charles Darwin (Figure 1.14) and Alfred Russel Wallace. By then geologists had shown that Earth had changed over millions of years, and many people were willing to think in terms of longer time spans. Thus, the presentation in the latter half of the nineteenth century of a well-documented and thoroughly scientific argument for evolution triggered a transformation of biology.

Darwin's Scientific Study of Evolution

Charles Darwin's approach to evolution incorporated the following hypotheses: (1) Earth is very old, and organisms have been changing steadily throughout the history of life. (2) All organisms are descendants of a common ancestor—that is, life arose only once on Earth. (3) Species multiply by splitting into daughter species, and such speciation has resulted in the great diversity of life found on Earth. (4) Evolution proceeds via gradual changes in populations, not by the sudden production of individuals of dramatically different types. (5) The major agent of evolutionary change is natural selection. Remarkably, these five hypotheses have all been supported by the mass of research that has been conducted since Darwin published his book *The Origin of Species* in 1859.

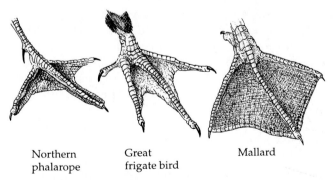

Northern   Great   Mallard
phalarope   frigate bird

1.13 Partially to Fully Webbed
Lamarck believed that the offspring of these birds, all of which stretch their feet while swimming, would inherit this stretched condition and eventually have webbed feet like the mallard's. Scientists today generally agree that the mechanism proposed by Lamarck does not explain evolutionary changes.

Darwin used some facts that were familiar to most of his fellow biologists. He knew that populations of all species, even those with very low reproductive rates, have the potential for exponential increases in numbers. Yet such rates of increase are rarely achieved in nature; populations of most species are relatively stable through time. Therefore, death rates in nature must be high (Figure 1.15). Without high death rates, even the most slowly reproducing forms would quickly reach enormous population sizes.

Darwin also noted that, although offspring tend to resemble their parents, the individuals of most types of organisms are not identical. He suggested that slight variations among individuals significantly affect the chance that a given individual will survive and reproduce. He called this differential reproductive success of individuals **natural selection**, probably because he was deeply interested in the artificial selection practices of animal and plant breeders. Indeed, many of Darwin's observations on the nature of variation came from domesticated plants and animals. Because Darwin himself was a pigeon breeder, he recognized close parallels between artificial selection by breeders and selection in nature.

Darwin argued his case for natural selection as follows:

> How can it be doubted, from the struggle each individual has to obtain subsistence, that any minute variation in structure, habits or instincts, adapting that individual better to the new conditions, would tell upon its vigour and health? In the struggle it would have a better chance of surviving; and those of its offspring which inherited the variation, be it ever so slight, would have a better chance. Yearly more are bred than can survive; the smallest gain in the balance, in the long run, must tell on which death shall fall, and which shall survive. Let this work of selection on the one hand, and death on the other, go on for a thousand generations, who will pretend to affirm that it would produce no effect, when we remember what, in a few years animal breeders effected in cattle, and . . . in sheep, by the identical principle of selection?

That statement, written more than 100 years ago, still stands as a good expression of the idea of evolution by natural selection.

The Importance of a World View

Biologists develop hypotheses, devise tests, and modify their ideas in the light of observational and experimental results, but they carry out these activities within a broader framework. All of us, whether scientists or not, operate within the framework of a general world view, which is sometimes called a **paradigm**. The paradigm determines which problems are interesting and which are not, and it strongly influences the responses we make to information that seems to oppose it.

Biology began a major paradigm shift a little over a century ago with the general acceptance of long-

1.15 Many Organisms Have High Reproductive Rates
A frog surrounded by the eggs it has laid. If all the offspring of frogs such as this one grew to adulthood, the world's frog population would be overwhelming. Many of these eggs will not survive, however. A rate of reproduction as high as this frog's is usually accompanied by a high mortality rate of eggs and young.

term evolutionary change and the recognition that natural selection is the primary agent producing adaptation. The shift took a long time because it required abandoning many components of a different world view. In the pre-Darwinian view, the world was thought to be a young one in which organisms had been created in their current forms. In the Darwinian view, the world is an ancient one in which both Earth and its inhabitants have been evolving from forms very different from the ones they now have. It is a world in which we would not recognize many organisms if we were transported far back into the past or far forward into the future. Accepting this paradigm means accepting not only the processes of evolution, but also the view that the living world is constantly evolving, but without any "goals." The idea that evolutionary change is not directed toward a final goal or state has been more difficult for some people to accept than the process of evolution itself.

LIFE'S SIX KINGDOMS

Perhaps as many as 30 million species of organisms inhabit Earth today. Many times that number lived in the past but are now extinct. To classify this past and present diversity of organisms, biologists have devised systems that reflect the evolutionary history of life. The details of the criteria biologists use today to classify organisms are given in Chapter 21, but because some key terms are used in the intervening chapters, we introduce the broad categories of the system here.

In the classification system used in this text, organisms are grouped into six large categories called **kingdoms** (Figure 1.16). Organisms belonging to the kingdoms **Archaebacteria** and **Eubacteria** have distinctive cells referred to a **prokaryotic** ("prenuclear") because they lack a nucleus as well as some of the other internal structures found in the cells of members of other kingdoms. Prokaryotes are exceedingly abundant and are found everywhere on Earth where life can exist. Some are critical links in the biogeochemical cycles essential to all life (see Chapter 49).

More than a billion years ago, some prokaryotes invaded the cells of others. Over time, this relationship between hosts and parasites gave rise to organisms with structurally more complex **eukaryotic** cells. Eukaryotes are divided into four kingdoms, the first of which, the kingdom **Protista** (protists), includes many single-celled organisms. The remaining three kingdoms, whose members are all multicellular organisms, are believed to have arisen from ancestral protists. The kingdom **Fungi** includes molds, mushrooms, yeasts, and other similar organisms. Fungi absorb food substances from their surroundings and digest them within their cells. Many are important as decomposers of the dead bodies of other organisms. Most members of the kingdom **Plantae** (plants) convert light energy to the energy of chemical bonds by photosynthesis (see Chapter 8). The biological molecules they synthesize are the primary food for nearly all other living organisms. The kingdom **Animalia** (animals) consists of organisms that digest food outside their cells, and then absorb the products. Animals depend on other forms of life for most of their materials and energy.

Sometimes organisms are referred to as "primitive" or "advanced." These terms can be useful when we are comparing ancestral and derived traits of organisms in the same evolutionary lineage. But they, and similar terms such as "lower" and "higher," are inappropriate in contexts where they imply that some organisms "work better" than others. The abundance of prokaryotes—all of which are relatively simple—readily demonstrates that they are highly functional, despite their simplicity.

SCIENCE AND RELIGION

Understanding the methods of science enables us to distinguish science from that which is not science. Recently some people have claimed that "creation science," sometimes called "scientific creationism," is a legitimate science that deserves to be taught in schools together with the evolutionary view of the world presented in this book. In spite of these claims, creation science is not science. Science begins with observations and the formulation of testable hypotheses. Creation science begins with the unsubstantiated assertion that Earth is only about 4,000 years old and that all species of organisms were created in approximately their present forms. This assertion is not presented as a hypothesis from which testable predictions are derived. Advocates of creation science do not believe that tests are needed because they assume the assertion to be true.

In this book we present evidence supporting the hypothesis that Earth is several billion years old, that today's living organisms have evolved from single-celled ancestors, and that many organisms dramatically different from those we see today lived on Earth in the remote past. All of this extensive scientific evidence is rejected by proponents of creation science in favor of a religious belief held by a very small minority of the world's human population. Evidence gathered by scientific procedures does not diminish the value of the biblical account of creation. Religious beliefs are not based on falsifiable hypotheses, as science is; they serve different purposes, giving meaning and guidance to human lives. The legitimacy of both religion and science is undermined when a religious belief is called science.

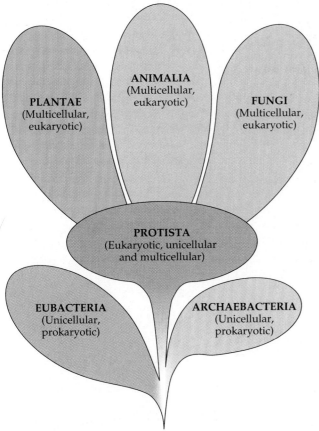

1.16 Six Kingdoms
In the classification system used in this book, Earth's organisms are first divided into these six groups, called kingdoms.

FOR STUDY

1. Why is it so important in science that we design and perform tests capable of rejecting a hypothesis?

2. Some philosophers and practitioners of science believe that it is impossible to prove any scientific hypothesis —that instead we can only fail to find a cause to reject it. Evaluate this view. Can you think of a reason that we can be more certain about rejecting a hypothesis than we can about accepting it?

3. One hypothesis about the conspicuous coloration of caterpillars was described in this chapter, and some tests were mentioned. Suggest some other plausible hypotheses for conspicuous coloration in these animals. Develop some critical tests for one of these alternatives. What are the appropriate associated null hypotheses?

4. According to the theory of adaptation, an organism evolves certain features because they improve the chances that the organism will survive and reproduce. There is no evidence, however, that evolutionary mechanisms have foresight or that organisms can anticipate future conditions. What, then, do biologists mean when they say, for example, that wings are "for flying"?

5. Consider a single-celled organism. Explain why it is not feasible for this organism to grow to a size of 10 cm in diameter. Cover the following topics: (1) surface-to-volume ratio; (2) transport of nutrients; (3) gas exchange; (4) excretion; and (5) support.

READINGS

Darwin, C. 1859. *The Origin of Species by Means of Natural Selection*. John Murray, London. The book that set the world to thinking about evolution; still well worth reading. Many reprinted versions are available.

Irvine, W. 1955. *Apes, Angels, and Victorians*. McGraw-Hill, New York. A delightful account of the reactions of English society to the theory of evolution by means of natural selection.

Kuhn, T. S. 1970. *The Structure of Scientific Revolutions*, 2nd Edition. University of Chicago Press, Chicago. A widely discussed book that developed a view of science as a succession of paradigms.

Margulis, L. and K. V. Schwartz. 1988. *Five Kingdoms: An Illustrated Guide to the Phyla of Life on Earth*, 2nd Edition. W. H. Freeman, New York. A good introduction to the kingdoms of organisms, in which the two kingdoms of prokaryotes are united into one. Excellent examples and illustrations.

Mayr, E. 1991. *One Long Argument: Charles Darwin and the Genesis of Modern Evolutionary Thought*. Harvard University Press, Cambridge, MA. An excellent, concise account of the history of evolutionary thinking during the past century.

National Academy of Sciences. 1984. *Science and Creationism. A View from the National Academy of Sciences*. National Academy Press, Washington, D.C. A good summary of the evidence demonstrating the great age of Earth and the evolutionary processes that generated the modern Earth.

Young, J. Z. 1951. *Doubt and Certainty in Science*. Clarendon Press, Oxford. A discussion of how scientists develop confidence in their theories.

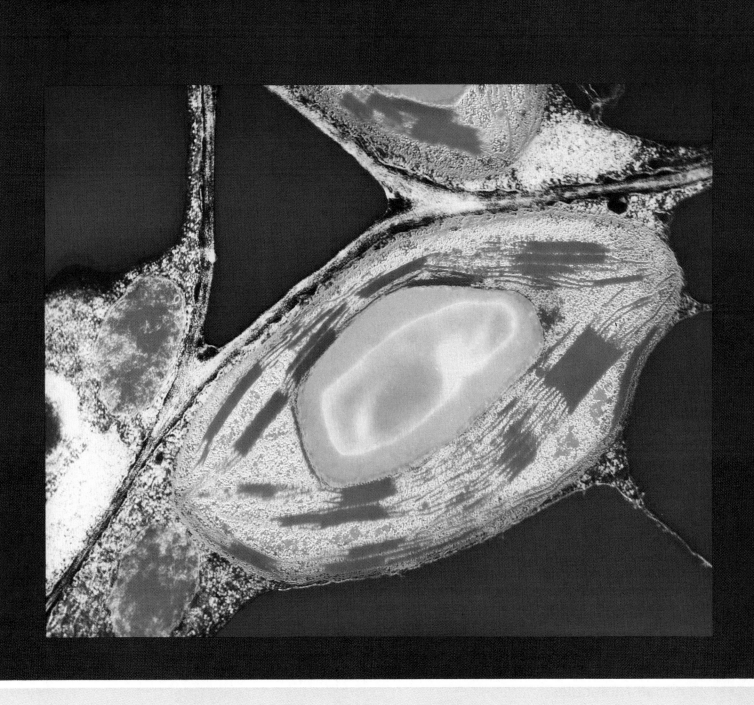

PART ONE
The Cell

2 Small Molecules
3 Large Molecules
4 Organization of the Cell
5 Membranes
6 Energy, Enzymes, and Catalysis
7 Pathways that Release Energy
   in Cells
8 Photosynthesis

What do you have in common with a rock? Like a rock, you have more oxygen in your molecules than anything else (you are about 65 percent oxygen, the rock about 47 percent). Inside both you and the rock, electrons spinning in orbits bind these oxygen atoms to other atoms. Electrons and other particles in the atoms determine their behavior and thus, ultimately, your behavior and that of the rock.

How do you differ from a rock (chemically, at least)? The rock may be about 28 percent silicon, but you have none, unless you have received a breast implant. Your body is 18 percent carbon—the second most common element in all living things—but the rock may contain no carbon. Probably the biggest difference between you and a rock is the number and speed of the chemical reactions going on inside.

Still, perhaps you have more in common with a rock than you think. What is Earth's matter—animal, vegetable, and mineral—composed of? What holds the units of matter together, what causes them to come apart, and how and why do they exchange parts?

In Part One of this book we begin with atoms (the lowest level of the hierarchical organization of biology) and travel up to the level of the cell. In this chapter we study the small molecules formed by the combination of atoms. Chapter 3 continues with the formation of large molecules from small molecules.

ATOMS

All matter, living and nonliving, is composed of **atoms**. More than a million million atoms could fit in a single layer over the period at the end of this sentence. Each atom consists of a dense, positively charged nucleus, around which one or more negatively charged electrons move. The nucleus contains one or more protons and may contain one or more neutrons. Electrons, protons, and neutrons are not indivisible particles. Each has a substructure, but that level of organization (the world of quarks) has no known consequence for biology.

The weight, or mass, of a proton serves as a standard unit: the atomic mass unit (amu) or dalton (after the English scientist John Dalton, who studied atoms two centuries ago). A single proton or neutron weighs one dalton, which is 1.7×10^{-24} gram (0.0000000000000000000000017 g), whereas the mass of an electron is 9×10^{-28} g (0.0005 dalton). Because they weigh so much less than protons and neutrons, electrons contribute only negligibly to the mass of an atom.

The positive electric charge on a proton is defined as a unit of charge. An electron has a charge equal and opposite to that of a proton. Thus the charge of

2

Small Molecules

a proton is +1 unit, that of an electron is −1. The neutron, as its name suggests, is electrically neutral, so its charge is 0. The number of protons in an atom equals the number of electrons, so the atom itself is electrically neutral.

ELEMENTS

An **element** is a substance made up of only one kind of atom. The element hydrogen consists only of hydrogen atoms; the element iron consists only of iron atoms. An element cannot be broken down into simpler substances. Although there are 92 or more different chemical elements in the world, more than 99 percent of the living matter of all organisms (including humans) is composed of just six elements—carbon, hydrogen, nitrogen, oxygen, phosphorus, and sulfur.

In addition to the 92 elements that occur naturally on Earth, physicists have produced other elements by using cyclotrons and other particle accelerators. Of the natural elements, some, such as silver, gold, and thulium, are extremely rare; others, such as hydrogen, nitrogen, and oxygen, are abundant on this planet. Table 2.1 compares the proportions of some representative elements in the human body, in Earth's crust, and in the universe as a whole. Some elements, such as silicon—a major component of rocks and soils—are abundant in Earth's crust but are not present in significant concentration in living things.

How Elements Differ

An atom is identified by the number of its protons, which is unchanging. This number is called the **atomic number**. An atom of hydrogen contains 1 proton; an atom with 2 protons would be helium; carbon has 6 protons and plutonium has 94. Their atomic numbers are thus 1, 2, 6, and 94, respectively.

Every atom except hydrogen has one or more neu-

TABLE 2.1
Abundance of Some Chemical Elements

ATOMIC NUMBER	SYM-BOL	ELEMENT	ATOMIC WEIGHT	ABUNDANCE, AS % OF			ATOMIC NUMBER	SYM-BOL	ELEMENT	ATOMIC WEIGHT	ABUNDANCE, AS % OF EARTH'S CRUST
				UNI-VERSE	EARTH'S CRUST	HUMAN BODY					
1	H	Hydrogen	1.01	87	0.14	9.5	28	Ni	Nickel	58.70	0.01
2	He	Helium	4.00	12			29	Cu	Copper	63.55	0.01
3	Li	Lithium	6.94		0.01		30	Zn	Zinc	65.38	0.01
4	Be	Beryllium	9.01				31	Ga	Gallium	69.72	
5	B	Boron	10.81				32	Ge	Germanium	72.59	
6	C	Carbon	12.01	0.03	0.03	18.5	33	As	Arsenic	74.92	
7	N	Nitrogen	14.01	0.01		3.3	34	Se	Selenium	78.96	
8	O	Oxygen	16.00	0.06	46.6	65.0	35	Br	Bromine	79.90	
9	F	Fluorine	19.00		0.03		36	Kr	Krypton	83.80	
10	Ne	Neon	20.18	0.02			37	Rb	Rubidium	85.47	0.03
11	Na	Sodium	22.99		2.83	0.2	38	Sr	Strontium	87.62	0.03
12	Mg	Magnesium	24.31		2.09	0.1	39	Y	Yttrium	88.91	
13	Al	Aluminum	26.98		8.13		40	Zr	Zirconium	91.22	0.02
14	Si	Silicon	28.09		27.7		41	Nb	Niobium	92.91	
15	P	Phosphorus	30.97		0.12	1.0	42	Mo	Molybdenum	95.94	
16	S	Sulfur	32.06		0.05	0.3	43	Tc	Technetium	98.91	
17	Cl	Chlorine	35.45		0.03	0.2	44	Ru	Ruthenium	101.07	
18	A	Argon	39.95				45	Rh	Rhodium	102.91	
19	K	Potassium	39.10		2.59	0.4	46	Pd	Palladium	106.40	
20	Ca	Calcium	40.08		3.63	1.5	47	Ag	Silver	107.87	
21	Sc	Scandium	44.96				48	Cd	Cadmium	112.41	
22	Ti	Titanium	47.90		0.44		49	In	Indium	114.82	
23	V	Vanadium	50.94		0.02		50	Sn	Tin	118.69	
24	Cr	Chromium	52.00		0.02		51	Sb	Antimony	121.75	
25	Mn	Manganese	54.94		0.1		52	Te	Tellurium	127.60	
26	Fe	Iron	55.85		5.0		53	I	Iodine	126.90	
27	Co	Cobalt	58.93				54	Xe	Xenon	131.30	

This list contains only the first 54 of the 92 natural elements. Elements found in the human body are divided into three categories on the basis of abundance: ■, most abundant; ▨, 0.1–3.3 percent; □, trace (less than 0.01 percent).

trons in its nucleus. The **mass number** of an atom equals the sum of the number of protons and neutrons in its nucleus. Because the mass of an electron is infinitesimal compared with that of a neutron or proton, electrons are ignored in calculating the mass number. The nucleus of a helium atom contains 2 protons and 2 neutrons; oxygen has 8 protons and 8 neutrons. Helium, therefore, has a mass number of 4 and oxygen a mass number of 16. The mass number may be thought of as the weight of the atom, in daltons.

Each element has its own one- or two-letter symbol. Thus H = hydrogen, He = helium, and O = oxygen. Some symbols come from other languages: Fe (from Latin *ferrum*) = iron, Na (Latin *natrium*) = sodium, and W (German *Wolfram*) = tungsten. Table 2.1 lists the symbols for 54 of the 92 natural elements. Sometimes the atomic and mass numbers of an element are written with the element's symbol, the atomic number at its lower left, and the mass number at its upper left. Thus hydrogen, carbon, and oxygen are written as ^1_1H, $^{12}_6C$, and $^{16}_8O$.

Isotopes

We have been speaking of hydrogen and oxygen as if each had only one form. To be precise, we should have said "the common form of" oxygen or hydrogen because not all atoms of the same element have the same mass number. The common form of hydrogen is ^1H, but about one out of every 6,500 hydrogen atoms on Earth has a neutron as well as a proton in its nucleus and is thus ^2H, called deuterium. Furthermore, it is possible to create ^3H, tritium, which has *two* neutrons and a proton in its nucleus. Because all three types of hydrogen atoms have just one proton, however, they all have the atomic number 1. Deuterium, tritium, and common hydrogen have virtually identical properties, although ^2H is twice and ^3H three times as heavy as ^1H. Such multiple forms of a single element are called **isotopes** of the element (Figure 2.1). (The prefix *iso-*, encountered in many technical terms, means "same.")

Many elements exist in several isotopic forms in nature. For example, the natural isotopes of carbon are ^{12}C, ^{13}C, and ^{14}C. Unlike the hydrogen isotopes, the isotopes of most elements do not have distinct names but rather are written in the form shown here and are spoken of as "carbon-12," "carbon-13," and "carbon-14," respectively. Most carbon atoms are ^{12}C, but about 1.1 percent are ^{13}C, and a tiny fraction are ^{14}C. An element's **atomic weight** is the average of the mass numbers of a representative sample of atoms of the element, with all isotopes in their normal proportions. For example, the atomic weight of carbon is 12.011.

Some isotopes are radioactive: they spontaneously

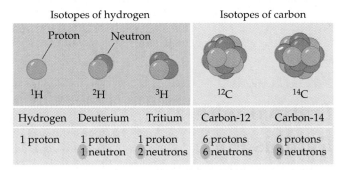

Isotopes of hydrogen			Isotopes of carbon	
^1H	^2H	^3H	^{12}C	^{14}C
Hydrogen	Deuterium	Tritium	Carbon-12	Carbon-14
1 proton	1 proton 1 neutron	1 proton 2 neutrons	6 protons 6 neutrons	6 protons 8 neutrons

2.1 Isotopes Have Different Numbers of Neutrons

give off energy or subatomic particles. Such isotopes are called **radioisotopes**. Radioactive decay transforms the original atom into another type, usually of another element. The radioisotope $^{14}_6C$, for example, is converted to $^{14}_7N$: Carbon becomes nitrogen by the emission of an electron as a neutron becomes a proton. In biological research, radioisotopes are employed as tracers of biochemical reactions. Some commonly used radioisotopes are tritium (^3H), ^{14}C, and ^{32}P (phosphorus-32).

Liquid scintillation counting is the most common method of measuring the amount of radioactivity in a sample. The sample is added to a solution containing a substance that emits light (scintillates) when it absorbs the products of radioactive decay. Like a light meter, the liquid scintillation counter measures the amount of light emitted (Figure 2.2). The amount of light detected corresponds directly to the amount of radioactive decay in the sample. Emitted particles can also be detected by imaging techniques such as **autoradiography**, which is described in Box 2.A.

Rare, nonradioactive isotopes, such as deuterium, do not decay and can be detected only by their dif-

2.2 Measuring Radioactivity by Liquid Scintillation Counting
The machine shown here accepts hundreds of samples, each in its radiation-sensitive "cocktail," and lowers each sample in turn into a darkened well, where its emitted light is measured.

BOX 2.A

Probing an Embryo with a Radioisotope

One of the most exciting and puzzling of biology's phenomena is development—the complex series of progressive changes in an organism from its beginning as a single cell to its adult form. Radioisotopes are useful tools with which to tackle questions about development. They have been extensively used in experiments on the embryos of fruit flies.

An adult fruit fly's body is made up of 13 segments, each with a specific function. How do the different segments form and take on their dis-

tinct functions? We suspect that different chemical substances act during the course of the fly's development to organize the segments. To test this idea, biologists can label a particular chemical substance with radioisotopes and add it to the fly embryo; then they see where in the embryo the substance accumulates. They wash the embryo, press it against X-ray film, and leave the embyro and film in the dark for several days. Wherever a radioactive atom decays, the emitted particle exposes the film. When the film is developed, silver grains—seen in the resulting **autoradiograph** as intense black dots— appear over the parts of the embryo that have the radioactive substance.

The striking bands of silver grains in this autoradiograph of a fly embryo show that the particular substance being tested accumulated in seven segments of the developing fly and not in the others. This knowl-

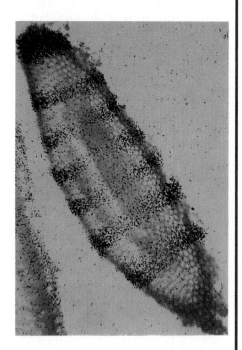

edge, when added to other similar discoveries, advances our still incomplete knowledge of what goes on in the developing insect body.

---

ferent mass. Measuring mass is more difficult than measuring radioactivity and usually requires the use of an expensive instrument called a mass spectrometer. Nevertheless, heavy water, which contains deuterium, is useful in studying biochemical reactions.

The decay of any radioisotope is regular. In successive, equal periods of time, the same fraction of the remaining radioactive material decays. The rate of decay of an isotope is its **half-life**. For example, in 14.3 days, one-half of any sample of ^{32}P decays. In the next 14.3 days, one-half of the remaining half decays, leaving one-fourth of the original sample of ^{32}P, and so on. Thus the half-life of ^{32}P is 14.3 days. Tritium has a half-life of 12.3 years; the half-life of ^{14}C is about 5,700 years. This regularity of decay allows us to use the radioactive isotopes present in nature to determine the ages of ancient bones, rocks, wood, and other materials (see Chapter 18).

THE BEHAVIOR OF ELECTRONS

The part of the atom of greatest biological interest is the electron. What happens in cells happens because of the way electrons behave. The characteristic number of electrons in each atom of an element determines how the atom reacts with other atoms. All **chemical reactions**, in cells or anywhere else, are

exchanges of electrons, or changes in the sharing of electrons between atoms.

Where a given electron in an atom is at any given time is impossible to say. We can only describe a volume of space within the atom where the electron is likely to be. The region of space within which the electron is found at least 90 percent of the time is the electron's **orbital** (Figure 2.3). An electron spins like a top—or like Earth on its axis—and, like a top, may spin in one of two directions: clockwise or counterclockwise. In an atom, a given orbital can be occupied by at most two electrons, which must spin in opposite directions. Thus any atom larger than helium (atomic number = 2) must have electrons in two or more orbitals. As shown in Figure 2.3, the different orbitals have characteristic forms. (Some atoms have more orbitals than are shown in the figure.)

The orbitals constitute a series of **shells** around the nucleus. The innermost shell, called the K shell, consists of only one orbital, an s orbital. The s orbital fills first, and its electrons have the lowest energy. Hydrogen ($_1$H) has one K-shell electron; helium ($_2$He) has two; all other atoms have two K-shell electrons and electrons in other shells as well. The L shell is made up of four orbitals (an s orbital and three p orbitals) and hence can hold up to eight electrons. The M, N, O, P, and Q shells have different numbers of orbitals.

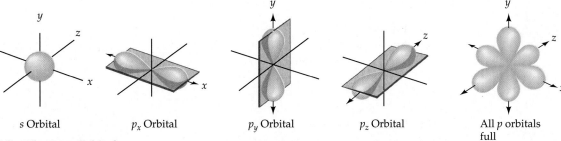

s Orbital p_x Orbital p_y Orbital p_z Orbital   All p orbitals full

2.3 Electron Orbitals

Orbitals are the regions around an atom's nucleus where electrons are most likely to be found. The movements of the two electrons closest to the nucleus form a spherical *s* orbital. The next two electrons form a larger, spherical *s* orbital (not shown). The next six electrons fill three dumbbell-shaped *p* orbitals, one pair of electrons per orbital, oriented on the *x*, *y*, and *z* axes through a point in the center of the atom.

In any atom, it is the outermost shell of electrons that determines what the atom can do chemically. When the outer shell is full, the atom will not react with other atoms. Examples of some chemically inert elements (elements with full outer shells) are helium, neon, and argon. Other elements are reactive in various degrees, depending on the number of electrons in the outermost shell. They are, in a sense, seeking ways to fill their outer shells with electrons by combining with other atoms (Figure 2.4).

CHEMICAL BONDS

Two atoms can cooperate to give each atom a full outer shell of electrons. In the process the two atoms become joined. **Molecules** consist of two or more joined atoms.

Covalent Bonds

A hydrogen atom has one electron in its only shell. Picture two hydrogen atoms, initially far apart but coming closer and closer, until they begin to interact. The negatively charged electron of atom A is attracted by the positively charged proton in nucleus B, as well as by its own nucleus; electron B experiences similar attractions. So the two electrons spend time between the two nuclei. The atoms do not get *too* close to-

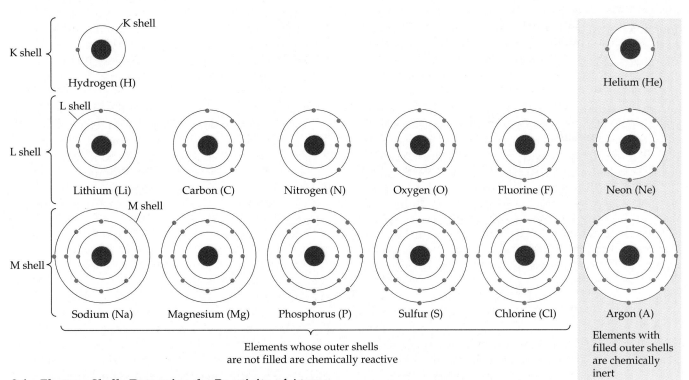

Elements whose outer shells are not filled are chemically reactive

Elements with filled outer shells are chemically inert

2.4 Electron Shells Determine the Reactivity of Atoms

Each shell can hold a specific maximum number of electrons: the K shell holds two; the L and M shells each hold eight. An atom with room for more electrons in its outermost shell may react with other atoms.

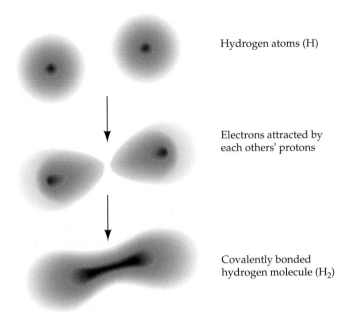

Hydrogen atoms (H)

Electrons attracted by
each others' protons

Covalently bonded
hydrogen molecule (H₂)

2.5 Electrons Are Shared in Covalent Bonds

Two hydrogen atoms combine to form a hydrogen molecule. Each electron is attracted to both protons, but the two protons cannot come so close together that they repel each other. A covalent bond forms when the electron orbitals in the K shells of the two atoms overlap.

gether, because their positively charged nuclei would then repel each other strongly. A certain distance between the coupled atoms, however, gives them a minimum amount of energy, and this is the most stable arrangement. (Pulling the atoms slightly farther apart would require an input of energy because of the "gluing" effect of the shared electrons; pushing the atoms closer together would require energy because of the mutual repulsion of the protons.) The two hydrogen nuclei share the two electrons completely. A **chemical bond** joins the two atoms, forming a molecule of hydrogen gas. A chemical bond is an attractive force that links two atoms. This type of chemical bond, consisting of a shared pair of electrons, is called a **covalent bond** (Figure 2.5). Because of the shared electrons, each hydrogen atom now has, in a sense, a full outer shell containing two electrons. The covalently bonded pair of hydrogen

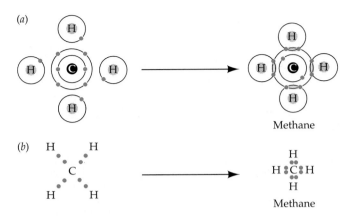

Methane

Methane

2.6 Covalent Bonding with Carbon

Carbon can complete its outer shell by sharing the electrons of four hydrogen atoms, forming methane. The drawings show two methods of representing the formation of the covalent bonds. Note that (b) depicts only the electrons from the initially unfilled shells (the two electrons in carbon's K shell are not shown).

atoms, with completely filled outer shells, is less reactive than the individual atoms, which have incomplete K shells.

A carbon atom has a total of six electrons: two in its (full) K shell and four in its (outer) L shell. Because the L shell can hold eight electrons, this atom can share electrons with up to four other atoms. Thus it can form four covalent bonds. When an atom of carbon reacts with four hydrogen atoms, a substance called methane forms. The carbon atom of methane has eight electrons in its (full) L shell, and each of the hydrogen atoms has a full K shell, thanks to the sharing. Thus four covalent bonds—each bond being a shared pair of electrons—hold methane together (Figure 2.6).

Bonds in which a single pair of electrons is shared are called **single bonds**. When four electrons (two pairs) are shared, the link is a **double bond**. Two oxygen atoms joined by a double bond make up a molecule of oxygen gas. **Triple bonds** (six shared electrons) are rare in biological molecules, but there is one in nitrogen gas, the chief component of the air we breathe (Figure 2.7).

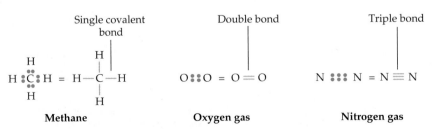

2.7 Single, Double, and Triple Bonds

Bonds may be shown by dots representing electron pairs (on the left in each pair) or by solid lines (on the right).

The Covalent Bonds of Different Elements

The atoms of a given element tend to form a specific number of covalent bonds with other atoms. The number of electrons in the outer shell determines this number. For example, carbon tends to form four covalent bonds, nitrogen three, oxygen two, and hydrogen one. The Harvard biologist George Wald once suggested that one reason these four elements are so important to living things is that their atoms are the smallest ones that can fill their outer electron shells by gaining one, two, three, or four electrons. In larger atoms that form covalent bonds with their M-shell electrons, the shared electrons are "screened" from the positively charged nucleus by the K-shell and L-shell electrons. Hence their shared electrons are held less tightly by the nucleus and form less stable covalent bonds. Hydrogen, carbon, oxygen, and nitrogen form the most stable covalent bonds. Wald also pointed out that carbon, oxygen, and nitrogen are biologically significant partly because they are among the few elements that can form double or triple bonds (two others are sulfur and phosphorus, which are also essential to living organisms). Because of double bonding, for example, carbon and oxygen can combine to form carbon dioxide ($O{=}C{=}O$), a water-soluble gas readily taken up by plants.

Ions and Ionic Bonds

When dissolved in water, many substances ionize: Their component atoms break apart, gaining or losing electrons in the process. Because the number of protons and electrons is no longer balanced, these atoms or groups of atoms become electrically charged and are known as **ions**. If an ion carries one or more positive electric charges, it is a cation; if it is negatively charged, it is an anion. Ions play major roles in many biological events.

Hydrochloric acid (HCl) offers a good example of ionization in action. Composed of hydrogen and chlorine atoms, HCl is a gas at room temperature. The hydrogen and the chlorine atoms share a pair of electrons (one from each atom) that form a single covalent bond. When HCl is dissolved in water, the atoms separate, but the chlorine atom keeps *both* the originally shared electrons (Figure 2.8). The chloride ion (Cl^-) thus has one more electron than elemental chlorine (Cl), giving its outer, L shell a full, stable load of eight electrons. The hydrogen ion (H^+) has a single positive charge because it has lost an electron—actually, H^+ is just a lonely proton. It is stable because it has no incomplete electron shells.

Some elements form ions with multiple charges by losing or gaining more than one electron to fill or completely empty out shells. Examples are Ca^{2+} (the calcium ion; a calcium atom has lost two electrons), Mg^{2+} (magnesium ion), and Al^{3+} (aluminum ion). Groups of atoms can also form ions: NH_4^+ (ammonium ion), SO_4^{2-} (sulfate ion), PO_4^{3-} (phosphate ion). Two biologically important elements each yield more than one stable ion: iron yields Fe^{2+} (ferrous ion) and Fe^{3+} (ferric ion), and copper yields Cu^+ (cuprous ion) and Cu^{2+} (cupric ion).

Oppositely charged ions attract one another. If ions are densely concentrated, as when water or another solvent evaporates, crystals form. Solid table salt (sodium chloride, or NaCl) consists of sodium ions (Na^+) and chloride ions (Cl^-) in a highly ordered crystalline array (Figure 2.9). The array is held together by **ionic bonds**, chemical bonds in which the attractive force is the electrical attraction between cations and anions. Like the covalently bonded HCl, NaCl dissolves readily in water. Actually, there is no sharp dividing line between covalent bonds and ionic bonds. In some covalent bonds, the bond partners share the electrons equally. Other pairs of atoms share electrons unequally. An ionic bond is simply a case in which one of the partners has the "shared" electron pair *all* the time. In solution, an ionic bond is less than one-tenth as strong as a covalent bond that shares electrons equally, so the ionic bond can be broken much more readily.

Covalent and ionic bonds are the strongest forces that hold atoms together in molecules. However, other, weaker interactions are also important in mol-

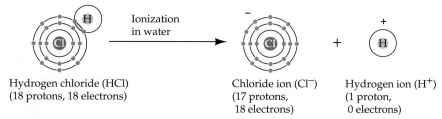

Hydrogen chloride (HCl)
(18 protons, 18 electrons)

Chloride ion (Cl^-)
(17 protons,
18 electrons)

Hydrogen ion (H^+)
(1 proton,
0 electrons)

Ionization
in water

2.8 Ionization of Hydrogen Chloride
When HCl—which has no electric charge—is dissolved in water, the chlorine atom retains *both* the shared electrons from the covalent bond and becomes a chloride ion (Cl^-). This ion is negatively charged because it contains one more electron than it does protons. The proton of the hydrogen atom, no longer balanced electrically by an electron, becomes a positively charged hydrogen ion (H^+).

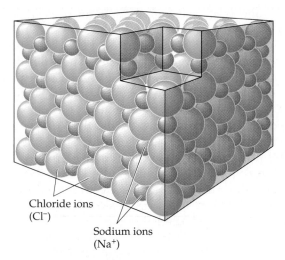

2.9 Ionic Bonding
A crystal of sodium chloride (NaCl) is held together by ionic bonds between the sodium cations (Na$^+$) and the chloride anions (Cl$^-$).

Chloride ions (Cl$^-$)

Sodium ions (Na$^+$)

ecules and cellular structures. The most important of these are hydrogen bonds and interactions between nonpolar molecules. Before describing them, we must introduce some chemical principles upon which they depend.

MOLECULES

A molecule consists of two or more atoms linked by chemical bonds. A substance whose molecules contain more than one kind of atom is a **compound**. Most biological substances are compounds. A substance (such as oxygen gas) that contains only one kind of atom is an **elemental substance**.

The **molecular formula** of a compound or an elemental substance shows how many atoms of each element are present in the molecule. This number is written to the lower right of the symbol. For example, the molecular formula for methane is CH_4 (each molecule contains one carbon atom and four hydrogen atoms), that for oxygen gas is O_2, that for nitric oxide is NO (Box 2.B), and that for sucrose (table sugar) is $C_{12}H_{22}O_{11}$. The hormone insulin is represented by the molecular formula $C_{254}H_{377}N_{65}O_{76}S_6$! Molecular formulas do not tell us anything about which atoms are linked to which. **Structural formulas**, as in Figures 2.6 and 2.7, give us this information.

Each compound has a **molecular weight**, which is simply the sum of the atomic weights of the atoms in the molecule. The atomic weights of hydrogen, carbon, and oxygen are, respectively, 1.008, 12.011, and 16.000. Thus the molecular weight of water (H_2O) is $(2 \times 1.008) + 16.000 = 18.016$, or about 18. What is the molecular weight of sucrose ($C_{12}H_{22}O_{11}$)? You can calculate this and find that the answer is

approximately 342. If you remember the molecular weights of a few representative biological compounds, you will be able to picture the relative sizes of molecules that interact with one another (Figure 2.10).

Suppose we want to compare how sodium chloride (NaCl), potassium chloride (KCl), and lithium chloride (LiCl) affect a biological process. You might at first think that we could simply give, say, 2 grams (g) of NaCl to one set of subjects, 2 g of KCl to another, and 2 g of LiCl to the third. But because the molecular weights of NaCl, KCl, and LiCl are different, 2-g samples of each of these substances contain different numbers of molecules. The comparison would thus not be legitimate. Instead, we want to give *equal numbers of molecules* of each substance so that we can compare the activity of one molecule of one substance with that of one molecule of another. But the weight of a single molecule of sodium chloride is 10^{-22} g—hardly a workable quantity.

Since we can neither weigh nor count individual molecules, we work with **moles** (also known as gram molecular weights). *One mole of a substance is an amount whose weight in grams is numerically equal to the molecular weight of the substance.* Potassium chloride (KCl) has a molecular weight of 74.55, so a mole of

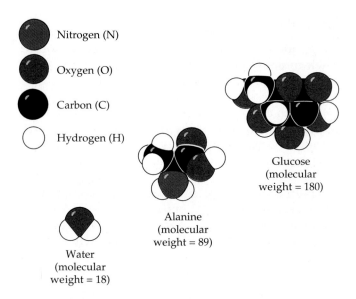

Nitrogen (N)

Oxygen (O)

Carbon (C)

Hydrogen (H)

Glucose (molecular weight = 180)

Alanine (molecular weight = 89)

Water (molecular weight = 18)

2.10 Molecular Weight and Size
The relative sizes of three common molecules and their molecular weights. Water is the solvent in which many biological reactions take place; alanine is one of the building blocks of proteins; glucose is a sugar, an important food substance in most cells. These space-filling models are the most realistic representations of molecules we can create; you will continue to see them in this and subsequent chapters. Space-filling models are particularly valuable in helping us understand how molecules interact. These color conventions for the atoms (along with yellow, which is used for sulfur and phosphorus atoms when they appear) are standard.

BOX 2.B

The Gas That Says NO

Nitrogen can react with oxygen to form any of three nitrogen oxides, depending on the reaction conditions. These oxides are nitric oxide (NO), nitrous oxide (N_2O, "laughing gas"), and nitrogen dioxide (NO_2). Until recently, we thought of NO primarily as a dangerous pollutant of the environment. It is a significant component of smog and of cigarette smoke. It contributes to the destruction of the ozone layer and to the formation of acid rain. We suspect that it can cause cancer.

What we did *not* suspect was that our bodies produce NO naturally and that NO plays a number of key roles in normal body functioning. In spite of its chemical simplicity, we began to recognize the biological functions of NO only in the late 1980s. Hundreds of technical papers on NO appeared in 1992 alone. What are the biological functions of NO? They are exceedingly diverse. NO helps defend the body against invading microorganisms: White blood cells called macrophages produce and release tiny quantities of NO, which destroys cells infected with bacteria. By releasing NO, macrophages also destroy tumor cells, thus perhaps contributing to the body's defenses against cancer.

In a dramatically different role, NO appears to participate in the formation of memories within the nervous system, as we will explain in Chapter 38. NO produced by cells lining blood vessels relaxes muscles in the vessel walls, helping regulate blood pressure. NO performs many other functions, among which is the translation of sexual excitement in male mammals into erection of the penis. In response to sexual stimulation, nerves in the pelvis release NO, which dilates certain blood vessels, allowing blood to rush in and produce the erection.

Natural regulators can wreak havoc if their production goes out of control. This is as true of NO as of other key regulators. Septic shock, a usually deadly condition that affects more than 300,000 people in the United States each year, results in part from a local excess of NO. NO is also implicated in some cases of diabetes and other diseases.

KCl weighs 74.55 g; a mole of NaCl weighs 58.45 g, and a mole of LiCl, 42.40 g. *A mole of one substance contains the same number of molecules as does a mole of any other substance.* This number, known as **Avogadro's number**, is 6.023×10^{23} molecules per mole. The concept of the mole is important for biology because it enables us to work easily with known numbers of molecules.

A solution containing one mole of solute per liter is a **molar** solution, 1 *M*. A solution containing one-half mole per liter is referred to as 0.5 *M*, or half-molar. How would you make 100 milliliters (ml) of a 0.02 *M* sucrose solution? The molecular weight of sucrose is 342, so one liter (1,000 ml) of a 1 *M* sucrose solution contains one mole, or 342 g, of sucrose. You were asked to make just 100 ml of 0.02 *M* solution. Since 34.2 g of sucrose would make 100 ml of 1 *M* sucrose, to make 100 ml of a 0.02 *M* sucrose solution you would use 0.02×34.2 g = 0.684 g of sucrose.

CHEMICAL REACTIONS

When atoms combine or change bonding partners, a chemical reaction is occurring. Consider the flame of a propane kitchen stove or water heater. When propane (C_3H_8) reacts with oxygen gas (O_2), the carbon atoms become bonded to oxygen atoms instead of to hydrogen atoms, and the hydrogen atoms become bonded to oxygen instead of carbon. This process is shown in Figure 2.11. As the covalently bonded atoms change bonding partners, the composition of the matter changes, and propane and oxygen gas become carbon dioxide and water. Chemical reactions in which covalently bonded atoms change partners are common and important in organisms.

The heat of the stove's flame and its blue light reveal that the reaction of propane and oxygen releases a great deal of energy. Changes in energy usually accompany chemical reactions: Energy may be given off to the environment, as in the reaction of propane with oxygen, or energy may be taken up from the environment (some substances will react only after being heated, for example).

We can measure the energy associated with chemical bonds. Work must be done to break a bond, and that work, or energy, can be expressed in calories. In most chemical reactions in which bond partners change, the total bond energies of the products differ from those of the reactants; these energies can also be expressed in calories. A calorie is the amount of heat energy needed to raise the temperature of 1 g of pure water (which contains no other substance) from 14.5°C to 15.5°C. (The nutritionist's Calorie, which biologists call a kilocalorie, is equal to 1,000 heat-energy calories.) Although defined in terms of heat, the calorie is a measure of any form of energy— mechanical, electric, or chemical.

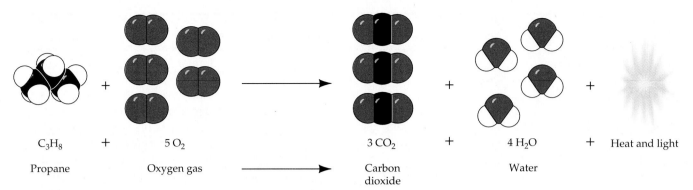

$$C_3H_8 \quad + \quad 5\,O_2 \quad \longrightarrow \quad 3\,CO_2 \quad + \quad 4\,H_2O \quad + \quad \text{Heat and light}$$

Propane   Oxygen gas   ⟶   Carbon dioxide   Water

2.11 Bonding Partners and Energy May Change in a Chemical Reaction
One molecule of propane reacts with five molecules of oxygen gas to give three molecules of carbon dioxide and four molecules of water. This particular reaction releases energy, in the form of heat and light.

WATER

One of the simplest compounds, yet one of the most biologically important and chemically interesting, is water (H_2O). Life originated in water; water covers three-fourths of present-day Earth, and somewhere between 45 and 95 percent of any active living organism's weight consists of water. We have all experienced the biological imperative of a raging thirst, and some organisms must live out their lives in water. No organism can remain biologically active without water.

More kinds of substances dissolve in water than in any other liquid, making it the most effective solvent known. The chemical reactions of interest to biologists take place in solution, many of them in watery or aqueous solutions (although many other reactions occur in solutions in which fatty substances are the solvent). Water itself takes part in a number of important reactions.

How is water different from molecularly similar substances (such as hydrogen sulfide, H_2S, a foul-smelling gas that is poisonous to humans)? Water has three different, temperature-dependent, physical states—solid, liquid, and gas. Its solid state, ice, is less dense than its liquid form, which is why ice floats in water (Figure 2.12). If ice sank in water, as almost all other solids do in their corresponding liquids, ponds and lakes would freeze from the bottom up, becoming solid blocks of ice in winter and killing most of the organisms living in them. Once the whole pond had frozen, its temperature could drop well below the freezing point of water. In fact, however,

2.12 Water: Solid and Liquid
Solid water from a glacier floats in its liquid form. The clouds are also water, but not in its gaseous phase: They are composed of fine drops of liquid water.

ice floats, forming a protective insulating layer at the top of a pond and reducing heat flow to the cold air above. Thus fish, plants, and other organisms in the pond can survive the winter without having to endure subfreezing temperatures. Unless the entire pond freezes, there will be a liquid portion no colder than 0°C, the freezing point of pure water.

As water changes from its liquid to its gaseous state, vapor, it uses heat. This is why sweating is a useful cooling device for humans—your body loses heat as the water in sweat evaporates. It also takes a relatively large amount of heat to raise the temperature of water. The temperature of a given quantity of water is raised only 1°C by an amount of heat that would increase the temperature of the same quantity of ethyl alcohol by 2°C, or of chloroform by 4°C. This important phenomenon contributes to the surprising constancy of the temperature of the oceans and other large bodies of water through the seasons of the year. This constancy is useful to the organisms living in lakes and oceans, for it means that they need not adapt to great variations in temperature. In addition, the relative constancy of water temperature helps to minimize variations in atmospheric temperature throughout the planet.

Water ionizes, but only to a limited extent. About one water molecule in 500 million is ionized at any one time. In a somewhat simplified form, the ionization of water may be represented as

$$H_2O \rightarrow H^+ + OH^-$$

H^+ is, of course, a hydrogen ion; the OH^- is known as a **hydroxide ion**. H^+ and OH^- ions participate in many important chemical reactions.

ACIDS, BASES, AND pH

In pure water, the concentration of hydrogen ions exactly equals that of OH^- ions, and this "solution" is said to be **neutral**. Now suppose we add some HCl (hydrochloric acid). As it dissolves, the HCl ionizes, releasing H^+ ions, so now there are more H^+ than OH^- ions. Such a solution is acidic. A basic, or alkaline, solution is one in which there are more OH^- than H^+ ions. A basic solution can be made from water by adding, for example, sodium hydroxide (NaOH), which ionizes to yield OH^- and Na^+ ions, thus making the concentration of OH^- ions greater than that of H^+ ions.

A compound that can *release* H^+ ions in solution is an **acid**. HCl is an acid, as is H_2SO_4 (sulfuric acid), one molecule of which may ionize to yield two H^+ ions and one SO_4^{2-} ion. Biological compounds such as acetic acid and pyruvic acid, which contain —COOH (the **carboxyl group**; see Figure 2.20) are also acids, because —COOH → —COO⁻ + H⁺.

Compounds that can *accept* H^+ ions are called **bases**. These include the bicarbonate ion (HCO_3^-), which can accept an H^+ ion and become carbonic acid (H_2CO_3); ammonia (NH_3), which can accept an H^+ ion and become an ammonium ion (NH_4^+); and many others.

Note that, although —COOH is an acid, —COO⁻ is a base, because —COO⁻ + H⁺ → —COOH. Acids and bases exist as pairs, such as —COOH and —COO⁻ because any acid becomes a base when it releases a proton, and any base becomes an acid when it gains a proton.

You may have noticed that the two reactions just discussed are the opposites of each other. The reaction that yields —COO⁻ and H⁺ is reversible and may be expressed as —COOH ⇌ —COO⁻ + H⁺. A **reversible reaction** is one that can proceed in either direction—left to right or right to left—depending on the relative starting concentrations of reacting substances and products. In principle, *all* chemical reactions are reversible. Some consequences of this reversibility will be discussed in Chapter 6.

The terms acid*ic* and bas*ic* refer only to *solutions*. How acidic or basic a solution is depends on the relative concentrations of H^+ and OH^- ions in it. *Acid* and *base* refer to *compounds* and *ions*. A compound or ion that is an acid can donate H^+ ions; one that is a base can accept H^+ ions.

How do we specify how acidic or basic a solution is? First, let's look at the H^+ ion concentrations of a few contrasting solutions. In pure water the H^+ concentration is $10^{-7}\,M$. In $1\,M$ hydrochloric acid the H^+ concentration is $1\,M$; and in $1\,M$ sodium hydroxide the H^+ concentration is $10^{-14}\,M$. With its values ranging so widely—from more than $1.0\,M$ to less than $10^{-14}\,M$—the H^+ concentration itself is an inconvenient quantity. It is easier to work with the logarithm of the concentration, because logarithms compress this range.

How acidic or basic a solution is is indicated by its **pH** (a term derived from *potential* of *Hydrogen*). The pH value is defined as the negative logarithm of the hydrogen ion concentration in moles per liter (molar concentration). In chemical notation, molar concentration is often indicated by putting brackets around the symbol for a substance: $[H^+]$ = the molar concentration of H^+. We can now write the equation

$$pH = -\log_{10}[H^+]$$

Since the H^+ concentration of pure water is $10^{-7}\,M$, its pH is $-\log(10^{-7}) = -(-7)$, or 7. A smaller negative logarithm means a larger number; in practical terms, a lower pH means a higher H^+ concentration, or greater acidity. In $1\,M$ HCl, the H^+ concentration is $1\,M$, so the pH is the negative logarithm of 1 ($-\log 10^0$), or 0. The pH of $1\,M$ NaOH is the negative logarithm of 10^{-14}, or 14. A solution with a pH of

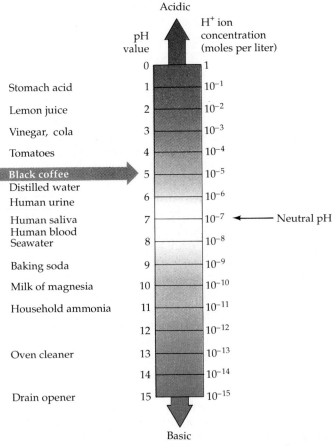

2.13 pH Values of Some Familiar Substances
A pH meter such as the one on the left tells us the pH of a solution. This scale reads from low (acidic) pH values at the top to high (basic) pH values at the bottom.

less than 7 is acidic: It contains more H^+ than OH^- ions. A solution with a pH of 7 is neutral, and a solution with a pH value greater than 7 is basic. Because the pH scale is logarithmic, the values are exponential: A solution with a pH of 5 is 10 times more acidic than one with a pH of 6 (it has 10 times as great a concentration of H^+); a solution with a pH of 4 is 100 times more acidic than one with a pH of 6. The pH values of some common substances are shown in Figure 2.13.

BUFFERS

An organism must control the chemistry of its cells—in particular, the pH of the separate compartments within cells. Animals must also control the pH of their blood. The normal pH of human blood is 7.4, and deviations of even a few tenths of a pH unit can be fatal. The control of pH is made possible in part by **buffers**, which are systems that maintain a relatively constant pH even when substantial amounts of acid or base are added. A buffer is a mixture of an acid that does not ionize completely in water and its corresponding base—for example, carbonic acid (H_2CO_3) and bicarbonate ions (HCO_3^-). If acid is added to this buffer, not all the H^+ ions from the acid stay in solution. Instead, many of the added H^+ ions combine with bicarbonate ions to produce more carbonic acid, thus using up some of the H^+ ions in the solution and decreasing the acidifying effect of the added acid:

$$HCO_3^- + H^+ \rightleftharpoons H_2CO_3$$

If base is added, the reaction reverses. Some of the carbonic acid ionizes to produce bicarbonate ions and more H^+, which counteracts some of the added base. In this way, the buffer minimizes the effects of added acid or base on pH. A given amount of acid or base causes a smaller change in pH in a buffered solution than in an unbuffered one (Figure 2.14). Buffers illustrate the reversibility of chemical reactions: The addition of acid drives the reaction in one direction, whereas addition of base drives it in the other direction.

POLARITY

In some molecules, called **polar molecules**, the electric charge is not distributed evenly in the covalent bonds. Water is an important polar molecule. In the O—H covalent bonds, the shared electrons are drawn more strongly to the oxygen nucleus, which has eight protons, than to the hydrogen nuclei, which have only one positive charge each. Because of this tendency of the electrons, the hydrogen atoms represent slightly positive regions of the water molecule. In addition, the two hydrogen atoms of water do not lie on directly opposite sides of the oxygen atom; rather, they are separated by an angle of only 104.5

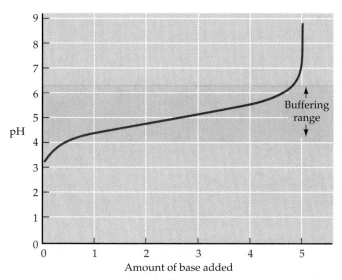

2.14 Buffers Minimize Changes in pH
Adding a base increases the pH of a solution. With increasing amounts of added base, the overall slope of a graph of pH is upward. In the buffering range, however, the slope is shallow. At high and low values of pH, where the buffer is ineffective, the slopes are much steeper.

degrees. Because the electrons are drawn away from the hydrogen nuclei, the electron cloud is most dense in the opposite region, which therefore has a slightly negative charge. In contrast, the electric charge in **nonpolar molecules** is evenly balanced from one end of the molecule to the other. Ethane is an example of a nonpolar molecule (Figure 2.15).

Much of the excellence of water as a solvent is due to its polarity. Substances such as sodium chloride dissolve easily in water because the Na^+ and Cl^- ions become hydrated—that is, surrounded by water molecules (Figure 2.16). Because the water molecules shield the ions from interacting with one another, the ions are prevented from dropping back out of solution as solid particles of NaCl.

Hydrogen Bonds

Because of their polarity, many water molecules may become loosely attracted to one another. Although

Water, a polar molecule   Ethane, a nonpolar molecule

2.15 Polarity of Molecules
The polar water molecule is slightly more positive near the hydrogen atoms and slightly more negative on the other side. The electrons are distributed evenly over the symmetrical surface of the nonpolar ethane molecule; there is no excess positive charge at any point.

this type of attraction—that between a slight positive charge on a hydrogen atom and a slight negative charge on a nearby atom—is only about one-tenth (or less) as strong as a covalent bond, it is strong enough to deserve a name: the **hydrogen bond**. It is also strong enough to be very important in biology (Figure 2.17). Hydrogen bonding plays major roles in determining the shapes of the giant molecules (for example, proteins and DNA; see Chapter 3) and in conserving and decoding genetic information (see Chapter 11). Compounds such as sugars dissolve readily in water because hydrogen bonds form between hydroxyl (—OH) groups on the sugars and the oxygen atoms of water. Hydrogen bonding accounts, too, for most of the unusual properties of water mentioned earlier.

Hydrogen bonding between the molecules of water in its liquid state give it a high **surface tension**. Surface tension creates an invisible "skin" that is so strong it permits some insects literally to walk on water (Figure 2.18). Hydrogen bonding causes **capillary action**—the rising of water and watery solu-

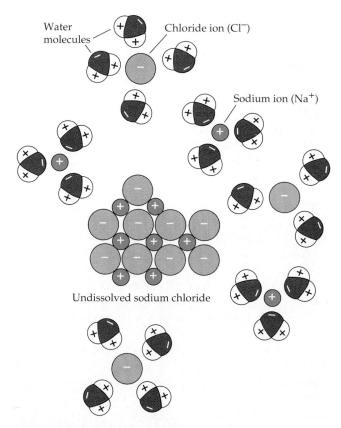

2.16 Water Hydrates Ions
Because water molecules are polar, they cluster around either cations or anions in solutions, blocking the reassociation of the dissolved ions. In this schematic representation of a solution, the negative ends of the water molecules are attracted to the sodium cations, whereas the positive hydrogen atoms in water are attracted to the chloride anions.

(a)

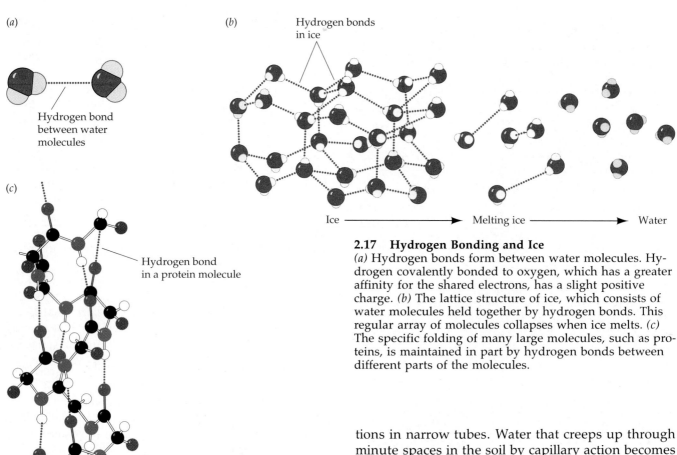

Hydrogen bond
between water
molecules

(b) Hydrogen bonds
in ice

Ice ⟶ Melting ice ⟶ Water

(c)

Hydrogen bond
in a protein molecule

2.17 Hydrogen Bonding and Ice
(a) Hydrogen bonds form between water molecules. Hydrogen covalently bonded to oxygen, which has a greater affinity for the shared electrons, has a slight positive charge. (b) The lattice structure of ice, which consists of water molecules held together by hydrogen bonds. This regular array of molecules collapses when ice melts. (c) The specific folding of many large molecules, such as proteins, is maintained in part by hydrogen bonds between different parts of the molecules.

tions in narrow tubes. Water that creeps up through minute spaces in the soil by capillary action becomes available to the roots of plants. The ability of water to be pulled up through conducting tissues up to the tops of trees as tall as 100 meters (see Chapter 30) is also a result of the interaction between polar molecules known as hydrogen bonding.

(a)

(b)

2.18 Surface Tension
(a) A water strider "skates" along, supported by the surface tension of the water that is its home. (b) Surface tension demonstrated by a soap bubble on a teacup.

**TABLE 2.2
Chemical Bonds**

TYPE OF BOND	BASIS OF BONDING	ENERGY	BOND LENGTH
Covalent bond	Sharing of electron pairs	50–110 kcal/mol[a]	<0.2 nm
Ionic bond	Attraction of opposite charges	3–7 kcal/mol	0.28 nm (optimal)
Hydrogen bond	Sharing of H atom	3–7 kcal/mol	0.26–0.31 nm (between atoms that share H)
van der Waals interaction	Interaction of electron clouds	~1 kcal/mol	0.24–0.4 nm

[a] kcal/mol = kilocalories per mole; for other abbreviations of units of measurement see the inside front cover.

Interactions between Nonpolar Molecules

Nonpolar molecules also interact to form weak chemical bonds. When uncharged molecules, or parts of molecules, come so close together that their electron clouds touch, the electrons of one molecule are weakly attracted by the nuclei of the atoms in the other molecule. The force of this attraction, which exceeds the repulsive force between the electron clouds, is called a **van der Waals interaction**. Table 2.2 compares these interactions with the other chemical bonds we have discussed. Although only one-fourth to one-third as strong as hydrogen bonds, van der Waals interactions do contribute to the maintenance of the specific structures of large molecules.

Another type of weak attraction, comparable in strength to van der Waals interactions, is the **hydrophobic interaction**, which occurs when nonpolar molecules, or parts of molecules, that come together in the presence of water associate with one another in such a way as to minimize their exposure to the water. For example, molecules of oil in water minimize the area of oil–water contact by aggregating into droplets. This configuration requires the least energy to maintain; work must be done for the area of contact between the oil and the water to increase.

SOME SIMPLE ORGANIC COMPOUNDS AND FUNCTIONAL GROUPS

Organic compounds are made of molecules that contain the element carbon. Organisms produce several classes of organic compounds. The simplest class is the **hydrocarbons**, compounds composed of only hydrogen and carbon atoms. The hydrocarbons include methane (CH_4), ethane (CH_3—CH_3), propane (CH_3—CH_2—CH_3), and ethylene (CH_2=CH_2) (Figure 2.19). Methane, ethane, and propane are called **saturated hydrocarbons** because they contain no carbon–carbon double bonds and are thus saturated with hydrogens. Ethylene, in contrast, is **unsaturated** because it can add more hydrogen to the carbon atoms connected by the double bond—thus becoming the saturated hydrocarbon ethane:

$$H_2C{=}CH_2 + H_2 \rightarrow H_3C{-}CH_3$$

We may write the formula for ethylene as CH_2=CH_2, which makes it easy to recognize that two identical parts are covalently bonded together, or, alternatively, as H_2C=CH_2, which reminds us it is the two carbon atoms that share the double bond.

Gasolines are hydrocarbons with six to ten carbon atoms arranged in a chain; a typical gasoline is octane, with eight carbon atoms. Motor oils have 12 to 20 carbon atoms, and waxy semisolids called paraffin waxes are longer-chain hydrocarbons. Polyethylene plastic is a large hydrocarbon, with chains thousands of carbon atoms long. Animal and plant fats also have long hydrocarbon chains (Chapter 3). Hydrocarbons are flammable, oily, and immiscible (they do not mix) with water. Things that dissolve in hydrocarbons ordinarily do not dissolve in water, and vice versa.

Several classes of biologically important compounds are formed by the replacement of a hydrogen atom on a hydrocarbon with any of several different functional groups (Figure 2.20). When the functional group is a **hydroxyl group** (—OH), for example, the product is an **alcohol**. Perhaps the most familiar alcohol is **ethanol** (ethyl alcohol). Small alcohols like ethanol are soluble in water, but larger alcohols are not soluble in water because of their long hydrocarbon chains.

Sugars contain both hydroxyl and carbonyl groups. The carbonyl group has a central carbon atom with a double bond to an oxygen atom. If one of the other two bonds on the carbon atom in a carbonyl group is to a hydrogen atom, the compound is an **aldehyde**; otherwise it is a **ketone**.

Molecules containing one or more carboxyl groups (—COOH) are acids because of the tendency of the carboxyl group to ionize. **Amines**, on the other hand, are organic bases. These compounds possess an **amino group** (—NH_2), which has a tendency to react

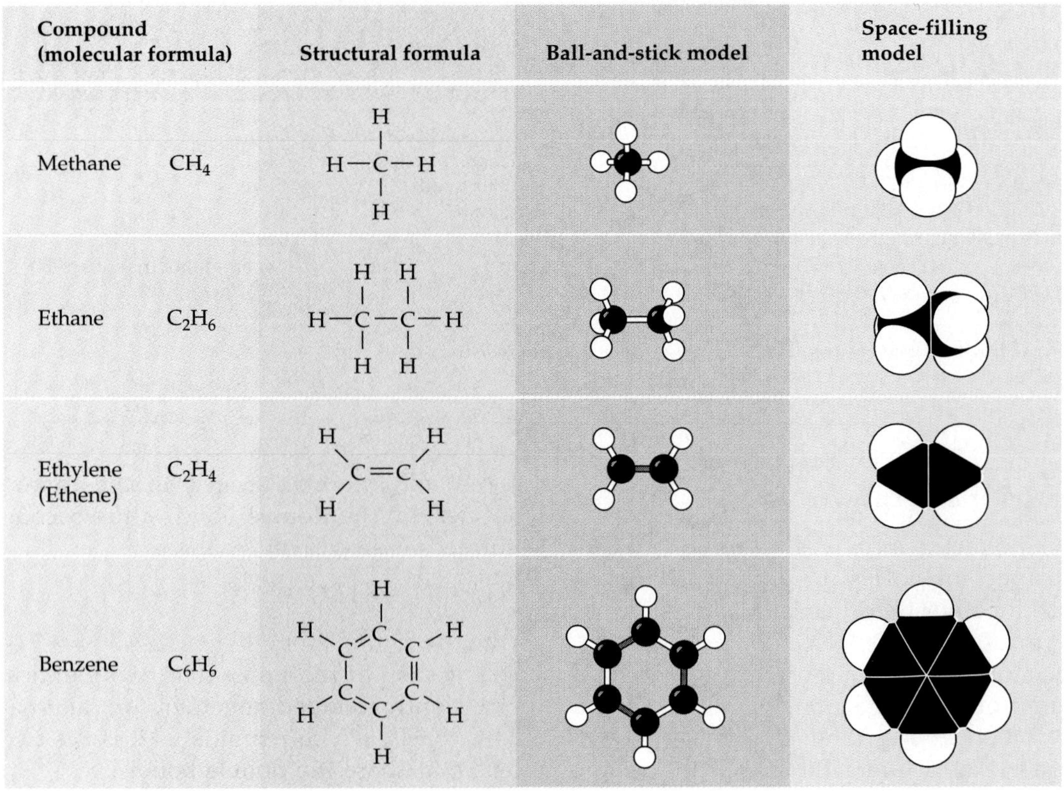

Compound (molecular formula)	Structural formula	Ball-and-stick model	Space-filling model
Methane CH_4			
Ethane C_2H_6			
Ethylene (Ethene) C_2H_4			
Benzene C_6H_6			

2.19 Some Small Hydrocarbons
Compare the structures of these hydrocarbons, noting which are saturated and which unsaturated. The molecules in the figure are represented in four different ways. The representations in the two right-hand columns emphasize the three-dimensional structure of the molecules; ball-and-stick models focus on bond angles, and space-filling models focus on the molecule's overall shape.

with H^+ to give the positively charged $—NH_3^+$ group. This H^+-accepting characteristic accounts for the classification of amines as bases.

Amino acids are important compounds that possess a carboxyl group *and* an amino group, both of which are attached to the same carbon atom, the α (alpha) carbon. Also attached to the α carbon atom are a hydrogen atom and a side chain (Figure 2.21). Twenty different amino acids constitute the building blocks of the giant protein molecules of living things. Each amino acid has a different side chain that gives it its distinctive chemical properties (Chapter 3). Because they possess both carboxyl and amino groups, amino acids are simultaneously acids and bases. At the pH values commonly found in cells, both the carboxyl and the amino groups are ionized: The carboxyl group has lost a proton, and the amino group has gained one.

Two other functional groups should be introduced here. The **sulfhydryl group** (—SH; Figure 2.20) is important in protein structure (Chapter 3) and in biochemical reactions (Chapter 7). The **phosphate group** ($—OPO_3^{2-}$) participates in many crucial reactions in which energy is transferred (Chapters 7, 8, 11, and 40). Phosphate groups are exchanged between sugars and many other compounds.

Isomers are compounds with the same chemical formula but different arrangements of the atoms. Whenever a carbon atom has four *different* atoms or groups attached to it, there are two different ways of making the attachments, each the mirror image of the other. Such a carbon atom is an asymmetric carbon, and the pair of compounds are optical isomers of each other (Figure 2.22). Your right and left hands are optical isomers. Just as a glove is specific for a particular hand, so some biochemical molecules can interact with a specific optical isomer of a compound but are unable to "fit" the other. The α carbon in an amino acid is an asymmetric carbon; hence, amino acids exist in two isomeric forms, called D- and L-amino acids (Figure 2.23). D- and L- are abbreviations for *dextro-* and *levorotatory*, referring to the directions (right or left, respectively) in which solutions of these compounds rotate the plane of polarized light; they

Functional group	Class of compounds	Structural formula	Example	Ball-and-stick model
Hydroxyl —OH	Alcohols	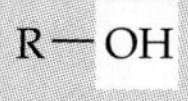 R—OH	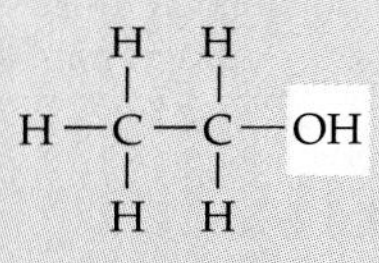 Ethanol	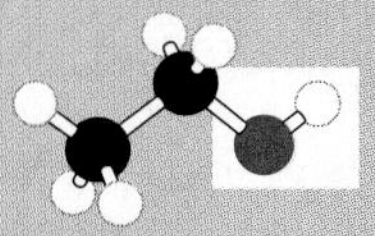
Carbonyl —CHO	Aldehydes	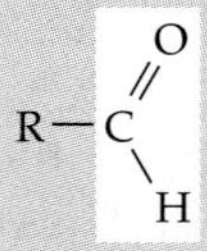 R—CHO	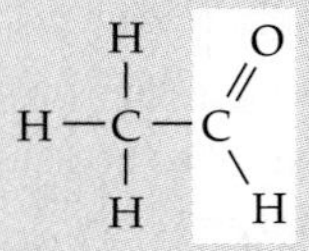 Acetaldehyde	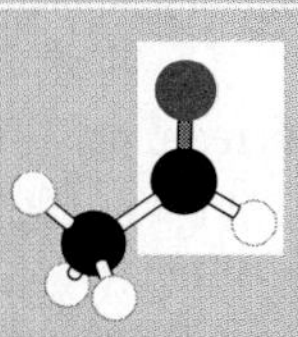
Carbonyl CO	Ketones	 R—CO—R	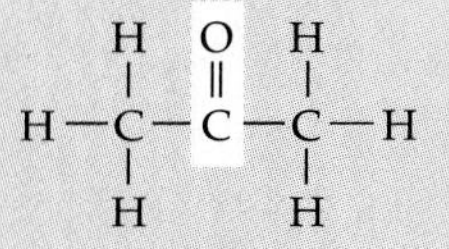 Acetone	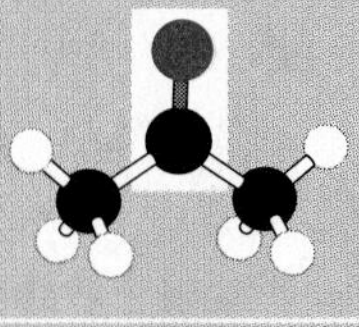
Carboxyl —COOH	Carboxylic acids	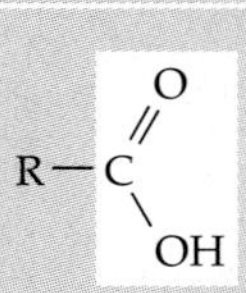 R—COOH	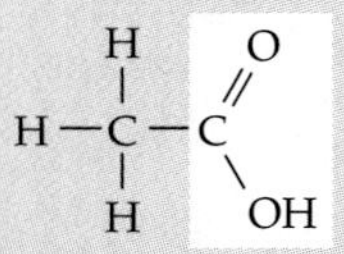 Acetic acid	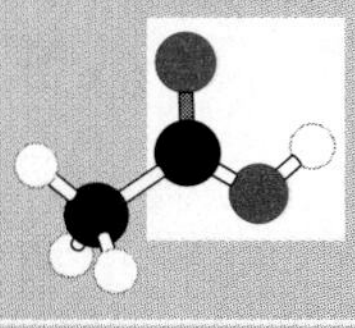
Amino —NH$_2$	Amines	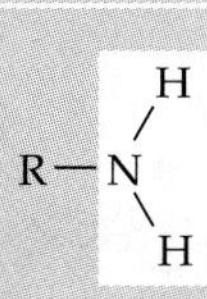 R—NH$_2$	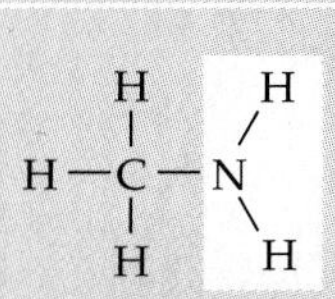 Methylamine	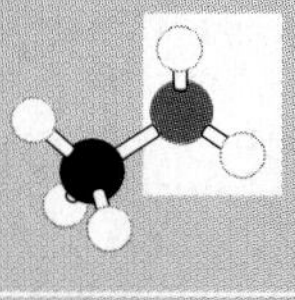
Phosphate —OPO$_3{}^{2-}$	Organic phosphates	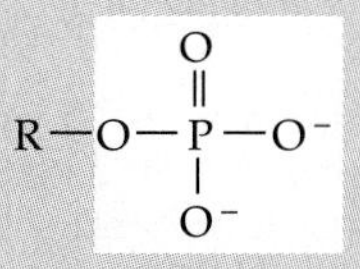 R—O—PO$_3{}^{2-}$	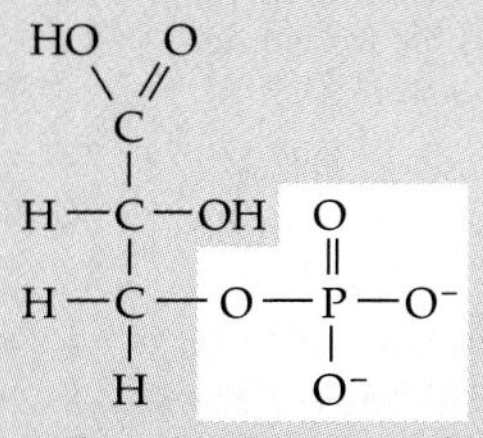 3-Phosphoglyceric acid	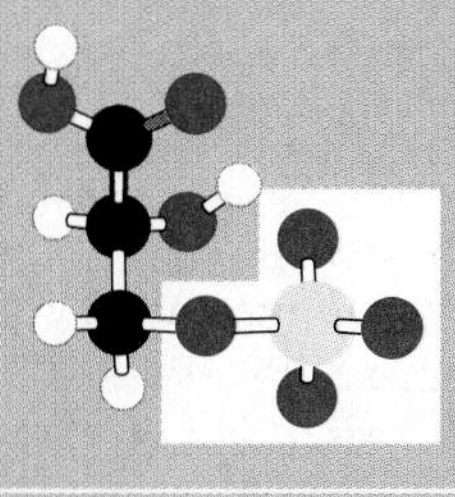
Sulfhydryl —SH	Thiols	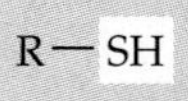 R—SH	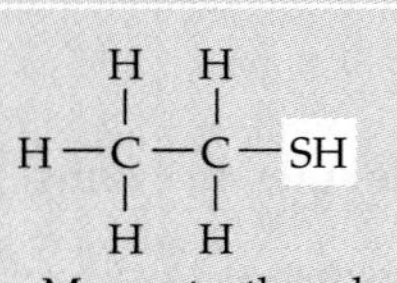 Mercaptoethanol	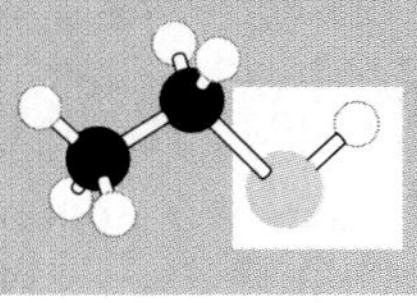

2.20 Simple Organic Compounds and Functional Groups
Compounds of the types shown here will appear throughout this book. The functional groups (highlighted) are the most common ones found in biologically important molecules. The term "R" represents the remainder of the molecule; it may be any of a large number of carbon skeletons or other chemical groupings.

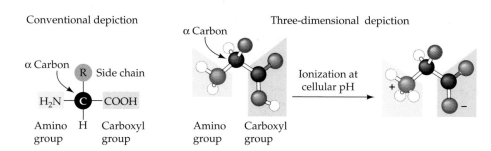

Conventional depiction

α Carbon

Side chain

H_2N — C — COOH

Amino   H   Carboxyl
group   group

Three-dimensional depiction

α Carbon

Ionization at
cellular pH

Amino   Carboxyl
group   group

2.21   Amino Acid Structure
Two depictions of the general structure of an amino acid. The side chain attached to the α carbon differs from one amino acid to another. At pH values found in living cells, both the carboxyl group and the amino group of an amino acid are ionized, as shown in the right-hand model.

refer to the "handedness" of a molecule with one or more asymmetric carbons. Only L-amino acids are commonly found in most proteins of living things.

The compounds discussed in this chapter include some of the more common ones found in organisms.

Between these small molecules and the world of the living stands another level, that of the giant macromolecules. These huge molecules—the proteins, lipids, carbohydrates, and nucleic acids—are the subject of the next chapter.

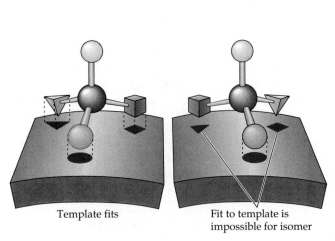

Template fits

Fit to template is impossible for isomer

2.22   Optical Isomers
Optical isomers are mirror images of each other. They result when four different groups are attached to a single carbon atom (the dark gray sphere in the center). If a template is laid out to match the groups on one carbon atom, there is no way the groups on the mirror-image isomer can be rotated to fit the same template.

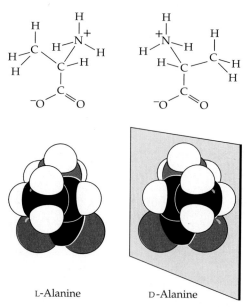

L-Alanine   D-Alanine

2.23   Isomers of Alanine
Structural formulas and space-filling models of the D- and L- forms of the amino acid alanine. Only L-alanine (on the left) is commonly found in living things.

SUMMARY of Main Ideas about Small Molecules in Living Things

Organisms, like nonliving matter, are composed of atoms bonded together as molecules.

Living things consist primarily of atoms of carbon, hydrogen, oxygen, nitrogen, sulfur, and phosphorus.

The numbers of electrons and protons they possess make each element unique.
  Review Figures 2.1, 2.3, and 2.4

A chemical reaction forms bonds between atoms, breaks such bonds, or does both.

The strongest chemical bonds are covalent bonds, in which two atoms share electrons.
  Review Figure 2.5

Carbon atoms form four covalent bonds, nitrogen three, oxygen two, and hydrogen one.
  Review Figures 2.6 and 2.7

Most biological compounds form by covalent bonding.

In a type of chemical reaction important in organisms, covalently bonded atoms change bonding

partners and the bond energies of the products differ from the bond energies of the reactants.
Review Figure 2.11

The molecules of many substances break apart in water into ions, which carry positive or negative electric charges.
Review Figures 2.8 and 2.16

Ions of opposite charge attract each other and may form ionic bonds.
Review Figure 2.9

Weaker bonds (hydrogen bonds, van der Waals interactions, and hydrophobic interactions) also help form large molecules and help molecules aggregate into larger structures.
Review Figure 2.17

Most of the chemistry of interest to biologists takes place in water.

The polarity of water makes it an exceptionally effective solvent.
Review Figures 2.15 and 2.16

Acids release hydrogen ions (protons), bases accept them.

The pH indicates how acidic or basic a solution is; buffers resist pH changes.
Review Figures 2.13 and 2.14

Various functional groups, such as hydroxyl, carbonyl, carboxyl, amino, sulfhydryl, and phosphate groups, give molecules some of their chemical properties.
Review Figure 2.20

SELF-QUIZ

1. The atomic number of an element
   a. equals the number of neutrons in an atom.
   b. equals the number of protons in an atom.
   c. equals the number of protons minus the number of electrons.
   d. equals the number of neutrons plus the number of protons.
   e. depends on the isotope.

2. The atomic weight of an element
   a. equals the number of neutrons in an atom.
   b. equals the number of protons in an atom.
   c. equals the number of electrons in an atom.
   d. equals the number of neutrons plus the number of protons.
   e. depends on the relative abundances of its isotopes.

3. Which of the following statements about all the isotopes of an element is *not* true?
   a. They have the same atomic number.
   b. They have the same number of protons.
   c. They have the same number of neutrons.
   d. They have the same number of electrons.
   e. They have identical chemical properties.

4. Which of the following statements about a covalent bond is *not* true?
   a. It is stronger than a hydrogen bond.
   b. One can form between atoms of the same element.
   c. Only a single covalent bond can form between two atoms.
   d. It results from the sharing of two electrons by two atoms.
   e. One can form between atoms of different elements.

5. Hydrophobic interactions
   a. Are stronger than hydrogen bonds.
   b. Are stronger than covalent bonds.
   c. Can hold two ions together.
   d. Can hold two nonpolar molecules together.
   e. Are responsible for the surface tension of water.

6. Which of the following statements about water is *not* true?
   a. It releases a large amount of heat when changing from liquid into vapor.
   b. Its solid form is less dense than its liquid form.
   c. It is the most effective solvent known.
   d. It is typically the most abundant substance in an active organism.
   e. It takes part in some important chemical reactions.

7. A solution with a pH of 9
   a. is acidic.
   b. is more basic than a solution with a pH of 10.
   c. has 10 times the hydrogen ion concentration of a solution with pH 10.
   d. has a hydrogen ion concentration of 9 molar.
   e. has a hydroxide ion concentration of 9 molar.

8. Which of the following compounds is an alcohol?
   a. O_2
   b. $CH_3CH_2CH_2OH$
   c. CH_3COOH
   d. C_3H_8
   e. CH_3COCH_3

9. Which of the following statements about the carboxyl group is *not* true?
   a. It has the chemical formula —COOH.
   b. It is an acidic group.
   c. It can ionize.
   d. It is found in amino acids.
   e. It has an atomic weight of 45.

10. Which of the following statements about amino acids is *not* true?
    a. They are the building blocks of proteins.
    b. They contain carboxyl groups.
    c. They contain amino groups.
    d. They do not ionize.
    e. They have both L- and D-isomers.

FOR STUDY

1. The elemental compositions of the universe, Earth's crust, and the human body differ sharply (Table 2.1). What factors might contribute to these differences?

2. Lithium (Li) is the element with atomic number 3. Draw the structures of the Li atom and of the Li^+ ion.

3. Draw the structure of a pair of water molecules held together by a hydrogen bond. Your drawing should indicate the covalent bonds.

4. The molecular weight of sodium chloride (NaCl) is 58.45. How many grams of NaCl are there in 1 l of a 0.1 M NaCl solution? How many in 0.5 l of a 0.5 M NaCl solution?

5. The side chain of the amino acid alanine is —CH_3 (see Figure 2.21). Draw the structures of the two optical isomers of alanine. The side chain of the amino acid glycine is simply a hydrogen atom (—H). Are there two optical isomers of glycine? Explain.

READINGS

Atkins, P. W. and J. A. Beran. 1992. *General Chemistry*, 2nd Edition. W. H. Freeman, New York. A first-rate textbook, beautifully illustrated.

Breed, A., T. Rodella and R. Basmajian. 1982. *Through the Molecular Maze*. William Kaufmann, Los Altos, CA. A short, inexpensive guide to the rudiments of chemical concepts and terminology needed by students in introductory courses on the life sciences.

Henderson, L. J. 1958. *The Fitness of the Environment*. Beacon Press, Boston. An essay written in 1912 about physical properties of water and carbon dioxide in relation to life. With a thought-provoking introduction.

Kotz, J. C. and K. F. Purcell. 1991. *Chemistry and Chemical Reactivity*, 2nd Edition. Saunders, Philadelphia. A well-illustrated modern textbook of general chemistry.

Lancaster, J. R. 1992. "Nitric Oxide in Cells." *American Scientist*, vol. 80, pages 248–259. A readable account of the multiple biological effects of this very small molecule.

Mertz, W. 1981. "The Essential Trace Elements." *Science*, vol. 213, pages 1332–1338. A review of the roles of more than a dozen elements needed in small amounts by animals if they are to function normally.

Zumdahl, S. 1992. *Chemical Principles*. D. C. Heath, Lexington, MA. A fine higher-level text for students with strong math backgrounds.

When we eat proteins, we are taking in molecules that were built up within the body of a plant or animal from smaller building blocks. In Chapter 2 we considered the structures of the smallest building blocks of organic molecules. In this chapter we see what organisms do with organic building blocks.

In consuming proteins, we first break them up into their constituent pieces and later reassemble the pieces into the chemically different proteins of our own bodies. This is not unlike picking up Lego toys that somebody else has made, taking them apart, and reassembling the parts to make the toys *we* want. One large molecule that our bodies must mass-produce (in kilogram amounts) is the hemoglobin that carries oxygen from lungs to working tissues via the bloodstream. Meat is mostly muscle, and muscle is mostly protein. We take apart the proteins in meat or vegetables and reassemble the pieces into our hemoglobin and other proteins.

In this chapter we take a brief look at the major classes of large molecules: lipids, carbohydrates, proteins, and nucleic acids. We begin to see that molecular structure (the way one piece fits with another) governs the way particular molecules function in the activities of living things.

LIPIDS

Our first class of large molecules is best understood by thinking about two behaviors that define these molecules: **Lipids** are insoluble in water but are readily soluble in organic (carbon-based) solvents such as ether, and they release large amounts of energy when they break down. Each of these properties is significant in the biology of these compounds. Because lipids do not dissolve in water and water does not dissolve in lipids, a mixture of water and lipids forms two distinct layers. Also, many biological materials that are soluble in water are much less soluble in lipids. Such materials include ions, sugars, and amino acids.

Suppose that you must design water-filled compartments, separated from each other and from their environment by barriers that limit the passage of materials. Based on the properties of lipids, a seemingly effective way to accomplish this is to use membranes containing lipids (Figure 3.1). This is, in fact, the system that has evolved in nature. Molecular traffic within an organism or into and out of its compartments is constrained by the properties of the lipid portion of the surrounding membrane. Compounds that dissolve readily in lipids can move rapidly through biological membranes, but compounds that are insoluble in lipids are prevented from passing, or

Different Plans for the Same Building Blocks

3

Large Molecules

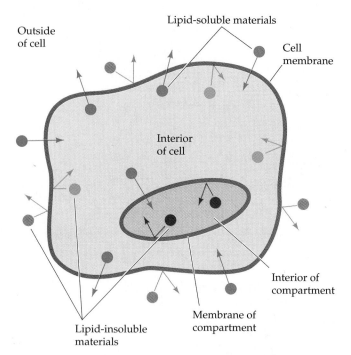

3.1 Lipids Assemble as Membranes Enclose Cells
Membranes made of lipids separate the cell from its environment; they also separate the contents of internal compartments from the rest of the cell. Materials that do not dissolve in lipids usually cannot pass through membranes, but lipid-soluble materials move through with relative ease.

must be transported across the membrane by specific proteins (see Chapter 5).

Lipids are marvelous storehouses for energy. By taking in excess food, many animal species deposit fat (lipid) droplets in their cells as a means for storing energy (Figure 3.2). Some plants, such as olives, avocados, sesame, castor beans, and all nuts, have substantial amounts of lipids in their seeds or fruits that serve as energy reserves for the next generation.

Triglycerides

One important group of lipids is the **triglycerides**, also known as *simple lipids*. Triglycerides that are solid at room temperature (20°C) are called **fats**; those that are liquid at room temperature are called **oils**. Triglycerides are composed of two types of building blocks: **fatty acids** and **glycerol**. Glycerol is a small alcohol with three hydroxyl (—OH) groups. Fatty acids are carboxylic acids with long hydrocarbon tails. Four typical fatty acids are shown in Figure 3.3. Palmitic acid is found in animal fats. Like palmitic acid, stearic acid is a **saturated** fatty acid because its hydrocarbon tail contains no double bonds. Oleic acid is **unsaturated**. Its double bond, which is near the middle of the hydrocarbon chain, causes a kink in the molecule. Fatty acids, such as linoleic acid, that

have more than one double bond are **polyunsaturated**. These molecules have multiple kinks. Unsaturated and polyunsaturated fatty acids can accept hydrogen atoms—that is, they can become hydrogenated. The addition of two hydrogen atoms across the double bond of oleic acid, for example, would produce stearic acid.

Three fatty acid molecules combine with a molecule of glycerol to form a molecule of a triglyceride (Figure 3.4). The three fatty acids in one triglyceride molecule are not always the same length, nor do they all have to be either saturated or unsaturated. The kinks associated with double bonds are important in determining the fluidity and melting point of a lipid. Triglycerides with short or unsaturated chains are usually oily liquids; those with long and saturated chains are waxy solids. Animal fats such as lard and tallow are usually solids with long-chain saturated or singly unsaturated fatty acids. In these fats, hydrocarbon chain lengths range from 10 to 20 carbon

3.2 Energy to Fight the Weather
These Alaskan walrus spend much of their time in frigid water. Lipids, deposited as layers of body fat, insulate their bodies against the cold and also store energy efficiently.

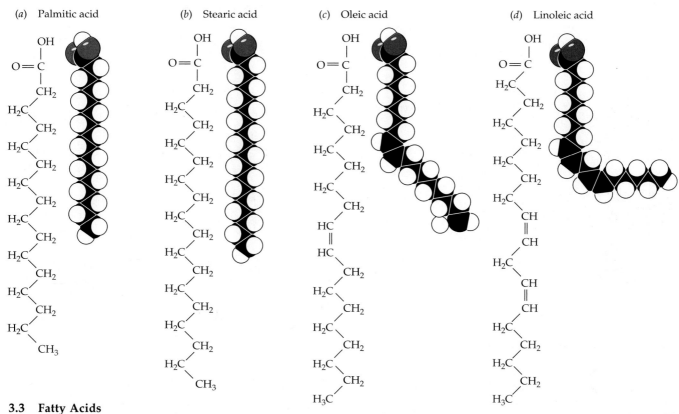

(a) Palmitic acid (b) Stearic acid (c) Oleic acid (d) Linoleic acid

3.3 Fatty Acids

(a) The absence of double bonds between carbon atoms in the chain means that palmitic acid is a saturated fatty acid. The straight-chain configuration seen in the model of the molecule is characteristic of saturated fatty acids. (b) Stearic acid has two more carbons and four more hydrogens than palmitic acid and is also saturated. (c) Oleic acid has a double bond between two carbons in the chain and is therefore unsaturated. The double bond causes the molecule to bend. (d) With two double bonds in its chain, linoleic acid is polyunsaturated.

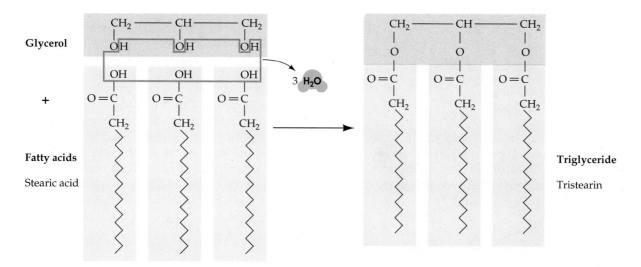

3.4 A Triglyceride and Its Components

In an example of triglyceride formation, tristearin forms from glycerol and three molecules of stearic acid by condensation. Condensations release water molecules. In living things the reaction is more complex, but the end result is as shown here. The jagged lines represent hydrocarbon chains.

Phosphatidate

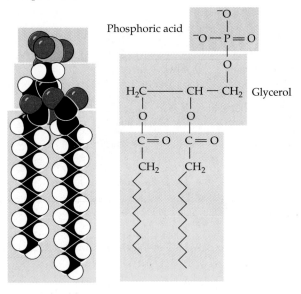

Phosphoric acid

Glycerol

Phosphatidyl ethanolamine

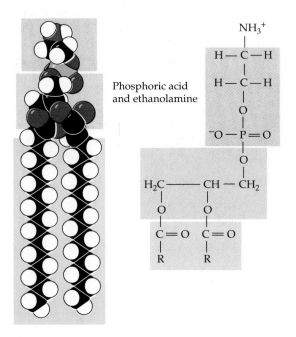

Phosphoric acid
and ethanolamine

Phosphatidyl choline

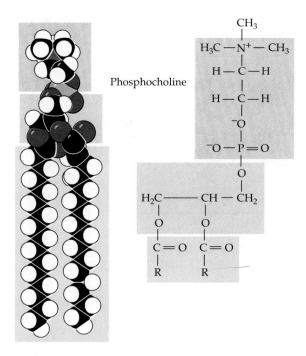

Phosphocholine

3.5 Some Phospholipids
Each phospholipid consists of a phosphorus-containing molecule (yellow shading), glycerol (red shading), and two molecules of fatty acid (blue shading). The fatty acid hydrocarbon chains may be shown as jagged lines or abbreviated to R (for "residue"). The hydrocarbon chains are nonpolar. The phosphorus-containing portions are electrically charged.

atoms. The triglycerides of plants tend to be less saturated, oily liquids. Natural peanut butter, for example, contains a great deal of oil. Peanut butter manufacturers often hydrogenate their product to reduce the number of double bonds and give a saturated, solid product.

Phospholipids

Like triglycerides, **phospholipids** have fatty acids bound to glycerol. In phospholipids, however, any of certain phosphorus-containing compounds replaces one of the fatty acids (Figure 3.5). Many phos-

pholipids are important constituents of biological membranes. If you examine the structure of phospholipids closely, you will find it easy to understand how they are oriented in membranes. Because the phosphorus-containing portion of the phospholipid molecule carries one or more electric charges, this portion is **hydrophilic** (water-loving; remember that water is a polar molecule). The two fatty acid regions, however, are **hydrophobic** (water-fearing). In a biological membrane, phospholipids line up in such a way that the nonpolar, hydrophobic "tails" pack tightly together to form the interior of the membrane, and the phosphorus-containing "heads" face outward (some to one side of the membrane and some to the other), where they interact with water, which is excluded from the interior of the membrane. The phospholipids thus form a bilayer, a sheet two molecules thick (Figure 3.6).

Other Lipids

The term *lipid* defines compounds not on the basis of structural similarity but in terms of their solubility. Remember that lipids are insoluble in water but are readily soluble in organic solvents such as ether, chlo-

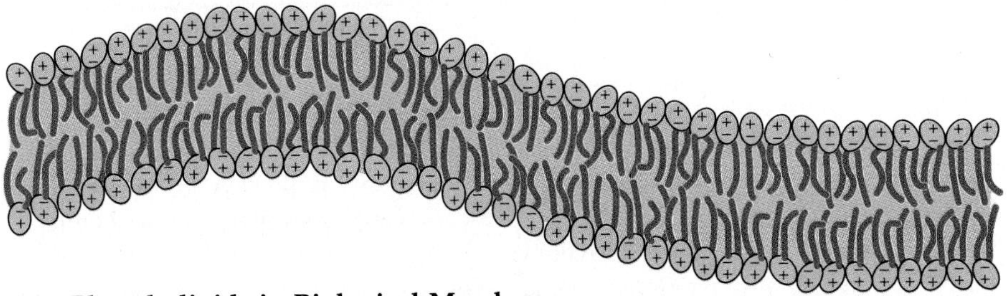

3.6 Phospholipids in Biological Membranes
Hydrophobic interactions bring the "tails" together in the interior of a phospholipid bilayer. Hydrophilic "heads" face outward on both sides of the membrane. The details of this important structure are the subject of Chapter 5.

roform, or benzene. Two more groups of compounds with these properties (and hence classifiable as lipids) are the carotenoids and the steroids.

The **carotenoids** are a family of light-absorbing pigments found in plants and animals (Figure 3.7). Beta-carotene (β-carotene) is one of the pigments that traps light energy in leaves as part of the process of photosynthesis (see Chapter 8). It is the β-carotene in plants that senses light and causes their parts to grow toward or away from the light (a behavior called phototropism, discussed in Chapter 34). In humans, a molecule of β-carotene can be broken down into two vitamin A molecules, from which we make the pigment rhodopsin that is required for vision (Chapter 39). Another derivative of vitamin A is used in treating a form of cancer (Box 3.A). Carotenoids are responsible for the color of carrots, tomatoes, pumpkins, egg yolks, and butter.

The **steroids** are a family of organic compounds whose multiple rings share carbons (Figure 3.8). Some steroids are important constituents of membranes. Others are hormones, chemical signals that carry messages from one part of the body to another (Chapter 34). Testosterone is a steroid hormone that regulates sexual development in male vertebrates (animals with backbones); the chemically related estrogens play a similar role in females. Cortisone and related hormones play a wide variety of regulatory

roles in the digestion of carbohydrates and proteins, salt and water balance, and sexual development. Vitamin D is a steroid that regulates the absorption of calcium from the intestines. It is necessary for the proper deposition of calcium in bones; a deficiency of vitamin D can lead to rickets, a bone-softening disease. Irradiation of certain other steroids with sunlight or ultraviolet light converts them to vitamin D.

Cholesterol is synthesized in the liver. It is the starting material for making testosterone and several other steroid hormones and for the bile salts that help to get fats into solution so they can be digested. We absorb cholesterol from foods such as milk, butter, and animal fats. When we have too much cholesterol in our blood, it is deposited in our arteries (along with other substances), a condition that may lead to arteriosclerosis and heart attack.

The chemical structures of lipids are quite varied (see Figures 3.3 through 3.8). Their diversity matches the variety of their functions in living things: energy storage, digestion, membrane structure, bone formation, vision, and chemical signaling. Most lipids can be synthesized in the bodies of animals; the synthesis and storage of fats is an important means of locking energy away until it is needed. The few lipids that cannot be synthesized must be obtained in small amounts from the diet. For humans, the diet must include three particular unsaturated fatty acids and the fat-soluble vitamins: A, D, E, and K.

Although lipids are structurally and functionally diverse, there is one thing they share that sets them apart from the large molecules we will consider

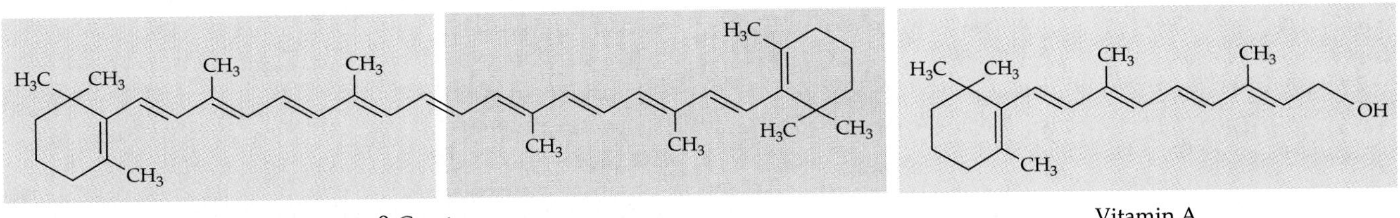

β-Carotene

Vitamin A

3.7 Carotenoids
These carotenoids have carbon atoms at each angle of the rings and chains. (Omitting the Cs that represent carbon atoms is standard chemical shorthand.) Methyl groups are indicated. Notice that β-carotene is symmetrical around the central double bond. Splitting β-carotene in the middle produces two vitamin A molecules.

BOX 3.A

Making a Leukemia Grow Up

Among the forms of cancer, leukemia has been one of the hardest to treat. In general, we treat cancers by radiation therapy or chemotherapy. Radiation therapy is most effective against localized cancers because only the particular part of the body affected needs to be subjected to radiation. Leukemia, however, is a cancer of the white blood cells, which travel throughout the body.

Recently, biomedical scientists have achieved great success with a novel treatment for a particular type of leukemia called acute promyelocytic leukemia. A substance derived from vitamin A, all-*trans*-retinoic acid, eliminates all symptoms of this leukemia in a high percentage of patients who take the acid orally. Medical researchers have learned that many victims of acute promyelocytic leukemia have a genetic defect. People with or without the defect produce retinoic acid naturally in their bodies, but people with the defect cannot process the acid normally. By ingesting additional retinoic acid as a supplement, patients with the defect can combat their cancer.

Why should this lipid, retinoic acid, act against certain cancer cells?

Does it kill them? In fact, it does *not* kill them. Rather, it makes them "grow up." Like several other cancers, acute promyelocytic leukemia results from the failure of the cancerous cells to complete normal cell development. Biologists have speculated for some time that retinoic acid promotes the normal development of cells in bird wings. Now, medical researchers propose that when it is given to patients with this type of leukemia, retinoic acid causes the cancerous cells to complete normal development and cease dividing.

Like many biologically active compounds, retinoic acid has multiple effects. For example, it is also a treatment for acne, in the skin cream Retin-A.

---

throughout the rest of the chapter: Lipid molecules do not form polymers in reactions of the type discussed in the next section.

FROM MONOMERS TO POLYMERS

The largest molecules covered in this chapter are called **macromolecules**—giant molecules, or aggregates of molecules, with molecular weights in excess of 1,000 daltons. All macromolecules are **polymers**: molecules made by the combination of many smaller molecules, called **monomers**. An **oligomer** contains only a few monomers. A cell can combine a limited variety of monomers into a near-infinite variety of polymers.

Macromolecules perform many essential functions in organisms. These functions arise directly from the structures of the molecules. Some macromolecules fold into globular forms with surface features that enable them to recognize and interact with certain other molecules. Others form long, fibrous systems that provide strength and rigidity to parts of an organism. Still others contract and allow the organism to generate the force to move itself. Some macromolecules aggregate to form structures that determine what materials enter or leave the compartments within an organism; others accelerate chemical reactions in cells. There is a flow of *information* among all the classes of macromolecules, but one class (the nucleic acids, discussed later in this chapter) specializes in information.

(a) Testosterone (b) Cortisone (c) Vitamin D (d) Cholesterol

3.8 Steroids Share a Common Ring Structure
Among the important steroids in vertebrates are (a) the male sex hormone testosterone, (b) the hormone cortisone, (c) vitamin D, and (d) cholesterol.

Macromolecules in living things are polymers built from simpler monomers through a series of reactions called **condensations** or **dehydrations**. These reactions are of the general type

$$A—H + B—OH \rightarrow A—B + H_2O$$

(A—H is a molecule consisting of a hydrogen atom attached to another part, A; B—OH is a molecule consisting of an —OH group attached to another part, B.) The product A—B is formed, along with a molecule of water; the atoms of water are derived from the reactants—one hydrogen atom from one reactant, and an oxygen atom and the other hydrogen atom from the other reactant. **Reactants** are the molecules undergoing a chemical reaction.

The polymerization reactions that produce the different kinds of macromolecules differ in detail. In all cases, energy must be added to the system for polymers to form. Other kinds of specific molecules participate; their function is to activate the reactants—to provide the necessary energy for the reactions to be carried out. Large molecules are assembled through the repeated condensations of activated monomers.

CARBOHYDRATES

A **carbohydrate** is a compound based on the general formula CH_2O. Carbohydrates are a diverse group of compounds with molecular weights ranging from less than 100 to hundreds of thousands. They fall into three categories: the **monosaccharides**, or *simple sugars*, which are monomers; the **oligosaccharides**, made up of a few monosaccharides linked together; and the **polysaccharides**, polymeric carbohydrates that include starches, glycogen, cellulose, and many other important biological materials. (*Mono-* means "single," *oligo-* means "few," and *poly-* means "many"; *saccharide* means "sugar.") There is no clear dividing line between a large oligosaccharide and a small polysaccharide, for these are simply terms of convenience used to separate "classes" within what is really a continuum of compounds of various sizes.

The general formula for carbohydrates, CH_2O, is true for monosaccharides, all of which have the same number of carbon and oxygen atoms but twice as many hydrogen atoms as oxygen atoms. The formulas of oligosaccharides and polysaccharides, however, differ slightly from this general formula.

Monosaccharides

All living cells contain the monosaccharide **glucose**, $C_6H_{12}O_6$. Green plants produce it by photosynthesis (see Chapter 8). Cells metabolize it to yield energy during cellular respiration (see Chapter 7). Glucose exists in both straight-chain and ring forms, in equilibrium with each other (Figure 3.9). The two distinct ring forms (α- and β-glucose) differ only in the placement of the —H and —OH attached to carbon 1. (The convention for numbering carbons shown in Figure 3.9 is used throughout this book). Although chemically and physically distinct substances, α- and β-glucose interconvert constantly in aqueous solution.

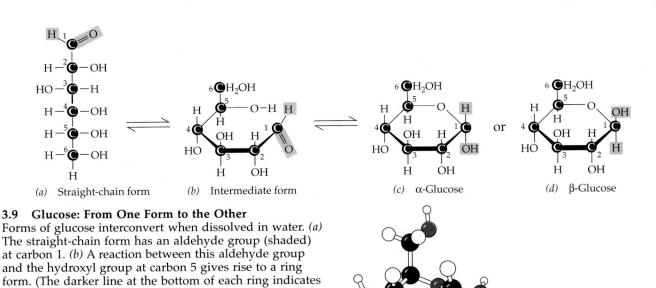

(a) Straight-chain form   (b) Intermediate form   (c) α-Glucose   (d) β-Glucose

3.9 Glucose: From One Form to the Other
Forms of glucose interconvert when dissolved in water. (*a*) The straight-chain form has an aldehyde group (shaded) at carbon 1. (*b*) A reaction between this aldehyde group and the hydroxyl group at carbon 5 gives rise to a ring form. (The darker line at the bottom of each ring indicates that that edge of the molecule extends toward you, while the upper edge extends back into the page.) (*c,d*) Depending on the orientation of the aldehyde group when the ring closes, either of two rapidly and spontaneously interconverting molecules—α-glucose or β-glucose—forms. The ball-and-stick model depicts α-glucose; it should be easy to visualize a comparable model of β-glucose.

Three-carbon sugar

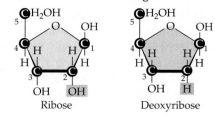

Glyceraldehyde

Five-carbon sugars

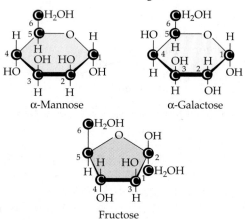

Ribose   Deoxyribose

Six-carbon sugars

α-Mannose   α-Galactose

Fructose

3.10   Monosaccharides

The three-carbon sugar glyceraldehyde has the formula $C_3H_6O_3$; its common form is the straight chain shown. Ribose and deoxyribose each have five carbons; their formulas are $C_5H_{10}O_5$ and $C_5H_{10}O_4$, respectively. All of the six-carbon sugars shown here have the formula $C_6H_{12}O_6$, but they are chemically and biologically distinct from one another.

Some other simple sugars are illustrated in Figure 3.10. Many monosaccharides have the same formula as glucose, $C_6H_{12}O_6$, including fructose ("fruit sugar"), mannose, and galactose. These compounds are isomers of each other: They are composed of the same kinds and numbers of atoms, but the atoms combined differently in each to yield different arrangements. Five-carbon sugars are referred to collectively as **pentoses**, and six-carbon sugars are called **hexoses**. Some pentoses are found primarily in the cell walls of plants, as are several of the hexoses. Two pentoses are of particular importance: **ribose** and **deoxyribose**, which form part of the backbones of RNA and of DNA, respectively. Ribose and deoxyribose differ by only one oxygen atom associated with one of the carbon atoms, carbon 2.

Disaccharides

Larger carbohydrates form by the bonding together of two or more monosaccharide molecules. The monosaccharides become covalently coupled by condensation reactions to form specific oligosaccharides and polysaccharides. The linkages between the monosaccharides are **glycosidic linkages**. The smallest oligosaccharides are the disaccharides and the trisaccharides, made up of two and three simple sugars, respectively. If one glucose molecule combines with another, as shown in Figure 3.11, the disaccharide product must be one of two types: α-linked or β-linked, depending on whether the molecule that bonds at the carbon 1 position is α-glucose or β-glucose. An α linkage with carbon 4 of a second glucose molecule gives maltose, whereas a β linkage gives cellobiose. Maltose and cellobiose are disaccharides, both with the formula $C_{12}H_{22}O_{11}$. Both are composed of two glucose molecules (minus one molecule of water), but they are different compounds; they are recognized by different enzymes and undergo different chemical reactions. Two other common disaccharides are sucrose and lactose. Sucrose (common table sugar; also $C_{12}H_{22}O_{11}$) is made from one molecule of glucose and one of fructose. Lactose (milk sugar) consists of glucose and galactose.

Polysaccharides

As we saw in Figure 3.11, maltose consists of two glucose units connected by an α linkage. Imagine a trisaccharide (three glucose units), a tetrasaccharide (four glucose units), and finally a giant polysaccharide consisting of hundreds or thousands of glucose units, each connected to the next by an α glycosidic linkage from carbon 1 of one unit to carbon 4 of the next. This polymer is **starch**, an important storage compound.

A similar giant polysaccharide, made up solely of glucose but with the individual units connected by β linkages instead of α linkages, is **cellulose** (Figure 3.12a). Cellulose is the predominant component of plant cell walls and by far the most abundant organic compound on this planet. Both starch and cellulose are composed of nothing but glucose; yet their biological functions and chemical and physical properties are entirely different. Enzymes that digest one do not affect the other at all.

Starch is not a single chemical substance; rather, the term denotes a large family of giant molecules of broadly similar structure. All starches are polymers of glucose with α linkages. All are large, but some are enormous, containing tens of thousands of glucose units. An important distinguishing characteristic of starches is the degree of branching. The starches that store glucose in plants, called **amylose**, are not highly branched (Figure 3.12b). The highly branched

α-Glucose + β-Glucose → (Formation of α-linkage) + H_2O → **β-Maltose** (α-Glucose + β-Glucose)

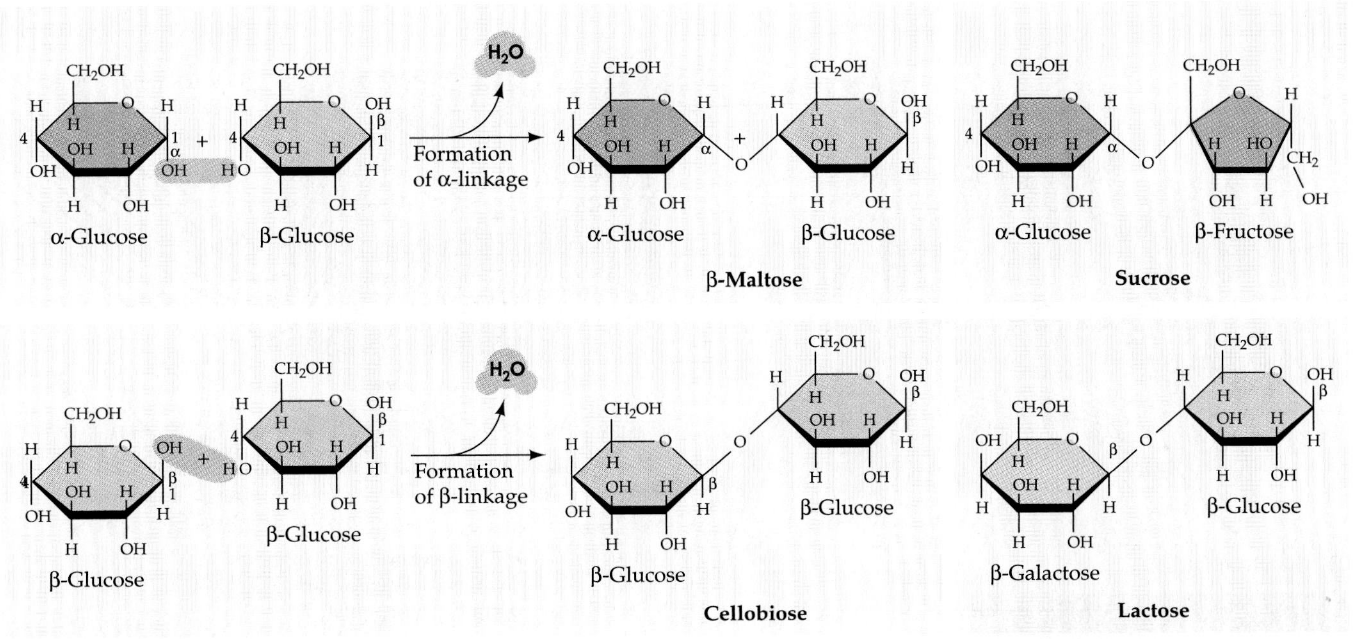

α-Glucose + β-Fructose → **Sucrose**

β-Glucose + β-Glucose → (Formation of β-linkage) + H_2O → **Cellobiose** (β-Glucose + β-Glucose)

β-Galactose + β-Glucose → **Lactose**

3.11 Disaccharides Are Composed of Two Monosaccharides

In the reaction shown at the top left (a simplified version of the reaction in nature), maltose is produced when an α-1,4 linkage forms between two glucose molecules, with the hydroxyl group on carbon 1 of one glucose in the α (down) position as it reacts with the 4-hydroxyl group of the other. In cellobiose (bottom left), the two glucoses are linked by a β-1,4 linkage. Lactose (bottom right) is made by a β linkage between carbon 1 of galactose and carbon 4 of glucose. In sucrose (top right), carbon 1 of glucose is joined by an α-1,2 linkage to carbon 2 of fructose.

polysaccharide that stores glucose in animals is **glycogen** (Figure 3.12c). Animals use glycogen to store energy in liver and muscle.

What do we mean when we say that starch and glycogen are storage compounds for energy? Very simply, these compounds can readily be depolymerized to yield glucose monomers. Glucose, in turn, can be further digested, or metabolized—that is, it can undergo chemical reactions—to yield energy for cellular work. Alternatively, glucose can be metabolized so that its carbon atoms are rearranged to form the skeletons of other compounds. Glycogen and starch are thus storage depots for carbon atoms as well as for energy. Each is chemically stable but is readily mobilized by digestion and further metabolism.

Derivative Carbohydrates

Derivative carbohydrates deviate from the general formula for carbohydrates by containing elements other than C, H, and O. Examples include sugar phosphates, amino sugars, and chitin (Figure 3.13). A number of sugar phosphates, such as fructose 1,6-bisphosphate, are important intermediates in cellular respiration (Chapter 7) and photosynthesis (Chapter 8). Sugar phosphates have phosphate groups attached to one or more —OH groups of the parent sugar. The two **amino sugars** shown in the figure, glucosamine and galactosamine, have an amino group in place of an —OH group. Galactosamine is a major component of cartilage, the material that forms caps on the ends of bones and stiffens the protruding parts of the ears and nose. The polymer chitin is made from a derivative of glucosamine. Chitin is the principal structural polysaccharide in the skeletons of insects and their relatives such as crabs and lobsters, as well as in the cell walls of fungi. Fungi and insects (and their relatives) constitute more than 80 percent of the species ever described, and chitin is another of the most abundant substances on Earth.

PROTEINS

In Chapter 2 we considered the amino acids (small molecules containing both carboxyl and amino groups). These are the monomers from which a fascinating set of polymers are formed—the **proteins**. Proteins account for many of the mechanical elements of living things, from parts of subcellular membranes to skin, bones, and tendons. In vertebrates, proteins called immunoglobulins form a major line of defense against foreign organisms. The specialized molecules needed to bring about all biochemical reactions make up a major class of proteins called enzymes. Our every movement results from the contraction and relaxation of muscles, resulting in turn from the delicately regulated sliding of particular pro-

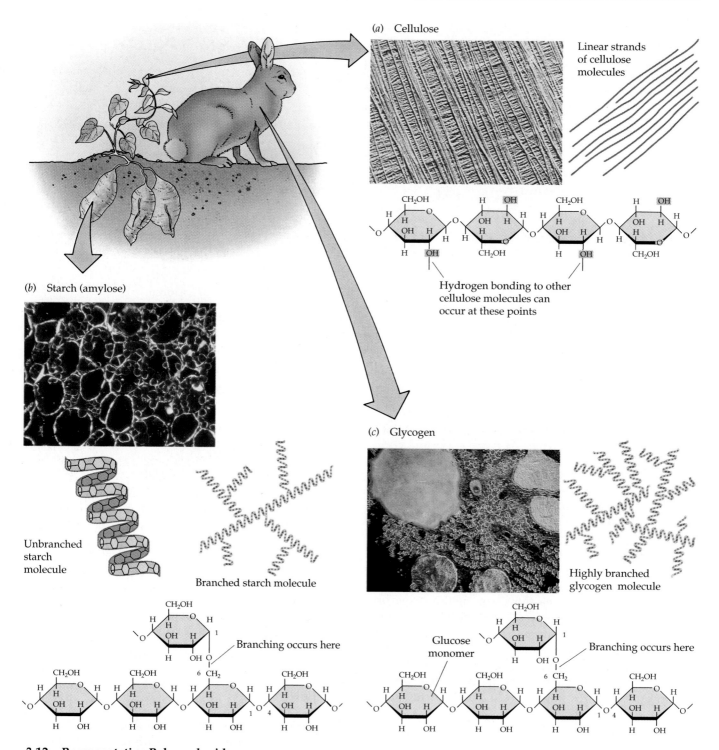

(a) Cellulose

Linear strands of cellulose molecules

Hydrogen bonding to other cellulose molecules can occur at these points

(b) Starch (amylose)

Unbranched starch molecule

Branched starch molecule

Branching occurs here

(c) Glycogen

Highly branched glycogen molecule

Glucose monomer

Branching occurs here

3.12 Representative Polysaccharides
(a) Cellulose is an unbranched polymer of glucose. Many adjacent cellulose molecules form the cellulose fibrils in photosynthetic cells. *(b)* In starches such as amylose, branching may occur. In the micrograph, a red dye stains the amylose grains in sweet potato cells. *(c)* Glycogen from animal cells differs from plant starch only in being more extensively branched. The tinted electron micrograph shows part of a human liver cell; the pink-tinted bodies are glycogen granules.

teins in muscle cells past one another. Still other proteins act as adjustable channels through which sodium ions (Na$^+$), potassium ions (K$^+$), and other ions are passed from one side of a nerve-cell membrane to the other, resulting in phenomena such as the transmission of electric signals along a nerve. To understand this stunning variety of functions, we must first explore the structure of these molecules.

Amino Acids

Twenty different amino acids are found in proteins. The **side chains** of amino acids show a wide variety of chemical properties. Side chains control the function of a protein—they are the reactive groups in proteins. The order of a protein's side chains determines how it folds into a three-dimensional configuration (discussed later in this chapter). Despite their importance, side chains are commonly left out of structural formulas, where they are represented simply by "R" (for "residue"); thus they are sometimes called R groups. Side chains are highlighted in the amino acid structural formulas in Table 3.1.

One useful classification of amino acids is based on whether their side chains are electrically charged, polar but uncharged, or hydrophobic. There are two groups of amino acids with electrically charged side

chains: those with positive charges and those with negative charges. All five charged amino acids are very hydrophilic. The four amino acids with polar but uncharged side chains tend to form hydrogen bonds readily, both with water and with other molecules. They, too, are hydrophilic. The side chains of eight other amino acids either are hydrocarbon or are very slightly modified from hydrocarbons; hence they are hydrophobic.

Three amino acids—cysteine, glycine, and proline—are special cases, although their side chains are generally hydrophobic. Two cysteine side chains can lose hydrogen atoms so that their sulfur atoms are joined by a covalent bond in a **disulfide bridge** (Figure 3.14). Hydrogen bonds and disulfide bridges help determine how a protein chain folds. When cysteine is not part of a disulfide bridge, its side chain is very hydrophobic. The glycine side chain is just a hydrogen atom; thus glycines may fit into tight corners in the interior of a protein molecule, where a larger side chain could not fit. Proline differs from other amino acids because it possesses a modified amino group (see Table 3.1).

Peptide Linkages

When amino acids polymerize, the carboxyl group of one amino acid reacts with the amino group of another, undergoing a condensation reaction and forming a **peptide linkage**. Figure 3.15 gives a simplified description of the reaction (actually, other molecules must activate the reactants, and there are intermediate steps). A linear polymer of amino acids connected by peptide linkages is a polypeptide. A protein is made up of one or more polypeptides. At one end of the polypeptide molecules is a free amino group. This end is the N-terminus, named for the nitrogen atom in the amino group. At the other end of the polypeptide—the C-terminus—there is a free carboxyl group. The other amino and carboxyl groups are bound in peptide linkages. Thus a protein has direction. For one example, the dipeptide glycine–alanine, in which glycine has the free amino group, differs from alanine–glycine, in which alanine has the free amino group.

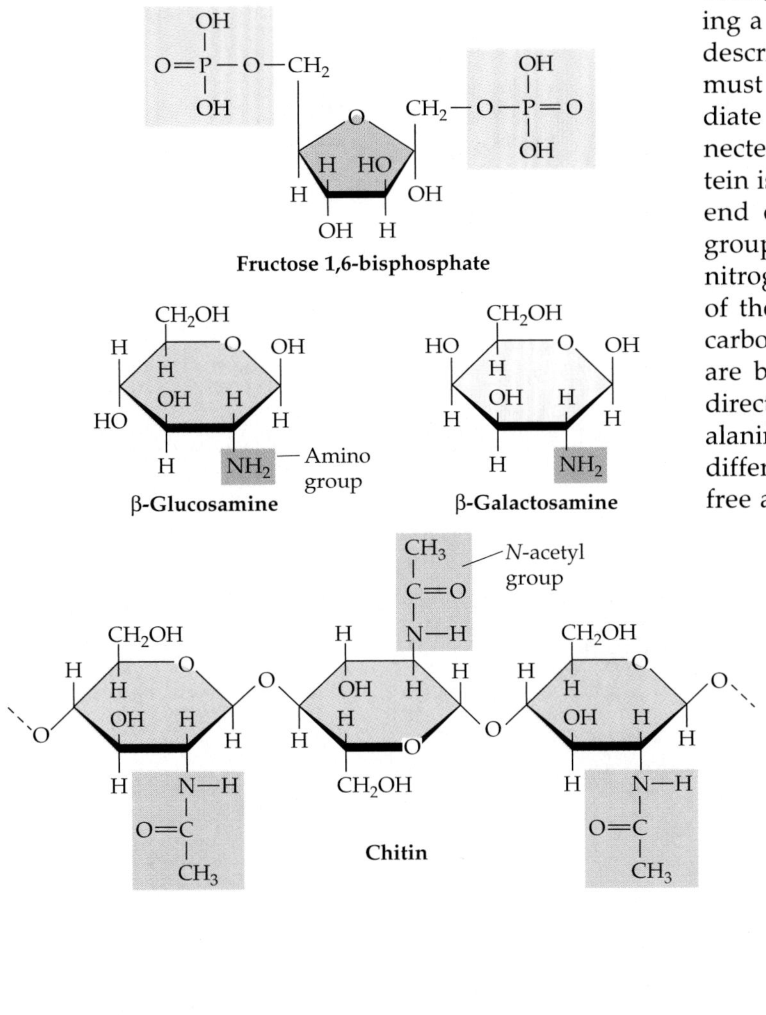

3.13 Derivative Carbohydrates
(a) Fructose 1,6-bisphosphate is a sugar phosphate; the numbers in its name refer to the bonding of the phosphate groups (shaded yellow) to carbons 1 and 6 of the sugar. (b) The amino groups on the amino sugars β-glucosamine and β-galactosamine are shown in green; recall that the β refers to the position of the –OH group on carbon 1. Chitin is a polymer of *N*-acetylglucosamine; *N*-acetyl groups are shown in pale green.

TABLE 3.1
Twenty amino acids found in proteins

A. *Amino acids with electrically charged side chains*

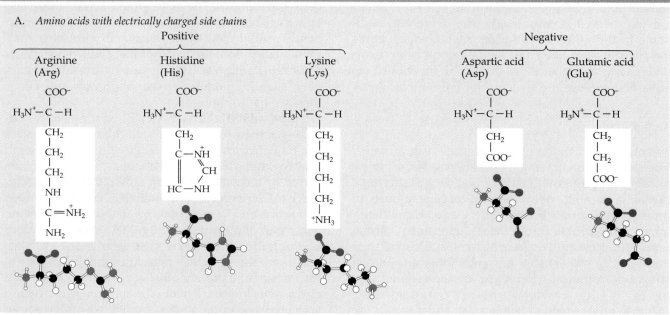

B. *Amino acids with polar but uncharged side chains*

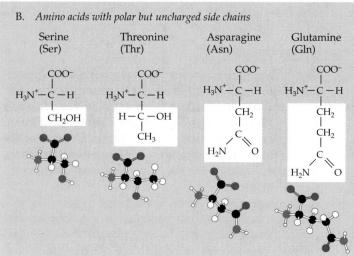

C. *Special cases*

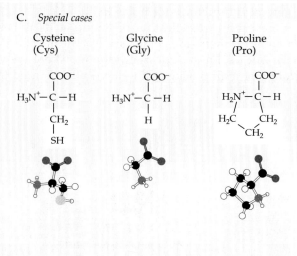

D. *Amino acids with hydrophobic side chains*

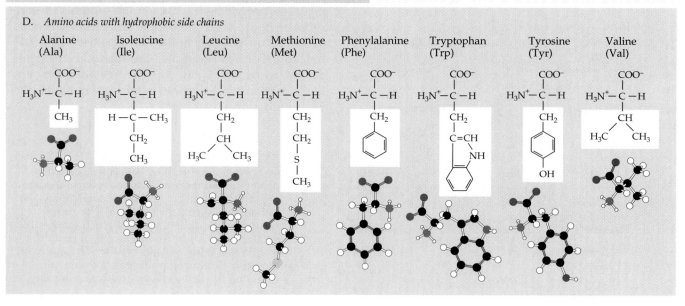

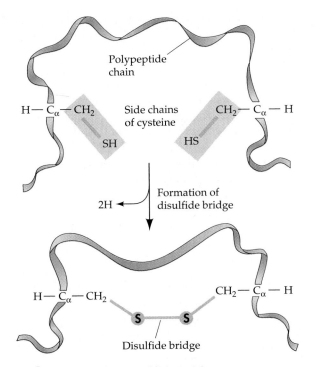

3.14 Formation of a Disulfide Bridge
The –SH groups on two cysteine side chains in a chain of amino acids can react to form a linkage between the two sulfur atoms. Such disulfide bridges are important in maintaining the proper three-dimensional shapes of protein molecules.

In the peptide linkage, the C=O oxygen carries a slight negative charge, whereas the N—H hydrogen is slightly positive. This asymmetry of charge favors hydrogen bonding (see Chapter 2) within the protein molecule itself and with other molecules, contributing to both the structure and the function of many proteins.

LEVELS OF PROTEIN STRUCTURE

Primary Structure

Protein structure is elegant and complex—so complex that it is described as consisting of several levels. The precise sequence of amino acids in the unbranched chain of a polypeptide (which is a linear polymer) constitutes the **primary structure** of a protein (Figure 3.16). This sequence is dictated by the precise sequence of monomers (which are called nucleotides) in a linear segment of a DNA molecule. The elucidation of this relationship between DNA primary structure and protein primary structure was one of the triumphs of molecular biology (it will be described in Chapter 11).

The theoretical number of different proteins is enormous. There are 20 different amino acids; this means there are $20 \times 20 = 400$ distinct dipeptides, and $20 \times 20 \times 20 = 8,000$ different tripeptides. Imagine this process of multiplying by 20 extended to a protein made up of 100 monomers (considered a small protein): There could be 20^{100} of these small proteins, each with its own distinctive primary structure. The higher levels of protein structure—from

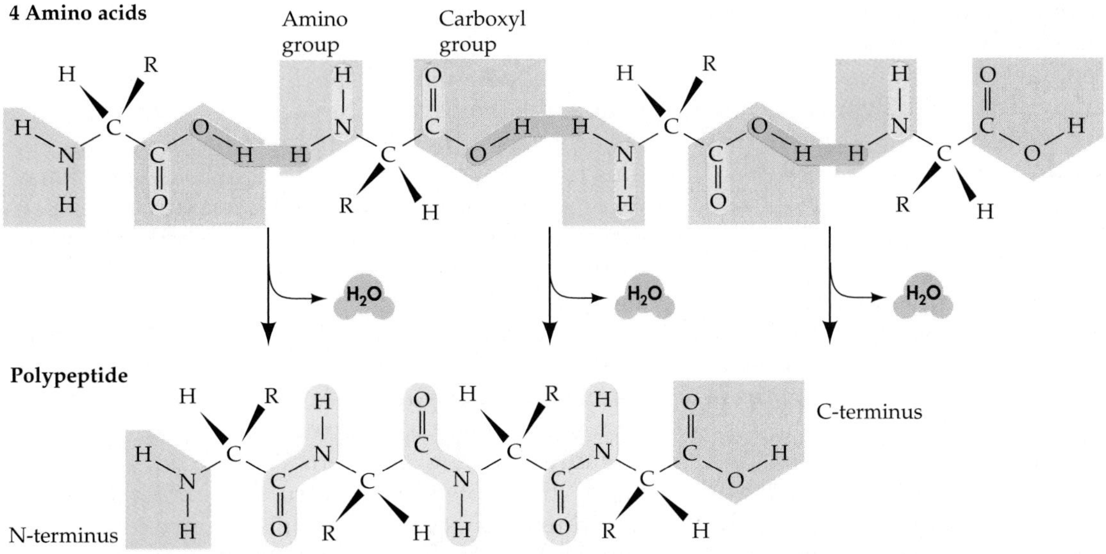

4 Amino acids

Amino group
Carboxyl group

H_2O H_2O H_2O

Polypeptide

C-terminus

N-terminus

3.15 Formation of Peptide Linkages
In this depiction, four amino acids combine across their amino and carboxyl groups, as indicated in yellow; the resulting peptide linkages create a polypeptide. A molecule of water (blue) is lost as each peptide linkage forms. (In living things the reaction is substantially more complex, but the end result is as shown here.)

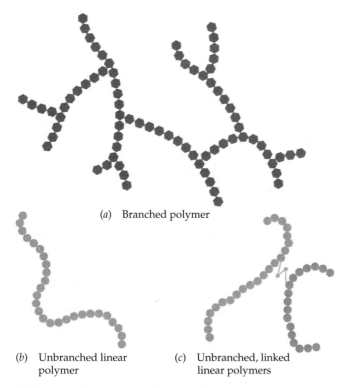

(a) Branched polymer

(b) Unbranched linear polymer

(c) Unbranched, linked linear polymers

3.16 Branched versus Linear Polymers
(a) Some biological polymers, such as the carbohydrate glycogen (see Figure 3.12c) are highly branched, as depicted in the generalized molecule here. Proteins, however, are unbranched (b), although the chains of amino acids may be linked together, as shown in (c).

local coiling and folding to the overall shape of the molecule—all derive from the primary structure. By presenting side chains of differing character (hydrophilic or hydrophobic, for example) in a specific and unique order, the precise sequence of amino acids in a given protein determines the ways in which the polypeptide chain can twist and fold. By twisting and folding, each protein adopts a specific structure that distinguishes it from every other protein.

Secondary Structure

Although the primary structure of each type of protein is unique, the secondary structure of many different proteins may be the same. A protein's **secondary structure** consists of regular, repeated patterns of orientation of parts of a polypeptide chain. One type of secondary structure, the **α helix** (alpha helix), is a right-handed coil "threaded" in the same direction as a standard wood screw. The twisting of a single polypeptide chain about its axis often allows hydrogen bonds to form between amino acids that are four monomers apart along the chain (Figure 3.17, top left). When this pattern of hydrogen bonding is established repeatedly over a segment of the protein, it stabilizes the twisted form, resulting in an α helix.

The ability of a protein to form an α helix depends on its primary structure: Certain amino acids have side chains that distort the coil or otherwise prevent the formation of hydrogen bonds.

Alpha helical secondary structure is particularly evident in the fibrous structural proteins called keratins. These include most of the protective tissues found in animals, such as fingernails and claws, skin, hair, and wool. Hair can be stretched because this requires breaking only hydrogen bonds in an α helix, rather than breaking covalent bonds; when the tension on the hair is released, both the helix and the hydrogen bonds re-form.

Silk is an example of a protein with another type of secondary structure, the **β-pleated sheet**. Here the protein chains are almost completely extended and are bound into sheets by hydrogen bonds connecting one chain to another (Figure 3.17, right). In many proteins, regions of β-pleated sheet are formed by bonding between different parts of the same polypeptide chain.

A third type of secondary structure, the triple helix, is found in collagen (Figure 3.17, bottom left). Collagen, an important protein found in cartilage, tendons, the underlayers of skin, and the cornea of the eye, consists of three polypeptide chains twisted around each other like the strands of a cable. Hydrogen bonds connect the chains, resulting in a structure that is strong, rigid, and unstretchable. The tail of a rat, under the skin, is almost pure collagen.

Tertiary Structure

The overall shape of a whole polypeptide molecule is its **tertiary structure**. The α helices and β-pleated sheets sometimes predominate throughout a protein molecule, determining the tertiary structure. More frequently, however, only limited portions of the molecule have these secondary structures, and they thus make only minor contributions to overall shape. A complete description of the tertiary structure specifies the location of every atom in the molecule in three-dimensional space, in relation to all the other atoms. The tertiary structure of the protein lysozyme is represented in Figure 3.18. Bear in mind that both this tertiary structure and the secondary structure emphasized in Figure 3.18c derive entirely from the protein's primary structure. If lysozyme is heated carefully, causing the tertiary structure to break down, the protein will return to its normal tertiary structure when it cools. The only information needed to specify the unique shape of the lysozyme molecule is the information contained in its primary structure.

Myoglobin is an important protein (Figure 3.19). Its function—to store oxygen in certain animal tissues—is discussed in Chapter 41. Myoglobin has 153 amino acids in its single polypeptide chain; there are

The α helix

The α helix is a secondary structure found in many proteins. The atoms in the relatively rigid plane of the peptide linkages are in color, and the hydrogen bonds that stabilize the helix are shown as red dotted lines.

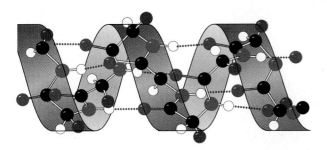

This computer drawing gives a three-dimensional sense of the relative positions of the atoms. Carbons are black, oxygens red, hydrogens white, and nitrogens blue. R groups are shown in purple.

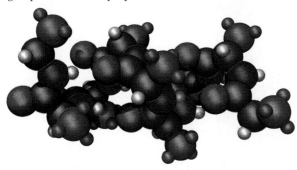

The triple helix

The secondary structure of the protein collagen is a triple helix of helical polypeptide chains (left). Such a triple helix is called tropocollagen. On the right, a number of triple helices of tropocollagen join in parallel fashion to create a strong, flexible collagen fibril; several collagen fibrils are shown here, magnified about 20,000 times. The spacing between black bands corresponds to the length of a single tropocollagen molecule.

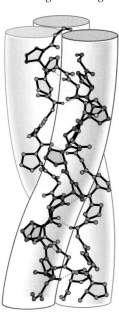
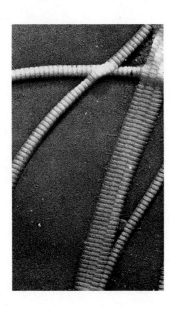

The β–pleated sheet

In the β-pleated sheet, polypeptide chains run side by side, linked by hydrogen bonds. The polypeptides are extended rather than coiling into a helix.

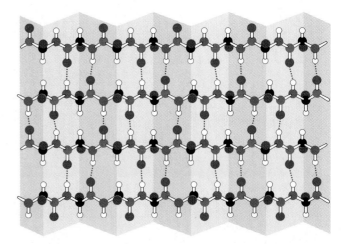

This computer drawing views a β-pleated sheet from the top. Four parallel strands of the polypeptide are joined by hydrogen bonds to form the sheet.

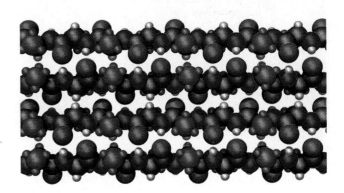

The computer drawing below shows the same material as the above drawing but viewed from the bottom edge to emphasize the "pleats" in the sheet.

3.17 Forms of Protein Secondary Structure
Many proteins share local regions of similarity. These recognizable motifs are called secondary structure.

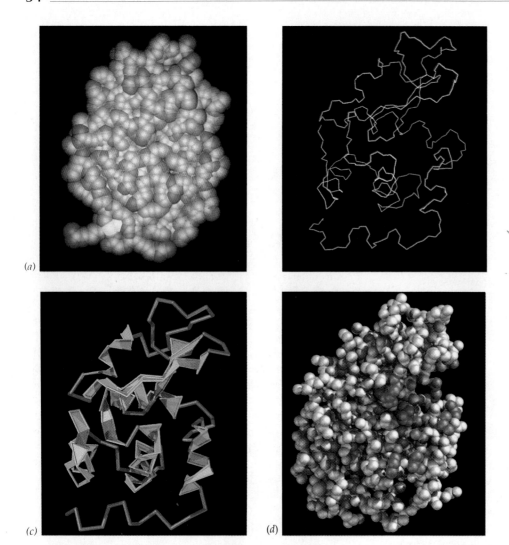

(a)

(c)　　　(d)

3.18 Four Representations of Lysozyme

Different molecular representations emphasize different aspects of tertiary structure. These four representations of lysozyme are similarly oriented. (a) This computer drawing gives the most realistic impression of lysozyme's tertiary structure, which is densely packed. (b) Another computer drawing emphasizes the backbone of the folded polypeptide. Regions in green have α-helical secondary structure; those in red constitute a β-pleated sheet. (c) The green coils here represent the α helices, and orange arrows represent the β-pleated sheet. (d) Another space-filling model, this one emphasizing the position (shown in purple) of the active site—the part of the enzyme molecule that binds reactant molecules. From its position here, you can infer the position of the active site in the other three representations.

3.19 Tertiary Structure of a Protein

Tertiary structure (the exact three-dimensional folding of a protein molecule) is illustrated here for the protein myoglobin. The individual atoms are not shown, nor are the individual amino acids, which form the coiled polypeptide chain. The chain is helical throughout most of its length. The blue shading shows the overall tertiary configuration of the molecule. The red structure in the upper part of the drawing is an iron-containing heme group.

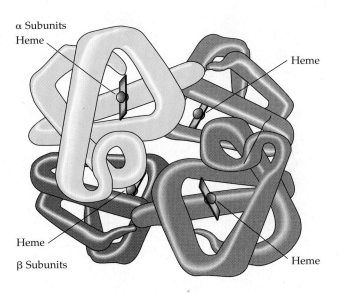

3.20 Quaternary Structure of a Protein
Hemoglobin consists of four folded polypeptide subunits
that assemble themselves into the quaternary structure
shown here. In these two analogous representations, each
subunit is a different color. Note the heme groups, which
are red.

no disulfide bridges, and the molecule is unusual in
that it consists almost entirely of helices. The eight
helical segments bend relative to each other and form
a pocket that encloses a **heme group**: an iron-containing ring structure that binds O_2 (oxygen gas).
Hydrophobic side chains on the inner sides of the
helices help to ensure that the helices fold against
one another correctly as the molecule is formed.
Myoglobin and most other proteins are made from a
single polypeptide chain.

Quaternary Structure

Some, although not most, proteins are made from
two or more polypeptide chains. The overall shape
of such a molecule, its **quaternary structure**, results
both from how its subunits fit together and from how
each subunit folds. Hemoglobin, a protein that brings
oxygen from the lungs to the tissues and delivers it
to myoglobin for storage, illustrates quaternary structure (Figure 3.20). Hydrophobic interactions, hydrogen bonds, and ionic bonds hold together four polypeptide chains (two each of two types). As the
hemoglobin molecule takes up or releases oxygen, its
four subunits shift their relative positions slightly,
changing the quaternary structure. Ionic bonds are
broken, exposing buried side chains that enhance the
binding of molecular oxygen.

Each subunit of hemoglobin is folded like a myoglobin molecule, suggesting that both hemoglobin
and myoglobin are evolutionary descendants of the
same oxygen-binding ancestral protein. But on the

surfaces where its subunits come in contact with each
other—regions that on myoglobin are exposed to
aqueous surroundings—hemoglobin has hydrophobic side chains where myoglobin has hydrophilic
ones. Again, the chemical nature of side chains on
individual amino acids determines how the molecule
folds and packs in three dimensions.

The four levels of protein structure are summarized in Figure 3.21.

NUCLEIC ACIDS

The proteins of today exist because of the structures
and activities of various **nucleic acids**. One group of
these, the **DNAs**, or deoxyribonucleic acids, are giant
polymers that carry the instructions for making proteins; another group, the **RNAs**, or ribonucleic acids,
interpret and carry out the instructions coded in the
DNAs.

Nucleic acids form from monomers called **nucleotides**, each of which consists of a pentose sugar, a
phosphate group, and a nitrogenous (nitrogen-containing) base (Figure 3.22*a*,*b*). Molecules consisting of
a pentose sugar and a nitrogenous base, but no phosphate group, are called nucleo*sides* (Figure 3.22*c*,*d*).
In DNA, the pentose sugar is deoxyribose, which
differs from the ribose found in RNA by one oxygen
atom (see Figure 3.10).

The "backbones" of both RNA and DNA consist
of alternating sugars and phosphates; the bases,
which are attached to the sugars, project from the
chain (Figure 3.23). The nucleotides are joined by
phosphodiester linkages between the sugar of one
nucleotide and the phosphate of the next. (The name
phosphodiester comes from the fact that each phosphate is connected to two sugars.) Most RNA mole-

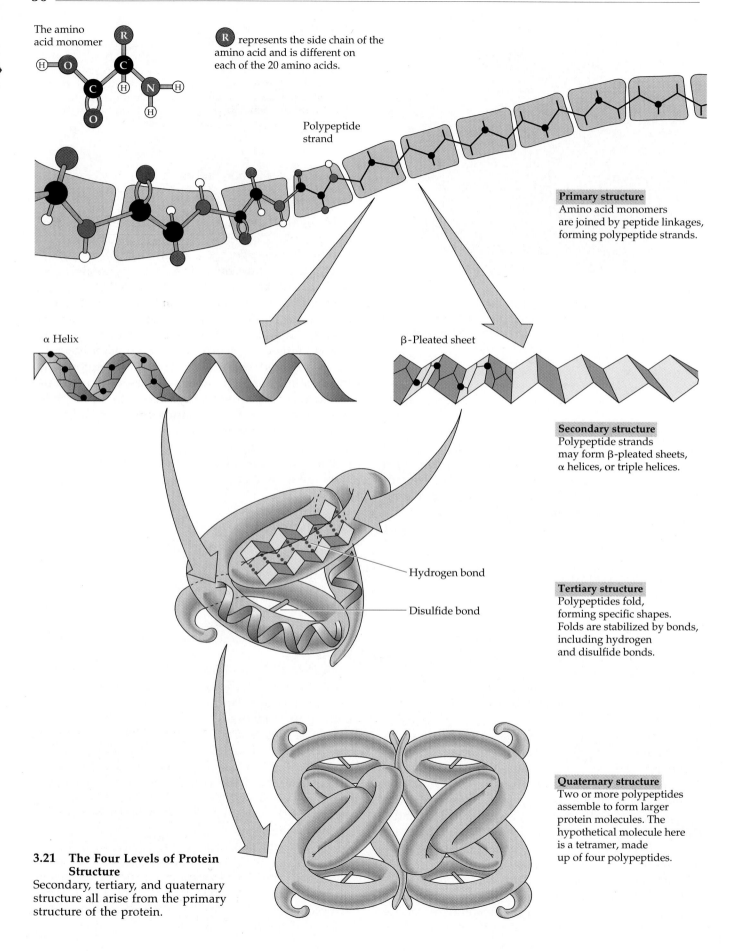

The amino acid monomer

(R) represents the side chain of the amino acid and is different on each of the 20 amino acids.

Polypeptide strand

Primary structure
Amino acid monomers are joined by peptide linkages, forming polypeptide strands.

α Helix

β-Pleated sheet

Secondary structure
Polypeptide strands may form β-pleated sheets, α helices, or triple helices.

Hydrogen bond

Disulfide bond

Tertiary structure
Polypeptides fold, forming specific shapes. Folds are stabilized by bonds, including hydrogen and disulfide bonds.

Quaternary structure
Two or more polypeptides assemble to form larger protein molecules. The hypothetical molecule here is a tetramer, made up of four polypeptides.

3.21 The Four Levels of Protein Structure
Secondary, tertiary, and quaternary structure all arise from the primary structure of the protein.

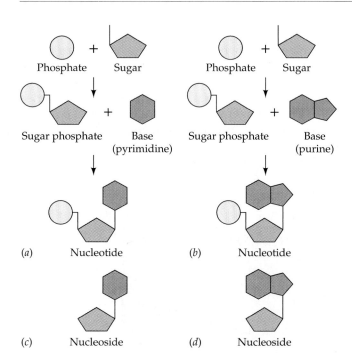

3.22 Components of a Nucleotide

(*a,b*) A sugar and a phosphate form a sugar–phosphate. In nucleotide synthesis, a nitrogen-containing base is then built on the sugar–phosphate in several steps not depicted here, forming the complete nucleotide monomer. The nitrogenous bases fall into two categories, as indicated here by their shapes. (*c,d*) A nucle*oside* consists of a sugar (*not* a sugar–phosphate) and a base.

cules are single-stranded: Each molecule consists of one polynucleotide chain. DNA, however, is usually double-stranded, with two polynucleotide chains held together by hydrogen bonding between their nitrogenous bases. The two strands are antiparallel—that is, they run in opposite directions.

Only four nitrogenous bases—and thus only four nucleotides—are found in DNA. The DNA bases are adenine, cytosine, guanine, and thymine. A key to understanding the structures and functions of nucleic acids is the principle of **complementary base pairing**: Particular bases pair only with certain other bases. In DNA, wherever one strand carries adenine, the other must carry thymine at the corresponding point. Wherever one chain has cytosine, the other must have guanine. The base-pairing rules for DNA and

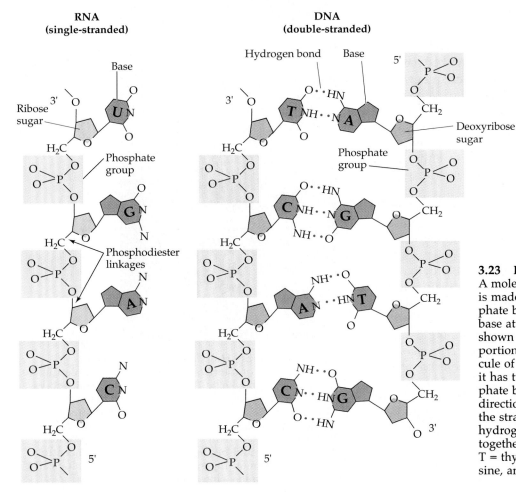

3.23 RNA versus DNA

A molecule of ribonucleic acid (RNA) is made up of a ribose sugar–phosphate backbone with a nitrogenous base attached to each sugar, as shown on the left. On the right is a portion of a double-stranded molecule of deoxyribonucleic acid (DNA); it has two deoxyribose sugar–phosphate backbones, running in opposite directions, with the bases between the strands. Red dots represent the hydrogen bonds that hold the strands together. For the bases, A = adenine, T = thymine, G = guanine, C = cytosine, and U = uracil.

TABLE 3.2
Base-Pairing Rules for the Nucleic Acids

IN DNA	WHEN RNA AND DNA INTERACT		WHEN RNA PAIRS WITH RNA
	RNA	DNA	
A pairs with T	A pairs with T		A pairs with U
T pairs with A	U pairs with A		U pairs with A
G pairs with C	G pairs with C		G pairs with C
C pairs with G	C pairs with G		C pairs with G

differ so greatly. In particular, enzymes must each recognize their own specific "target" molecules. They do this by having a unique three-dimensional form that can match at least a portion of the surface of their targets; structural diversity in the molecules with which enzymes react calls for corresponding diversity in the structure of the enzymes themselves. DNAs are similar and uniform and are read by simple machinery; proteins are diverse and interact with many other compounds.

RNA are shown in Table 3.2. The pairing scheme maximizes hydrogen bonding between the two strands of DNA. Because one of the **purine** bases (adenine or guanine), which are large, always pairs with one of the **pyrimidine** bases (thymine or cytosine), which are small, all base pairs are the same size. Complementary base pairing between the two strands of the DNA molecule makes it possible to copy DNA molecules very faithfully (see Chapter 11).

Ribonucleic acids also have four different nucleotides, but the nucleotides differ from those of DNA. The **ribonucleotides** contain ribose rather than deoxyribose, and one of their four bases is different from that in DNA. The four principal bases in RNA are adenine, cytosine, guanine, and uracil (instead of thymine). Although RNA is generally single-stranded, complementary associations between nucleotides are important in the formation of new RNA strands, in determining the shapes of some RNA molecules, and in associations between RNA molecules during protein synthesis. Guanine and cytosine pair as in DNA, and uracil pairs with adenine. Adenine in an RNA strand can pair with either uracil (in an RNA strand) or with thymine (in a DNA strand).

The three-dimensional appearance of DNA is strikingly regular. The segment shown in Figure 3.24 could be from any DNA molecule. Through hydrogen bonding, the two complementary polynucleotide strands pair and twist to form a **double helix**. How regular this formation seems in comparison with the complex and varied structures of proteins! But this structural difference makes sense in terms of the functions of these two classes of compounds. DNA is a purely informational molecule. *The information carried by DNA resides simply in the sequence of bases carried in its chains.* This is, in a sense, like the tape of a tape recorder. The message must be read easily and reliably. A uniform molecule like DNA can be interpreted by standard molecular machinery, and the machinery can read any molecule of DNA—just as a tape player can play any tape of the right size. Proteins, on the other hand, have good reason to

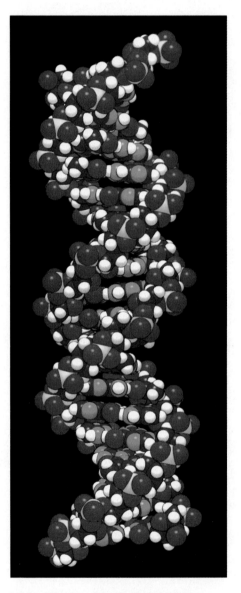

3.24   The Double Helix of DNA
The backbones of the two strands in a DNA molecule are coiled in a double helix. This computer drawing shows the atoms in a length of DNA containing 20 base pairs. Allow your eye to follow the yellow phosphorus atoms and their attached red oxygen atoms in the two helical backbones. The paired bases are stacked in the center of the coil and should become apparent if you concentrate on the light blue nitrogen atoms and the dark blue carbon atoms. The smaller white atoms are hydrogens.

GLYCOLIPIDS, GLYCOPROTEINS, LIPOPROTEINS, AND NUCLEOPROTEINS

We have been treating the classes of macromolecules as if each were completely separate from the others. In fact, certain macromolecules of different classes attach to one another to form covalently bonded products of great importance to cells. Many proteins have attached carbohydrate chains; this combination is called a glycoprotein. The carbohydrate chain often determines the placement of the glycoprotein within the cell. Other carbohydrate chains bind to lipids, resulting in glycolipids. Glycolipids usually reside in the membrane surrounding a cell, with the carbo- hydrate chain extending out into the cell's environment. The human blood group antigens, which determine the ABO blood types, are carbohydrates that combine with either proteins or lipids. When a cell becomes cancerous, both glycolipids and glycoproteins are modified.

Proteins bind DNA, forming nucleoproteins that regulate the activities of DNA. Still other proteins, in combination with cholesterol and other lipids, form lipoproteins. The lipoproteins make it possible to move very hydrophobic lipids through the predominantly hydrophilic environment of the human body, and they deliver cholesterol and certain other lipids to the appropriate cells and tissues within the body.

SUMMARY of Main Ideas about Large Molecules in Living Things

Organisms take in large molecules, break them down into their constituent parts, and from these parts assemble other necessary large molecules.

Proteins, nucleic acids, and large carbohydrates are polymers formed from monomers by condensation reactions, which release water.
Review Figures 3.4 and 3.15

Lipids are compounds that are insoluble in water but readily soluble in organic solvents.

Lipids differ from other molecules discussed in this chapter in that they do not form true polymers.

Triglycerides are composed of glycerol and fatty acids and serve as stored fuel.
Review Figures 3.3 and 3.4

Lipid molecules have large hydrophobic regions, so lipids present in an aqueous system tend to aggregate like oil in water.

Phospholipids tend to aggregate to form a continuous bilayer, as in biological membranes, and they control molecular traffic through the membranes.
Review Figures 3.1, 3.5, and 3.6

Carbohydrates (monosaccharides, oligosaccharides, and polysacchrides) have formulas based on CH_2O.
Review Figures 3.10, 3.11, and 3.12

Monosaccharides serve as fuel and as building blocks for polysaccharides.

The polysaccharides starch and glycogen are storage compounds.

Cellulose imparts strength to plant cell walls.

A protein is composed of one or more polypeptide chains. Amino acids are the monomers of polypeptides.
Review Figure 3.15 and Table 3.1

A protein's primary structure is the sequence of amino acids in its polypeptide chains.
Review Figure 3.21

Regular, repeated patterns of orientation such as α helices and β-pleated sheets are a protein's secondary structure.
Review Figures 3.17 and 3.21

How a polypeptide is folded, its overall shape, is its tertiary structure.
Review Figures 3.18, 3.19, and 3.21

The tertiary structure of a polypeptide arises spontaneously from its primary structure.

For a protein consisting of more than one polypeptide chain, the spatial arrangement of the polypeptides relative to each other is the quaternary structure of the protein.
Review Figures 3.20 and 3.21

DNA molecules are composed of only four different monomers joined by complementary base pairing.
Review Figures 3.22, 3.24, and Table 3.2

Although the structure of DNA is very uniform, the sequence of monomers within the polymer contains an enormous amount of information.

RNA differs from DNA in the sugar component, in one of the bases, and sometimes in the number of polynucleotide stands.
Review Figure 3.23

The different classes of large molecules aggregate with one another to form important compounds such as glycolipids, glycoproteins, lipoproteins, and nucleoproteins.

The structures of large molecules are the keys to how they perform their functions in organisms.

SELF-QUIZ

1. All lipids
   a. are triglycerides.
   b. are polar.
   c. are hydrophilic.
   d. are polymers.
   e. are more soluble in nonpolar solvents than in water.

2. Which of the following is *not* a lipid?
   a. A steroid
   b. A fat
   c. A triglyceride
   d. A biological membrane
   e. A carotenoid

3. All carbohydrates
   a. are polymers.
   b. are simple sugars.
   c. consist of one or more simple sugars.
   d. are found in biological membranes.
   e. are more soluble in nonpolar solvents than in water.

4. Which of the following is *not* a carbohydrate?
   a. Glucose
   b. Starch
   c. Cellulose
   d. Hemoglobin
   e. Deoxyribose

5. All proteins
   a. are enzymes.
   b. consist of one or more polypeptides.
   c. are amino acids.
   d. have quaternary structures.
   e. have prosthetic groups.

6. Which statement is *not* true of the primary structure of a protein?
   a. It may be branched.
   b. It is determined by the structure of the corresponding DNA.
   c. It is unique to that protein.
   d. It determines the tertiary structure of the protein.
   e. It is the sequence of amino acids in the protein.

7. The amino acid leucine (Table 3.1)
   a. is found in all proteins.
   b. cannot form peptide linkages.
   c. is likely to appear in the part of a membrane protein that lies within the phospholipid bilayer.
   d. is likely to appear in the part of a membrane protein that lies outside the phospholipid bilayer.
   e. is identical to the amino acid lysine.

8. The quaternary structure of a protein
   a. consists of four subunits—hence the name *quaternary*.
   b. is unrelated to the function of the protein.
   c. may be α, β, or γ.
   d. depends on covalent bonding among the subunits.
   e. depends on the primary structures of the subunits.

9. All nucleic acids
   a. are polymers of nucleotides.
   b. are polymers of amino acids.
   c. are double-stranded.
   d. are double-helical.
   e. contain deoxyribose.

10. Which statement is *not* true of condensation reactions?
    a. Protein synthesis results from them.
    b. Polysaccharide synthesis results from them.
    c. Nucleic acid synthesis results from them.
    d. They consume water as a reactant.
    e. Different ones produce different kinds of macromolecules.

FOR STUDY

1. Phospholipids make up a major part of every biological membrane; cellulose is the major constituent of the cell walls of plants. How do the chemical structures and physical properties of phospholipids and cellulose relate to their functions in cells?

2. Suppose that, in a given protein, one lysine is replaced by aspartic acid (Table 3.1). Is this a change in primary or secondary structure? How might it result in a change in tertiary structure? In quaternary structure.?

3. If there are 20 different amino acids commonly found in proteins, how many different dipeptides are there? How many different tripeptides? How many different polypeptides composed of 200 amino acid subunits? If there are four different nitrogenous bases commonly found in RNA, how many different dinucleotides are there? How many different trinucleotides? How many different single-stranded RNAs composed of 200 nucleotides?

4. Contrast the structures of hemoglobin, a DNA molecule, and a protein that spans a biological membrane.

READINGS

Branden, C. and J. Tooze. 1991. *Introduction to Protein Structure*. Garland, New York. A well-illustrated book suitable for undergraduates.

Doolittle, R. F. 1985. "Proteins." *Scientific American*, October. A strikingly illustrated treatment of protein structure and evolution.

Stryer, L. 1995. *Biochemistry*, 5th Edition. W. H. Freeman, New York. A relatively advanced but beautiful reference on the subjects of this chapter; outstanding illustrations, concise descriptions, clear prose.

Voet, D. and J. G. Voet. 1990. *Biochemistry*. John Wiley & Sons, New York. A fine advanced textbook with outstanding illustrations.

Some free-living organisms are single cells. Even for multicellular organisms, the single cell, not the organism itself, is the basic unit of life. With a geranium from your garden you can demonstrate for yourself that the unit of life is something smaller than the whole organism. You can take a cutting consisting of a bit of stem and a leaf or two, put it in soil and care for it, and end up with an entire plant. Going a step further, scientists can isolate single cells from many plant species and induce them, by treatment with natural substances that control plant growth, to develop into whole plants.

If we try to go to a level below an entire cell, however, we come to the end of the line. Subcellular structures—organelles—such as nuclei and chloroplasts may be isolated from cells and induced to carry out their normal functions, but they can never be induced to regenerate whole cells, let alone an entire organism. The inability of even the nucleus to produce any sort of life form verifies that it is the whole cell that is the basic unit of life.

CELLS AND THE CELL THEORY

The basic unit of organization in living things is the cell. All organisms are composed of cells, and all cells come from preexisting cells—these two statements constitute the **cell theory** (Chapter 1). Even viruses, which are not cells themselves, are entirely dependent on the presence and chemical machinery of cells for their reproduction. Cells from each of the six kingdoms are shown in Figure 4.1.

Most cells are tiny. They have volumes from 1 to 1,000 μm^3. The eggs of some birds are enormous exceptions, to be sure, and individual cells of several types of algae are large enough to be viewed with the unaided eye. Neurons (nerve cells) have volumes that fit within the "normal" range, but they often have fine projections that may extend for meters, carrying signals from one part of a large animal to another. In spite of these special cases, we may generalize and say that cells are very small objects (Box 4.A). What else do cells have in common?

COMMON CHARACTERISTICS OF CELLS

Cells must do many things in order to survive. They must obtain and process energy, they must convert the genetic information of DNA into protein, and they must keep certain biochemical reactions separate from other, incompatible reactions that must occur simultaneously. Structures within cells carry out these functions, as we will examine in detail in the remaining chapters of Parts One and Two. We begin

Some Cells Can Give Rise to Whole Organisms
Each of these young carrot plants began as one or a few cells in culture.

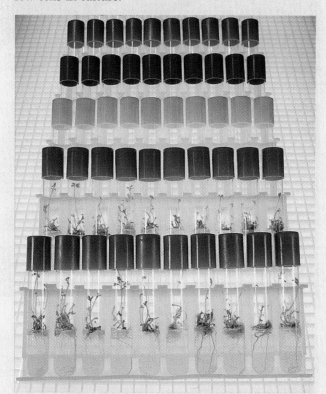

4

Organization of the Cell

4.1 Cells Come in Many Shapes

In these micrographs we see *(a)* a filamentous bacterium in the kingdom Archaebacteria; *(b)* two species of filamentous cyanobacteria (kingdom Eubacteria); *(c) Euglena,* a plantlike protist; *(d) Paramecium,* an animallike protist; *(e)* brewer's yeast, a fungus; *(f)* "leaf" cells of a moss, packed with green, photosynthetic chloroplasts; *(g)* blood cells of a frog; and *(h)* mammalian cells grown in culture in the laboratory.

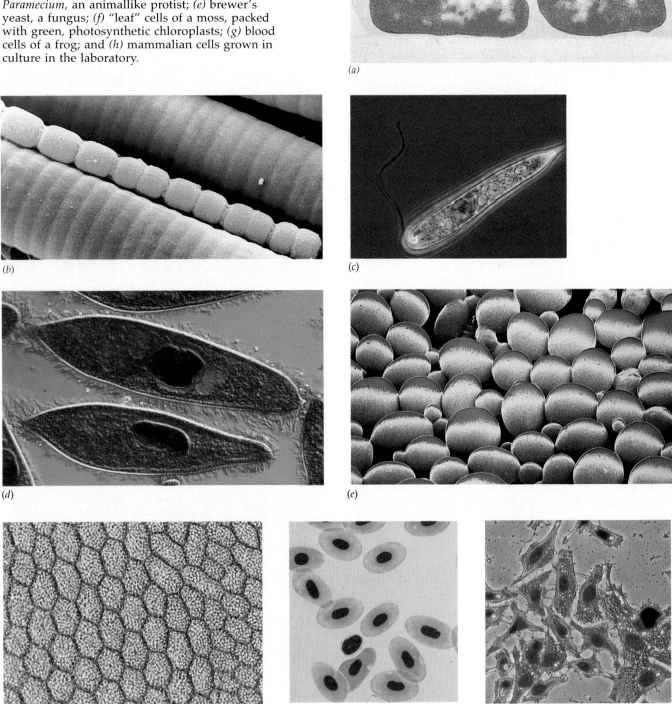

(a)

(b)

(c)

(d)

(e)

(f)

(g)

(h)

in this chapter by considering the component parts of cells.

Among all the kinds of cells there are two distinct general arrangements, with only a few intermediate forms in evidence. One general arrangement, the simpler, is the **prokaryotic cell,** characteristic of the kingdoms Eubacteria and Archaebacteria. Organisms in these kingdoms are called prokaryotes. Their single cells lack nuclear compartments and membrane-bounded internal compartments. The rest of the living world has **eukaryotic cells**—cells that contain membrane-bounded nuclei. Eukaryotic cells have

BOX 4.A

The Sizes of Things

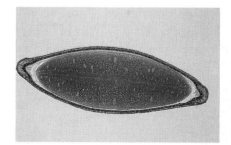
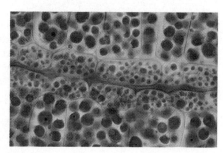

On the left, a pumpkin seed is shown magnified 10 times. When the same seed is magnified 1,000 times, we see the starch grains in the seed's cells as red balls.

Biologists study objects of very different sizes, ranging from molecules with diameters of less than one nanometer to organisms that are many meters long. You need a sense of the sizes of things to appreciate how they function and how their parts interact. The sizes of some objects are compared in the diagram, which also indicates the methods by which they are usually viewed. To help you develop a sense of sizes we will sometimes identify the magnifications of figures, as in the pictures above, taken with a light microscope. However, you will need to remember the approximate sizes of cells and their parts.

Inside the front cover of this book, there is a table of measurement units and their symbols. You may want to refer to it several times as you read this chapter. What is the size range 1 to 1,000 μm^3? If we assume that many cells are almost spherical, this range of volumes corresponds with a range of diameters of about 1.2 to 12 μm.

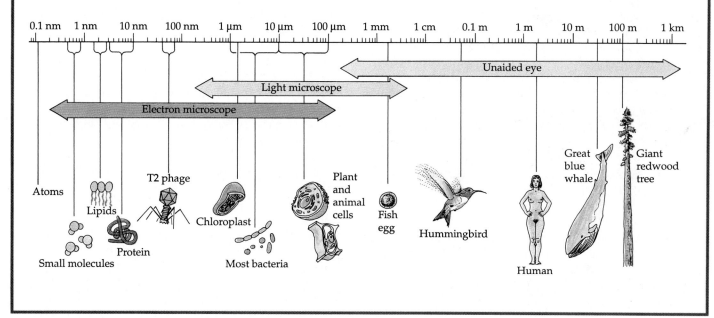

other internal compartments (called organelles) that, like the nucleus, are surrounded by membranes. Organisms with this type of cell are known as eukaryotes. Both prokaryotes and eukaryotes have prospered through many hundreds of millions of years of evolution, and both are great success stories.

PROKARYOTIC CELLS

We will first consider the characteristics that cells throughout the kingdoms Eubacteria and Archaebacteria have in common. Then, in the next section, we will discuss structural features that typify many, but not all, prokaryotes.

Features Shared by All Prokaryotic Cells

All prokaryotic cells have, without exception, a plasma membrane, a nucleoid, and cytoplasm filled with ribosomes. The **plasma membrane** separates the cell from its environment and regulates the traffic of materials in and out of the cell. The **nucleoid** is a relatively clear region (as seen under the electron microscope) that contains the hereditary material (DNA) of the cell. Each prokaryotic cell has at least

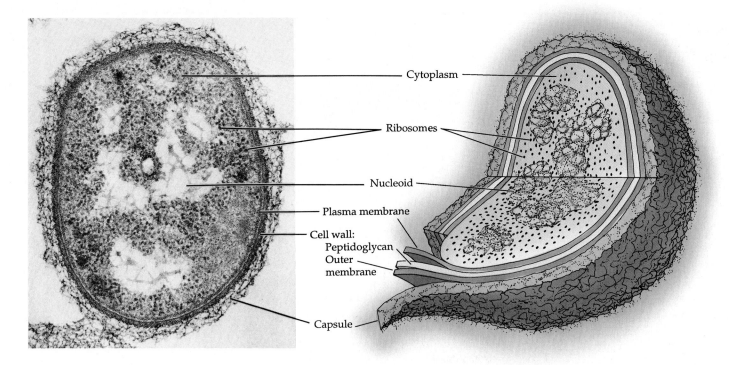

4.2 A Bacterial Cell
The eubacterium *Pseudomonas aeruginosa* illustrates typical prokaryotic cell structures. The electron micrograph on the left is magnified about 80,000 times.

one nucleoid; some cells have more than one. The rest of the material in the cell is called the **cytoplasm.** At high magnification, the cytoplasm appears full of minute, spherical structures called **ribosomes.** Prokaryotic ribosomes are approximately 15 to 20 nm in diameter and consist of three molecules of RNA and about 50 different protein molecules. Ribosomes coordinate the synthesis of proteins (see Chapter 11). In addition to ribosomes, the cytoplasm contains many kinds of enzymes and other chemical constituents of the cell.

Although structurally simple (consisting primarily of plasma membrane, nucleoid, and ribosome-filled cytoplasm), the prokaryotic cell is functionally complex. Enzymes in prokaryotic cells direct thousands of chemical reactions, with the cell's DNA serving as the molecular memory that allows successive generations of a given cell to be very much alike.

Other Features of Prokaryotic Cells

All but the simplest prokaryotic cells have at least a few more structural complexities. Most, for example, have a **cell wall** outside the plasma membrane (Figure 4.2). The rigidity of the wall lends support to the cell and determines its shape. The cell walls of most eubacteria consist primarily of **peptidoglycan,** a polymer of amino sugars, cross-linked to form a single molecule around the entire cell! Outside the cell wall there is often a layer of slime (composed mostly of polysaccharide), referred to as a **capsule.** The capsules of some bacteria may protect them from attack by white blood cells within the bodies of animals they

infect. The capsule provides protection against drying of the cell, and in some cases it may trap other cells for attack by the bacterium. Many prokaryotes produce no capsule at all, and even those that have capsules will not die if they lose them.

The cyanobacteria (Figure 4.1*b*) and some other bacteria carry on photosynthesis. They convert the energy of sunlight to chemical energy to produce food and to drive other energy-requiring reactions. In these photosynthetic prokaryotes, the plasma membrane folds into the cytoplasm—often very extensively—to form an internal membrane system containing chlorophyll and other compounds needed for photosynthesis (Figure 4.3). Other bacteria possess different sorts of membranous structures called **mesosomes,** which may function in cell division or in various energy-releasing reactions. Like the photosynthetic membrane systems, mesosomes are formed by infolding of the plasma membrane. They remain attached to the plasma membrane and never form the free-floating, isolated, membranous organelles that are characteristic of eukaryotic cells (Figure 4.4). The fluid portion of a cell's cytoplasm, in which ribosomes and membranous structures are found, is called the **cytosol.**

Some prokaryotes swim by using appendages called **flagella** (Figure 4.5*a*). A single flagellum, made of a protein called flagellin, looks something like a tiny corkscrew. It spins about its axis like a propeller,

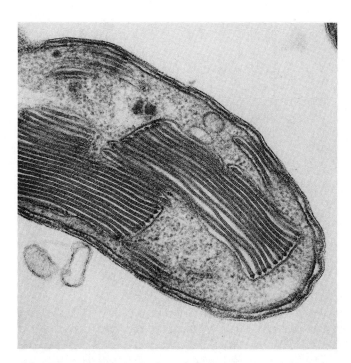

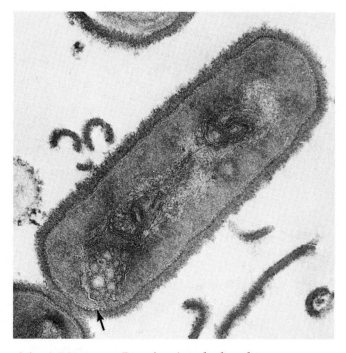

4.3 Photosynthetic Prokaryotes Have Extensive Internal Membranes

Photosynthetic membranes fold into "stacks" inside a bacterial cell; such organized collections of internal membranes contradict the mistaken notion that bacteria are nothing more than tiny bags of molecules.

4.4 A Mesosome Remains Attached to the Plasma Membrane

Convoluted membranes of a large mesosome extend throughout this bacterial cell. At the lower left (arrow) you can see that the membrane of the mesosome is continuous with the plasma membrane.

driving the bacterium along. Ring structures anchor the flagellum to the plasma membrane and, in some bacteria, to the outer membrane of the cell wall (Figure 4.5b). The fact that flagella actually cause the motion of the cell can be shown by removing these

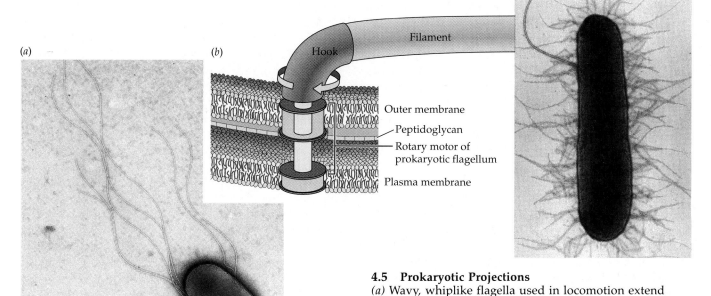

4.5 Prokaryotic Projections

(a) Wavy, whiplike flagella used in locomotion extend from this bacterium. (b) The basal ends of these tiny structures are complex; the mechanism by which bacterial flagella rotate is still under investigation. (c) The small, hairlike pili bristling from the surface of this cell of *Escherichia coli* help it adhere to other cells.

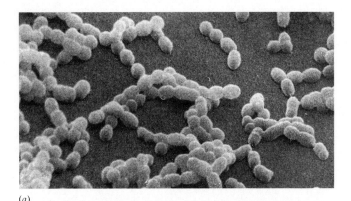

(a)

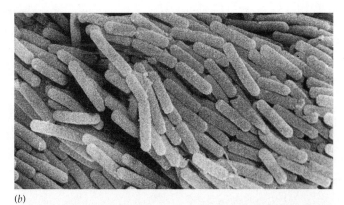

(b)

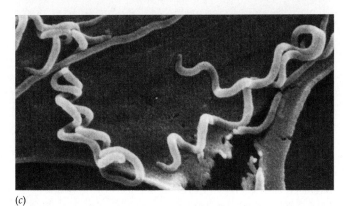

(c)

4.6   Bacterial Shapes and Aggregations
(a) These spherical cocci of an acid-producing bacterium grow on tooth enamel and cause decay. *(b)* Rod-shaped bacteria. *(c)* These spirochetes (*Treponema pallidum*) cause syphilis in humans.

from a cell. When this is done, the cell no longer moves. In addition, if the tip of a flagellum is attached to an immovable cell, the spinning of the flagellum causes the entire cell to rotate.

Pili project from the surface of some groups of bacteria (Figure 4.5c). Shorter than flagella, these threadlike structures seem to help bacteria adhere to other cells—to one another during mating, and also to animal cells.

Kinds of Prokaryotes

Bacteria are often categorized by the shapes of their cells. Spherical cells are **cocci;** rod-shaped cells are **bacilli;** and helical cells, sometimes coiled like corkscrews, are **spirilla** or **spirochetes** (Figure 4.6). Although each prokaryote is a single cell, some types are seldom seen singly. Many types are usually seen in chains, small clusters, or even colonies with hundreds of individuals. The diversity within the bacterial kingdoms is the subject of Chapter 22.

Many prokaryotes have been used extensively in biological research. The most familiar bacterium is *Escherichia coli* (Box 4.B).

PROBING THE SUBCELLULAR WORLD: MICROSCOPY

Many significant advances in our knowledge of cells have depended upon the **resolution,** or **resolving power,** of the instruments we use for magnifying tiny objects. We define the resolving power of a lens or microscope as the smallest distance separating two objects that allows them to be seen as two distinct things rather than as a single entity. For example, most humans can see two fine parallel lines as dis-

tinct markings if they are separated by at least 0.1 millimeter (0.1 mm); if they are any closer together, we see them as a single line. Thus, the resolving power of the human eye is about 0.1 mm, which is the approximate diameter of the human egg. To see anything smaller, we must use some form of microscope.

The invention of the **light microscope** (Figure 4.7a)—so called because it allows objects to be viewed in visible light—made the study of cells possible. In its contemporary form, the light microscope has a resolving power of about 200 nanometers (200 nm = 0.2 μm = 0.0002 mm), so it gives a useful view of cells and can reveal features of some of the subcellular organelles. Today, half a century after the invention of the more powerful electron microscope, the light microscope remains an important tool for the biologist. Many of the illustrations in this book are photographs taken through the light microscope; they are called photomicrographs. The light microscope has its limitations, however—principally its 200-nm resolving power. This resolving power cannot be improved by adding more lenses or by taking photomicrographs and then enlarging them. Such enlargements can be made, but they do not increase the resolution; as the images become larger, they simply become fuzzier.

Figure 4.2 and many other figures throughout this book show cellular structures that are far too small

BOX 4.B

The Best-Known Prokaryote: *Escherichia coli*

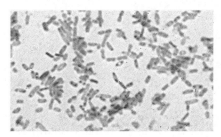

Cells of *E. coli* appear as red rods in this light micrograph of a stained preparation.

Without a doubt, the best-understood living creature is a humble bacterium living in our intestinal tracts: *Escherichia coli*—or, as it is commonly known, *E. coli*. This rod-shaped bacterium (shown in the figure) is about 2 μm in length and 0.8 μm in diameter, giving it a volume of about 1 μm^3 and a weight of approximately 10^{-12} g (one-millionth of one-millionth of a gram). *E. coli* are about 100 times larger than the smallest living cells, the mycoplasmas (see Chapter 22). Immediately outside the *E. coli* plasma membrane is a cell wall about 10 nm thick, and projecting from the cell are flagella and pili. The flagella gather into a bundle and push the bacterium at a speed such that if the bacterium were magnified to human dimensions it would move at 30 miles per hour! Every second or so, the bundle of flagella separates and re-forms, causing the cell to change its direction.

An *E. coli* cell is approximately 70 percent water, 15 percent protein, 1 percent DNA, 6 percent RNA, 3 percent carbohydrate, 2 percent lipid, and 1 percent simple ions such as K$^+$ (potassium ions), as well as small amounts of other substances. It has 15,000 to 30,000 ribosomes and contains from one to four identical molecules of DNA. This genetic material is only about 1/500 as much DNA as is contained in a single cell of a human being. Nonetheless, as simple as it is, each prokaryotic cell of *E. coli* makes thousands of specific proteins.

E. coli is favorable for biological experimentation for several reasons. As noted, it is very small. Under the best conditions, it can divide once every 20 minutes, whereas most animal cells require about a day to go through a division cycle. Because of this rapid division, immense populations of *E. coli* can be grown very quickly. One cell can become 8 in an hour, 512 in 3 hours, over a billion in 10 hours, and more than 10^{21} in a day (in principle, that is, and with unlimited food and space). Its nutritional requirements are simple: water, some mineral ions, and an energy source such as glucose. Unlike some bacteria, most varieties of *E. coli* do not present a great health hazard, so they can be grown without extensive precautions. Many genetic strains with different, known characteristics are readily available. As a result of these and other advantages, *E. coli* has been used in countless investigations of genetics, biochemistry, and other areas of biology. It is extensively used in research on recombinant DNA ("genetic engineering"), a topic considered in detail in Chapter 14.

(a)   *(b)*

4.7 Exploring Cells with Microscopes
(a) A research-quality light microscope. A camera is mounted at the top of the instrument for making photomicrographs. *(b)* A transmission electron microscope. The magnets that focus the electron beam are in the tall cylinder.

to be resolved with the light microscope. Ribosomes, for example, being 20 nm or less in diameter, cannot be resolved as individual objects under the light microscope. On the other hand, ribosomes are readily resolved with the **electron microscope** (Figure 4.7b). An electron microscope uses powerful magnets as lenses to focus an electron beam, much as the light microscope employs glass lenses to focus a beam of light. We cannot see electrons, however, so they are directed at a fluorescent screen or a photographic film to create an image we can see. The resulting images are called electron micrographs, many of which appear in this and later chapters. The resolving power of modern electron microscopes is about 0.2 nm, but no biological specimen has yet been seen in such detail with an electron microscope. One reason is that the energy of the electron beam at that power is so great that it destroys biological molecules before they can be seen. Because of this and other technical limitations, most electron micrographs resolve detail no finer than 2 nm, and even the best micrographs rarely resolve detail as fine as 1 nm. This corresponds to a resolving power about 100,000 times finer than that of the human eye.

There are two types of electron microscopy. In **transmission electron microscopy,** which produced Figures 4.3 and 4.4, electrons pass *through* a sample. Transmission electron microscopy is used to examine thin slices of objects—similar to the sections shaved off a material and placed on the slides used with a light microscope, but much thinner.

In **scanning electron microscopy,** electrons are directed at the surface of the sample, where they cause other electrons to be emitted; the scanning electron microscope focuses these secondary electrons on a viewing screen. Scanning electron microscopy reveals the *surface* structures of three-dimensional objects, such as the bacteria shown in Figure 4.6. A scanning electron microscope has a resolving power no better than 10 nm, so scanning electron micrographs are usually at a somewhat lower magnification than that of transmission electron micrographs.

You might think that with such a resolving power the electron microscope would be used for all microscopic studies, but this is not so. For some applications it would be sheer overkill, like using a magnifying glass to get an overall view of an elephant. A more important limitation is that biological samples have to be killed and dehydrated before they can be examined with an electron microscope. Light microscopy, on the other hand, allows us to observe *living* cells.

Samples for transmission electron microscopy have to be thinly sliced. Samples are also often sliced before examination under a light microscope. To get a reasonable three-dimensional view of large cells or tissues with a microscope, therefore, one looks at many successive slices, rather like examining successive slices of Swiss cheese to "see" one of the holes.

THE EUKARYOTIC CELL

The vast majority of living species, including all animals, plants, fungi, and protists, have cells that are structurally more complex than those of the prokaryotes. Compare Figures 4.8 and 4.9 with Figure 4.2 for a quick sense of the prominent differences. Eukaryotic cells are full of membranous structures of wondrous diversity. One or two membranes enclose each of many of these structures, which carry on particular biochemical functions. These structures are neatly packaged subsystems, with membranes to control their functions and to regulate what gets in and out. Some of the subsystems are like little factories that make specific products. Others are like power plants that take energy in one form and convert it to a more useful form (see Chapter 6). These membranous subsystems, as well as other structures (such as ribosomes) that lack membranes but possess distinctive shapes and functions, are called **organelles.** Like prokaryotic cells, eukaryotic cells have a plasma membrane, cytoplasm, and ribosomes.

Roles of Membranes in Eukaryotic Cells

In 1952 the first people to look at reasonably clear electron micrographs of eukaryotic cells were stunned by the complexity of what they saw. Based on chemical and biological observations, scientists had expected to find cells surrounded by plasma membranes, even though these structures were not resolved with the light microscope. It was also known that cells teemed with organelles, and it was suspected that at least some of these organelles were bounded by membranes. It is doubtful, however, that anyone expected membranes to be as profuse in the eukaryotic cell as they actually are. What are all those membranes for? How do they function? What is their structure—or structures, if all membranes are not alike? We will deal with these questions in Chapter 5, but will note a few of the most basic ideas here.

Biological membranes regulate molecular traffic from one side of the membrane to the other. The hydrophobic interior of the membrane is a barrier to the passage of many materials, especially polar materials that are readily soluble in water (see Figure 3.6). Many materials are transported through the membrane with the help of highly specific protein molecules. As discussed later in this chapter, the plasma membrane of some cells can fold inward and form compartments called vesicles in the cell to trap a portion of the cell's environment, as if taking a bite out of it.

Membranes participate in many activities besides transport. They are staging areas for interactions between cells. For example, immunologically active white blood cells recognize and interact with their targets by means of specific protein molecules built into their plasma membranes (see Chapter 16). The proper development and organization of multicellular animals depends upon recognition reactions between cells, which are mediated by the plasma membrane (see Chapter 17). Many intracellular membranes carry the components responsible for energy transformations in cells. Chlorophyll and other substances necessary for energy-capturing photosynthesis are bound in a specific way to membranes in chloroplasts, one type of organelle. The electron carriers that help transform food energy into a form the cell can use are organized as parts of the inner membrane of mitochondria, another organelle type. In many respects, a discussion of eukaryotic cells is a discussion of membranes that are specialized for various cellular activities.

INFORMATION-PROCESSING ORGANELLES

Living things depend on accurate, appropriate information. Information is *stored* as the sequence of bases in DNA molecules. The bulk of the DNA in eukaryotic cells resides in the nucleus. Information is *translated,* from the language of DNA into the language of proteins, on the surfaces of the ribosomes.

The Nucleus

Typically, the **nucleus** is the largest organelle in the eukaryotic cell (Figures 4.8 and 4.9). Most animal cells have a nucleus that is approximately 5 μm in diameter. The possession of a membrane-bounded nucleus is the defining property of the eukaryotic cell. (Remember that in prokaryotes there is no membrane separating the nucleoid from the surrounding cytoplasm.) As viewed under the electron microscope, a nucleus is surrounded by *two* membranes separated by a few tens of nanometers. The **nuclear envelope,** as this pair of membranes is called, is perforated by **nuclear pores** approximately 9 nm in diameter (Figure 4.10). Each pore is surrounded by eight large protein granules arranged in an octagon where the inner and outer membranes merge. RNA and water-soluble molecules pass through the pores to enter or leave the nucleus.

The outer membrane of the nuclear envelope sometimes folds outward into the cytoplasm and is continuous with the network called the endoplasmic reticulum (discussed later in this chapter). The endoplasmic reticulum and, to a lesser extent, the outer surface of the outer membrane of the nuclear envelope often carry great numbers of ribosomes. There are no ribosomes on the inner surface of the outer membrane, or on either surface of the nuclear envelope's inner membrane.

In the nucleus, DNA combines with proteins in a fibrous complex called **chromatin.** We will consider the structure of chromatin in Chapter 9. Throughout most of the life cycle of the cell, the chromatin exists as exceedingly long, fine threads that are so tangled that they cannot be seen clearly with any microscope. When the nucleus is about to divide (that is, to undergo mitosis or meiosis; see Chapter 9), the chromatin condenses and coils tightly to form a precise number of readily visible objects called **chromosomes** (Figure 4.11). Each chromosome contains one long molecule of DNA. The chromosomes are the bearers of hereditary instructions; their DNA carries the information required to carry out the synthetic functions of the cell and to endow the cell's descendants with the same instructions. Between nuclear divisions, the chromatin attaches to a protein meshwork, the nuclear lamina, on the inside of the nuclear envelope. The chromatin detaches when nuclear division commences, and the envelope breaks up into vesicles.

During most of the nuclear cycle, dense, roughly spherical bodies called **nucleoli** are visible in the nucleus (see Figure 4.10). Taken together, the nucleoli contain from 10 to 20 percent of a cell's RNA. Ribosomes are assembled in the nucleolus. Protein molecules move into the nucleus and then into the nucleoli, where they combine with RNA molecules to form the cell's ribosomes. The ribosomes then move out of the nucleus. Each nucleus must have at least one nucleolus, and those of some species have several. The exact number of nucleoli in its cells is characteristic of a species.

Chromosomes and nucleoli float in a fluid called **nucleoplasm.** This fluid is a suspension of various particles, fibers, proteins, and other compounds. (Whereas in a solution solid particles dissolve, in a suspension they disperse but remain solid.) Recall that the fluid portion of the cytoplasm, in which the various organelles, particles, and fibers are suspended, is called the cytosol.

Nucleus and Cytoplasm

The nucleus contains most of the cell's DNA. The DNA encodes the information needed to make the cell's macromolecules, which carry out the activities of the nucleus and of the cytoplasm. Beyond this, what can we say about the relationship between the nucleus and the cytoplasm?

To answer this question, consider experiments performed with a giant single-celled alga of the genus *Acetabularia.* Cells of *Acetabularia* reach lengths of a

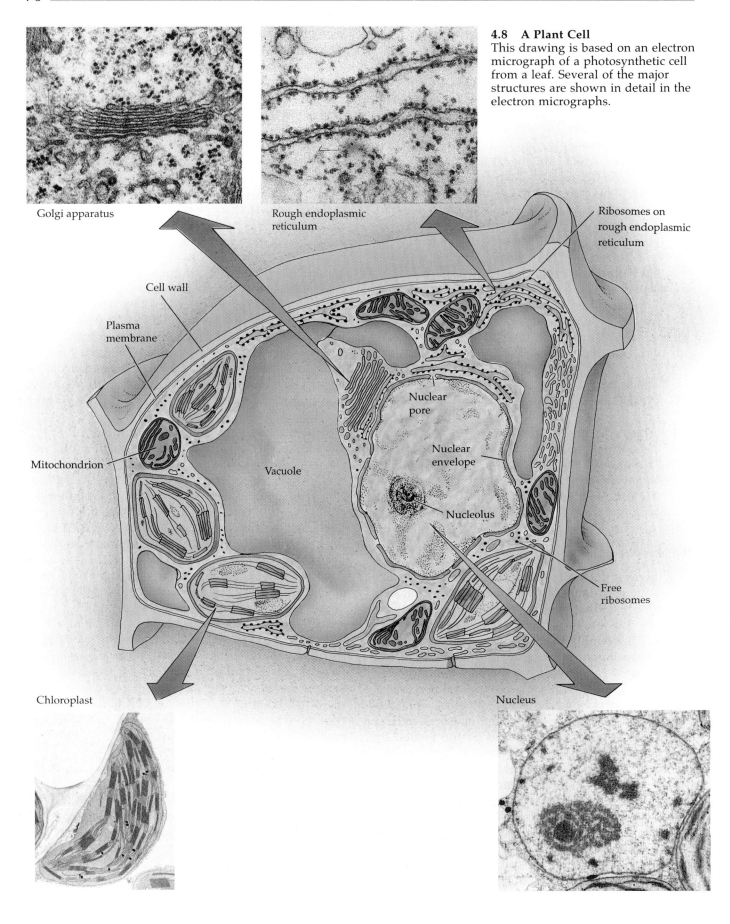

Golgi apparatus

Rough endoplasmic reticulum

4.8 A Plant Cell
This drawing is based on an electron micrograph of a photosynthetic cell from a leaf. Several of the major structures are shown in detail in the electron micrographs.

Ribosomes on rough endoplasmic reticulum

Cell wall

Plasma membrane

Nuclear pore

Nuclear envelope

Mitochondrion

Vacuole

Nucleolus

Free ribosomes

Chloroplast

Nucleus

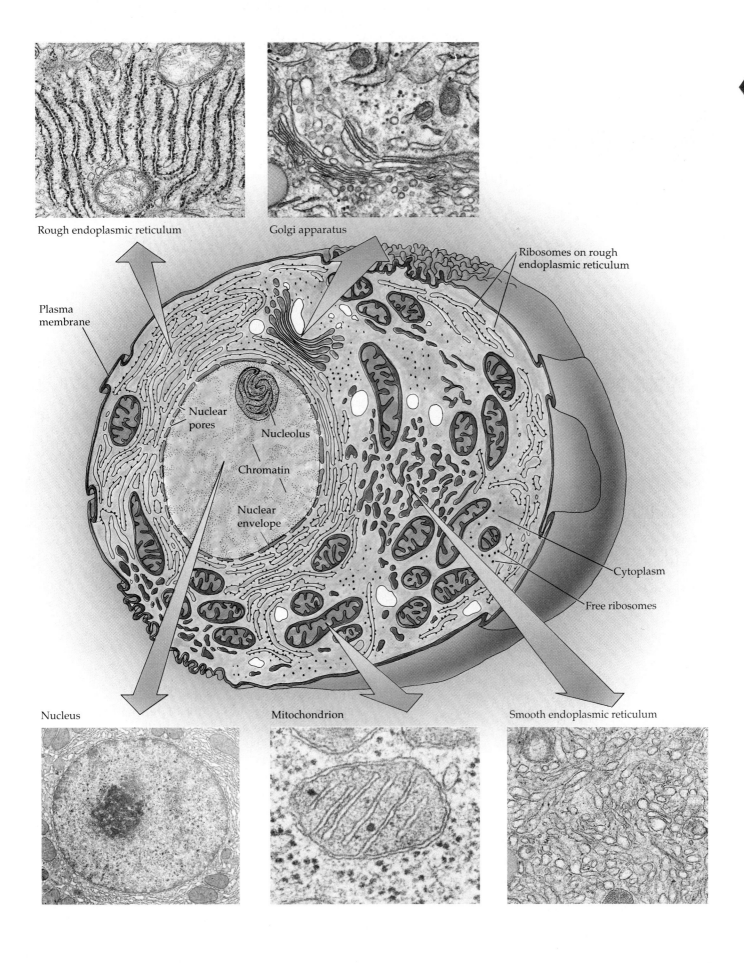

Rough endoplasmic reticulum

Golgi apparatus

Ribosomes on rough
endoplasmic reticulum

Plasma
membrane

Nuclear
pores

Nucleolus

Chromatin

Nuclear
envelope

Cytoplasm

Free ribosomes

Nucleus

Mitochondrion

Smooth endoplasmic reticulum

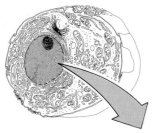

4.10 The Eukaryotic Nucleus Is Bounded by a Double Membrane
The electron micrograph shows the nucleus of an animal cell. Notice the double-membraned nuclear envelope, the nucleolus, and other common features of animal cell nuclei. The bottom two drawings illustrate pores in the nuclear envelope. Each nuclear pore complex comprises eight protein granules surrounding a pore.

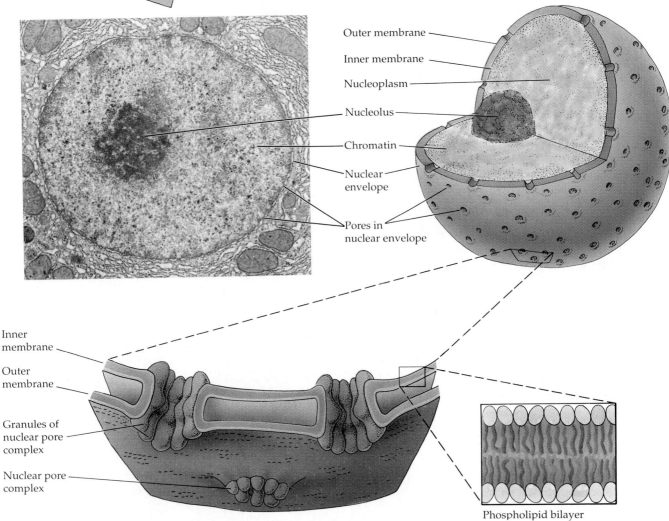

Outer membrane
Inner membrane
Nucleoplasm
Nucleolus
Chromatin
Nuclear envelope
Pores in nuclear envelope

Inner membrane
Outer membrane
Granules of nuclear pore complex
Nuclear pore complex

Phospholipid bilayer

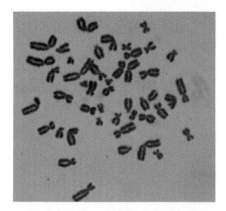

4.11 Humans Have 46 Chromosomes
The chromosome complement in a normal human cell. If a nucleus about to divide is ruptured and treated with certain stains, its chromosomes are readily visible under the light microscope, as shown here.

few centimeters and can readily be picked up and handled. A single cell of this organism is just large enough to be easy to use for dissecting and grafting experiments. The cells consist of three principal regions: the cap, the stalk, and the rhizoids. The rhizoids anchor the organism in its watery environment. The nucleus is within the rhizoid region throughout most of the life of the cell. Rhizoids have no ribosomes, and most of the cell's cytoplasm is in the stalk.

If the cap is cut off a cell of *Acetabularia*, a new cap forms over a period of several days, regenerated with proteins and lipids synthesized by the cytoplasm in the stalk. If this new cap is removed, still another forms, and so forth (Figure 4.12a,b).

Acetabularia cells of two different species can be grafted together to show the origin of the information for the cap structures. An *Acetabularia mediterranea* cap

4.12 Domination by the Nucleus

Grafting experiments with the giant protist *Acetabularia* point to the regulatory activity of the nucleus. The nucleus-containing rhizoids determine the type of cap produced by regenerating *Acetabularia*.

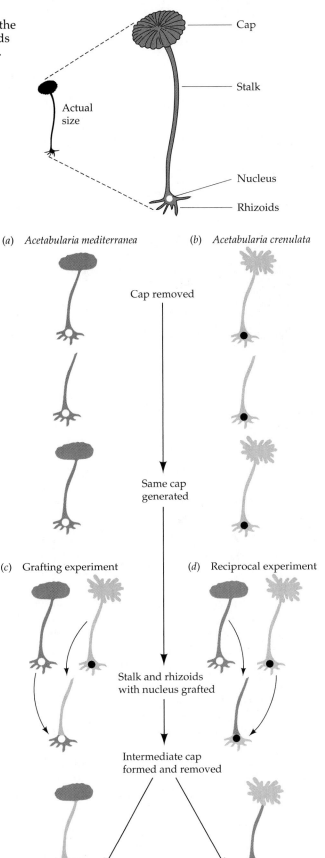

(a) *Acetabularia mediterranea*   (b) *Acetabularia crenulata*

Cap removed

Same cap generated

(c) Grafting experiment   (d) Reciprocal experiment

Stalk and rhizoids with nucleus grafted

Intermediate cap formed and removed

Cap of *Acetabularia mediterranea* forms   Cap of *Acetabularia crenulata* forms

looks like an umbrella, whereas an *Acetabularia crenulata* cap looks more like a bunch of bananas. If we cut up two *Acetabularia* and graft together the rhizoids from *A. mediterranea* and the stalk from *A. crenulata*, a new cap will form from the cytoplasm of the *A. crenulata* stalk. Which type will it be? The first new cap to form is of intermediate appearance. If we cut off this cap and wait for another to appear, the next cap looks like a typical *A. mediterranea* cap, even though it is made from the cytoplasm of the *A. crenulata* stalk (Figure 4.12c). With more refined experiments, we can show that it is the nucleus—which happens to lie in the rhizoids in both species—that provides the instructions for making a new cap. So the nucleus is the storehouse of information for the cell, even when the cell has been put together by grafting.

Can we be sure that we have interpreted the grafting experiment correctly? Perhaps *A. mediterranea* just happens to be somehow "dominant" over *A. crenulata*. We can test this possibility by turning the experiment around—by doing a reciprocal experiment. To do this, we combine *A. crenulata* rhizoids with an *A. mediterranea* stalk. Again, a cap of intermediate form is made first and is cut away; all subsequent caps regenerate as the *A. crenulata* type (Figure 4.12d). Now we can be more confident of our conclusion: The nucleus controls what the cytoplasm builds.

Ribosomes

In both eukaryotic and prokaryotic cells, ribosomes fill the need for a site where a crucial cellular activity—protein synthesis—can take place. Ribosomes reside in three places in eukaryotic cells: free in the cytoplasm; attached to the surface of endoplasmic reticulum, as will be described later in this chapter; and in the energy-processing organelles discussed in the next section. In each of these places the ribosomes provide the site where proteins are synthesized under the direction of nucleic acids (see Chapter 11).

The ribosomes of prokaryotes and of eukaryotes are similar in that they each consist of two different-sized subunits (Figure 4.13). Eukaryotic ribosomes are somewhat larger, but the structure of prokaryotic ribosomes is better understood. Chemically, ribosomes consist of a type of RNA to and around which are bound more than 50 different protein molecules. The ribosome temporarily binds two other types of RNA molecules (one large and one small) as it trans-

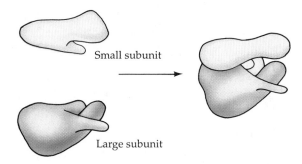

Small subunit

Large subunit

4.13 Tiny Ribosomes Have a Complex Structure
A ribosome consists of a large subunit and a small subunit; these subunits come together only when participating in protein synthesis. Ribosomes do not have membranes.

lates hereditary information into the structures of the cell's structural and regulatory proteins.

ENERGY-PROCESSING ORGANELLES

In addition to information, cells require energy. Eukaryotic cells have organelles for obtaining energy from food molecules, and some cells of plants and of some protists have organelles in which the energy of sunlight is captured. In prokaryotic cells energy is transformed on the plasma membrane, on membrane infoldings, and in the cytosol.

Mitochondria

Utilization of "fuels" for the eukaryotic cell begins in the cytosol, where chemicals from food become molecules that are then taken up by organelles called **mitochondria** (singular: mitochondrion). Mitochondria function primarily to convert energy from food into a form the cell can use. In mitochondria, energy-rich substances from the cytosol are oxidized—that is, electrons are removed from them (see Chapter 7). Some of the energy available from these electrons is used to make a substance—ATP—that stores energy in two special chemical bonds. The stored energy may be used either immediately or later to perform various kinds of work for the cell. The utilization of food in the mitochondria, with the associated formation of ATP, is called cellular respiration.

Typical mitochondria are small, somewhat less than 1.5 μm in diameter and 2 to 8 μm in length, about the size of many bacteria. Mitochondria are visible with a light microscope, but virtually nothing was known of their structure until they were examined with the electron microscope. Electron micrographs show that mitochondria have an outer membrane that is smooth and unfolded. Immediately inside this is an inner membrane that folds inward at many points, giving it a much greater surface area

than that of the outer membrane (Figure 4.14). In animal cells these folds tend to be quite regular, giving rise to shelflike structures called **cristae.** The mitochondria of plants also have cristae, but plant cristae tend to be much less regular in size and structure, and their inner membranes form both shelves and tubes. Special techniques and very high magnification have been used to show that the inner mitochondrial membrane contains large protein structures now known to participate in cellular respiration (see Chapter 7). The region enclosed by the inner membrane is referred to as the **mitochondrial matrix.** Within the matrix are ribosomes and DNA that make some of the proteins needed for the synthesis of mitochondria.

Almost all eukaryotes have mitochondria. The few exceptions are microscopic organisms that live in environments without oxygen, and parasites that exploit the energy resources of their hosts. The number

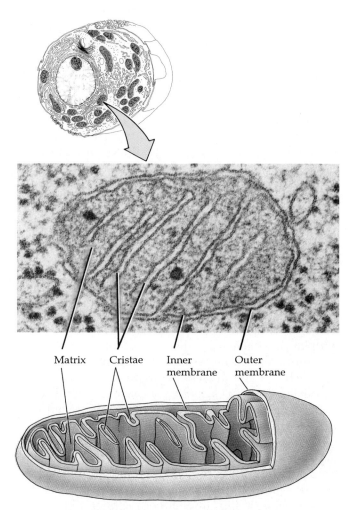

Matrix   Cristae   Inner membrane   Outer membrane

4.14 A Mitochondrion: Where Foods Yield Up Their Energy
A mitochondrion as seen by electron microscopy (top). The drawing shows the mitochondrion's surface cut away to expose internal structures.

of mitochondria per cell ranges from one contorted giant in some unicellular protists to a few hundred thousand in large egg cells. An average human liver cell contains more than a thousand mitochondria. The primary function of all mitochondria is cellular respiration, by which usable energy is derived from food materials and stored in ATP. The ATP is exported into the cytosol, where most of it is used; some also goes into the nucleus. The cells that require the most chemical energy tend to have more mitochondria per unit volume. In Chapter 7 we will see how different parts of the mitochondrion work together in the respiratory process.

Plastids

One class of organelles—the **plastids**—is produced only in plants and certain protists. The most familiar of the plastids is the **chloroplast,** which is the site of photosynthesis and contains all the chlorophyll in the plant or photosynthetic protist cell (Figure 4.15). Photosynthesis (see Chapter 8) is the process by which light energy is converted into the energy of chemical bonds. The molecules formed in photosynthesis provide food for the plant itself and for other organisms; directly or indirectly, photosynthesis is the energy source for much of the living world. The chloroplast has a number of other metabolic functions

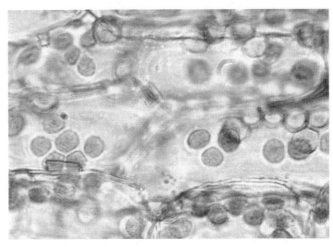

4.15 Chloroplasts in a Cell
Plant cells contains dozens of chloroplasts, the green organelles that carry on photosynthesis.

besides photosynthesis. For example, it helps make nitrogen available to the rest of the plant.

Like the mitochondrion, the chloroplast is surrounded by two membranes. However, both membranes are unfolded and surround the organelle as a smooth, closely fitting, double layer. Arising from the inner membrane is a series of discrete internal membranes, whose structure and arrangement vary from one group of photosynthetic organisms to another. As an introduction, we concentrate on the chloroplasts of the flowering plants. Even these show some variation, but the pattern in Figure 4.16 is reasonably typical.

4.16 The Organelle That Feeds the World
An electron micrograph of a section of a chloroplast from a leaf of corn. Note the stacks of thylakoids, called grana, and the membranous connections between the grana, making up an extensive network of photosynthetic membranes in the organelle. Only a thin layer of cytoplasm surrounds the chloroplast in this mature cell.

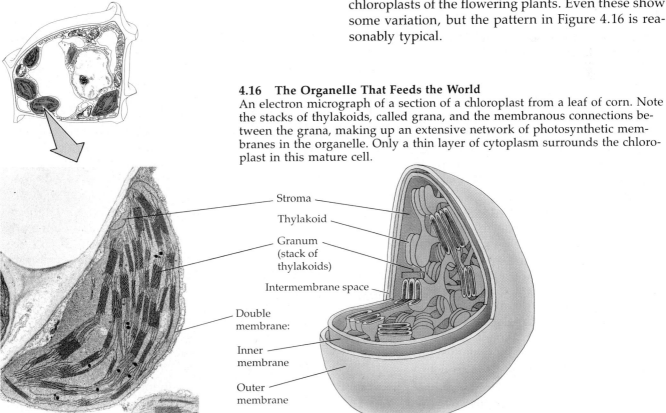

Stroma

Thylakoid

Granum
(stack of
thylakoids)

Intermembrane space

Double
membrane:

Inner
membrane

Outer
membrane

4.17 Living Together: Anemone–Alga Symbiosis
This giant sea anemone owes the intensity of its green color to the chloroplasts in a unicellular alga that lives and carries on photosynthesis within the tissues of the anemone.

As seen in electron micrographs, chloroplasts contain structures that look like stacks of pancakes. These stacks, called **grana** (singular: granum), consist of a series of flat, closely packed, circular sacs called **thylakoids.** Each thylakoid is a single membrane composed of the usual membrane components (lipids and proteins) to which have been added chlorophyll and other substances needed for trapping photosynthetic energy and producing food. All the cell's chlorophyll is contained in the thylakoid membranes. Thylakoids of one granum may be connected to those of other grana (see Figure 4.16), making the interior of the chloroplast a highly developed network of membranes. The fluid in which the grana are suspended is referred to as **stroma.** Like the mitochondrial matrix, the chloroplast stroma contains ribosomes and DNA. These ribosomes and this DNA provide some—but only some—of the proteins that make up the chloroplast.

Chloroplasts are what give plant leaves their green color. Looking at a thin slice of a leaf under the microscope reveals that most of the leaf is quite colorless—the only green to be seen is contained in the numerous chloroplasts within its cells. Not all plant cells contain chloroplasts, however. Most roots, for example, are colorless (or at least not green). This is just as well, because it would be a waste of energy and materials for the plant to provide chloroplasts for cells that reside in the dark, since photosynthesis requires light.

Animal cells do not *produce* chloroplasts, but some *contain* functional chloroplasts. These organelles are taken up either as free chloroplasts derived from plants eaten as food, or as bound chloroplasts contained within unicellular algae living within the animal tissues. The green color of some corals and sea anemones results from chloroplasts in algae that live within the animals (Figure 4.17). The animals derive some of their nutrition from photosynthesis carried out by their algal "guests."

Chloroplasts are not the only plastids found in plants. The red color of a ripe tomato results from the presence of legions of plastids called **chromoplasts.** Just as chloroplasts derive their color from chlorophyll, chromoplasts are red, orange, or yellow because of the pigments (called carotenoids; see Chapter 3) that they contain. Chromoplasts have no known chemical function in the cell, but the colors they give to some petals and fruits probably help attract animals that assist in pollination or seed dispersal. On the other hand, there is no apparent advantage in a carrot root being orange. Other types of plastids, called **leucoplasts,** serve as storage depots for starch and fats. All plastid types are related to one another. Chromoplasts, for example, are formed from chloroplasts by a loss of chlorophyll and some change in internal structure. All plastids develop from **proplastids,** which are very simple in structure.

The Origins of Plastids, Mitochondria, and Eukaryotes

In the past, biologists tried to grow chloroplasts or mitochondria in culture, outside the cells that they normally inhabit. These organelles are about the size of bacteria, they contain ribosomes and DNA, and they divide within the cell—might they not be treated like little cells in their own right? Although all such efforts at organelle culture failed because the organelles depend on the cell's nucleus and cytoplasm for some parts, the experiments helped nurture thoughts about another important question: How did the eukaryotic cell with its organelles arise in the first place? As we have seen, prokaryotic cells are generally much simpler in structure than eukaryotic cells because prokaryotes lack membrane-bounded organelles. Prokaryotic fossils can be found in sediments well over 3 billion years old, whereas the earliest known eukaryotic fossils date back to only 1.4 billion years ago. It is generally agreed that eukaryotes evolved from prokaryotes. But how?

One suggestion that has been alternately popular and scorned, over and over again for many years, is the **endosymbiotic theory** of the origin of mitochondria and chloroplasts. An important current champion of and contributor to this theory is Lynn Margulis of the University of Massachusetts, Amherst, who proposed the following idea: Picture a time, well over a billion years ago, when only prokaryotes inhabited Earth. Some of them got their food by absorbing it directly from the environment, others were photosynthetic, and still others fed by eating their

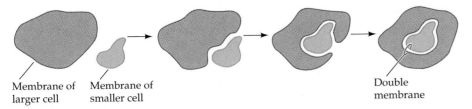

Membrane of larger cell   Membrane of smaller cell   Double membrane

4.18   Creation of a Double Membrane
Double membranes, such as those surrounding chloroplasts and mitochondria, might have been created when a larger cell (purple) engulfed a smaller cell (green) but did not digest the smaller cell. Thus the double membrane would consist of the portion of the larger cell's plasma membrane that enclosed the smaller cell and the plasma membrane of the smaller cell.

prokaryotic neighbors. Suppose that an occasional small, photosynthetic prokaryote was ingested by a larger one but did not get digested, so it sat trapped within the larger cell. Suppose further that the smaller prokaryote survived there and that it divided at about the same rate as the larger one, so successive generations of the larger prokaryote continued to be inhabited (or infected) by the offspring of the smaller one. We would call this endosymbiosis: "living within" another cell or organism, as, for example, certain algae live within sea anemones (see Figure 4.17).

Could the little green prokaryote "eaten" by the larger prokaryote have been the first chloroplast? A present-day chloroplast is surrounded by a double membrane. Such a structure might have arisen when, in the process of engulfing the photosynthetic cell, the membrane of the larger cell stretched around the plasma membrane of the smaller cell (Figure 4.18). The fact that chloroplasts contain ribosomes and DNA also fits the endosymbiotic theory because the chloroplast is proposed to have arisen from an engulfed prokaryote. In addition, whereas ribosomes in the eukaryotic cytosol are larger than those of prokaryotes, ribosomes in the chloroplast are similar in size to those of prokaryotes, further supporting the theory.

Arguments like these can be made for the proposition that mitochondria represent the descendants of respiring prokaryotes engulfed by, and ultimately endosymbiotic with, larger prokaryotes. Also, there are striking similarities between some functions of bacterial plasma membranes and mitochondrial inner membranes, and between the primary structures of certain bacterial and mitochondrial enzymes (see Chapter 3). Finally, a few modern cells do contain other, smaller cells as endosymbionts, suggesting that the endosymbiotic theory is plausible.

A spectacular example of endosymbiosis is found in the guts of certain Australian termites. We think of a termite as being able to digest wood; yet, strictly speaking, it cannot. Much of the digestive chemistry is accomplished by an endosymbiotic protist, *Mixo-tricha paradoxa*, that lives in the termite's gut. But that is far from the whole story. *Mixotricha* itself harbors an amazing colony of endosymbionts. *Mixotricha* swims around within the termite gut, apparently propelled by multitudes of flagella. Closer examination, however, shows that although there are a few true flagella in a tuft at one end of the organism, the hundreds of other flagellumlike propellers are not flagella at all—they are long, motile bacteria (spirochetes) that are attached at regular intervals to the plasma membrane of the *Mixotricha* cell and that beat just like real flagella. Also covering the surface of the protist, organized in a precise pattern, are other, smaller bacteria. *And* inside the *Mixotricha* are numerous bacteria of a third species, which are thought to help with the digestion of the tiny wood particles ingested by the protist—which, in turn, obtains them from the gut of the termite that is carrying this strange menagerie around inside itself. Perhaps termites are not all that special. If we carried colonies of *Mixotricha* in our digestive tracts, we could eat wood too. We do carry colonies of *Escherichia coli* (see Box 4.B) that aid in the digestion of our food, as discussed in Chapter 43.

We should stress that mitochondria and chloroplasts are not enough to make a prokaryote into a eukaryote. The origin of the nuclear envelope, as well as that of other important structures, including those responsible for nuclear division, still needs to be illuminated. Thus far, the endosymbiotic theory is incomplete, although suggestions have been made for its extension to deal with the origin of other eukaryotic organelles. Is the endosymbiotic theory true? Certainly it has not yet been proved. A number of compelling objections to the theory have been raised, among them the fact that the DNA responsible for the synthesis of most of the enzymes in chloroplasts and mitochondria resides in the nucleus. However the matter ultimately is resolved, the endosymbiotic theory is a good example of creative biological thinking; it gives us a useful perspective on the structures, functions, and origins of mitochondria and chloroplasts.

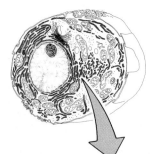

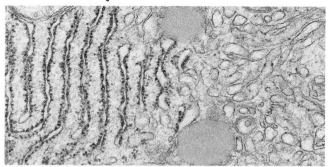

4.19   Endoplasmic Reticulum
The micrograph and drawing show rough ER on the left and smooth ER on the right. The "smoothness" is simply the absence of ribosomes.

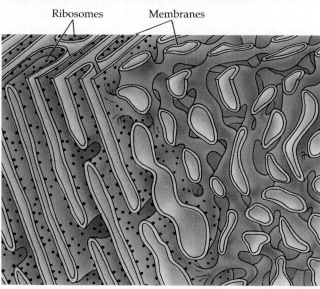

Ribosomes   Membranes

THE ENDOMEMBRANE SYSTEM

Much of the volume of a eukaryotic cell is taken up by extensive membrane systems that play numerous roles in the life of the cell. These membrane systems are closely interrelated and arise from one another. They are referred to collectively as the **endomembrane system.**

Endoplasmic Reticulum

Running here and there throughout the cytoplasm, branching and rejoining, is a network of tubes and flattened sacs called the **endoplasmic reticulum,** or **ER.** As we have noted, electron micrographs often show this membrane system to be continuous with the outer membrane of the nuclear envelope.

Parts of the ER are sprinkled liberally with ribosomes, which are attached to the outer faces of the flattened sacs. Because of their appearance in the electron microscope (Figure 4.19), these regions are called rough ER. The attached ribosomes are sites for the synthesis of proteins that function outside the cytosol—that is, proteins that are to be exported from the cell, incorporated into membranes, or moved into organelles of the endomembrane system. These proteins enter the lumen (interior) of the ER as they are synthesized, directed by a special sequence of amino acids known as the signal sequence (see Chapter 11). In the ER these proteins undergo a number of changes, including the formation of disulfide bridges (see Figure 3.14) and folding into their final, tertiary structures; some aggregate, forming quaternary structures (see Chapter 3). Some proteins gain carbohydrate groups in rough ER and thus become glycoproteins. The carbohydrate groups are part of an addressing system that ensures that the right proteins are directed to the right parts of the cell. Proteins that are to remain within the cytosol or to move into mitochondria and chloroplasts are synthesized on "free" ribosomes, that is, ribosomes that are not attached to the ER. These proteins lack the signal sequence that would otherwise direct them into the ER.

Other parts of the endoplasmic reticulum lack ribosomes and are referred to as smooth ER (see Figure 4.19). Smooth ER modifies proteins synthesized by rough ER. As you might guess from its function, the amount of ER in a cell is related to how busy the cell is in making proteins for export. Cells that are synthesizing a lot of proteins—glandular cells (see Chapter 36) or the immune system's plasma cells (see Chapter 16), for example—may be heavily packed with ER, whereas others with less work to do (such as food-storing cells) contain very little ER.

The Golgi Apparatus

In 1898 the Italian microscopist Camillo Golgi discovered a delicate structure in nerve cells. He reported that he always observed it near the nucleus. Unfortunately, his technique for staining this structure was tricky and often failed to work, so the **Golgi apparatus** was regarded by most biologists as a figment of Golgi's imagination. Work with the electron microscope in the late 1950s, however, showed clearly that these structures do exist—and not just in nerve cells, but in most eukaryotic cells.

The appearance of the Golgi apparatus varies from

(a)

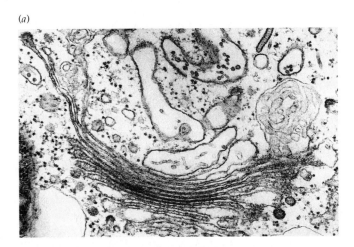

4.20 The Golgi Apparatus

(a) A Golgi apparatus in an alga. (b) Vesicles from the endoplasmic reticulum transfer substances to the apparatus by fusing with the *cis* region at the bottom of the stack, and vesicles with substances for export or transport to elsewhere in the cell leave the apparatus by budding off the *trans* region. The Golgi apparatus modifies incoming proteins and "targets" them to the correct addresses.

species to species, but a number of flattened sacs are always seen lying together like a stack of saucers (Figure 4.20a). In the cells of plants, protists, fungi, and many invertebrate animals these stacks are individual units scattered throughout the cytoplasm. In vertebrate cells, a few such stacks may form a more complex Golgi apparatus. The bottom saucers, constituting the *cis* region of the Golgi apparatus, lie nearest the nucleus or a patch of rough ER; the top saucers, constituting the *trans* region, lie closest to the surface of the cell. The saucers in the middle make up the *medial* region of the complex. These three parts of the Golgi apparatus contain different enzymes and perform different functions.

What are the functions of this organelle that Golgi discovered? The first clue comes from observing the relationships between the Golgi apparatus and other parts of the cell. Vesicles from the rough ER travel to and merge with the *cis* region of the Golgi apparatus. Other small vesicles move between the flattened sacs of the Golgi apparatus, always in the direction *cis* to *trans*, transporting proteins. Associated with the sacs, particularly those toward the *trans* region, are numerous tiny vesicles that pinch off from the sacs and then move away (Figure 4.20b). The vesicles sometimes merge with each other and finally merge with other organelles or with the plasma membrane, where they release their contents in a process called exocytosis (see next section). This behavior is the key to the important cellular functions of the Golgi apparatus. The Golgi apparatus serves as a sort of postal service depot in which some of the proteins synthesized on ribosomes on the rough ER are stored,

(b)

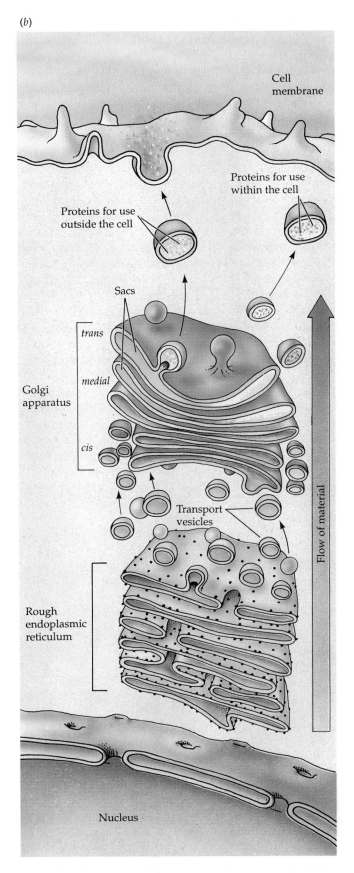

chemically modified, and packaged for delivery to the environment outside the cell or to other organelles of the cell.

How does the Golgi apparatus address the right proteins to the right destinations? The addressing consists of a series of chemical reactions in which proteins gain specific chemical "address tags." For example, a protein destined for use in an organelle called a lysosome (discussed later in this chapter) possesses a signal sequence that directs it into the lumen of the rough ER as the protein is synthesized. Enzymes in the ER add an initial carbohydrate tag—the same tag that directs other proteins to the plasma membrane to be secreted from the cell. However, other enzymes in the Golgi sacs modify the carbohydrate tags of the lysosome glycoproteins—proteins with carbohydrate residues (see Chapter 3)—changing the structure of the tag and adding a phosphate group in a process called phosphorylation. When the phosphorylated glycoprotein reaches the *trans* region of the Golgi apparatus, it binds to a receptor protein in the Golgi membrane. A vesicle containing the bound, phosphorylated glycoprotein separates from the Golgi apparatus and delivers its contents to the developing lysosome. Comparable mechanisms, most using addressing systems other than carbohydrates, deliver other proteins to other parts of the cell. The Golgi apparatus exports some proteins constantly but retains others, releasing them only at the appropriate time. In these ways, the Golgi apparatus directs the molecular mail of the cell.

Exocytosis and Endocytosis

Certain organelles such as the Golgi apparatus secrete materials to the environment; other organelles acquire materials by enveloping a portion of the environment. In eukaryotic cells there is an ongoing traffic—in and out—of membrane-bounded "packages." The plasma membrane and the endomembrane system participate actively in shipping these packages. Macromolecules for export are contained in membranous vesicles that move toward the exterior of the cell and ultimately meet the plasma membrane. The phospholipid regions of the two membranes merge, and an opening to the outside of the cell develops. The contents of the vesicle are released to the environment, and the vesicle membrane is smoothly incorporated into the plasma membrane. This entire process—export of material and transformation of the membrane—is called **exocytosis** (Figure 4.21a).

Materials may be brought into the eukaryotic cell by a related process known as **endocytosis** (Figure 4.21b). The cell surface folds to make a small pocket that is lined by the plasma membrane. The folding increases until the pocket seals off, forming a vesicle whose contents are materials from the environment. This vesicle, enclosed in membrane taken from the plasma membrane, separates from the cell surface and migrates to the interior. Usually the vesicle fuses

4.21 Exocytosis and Endocytosis
(a) In exocytosis, a membrane-enclosed secretory vesicle containing substances for export fuses with the plasma membrane. First one layer of the membrane phospholipids of the two membranes fuses, then the other. The contents of the vesicle scatter, and the vesicle membrane becomes part of the plasma membrane. (b) In endocytosis, the plasma membrane surrounds a part of the exterior environment and the whole unit buds off to the interior as a membrane-surrounded vesicle. (c) In phagocytosis, a form of endocytosis, whole particles are engulfed. Here, an amoeba is engulfing another protist.

Environment

(a) Exocytosis

(b) Endocytosis

Vesicle

Secretory vesicle

Cytoplasm

Phospholipid bilayer of plasma membrane

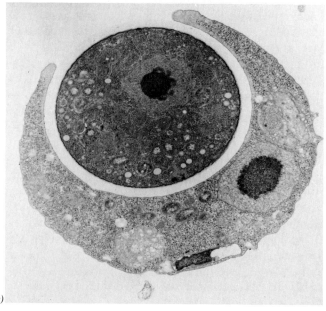

(c)

with a lysosome (see next section) and its contents are digested.

Endocytosis is a blanket term for several processes that have more specialized names. A distinction is often made between **pinocytosis** (cell drinking), in which tiny, liquid-containing vesicles are formed, and **phagocytosis** (cell eating), in which particles or even entire cells may be trapped in large vesicles (Figure 4.21c). Phagocytosis constitutes an important part of our immune system for defense against foreign cells (see Chapter 16). Many kinds of cells are able to engulf food materials from their environment, forming vesicles surrounded by pieces of the plasma membrane.

Lysosomes

Originating from the Golgi apparatus, organelles called **lysosomes** keep a cell from self-destructing by containing and transporting digestive enzymes that accelerate the breakdown of proteins, polysaccharides, nucleic acids, and lipids. Lysosomes are surrounded by a single membrane and have a densely staining, featureless interior (Figure 4.22a). The cells of animals, many protists, fungi, and a few plants have lysosomes.

Lysosomes are sites for the breakdown of food and foreign objects taken up by endocytosis. As Figure 4.22b shows, some of the vesicles that pinch off from the Golgi apparatus become primary lysosomes. After a primary lysosome fuses with a food-containing vesicle, the merged compartment is called a secondary lysosome. The effect of this fusion is rather like releasing hungry foxes into a chicken coop. The food particles are quickly digested by the enzymes in the secondary lysosome. The activity of the enzymes is enhanced by the mild acidity of the lysosome's interior, where the pH is lower than in the surrounding cytoplasm. The products of digestion exit through the membrane of the lysosome to be used by the rest of the cell. The "used" secondary lysosome then moves to the plasma membrane, fuses with it, and releases the remaining undigested contents to the environment by exocytosis.

This lysosomal compartmentalization is a very good arrangement. The digestive enzymes of the ly-

sosome would be highly destructive if they were released into the cell, where they would attack the contents of the cytosol and the other organelles. Instead, enzymes are sealed in the lysosome, from which they cannot escape. The raw materials for enzymic breakdown are delivered tightly packaged in a vesicle that fuses with the lysosome; the useful products of digestion leak out; finally the enzymes, along with the unusable products, are thrown out of the cell.

(a)

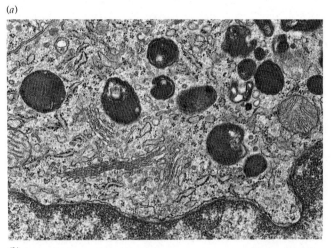

(b)

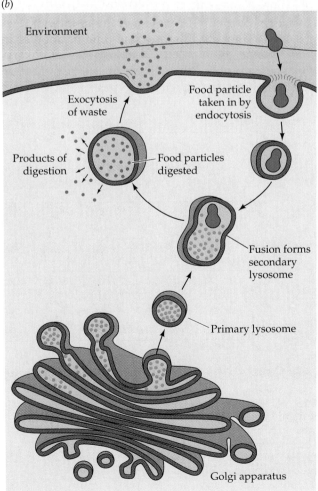

4.22 Lysosomes Keep Digestive Enzymes Where They Belong
(a) The darkly stained organelles in the upper half of this electron micrograph of a rat cell are secondary lysosomes with material being digested inside them. At the bottom is a portion of the cell's nucleus, above which are some of the flattened membrane sacs of the Golgi apparatus. (b) Food-containing vesicles fuse with enzyme-containing primary lysosomes. Digestive enzymes do not contact cell parts other than lysosome interiors.

Clearly, the consequences of digestive enzymes escaping from the lysosomes can be severe. At times, though, such digestive activity is appropriate, as during the development of a frog from a tadpole. The tadpole has a fleshy tail, whereas the mature frog has none. How does the tail disappear? Part of the job is accomplished by the breakdown of lysosomes within the tail cells of the tadpole, releasing enzymes that digest the cells themselves.

Why are the lysosomes not destroyed by these enzymes? It seems as if lysosomes themselves also should be subject to attack by the digestive enzymes they house, yet they survive and function. Despite much interest in this problem, no generally acceptable solution has yet been proposed.

OTHER ORGANELLES

In addition to the information-processing organelles (nucleus and ribosomes), the energy-processing organelles (mitochondria and plastids), and the organelles that form the endomembrane system of the cell (endoplasmic reticulum, Golgi apparatus, and lysosomes), there are several other organelles. **Microbodies** form by pinching off from rough endoplasmic reticulum. As seen with the electron microscope, these small organelles, 0.2 to 1.7 μm in diameter, have a single membrane and a granular interior (Figure 4.23). They are found at one time or another in at least some cells of almost every species of eukaryote. The same microbody structure serves different functions, and the functional types of microbodies

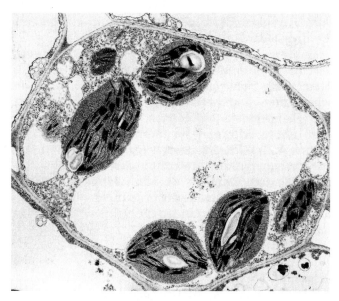

4.24   Vacuoles in Plant Cells Are Big
The large central vacuole is typical of mature plant cells. Smaller vacuoles are visible toward each end of the cell.

have their own names. Microbodies called **peroxisomes** house reactions in which toxic peroxides (such as hydrogen peroxide, H_2O_2) are formed as unavoidable side products of chemical reactions. Subsequently, the peroxides are broken down safely within the peroxisomes without mixing with other parts of the cell. Both plant and animal cells have peroxisomes, but another type of microbody, the **glyoxysome,** is found only in plants. Glyoxysomes, which are most prominent in young plants, are the sites at which stored lipids are converted into carbohydrates.

We think of organelles as self-contained compartments or "factories" for specific processes, but remember that materials are constantly being shuttled from one type of organelle to another. For example, there is considerable traffic of compounds among the organelles of plant cells in a process called photorespiration (see Chapter 8). The process of photorespiration begins in the chloroplasts. One intermediate product is transported from the chloroplasts to the peroxisomes for further chemical changes. Some of the products of peroxisome action are then passed to the mitochondria, whereas others are returned to the chloroplast for more changes.

In many eukaryotic cells, but particularly in those of plants and protists, are structures that look empty under the electron microscope. They are called **vacuoles** (Figure 4.24). They are not actually empty but are filled with aqueous solutions containing many dissolved substances. Each vacuole is surrounded by a single membrane.

For all their structural simplicity, vacuoles play a variety of crucial roles in the lives of cells. For example, consider one of the problems a plant faces.

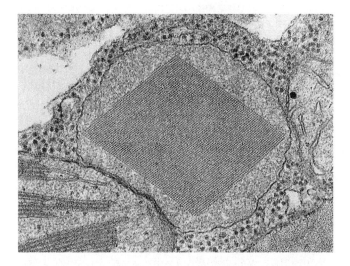

4.23   A Microbody
A diamond-shaped crystal, composed of an enzyme, almost entirely fills this rounded microbody in a leaf cell. The enzyme catalyzes one of the reactions fulfilling the special function of the microbody. The microbody is pressed against a chloroplast (lower left).

Like animals and other organisms, plants produce a number of by-products that would be toxic to the organism if not set aside. Animals have excretory mechanisms for getting rid of such wastes, but plants are not equipped in the same way. Although plants manage to secrete some wastes to their environment, many compounds must be stored within the cells. The solution to this storage problem is the vacuole. The vacuolar membrane can keep wastes away from the rest of the cell, preventing toxic reactions. The vacuoles of many plants store large amounts of chemicals that are poisonous or distasteful to herbivores (plant-eating animals), thus deterring animals from eating the plants.

In many plant cells, enormous vacuoles take up more than 90 percent of the total volume and grow as the cell grows. Vacuoles are by no means a waste of space, however, for the dissolved substances in the vacuole, working together with the vacuolar membrane, provide turgor, or stiffness, to the cell, which in turn supports the structure of nonwoody plants.

Some unicellular protists, sponges, and some of the more ancient invertebrate animals obtain nutrients by endocytosis. Particles trapped from the environment in this way end up in vesicles called **food vacuoles.** Many freshwater protists have a highly specialized **contractile vacuole** (see Chapter 23). Its function is to rid the cell of excess water that rushes in because of the imbalance in salt concentration between the relatively salty interior of the cell and its freshwater environment. Vacuoles even play a role in the sex life of plants. Some pigments (especially blue and pink ones) in petals and fruits (and sometimes in leaves) are contained in vacuoles. These pigments—the anthocyanins—serve as cues that encourage animals to visit flowers and thus aid in pollination, or to eat fruits and thus aid in seed dispersal.

THE CYTOSKELETON

Membranes divide the cytoplasm of eukaryotic cells into numerous compartments, as already described. But even the cytosol (the space and material inside the plasma membrane but outside the membrane-bounded organelles) is not a simple aqueous solution—far from it! In this compartment of the cell is a dynamic set of fibers—the **cytoskeleton**—that contributes to the cell's shape and physical texture. Some fibers also act as tracks for "motors" that help a cell to move. At least three components of the cytoskeleton are visible in electron micrographs: microtubules, microfilaments, and intermediate filaments.

Microfilaments are a common type of fiber in the cytosol. They drive many types of cellular movement.

Microfilaments are assembled from a protein called actin, often in combination with other proteins (Figure 4.25a). Each microfilament is 7 nm in diameter and several micrometers long. Microfilaments may be single or in bundles and networks. Often they are attached to the plasma membrane. In muscle cells the microfilaments are stable for days at a time, but in other cells they are more dynamic, forming and breaking down in minutes. Microfilaments help the cell to contract; many types of motion within the cell require their participation. Microfilaments take part in changes in cell shape (including cell length, as in the contraction of muscles), in the streaming of cytoplasm (a flowing movement observed in some cells), in movements of organelles and particles, and in pinching movements such as those that separate the daughter cells after an animal cell has undergone nuclear division. In muscles, filaments of a protein called myosin have "motor" projections that "row" the actin and myosin filaments past each other, pulling the ends of the cell together; the action of microfilaments in muscles will be discussed in detail in Chapter 40.

Filaments of another type, the **intermediate filaments,** play more static roles: They stabilize cell structure and resist tension. They are found only in multicellular organisms. Different kinds of cells contain different kinds of intermediate filaments. Although there are at least five distinct types of intermediate filaments, all share the same general structure (Figure 4.25b). Intermediate filaments are composed of fibrous proteins similar to those that make up hair and skin. In cells, these proteins are organized into tough, ropelike assemblages about 8 to 12 nm in diameter. In some cells, intermediate filaments end at the nuclear envelope and may maintain the positions of the nucleus and other organelles in the cell. Other types of intermediate filaments help hold a complex apparatus of microfilaments in place in muscle cells (see Chapter 40). Still other types help to stabilize and maintain rigidity in surface tissues by connecting "spot welds" called desmosomes (see Chapter 5). Rapidly growing or newly formed cells do not contain intermediate filaments.

Microtubules are long, hollow, unbranched cylinders about 25 nm in diameter and up to several micrometers long. Many of the microtubules in a cell radiate from a region called the microtubule organizing center. Microtubules are made up of many molecules of a protein called **tubulin.** Tubulin itself is a dimer consisting of two polypeptide subunits called α-tubulin and β-tubulin. Thirteen rows, or protofilaments, of tubulin dimers surround the central cavity of the microtubule (Figure 4.25c). Microtubules have a polarity—one end is called the + end and the other the − end. Tubulin dimers can be added to or subtracted from the + end, lengthening or shortening

(a) Microfilaments

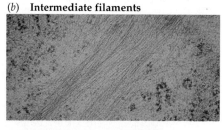

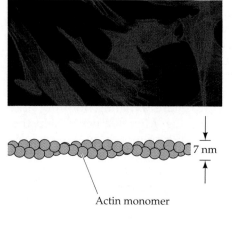

7 nm

Actin monomer

Microfilaments are made up of strands of the protein actin and often interact with strands of other proteins. Microfilaments may occur singly, or in bundles or networks. They change cell shape and drive cellular motion, including contraction, cytoplasmic streaming, and the "pinched" shape changes that occur during cell division. Microfilaments and myosin strands together drive muscle action.

(b) Intermediate filaments

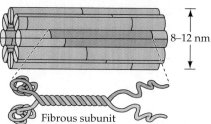

8–12 nm

Fibrous subunit

Intermediate filaments are made up of fibrous proteins organized into tough, ropelike assemblages that stabilize a cell's structure and help maintain its shape. Some intermediate filaments hold neighboring cells together.

(c) Microtubules

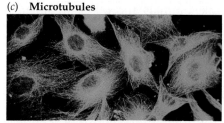

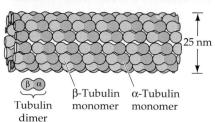

25 nm

β α

Tubulin dimer

β-Tubulin monomer   α-Tubulin monomer

Microtubules are long, hollow cylinders made up of many molecules of the protein tubulin. Tubulin consists of two subunits, α-tubulin and β-tubulin. Microtubules lengthen or shorten by adding or subtracting tubulin dimers. Microtubule shortening moves chromosomes. Interactions between microtubules drive the movement of cells. Microtubules serve as "tracks" for the movement of vesicles.

4.25   Components of the Cytoskeleton

the microtubule and thus affecting the cell in various ways. The capacity to change length rapidly makes microtubules dynamic structures.

In plants, microtubules help control the arrangement of the fibrous components of the cell wall. Electron micrographs of plants frequently show microtubules lying next to the cell wall, and disruption of the cell's microtubules leads to a disordered arrangement of newly synthesized fibers in the cell wall. In animal cells, microtubules are often found in the parts of the cell that are changing shape, but the mechanisms by which the microtubules function are not yet known. In some specialized cells, such as nerve cells, the cytoplasm contains microtubules running parallel to the length of long cellular projections. Here the microtubules contribute to the mechanical stability of the projections. In these and some other cells, microtubules serve as tracks along which tiny "motors" carry protein-laden vesicles from one part of the cell to another (Box 4.C).

Many eukaryotic cells possess whiplike appendages, the flagella and cilia that are built from specialized microtubules. These organelles push or pull the cell through its aqueous environment or promote movement of the surrounding liquid over the surface of the cell (Figure 4.26a). Cilia and eukaryotic flagella are identical in internal structure but differ in relative lengths and patterns of beating. Longer appendages, usually single or in pairs, are called flagella; these

propagate waves of bending from one end to the other in snakelike undulation. The shorter appendages, usually present in great numbers, are called cilia; they beat stiffly in one direction and recover flexibly in the other direction (like a swimmer's arm), so the recovery stroke does not undo the work of the power stroke (see Chapter 40).

In cross section, a typical cilium or eukaryotic flagellum is seen to be covered by the plasma membrane and to contain what is usually called a 9 + 2 array of microtubules (Figure 4.26b). As you can see from the figure, this name is somewhat misleading: There are actually nine fused pairs of microtubules, called **doublets**, forming a cylinder, and one pair of unfused microtubules running up the center. The motion of cilia and eukaryotic flagella results from these microtubules sliding past one another, as described in Chapter 40. But what is the "motor" that drives this sliding? It is a protein called **dynein**, which can undergo changes in tertiary structure driven by en-

4.26   Cilia Move Cells   ▶

(a) Cilia covering the surface of this protist propel the unicellular organism through its watery environment. (b) Longitudinal section through three cilia on a protist cell. If one of these cilia and its basal body were viewed in cross section, the structures diagrammed would be seen. Movement is powered by "arms" of the protein dynein sliding the microtubules along.

BOX 4.C

Molecular "Motors" Carry Vesicles along Microtubules

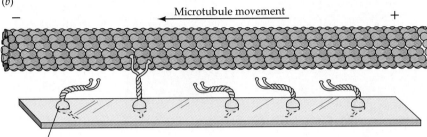

(a)

(a) Kinesin molecules "walk" along a microtubule, moving a vesicle along with them.
(b) Diagram of the experiment that established the mechanisms of kinesin and other "motor molecules."

A remarkable group of protein molecules serve as "motors" on the cytoskeleton. One such protein, kinesin, carries vesicles from the − ends to the + ends of microtubules. Other proteins carry other kinds of loads, such as mitochondria. Kinesin consists of four polypeptide chains: two identical heavy chains and two identical light chains. The "heads" of the heavy chains bind to microtubules and undergo changes in shape (tertiary structure). The shape changes cause the kinesin molecule to "walk" along the tubulin subunits of the microtubules. The light chains bind vesicles, which thus go along for the ride.

How were the functions of kinesin and the other motors discovered? Kinesin was isolated and purified by techniques that took advantage of its affinity for microtubules. The light chains of purified kinesin bind to glass surfaces, enabling the following elegant experiment: After binding kinesin to a glass slide and adding a solution containing microtubules and ATP, scientists discovered that the microtubules slithered along the slide, moving in the direction of their − ends. Because the kinesin molecules were bound tightly to the glass, as their heads walked toward the + ends of the microtubules, the microtubules were pushed in the direction of their − ends. In other studies, scientists showed that kinesin can also carry plastic beads toward the + ends of isolated microtubules.

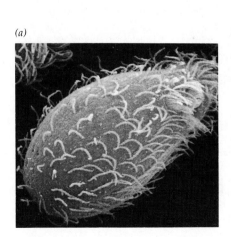

(a)

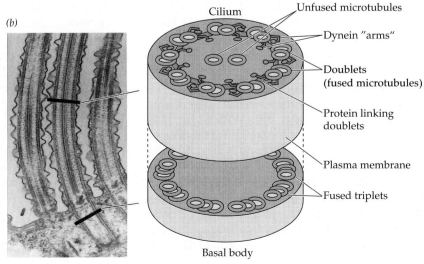

(b)

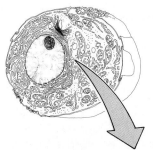

4.27 Centrioles Contain Triplets of Microtubules
Centrioles are found in the microtubule organizing center, a region near the nucleus. *(a)* This thin-section micrograph shows a pair of centrioles at right angles to each other. Nine sets of three microtubules are evident in the centriole on the right, which is seen in cross section. *(b)* The diagram emphasizes the three-dimensional structure of a centriole; compare with the diagram of the basal body in Figure 4.26.

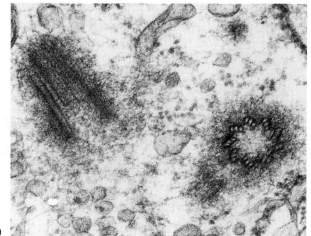

(a)

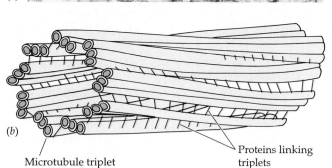

(b)

Microtubule triplet

Proteins linking triplets

accompanied by another microtubule, making nine sets of *three* microtubules. The central, unfused microtubules of the cilium or flagellum do not extend into the basal body.

Centrioles are organelles that are virtually identical in structure to basal bodies. Centrioles are found in all eukaryotes except those that never produce cells with cilia or flagella (that is, the flowering plants, the pines and their relatives, and some protists). Under the light microscope, a centriole looks like a small, featureless particle, but the electron microscope reveals that it is made up of a precise bundle of microtubules, arranged as nine sets of three fused microtubules each (Figure 4.27). When present, two centrioles lie at right angles to each other in the microtubule organizing center in cells about to undergo division (see Chapter 9).

THE OUTER "SKELETON"—THE CELL WALL

The eukaryotic **cell wall** is a semirigid structure outside the plasma membrane of cells of plants (Figure 4.28*a*), fungi, and some protists. (There is no com-

ergy from ATP. Dynein molecules on one microtubule bind to a neighboring microtubule; as the dynein molecules change shape, they "row" one microtubule past its neighbor.

Although some prokaryotes also have flagella, as described earlier in this chapter, prokaryotic flagella are very different from the 9 + 2 arrangement of the eukaryotic flagellum: They lack microtubules and dynein. There is no structural or evolutionary relationship between the flagella of prokaryotes and those of eukaryotes. The prokaryotic flagellum has a much simpler structure than that of the eukaryotic flagellum (see Figure 4.5) and a smaller diameter than that of a single microtubule. Prokaryotic flagella rotate, whereas eukaryotic flagella beat in a wavelike (or undulating) motion.

An organelle called a **basal body** is found at the base of every eukaryotic flagellum or cilium (see Figure 4.26*b*). The nine microtubule doublets extend into the basal body. In the basal body, each doublet is

(a)

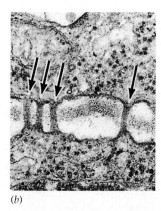

(b)

4.28 The Plant Cell Wall
(a) The brilliant gold and blue structures in this polarizing micrograph of seed tissue are the cell walls. Cell walls make up a substantial fraction of the tissue. (Compare with Hooke's drawing of cork in Figure 1.11.) *(b)* In this electron micrograph of a grass root, plasmodesmata are the gray channels (arrows) crossing the cell walls between the cytoplasm of the cell above and the cell below.

parable structure in animal cells.) The cell wall is made up primarily of polysaccharides. It provides support for the cell, limits the cell's volume, and in some instances restricts the flow of water into and out of the cell. In plants, modifications of the cell wall are important in determining the specific function of the cell.

Although it might seem that plant cells are completely isolated from one another by their cell walls, this is not the case. Plant cells are connected by channels, or **plasmodesmata,** each 20 to 40 nm in diameter, that extend through the walls of adjoining cells (Figure 4.28*b*). A strand of cytoplasm about 4 nm in diameter runs through most plasmodesmata. A plasmodesma allows relatively free passage of many molecules—not only is there no cell wall across the hole, but there is also no plasma membrane.

EUKARYOTES, PROKARYOTES, AND VIRUSES

Let's review the major differences between the cells of eukaryotes and those of prokaryotes. In contrast to prokaryotes, which have nucleoids, eukaryotes have a true, membrane-enveloped nucleus, and unlike the prokaryotes, eukaryotes have other membrane-bounded organelles that allow various cellular activities to be concentrated in specialized compartments. Eukaryotic cells, but not prokaryotic ones, can perform endocytosis and exocytosis. Eukaryotes also have many specialized molecules, such as tubulin and actin, that are used for movement and are not found in prokaryotes. The cell walls of prokaryotes differ structurally and chemically from those of eukaryotes.

Do such differences mean that eukaryotes are more advanced, higher, or more successful than prokaryotes? Not at all! Every surviving species is the product of eons of natural selection and is superbly adapted to its environmental niche. Each species has characteristics that enable it to live where and how it does and to compete successfully against other species. Both eukaryotic and prokaryotic cells are marvels of systematic organization to achieve complex function.

Where do the viruses fit into this picture? They are simpler in structure than the bacteria—so much simpler that they cannot be called cells. Viruses lack ribosomes and must use the ribosomes of a host cell (prokaryotic or eukaryotic) to synthesize the proteins they need. The hereditary material of a virus enters the host cell and there subverts the host's metabolic machinery to make new viruses. Except for a few complex forms, viruses are simply packets of hereditary material wrapped in coats of protein; in no way do they demonstrate the life processes of independent cells. Viruses are treated in detail in Chapter 22.

FRACTIONATING THE EUKARYOTIC CELL: ISOLATING ORGANELLES

Since organelles are so small (just 1 g of chloroplasts contains more than 10 billion individual chloroplasts of average size), isolating particular organelles for study is difficult. During the early days of cell biology in the nineteenth century, all that one could do with organelles—and only the largest ones at that—was look at them with a microscope as they sat within the cell. Later, with the electron microscope, it became possible to view cells at a higher magnification and to isolate and purify substantial quantities of specific organelles, enabling researchers to study their biochemical activities and physiological functions. Today, scientists who want to work with isolated organelles must accomplish two tasks: removing the organelles from cells, and separating the various types of organelles from one another. This process is called cell fractionation.

Rupturing the Cell

The first step in cell fractionation is opening up the cell; during this process, one must be careful not to burst the organelles. Various methods, most of which employ strong shearing forces, can be used to break cells open. The simplest methods use an old-fashioned mortar and pestle or a hand-operated glass homogenizer (which squeezes and shears cells between two tightly fitting, counter-rotating, ground-glass surfaces). Cells of some tissues can be opened by rapid chopping with a razor blade against a glass plate. Motor-driven homogenizers or blenders are also commonly used.

These methods rupture the cells by tearing their plasma membranes and (if present) cell walls. The rupturing of cells is conducted in a solution with a high concentration of solutes, which prevents the organelles from bursting. In a more dilute solution or pure water, the organelles would take up water and explode, as described in Chapter 5.

Separating the Organelles

Using techniques like those described in the previous section, one may reduce biological tissue to a crude suspension of mixed organelles, unbroken cells, and debris. But how can the components of such a suspension be separated from one another? The methods of choice are two types of centrifugation. The **centrifuge** is a laboratory instrument that can spin materials extremely rapidly about a fixed axis, which causes the particles in suspension to fall or rise more rapidly than they would under the force of gravity alone (Figure 4.29). Different classes of organelles sediment (settle out of a suspension) at different rates

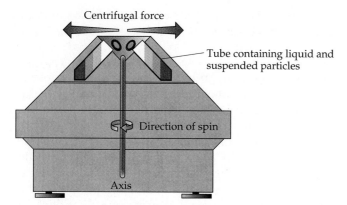

4.29   Separation by Centrifugation
Extremely rapid rotation of liquids in centrifuge tubes around an axis produces a force in the tubes analogous to that of gravity, but much more powerful. Suspended particles denser than the liquid in the tubes sink rapidly. Following centrifugation, the fluid can be poured off, leaving a dense pellet of particles in the bottom of the tube.

in a centrifuge. Factors determining the rate of sedimentation include the organelle's size (radius) and density (weight per unit volume). For example, an organelle that is more dense than the liquid in which it is suspended settles toward the bottom of the centrifuge tube. An organelle less dense than the liquid floats toward the surface. This is the same principle that explains why oil rises to the top of a mixture of oil and water that is shaken and then allowed to separate.

Equilibrium centrifugation can be used to separate two or more types of organelles that differ in density (Figure 4.30). First a centrifuge tube is filled with liquid whose density varies from the top to the bottom of the tube. This could be done, for example, by first putting in a small amount of a highly concentrated (and hence very dense) sugar solution—say, 60 percent sucrose. Next a small amount of a 50 percent sucrose solution is added, then 40 percent, and so on. Such a **density gradient** is usually con-

structed by an automatic device so that the gradient is smooth rather than erratic and jumpy. After the gradient is established, some of the mixture of, say, two organelles to be separated is carefully added to the top and the tube is centrifuged. Both populations of organelles sediment into the gradient as long as they are more dense than the surrounding liquid. When an organelle reaches the part of the gradient where its density matches that of the liquid, it stops sedimenting—it has reached buoyant equilibrium. If the organelle were pushed farther down, it would float back up to this point. If the two types of organelles have different average densities, they form separate bands in the liquid. After the tube is removed from the centrifuge, the organelles can be collected separately by the careful use of a pipette or by punching a hole in the bottom of the tube and collecting samples as the liquid slowly drips out.

A second approach, called **differential centrifugation,** depends on differences in either the radius or the density of the particles being centrifuged. No density gradient is used, and the liquid in the tube is usually of low density. A mixture containing various organelles is centrifuged briefly at low speed (and hence low relative centrifugal force). The largest and densest particles sediment, forming a pellet in the bottom of the tube; this pellet is left behind when the remaining liquid and its suspended contents (together called the supernatant) are poured off. The liquid and its remaining contents are then spun at a higher speed and for a longer time, causing other organelles to sediment. By repeating this procedure with ever-increasing centrifugal force, one can separate many organelles. Nuclei sediment into a pellet even when the centrifuge is spun slowly. The tiny ribosomes, on the other hand, require considerably longer centrifugation at extremely high speeds before they will sediment—in most rotors, the spinning rate must be about 40,000 revolutions per minute, giving forces as great as 100,000 times the force of gravity.

Partially purified organelles are obtained by equilibrium or differential centrifugation. Further purification is achieved by repeating the centrifugation

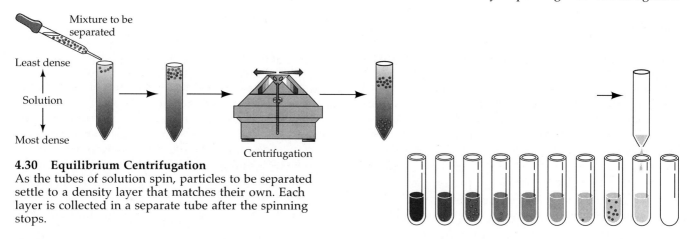

4.30   Equilibrium Centrifugation
As the tubes of solution spin, particles to be separated settle to a density layer that matches their own. Each layer is collected in a separate tube after the spinning stops.

routine. The purity may be determined by examining the final sample under the microscope—light or electron—or by testing the chemical activities of the sample in comparison with the known behavior of various organelles. We actually *can* get a 1-g sample of chloroplasts (or other organelles) in this way and study them to our heart's content.

Early work with isolated organelles focused on their chemical composition and then on the biochemical reactions that take place within them. Gradually we have come to have an extensive, but still partial, knowledge of the activities of the different organelles (as discussed in this and the next few chapters) and hence of the cell itself.

SUMMARY of Main Ideas about Cells

All organisms are composed of cells, and all cells come from preexisting cells.

In the kingdoms Eubacteria and Archaebacteria, each organism is a single, uncompartmented prokaryotic cell.

In the other kingdoms, each organism consists of one or more nucleated, compartmented eukaryotic cells.

All prokaryotic cells have a plasma membrane, ribosomes, and DNA in a nucleoid, and many have flagella, cell walls, capsules, and internal membranes formed from the plasma membrane.
  Review Figures 4.2, 4.3, 4.4, and 4.5

The nucleus of a eukaryotic cell contains chromosomes, nucleoli, and nucleoplasm, all surrounded by a nuclear envelope.
  Review Figures 4.8, 4.9, and 4.10

A cell's genetic information is stored in its nucleoid (prokaryotes) or nucleus (eukaryotes).

All cells synthesize proteins on ribosomes consisting of two subunits.
  Review Figure 4.13

Mitochondria convert energy into a usable form in eukaryotic cells.
  Review Figures 4.8, 4.9, and 4.14

Chloroplasts in photosynthetic eukaryotic cells capture light energy.
  Review Figures 4.8, 4.15, and 4.16

If the endosymbiotic theory is correct, mitochondria and chloroplasts evolved by the incorporation of smaller prokaryotic cells into larger ones.
  Review Figure 4.18

Endocytosis brings substances into eukaryotic cells, and exocytosis ejects substances from them.
  Review Figures 4.21 and 4.22

The endoplasmic reticulum and the Golgi apparatus sort, store, modify, and move proteins in eukaryotic cells.
  Review Figures 4.8, 4.9, 4.19, and 4.20

Reactions that either form toxic products or convert lipids to carbohydrates are housed, respectively, in peroxisomes or glyoxysomes; these are examples of microbodies.
  Review Figure 4.23

Vacuoles in eukaryotic cells store wastes and provide turgor.
  Review Figures 4.8 and 4.24

The cytoskeleton gives a eukaryotic cell shape and may help it move.
  Review Figure 4.25

Cilia and flagella propel eukaryotic cells or move substances past them.
  Review Figure 4.26

A cell wall surrounds the cells of plants, fungi, and some protists.

SELF-QUIZ

1. Which statement is true of both prokaryotic and eukaryotic cells?
   a. They contain ribosomes.
   b. They have peptidoglycan cell walls.
   c. They contain membrane-bounded organelles.
   d. They contain true nuclei.
   e. Their flagella have the 9 + 2 structure.

2. Which statement is *not* true of the nuclear envelope?
   a. It is continuous with the endoplasmic reticulum.
   b. It has pores.
   c. It consists of two membranes.
   d. RNA and some proteins pass through it to move in and out of the nucleus.
   e. Its inner membrane bears ribosomes.

3. Which statement is *not* true of mitochondria?
   a. Their inner membrane folds to form cristae.
   b. They are usually one micrometer or less in diameter.
   c. They are green because of the chlorophyll they contain.
   d. Energy-rich substances from the cytosol are oxidized in them.
   e. Much ATP is synthesized in them.

4. Which statement is true of plastids?
   a. They are found in prokaryotes.
   b. They are surrounded by a single membrane.
   c. They are the sites of cellular respiration.
   d. They are found in fungi.
   e. They are of several types, with different functions.

5. Which statement is *not* true of the endoplasmic reticulum?
   a. It is of two types: rough and smooth.
   b. It is a network of tubes and flattened sacs.
   c. It is found in all living cells.
   d. Some of it is sprinkled with ribosomes.
   e. Parts of it modify proteins.

6. The Golgi apparatus
   a. is found only in animals.
   b. is found in prokaryotes.
   c. is the appendage that moves a cell around in its environment.
   d. is a site of rapid ATP production.
   e. packages and modifies proteins.

7. Which of the following organelles is *not* surrounded by one or more membranes?
   a. Ribosome
   b. Chloroplast
   c. Mitochondrion
   d. Microbody
   e. Vacuole

8. Eukaryotic flagella
   a. are composed of a protein called flagellin.
   b. rotate like propellers.
   c. cause the cell to contract.
   d. have the same internal structure as cilia.
   e. cause the movement of chromosomes.

9. Microfilaments
   a. are composed of polysaccharides.
   b. are composed of actin.
   c. provide the motive force for cilia and flagella.
   d. make up the spindle that aids movement of chromosomes.
   e. maintain the position of the nucleus in the cell.

10. Which statement is *not* true of the plant cell wall?
    a. Its principal chemical components are polysaccharides.
    b. It lies outside the plasma membrane.
    c. It provides support for the cell.
    d. It completely isolates adjacent cells from one another.
    e. It is semirigid.

FOR STUDY

1. Which organelles and other structures are found in both plant and animal cells? Which are found in plant but not animal cells? Which in animal but not plant cells? Discuss, in relation to the activities of plants and animals.

2. Through how many membranes would a molecule have to pass in going from the interior of a chloroplast to the interior of a mitochondrion? from the interior of a lysosome to the outside of a cell? from one ribosome to another?

3. How does the possession of double membranes by chloroplasts and mitochondria relate to the endosymbiotic theory of the origins of these organelles? What other evidence supports the theory?

4. What sorts of cells and subcellular structures would you choose to examine by transmission electron microscopy? by scanning electron microscopy? by light microscopy? What are the advantages and disadvantages of each of these modes of microscopy?

5. Some organelles that cannot be separated from one another by equilibrium centrifugation can be separated by differential centrifugation. Other organelles cannot be separated from one another by differential centrifugation but can be separated by equilibrium centrifugation. Explain these observations.

READINGS

Alberts, B., D. Bray, J. Lewis, M. Raff, K. Roberts and J. D. Watson. 1994. *Molecular Biology of the Cell*, 3rd Edition. Garland Publishing, New York. An outstanding book in which to pursue the topics of this chapter in greater detail; authoritative treatment of modern cell biology and its experimental basis.

Allen, R. D. 1987. "The Microtubule as an Intracellular Engine." *Scientific American*, February. How microtubules cause two-way transport of materials in cells.

Brandt, W. H. 1975. *The Student's Guide to Optical Microscopes*. William Kaufmann, Los Altos, CA. A short, programmed guide for those interested in learning how to use a light microscope.

De Duve, C. 1975. "Exploring Cells with a Centrifuge." *Science*, vol. 189, pages 186–194. A discussion by a Nobel laureate of the uses of centrifugation in studies of cells.

Fawcett, D. W. 1981. *The Cell*, 2nd Edition. Saunders, Philadelphia. Beautiful electron micrographs of subcellular structures in animal cells.

Glover, D. M., C. Gonzalez and J. W. Raff. 1993. "The Centrosome." *Scientific American*, June. New findings about the structure and function of the organelle that directs the assembly of the cytoskeleton and controls cell division—when it is present.

Howells, M. R., J. Kirz and D. Sayre. 1991. "X-Ray Microscopes." *Scientific American*, February. A description of novel methods of microscopy afford striking improvements in resolution.

Lodish, H., D. Baltimore, A. Berk, L. Zipursky, P. Matsudaira and J. Darnell. 1995. *Molecular Cell Biology*, 3rd Edition. Scientific American Books, New York. Another excellent middle-level book; fine illustrations.

Margulis, L. 1993. *Symbiosis in Cell Evolution*, 2nd Edition. W. H. Freeman, New York. An authoritative and thought-provoking reference on the origin and evolution of eukaryotic cells by a leading student of the problem.

Rothman, J. E. 1985. "The Compartmental Organization of the Golgi Apparatus." *Scientific American*, September. Discusses the structure and function of the Golgi apparatus.

Weber, K. and M. Osborn. 1985. "The Molecules of the Cell Matrix." *Scientific American*, October. A clear treatment of microfilaments, intermediate filaments, tubulin, and the ways in which they are studied.

The Plasma Membrane
The edge of a red blood cell, magnified about 300,000 times by a transmission electron microscope, shows the bilayered phospholipid membrane as two dark lines reminiscent of railroad tracks.

5

Membranes

Poised between every cell and its environment is a filmy sheet so thin that it can be seen only with the aid of an electron microscope. Similar sheets within the cytoplasm divide many cells into compartments. The sheets are dynamic—forming, changing, merging, moving. Filmy and thin as they are, these sheets perform a sweeping array of vital functions. They recognize specific chemical substances as well as other cells, hold groups of molecules in place, and transmit signals along nerves. They let some substances pass through and keep other substances out.

How are such diverse tasks performed? How can some things pass through while others are stopped? These are the subjects of this chapter. The story begins with a close look at the filmy sheets and with consideration of the fundamental process of diffusion, followed by an explanation of other ways that particular substances move through and a description of more of the diverse tasks.

MEMBRANE STRUCTURE AND COMPOSITION

The filmy sheets described in our opening paragraphs are, of course, the plasma membrane and other **membranes,** all of which are thin, pliable bilayers of phospholipids with embedded proteins. (The phospholipid bilayer was introduced in Chapter 3, and membranous organelles are described in Chapter 4.) As we consider membranes, the relationship between the function and the physical structure of the membrane is particularly obvious; this relationship is evident at all levels, from the overall shape of the membrane down to its individual chemical components.

The chemical makeup, physical organization, and functioning of a biological membrane depend upon three classes of biochemical compounds: lipids, proteins, and carbohydrates. The lipids are an effective barrier to the passage of many materials between the inside and the outside of a cell or organelle. Many functions of membranes result from this **selective permeability**—some materials move through them more readily than others. The lipids also account for much of the physical integrity of the membrane. The lipids are present as a double layer that constitutes the continuous portion of the membrane. In this lipid "lake" floats a variety of proteins.

Some proteins reach from one side of a membrane to the other; others reside primarily on one side. The proteins are responsible for many of the specific tasks performed by membranes. Certain membrane proteins allow materials to pass through the membrane that cannot pass through the pure lipid bilayer. Other proteins receive chemical signals from the cell's ex-

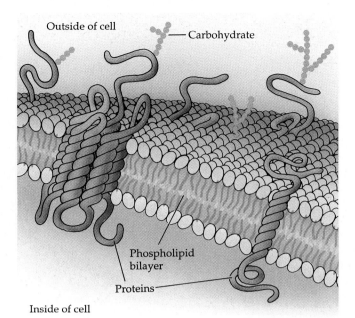

Outside of cell

Carbohydrate

Phospholipid
bilayer

Proteins

Inside of cell

5.1 A Model of a Biological Membrane
The general molecular structure of biological membranes, sometimes called the fluid mosaic model, is a continuous phospholipid bilayer. Proteins are embedded in the bilayer. Carbohydrates may be attached to the proteins or to the phospholipids.

ternal environment and respond by regulating certain processes within the cell. Still other proteins accelerate chemical reactions on the membrane surface.

Like some proteins, the carbohydrates, the third class of compounds important in membranes, are crucial in recognizing specific molecules. The carbohydrates are attached to lipid or protein molecules, mostly on the outside of the plasma membrane, where they protrude into the environment, away from the cell.

Generalized membrane architecture is shown in Figure 5.1. The two sides of the membrane are not identical—in fact, the membrane is decidedly asymmetric. Not even all the lipids are the same: Those on the inward-facing half are different from those on the outward-facing half.

Membrane Lipids

Nearly all lipids in biological membranes are **phospholipids.** Recall that some compounds are hydrophilic ("water-loving") and others are hydrophobic ("water-fearing"). Phospholipids are both: They have hydrophilic regions and hydrophobic regions. The large, nonpolar fatty acid parts of phospholipid molecules associate easily with other fatty materials but do not dissolve in water. The phosphorus-containing region of the phospholipid is electrically charged and hence very hydrophilic. As a consequence, one way for phospholipids and water to coexist is for the

phospholipids to form a double layer with the fatty acids of the two layers pointing toward each other and the polar regions facing the outside (Figure 5.2). Artificial membranes with the same two-layered arrangement are made easily in the laboratory.

Both artificial and natural membranes form continuous sheets. Because of the tendency of the fatty acids to associate with one another and exclude water, small holes or rips in a membrane seal themselves spontaneously. This property helps membranes fuse during endocytosis, exocytosis, and cell fusion.

The phospholipid bilayer stabilizes the entire structure. At the same time, it makes the membrane fluid—about as fluid as lightweight machine oil—so materials can move laterally within the membrane. As we will see, some membrane proteins are relatively free to migrate about, and individual phospholipid molecules may also move. A phospholipid molecule in the plasma membrane of a bacterium may travel from one end of the bacterium to the other in a little over a second. On the other hand, it is *not* common for a phospholipid molecule in one half of the bilayer to flop over to the other side and trade places with another phospholipid molecule. For this to happen, the polar part of each molecule would

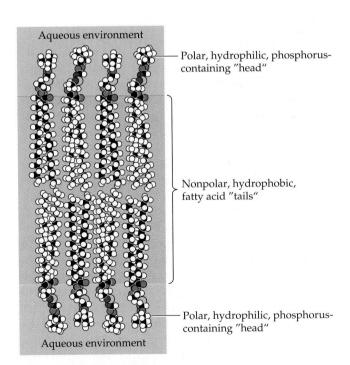

Aqueous environment

Polar, hydrophilic, phosphorus-containing "head"

Nonpolar, hydrophobic, fatty acid "tails"

Polar, hydrophilic, phosphorus-containing "head"

Aqueous environment

5.2 A Phospholipid Bilayer Separates Two Aqueous Regions
The six phospholipid molecules shown here represent a small section of a membrane bilayer. The charged, hydrophilic heads of the molecules orient toward the surfaces of the membrane. The hydrophobic fatty-acid tails mingle with one another in the interior.

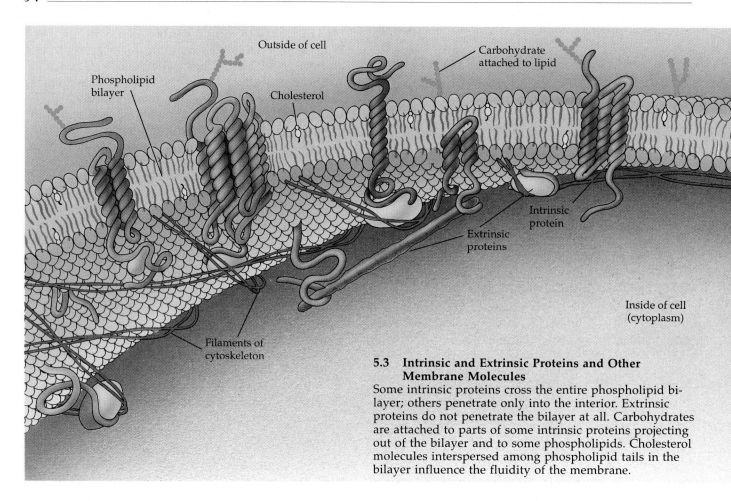

Outside of cell

Phospholipid bilayer

Cholesterol

Carbohydrate attached to lipid

Intrinsic protein

Extrinsic proteins

Inside of cell (cytoplasm)

Filaments of cytoskeleton

5.3 Intrinsic and Extrinsic Proteins and Other Membrane Molecules
Some intrinsic proteins cross the entire phospholipid bilayer; others penetrate only into the interior. Extrinsic proteins do not penetrate the bilayer at all. Carbohydrates are attached to parts of some intrinsic proteins projecting out of the bilayer and to some phospholipids. Cholesterol molecules interspersed among phospholipid tails in the bilayer influence the fluidity of the membrane.

have to move through the hydrophobic interior of the membrane. Since phospholipid flip-flops are rare, the two halves of the bilayer may be quite different.

All biological membranes have similarities, but different membranes may differ greatly in the details of their composition—even within the same cell. One big area of difference is lipid composition. The proportions of different types of lipids vary from one type of membrane to another. For example, 25 percent of the lipid in many membranes is cholesterol, but some membranes have no cholesterol at all. In a membrane, cholesterol is commonly next to an unsaturated fatty acid, and its polar region extends into the surrounding aqueous layer. Cholesterol plays more than one role in determining the fluidity of the membrane. Under some circumstances, cholesterol increases membrane fluidity; under others, it decreases membrane fluidity. Shorter fatty acid chains make for a more fluid membrane, as do unsaturated fatty acids. Organisms can modify their membrane lipid composition, thus changing membrane fluidity, to compensate for changes in temperature that are not too sudden. House plants adapted to indoor temperatures may die when accidentally left outdoors overnight. The sudden change in temperature does not give them time to adapt by adjusting their membrane lipid composition. Lipids constitute a major fraction of all membranes, and they always form the continuous matrix into which the other chemical components become inserted.

Membrane Proteins

Each protein in a biological membrane is either an **intrinsic protein** or an **extrinsic protein** (Figure 5.3). Intrinsic proteins penetrate the phospholipid bilayer; many extend from one side of the membrane to the other. Extrinsic proteins are entirely outside the bilayer; they are attached to the surface of the membrane by weak (noncovalent) bonds with the exposed parts of the intrinsic proteins or with the hydrophilic parts of phospholipid molecules. These two types of proteins play different roles in membrane function.

The membranes of the various organelles differ sharply in protein composition. Different biochemical reactions (many of them requiring membrane-bound enzymes) occur in different organelles. In many of the important reactions of cellular respiration and photosynthesis, membrane-bound enzymes carry electrons from a donor to an acceptor molecule. Accordingly, both mitochondria and chloroplasts have highly specialized internal membranes, and these differ markedly (Figure 5.4).

Many membrane proteins move relatively freely

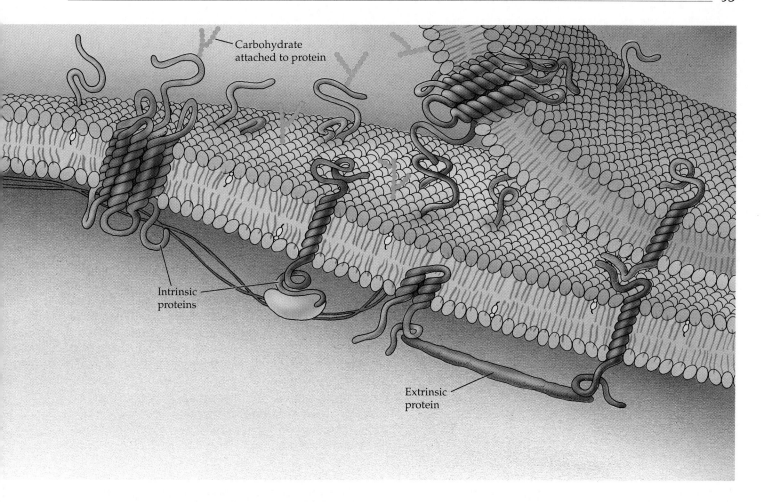

Carbohydrate
attached to protein

Intrinsic
proteins

Extrinsic
protein

within the phospholipid bilayer. Evidence of this migration is dramatically illustrated by experiments using the technique of cell fusion. Specially treated cells of two different species, such as human and mouse, can be fused so that one continuous membrane surrounds the combined cytoplasm and both nuclei. Ini-

(a)

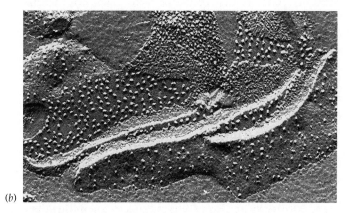

(b)

5.4 Proteins in Specialized Membranes

(a) The outer membrane of this mitochondrion has been fractured away, exposing the inner membrane. The image has been magnified about 65,000 times. The particles giving the inner membrane a grainy appearance are proteins necessary for cellular respiration. (b) The distinct proteins embedded in these thylakoids from a spinach chloroplast, magnified about 70,000 times, are necessary for photosynthesis.

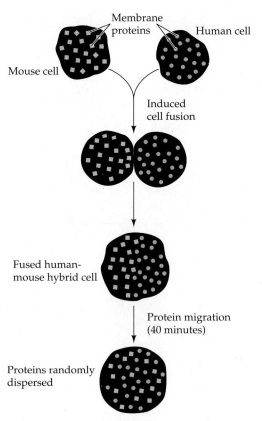

5.5 Proteins Move Around in Membranes

When specially treated mouse and human cells are joined, one continuous membrane surrounds the fused hybrid cell. Distinguished by different fluorescent dyes, many of the mouse and human membrane proteins mingle over time, demonstrating that some proteins move in the fluid phospholipid bilayer.

tially the experimenter can tell by the protein content which part of the plasma membrane came from which species, but the membrane proteins of the two cells migrate in the joint membrane until, after about 40 minutes, they are uniformly dispersed (Figure 5.5). On the other hand, there is also good evidence that other membrane proteins are *not* free to migrate at will—that they are, to an extent, held in place. These proteins are anchored by microfilaments, as described later in this chapter. Microtubules may also play a role.

Proteins are asymmetrically distributed in membranes. Many intrinsic proteins extend completely through a membrane and have specific parts of their primary structures on one side of the membrane, specific parts within the membrane, and specific parts on the other side of the membrane. All the other membrane proteins (intrinsic and extrinsic) are localized on one side of the membrane or the other, but not both.

The fact that molecules of one type of protein may be confined to one part of the cell surface, rather than scattered evenly about also contributes to asymmetry. In certain muscle cells, for example, the membrane

protein that receives the chemical signal from nerve cells is normally found only where a nerve cell meets the muscle cell. None of this protein is found elsewhere on the muscle cell plasma membrane. If the nerve is severed, however, the protein molecules may later be found evenly distributed over the entire membrane. If the nerve regenerates its attachment to the muscle, then the protein is once again limited to the junction area.

What determines whether a particular membrane protein is intrinsic or extrinsic? If it is intrinsic, what controls whether it reaches all the way through the membrane or is limited to one side? What keeps it in the bilayer, and what determines just how far in it reaches? All these questions are answered in terms of the tertiary structure of the protein (see Chapter 3). Recall that the side chains of the various amino acids in a protein differ chemically. What matters here is that some of the side chains are hydrophilic and others hydrophobic. After a polypeptide chain folds into the final tertiary structure, the protein may have both hydrophilic and hydrophobic surfaces. If one end of a folded protein is hydrophilic and the other hydrophobic, it will be an intrinsic protein, sticking out of one side of the membrane. Many intrinsic proteins that reach from one side of the membrane to the other have hydrophobic α-helical regions large enough to penetrate the entire depth of the phospholipid bilayer. They also have hydrophilic ends that protrude into the aqueous environments on either side of the membrane (Figure 5.6). Proteins like this resist being removed from the membrane. If the hydrophobic surface of such a protein is pulled partway out of the phospholipid bilayer, it is repelled by the aqueous environment. If an intrinsic protein is pushed farther into the membrane, its hydrophilic end is pushed back by the hydrophobic fatty acid region of the lipids. Thus such a protein may migrate

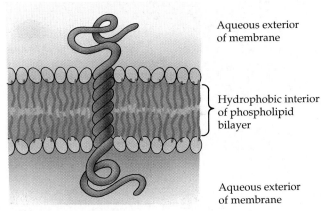

5.6 Controlling Surfaces of Intrinsic Proteins

This intrinsic membrane protein has hydrophilic surfaces (purple) made up of the hydrophilic side chains of some of its amino acids. A region consisting of hydrophobic side chains (brown) buries itself among the fatty acid tails in the bilayer's interior.

laterally in the membrane sheet, but it may not push through the membrane or pop out of it.

How does an intrinsic protein penetrate the membrane in the first place? Clearly, it cannot fold into its final shape and then be pushed into the membrane—for the same reasons that it cannot leave the membrane once in place. In fact, its insertion into the membrane usually proceeds while the protein is being synthesized. Targeting of the sort discussed in Chapter 4 plays a crucial role—specific sequences of amino acids in the growing protein interact with receptors on the membrane, causing the protein to enter the membrane (or not) and to exit on the other side (or not). Final folding is achieved after the different regions of the protein have been appropriately situated.

Membrane Carbohydrates

All plasma membranes and some other membranes contain substantial amounts of carbohydrate along with the lipids and proteins. For example, the plasma membrane of a red blood cell consists of approximately 40 percent lipid, 52 percent protein, and 8 percent carbohydrate by weight. The carbohydrates are recognition sites. Some of the membrane carbohydrates are bound to lipids, forming glycolipids. Glycolipids enable a cell to recognize other cells. Although there is still much to learn about glycolipids, we know that they are important—the structures of some glycolipids change when a cell becomes cancerous, for example.

Most of the carbohydrates in membranes are bound to proteins, forming glycoproteins. These bound carbohydrates are oligosaccharide chains, usually not exceeding 15 monosaccharide units in length. The glycoproteins enable a cell to recognize foreign substances. The oligosaccharide chains are added to the membrane proteins inside the endoplasmic reticulum and are modified in the Golgi apparatus.

The carbohydrates linked to membrane proteins and lipids are relatively small and made from only nine building blocks—the monosaccharides. Nevertheless, these carbohydrates are exceedingly diverse. To understand how this can be, recall that monosaccharides may join to form branched oligomers (see Figure 3.16). The possibility of different branching patterns greatly increases the diversity that can be achieved by the monosaccharide sequence of a carbohydrate alone. Also, monosaccharides may link together at any of several different carbons. All in all, membrane carbohydrates have great specificity and diversity. This structural diversity is important in all sorts of reactions at the cell surface in which different membrane carbohydrates recognize and react with specific foreign substances. All plasma membrane carbohydrates are on the *outside* of the plasma membrane, as befits their role as recognition sites for foreign substances and cells; none face into the cell (see Figure 5.3).

MICROSCOPIC VIEWS OF BIOLOGICAL MEMBRANES

Light microscopes cannot resolve anything as thin as the plasma membrane and the membranes within cells. Electron microscopes, however, offer various ways to examine membranes. The first visualizations were produced when very thin slices of tissue were made with diamond knives and the resulting sections were examined by transmission electron microscopy. The cuts were very clean, and membranes were often seen in cross section in electron micrographs of the slices. (Look back at the micrograph on page 92, noticing that the plasma membrane's phospholipid bilayer appears as two dark lines separated by a light region.) A more detailed understanding of membrane architecture, however, had to await methods for seeing surface views and for revealing the hydrophobic interior of the membrane.

The first successes were achieved by a technique known as **freeze-fracture.** In this procedure the tissue to be examined is frozen solid and then broken rather than cut. To visualize the consequences of this fracture, picture a chocolate bar with almonds. If the bar is cut carefully with a very sharp knife, the cut will pass cleanly through the nuts as well as the chocolate. If, on the other hand, the bar is broken, the break will pass around any almonds in its path and reveal them where they protrude from the chocolate. Similarly, in the freeze-fracture technique, the break tends to pass around membrane-encased organelles and to pass between the two halves of the phospholipid bilayer but around the proteins within it. When the exposed surfaces of the fractured bilayer are examined with the electron microscope, they appear bumpy—the intrinsic proteins are revealed (Figure 5.7a).

Further clarification of membrane structure is obtained by **freeze-etch.** In this method, a freeze-fractured sample is kept cold and put under a high vacuum for a minute or so, allowing some of the frozen water to evaporate, wherever it is exposed to the vacuum. The water evaporation reveals more of the texture of the membrane by uncovering surfaces that were covered by ice (Figure 5.7b).

Contrast in both freeze-fractured and freeze-etched preparations is enhanced by **shadowcasting** with platinum. The metal is sprayed on from an angle so that bumps and dips on the surface give shadow patterns that make them easier to see and interpret. Figure 5.7c is an example of a platinum-shadowed micrograph of a freeze-etched sample.

Freeze-fracture and freeze-etch were instrumental

(a) Freeze-fracture. Frozen cells are broken open, exposing membrane faces and interiors

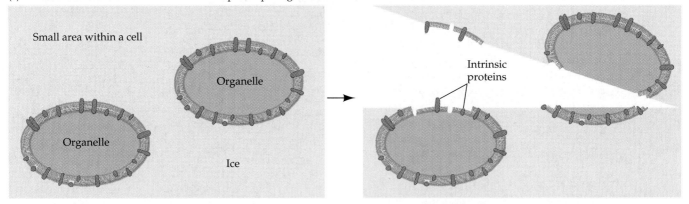

Small area within a cell

Organelle

Organelle

Ice

Intrinsic proteins

(b) Freeze-etch. Water is evaporated from surface of a freeze-fractured sample, exposing more structures

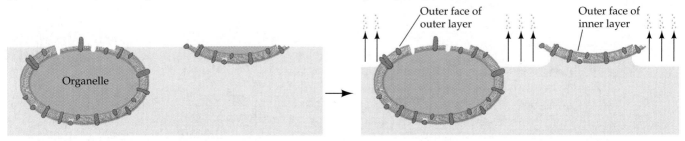

Organelle

Outer face of outer layer

Outer face of inner layer

5.7 Freeze-Fracture and Freeze-Etch

(a) Freeze-fracture of a cell often splits the lipid bilayer of a membrane, exposing membrane faces and intrinsic proteins. *(b)* Freeze-etch technique enhances the sample, revealing even more detail. *(c)* A freeze-etched sample of part of a photosynthetic protist cell, shadowed with platinum. We can see the Golgi apparatus (lower left), a chloroplast (upper right), and an inside view of the nuclear envelope, dotted with pores (upper left). Notice the abundant vesicles between the nucleus and the *cis* region of the Golgi apparatus. More examples of electron micrographs of freeze fractured preparations are found in Figures 5.4 and 5.8 (top photo).

(c) Freeze-etched sample shadowed with platinum reveals cellular structures in electron micrograph

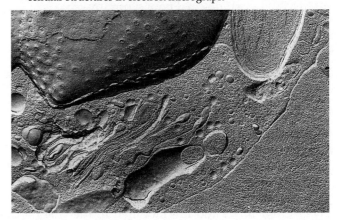

in formulating our current picture of membrane structure. In particular, they made it clear that the membrane is not just a continuous bilayer of phospholipids with the proteins spread on its surface, but that many of the proteins are embedded in the bilayer to various depths.

You now know quite a bit about the chemistry and appearance of plasma membranes. What happens when the plasma membranes of two cells meet? What can microscopic views tell us about this?

WHERE ANIMAL CELLS MEET

Plasma membranes of adjacent animal cells sometimes touch, but usually there is a space between the membranes. What occupies that space? Electron microscopic studies reveal that animal cells are surrounded by an **extracellular matrix** consisting of a fine meshwork of polysaccharides permeated by fibrous proteins. The most abundant protein in the

extracellular matrix is collagen, whose triple-helical structure was described in Chapter 3. One major function of the matrix is to hold the cells together as tissues. In addition, the matrix helps to direct the ways in which cells interact and to orient individual cells. This relationship is reciprocal because the cells themselves secrete the extracellular matrix and establish its orientation. Thus the organizational pattern of a tissue, once established, tends to be maintained as the tissue grows—the first cells orient the matrix, which in turn orients the new cells, and so forth.

Some epithelial cells (cells that line a body cavity or an exterior body surface) come in direct physical contact with one another and form special links, called **junctions.** By their structure and function, junctions fall into three categories: desmosomes,

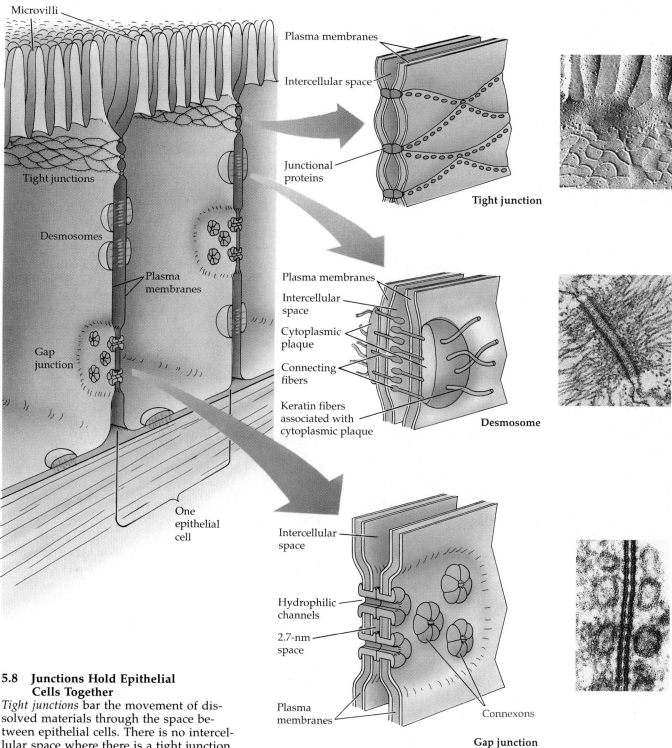

Labels on the illustration:

Microvilli

Plasma membranes

Intercellular space

Junctional proteins

Tight junction

Tight junctions

Desmosomes

Plasma membranes

Intercellular space

Cytoplasmic plaque

Connecting fibers

Keratin fibers associated with cytoplasmic plaque

Desmosome

Plasma membranes

Gap junction

Intercellular space

Hydrophilic channels

2.7-nm space

One epithelial cell

Plasma membranes

Connexons

Gap junction

5.8 Junctions Hold Epithelial Cells Together

Tight junctions bar the movement of dissolved materials through the space between epithelial cells. There is no intercellular space where there is a tight junction. Long rows of the tight-junction proteins form a complex meshwork, seen at the bottom of the freeze-etch micrograph. *Desmosomes* link adjacent cells but permit materials to move between them. Connecting fibers bind the neighboring cells to each other; cytoplasmic plaques anchor these fibers in both cells. *Gap junctions* let adjacent cells communicate. Dissolved molecules and electrical signals may pass from one cell to the other through the channels. Each channel is made of two connexons. A connexon reaches through the phospholipid bilayer of the membrane and extends into the cytoplasm on one side and into the external environment on the other side, where it abuts a connexon on an adjacent cell.

tight junctions, and gap junctions. As you read about each type of junction in the sections that follow, refer to the illustrations in Figure 5.8.

Tight Junctions

Tight junctions bind cells so closely that materials do not move between the joined cells; there is no space at all between cells at a tight junction. Tight junctions result from the mutual binding of strands of specific proteins in the plasma membranes of the two cells, forming belts that virtually fuse the two membranes. Epithelial cells are so extensively linked by tight junctions between adjacent cells that substances on one side of an epithelium (flat tissue composed of epithelial cells) cannot seep through to the other side. Thus, for example, the contents of the gut (digestive tract) cannot seep between the epithelial cells of the gut lining—the contents can pass through the epithelium only if they enter the epithelial cells from one side of the tight junction and are then released from the other ends of the cells. The selective permeability of the plasma membranes ensures that food substances from the digestive tract pass through the epithelial cells to the bloodstream and unwanted substances are unable to pass through the membrane. So tight are the tight junctions that membrane proteins and phospholipids on one side of a tight junction cannot flow through the junction to the other part of the plasma membrane. By forcing materials to enter some cells, and by allowing different ends of cells to have different membrane proteins, tight junctions help direct the transport of materials in the body.

Desmosomes

Unlike tight junctions, which cement cells so closely that the cells are sealed off from each other, **desmosomes** simply cause neighboring cells to adhere tightly. Although a desmosome holds adjacent epithelial cells firmly together, there is a 24-nm space between the two plasma membranes. Some desmosomes act like spot welds or rivets at individual points, while others form continuous belts around the outer ends of adjacent epithelial cells. At each weld, in each of the two cells, is a protein-containing mass called a cytoplasmic plaque. Desmosomes are easily recognized in electron micrographs by the plaques and their associated dense networks of keratin fibers, which extend into each of the cells. The fibers connect the cells through the space between the two plasma membranes. (Keratin is a fibrous protein that is classified as an intermediate filament; see Chapters 3 and 4. It makes up the bulk of our fingernails and hair.) Epithelial tissue is strengthened by the many desmosomes holding its cells together.

Gap Junctions

Adjacent cells in some animal tissues communicate through the third type of junction, called a **gap junction** because of the gap of 2.7 nm between the plasma membranes of two cells spanned by its many pipelike channels. The channels, called connexons, are made up of a specific protein. Connexons provide a cyto-plasmic connection between cells, through which chemical substances or electric signals may pass.

In Chapters 38 and 40 we will see that the muscle cells of the vertebrate heart, many smooth muscles, and some nerve cells are connected by gap junctions, allowing the direct passage of an electric signal—the nerve impulse—from one cell to the next. Gap junctions are also important in embryonic development, for they appear at a specific developmental stage. If animal tissues are experimentally disrupted to dissociate the cells, the cells quickly form new gap junctions as they reassociate. In fact, cells isolated from one species of vertebrate readily form gap junctions with cells from other vertebrate species. Cancer cells, however, never develop gap junctions; presumably this means that they do not communicate with other cells as normal cells do.

Review the differences among the three types of junctions. Desmosomes allow cells to *adhere* strongly to one another. Tight junctions are *barriers* to the passage of molecules through the space between cells. Gap junctions provide channels for chemical and electric *communication* between the cells on the opposite sides of the junctions.

We have examined the structures of membranes and their junctions. Now it is time to consider their functions. Some of the most important functions of membranes depend on their selective permeability—their ability to allow some substances, but not others, to pass through.

DIFFUSION

How do molecules move in an aqueous environment? Before we discuss further the movements of molecules across membranes, it is important to consider this: Nothing in this world is ever absolutely at rest. Everything is in motion, though the motions may be very small. Molecules and ions in solution are constantly jiggling. An immediate consequence of this random jiggling is that all the components of a solution tend eventually to become evenly distributed throughout the system. If, for example, a drop of ink falls into a container of water, the pigment molecules of the ink will move about at random, spreading through the system until the concentration of pigment—and thus the intensity of color—is exactly the same in every drop of liquid in the container. A solution in which the particles are uniformly distributed is said to be at equilibrium, and the process of random movement toward a state of equilibrium is called **diffusion.**

In diffusion, the motion of each individual particle is absolutely random, even though the net movement of particles is directional until equilibrium is reached. Diffusion is this net movement—always in the direc-

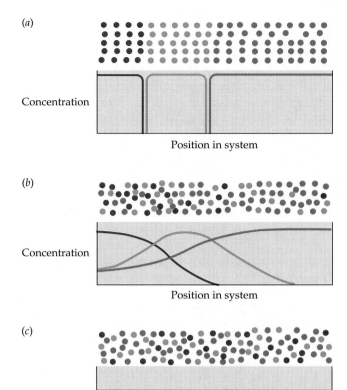

(a)

Concentration

Position in system

(b)

Concentration

Position in system

(c)

Concentration

Position in system

5.9 Diffusion Leads to Uniform Distribution
(a) Initially, each of three dissolved substances is highly concentrated in part of the solution in which they have been combined, and absent from the rest of it. *(b)* As the substances mix, each substance is at a higher concentration—its peak on the graph—in part of the solution but is also present elsewhere. *(c)* At equilibrium, all three substances are uniformly distributed throughout the solution, as shown by the lack of peaks on the graph.

tion *from greater* concentration *to lesser* concentration (Figure 5.9). In addition, in a complex solution the diffusion of each substance is independent of that of the other substances.

How fast substances diffuse depends on three physical properties—the diameter of the molecules or ions, the temperature, and the electric charge, if any, of the diffusing material—as well as on the **concentration gradient** in the system. The concentration gradient is the change in concentration with distance. The greater the concentration gradient, the more rapidly substances diffuse.

Within cells, where distances are very short, solutes distribute rapidly by diffusion. Small molecules and ions may move from one end of an organelle to another in a fraction of a millisecond, or from the center of a cell to its surface almost as fast. Diffusion of a chemical signal from one nerve cell to another takes less than a millionth of a second. On the other

hand, the usefulness of diffusion as a transport mechanism is diminished drastically as distances become greater. Diffusion over a centimeter may take an hour or more; over meters it takes years (it is assumed throughout this discussion that the fluid is not stirred or moved in any other way). Diffusion is not enough to distribute materials over the length of the human body, but within our cells or across layers of one or two cells it is rapid enough to distribute small molecules and ions almost instantaneously.

CROSSING THE MEMBRANE BARRIER

In principle, *all* substances diffuse, although the rates of diffusion vary. In a solution without barriers, all the solutes diffuse at rates determined by their physical properties and in directions determined by the concentration gradient of each solute. If a biological membrane (which is selectively permeable) is introduced as a barrier, the movement of the different solutes can be affected. Some solutes move fairly readily through the membrane, whereas others are prevented from crossing it. Molecules that can move through the phospholipid barrier diffuse from one compartment to the other until their concentrations are equal on both sides of the membrane. Molecules that cannot cross the membrane diffuse only within their own compartments, so their concentrations remain different on the two sides of the membrane.

Substances move through biological membranes in three ways: simple diffusion, facilitated diffusion, and active transport (Figure 5.10). In **simple diffusion,** small, nonpolar molecules pass through the lipid bilayer of the membrane. Equilibrium is reached when the concentrations of the diffusing substance are identical on both sides of the membrane. At equilibrium individual molecules are still passing through the membrane, but equal numbers of molecules are moving in each direction, so there is no change in concentration.

Facilitated diffusion (sometimes called carrier-mediated diffusion) also involves movement down a concentration gradient to produce equal concentrations of solute on the two sides of a membrane, but in contrast to simple diffusion, the solute molecules do not diffuse through the membrane on their own. Rather, they combine with **carrier molecules** in the membrane. By a mechanism that is still not understood, these carriers enable the solute molecules to pass through the membrane.

Both simple diffusion and facilitated diffusion permit the passage of a solute across a membrane down a concentration gradient—that is, from the side of higher concentration to the side of lower concentration. Neither of these mechanisms, however, allows for the transport of a solute *against* a concentration

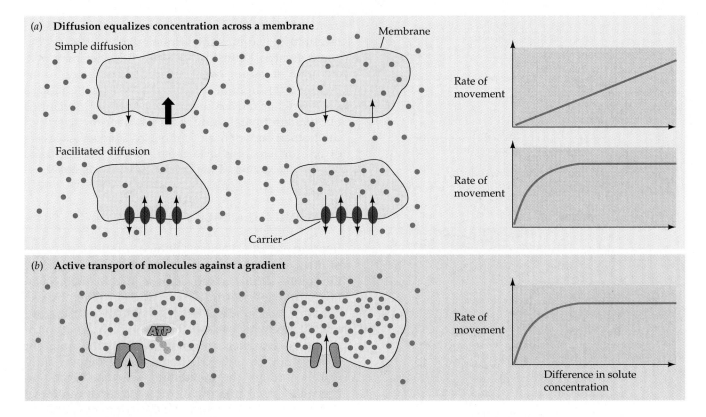

(a) **Diffusion equalizes concentration across a membrane**

Simple diffusion

Membrane

Rate of movement

Facilitated diffusion

Rate of movement

Carrier

(b) **Active transport of molecules against a gradient**

ATP

Rate of movement

Difference in solute concentration

5.10 Crossing Biological Membranes
(a) Both simple and facilitated diffusion equalize concentrations of a solute across a membrane. The rate of simple diffusion is directly proportional to the concentration difference, as the top graph shows. At equilibrium the concentration of solute inside the membrane equals that outside. With facilitated diffusion, equal concentrations are also reached, but a protein carrier in the membrane allows the rate of solute crossing to be greater. This rate reaches a maximum when all carriers are saturated with solute. *(b)* Active transport employs energy (ATP) and can move solutes against a concentration gradient. The rate of movement is similar to that of facilitated diffusion and reaches a maximum when membrane carriers are saturated, but the final concentrations of solute on either side of the membrane can be quite different due to the expenditure of energy.

gradient—that is, from the side of the membrane where the concentration is lower to the side where it is higher. Exactly this phenomenon, called **active transport,** is of extreme importance to living things. Like facilitated diffusion, active transport relies on carrier molecules. Unlike facilitated diffusion, this active process by which ions or molecules are moved against their own concentration difference requires a great deal of energy, which is provided by ATP obtained through cellular respiration (see Chapter 7).

In all three mechanisms of movement through membranes, the *rate* of movement depends on the concentration difference across the membrane. In simple diffusion, the net rate of movement is directly proportional to the concentration difference across the membrane. In facilitated diffusion, the rate of

movement also increases with the difference in solute concentration across the membrane, but a point is reached at which further increases in concentration difference are not accompanied by an increased rate (see Figure 5.10*a*). The facilitated diffusion system is said to be saturated at this high concentration. If there are only so many carrier molecules per unit area of membrane, then the rate of movement reaches a maximum when all the carrier molecules are fully engaged in moving solute molecules. In other words, at high solute concentration differences across the membrane, there are not enough carrier molecules free at a given moment to handle all the solute molecules. Like facilitated diffusion, the rate of active transport stops increasing at high solute concentrations (see Figure 5.10*b*).

Simple Diffusion

Whether a substance can pass through biological membranes depends on how soluble it is in lipids. The more lipid-soluble the compound, the more rapidly it diffuses (Figure 5.11). This statement holds true over a wide range of molecular weights. Only certain ions and the smallest molecules seem to deviate from this rule: Materials such as water, potassium ions (K^+), and chloride ions (Cl^-) pass through membranes much more rapidly than their solubilities in lipid would predict. Let's consider, then, how a membrane's chemical structure affects diffusion.

The key feature of membrane architecture, as we

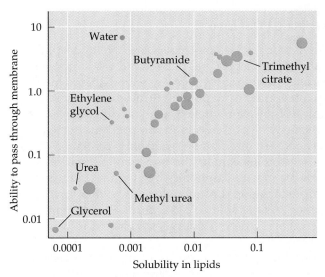

5.11 Membrane Permeability
Most substances cross membranes at rates relative to their solubility in lipids, as shown by the data points on this graph: The more lipid-soluble the molecule, the greater its ability to pass through the membrane. The sizes of the points correspond roughly to the sizes of the molecules studied; the assortment of molecules of all sizes along the curve indicates that size alone is not an important factor for permeability. The rate of diffusion of a few molecules, such as water, does not seem to be related to their solubility in lipids.

have seen, is the phospholipid bilayer that forms its framework (see Figure 5.1). The inner portion of the bilayer consists of the fatty acid chains of the phospholipids, along with cholesterol and other highly hydrophobic, nonpolar materials. When a hydrophilic molecule or ion moves into such a hydrophobic region, it is rejected by the lipid layer and forced out. Such a molecule enters the hydrophobic region only when energy is available to push it in. On the other hand, a molecule that is itself hydrophobic, and hence soluble in lipids, enters the membrane readily and is thus able to pass through it.

This explanation accounts for most of the information in Figure 5.11, but it does not explain how water itself can move so rapidly through biological membranes—water is not hydrophobic enough to account for this flow. The diffusion of water into and out of cells is still under debate; many workers feel that the rapid movement of water through membranes can be explained in several ways.

We must also account for the rapid movement of certain ions through biological membranes. Ions pass through water-filled pores, or **channels,** in the membranes of all eukaryotic cells. There are specific channels for potassium, sodium, calcium, and chloride ions, allowing these ions to diffuse through membranes in spite of their hydrophilic character.

Membrane Transport Proteins

Both the channels through which certain ions diffuse across membranes and the carriers for facilitated diffusion and active transport are **membrane transport proteins:** intrinsic proteins that reach from one side of the membrane to the other. Different membrane transport proteins allow only specific substances to pass through, employing one of various methods of transport (Figure 5.12). The channels, for example, have an aqueous region through which ions can diffuse.

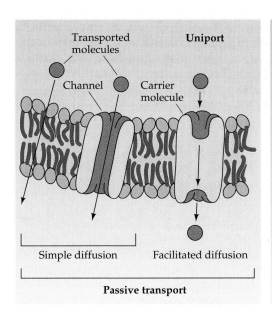
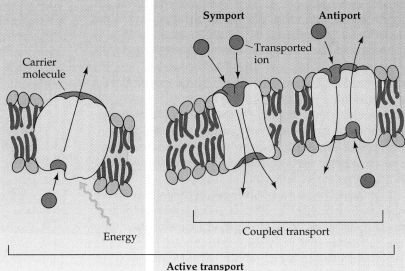

5.12 Getting Across the Membrane
Substances pass through biological membranes by many mechanisms, as illustrated here. Most of the mechanisms—all except for simple diffusion through the phospholipid bilayer, shown at the far left—make use of proteins that act as either channels or carriers.

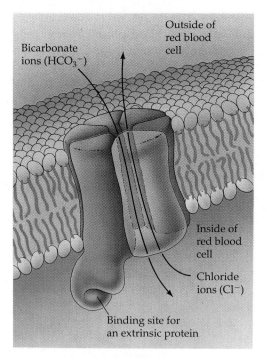

5.13 An Anion Channel
Red blood cell membranes contain anion channels
through which chloride and bicarbonate ions are exchanged. One end of the channel protein binds an extrinsic protein inside the red blood cell, anchoring the channel to a network of other proteins that stabilizes the
positions of intrinsic proteins.

There are two classes of membrane transport proteins: uniports and coupled transport systems. **Uniports** transport a single type of solute. **Coupled transport systems** transport two or more different solutes,
but neither solute can be moved by the system unless
the other solute is also present. If the two coupled
solutes are transported in the same direction, the
system is a **symport.** If they are transported in opposite directions, the system is an **antiport.** One
example of an antiport is an anion channel that is
abundant in the plasma membrane of red blood cells
(Figure 5.13). Each red blood cell contains about 1
million of these channels, which allow the exchange
of bicarbonate (HCO_3^-) and chloride (Cl^-) ions. As
you will learn in Chapter 41, this particular exchange
is important in transporting carbon dioxide (present
in the bloodstream as bicarbonate ions) from working
tissues where carbon dioxide is produced to the
lungs, where it is released.

The structures of membrane transport proteins
make the proteins specific to certain substances. A
given membrane transport protein thus carries only
one particular solute, or solutes that have very similar
structures. The great diversity of protein structures
allows this high specificity for different transported
solutes, just as it allows an enzyme to be highly
specific for accelerating a particular chemical reaction.

Facilitated Diffusion

Most biochemical molecules are too hydrophilic to
enter the phospholipid bilayer and too large to move
through membranes the way water does. Thus they
are prevented from passing through the membrane—
unless they can interact with a carrier of suitable
specificity. As discussed in the previous section, the
carriers are membrane transport proteins. Where
these proteins contact the phospholipid bilayer, their
surfaces are hydrophobic, but within them is a hydrophilic region through which the diffusing material
passes. As the solute passes through the membrane,
the carrier protein undergoes a change in tertiary or
quaternary structure.

In facilitated diffusion, the carrier proteins enable
the solutes to pass in *both* directions. The net movement is toward the side where the solute concentration is lowest simply because on the side where the
concentration is greater, the carriers encounter more
solute molecules to transport.

Active Transport

Things are different in active transport. Active-trans-
port carriers operate in one direction only. Typically,
this direction is the one in which the transported
substance is moved against a concentration difference—that is, from a region of low concentration to
one of higher concentration. This "uphill" process
requires an input of energy (see Chapter 6). There
are two basic types of active transport: primary and
secondary.

Primary active transport requires the direct participation of adenosine triphosphate (ATP), an energy-
storing compound found in all cells (see Chapter 7).
In primary active transport, energy released from
ATP drives the movement of specific ions against a
concentration difference. For example, compare the
concentrations of potassium ions (K^+) and sodium
ions (Na^+) inside a nerve cell and in the fluid bathing
the nerve (Table 5.1). The K^+ concentration is much
higher inside the cell, whereas the Na^+ concentration

**TABLE 5.1
Concentration of Major Ions
Inside and Outside the Nerve Cell
of a Squid**

| ION | CONCENTRATION (MOLAR) | |
	IN NEURON	IN BLOOD
K^+	0.400	0.020
Na^+	0.050	0.440
Cl^-	0.120	0.560

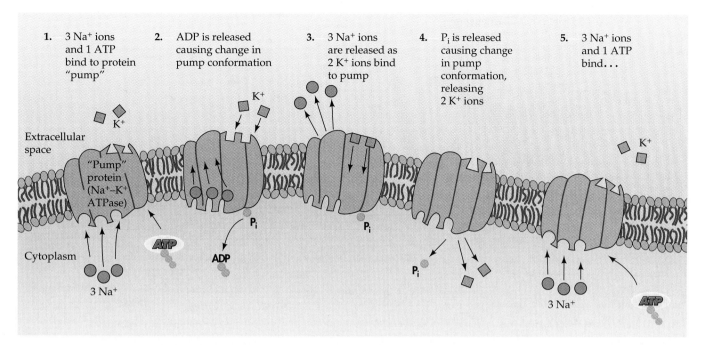

1. 3 Na⁺ ions and 1 ATP bind to protein "pump"

2. ADP is released causing change in pump conformation

3. 3 Na⁺ ions are released as 2 K⁺ ions bind to pump

4. P$_i$ is released causing change in pump conformation, releasing 2 K⁺ ions

5. 3 Na⁺ ions and 1 ATP bind...

5.14 The Work of the Sodium–Potassium Pump Is Primary Active Transport
For each molecule of ATP used, 2 K⁺ are pumped into the cell and 3 Na⁺ are pumped out of the cell. The transport protein molecule—the pump—extends all the way through the phospholipid bilayer of the membrane; thus the highly hydrophilic Na⁺ and K⁺ ions need not interact with the hydrophobic center of the membrane.

is much higher outside. In spite of this, the nerve cells continue to pump Na⁺ out and K⁺ in, against these concentration differences, ensuring that the differences are maintained. This **sodium–potassium pump** is found in all animal cells and is an intrinsic membrane glycoprotein. It repeatedly breaks down one molecule of ATP to ADP (adenosine diphosphate), brings two K⁺ ions into the cell, and exports three Na⁺ ions (Figure 5.14). The sodium–potassium pump is thus an antiport. There are pumps for the transport of several other ions, but only cations are transported directly by pumps in primary active transport. The transport of other solutes is achieved by secondary active transport.

Unlike primary active transport, **secondary active transport** does not use ATP directly; rather, transport of the solute is tightly coupled to the difference in ion concentration established by primary active transport. The movement of particular solutes, such as sugars and amino acids, is regulated by coupled transport systems (some symports, some antiports) that move these specific solutes against their concentration difference, using energy "regained" by letting Na⁺ or other ions move *with* their concentration difference. Putting the two forms of active transport together, we see that energy from ATP is used in one example of primary active transport to establish concentration differences of potassium and sodium ions; the movement of some sodium ions in the opposite

direction provides energy for the secondary active transport of the sugar glucose (Figure 5.15). Other secondary active transporters are used for the uptake of amino acids and other solutes.

Osmosis

For years scientists disagreed about whether water only diffuses through biological membranes or is sometimes actively transported as well. It is now clear that water moves through membranes only by **osmosis,** the movement of a solvent through a membrane in accordance with the laws of diffusion. This process, in which no metabolic energy is expended, can be understood in terms of a very few principles, which we will develop here using two simple examples.

Red blood cells are suspended in a fluid called plasma that contains salts, proteins, and other solutes. If a drop of blood is examined under the light microscope, the characteristic shape of the red cells is evident. If pure water is added to the drop of blood, however, the cells quickly swell and burst (Figure 5.16). Similarly, if slightly wilted lettuce is put in pure water, it soon becomes crisp; by weighing it before and after, we can show that it has taken up water (Figure 5.17). If, on the other hand, the red blood cells and crisp lettuce leaves are placed in a relatively concentrated solution of salt or sugar, the

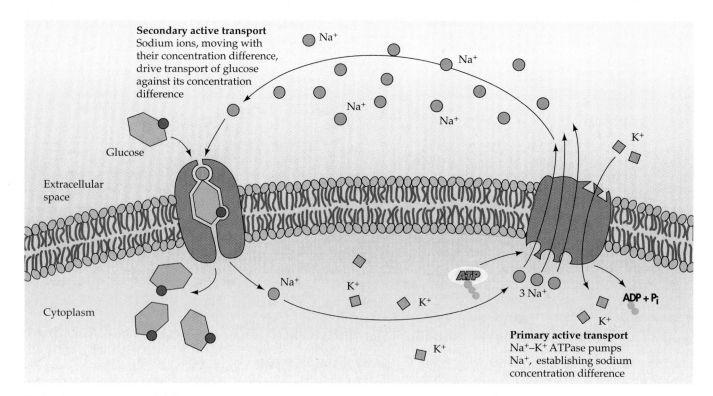

Secondary active transport
Sodium ions, moving with their concentration difference, drive transport of glucose against its concentration difference

Glucose

Extracellular space

Cytoplasm

Primary active transport
Na⁺–K⁺ ATPase pumps Na⁺, establishing sodium concentration difference

5.15 Secondary Active Transport
The Na^+ concentration difference established by primary active transport (right) powers the secondary active transport of glucose and some other substances. Glucose moves through the membrane against its concentration difference, accompanied by Na^+ ions that are moving with their concentration difference (left).

leaves become limp and the red blood cells pucker and shrink.

From these and other observations, we know that solute concentration is the principal factor in what is called the **osmotic potential** of a solution. The greater the solute concentration, the more negative the osmotic potential of the solution. Since pure water has nothing dissolved in it, its osmotic potential equals zero. Other things being equal, if two unlike solutions are separated by a differentially permeable membrane (one that allows water to pass through but not solutes), osmosis—movement of the water—proceeds toward the solution with the more negative osmotic potential.

If two solutions have identical osmotic potentials, they are **isotonic** to one another. (This is true even if their chemical compositions are very different). If they are not isotonic, then solution A with a more negative osmotic potential (that is, with a higher concentration of solutes) is said to be **hypertonic** to solution B; solution B is **hypotonic** to solution A. All three of these terms are strictly relative. They can be used only in comparing the osmotic potentials of two solutions; no solution can be called hypertonic, for example, except in comparison with another solution that has a less negative osmotic potential.

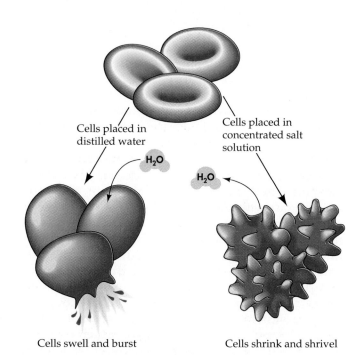

Cells placed in distilled water

Cells placed in concentrated salt solution

H_2O

H_2O

Cells swell and burst

Cells shrink and shrivel

5.16 Osmosis Modifies Cell Shape
A mammalian red blood cell suspended in plasma has a biconcave shape (indented on both sides). If the cell is placed in distilled water, water enters by osmosis and the cell swells and bursts. If the cell is placed in a solution in which salts are more concentrated than they are in plasma, water leaves the cell by osmosis, causing it to shrivel.

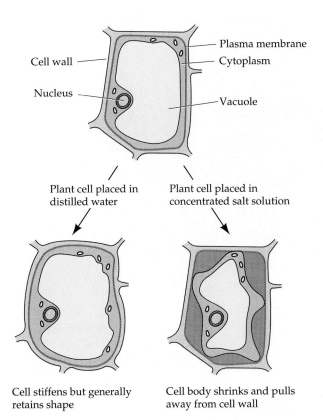

Cell wall

Plasma membrane

Cytoplasm

Nucleus

Vacuole

Plant cell placed in distilled water

Plant cell placed in concentrated salt solution

Cell stiffens but generally retains shape

Cell body shrinks and pulls away from cell wall

5.17 Osmosis and Cells with Walls
Water enters the vacuole of a plant cell placed in distilled water, but the cell retains its shape due to the rigid cell wall. When a plant cell is placed in a concentrated salt solution, the cell loses water to its surroundings; the cell wall retains its shape, but the plasma membrane shrinks away from the inside of the wall as the cell's volume is reduced.

Osmotic potentials determine the direction of osmosis in animal cells. A red blood cell takes up water from a solution that is hypotonic to the cell's contents. The cell bursts because its delicate plasma membrane cannot resist the swelling of the cell. The integrity of red blood cells and other blood cells is absolutely dependent upon the maintenance of a constant osmotic potential in the plasma in which they are suspended; the plasma must be isotonic with the cells if the cells are not to burst or shrink.

By contrast, the cells of plants, monerans, fungi, and some protists have cell walls that limit the volume of the cells and keep them from bursting. Cells with sturdy cell walls take up a limited amount of water and, in so doing, build up a pressure against the cell wall that prevents further water from entering. This pressure is the driving force for growth in plant cells—it is a normal and essential component of plant development. In cells with walls, then, osmosis is regulated not only by osmotic potentials but also by this opposed **pressure potential**. Osmotic phenomena in plants are discussed in Chapter 30.

MORE ACTIVITIES OF MEMBRANES

We have discussed some functions of membranes: the compartmentalization of cells, the regulation of traffic between compartments, and the active pumping of solutes. Membranes have many more functions. As discussed in Chapter 4, the membrane of rough endoplasmic reticulum is a site for ribosome attachment. Newly formed proteins are passed from the ribosomes through the membrane and into the interior of the endoplasmic reticulum for delivery to other parts of the cell. In nerve cells, the plasma membrane conducts the nerve impulse from one end of the cell to the other. The membranes of muscle cells, some eggs, and other cells are also electrically excitable. Many other biological activities and properties discussed in the chapters to follow are integrally associated with membranes.

To show the versatility of biological membranes, we will discuss here three more of their functions: mediating energy-trapping and energy-releasing reactions, recognizing materials at the cell surface, and helping cells associate as tissues.

Energy Transformations

Certain biological membranes are specialized to process energy—to convert light energy to the energy of chemical bonds, to trap energy released in oxidation–reduction reactions, and other vital activities. Why should membranes be involved in these activities? There are two reasons: structural organization and the separation of electric charges.

STRUCTURAL ORGANIZATION. Many processes in cells require various substances and take place step-by-step, with the products of one step being the reactants for the next step. If the necessary chemical substances for these reactions are all moving about at random, only chance collisions will bring them together and the processes will go forward slowly, if at all. If, on the other hand, the different substances are bound to a membrane (and especially if they are arranged in an orderly fashion), the product of one reaction can be released in close proximity to where it is needed for the next step in the pathway, and so forth—a virtual assembly line is established. In this sense, the membrane is a pegboard for the orderly attachment of specific proteins and other molecules.

SEPARATION OF CHARGES. A biological membrane can act like an electric battery. Work can be obtained from a battery by letting electrons flow from one of its terminals to the other by way of some device, like a motor or a light bulb, that uses the electric current. A similar process takes place in both photosynthesis and cellular respiration (see Chapters 7 and 8). Briefly, because of the limited permeability of the membranes in mitochondria and chloroplasts and because of the activities of certain electron carriers in those membranes, a substantial gradient of both electric charge and pH can be established across them. When these gradients are discharged by letting electric charge flow back through the membrane, the flow of charge can be used to do work or to form the energy-rich bonds of ATP. Without a membrane to enable the separation of charge, these reactions could not proceed.

Both the structure of membranes and their ability to separate electric charges relate to the properties of the two bulk components of membranes: lipids and proteins. The pegboard effect of the membrane's structure comes from the ability of the phospholipid bilayer to hold certain proteins in a defined plane so that they do not diffuse freely throughout the cell. The separation of charges is accomplished by certain membrane proteins and maintained by the insulating effect of the phospholipid bilayer.

Recognition and Binding

Membrane proteins and carbohydrates recognize and bind a variety of things to the outer surface of the plasma membrane. Antibodies (see Chapter 16) recognize target cells by virtue of specific proteins or carbohydrates on the surfaces of those cells. Viruses may begin their attack on intended host cells by attaching to carbohydrates on the host surface. Many hormones, including insulin, are recognized by membrane proteins that serve as receptors (see Chapter 36). Many nerve cells pass information to other nerve cells or to muscle fibers by means of a substance called acetylcholine (see Chapter 38). Acetylcholine activates these cells by attaching to a membrane protein—the acetylcholine receptor (Figure 5.18)—and changing the permeability of that membrane to ions. One of the most important classes of receptors binds substances, called growth factors, that regulate cell reproduction and differentiation (see Chapter 17).

Most animal cells have a mechanism, known as **receptor-mediated endocytosis,** that captures specific macromolecules from the cell's environment. The uptake is similar to endocytosis as described in Chapter 4, except that in receptor-mediated endocytosis parts of the plasma membrane contain specific receptor protein molecules. The parts of the membrane that contain receptor molecules are called coated pits because the inner surfaces of the membrane at these points are coated with several other, fibrous proteins, the best known of which is clathrin. When the receptor proteins bind the appropriate macromolecules

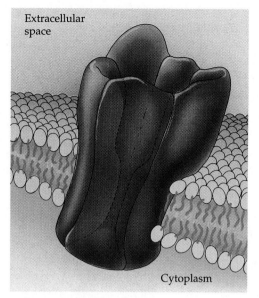

5.18 The Acetylcholine Receptor Protein
The acetylcholine receptor extends through the phospholipid bilayer and protrudes well beyond the bilayer into the extracellular space of nerve cells. There is a channel through the interior of the receptor protein (outlined by dashed lines); when molecules of acetylcholine bind to the receptor protein, the passage of ions through the channel is affected.

from the environment, the associated coated pit invaginates and forms a **coated vesicle** around the bound macromolecules. Strengthened and stabilized by clathrin molecules, this vesicle carries the macromolecules into the cell (Figure 5.19). Once inside, the vesicle loses its coat, making the macromolecules available to the cell. Because it is specific for particular macromolecules and can transport a group of macromolecules in one vesicle, receptor-mediated endocytosis is more rapid and efficient for taking up specific molecules than simple endocytosis is.

Receptor proteins must be intrinsic membrane proteins that span the entire thickness of the plasma membrane. Receptor carbohydrates are bound to such proteins or to phospholipid molecules. When a substance (hormone, virus, or other "visitor"), called a **ligand,** binds to its specific receptor, changes occur within the cell, on the cytoplasmic side of the membrane. Generally the visitor does not even cross the membrane to enter the cell. How can such a visitor produce an effect inside the cell if it does not enter it? Apparently the receptor protein undergoes structural changes. The portion of the protein inside the cell is altered, as is the way the protein functions in the cytoplasm.

The specificity of receptors (both proteins and carbohydrates) resides in their particular tertiary structures—that is, in their three-dimensional shapes. Some portion of the receptor protein or carbohydrate

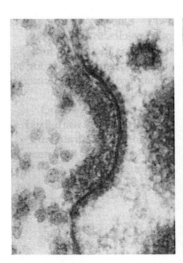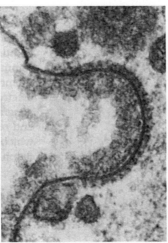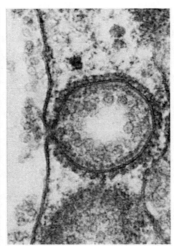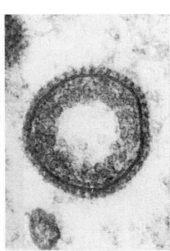

5.19 Coated Vesicle Formation
The micrograph at the far left shows a coated pit, with clathrin and other proteins concentrated against its cytoplasmic surface (arrow). The remainder of the sequence illustrates the development of a coated vesicle (far right) from the coated pit.

fits—hand in glove, as it were—the hormone, growth factor, part of a virus, or other ligand it is meant to recognize. We will discuss this sort of specificity in detail when we talk about enzymes in Chapter 6 and again in Chapter 16 when we talk about immune systems.

Cell Adhesion

During the growth of an animal embryo, specific membrane proteins known as **cell adhesion molecules** are crucial to the formation of the adult organism. As the embryo develops, its cells move about and associate with other specific cell types. This behavior is mediated by cell adhesion molecules in the membranes, which play roles in organizing the cells into tissues. One type of cell adhesion molecule, for example, organizes individual nerve cells into nerve cell bundles. Groups of cells that are about to migrate within the embryo lose their specific cell adhesion molecules; when they reach their new location, these cells regain their cell adhesion molecules and reorganize into tissue.

MEMBRANE INTEGRITY UNDER STRESS

Red blood cells appear fragile, yet they survive repeated compression and deformation as they squeeze through the finest of capillaries. This surprising resilience comes from certain intrinsic and extrinsic proteins in the plasma membrane of the cell (Figure 5.20). The extrinsic protein spectrin forms a meshwork of microfibrils on the cytoplasmic surface of the plasma membrane. This meshwork provides structural support. Another extrinsic protein, ankyrin, anchors the spectrin to the membrane at many points by binding both the spectrin and an anion transporter that is an intrinsic membrane protein. (This anion transporter exchanges Cl^- for HCO_3^- as described earlier in this chapter.)

Genetic defects in spectrin and others of these proteins result in abnormal red blood cells and thus in various diseases. Mice with hemolytic anemia have spherical, fragile red blood cells. Their red blood cells have very little spectrin, but the cells take on a normal shape if spectrin is provided.

MEMBRANE FORMATION AND CONTINUITY

As we have seen, membranes are dynamic in that they participate in numerous physiological and biochemical processes. Membranes are dynamic in another sense as well: They are constantly being formed, transformed from one type to another, and broken down. In eukaryotes, phospholipids are synthesized within the sacs of the rough endoplasmic reticulum and are rapidly distributed to membranes throughout the cell. Membrane proteins are inserted into the sacs of the rough endoplasmic reticulum as they are formed on ribosomes. Sugars may be added to the proteins while they are in the endoplasmic reticulum. Next the proteins go to the Golgi apparatus, where some have carbohydrates added to them. The proteins then travel in Golgi-derived vesicles to the plasma membrane and are incorporated into it (see Chapter 4).

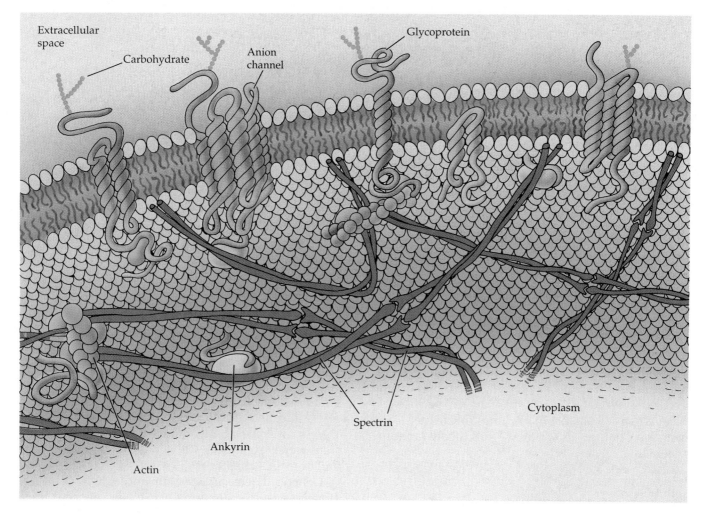

5.20 Some Proteins of the Red Blood Cell Membrane
Many intrinsic and extrinsic proteins contribute to this structure. Spectrin (purple) acts as a skeleton, binding the other proteins and thus strengthening the membrane that encloses the highly flexible red blood cell. Spectrin does not bind the membrane directly but connects by way of linker proteins such as ankyrin (green), which in turn binds the anion channel. Actin (yellow) also is a linker protein.

Functioning membranes move about within eukaryotic cells. For example, portions of the rough endoplasmic reticulum bud away from the endoplasmic reticulum and join the *cis* region of the Golgi apparatus (see Chapter 4). Rapidly—often in less than an hour—these segments of membrane find themselves in the *trans* regions of the apparatus, from which they bud away to join the plasma membrane (Figure 5.21). Bits of membrane are constantly merging with the plasma membrane in the process of exocytosis, but this is largely balanced by the removal of membrane in endocytosis. This removal by endocytosis affords a recovery path by which internal membranes are replenished. In sum, there is a steady flux of membranes as well as membrane components in cells.

Because we know about the constant interconversion of membranes, we might expect all subcellular membranes to be chemically identical. As you already know, this is not the case. There are major chemical differences among the membranes of even a single cell. Apparently membranes are changed chemically when they form parts of certain organelles. In the Golgi apparatus, for example, the membranes of the *cis* region are very similar to those of the endoplasmic reticulum, but the membranes of the *trans* region are more similar in composition to the plasma membrane. Ceaselessly moving, constantly carrying out functions vital to the life of the cell, biological membranes certainly are not the static, stodgy structures they once were thought to be.

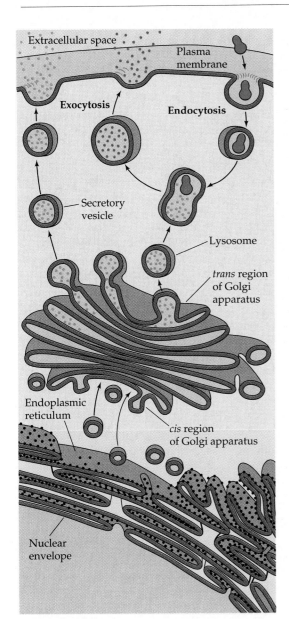

Extracellular space

Plasma membrane

Exocytosis

Endocytosis

Secretory vesicle

Lysosome

trans region of Golgi apparatus

Endoplasmic reticulum

cis region of Golgi apparatus

Nuclear envelope

5.21 Membrane Continuity in Cells

Arrows trace how membranes form, move, and fuse in cells. New stretches of membranes may be generated at certain locations, such as the outer membrane of the nuclear envelope. Vesicles budding from the *trans* region of the Golgi apparatus are also membrane-bounded. The vesicles may remain inside the cell as organelles, such as lysosomes, or they may fuse with the plasma membrane, delivering their contents to the exterior of the cell (exocytosis) and adding their membranes to the plasma membrane. Membrane is subtracted from the plasma membrane in the process of endocytosis.

SUMMARY of Main Ideas about Membranes

The membranes around and within cells are thin, pliable bilayers of phospholipids with embedded proteins and attached carbohydrates.
  Review Figures 5.1, 5.2, 5.3, and 5.6

Membrane lipids restrict the rates at which many solutes pass through membranes.

In a solution without barriers, solutes diffuse at rates determined by their physical properties and in directions determined by their concentration gradients.
  Review Figure 5.9

Solutes that can pass through a membrane do so by simple diffusion, facilitated diffusion, or active transport.
  Review Figure 5.10

Different membrane transport proteins allow specific substances to pass through in various ways.
  Review Figures 5.12 and 5.13

Only active transport can move solutes against a concentration difference.
  Review Figures 5.14 and 5.15

Water moves through membranes only by osmosis.

Water moves across the plasma membrane of an animal cell toward the side where the osmotic potential is more negative.
  Review Figure 5.16

Pressure potential as well as osmotic potential regulates osmosis in cells with walls.
  Review Figure 5.17

Carbohydrates attached to lipids or proteins on the outside surface of a plasma membrane are sites where the cell recognizes other cells or foreign substances.

Particular membrane proteins recognize other cells or molecules, regulate specific chemical reactions, or transmit information across the membrane.

Membranes within cells continually form, move, and fuse.
Review Figure 5.21

Where animal cells meet, junctions form, linking the plasma membranes: desmosomes for adherence, tight junctions that limit flow between cells, and gap junctions for communication.
Review Figure 5.8

SELF-QUIZ

1. Which statement is *not* true of membrane phospholipids?
   a. They associate to form bilayers.
   b. They have hydrophobic "tails."
   c. They have hydrophilic "heads."
   d. They give the membrane fluidity.
   e. They flop readily from one side of the membrane to the other.

2. The phospholipid bilayer
   a. is readily permeable to large, polar molecules.
   b. is entirely hydrophobic.
   c. is entirely hydrophilic.
   d. has different lipids in the two layers.
   e. is made up of polymerized amino acids.

3. Which statement is *not* true of membrane proteins?
   a. They all extend from one side of the membrane to the other.
   b. Some serve as channels for ions to cross the membrane.
   c. Many are free to migrate laterally within the membrane.
   d. Their position in the membrane is determined by their tertiary structure.
   e. Some play roles in photosynthesis.

4. Which statement is *not* true of membrane carbohydrates?
   a. Most are bound to proteins.
   b. Some are bound to lipids.
   c. Carbohydrates are added to proteins in the Golgi apparatus.
   d. They show little diversity.
   e. They are important in recognition reactions at the cell surface.

5. Which statement about animal membrane junctions is *not* true?
   a. Tight junctions are barriers to the passage of molecules between cells.
   b. Desmosomes allow cells to adhere strongly to one another.
   c. Gap junctions block communication between adjacent cells.
   d. Connexons are made of protein.
   e. The fibers associated with desmosomes are made of protein.

6. Which statement is *not* true of diffusion?
   a. It is the movement of molecules or ions to a state of even distribution.
   b. At the subcellular level it is a slow process.
   c. The motion of each molecule or ion is random.
   d. The diffusion of each substance is independent of that of other substances.
   e. Diffusion over meters takes years.

7. Which statement is *not* true of channels in membranes?
   a. They are pores in the membrane.
   b. They are proteins.
   c. All ions pass through the same type of channel.
   d. Some channels are gated.
   e. Movement through channels is by simple diffusion.

8. Facilitated diffusion and active transport
   a. both require ATP.
   b. both require the use of proteins as carriers.
   c. both carry solutes in only one direction.
   d. both increase without limit as the solute concentration increases.
   e. both depend on the solubility of the solute in lipid.

9. Primary and secondary active transport
   a. both generate ATP.
   b. both are based on passive movement of sodium ions.
   c. both include the passive movement of glucose molecules.
   d. both use ATP directly.
   e. both can move solutes against their concentration gradients.

10. Which statement is *not* true of osmosis?
    a. It obeys the laws of diffusion.
    b. In animal tissues, water moves to the cell with the most negative osmotic potential.
    c. Red blood cells must be kept in a plasma that is hypotonic to the cells.
    d. Two cells with identical osmotic potentials are isotonic to each other.
    e. Solute concentration is the principal factor in the osmotic potential.

FOR STUDY

1. How do freeze-fracture and freeze-etch techniques reveal aspects of membrane structure not revealed by transmission electron microscopy of thin sections of biological material?

2. In Chapter 40 we will see that the functioning of muscles requires calcium ions to be pumped into a subcellular compartment against a calcium concentration gradient. What types of chemical substances are required for this to happen?

3. Some algae have complex glassy structures in their cell walls. The structures form within the Golgi apparatus. How do these structures get to the cell wall without having to pass through a membrane?

4. Organisms that live in fresh water are almost always hypertonic to their environment. In what way is this a serious problem? How do some organisms cope with this problem?

5. Contrast simple endocytosis (see Chapter 4) and receptor-mediated endocytosis (this chapter) with respect to mechanism and to performance.

READINGS

Alberts, B., D. Bray, J. Lewis, M. Raff, K. Roberts and J. D. Watson. 1994. *Molecular Biology of the Cell*, 3rd Edition. Garland Publishing, New York. An outstanding general text in modern cell and molecular biology; Chapters 10 and 12 are particularly suitable for further study of biological membranes, and Chapters 17 and 19 are also useful.

Bretscher, M. S. 1985. "The Molecules of the Cell Membrane." *Scientific American*, October. A fine treatment of membrane chemistry, cell junctions, endocytosis, and other topics.

Lodish, H., D. Baltimore, A. Berk, L. Zipursky, P. Matsudaira and J. Darnell. 1995. *Molecular Cell Biology*, 3rd Edition. Scientific American Books, New York. Another fine general text; see Chapters 14, 15, and 16.

Lodish, H. F. and J. E. Rothman. 1979. "The Assembly of Cell Membranes." Scientific American, January. Good information on how membranes grow and why the two sides of a membrane differ.

Stryer, L. 1995. *Biochemistry*, 4th Edition. W. H. Freeman, New York. A reference for the major types of molecules found in membranes, with outstanding illustrations of phospholipids.

Unwin, N. and R. Henderson. 1984. "The Structure of Proteins in Biological Membranes." *Scientific American*, February. A clear presentation of how the proteins in membranes transport molecules.

Biochemical Self-Renewal
From biochemical reactions carried out in their cells, living things like this gecko replace worn out or lost materials with new ones.

6

Energy, Enzymes, and Catalysis

In a single second, a typical cell carries out thousands of biochemical reactions. Why must it do this? Perhaps the main answer is that life depends on energy. Energy is available in the environment, and, as we will see in this chapter, energy can be released and converted among many forms by biochemical reactions. In certain reactions, cells break molecules apart, releasing energy that then fuels other reactions.

Via biochemical reactions, cells convert raw materials obtained from the environment into the building blocks of proteins and other compounds unique to living things. To maintain themselves, organisms must replace lost materials with new ones; they also grow and reproduce. All this is accomplished because cells carry out thousands of reactions per second.

Cells have a big advantage over nonliving matter when it comes to carrying out reactions. They have evolved the means to produce special substances, called enzymes, that enhance how reactions proceed. Enzymes and how they speed up energy conversions within cells are the main topic of this chapter.

ENERGY AND THE LAWS OF THERMODYNAMICS

The sum total of all uses of energy by a cell or by an organism is its **metabolism.** Metabolism consists of thousands of individual biochemical reactions that take place every second in most cells. Some of these reactions form more complex molecules from simpler ones (anabolism), and others break down complex molecules into simpler ones (catabolism). Some reactions release energy that is used to do physical work, such as the contraction of a muscle against a resisting load. Many other biochemical reactions, including those that synthesize macromolecules, proceed very slowly unless a source of energy is provided.

Energy has been defined both as the "capacity to do work" and as "heat or anything that can be transformed into heat." Energy comes in many forms: heat, light, electric, mechanical, chemical, nuclear, and others. Matter itself is a form of energy. An atomic explosion or a reaction in a nuclear power plant converts a small amount of matter into enormous amounts of energy. When all forms of energy are accounted for, the total amount of energy in the universe is unchanging: Energy can neither be created nor destroyed. This is the **first law of thermodynamics.** The first law holds for the universe as a whole or for any closed system within the universe. A **closed system** is one that is not exchanging energy with its surroundings.

Energy in one form can be converted into energy in another form (Figure 6.1). In solar batteries, light

6.1 Biological Energy Transformations
The leaf traps light energy and produces food by photosynthesis. Sawfly larvae obtain energy by eating the leaf. As the larvae crawl and chew, they expend mechanical energy obtained from the chemical energy of food.

energy is converted into electric energy. Electric energy can be converted into light, heat, motion, and other forms of energy. Green plants convert light into chemical energy; in muscles, chemical energy is transformed into the energy of motion. None of these energy conversions is 100 percent efficient—some energy is always lost as heat. Heat, too, can be used to do work (think of a steam engine, for example), but here we run into limitations. The conversion of any other form of energy into heat is not fully reversible; that is, not all the heat can be converted back into the other forms of energy. Biological, chemical, and physical processes are often accompanied by the production of heat, not all of which can be made to do work.

Unusable heat is associated with an increase in disorder. Chemical changes, physical changes, biological processes, and anything else you can think of tend toward disorder, or randomness. A crystal of sodium chloride, which is highly ordered, will dissolve spontaneously in water to form a more random solution of sodium chloride. A sodium chloride solution, however, will not spontaneously reorder itself into a crystal of salt and pure water. Has your room become more or less orderly since the last time you expended energy to straighten it up?

Disorder can be discussed in quantitative terms; its measure is a quantity called **entropy.** Greater entropy implies greater disorder in any system. Not all energy conversion processes result in the same ability to do work. Various amounts of useful energy may be lost as the original forms are converted to unusable forms of energy—the heat associated with disorder. In the universe as a whole, or in any closed system, the amount of entropy increases; this is the principle of degradation of energy, also known as the **second law of thermodynamics.** Other ways of stating the second law will appear in this chapter.

CHEMICAL EQUILIBRIUM

In principle, all chemical reactions can run both forward and backward. For example, if compound A can be converted into compound B (A → B), then B can in principle be converted into A (B → A)—although at given concentrations of A and B only one of these directions will be favored. Think of the overall reaction as a result of competition between forward and reverse reactions. Increasing the concentration of the reactants (A) speeds up the forward reaction; increasing the concentration of the products (B) favors the reverse. At some point the forward and reverse reactions take place at the same rate. At this point no further change in the system is observable, although individual molecules are still forming and breaking apart. This balance between forward and reverse reactions is known as **chemical equilibrium.**

When a reaction goes more than halfway to completion, we say that the reaction A → B is a **spontaneous reaction;** in this case the reaction B → A is not spontaneous. A spontaneous reaction is one that, given enough time, goes largely to completion by itself, without the addition of energy. In fact, as spontaneous reactions proceed, they release energy (called free energy). If a reaction runs spontaneously in one direction (from reactant A to product B, for example), then the reverse reaction (from B to A) requires a steady supply of energy to drive it (Figure 6.2). For example, starch slowly but spontaneously breaks down in water, producing the disaccharide maltose; maltose, however, does not form starch spontaneously.

Entire chemical pathways also have a spontaneous direction. The complete oxidation of glucose in the processes that provide energy for cellular work is spontaneous. On the other hand, the synthesis of glucose from carbon dioxide and water by plants is

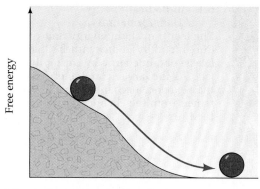

6.2 A Spontaneous Reaction
The reactants in a spontaneous reaction are like a ball on the side of a hill. The products form as the "reactant ball" rolls down the hill with a release of free energy. But the ball will not roll *up* the hill spontaneously; it would require work (the input of energy) to move the ball back up the hill.

a nonspontaneous reaction that must be driven by energy from the absorption of light in photosynthesis. (The important processes of fermentation, cellular respiration, and photosynthesis are discussed at length in Chapters 7 and 8).

Spontaneity has nothing to do with time. A reaction may be extremely slow, yet still be spontaneous. The burning of a newspaper and the slow browning of pages in old library files are both spontaneous processes, but they have different time scales. A spontaneous reaction moves toward equilibrium, using up reactants and making products. Given enough time, every spontaneous reaction eventually reaches equilibrium concentrations of reactants and products. By supplying energy to reactions, cells *prevent* the attainment of equilibrium as long as they live. In this way, cells keep their chemical composition different from that of the environment around them.

The Equilibrium Constant

Each reaction has a specific point of equilibrium—the point at which certain concentrations of reactants and products have been reached. Consider the following example. Every living cell contains glucose 1-phosphate, and its conversion to glucose 6-phosphate, accelerated by a particular enzyme, is a common event in cells. (Later in this chapter we explain how enzymes work.) In our example, the necessary enzyme is added to a solution of glucose 1-phosphate, at an initial concentration of 0.02 *M* (0.02 molar; see Chapter 2). As the reaction comes to equilibrium, the concentration of the product, glucose 6-phosphate, rises from 0 to 0.019 *M*, while the concentration of glucose 1-phosphate falls to 0.001 *M* (Figure 6.3). The reaction proceeds until equilibrium is reached at these concentrations. From then on, the reverse re-

action to glucose 1-phosphate is progressing at the same rate as the forward reaction to glucose 6-phosphate. At equilibrium, then, the forward reaction has gone 95 percent of the way to completion—the forward reaction is a spontaneous reaction. This result is obtained every time the experiment is run under the same conditions: at 25°C and pH 7.

The **equilibrium constant,** K_{eq}, is defined as the ratio of the concentrations of products and reactants *at equilibrium:*

$$K_{eq} = \frac{[\text{product}]}{[\text{reactant}]}$$

where the brackets indicate concentrations in, for example, moles per liter. By convention, reaction products are always shown in the numerator and reactants in the denominator. In our example,

$$K_{eq} = \frac{[\text{glucose 6-phosphate}]}{[\text{glucose 1-phosphate}]}$$

$$= 0.019\ M/0.001\ M = 19$$

Suppose that we run the experiment again, this time starting with only the product—with, say, 0.02 *M* glucose 6-phosphate and no glucose 1-phosphate. A reaction occurs, with equilibrium being reached at 0.001 *M* glucose 1-phosphate and 0.019 *M* glucose 6-phosphate—the same as before. As this result shows, the starting proportions of reactants and products do not affect equilibrium concentrations. Equilibrium concentrations are defined by K_{eq}—that is, the equilibrium constant *is constant* for a given chemical reaction as long as other conditions remain unchanged.

In general, for a reaction of the type A $\rightleftharpoons$ B, the equilibrium constant K_{eq} is defined by the equation

$$K_{eq} = \frac{[\text{B}]}{[\text{A}]}$$

where [B] and [A] are the concentrations of product B and reactant A, in moles per liter, *at equilibrium*. For the more complex reaction C + D $\rightleftharpoons$ E + F, the equilibrium constant is

$$K_{eq} = \frac{[\text{E}][\text{F}]}{[\text{C}][\text{D}]}$$

For the ionization of acetic acid in water at 25°C,

$$CH_3COOH \rightarrow CH_3COO^- + H^+$$

$$K_{eq} = \frac{[CH_3COO^-][H^+]}{[CH_3COOH]} = 2 \times 10^{-5}\ M$$

The smallness of this equilibrium constant tells us that the ionization of acetic acid is quite limited—less than one-half of one percent of the acetic acid molecules are ionized, and the ionization is not a spontaneous reaction. Acetic acid is a weak acid. When a strong acid such as hydrochloric acid dissolves in

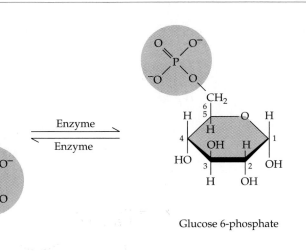

Glucose 1-phosphate

Glucose 6-phosphate

6.3 Concentration at Equilibrium

An enzyme speeds both the conversion of glucose 1-phosphate to glucose 6-phosphate and the reverse reaction. At 25°C and a pH of 7, there will always be 95 percent glucose 6-phosphate and 5 percent glucose 1-phosphate at equilibrium, no matter what the starting percentages of the two compounds are. Notice from the structural formulas that the compounds get their names from the position of attachment of the phosphate group (orange) to the carbon ring.

100% Glucose 1-phosphate

95% Glucose 6-phosphate
5% Glucose 1-phosphate

100% Glucose 6-phosphate

water, almost all the molecules ionize; thus the equilibrium constant is large, showing this reaction is spontaneous.

Table 6.1 gives examples of equilibrium constants. Notice that the values vary widely. A high value of K_{eq} means that the reaction goes far toward completion; a very low value means that the reaction scarcely goes at all—in fact, the reverse reaction predominates. The table also gives a related measure of a reaction's spontaneity, its free energy change, which we discuss in the next section.

Free Energy and Equilibria

What determines the point of chemical equilibrium? What distinguishes a spontaneous reaction from one that can proceed only with a considerable input of energy? The second question answers the first. A spontaneous reaction, one that goes far toward completion, is a reaction that gives off a great deal of **free energy,** energy that can be used to do work. Such a reaction is said to be **exergonic.** A reaction with a tiny equilibrium constant, on the other hand, is **endergonic** because it can be made to go only by the addition of free energy (Figure 6.4). Free energy is symbolized by G (for "Gibbs free energy," named after the nineteenth-century Yale thermodynamicist Josiah Willard Gibbs). It cannot be measured absolutely; however, the *change* in free energy, ΔG, of a reaction can be determined readily. It is related directly to the value of K_{eq} for the reaction: The greater the value of K_{eq}, the more free energy is given off. Values of ΔG are given in Table 6.1.

In the universe as a whole, or in any closed sys-

TABLE 6.1
Equilibrium Constants and Standard Free Energy Changes of Selected Reactions

REACTION[a]	EQUILIBRIUM CONSTANT	STANDARD FREE ENERGY CHANGE (KCAL/MOL)
Acetic acid + H_2O → acetate + H_3O^+	0.00002	+6.3
Malate → fumarate + H_2O	0.28	+0.75
Fructose 6-phosphate → glucose 6-phosphate	2.0	−0.4
Glucose 1-phosphate → glucose 6-phosphate	19	−1.7
Glucose 6-phosphate + H_2O → glucose + phosphate	260	−3.3
Sucrose + H_2O → glucose + fructose	140,000	−7.0

[a]The reactions are arranged from top to bottom in order of increasing tendency to go to completion as written ("go to the right").

Exergonic reaction
(spontaneous; energy-releasing)

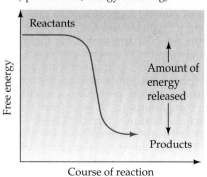

Endergonic reaction
(not spontaneous; energy-requiring)

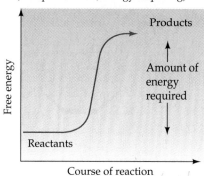

6.4 Exergonic and Endergonic Reactions

In an exergonic reaction, energy is *released* as reactants with a relatively high energy content form products with a lower amount of energy. Energy is *required* for an endergonic reaction, in which reactants with a low energy content are converted to products with a higher energy level.

tem, the quantity of free energy is always decreasing and entropy is always increasing. (This is another way of stating the second law of thermodynamics.) In a spontaneous reaction, the reactants possess more free energy than do the products. A reaction that goes nearly to completion ("to the right" as written) has a large, negative ΔG, indicating that it releases a large amount of free energy. In an exergonic reaction such as the conversion of glucose 1-phosphate to glucose 6-phosphate, ΔG is a negative number (in this example, $\Delta G = -1.7$ kcal/mol). Recall that equilibrium for this reaction has a product-to-reactant ratio of 19:1—that is, the reaction goes nearly to completion. A large, positive ΔG means that the reaction hardly proceeds at all as written; if the products are present, such a reaction runs backward ("to the left") to near completion. A ΔG value near zero is characteristic of a readily reversible reaction: Reactants and products have almost the same free energies.

Free Energy, Heat, and Entropy

We have seen how free energy is related to chemical equilibrium: Where equilibrium is attained, free energy is at a minimum. This relationship makes it clear that ΔG measures the useful chemical energy obtainable from a reaction. As an exergonic reaction proceeds, free energy is released and may be used to do chemical work. The cell harvests free energy from exergonic reactions such as the oxidation of foodstuffs, or from sunlight, and uses the free energy to drive vital endergonic reactions (such as those of photosynthesis).

Free energy is related to two other forms of energy: heat and a form associated with the entropy of the system. Both forms can be discussed in the context of spontaneous reactions: As a spontaneous reaction proceeds, entropy increases and heat is usually released. In any chemical reaction or physical process, each of these three forms of energy—free energy, heat, and the energy associated with entropy—may change.

Entropy, the measure of disorder, is expressed in kilocalories per degree (kcal/deg). To relate entropy to free energy, multiply entropy by the temperature at which the reaction occurs to obtain energy in kilocalories. For this calculation we use the Kelvin temperature scale, or absolute temperature. (In Kelvin units, 0 K is equal to about $-273°$C. Since one Kelvin unit is equivalent to one Celsius degree, to convert from Celsius to Kelvin, simply add 273 to the Celsius temperature.) As an example, the increase in entropy as 1 mole of ice melts to form water is 5.26 kcal/deg. When this reaction occurs at the freezing point (0°C, or 273 K), the energy lost to disorder is 5.26 kcal/deg $\times$ 273 K = 1436 kcal.

The change in free energy (ΔG) of any reaction is defined in terms of the change in heat (ΔH) and the change in entropy (ΔS):

$$\Delta G = \Delta H - T\Delta S$$

where T is the absolute temperature at which the reaction is occurring. The $T\Delta S$ term shows that the effect of a change in disorder is greater at high temperatures than at low. (Multiplying a given value of ΔS by the larger number that represents a higher temperature gives a larger energy change.)

Consider the relative importance of these factors in the following example: The combustion of 1 mole of glucose gives off 673 kcal of heat and the disorder increases by 0.0433 kcal/deg. At 25°C (298 K), $T\Delta S$ = 298 K (0.0433 kcal/deg) = 12.9 kcal, which can be rounded off to 13 kcal. Both heat and disorder contribute to the spontaneity of glucose combustion. Thus for the complete oxidation of 1 mole of glucose, the change in free energy is calculated as follows:

$$\Delta G = -673 \text{ kcal} - 13 \text{ kcal} = -686 \text{ kcal}$$

Although the entropy factor (13 kcal/mol) is much smaller than the heat factor (673 kcal/mol), it does contribute to the total change in free energy. Some other spontaneous reactions have large negative ΔG values because their changes in entropy are large.

Recall that the change in free energy determines

the equilibrium constant for the reaction. Each chemical reaction is characterized by a particular equilibrium constant (K_{eq}) and by another constant, which we discuss in the next section.

REACTION RATES

When we know a reaction's change in free energy (ΔG), we know where the equilibrium of the reaction lies. The more negative ΔG is, the further the reaction proceeds toward completion. ΔG does not tell us anything, however, about the **rate of a reaction**—the speed at which it moves toward equilibrium.

Getting Over the Energy Barrier

A key to understanding reaction rates lies in recognizing that there is an energy barrier between reactants and products. Think about a butane lighter. The burning of the fuel (that is, the reaction of butane with oxygen to release carbon dioxide and water vapor) is obviously exergonic—once started, the reaction goes to completion, which occurs when all the butane has been burned. Since burning butane liberates free energy as light and heat, you might expect this reaction to proceed on its own. If you simply allow butane to flow and mix with air, however, nothing happens. To start the burning of butane, you have to provide a spark. Even though the burning of butane is spontaneous because it releases energy, the need for a spark to start the reaction shows that there is some sort of energy barrier between reactants and products.

In general, reactions proceed only after they are pushed over the energy barrier by bits of energy (such as the heat from a spark), which are called **activation energy** (E_a; Figure 6.5a). Recall the ball rolling down the hill in Figure 6.2. The ball has more free energy at the top of the hill because somebody expended energy carrying or throwing it up there. As the ball rolls down the hill, the reaction is exergonic: The ball is losing energy. Rolling the ball back up the hill requires energy and is thus an endergonic reaction, which cannot happen spontaneously.

Suppose now that the ball on the hillside is in a little depression (Figure 6.5b). Rolling down the hill is still an exergonic process, but to start the ball rolling, a small amount of activation energy must first be exerted. In a chemical reaction, the energy barrier—the "hump" over which reactants must "roll" before they can proceed spontaneously to form products—is energy needed to change reactants into intermediate molecular forms called transition-state species. Transition-state species have higher free energies than either the products or the reactants. A transition-state species corresponds to a ball that has

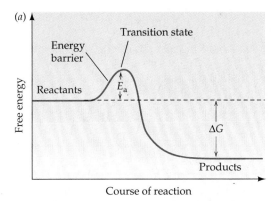

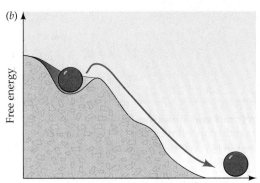

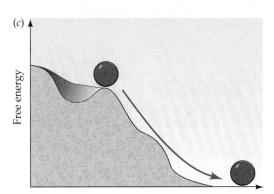

6.5 Reactions Need Activation Energy to Get Them Going
(a) If supplied with activation energy (E_a), the reactants surmount an energy barrier. Then an energy-releasing reaction proceeds spontaneously, releasing free energy (ΔG). (b) If a ball on the side of a hill is in a depression, it requires a push (activation energy) to get it out of the depression before it can roll to the bottom. (c) Here the reactant ball on the hill has received an input of activation energy and is poised (activated) for a spontaneous journey down to the product level, releasing free energy as it goes.

just been rolled up from the depression (Figure 6.5c). The activation energy that starts a reaction is recovered during the ensuing "downhill" phase of the reaction, so it is not a part of the drop in free energy, ΔG (see Figure 6.5a).

In any situation, some molecules have more energy than others. Picture a mixture of reactant molecules with various energies. Some of the molecules

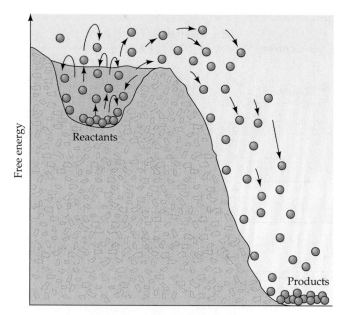

6.6 Energy Hump and Reaction Rate
Some reactant molecules have enough activation energy to get over the energy hump; others fall back into the depression, remaining there as reactants. The higher the hump, the fewer the molecules that can get over it, and the lower the reaction rate.

in the mixture have enough energy to get over the activation "hump" and enter the transition state, and some of these molecules react to yield products (Figure 6.6). A reaction with a low activation energy goes more rapidly because more of the reactant molecules

have enough energy to get over the hump. When activation energy is high, the reaction is slow unless more energy is provided, usually as heat. If the system is heated, all the reactant molecules become more energetic, and since more of them thus have energy in excess of the required activation energy, the reaction speeds up.

Biochemical reactions in our own mouths confront an energy barrier. Our saliva acts on the starch in food. As we saw in Chapter 3, starch is a polymer consisting of many glucose monomers. Starch reacts with water in the presence of saliva. The water cleaves the starch polymer into oligomers by breaking some of the bonds connecting the glucose units (Figure 6.7a). Starch in pure water, however, is quite stable because the activation energy of the reaction is high. (When we say "stable" we mean that a reaction does not proceed at all or proceeds so slowly that it would take a very long time to detect any change.) For the reaction to proceed, water and starch molecules must first collide; then the bonds connecting the glucose units must stretch and break. New bonds must then form between the broken ends and either H or OH derived from water. As shown in Figure 6.7b, there is an intermediate stage in which the bond between two glucose units is longer than normal and new bonds are in the process of forming. This is the transition-state species; it would be at the top of the energy barrier in a diagram like Figure 6.5. What is the difference between saliva and pure water that causes starch to break down in saliva but not in water? We'll explain in a moment.

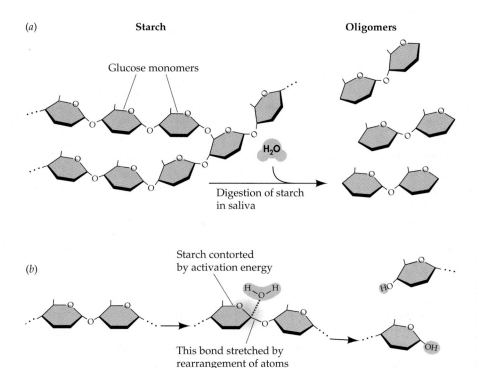

(a) **Starch** → **Oligomers**

Glucose monomers

H_2O

Digestion of starch in saliva

(b)

Starch contorted by activation energy

This bond stretched by rearrangement of atoms

6.7 The Breakdown of Starch
(a) In saliva, long starch molecules are digested into oligomers. (b) Before the bonds between glucose monomers in starch can break, an activated intermediate stage (center) must form, as. In the yellow area, activation energy contorts the starch molecule, stressing an otherwise stable bond and preparing the molecule for reaction with water. On the right, the reaction has been completed with the breaking of the stressed bond and the addition of atoms from a water molecule.

Rate Constants

Any chemical reaction proceeds at a rate directly proportional to the concentration of the reactants. The rate—in, for example, micromoles per minute—simply equals the reactant concentration times a **rate constant,** k, related to activation energy. At any given temperature, each specific reaction has its own characteristic rate constant. For the simple case,

$$A \xrightarrow{k} B$$

the rate of formation of product B is equal to k times the concentration of A, or $r_B = k[A]$, where r_B is the rate of formation of B. The rate constant is affected by temperature: As the temperature increases, so does the rate constant k. Recall from the previous section that increased temperature makes the reactant molecules more energetic, enabling more of them to get over the E_a hump and thus speeding up the reaction.

When starch dissolves in pure water, the rate of its conversion to smaller carbohydrate molecules is extremely slow, as we have already noted. Even when the starch solution is heated to the boiling point, essentially none of the water and starch molecules gain enough energy to exceed the activation energy for the reaction. Nonetheless, we know that starch does get digested at a significant rate in the presence of saliva (which is fortunate, for we would starve to death if all our digestive processes took place as slowly as the spontaneous breakdown of starch in pure water). How can a reaction such as that shown in Figure 6.7 be sped up by the presence of something such as saliva? The answer is that saliva is a catalyst.

THE HIGHLY SPECIFIC CATALYSIS OF ENZYMES

Although its function is to speed up a reaction, a **catalyst** does not cause anything to take place that would not take place eventually without it, and it does not become part of the reaction products. A catalyst merely lowers the activation energy of the reaction, allowing equilibrium to be approached at a faster rate. Most nonbiological catalysts are nonspecific. Platinum black (a soft powder), for example, catalyzes virtually any reaction in which molecular hydrogen is a reactant because it weakens the bond between the atoms in the H_2 molecule. By contrast, most biological catalysts, called **enzymes,** are highly specific. An enzyme usually catalyzes only a single chemical reaction or, at most, a very few closely related reactions. Saliva contains an enzyme that catalyzes the breakdown of starch, but that enzyme does

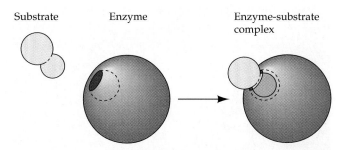

6.8 Enzyme and Substrate
An enzyme is a protein catalyst with an active site capable of binding a particular type of substrate molecule, forming an enzyme–substrate complex.

not catalyze the breakdown of fats, of proteins, or even of other polymers of glucose.

The molecules that are acted on catalytically are the enzyme's **substrates.** Substrate molecules bind themselves to the surface of the enzyme at the enzyme's **active site,** where catalysis takes place (Figure 6.8). The binding of a substrate to the active site forms an **enzyme–substrate complex** held together by one or more means, such as hydrogen bonding, ionic attraction, or covalent bonding. The enzyme–substrate complex may form product and free enzyme:

$$E + S \rightleftharpoons ES \rightleftharpoons E + P$$

where E is the enzyme, S is the substrate, P is the product, and ES the enzyme–substrate complex. An enzyme present in saliva acts on its substrate, starch, forming oligosaccharides, the product. Note that E, the free enzyme, is in the same chemical form at the end of the reaction as at the beginning. During the reaction it may change chemically, but by the end of the reaction it is restored to its initial form.

An enzyme gets its specificity from the exact structure of its active site. The tertiary structure of the enzyme lysozyme is shown in Figure 6.9. Lysozyme is an enzyme that protects the animals that produce it by destroying invading bacteria, which it accomplishes by cleaving certain polysaccharide chains in the cell walls of the bacteria. Lysozyme is found in tears and other bodily secretions and is particularly abundant in the whites of bird eggs. In Figure 6.9, the active site of lysozyme appears as a large indentation filled with the substrate (shown in green). The substrate fits precisely into the active site, whereas other molecules—with different shapes or different chemical groups on their surfaces—cannot form a complex with the enzyme.

Some enzymes change shape after binding with their substrate (Figure 6.10). These shape changes improve the alignment and result in an **induced fit** between the enzyme and the substrate. The enzyme α-amylase, which is present in saliva and specific for

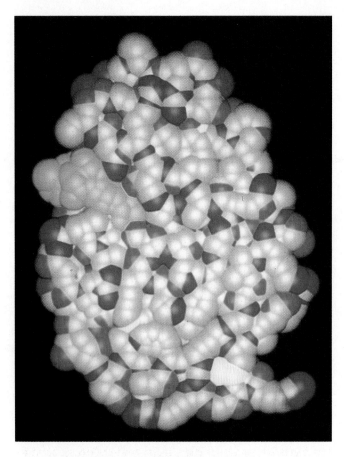

6.9 Tertiary Structure of Lysozyme
A substrate, shown in green, is bound to a lysozyme molecule. In the enzyme, the carbons are shown in gray, oxygens in red, nitrogens in blue, and sulfurs in yellow. Hydrogen atoms have been omitted. Lysozyme attaches precisely to the polysaccharide substrate, stressing particular bonds and allowing the usually stable polymer to be broken.

digesting starch to oligosaccharides, undergoes such a change as it catalyzes the reaction shown in Figure 6.7.

When an enzyme lowers the activation energy, both the forward and the reverse reactions speed up, so the enzyme-catalyzed reaction proceeds toward equilibrium more rapidly than the uncatalyzed one. Remember that the final equilibrium is the same with or without the enzyme. Adding an enzyme to a reaction does not change the difference in free energy (ΔG) between the reactants and the products; it changes only the activation energy (Figure 6.11), and consequently the rate constant.

Substrate Concentration and Reaction Rate

For a reaction of the type A → B, the rate of the uncatalyzed reaction is directly proportional to the concentration of A. Most enzyme-catalyzed reactions, on the other hand, generate plots like the one in Figure 6.12. At first the reaction rate increases as the substrate concentration increases, but then it levels off. Further increases in the substrate concentration do not increase the reaction rate. Since the concentration of the enzyme is usually much lower than that of the substrate, what we are seeing is a saturation phenomenon like the ones that occur in facilitated diffusion and active transport across membranes (see Figure 5.10). As more substrate is added,

6.10 Some Enzymes Change Shape When Substrate Binds
The deep cleft on the left side of the enzyme hexokinase divides the molecule into upper (darker shading) and lower lobes and contains the site where the substrate, glucose, binds.

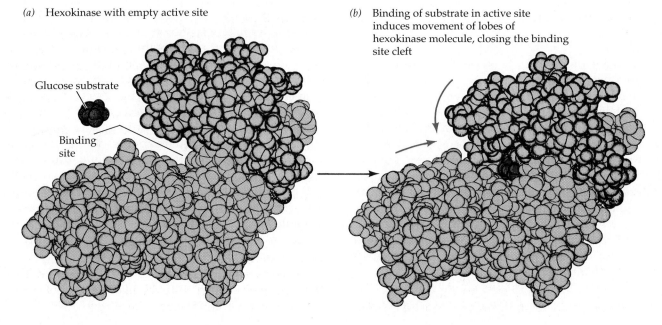

(a) Hexokinase with empty active site

Glucose substrate

Binding site

(b) Binding of substrate in active site induces movement of lobes of hexokinase molecule, closing the binding site cleft

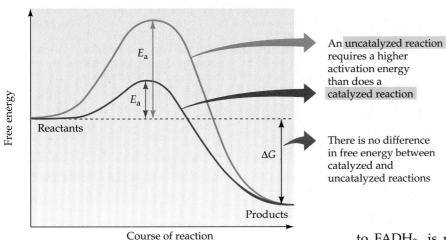

An enzyme-catalyzed reaction (red curve) has a lower activation energy (E_a) than does an uncatalyzed reaction (blue curve). There is no difference in the amount of free energy released (ΔG).

An uncatalyzed reaction requires a higher activation energy than does a catalyzed reaction

There is no difference in free energy between catalyzed and uncatalyzed reactions

more of the enzyme molecules are tied up as enzyme–substrate complexes. When all the enzyme molecules are bound to substrate molecules, nothing is gained by adding more substrate because no enzyme molecules are available to act as catalysts.

The study of the rates of enzyme-catalyzed reactions is called *enzyme kinetics.* As we will see later in this chapter, some graphs of rate as a function of substrate concentration are quite different from that in Figure 6.12. Such graphs tell us a great deal about the nature of the enzyme-catalyzed reaction.

Coupling of Reactions

Some of the most important reactions in living organisms are not spontaneous but proceed because specific enzymes **couple** them with other reactions that *are* spontaneous. Consider, for example, a pair of coupled reactions that occur in mitochondria. The first reaction, which converts succinate to fumarate, is highly spontaneous, with a large drop in free energy. The second reaction, the hydrogenation of FAD

to $FADH_2$, is nonspontaneous and requires a large input of free energy. The catalyst that couples these two reactions is the enzyme succinate dehydrogenase (Figure 6.13).

When the first reaction takes place in a mitochondrion, the two hydrogen atoms that are removed from succinate are transferred to a molecule of a carrier substance, FAD (flavin adenine dinucleotide). Succinate dehydrogenase couples the exergonic reaction to the endergonic one by ensuring that hydrogen atoms liberated by succinate are used to make $FADH_2$. One site on the enzyme surface binds succinate; another site nearby binds FAD. Every time a succinate ion reacts with succinate dehydrogenase to form fumarate, much of the free energy that is released by this highly exergonic process is immediately trapped and used to synthesize $FADH_2$. $FADH_2$ acts as a carrier of the hydrogen and the chemical free energy until another enzyme couples the exergonic dehydrogenation of $FADH_2$ with the endergonic hydrogenation of some other compound (see Chapter 7).

We have already seen, in Chapter 5, other examples of coupled reactions (see Figures 5.14 and 5.15). In animals the sodium–potassium pump (for primary active transport) is an enzyme that couples the exergonic breakdown of ATP with the endergonic pumping of Na^+ and K^+ against their concentration differences. In secondary active transport, another protein couples the exergonic influx of Na^+ with the endergonic influx of glucose. We will see in Chapter 40 how the contractile proteins of muscle couple the exergonic breakdown of ATP with the performance of mechanical work against a load. In metabolic pathways, there is another type of coupling, in which successive enzyme-catalyzed steps share compounds. A reaction $A + B \rightleftharpoons C$ may be endergonic but still proceed rapidly if the next step, $C + D \rightleftharpoons E$, is so exergonic that the overall reaction ($A + B + D \rightleftharpoons E$) is exergonic.

These examples illustrate an important generalization: Coupled reactions are the major means of carrying out energy-requiring reactions in cells.

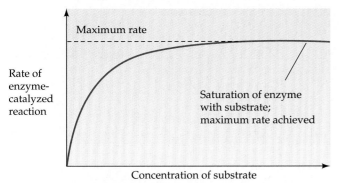

Maximum rate

Saturation of enzyme with substrate; maximum rate achieved

Rate of enzyme-catalyzed reaction

Concentration of substrate

6.12 Enzymes and Reaction Rate
At the maximum reaction rate, all enzyme molecules are tied up with substrate molecules; at this point, adding more substrate would not make the reaction go any faster.

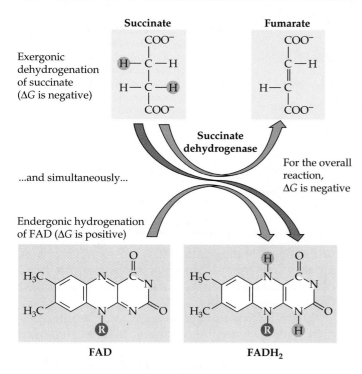

Exergonic dehydrogenation of succinate (ΔG is negative)

...and simultaneously...

Endergonic hydrogenation of FAD (ΔG is positive)

Succinate

Fumarate

Succinate dehydrogenase

For the overall reaction, ΔG is negative

FAD

FADH₂

6.13 Succinate Dehydrogenase Couples Two Reactions

In mitochondria, the enzyme succinate dehydrogenase couples the energy-producing reaction that converts succinate to fumarate with the energy-requiring hydrogenation of FAD (blue arrows). The enzyme enables the transfer of the two hydrogens lost by succinate to FAD, producing $FADH_2$ (red arrow).

MOLECULAR STRUCTURE OF ENZYMES

Until the 1960s biochemists knew little about the behavior of enzymes at the molecular level. It was generally agreed that the substrates of enzymes bind to active sites on the surface of the enzyme molecule, but the actual structure of an active site was not understood. The remarkable ability of an enzyme to select exactly the right substrate was explained by the assumption that the binding of the substrate to the site depends on a precise interlocking of molecular shapes. In 1894 the great German chemist Emil Fischer compared the fit between an enzyme and substrate to that of a lock and key. Fischer's model persisted for more than half a century with only indirect evidence to support it.

The first direct evidence came in 1965, when David Phillips and his colleagues at the Royal Institution in London succeeded in crystallizing the enzyme lysozyme and, using the techniques of X-ray crystallography, determined its structure. Since then, the structures of many other enzymes have been determined by such X-ray diffraction studies. Computers are now programmed to draw proteins from X-ray crystallographic data. This work has revealed a great deal about how the enzyme molecule is designed, how it works, and how it is controlled. Small enzymes consist of a single folded polypeptide chain; large en-

6.14 Some Enzymes Cleave Proteins

The hypothetical polypeptide in this figure is eight amino acids long; its backbone is shaded gray and the names of amino acids are indicated at the bottom of the figure. Arrows point to the peptide linkages cleaved by four different enzymes.

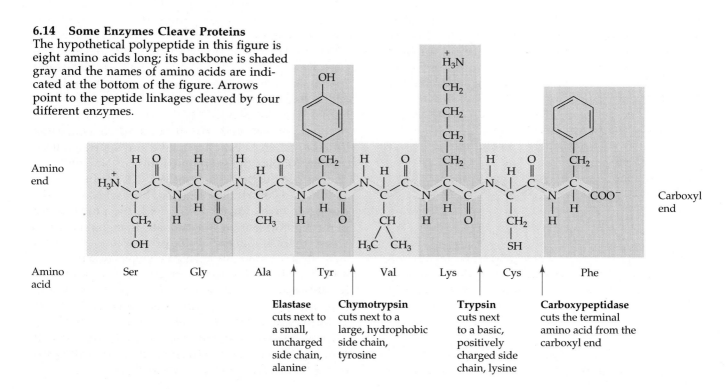

Amino end

Carboxyl end

| Amino acid | Ser | Gly | Ala | Tyr | Val | Lys | Cys | Phe |

Elastase cuts next to a small, uncharged side chain, alanine

Chymotrypsin cuts next to a large, hydrophobic side chain, tyrosine

Trypsin cuts next to a basic, positively charged side chain, lysine

Carboxypeptidase cuts the terminal amino acid from the carboxyl end

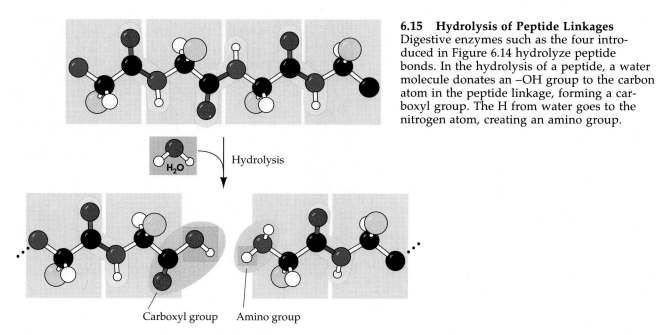

6.15 Hydrolysis of Peptide Linkages
Digestive enzymes such as the four introduced in Figure 6.14 hydrolyze peptide bonds. In the hydrolysis of a peptide, a water molecule donates an –OH group to the carbon atom in the peptide linkage, forming a carboxyl group. The H from water goes to the nitrogen atom, creating an amino group.

zymes may contain several, often identical, polypeptide chains. Frequently the active site contains a metal ion that enhances the reaction, or a small, nonprotein molecule may be attached to the active site. These metals and small molecules are called prosthetic groups and will be discussed later in this chapter.

Structures and Actions of Protein-Digesting Enzymes

Different enzymes may behave generally in the same way and yet have very specific differences in behavior. Consider four enzymes that allow animals to digest proteins: carboxypeptidase, chymotrypsin, trypsin, and elastase. All four break the peptide linkages that connect amino acids in polypeptide chains, but each attacks only a very specific linkage. Carboxypeptidase snips one amino acid at a time from the carboxyl (COOH) end of a chain; the other three enzymes cleave chains at particular places in the middle, as shown in Figure 6.14.

The explanation for this fastidious specificity, which underlies the entire biochemistry of living organisms, lies in the architecture of the enzyme molecules. As Fischer suggested, the structure of the active site is molded to fit the substrate molecule. The binding of the substrate to the active site of the enzyme depends on the same forces that maintain the folded tertiary structure of the enzyme itself: hydrogen bonds, the electrostatic attraction and repulsion of charged chemical groups, and the interaction of hydrophobic groups. Substrates may also be covalently bonded to enzymes. (These forces are described in Chapter 2.)

Protein-digesting enzymes **hydrolyze** a peptide bond—that is, they break the polypeptide chain by adding a water molecule across the peptide bond (Figure 6.15). Carboxypeptidase lowers the energy barrier for this reaction, thus sharply increasing the reaction rate. Because of the specificity of the enzyme, only certain peptide bonds are hydrolyzed at this rapid rate—only those carboxyl-terminal bonds next to bulky hydrophobic side chains that fit comfortably into a hydrophobic pocket in the active site of carboxypeptidase. The other protein-digesting enzymes have different active sites that bind different side chains on the substrate.

Prosthetic Groups and Coenzymes

Although some enzymes consist entirely of one or more polypeptide chains, others possess a tightly bound nonprotein portion called a **prosthetic group.** This may be a single metal ion, a metal ion contained in a small organic molecule, or a **coenzyme,** a complex organic molecule required in some way for the action of one or more enzymes (Figure 6.16). Not all coenzymes are bound as a prosthetic group, however; some are separate and move from enzyme molecule to enzyme molecule.

Some coenzymes assist the catalytic activities of enzymes by accepting or donating electrons or hydrogen atoms (see Chapter 7). Other coenzymes alter the structure of a substrate in such a way that the substrate becomes more reactive. In animals, coenzymes of these two types often are produced from vitamins in the diet. Another group of coenzymes transfers phosphate groups, along with a great deal of free energy, from molecule to molecule. Metal ions attached to enzyme proteins generally function by binding the substrate to the enzyme or by withdrawing electrons from the substrate (Figure 6.17).

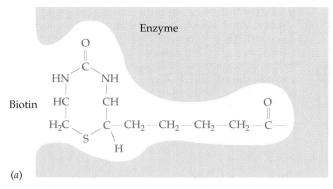

(a)

6.16 Coenzymes

Some enzymes require coenzymes in order to function. (a) The coenzyme biotin bonds covalently to any of several different enzymes that carry a carboxyl group. Because it bonds to its enzyme, this coenzyme is a prosthetic group. (b) The coenzyme NAD (nicotinamide adenine dinucleotide), shown in green, bonds to the enzyme G3PD (glyceraldehyde 3-phosphate dehydrogenase). This illustration shows more realistically than the representation in (a) the relative sizes of enzyme and coenzyme. Hydrogen atoms have been omitted in this drawing, which is colored in the same way as Figure 6.9.

REGULATION OF ENZYME ACTIVITY

Various substances, some occurring naturally in cells and others artificial, act upon enzymes to increase or decrease the rates of enzyme-catalyzed reactions. Those that occur naturally regulate metabolism; the

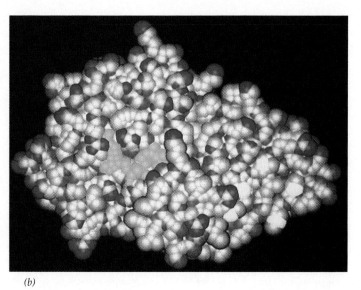

(b)

artificial ones are used either to treat disease or to study how enzymes work. Some substances, called **inhibitors,** that inhibit enzyme-catalyzed reactions produce irreversible effects. Other inhibitors produce reversible effects—that is, these inhibitors can become unbound. Enzymes consisting of multiple subunits are subject to another type of control called allosteric regulation. We will discuss all these types

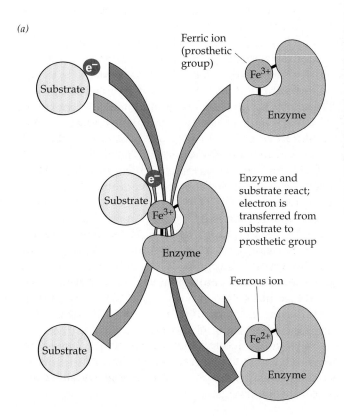

(a)

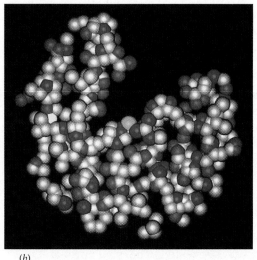

(b)

6.17 Metal Ions as Prosthetic Groups

(a) A ferric ion attached to an enzyme as a prosthetic group. In this reaction the ferric ion (Fe^{3+}) withdraws the electron from the substrate and becomes a ferrous ion (Fe^{2+}); the substrate moves on, altered by the loss of an electron. (b) A tiny part of the enzyme thermolysin, with a zinc ion bound as a prosthetic group (green). Close to the zinc ion is part of a carboxyl group of a glutamic acid. This carboxyl group and the zinc ion collaborate in binding the substrate. The side chain of the glutamic acid having this carboxyl group is thus the part of the enzyme molecule most important for catalysis.

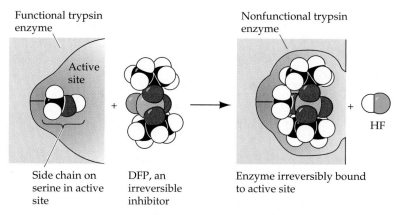

Functional trypsin enzyme

Nonfunctional trypsin enzyme

Active site

+

→

+

HF

Side chain on serine in active site

DFP, an irreversible inhibitor

Enzyme irreversibly bound to active site

6.18 Irreversible Inhibition
DFP disables the digestive enzyme trypsin by reacting with a hydroxyl group belonging to the amino acid serine. After this reaction, trypsin can no longer act on peptide linkages as described in Figure 6.14.

of regulation and conclude this section with a look at the effects of pH and temperature on enzyme activity.

Irreversible Inhibition

Some inhibitors irreversibly modify certain side chains at active sites, ruining the enzymes by destroying their capacity to function as catalysts. An example of such an **irreversible inhibitor** is DFP (diisopropylphosphorofluoridate). DFP reacts with a hydroxyl group belonging to the amino acid serine (see Table 3.1) at the active site of the enzyme trypsin, preventing the use of this side chain in the catalytic mechanism (Figure 6.18).

DFP is an irreversible inhibitor not only for the protein-digesting enzyme trypsin but also for many other enzymes whose active sites contain serine. Among these is acetylcholinesterase, an enzyme that is essential for the orderly propagation of signals from one nerve cell to another (see Chapter 38). Because of their effect on acetylcholinesterase, DFP and other similar compounds are classified as nerve gases.

Reversible Inhibition

Not all inhibitory action is irreversible. Some inhibitor molecules are similar enough to a particular enzyme's natural substrate to bind to the active site, yet different enough that the enzyme catalyzes no chemical reaction. When such a molecule is bound to the enzyme, the natural substrate cannot enter the active site; thus the intruder effectively wastes the enzyme's time, inhibiting its catalytic action. These are called **competitive inhibitors** because they compete with the natural substrate for the active site and block it (Figure 6.19a). The blockage is reversible, however; a competitive inhibitor may become unbound, leaving the active site unchanged. If enough natural sub-

strate molecules are present, they can compete successfully with inhibitor molecules for empty active sites.

Consider the enzyme succinate dehydrogenase, which is subject to competitive inhibition. Recall that this enzyme, found in all mitochondria, removes two hydrogen atoms from succinate to produce fumarate (see Figure 6.13); it then transfers the hydrogens to another molecule, as shown in Figure 6.19b. The other molecules shown in the figure are competitive inhibitors of succinate dehydrogenase. They resemble succinate enough that the enzyme is fooled into binding them. Having bound them, however, the enzyme can do nothing more with them, because the inhibitors are the wrong size and shape or have key chemical groups in the wrong places. The enzyme molecule cannot bind a succinate molecule until the inhibitor molecule has moved out of the active site.

Dissociation of the inhibitor does occur because binding of a competitive inhibitor is reversible, *as is binding of the substrate.* For example, when the competitive inhibitor malonate is added to a solution containing succinate and succinate dehydrogenase, the reaction of succinate to fumarate is slowed. The effect of malonate can be overcome, however, if enough succinate is added. The relative concentrations of substrate and inhibitor determine which of these is more likely to bind to the active site.

Inhibitors that do not react with the active site are called **noncompetitive inhibitors.** Noncompetitive inhibitors bind to the enzyme at a site away from the active site. Their binding causes a conformational change in the protein that alters the active site (Figure 6.19c). The active site still binds substrate molecules, but the rate of product formation is reduced. Noncompetitive inhibitors can become unbound, so their effects are reversible. Because they do not bind to the active site, their effects do not change as substrate concentration changes.

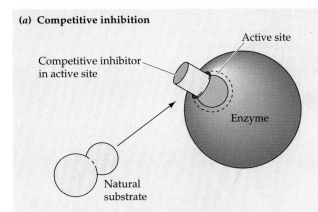

(a) Competitive inhibition

Active site

Competitive inhibitor
in active site

Enzyme

Natural
substrate

The enzyme molecule's function is disabled as long as
the inhibitor remains bound but if the inhibitor becomes
unbound, a substrate molecule may bind to the active site.

(b) Competitive inhibition of succinate dehydrogenase

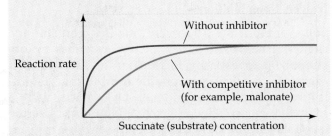

Succinate
(substrate)   + A ⇌   Fumarate   + AH$_2$

Catalyzed by
succinate
dehydrogenase

Competitive inhibitors

Oxalate   Malonate

Oxaloacetate   Glutarate

The above series of molecules of increasing length compete with
succinate for the enzyme's active site. The similarity that fits
them all to the same active site is the presence of two negatively
charged carboxyl groups, one at each end of the molecule.

Reaction rate

Without inhibitor

With competitive inhibitor
(for example, malonate)

Succinate (substrate) concentration

In the absence of a competitive inhibitor, the enzyme increases
reaction rate more than it does when one is present. As substrate
concentrations increase, however, competitive inhibition
becomes less effective and, eventually, completely ineffective.

(c) Noncompetitive inhibition

Substrate

Enzyme

Noncompetitive
inhibitor

Enzyme

A noncompetitive inhibitor may not prevent the substrate from
binding to the active site, but it modifies the active site.

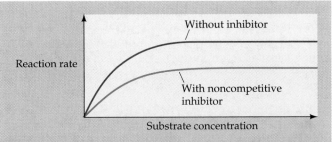

Reaction rate

Without inhibitor

With noncompetitive
inhibitor

Substrate concentration

The modification of the active site slows down the rate at which
the enzyme catalyzes the reaction. Even high concentrations of
substrate do not overcome the effect of the inhibitor; compare
with graph in (b).

6.19   Reversible Inhibition

(a) A competitive inhibitor, in competition with the substrate, can bind the active site and thus can disable the
enzyme. This binding, like that of the substrate, is reversible. (b) Succinate dehydrogenase (see Figure 6.13) is
an example of an enzyme that is subject to competitive
inhibition. Competitive inhibitors resemble the substrate
chemically. (c) A noncompetitive inhibitor binds at a site
away from the active site. The effect may not disable the
enzyme but may slow down the reaction.

Allosteric Enzymes

Many important enzymes are larger and more complex than the ones we have discussed so far, which are individual polypeptides. These complex enzymes have quaternary structures (see Chapter 3) consisting of two or more polypeptide subunits, each with a molecular weight in the tens of thousands.

The activity of these complex enzymes is controlled by molecules, called **effectors,** that may have no similarity either to the reactants or to the products of the reaction being catalyzed. Effectors operate by binding to a site on the enzyme other than the active site. Binding at this **allosteric site** enhances or diminishes reactions at the active site; effectors thus can be activators or inhibitors of enzymes. Because many effector–substrate pairs differ structurally, this phenomenon is called allostery, meaning "different shape." Enzymes subject to allosteric control are called **allosteric enzymes;** all have two or more subunits.

Allosteric enzymes and single-subunit enzymes differ greatly in their effects on reaction rates when the substrate concentration is low. Graphs of reaction rates plotted against substrate concentration show this difference. For an enzyme with a single subunit, the plot looks like that in Figure 6.20*a* (see also Figure 6.12). The reaction rate first increases very sharply with increasing substrate concentration, then tapers off to a constant maximum rate as the supply of enzyme becomes saturated with substrate. The plot for an allosteric enzyme is radically different (Figure 6.20*b*), with a sigmoidal (S-shaped) appearance. The increase in reaction rate with increasing substrate concentration is slight at low substrate concentrations, but there is a range over which the reaction rate is extremely sensitive to relatively small changes in the substrate concentration. Because of this sensitivity, allosteric enzymes are important in fine-tuning the activities of a cell. We can understand this behavior in terms of the structure of an allosteric enzyme.

Mechanism of Allosteric Effects

An allosteric enzyme has not only more than one subunit, but more than one *type* of subunit: The **catalytic subunit** has an active site that binds the enzyme's substrate; the **regulatory subunit** has one or more allosteric sites that bind specific effector molecules. A molecule of an allosteric enzyme usually consists of two or more catalytic subunits and two or more regulatory subunits. An allosteric enzyme exists in two or more distinct forms with different catalytic efficiencies, and these forms are in equilibrium with each other. In the simple cases we will examine, the **active form** has full catalytic activity, whereas the **inactive form** is totally without activity (Figure 6.21). In the active form, the active sites on the catalytic subunits can bind substrate and convert it to product. In the inactive form, the active sites have been distorted in such a way that they cannot bind substrate; however, the allosteric sites are able to bind an effector, which we will consider in this example to be an inhibitor. The regulatory subunits of the active form of the enzyme have deformed allosteric sites and cannot bind effector. When neither substrate nor inhibitor is present, the active and inactive forms convert rapidly back and forth in equilibrium (column 1 in Figure 6.21), the equilibrium constant being characteristic of the given enzyme.

What happens when inhibitor or substrate is present? If substrate is present (top of column 2 in Figure 6.21), some of the substrate binds to the active sites of active enzyme molecules; while the enzyme–substrate complex exists, those enzyme molecules cannot be converted to the inactive form. (The presence of a substrate molecule in either active site prevents the enzyme molecule from being converted to the inactive form.) Inactive molecules, however, are being converted to active ones at the same rate as before, so an increase in the concentration of active enzyme results from the presence of substrate! In addition, because each active enzyme molecule (in this example) has two active sites, one enzyme molecule can bind two substrate molecules and simultaneously catalyze reactions of both of them. This explains the upward curvature at the lower left of a plot of reaction rate versus substrate concentration for an alloste-

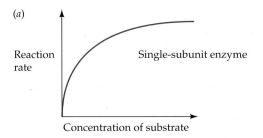

(a)

Reaction rate

Single-subunit enzyme

Concentration of substrate

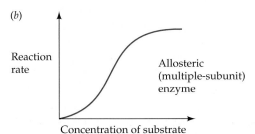

(b)

Reaction rate

Allosteric (multiple-subunit) enzyme

Concentration of substrate

6.20 Allostery and Reaction Rate
How the rate of an enzyme-catalyzed reaction changes with increasing substrate concentration depends on whether the enzyme consists of one or more than one polypeptide subunit.

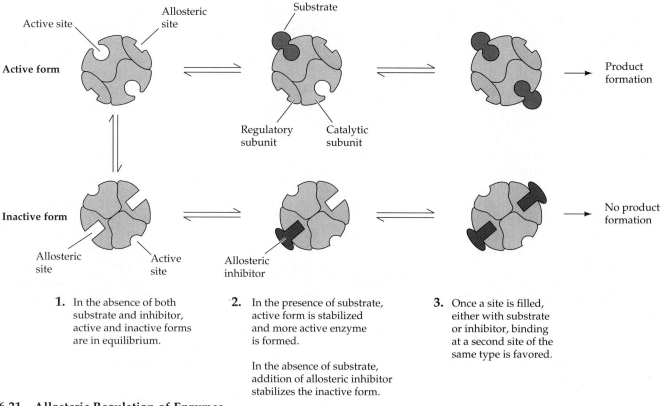

1. In the absence of both substrate and inhibitor, active and inactive forms are in equilibrium.

2. In the presence of substrate, active form is stabilized and more active enzyme is formed.

   In the absence of substrate, addition of allosteric inhibitor stabilizes the inactive form.

3. Once a site is filled, either with substrate or inhibitor, binding at a second site of the same type is favored.

6.21 Allosteric Regulation of Enzymes
The hypothetical enzyme shown here has four subunits, two catalytic (blue) and the other two regulatory. When the enzyme is in its active form, the active sites on the catalytic subunits can accept substrate. When the enzyme is in its inactive form, the allosteric sites on the regulatory subunits can accept inhibitor.

ric enzyme (see Figure 6.20): Increasing the substrate concentration increases the availability of active enzyme and of active sites and hence rapidly accelerates the reaction rate.

If inhibitor is present (bottom of column 2 in Figure 6.21), the concentration of active enzyme *decreases* and the reaction is thus inhibited. The inhibitor binds to the allosteric site of the *inactive* form of the enzyme, preventing the conversion of the inactive to the active form; but the conversion of the active to the inactive enzyme is not affected by the inhibitor. The overall effect is to decrease the concentration of the active form and thus inhibit the enzymatic reaction. An allosteric activator works in a similar way, except that it binds to the regulatory subunit of the *active* form of the enzyme and holds it in the active configuration. Note that allosteric inhibitors and activators do not modify the structure of the enzyme; rather, they interfere with its conversion to another form with which it is normally in equilibrium.

This mechanism of allosteric inhibition and activation was proposed by the French molecular biologist Jacques Monod and his colleagues in 1965. Yet another of Monod's many contributions to biology is described in Chapter 12.

Control of Metabolism through Allosteric Effects

An organism's metabolism is the sum total of the biochemical reactions that take place within it. These reactions proceed along **metabolic pathways,** which are sequences of enzyme-catalyzed reactions. In the sequences, the product of one reaction is the substrate for the next. Some pathways synthesize, step-by-step, the important chemical building blocks from which macromolecules are built; others trap energy from the environment; still others have different functions. Some metabolic pathways are branched. In a branched pathway, one or more of the intermediate substances are acted upon by more than one enzyme and thus are sent through more than one metabolic branch.

At the branching points where metabolic pathways diverge, **regulatory enzymes** catalyze reactions. These regulatory enzymes are like switches. What flips such a switch? The end product of a branch pathway may block the initial step in that branch pathway, reducing the formation of the end product (Figure 6.22). This illustrates the principle of **negative feedback,** also evident, for example, in thermostats on furnaces. (Negative feedback is discussed in detail

6.22 Feedback in Metabolic Pathways

The reactions C to D and C to H are the first committed steps in the branch pathways leading to end products G and J, respectively. These end products can block the first committed steps by acting as allosteric inhibitors of the enzymes catalyzing the reactions. Thus, if levels of products G and J build up beyond what the cell needs for other reactions, the extra molecules provide the negative feedback that turns off the synthesis of G and J.

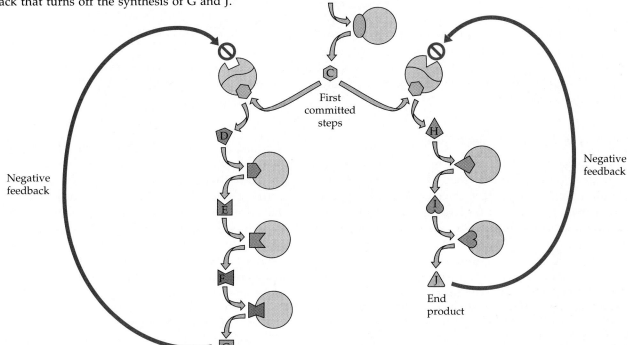

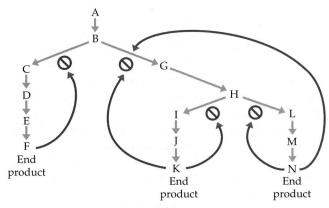

6.23 Concerted Feedback Inhibition

A variety of feedback controls come into play in branching metabolic pathways. Inhibiting the B → C reaction, for example, shunts reactant B into the B → G step. In some cases in which a single compound (H) gives rise to more than one end product (K and N), both end products can join in concerted inhibition of the enzyme catalyzing the first reaction committed to the production of H (B → G); each end product may also inhibit its own branch pathway.

in Chapters 35 and 36.) The end product of a particular pathway typically is an allosteric inhibitor of the regulatory enzyme catalyzing the first **committed step** in its own synthesis—that is, the earliest step in the branched pathway that leads only to the synthesis of that end product and no other. The first committed steps in metabolic pathways are particularly effective points for feedback control. For instance, inhibition of the B-to-C step in Figure 6.23 shunts all the reactants into the other branch of the pathway, whereas inhibition of the C-to-D reaction, one step later, leads only to a possibly harmful and certainly wasteful buildup of substance C. Living things generally do not accumulate unneeded intermediates.

When two different end products, produced by different branches of a pathway, are both present in excessive concentrations, they may act together to inhibit an earlier branch-point enzyme—one that catalyzes the first committed step for formation of these two products (see the B-to-G step in Figure 6.23). **Concerted feedback inhibition** like this results in further efficiency: In this example intermediates G and H do not build up as they would if only the steps to I and L were inhibited by their individual end products. Concerted feedback inhibition requires that the enzyme have two allosteric sites, both of which must be bound to inhibitors to stop the enzyme's activity.

Allosteric regulation is very effective: It allows rapid adjustment to short-term changes in metabolism or in the environment. The activities of enzyme molecules are adjusted by their interactions with small molecules, the end products. If a particular enzyme is not needed, might it not be a good idea simply to stop making it until it is needed? Wouldn't it be advantageous to be able to regulate *production* as well enzyme *activity*? This is indeed the case, and the regulation of enzyme synthesis plays an important role in controlling development and metabolism (see Chapters 12 and 17).

Sensitivity of Enzymes to the Environment

Enzymes enable cells to perform reactions under mild conditions, unlike the extremes of temperature and pH employed by chemists in the laboratory. Enzymes are extremely sensitive to changes in the medium around them. For example, the rates of most enzyme-catalyzed reactions depend on the pH of the medium in which they occur. Each enzyme is most active at a particular pH; its activity decreases as the solution is made more acidic or more basic (Figure 6.24).

Several factors contribute to this effect. One is the ionization of carboxyl, amino, and other groups on either the substrate or the enzyme. Carboxyl groups (—COOH) ionize to become negatively charged carboxylate ions (—COO$^-$) in neutral or basic solutions. Similarly, amino groups (—NH$_2$) accept H$^+$ ions in neutral or acidic solutions, becoming positively charged ammonium ions (—NH$_3{}^+$). This means, for

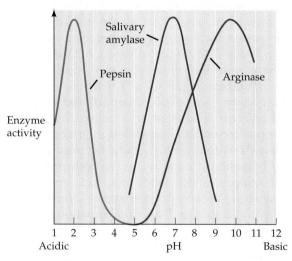

6.24 pH Affects Enzyme Activity
Each enzyme catalyzes reactions most efficiently at a particular pH, as shown by the peaks of the activity curves for three enzymes.

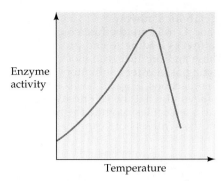

6.25 Temperature Affects Enzyme Activity
An enzyme is most active at a particular temperature.

example, that in a neutral solution a molecule with an amino group is attracted electrically to another molecule that has a carboxyl group because both groups are ionized and they have opposite charges. The attraction does not occur in acidic solution (in which the carboxyl group is not ionized) or in basic solution (in which the amino group is not ionized). Different enzymes function best at different pH values. Evolution has matched enzymes to their environment—for example, digestive enzymes that act in the stomach work best at the very low pH values that prevail in the stomach after a meal.

Temperature also has a profound effect on enzyme activity (Figure 6.25). At low temperatures, warming increases the rate of an enzyme-catalyzed reaction because at higher temperatures a greater fraction of the reactant molecules have enough energy to provide the activation energy of the reaction. Temperatures that are too high, however, inactivate enzymes because at high temperatures enzyme molecules vibrate and twist so rapidly that some of their noncovalent bonds break. The heat destroys their tertiary structure, and the enzyme molecules lose their activity. Enzymes become permanently inactivated, or **denatured,** at certain temperatures. Some enzymes denature at temperatures only slightly above that of the human body; a very few are stable even at the boiling point of water. A graph of enzyme activity versus temperature peaks at the **optimal temperature** for the enzyme. Above the optimal temperature, inactivation of enzyme molecules predominates.

Organisms adapt to changes in the environment in many ways, one of which is based on groups of enzymes, called **isozymes,** that catalyze the same reaction but have differing physical properties. Isozymes may be chemically similar to one another (made from different combinations of the same subunits, for example) or totally unrelated. Within a given set, different isozymes may have different optimal temperatures. An example is the enzyme acetylcholinesterase in the rainbow trout. If a rainbow trout is transferred from relatively warm water to near-freezing water (2°C), the fish produces an isozyme of acetylcholinesterase that is different from the acetylcholinesterase isozyme produced at the higher temperature. The new isozyme has a lower optimal temperature than does the previously formed one, which helps the fish to perform normally in the colder water.

ENZYMES, RIBOZYMES, AND ABZYMES

This chapter is about enzymes: biological catalysts that are proteins. However, enzymes may not have been the first catalytic macromolecules to evolve. As we will see in Chapters 13 and 18, the first biological catalysts may have been RNA molecules. Catalytic RNAs, or **ribozymes,** still function today. Also, by immunizing mice with synthetic analogs of substrates, we can cause them to produce novel antibodies (antibodies are another class of proteins; see Chapter 16) that have modest catalytic activity. Catalytic antibodies, or abzymes, have potential as "designer" catalysts.

In Chapters 7 and 8 we will see how enzymes that catalyze the reactions of two crucial sets of pathways—cellular respiration and photosynthesis—provide cells and organisms with the energy they need to live, grow, and reproduce. Enzymes and other biological catalysts will appear again and again as we continue our examination of the living world.

SUMMARY of Main Ideas about Enzymes, Catalysis, and Energy

Enzymes are biological catalysts that speed up chemical reactions in cells.

An enzyme's active site binds its substrate.
> **Review Figure 6.8**

The tertiary structure of some enzymes changes as they bind their substrates.
> **Review Figure 6.10**

Inhibitors block the functioning of enzymes.
> **Review Figures 6.18 and 6.19**

The rate of an enzyme-catalyzed reaction is affected by the concentrations of substrates and by the temperature and pH of the medium.
> **Review Figures 6.12, 6.24, and 6.25**

Forward and reverse reactions reach a point of chemical equilibrium at which they both proceed at the same rate.
> **Review Figure 6.3**

Every chemical reaction has its own equilibrium constant, K_{eq}, which is related to the change in free energy, ΔG.
> **Review Table 6.1**

Enzymes speed up reactions by lowering the energy barrier between reactants and products.
> **Review Figures 6.5, 6.6, and 6.11**

Exergonic reactions are spontaneous, but endergonic reactions proceed only if free energy is provided.
> **Review Figure 6.4.**

Many energy-requiring reactions in cells proceed because some enzymes couple exergonic and endergonic reactions.
> **Review Figure 6.13**

Prosthetic groups, such as coenzymes or metal ions, help some enzymes catalyze reactions.

Allosteric (multiple-subunit) enzymes and single-subunit enzymes behave differently.
> **Review Figures 6.20 and 6.21**

Feedback regulation is one type of control for regulating metabolic pathways.
> **Review Figures 6.22 and 6.23**

Two laws of thermodynamics, describing energy in any closed system, underlie what happens in a chemical reaction.

The first law of thermodynamics states that the quantity of energy in the system remains constant: energy cannot be created or destroyed.

The second law of thermodynamics states that the quantity of free (usable) energy in the system decreases and the quantity of entropy (disorder) increases.

Free energy, heat, temperature, and entropy are related by the equation

$$\Delta G = \Delta H - T\Delta S$$

SELF-QUIZ

1. Which statement about energy is incorrect?
   a. It can neither be created nor destroyed.
   b. It is the capacity to do work.
   c. All its conversions are fully reversible.
   d. In the universe as a whole, the amount of free energy decreases.
   e. In the universe as a whole, the amount of entropy increases.

2. Which statement about thermodynamics is incorrect?
   a. Free energy is given off in an exergonic reaction.
   b. Free energy can be used to do work.
   c. A spontaneous reaction is exergonic.
   d. Free energy tends always to a minimum.
   e. Entropy tends always to a minimum.

3. In a chemical reaction,
   a. the rate depends on the equilibrium constant.
   b. the rate depends on the activation energy.
   c. the entropy change depends on the activation energy.
   d. the activation energy depends on the equilibrium constant.
   e. the change in free energy depends on the activation energy.

4. Which statement is *not* true of enzymes?
   a. They consist of proteins, with or without a nonprotein part.
   b. They change the rate constant of the catalyzed reaction.
   c. They change the equilibrium constant of the catalyzed reaction.
   d. They are sensitive to heat.
   e. They are sensitive to pH.

5. The active site of an enzyme
   a. never changes shape.
   b. forms no chemical bonds with substrates.
   c. determines, by its structure, the specificity of the enzyme.
   d. looks like a lump projecting from the surface of the enzyme.
   e. changes the equilibrium constant of the reaction.

6. A prosthetic group
   a. is a tightly bound, nonprotein part of an enzyme.
   b. is composed of protein.
   c. does not participate in chemical reactions.
   d. is present in all enzymes.
   e. is an artificial enzyme.

7. The rate of an enzyme-catalyzed reaction
   a. is constant under all conditions.
   b. decreases with an increase in substrate concentration.
   c. cannot be measured.
   d. depends on the equilibrium constant.
   e. can be reduced by inhibitors.

8. Which statement is *not* true of enzyme inhibitors?
   a. A competitive inhibitor binds the active site of the enzyme.
   b. An allosteric inhibitor binds a site on the active form of the enzyme.
   c. A noncompetitive inhibitor binds elsewhere than the active site.
   d. Noncompetitive inhibition cannot be completely overcome by adding more substrate.
   e. Competitive inhibition can be completely overcome by adding more substrate.

9. Which statement is *not* true of feedback inhibition of enzymes?
   a. It is exerted through allosteric effects.
   b. It is directed at the enzyme catalyzing the first committed step in a branch of a pathway.
   c. Concerted feedback inhibition is based on two or more end products.
   d. It acts very slowly.
   e. It is an example of negative feedback.

10. Which statement is *not* true of temperature effects?
    a. Raising the temperature may reduce the activity of an enzyme.
    b. Raising the temperature may increase the activity of an enzyme.
    c. Raising the temperature may denature an enzyme.
    d. Some enzymes are stable at the boiling point of water.
    e. The isozymes of an enzyme have the same optimal temperature.

FOR STUDY

1. How is it possible for endergonic reactions to occur in organisms?

2. Consider two proteins: One is an enzyme dissolved in the cytosol; the other is an ion channel in a membrane. Contrast the structures of the two proteins, indicating at least two important differences.

3. Plot free energy versus the course of a reaction for an endergonic reaction and for an exergonic reaction. Include the activation energy in both plots. Label E_a and ΔG in both graphs.

4. Consider an enzyme that is subject to allosteric regulation. If a competitive inhibitor (not an allosteric inhibitor) is added to a solution of such an enzyme, the ratio of enzyme molecules in the active form to those in the inactive form increases. Explain this observation.

READINGS

Dressler, D. and H. Potter. 1991. *Discovering Enzymes*. W. H. Freeman, New York. How enzymes are studied; nicely illustrated.

Harold, F. 1986. *The Vital Force: A Study of Bioenergetics*. W. H. Freeman, New York. A detailed introduction to the study of energy and life.

Karplus, M. and J. A. MacCammon. 1986. "The Dynamics of Proteins." *Scientific American*, April. This article will correct any misconception of proteins as rigid molecules; it describes the constant, rapid changes in local shape that underlie the functioning of proteins.

Koshland, D. E., Jr. 1973. "Protein Shape and Biological Control." *Scientific American*, October. This paper shows that the ability of proteins to change shape in specific circumstances underlies the control and coordination of biological processes.

Newsholme, E. A. and C. Start. 1973. *Regulation of Metabolism*. John Wiley & Sons, New York. A rigorous treatment of regulation of enzyme activity, with emphasis on feedback control.

Stryer, L. 1995. *Biochemistry*, 4th Edition. W. H. Freeman, New York. Good discussion of the structure of proteins.

Cells convert energy from one form to another as they carry out the business of life. Plants, as we will see in the next chapter, convert light energy into chemical energy—food for themselves and for the animals that eat them. Some organisms convert the chemical energy of food into light, as does this small deep-sea animal, called a sea walnut. Instead of using chemical energy from food only to grow or reproduce, the sea walnut illuminates its environment. Why does it spend energy in such an apparently frivolous way? In fact, nobody knows for sure—it has been suggested that the animal uses its light to frighten off potential predators, but not many biologists believe this. The function of light emission by sea walnuts remains a mystery. We do know a lot about how the energy of food is made available, however.

GLYCOLYSIS, CELLULAR RESPIRATION, AND FERMENTATION

Organisms must have energy to live. How do they get it? Organisms draw their energy from four biochemical processes that power the machinery of life: photosynthesis, glycolysis, cellular respiration, and fermentation (Figure 7.1). These are the biochemical pathways. These processes consist of metabolic pathways made up of many small chemical steps. In photosynthesis, plants and some other autotrophs use light energy to synthesize food compounds, as we will discuss in Chapter 8. In this chapter we are concerned with the extraction of energy from food molecules as practiced by both heterotrophs *and* autotrophs.

Glycolysis prepares food for cellular respiration. A series of preparatory reactions convert food molecules into a compound in cellular respiration. The glycolytic pathway is also the first part of fermentation. Whether the process that follows glycolysis is cellular respiration or the rest of fermentation depends on the type of organism extracting the energy and on whether the environment is **aerobic** (containing oxygen gas, O_2) or **anaerobic** (lacking oxygen gas). Glycolysis takes place in either case. Fermentation, which occurs primarily in cells where the oxygen supply is limited or depleted, converts food molecules to waste products such as lactic acid or ethanol, releasing some energy. Cellular respiration, which requires oxygen, releases much more energy from a given amount of food than does fermentation. The pathways that release energy in cells are regulated by a system of allosteric feedback control.

The combined operation of glycolysis and cellular respiration is the biological equivalent of burning the sugar glucose. When glucose is burned with a match, it yields carbon dioxide, water, and energy in the form of heat and light. In cells, glucose is broken

Releasing Energy for a Novel Purpose
This sea walnut uses some of its energy to generate its own light in the deep ocean.

7

Pathways that Release Energy in Cells

136

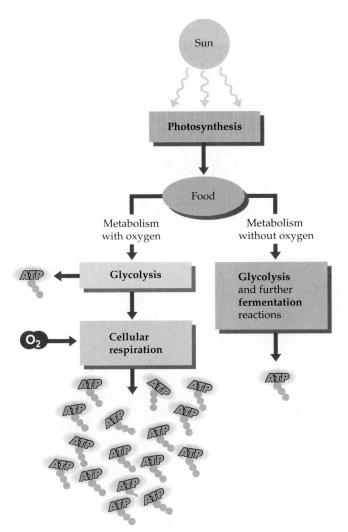

ATP = Energy for work

7.1 Energy for Life

Organisms obtain energy through four types of metabolism. In photosynthesis, autotrophic organisms use light energy to synthesize food compounds. Heterotrophic organisms process these food compounds by glycolysis, fermentation, and cellular respiration. Fermentation is particularly important to cells in which oxygen content is low or depleted, but it releases much less energy from a given amount of food than does cellular respiration. Glycolysis precedes cellular respiration and is the first part of fermentation.

down to the same products, but much of the released energy, rather than being lost as light and large amounts of heat, is trapped in the energy-storage compound **ATP** (adenosine triphosphate). In this chapter we show how cells use enzymes to "burn" glucose and then harness the released energy as ATP—the "spendable energy" referred to earlier.

The complete biological "combustion" of glucose takes place as coupled reactions. To understand these events we must first learn the biochemistry of ATP, which is the cell's principal compound for short-term energy storage and for energy transfers within the cell.

ATP

All living cells rely on the ATP molecule for the short-term storage of energy and the transfer or application of energy to do work. Bioluminescence is a visually dramatic example of the use of ATP (Box 7.A). ATP is a sort of energy currency. How do cells spend, make, and save this currency in the course of their activities?

Spending, Making, and Saving ATP

To "spend" a molecule of its ATP, a cell must break the molecule, releasing the energy of one of its bonds. Many different enzymes can catalyze the breakdown of ATP, whose structure is shown in Figure 7.2. The breakdown (hydrolysis) of ATP yields **ADP** (adenosine diphosphate) and an inorganic phosphate ion. This breakdown is an exergonic reaction, yielding approximately 12 kcal of free energy per mole of ATP under biological conditions. This is enough energy to drive typical endergonic reactions in the cell.

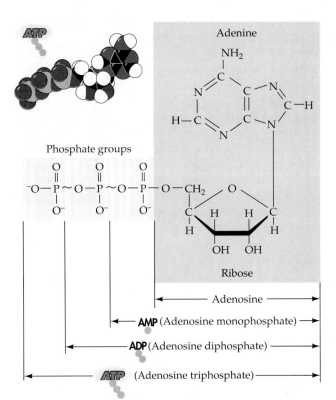

7.2 Structure of ATP

An ATP molecule consists of the base adenine bonded to ribose (a sugar), with three phosphate groups bonded to another carbon on the ribose. The compounds ADP and AMP are like ATP but have one and two fewer phosphate groups, respectively. Adenosine consists of adenine bonded to ribose, without phosphate groups. When the high-energy bonds (color) between the phosphate groups are broken, energy is released.

BOX 7.A

Some Organisms Use ATP to Make Light

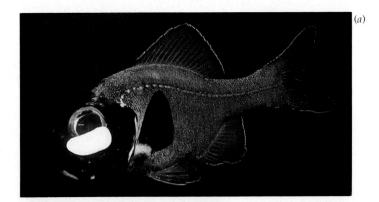
(a)

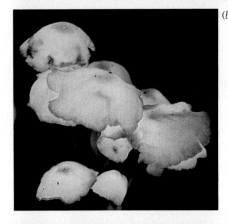
(b)

(c)

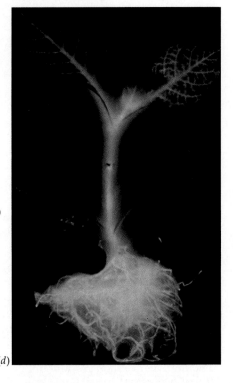
(d)

Bioluminescence—the production of light by living organisms—always requires ATP as an energy source. Although we know that a firefly uses its bioluminescence to communicate with fireflies of the opposite sex, we are in the dark as to the role of light in most bioluminescent species. Bioluminescence has evolved independently in many kinds of organisms, from bacteria to vertebrates. The comb jelly shown at the beginning of this chapter is a bioluminescent ctenophore. The glowing mushrooms in (b) are a bioluminescent fungus growing in the Costa Rican rainforest.

Many bioluminescent organisms live within the cells or tissues of other organisms. Their hosts then appear to emit light. For example, bioluminescent bacteria populate the kidney-shaped organ below the eye of the "flashlight" fish in (a). Another type of bioluminescent bacterium lives within the nematode worms that, in turn, infest the tissues of the caterpillars in (c).

In an example of bioluminescence "engineered" by scientists, the insertion of a gene from a firefly into tobacco plants produced a dazzling display when the plants were watered

with an appropriate substrate (d). The plant's ATP provides the energy for the reaction catalyzed by an enzyme encoded in the firefly gene. This triumph of recombinant DNA

technology (Chapter 14) affords a powerful tool for studying such diverse topics as development, gene expression, cellular energetics, and the movement of proteins within

---

ADP, which possesses less free energy than does ATP, can combine with a phosphate ion to make a new molecule of ATP if enough energy is provided by an exergonic reaction. The formation of ATP from ADP and a phosphate ion is endergonic and consumes as much free energy as is released by the breakdown of ATP. Many different enzyme-catalyzed reactions in the cell can provide the energy to convert ADP to ATP. The main ones in eukaryotes, however,

are the reactions of cellular respiration, in which the maximum amount of energy is released from food molecules and trapped as the stored energy of ATP. Later in this chapter we will see in detail exactly how cells produce ATP.

Figure 7.3 summarizes how ATP traps and releases energy. An exergonic reaction is coupled with the formation of ATP, and at a later time or in another part of the cell, an endergonic reaction is coupled

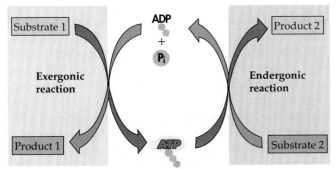

7.3 Formation and Use of ATP
Energy released by an exergonic reaction can be contained when it is used to meld ADP and an inorganic phosphate ion (P_i) into a molecule of ATP. The ATP molecule can then drive an endergonic reaction by splitting into ADP and P_i, releasing energy in the process.

with the splitting of the ATP to ADP and a phosphate ion.

An active cell requires millions of molecules of ATP per second to drive its biochemical machinery. Even so, the cell diverts some of its ATP into synthesizing long-term energy-storage compounds. For this purpose, plants synthesize starch, a long-chain polymer of glucose (see Chapter 3), and sometimes fats. Animals store energy in glycogen (another glucose polymer) and fats. Of course, any large molecule synthesized by the cell (a protein, for example) is a storehouse of energy, but energy storage may not be its primary function. ATP can be considered the circulating currency of energy exchange in living organisms, and starches and fats the savings accounts. When animals need energy, they draw on their deposits of fat and carbohydrates. They break down these deposits into carbon dioxide and water, forming ATP from the energy released in the process. Similarly, plants draw on their stored fats or on deposits of starch, which they convert to glucose; plants then break the glucose down to carbon dioxide and water while forming ATP.

The Energy Content of ATP

To understand the respiratory pathways, you need to have a feeling for why such large changes in free energy accompany the formation and hydrolysis of ATP. ATP is a phosphate ester whose hydrolysis releases much more free energy than most other esters release when hydrolyzed. (An **ester** is an organic compound produced by the reaction of an alcohol with an acid.)

For the hydrolysis of ATP to ADP and phosphate, the change in free energy, ΔG, is about -12 kcal/mol at the temperature, pH, and substrate concentrations typical of living cells. (The "standard" ΔG for ATP hydrolysis is generally given as -7.3 kcal/mol, but that value is valid only at pH 7 and with ATP, ADP,

and phosphate present at concentrations of 1 M—conditions that differ greatly from those found in cells.) The hydrolysis of most other phosphate esters produces considerably less than half as much free energy as the hydrolysis of ATP. By transferring phosphate groups, ATP can "prime" other compounds for future chemical reactions.

Part of the unusually large free energy of the hydrolysis of ATP comes from the large number of negative charges near each other on its neighboring phosphate groups. When ATP is hydrolyzed, the charges are spread over two molecules—ADP and inorganic phosphate—which can get far apart; these products are thus more stable than the ATP molecule with its cluster of negative charges. The hydrolysis of ADP to AMP (adenosine monophosphate) and inorganic phosphate liberates an even slightly greater amount of free energy than does that of ATP to ADP, although this energy is not often harnessed for work. Hydrolyzing the last phosphate group (that is, converting AMP to adenosine) does not spread out the negative charges any further, so the free-energy change is low, similar to that for any other ester hydrolysis. Because of their larger free energies of hydrolysis, the first and second bonds broken in ATP are sometimes called high-energy bonds, although "high-energy" refers to the energy of hydrolysis and not to any intrinsic energy of the bond itself. The high-energy bond is sometimes symbolized by $\sim$, and ATP can be written A—P$\sim$P$\sim$P, where P represents an entire phosphate group. The phosphate ion, HPO_4^{2-}, is often abbreviated P_i, meaning inorganic phosphate.

As one might expect, given the large amount of free energy released in the hydrolysis of ATP, the reverse reaction—making ATP from ADP and P_i—requires a substantial input of energy. The two major mechanisms of ATP production arise from reactions in which molecules transfer electrons or hydrogen atoms to other molecules. How do cells manage these transfers?

THE TRANSFER OF HYDROGEN ATOMS AND ELECTRONS

Certain pathways of energy metabolism release hydrogen atoms that must be captured and passed on to other reactions. A hydrogen atom consists of a proton and an electron. The transfer of electrons is an oxidation–reduction reaction, or **redox reaction**. The *gain* of one or more electrons by an atom, ion, or molecule is called **reduction**. The *loss* of one or more electrons is called **oxidation**. Although oxidation and reduction are always defined in terms of traffic in *electrons*, we must also think in these terms when hydrogen atoms (not hydrogen ions) are gained or lost, because transfers of hydrogen atoms

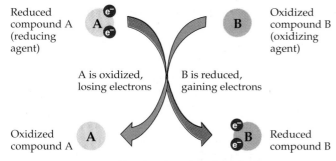

Reduced compound A (reducing agent)

Oxidized compound B (oxidizing agent)

A is oxidized, losing electrons

B is reduced, gaining electrons

Oxidized compound A

Reduced compound B

7.4 Oxidation and Reduction Are Coupled

As compound A is oxidized, compound B is reduced. In the process A loses electrons and B gains electrons. In this redox reaction, A is the reducing agent because it donates electrons and B is the oxidizing agent because it accepts electrons.

involve transfers of electrons. Thus when a molecule loses hydrogen atoms, the molecule becomes oxidized.

Oxidation and reduction *always* occur together: As one material is oxidized, the electrons it loses are transferred to another material, reducing it. In a redox reaction, we call the reactant that becomes reduced an **oxidizing agent** and the one that becomes oxidized a **reducing agent** (Figure 7.4). An oxidizing agent accepts electrons; a reducing agent gives up electrons. In the process of oxidizing the reducing agent, the oxidizing agent itself becomes reduced. Conversely, the reducing agent becomes oxidized as it reduces the oxidizing agent. Energy is transferred in the reaction, with energy originally present in the reducing agent becoming associated with the reduced product. ΔG is negative as long as the overall redox reaction is spontaneous. As we will see, some of the key reactions of cellular respiration are highly exergonic redox reactions.

At some early stage in evolution, organisms began to form reducing agents and oxidizing agents. Natural selection favored the use of certain of these

agents (the ones whose redox reactions have suitable values of ΔG) as a system for the orderly exchange of electrons, analogous to the use of the ATP–ADP system for the orderly transfer of energy. We have already encountered an example of such agents at work in cells: In Chapter 6 we saw that FAD accepts hydrogens during the respiratory conversion of succinate to fumarate; in that reaction, FAD is an oxidizing agent and $FADH_2$ is a reducing agent.

The main electron banking system is based on the compound **NAD** (nicotinamide adenine dinucleotide; Figure 7.5), which exists in two chemically distinct forms: one oxidized (NAD^+) and the other reduced ($NADH + H^+$). The function of NAD is to carry hydrogen atoms (with their high-energy electrons) and free energy from compounds being oxidized and to give up hydrogen atoms (with their electrons) and free energy to compounds being reduced (Figure 7.6). The reduction

$$NAD^+ + 2\ H \rightarrow NADH + H^+$$

is accompanied by a free energy increase of 52.4 kcal/mol if oxygen gas is the final oxidizing agent:

$$NADH + H^+ + \tfrac{1}{2}\ O_2 \rightarrow NAD^+ + H_2O$$

$$\Delta G = -52.4\ \text{kcal/mol}$$

(Note that the oxidizing agent appears here as "½ O_2" instead of "O." This is to emphasize that it is oxygen gas, O_2, that takes part in the reaction.) In the same way that ATP can be thought of as a means of packaging free energy in bundles of about 12 kcal/mol, NAD can be thought of as a means of packaging approximately 50-kcal/mol bundles.

Various energy carriers are chemically related. Be-

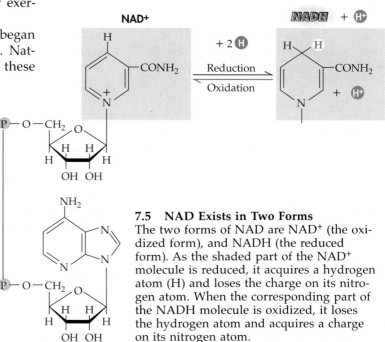

7.5 NAD Exists in Two Forms

The two forms of NAD are NAD^+ (the oxidized form), and NADH (the reduced form). As the shaded part of the NAD^+ molecule is reduced, it acquires a hydrogen atom (H) and loses the charge on its nitrogen atom. When the corresponding part of the NADH molecule is oxidized, it loses the hydrogen atom and acquires a charge on its nitrogen atom.

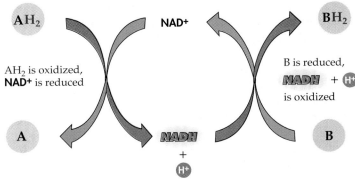

7.6 NAD Is an Energy Carrier

As compound AH_2 is oxidized, it releases its hydrogen atoms with their electrons to NAD^+, reducing NAD^+ to $NADH + H^+$. Elsewhere $NADH + H^+$ may reduce compound B to BH_2, at which time NADH is oxidized to NAD^+. Thanks to its ability to carry electrons and free energy, NAD is a major and universal energy intermediary in cells.

cause part of the NAD molecule looks very much like a molecule of ATP (see Figures 7.2 and 7.5), it is easy to imagine that several energy carriers made up of adenine, ribose, phosphates, and other groups have evolved over time from a common (and less efficient) precursor molecule.

The structures of NAD and some other carrier molecules include compounds that we humans need but cannot synthesize for ourselves; these are classified as vitamins. Nicotinamide, which is part of NAD, forms directly from nicotinic acid, or niacin, a member of the vitamin B complex. Another member of this same vitamin complex is riboflavin, which is part of a carrier called **FAD** (flavin adenine dinucleotide), which we will encounter frequently. We need only small amounts of vitamins because these carrier molecules are recycled through the metabolic machinery. Vitamins are discussed more fully in Chapter 43.

HOW DO CELLS PRODUCE ATP?

In many contexts in this and the next chapter we will be discussing the production of ATP by cells. Two basic mechanisms accomplish the production of ATP: substrate-level phosphorylation and the chemiosmotic mechanism. Every living cell produces at least part of its ATP by **substrate-level phosphorylation**, the enzyme-catalyzed transfer of phosphate groups from donor molecules to ADP molecules, driven by

energy obtained from oxidation. Consider, for example, the following pair of reactions in glycolysis. One intermediate in the glycolytic pathway, glyceraldehyde 3-phosphate, reacts with P_i and NAD^+, becoming 1,3-bisphosphoglycerate. In this enzyme-catalyzed reaction, an aldehyde is oxidized to a carboxylic acid, with NAD^+ acting as the oxidizing agent. The oxidation provides so much energy that the newly added phosphate group is linked to the rest of the molecule by a bond with even higher energy than the high-energy bond of ATP (Figure 7.7a). A second enzyme catalyzes the transfer of this phosphate group from 1,3-bisphosphoglycerate to ADP, forming ATP (Figure 7.7b). Both reactions are exergonic, even though a substantial amount of energy is consumed in the formation of ATP.

Free-floating enzymes catalyze phosphorylation at the substrate level. By contrast, the **chemiosmotic mechanism** requires the participation of molecules that are embedded in membranes. In the chemiosmotic mechanism, protons are pumped across a membrane, effectively charging a "battery." The pump causes a difference in proton concentration (pH) across the membrane. Because the protons are positively charged and the membrane has a low permeability to anions, a difference in electric charge also builds up across the membrane. One side of the membrane is then electrically negative to the other side. The proton concentration gradient and the charge difference together constitute a **proton-motive force** that tends to drive the protons back across the membrane, just as the charge on a battery drives the flow of electrons, discharging the battery. The dis-

(a) **Oxidation of substrate**

(b) **Transfer of phosphate to ADP**

7.7 Enzymes Catalyze Substrate-Level Phosphorylation

(a) Enzyme I catalyzes the oxidation of a substrate; in this example, glyceraldehyde 3-phosphate is oxidized, becoming 1,3-bisphosphoglycerate. 1,3-Bisphosphoglycerate contains a phosphate group linked to the molecule by a high-energy bond. (b) A second enzyme catalyzes the transfer of this phosphate group to ADP, forming ATP.

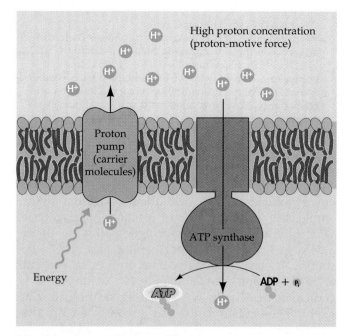

High proton concentration
(proton-motive force)

Proton
pump
(carrier
molecules)

ATP synthase

Energy

ADP + P$_i$

ATP

7.8 Membranes Support Chemiosmotic ATP Production

The energy-driven transport of electrons from one carrier in a membrane to another pumps protons across the membrane, establishing a proton-motive force that tends to drive the protons back. Protons return across the membrane only by using ATP synthases. These membrane proteins operate only when ADP is present; they couple energy from the discharge of the proton-motive force with the formation of ATP from ADP and P$_i$.

charge of the proton-motive force is prevented by the impermeability of most of the membrane to protons. Protons can return across the membrane only by passing through **ATP synthases**, membrane protein complexes that permit protons to flow only if ADP is present. The energy released by the discharge of the proton-motive force is coupled with the endergonic formation of ATP.

What pumps the protons across the membrane in the first place? High-energy electrons are passed from one carrier in the membrane to another. The individual redox reactions are exergonic, and some of the energy is used to pump protons against the gradient of pH and electric charge (Figure 7.8). In cellular respiration, the electrons derive from food molecules; in photosynthesis, the electron transfers are initiated by the energy of sunlight. We will consider these electron transfers in more detail later in this chapter and in Chapter 8.

The chemiosmotic mechanism plays several key roles in the living world. It is responsible for most of the production of ATP in cellular respiration; it participates in the trapping of light energy in ATP formation in photosynthesis (Chapter 8); it provides the energy for driving the propeller-like motion of bacterial flagella (Chapter 4). These three systems differ in detail, but each includes the pumping of protons

across a membrane and the return of the protons across the membrane, tightly coupled with the formation of ATP.

THE RELEASE OF ENERGY FROM GLUCOSE

The energy-extracting processes of cells—glycolysis, cellular respiration, and fermentation—may be divided into pathways that we will consider one at a time. These pathways also tend to be separated physically in the cell (Figure 7.9); they evolved separately, and perhaps at different times.

Glycolysis consists of the glycolytic pathway. This near-universal process was probably the first energy-releasing pathway to evolve; if any earlier pathway existed, it has disappeared from Earth. Today virtually all living cells, even the most evolutionarily ancient, use glycolysis. It is this pathway in which glucose is metabolized to **pyruvate** (pyruvic acid). The glycolytic pathway contains an oxidative step in which an electron carrier, NAD$^+$, becomes reduced, acquiring electrons. Each molecule of glucose processed through glycolysis produces a net yield of two molecules of ATP. The major products of glycolysis are ATP (which the cell will use to drive endergonic reactions), pyruvate, and the two electrons acquired by NAD. Both the pyruvate and the electrons must be processed further.

Cellular respiration is made of up two pathways: the citric acid cycle and the respiratory chain. In eukaryotes in the presence of oxygen and in some bacteria, the pyruvate from glycolysis is oxidized in a cyclical series of respiratory reactions known as the **citric acid cycle** (also called the Krebs cycle or the tricarboxylic acid cycle). In eukaryotes, the reactions of the citric acid cycle are catalyzed by enzymes present in the liquid matrix inside the mitochondrion. Several steps release the carbon atoms of pyruvate (originally the carbon atoms of glucose) as carbon dioxide (CO$_2$) molecules and transfer more electrons to carriers. The products of the citric acid cycle are, then, carbon dioxide, which must be eliminated in some way from the organism, and many more stored electrons (along with accompanying hydrogen nuclei) than are produced in glycolysis. As we are about to see, more stored electrons means a greater ultimate harvest of ATP.

Hydrogen is an outstanding fuel. When it reacts with oxygen, a great deal of free energy is released; better still, the "waste" product of this reaction—water—is no problem either to the environment or to any organism that produces it. In both glycolysis and the citric acid cycle, hydrogen atoms are acquired by the molecules they reduce. Through these pathways most of the energy originally present as the covalent bonds of glucose becomes associated with reduced NAD (NADH + H$^+$).

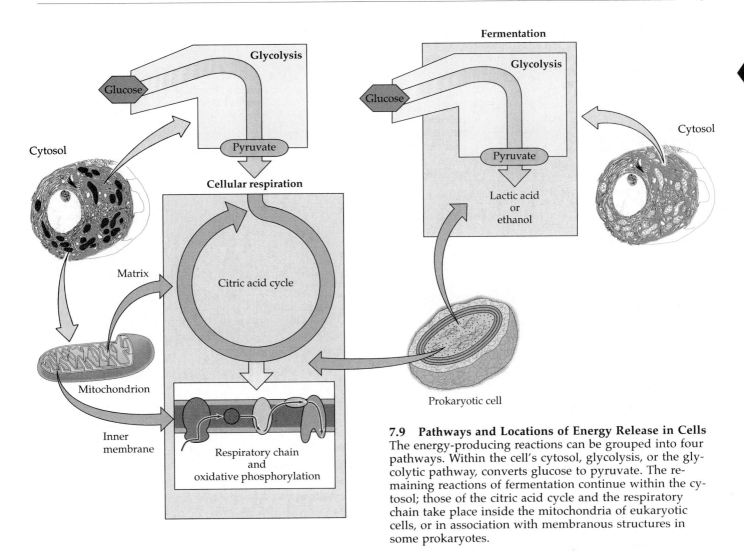

7.9 Pathways and Locations of Energy Release in Cells

The energy-producing reactions can be grouped into four pathways. Within the cell's cytosol, glycolysis, or the glycolytic pathway, converts glucose to pyruvate. The remaining reactions of fermentation continue within the cytosol; those of the citric acid cycle and the respiratory chain take place inside the mitochondria of eukaryotic cells, or in association with membranous structures in some prokaryotes.

The principal role of the **respiratory chain** is to release energy from reduced NAD in such a way that it may be used to form ATP. This pathway is a series of successive redox reactions in which hydrogen atoms—or, in the later steps, electrons derived from hydrogen atoms—are passed from one type of membrane carrier to another and finally are allowed to react with oxygen gas to produce water. In eukaryotes, the carriers (and associated enzymes) are bound to the folds of the inner mitochondrial membranes, the **cristae** (Figure 7.10 and Chapter 4). In both prokaryotes and eukaryotes, the transfer of electrons along the respiratory chain drives a chemiosmotic mechanism (see Figure 7.8). This is the way in which the vast majority of the ATP in animals is formed.

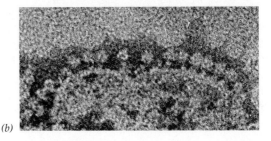

7.10 Cellular Powerhouses

(a) Numerous cristae, arising from the inner of the two mitochondrial membranes, reach into the matrix of these mitochondria. The cristae are the sites of the ATP-producing reactions of cellular respiration. (b) A high-magnification view of the inner mitochondrial membrane. Spherical "knobs" project into the mitochondrial matrix; these protein knobs catalyze the synthesis of ATP.

7.11 Glycolysis Converts Glucose to Pyruvate

Starting with hexokinase, ten enzymes catalyze ten reactions in turn. Along the way, ATP is produced (reactions 7 and 10), and two molecules of NAD^+ are reduced to $2\ NADH + 2\ H^+$ (reaction 6).

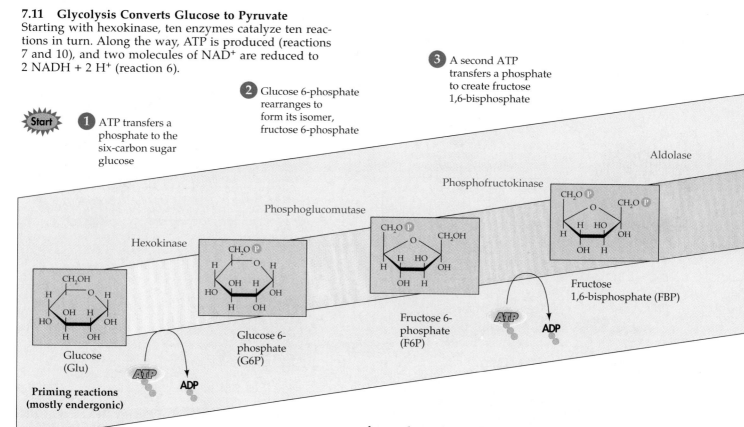

Start

1 ATP transfers a phosphate to the six-carbon sugar glucose

2 Glucose 6-phosphate rearranges to form its isomer, fructose 6-phosphate

3 A second ATP transfers a phosphate to create fructose 1,6-bisphosphate

Aldolase

Phosphofructokinase

Phosphoglucomutase

Hexokinase

Glucose (Glu)

Glucose 6-phosphate (G6P)

Fructose 6-phosphate (F6P)

Fructose 1,6-bisphosphate (FBP)

Priming reactions (mostly endergonic)

The chemiosmotic formation of ATP during the operation of the respiratory chain is called **oxidative phosphorylation**.

As the energy is released, the reduced NAD ($NADH + H^+$) and other agents of electron transfer are oxidized. They may then be reused in glycolysis and the citric acid cycle, steadily draining off hydrogen atoms and allowing those pathways to continue operating. This oxidation of $NADH + H^+$ is thus another consequence of the respiratory chain. Overall, the inputs to the respiratory chain are stored hydrogen atoms and oxygen gas (O_2), and the outputs are water and stored energy in the form of ATP.

If we are deprived of oxygen for too long, we die because the respiratory chain cannot function. Without oxygen molecules as receptors, the carriers in our mitochondrial cristae are unable to jettison the hydrogen atoms or electrons bound to them. Soon no oxidized carriers are available; when that happens, glycolysis and the citric acid cycle stop. With no glycolysis, no citric acid cycle, and no respiratory chain activity, we have insufficient ATP. Without ATP, our cells cannot maintain their structure and metabolism, and we die.

Our muscle cells have an alternative way to rid themselves of the hydrogen atoms produced during glycolysis: The hydrogens are passed right back to the end product of glycolysis (pyruvate), and lactic acid is formed. Because the hydrogen atoms are removed, the electron carriers that held them are oxidized and can be used again to process more glucose. Thus even in the absence of oxygen, glycolysis continues (sometimes at an increased rate), without the activity of the citric acid cycle or the respiratory chain, and ATP continues to be produced. The cells that have in their cytosol the enzymatic machinery necessary for this reaction are thereby enabled to function for a time in the absence of oxygen. Eventually, however, the concentration of lactic acid in muscles reaches a toxic level. This anaerobic production of ATP, which releases only a small part of the energy for organisms that require oxygen, is called **fermentation**. For some organisms that live entirely without oxygen, fermentation is the sole pathway that can combine with glycolysis to release energy. We will examine fermentation in more detail later in this chapter.

GLYCOLYSIS

A molecule of glucose taken in by a cell enters the glycolytic pathway, which consists of ten reactions that gradually convert the six-carbon glucose molecule into two molecules of the three-carbon compound pyruvic acid (Figure 7.11). These reactions are accompanied by the *net* formation of two molecules of ATP and by the reduction of two molecules of

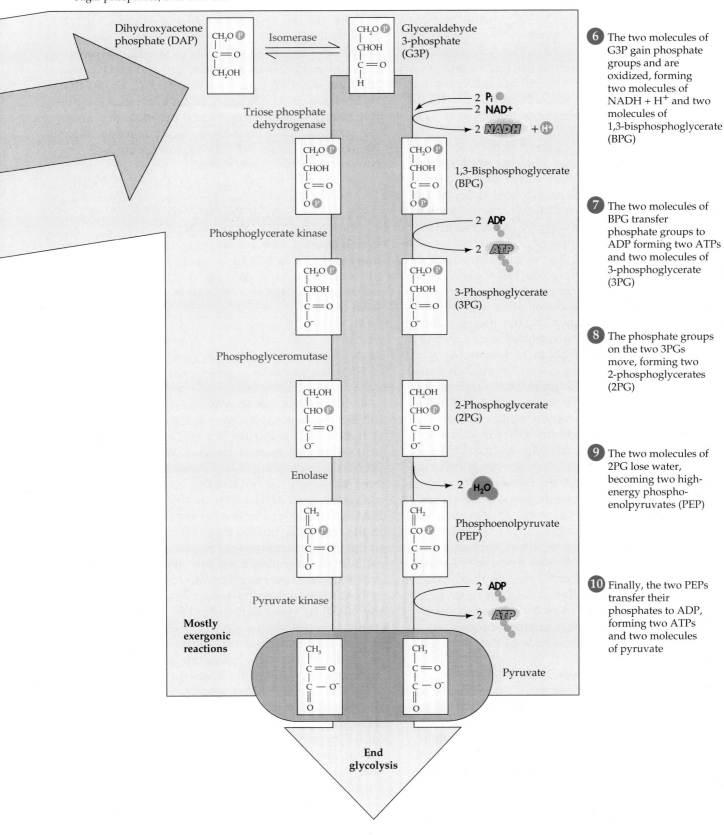

4 The fructose ring opens, and the six-carbon fructose 1,6-bisphosphate breaks into two different three-carbon sugar phosphates, DAP and G3P

5 Dihydroxyacetone phosphate rearranges to form its isomer, glyceraldehyde 3-phosphate (G3P)

Dihydroxyacetone phosphate (DAP)

Isomerase

Glyceraldehyde 3-phosphate (G3P)

Triose phosphate dehydrogenase

2 P_i
2 **NAD+**
2 *NADH* + H^+

6 The two molecules of G3P gain phosphate groups and are oxidized, forming two molecules of NADH + H^+ and two molecules of 1,3-bisphosphoglycerate (BPG)

1,3-Bisphosphoglycerate (BPG)

Phosphoglycerate kinase

2 **ADP**
2 *ATP*

7 The two molecules of BPG transfer phosphate groups to ADP forming two ATPs and two molecules of 3-phosphoglycerate (3PG)

3-Phosphoglycerate (3PG)

Phosphoglyceromutase

8 The phosphate groups on the two 3PGs move, forming two 2-phosphoglycerates (2PG)

2-Phosphoglycerate (2PG)

Enolase

2 H_2O

9 The two molecules of 2PG lose water, becoming two high-energy phospho-enolpyruvates (PEP)

Phosphoenolpyruvate (PEP)

Pyruvate kinase

2 **ADP**
2 *ATP*

10 Finally, the two PEPs transfer their phosphates to ADP, forming two ATPs and two molecules of pyruvate

Mostly exergonic reactions

Pyruvate

End glycolysis

NAD$^+$ to two molecules of NADH + H$^+$. At the end of the pathway, energy is located in ATP, and four hydrogen atoms are passed on in a reducing agent. The fate of the pyruvic acid depends on the type of cell carrying out glycolysis and on whether the environment is aerobic or anaerobic. The fate of the NADH + H$^+$, too, is variable. In most cases, NADH + H$^+$ will be oxidized through the respiratory chain to yield water and NAD$^+$—a chain of reactions that results in the formation of much more ATP (three molecules of ATP per molecule of NADH + H$^+$). In fermentation, however, NADH + H$^+$ is reoxidized to NAD$^+$ either by pyruvic acid itself or by one of its metabolites, with no further storage of free energy. In either case, glycolysis may be regarded as a series of *preparatory* reactions, to be followed either by the citric acid cycle or by the remainder of fermentation.

With the help of Figure 7.11, we can trace our way through the glycolytic pathway. The first five reactions may be viewed as "pump-priming." Each of these five reactions is endergonic, taking up free energy; the cell is *investing* free energy rather than gaining it during the early reactions of glycolysis. Two molecules of ATP are invested in attaching two phosphate groups to the sugar (reactions 1 and 3), thereby raising its free energy by about 15 kcal/mol (Figure 7.12). The phosphate groups will be used to make new molecules of ATP. Although both of these first steps of glycolysis use ATP as one of the substrates, each is catalyzed by a different, specific enzyme. The enzyme hexokinase catalyzes the reaction 1, in which glucose receives a phosphate group from ATP. (A *kinase* is any enzyme that catalyzes the transfer of phosphate group from ATP to another substrate.) In reaction 2, the six-membered glucose ring is rearranged to a five-membered fructose ring; then the enzyme phosphofructokinase adds a second phosphate (taken from another ATP) to the sugar ring (reaction 3). The sugar ring with its two phosphates is opened, and the six-carbon sugar bisphosphate is cleaved to give two different three-carbon sugar phosphates (reaction 4). In reaction 5, one of these sugar phosphates (dihydroxyacetone phosphate) is converted into a second molecule of the other (glyceraldehyde 3-phosphate).

By this time, the halfway point in glycolysis, the following things have happened: Two molecules of ATP have been *used* in the priming reactions, and the six-carbon glucose molecule has been converted into two molecules of a three-carbon sugar phosphate. No ATP has been gained, and nothing has been oxidized; in short, it looks as if we are going determinedly backward.

Now, however, the pump is primed and things begin to happen rapidly. In what follows, remember that each step occurs twice for each glucose molecule going through glycolysis, because each molecule has

by now been split into two molecules of three-carbon sugar phosphate, both of which go through the remaining steps of glycolysis. Reaction 6 is a two-step reaction catalyzed by the enzyme triose phosphate dehydrogenase. The end product of reaction 6 is 1,3-bisphosphoglycerate (1,3-bisphosphoglyceric acid). A phosphate ion has been snatched from the surroundings (but not, this time, from ATP) and tacked onto the three-carbon compound. Figure 7.12 shows that this reaction is accompanied by an enormous drop in free energy—more than 100 kcal per mole of glucose is released in this extremely exergonic reaction. What has happened here? Why the big energy change? The conversion of a sugar to an acid

$$R-\overset{\overset{\textstyle O}{\|}}{C}-H + (O) \rightarrow R-\overset{\overset{\textstyle O}{\|}}{C}-OH$$

is an oxidation and, as you know, oxidations are very exergonic. (Note that here we do *not* write $\frac{1}{2}O_2$ because in this case oxygen gas does not participate in the reaction.) The formation of the phosphate ester from the acid

$$R-\overset{\overset{\textstyle O}{\|}}{C}-OH + HPO_4^{2-} \rightarrow R-\overset{\overset{\textstyle O}{\|}}{C}-O-\overset{\overset{\textstyle O}{\|}}{\underset{\underset{\textstyle O^-}{|}}{P}}-O^- + H_2O$$

is slightly endergonic, but not nearly enough to offset the drop in free energy from the oxidation.

If this big energy drop were simply the loss of heat, glycolysis would be an extremely inefficient process for providing useful energy to the cell. However, this energy is not lost but is used to make two molecules of NADH + H$^+$ from two molecules of NAD$^+$. This stored energy is regained later—either in the respiratory chain, by the formation of ATP, or in the last step of fermentation when pyruvate or its product is reduced and the two molecules of NADH + H$^+$ are restored once again to NAD$^+$. This cycling of NAD is necessary to keep glycolysis going; if all the NAD$^+$ is converted to NADH + H$^+$, glycolysis comes to a halt.

The remaining steps of glycolysis in Figure 7.11 are simpler. The two phosphate groups of 1,3-bisphosphoglycerate are transferred, one at a time, to molecules of ADP, with a rearrangement in between. As a result of this substrate-level phosphorylation, more than 20 kcal of free energy are stored in ATP for every mole of 1,3-bisphosphoglycerate broken down. Finally, we are left with pyruvic acid (pyruvate)—2 mol for each mole of glucose that entered glycolysis.

A review of the reactions shows that at the beginning of glycolysis two molecules of ATP are used per molecule of glucose, but that ultimately four are produced (two for each of the two 1,3-bisphosphoglycer-

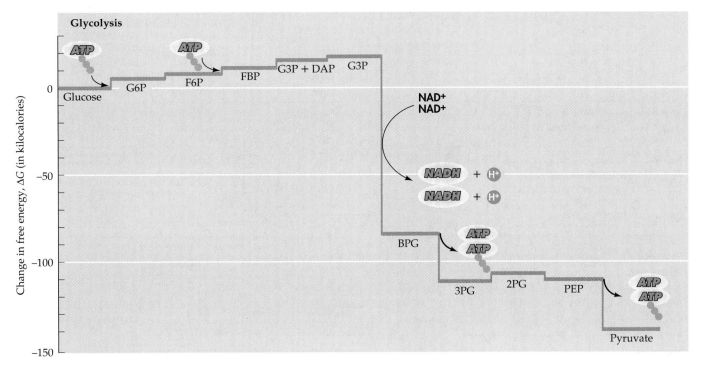

7.12 Free Energy Changes during Glycolysis
Each reaction of glycolysis changes the free energy available, as shown by the differing energy levels of the series of reactants and products from glucose to pyruvate. Note at the upper left that the investment of each of two ATPs in the priming reactions raises the free energy content of the sugar. Key energy-releasing reactions account for the largest drops in free energy. The quantities of NADH + H$^+$ and ATP in the diagram reflect those produced from one molecule of glucose by the reactions of glycolysis. The net gain is 2 ATP, and 2 NADH + 2 H$^+$ are released.

ates)—a net gain of two ATP molecules and two NADH + H$^+$. Under anaerobic conditions, the total usable energy yield from the metabolism of glucose is usually two ATP molecules. The NADH + H$^+$ is rapidly recycled to NAD$^+$ by fermentation for reuse by the triose phosphate dehydrogenase of glycolysis. In the presence of oxygen, on the other hand, eukaryotes and some bacteria are able to reap far more energy by the further metabolism of pyruvate and by reoxidizing the reduced NAD of glycolysis through the respiratory chain.

[By now, you might be wondering why we are using words like *pyruvate* and *pyruvic acid* interchangeably. At pH values commonly found in cells, the ionized form—pyruvate—is present rather than the acid—pyruvic acid. Similarly, all carboxylic acids are present as ions (the *–ate* forms) at these pHs. Thus on grounds of chemical accuracy and simplicity, it is better to name the negative ion than the acid. However, custom often prevails over accuracy, and the acids are often named instead; for example, nobody seems to want to change the name *citric acid cycle* to the more accurate *citrate cycle*.]

THE BEGINNING OF CELLULAR RESPIRATION: THE CITRIC ACID CYCLE

The end product of glycolysis, pyruvate, is the starting point of two different pathways: the citric acid cycle of cellular respiration, and the fermentation pathway (see Figure 7.9). In this section we take a look at the citric acid cycle, in which pyruvate is incinerated to carbon dioxide (CO_2).

Figure 7.12 shows that the metabolism of glucose to pyruvate is accompanied by a drop in free energy of about 140 kcal/mol. About one-third of this energy is captured in the formation of ATP and reduced NAD. Oxidizing the pyruvate yields additional free energy for biological work. The citric acid cycle takes pyruvate and breaks it down to CO_2, using the hydrogen atoms to reduce carrier molecules and to pass chemical free energy to those carriers. The reduced carriers are later oxidized in the respiratory chain, which we discuss in the next section; and an enormous amount of free energy is transferred from the reduced carriers to ATP in the process. The principal inputs to the citric acid cycle are pyruvic acid, water, and oxidized electron carriers; the principal outputs are carbon dioxide and reduced electron carriers:

pyruvic acid ($C_3H_4O_3$) + 3 H_2O + 5 carrier →
3 CO_2 + 5 carrier · (2 H)

Overall, then, for each molecule of pyruvate, during the citric acid cycle three carbons are removed as CO_2 and five pairs of hydrogen atoms are used to reduce carrier molecules, with the simultaneous storage of

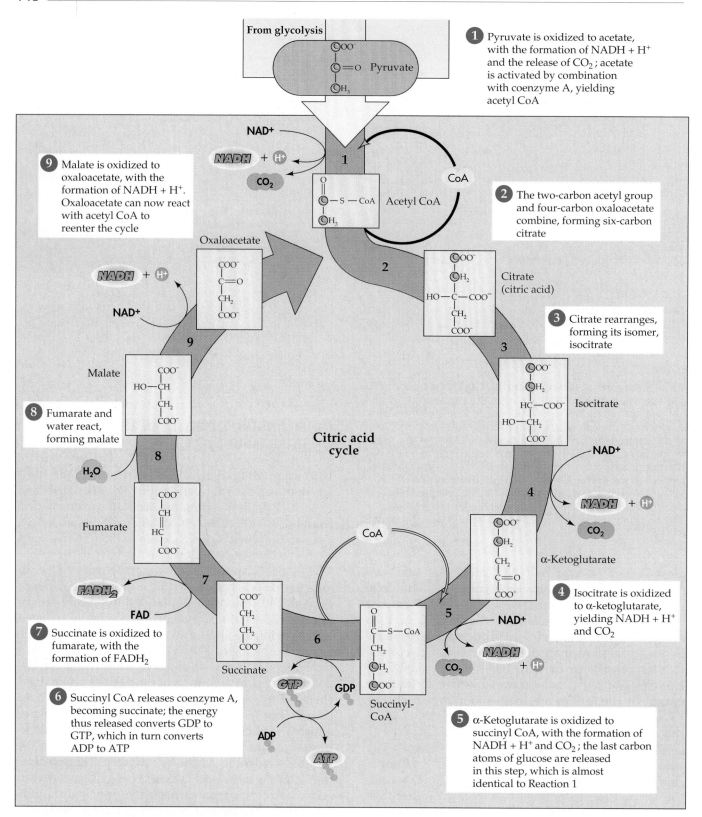

From glycolysis

Pyruvate

1 Pyruvate is oxidized to acetate, with the formation of NADH + H$^+$ and the release of CO_2; acetate is activated by combination with coenzyme A, yielding acetyl CoA

9 Malate is oxidized to oxaloacetate, with the formation of NADH + H$^+$. Oxaloacetate can now react with acetyl CoA to reenter the cycle

Acetyl CoA

CoA

2 The two-carbon acetyl group and four-carbon oxaloacetate combine, forming six-carbon citrate

Oxaloacetate

Citrate (citric acid)

3 Citrate rearranges, forming its isomer, isocitrate

Malate

Isocitrate

8 Fumarate and water react, forming malate

Citric acid cycle

4 Isocitrate is oxidized to α-ketoglutarate, yielding NADH + H$^+$ and CO_2

Fumarate

CoA

α-Ketoglutarate

7 Succinate is oxidized to fumarate, with the formation of FADH$_2$

Succinate

Succinyl-CoA

6 Succinyl CoA releases coenzyme A, becoming succinate; the energy thus released converts GDP to GTP, which in turn converts ADP to ATP

5 α-Ketoglutarate is oxidized to succinyl CoA, with the formation of NADH + H$^+$ and CO_2; the last carbon atoms of glucose are released in this step, which is almost identical to Reaction 1

7.13 The Citric Acid Cycle

The first reaction produces the two-carbon acetyl CoA. Notice that the two carbons from acetyl CoA are traced with color through reaction 5, after which they may be at either end of the molecule (note the symmetry of succinate and fumarate). Reactions 1, 4, 5, 7, and 9 accomplish the major overall effect of the cycle—the storing of energy—by passing electrons to the carrier molecule NAD. Reaction 6 also stores energy.

energy. The energy-storing reactions of the cycle are a major reason for its existence.

At the beginning of the citric acid cycle, pyruvate is oxidized (yielding useful free energy and CO_2) and converted into an activated form of acetic acid, CH_3COOH, called **acetyl CoA** (acetyl coenzyme A). Then acetyl CoA, with two carbon atoms in its acetate group, reacts with a four-carbon acid (oxaloacetate) to form the six-carbon compound citric acid (citrate). The remainder of the cycle consists of a series of enzyme-catalyzed reactions in which citric acid is degraded, leading to the production of useful free energy from redox reactions, to the release of two of the carbons as CO_2, and to the production of a new four-carbon molecule of oxaloacetate from the other four carbons. This new oxaloacetate can react with a second acetyl CoA, producing a second molecule of citrate, and so forth. Acetyl CoA comes into the cycle from pyruvate, CO_2 goes out, the rest of the compounds in the cycle are used and replaced, and energy from redox reactions is stored. As we describe the citric acid cycle in detail, concentrate on how it is maintained in a steady state—that is, with material entering and leaving and with intermediate compounds like succinate and malate turning over constantly but without changing concentration. (The concept of steady state is important; be sure it is clear to you.)

As you read the next paragraphs, you can follow the sequence of steps in the citric acid cycle in Figure 7.13. The product of reaction 1, acetyl CoA, is 7.5 kcal/mol higher in energy than simple acetate. (Acetyl CoA can donate acetate to acceptors such as oxaloacetate much as ATP can donate phosphate to various acceptors.) There are three steps in this first reaction: (1) pyruvate is oxidized to acetate, with the release of CO_2; (2) part of the energy from this oxidation is saved by reducing NAD^+ to $NADH + H^+$; and (3) some of the remaining energy is stored temporarily by combining the acetate with CoA. An analogous three-step reaction occurs in glycolysis when glyceraldehyde 3-phosphate is converted to 1,3-bisphosphoglycerate (see Figure 7.11, reaction 6). In that reaction, a sugar is oxidized to an acid, some of the energy released by oxidation is stored in $NADH + H^+$, and some of the remaining energy is preserved in a second phosphate bond in the molecule. A good metabolic idea is likely to appear more than once; we will see this one again, later in the citric acid cycle. As you might guess, a complex set of steps such as those in the reaction from pyruvate to acetyl CoA requires more than one type of catalytic protein. In fact, this reaction is catalyzed by the *pyruvate dehydrogenase complex*, which consists of 72 subunits—24 each of three different types of protein, for a total molecular weight of 4.6 million. This complex is an impressive example of biological organization.

The energy temporarily stored in acetyl CoA helps to drive the formation of citrate from oxaloacetate (reaction 2). During this reaction, the coenzyme molecule falls away, to be recycled and bound to another acetate by the pyruvate dehydrogenase complex. Citrate is rearranged to isocitrate (reaction 3). In reaction 4, a CO_2 molecule and two hydrogen atoms are removed in the conversion of isocitrate to α-ketoglutarate. As Figure 7.14 indicates, this reaction produces a large drop in free energy. The released energy is stored in $NADH + H^+$ and can be recovered later in the respiratory chain, when the $NADH + H^+$ is reoxidized.

Like reaction 1 (the oxidation of pyruvate to acetyl CoA), reaction 5 of the citric acid cycle is complex. The α-ketoglutarate molecule is oxidized to succinate, CO_2 is given off, some of the oxidation energy is stored in $NADH + H^+$, and some is preserved temporarily by combining succinate with CoA. This temporarily stored energy is saved in reaction 6, in which GTP (guanosine triphosphate) is first made and then used to make ATP—another example of substrate-level phosphorylation. A smaller amount of free energy is released in reaction 7, when two hydrogens are transferred to an enzyme containing FAD (an oxidizing agent similar to NAD^+ that we discussed earlier in this chapter); one more NAD^+ reduction (reaction 9) occurs after a molecular rearrangement (reaction 8). The oxaloacetate that remains after all these reactions is ready to combine with another acetyl CoA molecule and go around the cycle again. Bear in mind that the citric acid cycle operates twice for each glucose molecule that enters glycolysis.

CONTINUATION OF CELLULAR RESPIRATION: THE RESPIRATORY CHAIN

Without oxidizing agents to be reduced and act as electron carriers, the oxidative steps of glycolysis and the citric acid cycle could not occur. Two such agents—the crucial molecules NAD^+ and FAD—are regenerated by the respiratory chain. The reaction of substrate and oxidizing agent can be represented as follows

$$substrate \cdot H_2 \qquad NAD^+$$
$$Substrate \qquad NADH + H^+$$

with the hydrogen and its high-energy electrons being passed from the originally reduced substrate (such as malate in the citric acid cycle) to the oxidizing agent NAD^+. We see that the presence of the oxidizing agent is critical, because without it to accept electrons the substrates could not be oxidized, and there would be no respiratory metabolism.

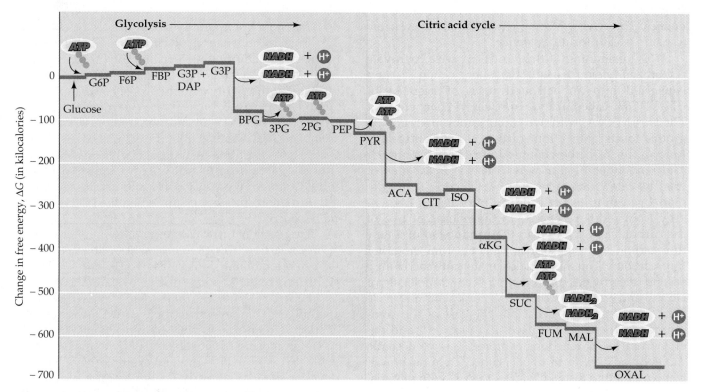

7.14 The Citric Acid Cycle Releases Much More Free Energy Than Glycolysis Does
Electron carriers (NAD in glycolysis; NAD and FAD in the citric acid cycle) are reduced and ATP is generated by reactions coupled to reactions producing major drops in free energy.

But what about all that NADH? If this reaction continued, it appears that all the cell's NAD⁺ would become reduced, leaving none to act as an oxidizing agent. Fortunately, there is something in most cells—a specific oxidizing agent—that can reoxidize the NADH + H⁺. This agent is a carrier called ubiquinone (Q). Q acts as follows to oxidize NADH + H⁺:

$$NADH + H^+ \longrightarrow Q$$
$$NAD^+ \longleftarrow QH_2$$

NAD⁺ is once again available, so glycolysis and the citric acid cycle may continue. But won't we run out of oxidized Q now? No, because there is another carrier, **cytochrome c**, a small protein that can reoxidize the QH₂. Does this respiratory chain have an end? Yes—cytochrome c is reoxidized by molecular oxygen, the final oxidizing agent:

$$cyt\ c\ (red) \longrightarrow \tfrac{1}{2} O_2$$
$$cyt\ c\ (ox) \longleftarrow H_2O$$

This is very satisfactory indeed. Oxygen gas is abundant in most places where life is found on Earth today, so there is no worry about running out of oxidizing agent. In addition, the "waste" product of the respiratory chain, water, is nontoxic and presents no disposal problem. The two hydrogens in the water, by the way, may be thought of as the hydrogens abstracted from a substrate back in the citric acid cycle or in glycolysis. They have been passed from one carrier to another and finally used to reduce molecular oxygen, reoxidizing cytochrome c and allowing the various respiratory pathways to continue.

The respiratory chain is more complicated than we have just indicated, for the chain also contains three large protein complexes through which electrons are passed (Figure 7.15). Between NADH + H⁺ and Q lies **NADH-Q reductase**, a complex of 25 polypeptide subunits, with a total molecular weight of 850,000. **Cytochrome reductase**, with 9 subunits and a molecular weight of 250,000, lies between Q and cytochrome c. **Cytochrome oxidase**, with 8 subunits and a molecular weight of 160,000, lies between cytochrome c and oxygen. Different subunits within each of the complexes bear different electron carriers, so electrons are transported *within* each complex. All the components of the respiratory chain are proteins or are attached to proteins, except for Q, which is a smaller molecule.

Why should the respiratory chain have so many links? Why, for example, do we not just use the following single step?

$$NADH + H^+ + \tfrac{1}{2} O_2 \rightarrow NAD^+ + H_2O$$

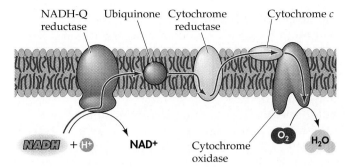

7.15 The Oxidation of NADH + H⁺

Oxidation of NADH + H⁺ by the respiratory chain produces a great deal of ATP. As traced by the purple arrows, electrons from NADH + H⁺ are passed through a series of carrier molecules in the inner mitochondrial membrane (or the plasma membrane of an aerobic prokaryote), releasing enough energy to produce ATP from ADP + P_i along the way. The carriers gain free energy and become reduced as electrons are passed to them and release free energy when they are oxidized, passing the electrons to the next carrier in the chain.

Would this not accomplish the same thing, and more efficiently? To begin with, there is no enzyme that will catalyze the direct oxidation of NADH by oxygen. More fundamentally, this would be an untamable reaction. It would be terrifically exergonic—rather like setting off a stick of dynamite in the cell. There is no biochemical way to harvest that burst of energy efficiently and put it to physiological use (no metabolic reaction is so endergonic as to consume a significant fraction of that energy in a single step).

Instead, evolution has led to the lengthy chain we observe today: a *series* of reactions, each releasing a smaller, relatively manageable amount of energy. Electron transport within each of the three protein complexes results, as we shall see, in the formation of ATP. Thus the vast energy supply originally contained in glucose and other foods is finally tucked into the cellular energy currency, ATP. For each pair of electrons passed along the respiratory chain from NADH + H⁺ to oxygen, three molecules of ATP are formed.

The carriers of the respiratory chain (including those contained in the three protein complexes) differ as to how they change upon reduction. NAD⁺, for example, accepts one proton and two electrons, leaving the proton from the other hydrogen atom to float free: NADH + H⁺. Others, including Q, bind both protons and both electrons in becoming, for example, QH_2. The remainder of the chain, however, is only an electron-transport process. Electrons, but not protons, are passed from Q to cytochrome *c*. The cytochromes contain iron atoms that in their oxidized (ferric) states are Fe^{3+} and in their reduced (ferrous) states Fe^{2+}. The iron atoms are held in place by **heme groups** like that found in hemoglobin.

Electrons pour into the pool of Q molecules from the NADH + H⁺ pathway, or they can come from another source: the succinate-to-fumarate reaction of the citric acid cycle (see Figure 7.13, reaction 7). Another protein complex, **succinate-Q reductase**, links the oxidation of succinate to the reduction of Q (Figure 7.16). The enzyme that constitutes the first part

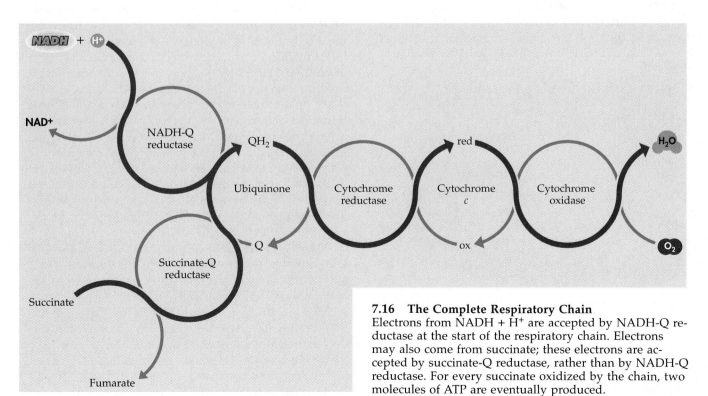

7.16 The Complete Respiratory Chain

Electrons from NADH + H⁺ are accepted by NADH-Q reductase at the start of the respiratory chain. Electrons may also come from succinate; these electrons are accepted by succinate-Q reductase, rather than by NADH-Q reductase. For every succinate oxidized by the chain, two molecules of ATP are eventually produced.

of succinate-Q reductase has attached to it an FAD carrier molecule, which is reduced by succinate to $FADH_2$. Later, hydrogen atoms are transferred to the Q molecules. No ATP is generated in the succinate-to-Q branch of the respiratory chain. Hence the pathway from the oxidation of succinate forms only two ATP molecules, compared with the three obtained when NAD^+ is the first oxidizing agent.

Oxidative Phosphorylation and Mitochondrial Structure

For many years, scientists struggled to understand how the operation of the respiratory chain caused oxidative phosphorylation—the formation of ATP in the mitochondrion. The problem was solved in 1961, when the British biochemist Peter Mitchell proposed the chemiosmotic theory. This elegant model illustrates once again the intimate relationship between structure and function in biology, so let's begin by reviewing the placement of the various components of respiratory metabolism within the cell.

The reactions of glycolysis are older than those of the citric acid cycle, having evolved before the most ancient of today's prokaryotes. It is not surprising, therefore, that the enzymes for glycolysis float free in the cytosol of the cell or are bound to the cytoskeleton. They are not enclosed within any organelle and are even found in most cells that lack organelles. By contrast, the enzymes of the citric acid cycle and the respiratory chain are isolated in the mitochondria.

(Some aerobic bacteria also carry out these reactions. Although they lack mitochondria, the bacteria have membrane systems with which these enzymes are closely associated.)

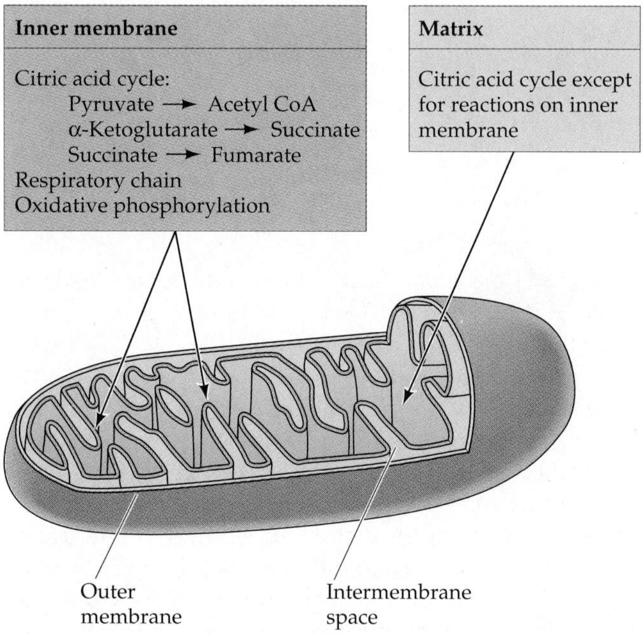

A typical mitochondrion from a mammalian cell is diagrammed in Figure 7.17. It has a relatively smooth outer membrane and an inner membrane that is folded back and forth deep into the interior of the organelle, giving the inner membrane an enormous surface area in relation to the volume that it encloses. That enclosed volume is filled with a protein-rich fluid, the mitochondrial matrix. The enzymes of the citric acid cycle are dissolved in the mitochondrial matrix, with three exceptions: succinate dehydrogenase, which catalyzes reaction 7, and the two enormous complexes that catalyze reactions 1 and 5 (see Figure 7.13). These enzymes are buried in the inner membrane (Box 7.B). The carriers and enzymes of the respiratory chain (other than cytochrome *c*) are also embedded in the inner mitochondrial membrane. Cytochrome *c* is an extrinsic protein (see Chapter 5) and lies in the space between the inner and outer mitochondrial membranes, loosely attached to the inner membrane. Ubiquinone, a small, nonprotein molecule, is free to move within the hydrophobic interior of the phospholipid bilayer of the inner membrane (Figure 7.18).

Mitchell proposed, and then showed, that operation of the respiratory chain results in the transport of hydrogen ions, against their concentration difference, through the inner membrane of the mitochondrion from inside to outside ("outside" being the space between the two mitochondrial membranes). This movement of H^+ appears to result from the particular location of the various respiratory chain intermediates in the membrane, and it acts as a sort of battery charger by establishing and maintaining a difference in pH across the inner membrane (see Figure 7.8). Because of the charge on the proton, this transport also causes a difference in electric charge across the membrane, further aiding the battery effect. The mechanisms by which these protons are transported are not yet known. As indicated in Figure 7.18, however, the protons travel through the membrane in conjunction with electron transport within the three protein complexes (NADH-Q reductase, cytochrome reductase, and cytochrome oxidase).

Figure 7.18 also shows that in the inner mitochondrial membrane is an enzyme (an ATP synthase) that allows the flow of protons through the membrane and catalyzes the production of ATP from ADP and P_i. This enzyme is perpendicular to the surface of the membrane and possesses a specific channel, embedded in the inner mitochondrial membrane, through which the excess protons on the outside of the membrane may flow back into the matrix. The ATP synthase part of this enzyme, which sticks out of the

Inner membrane	**Matrix**

Citric acid cycle:
    Pyruvate ⟶ Acetyl CoA
    α-Ketoglutarate ⟶ Succinate
    Succinate ⟶ Fumarate
Respiratory chain
Oxidative phosphorylation
	Citric acid cycle except for reactions on inner membrane

Outer membrane   Intermembrane space

7.17 Reactions in the Mitochondrion
Most of the important reactions of cellular respiration in eukaryotic cells take place in the mitochondrion's matrix or inner membrane.

BOX 7.B

Dissecting the Mitochondrion

We have said that certain enzymes of the citric acid cycle are in the mitochondrial matrix, whereas others are embedded in mitochondrial membranes. How was this learned? Think for a moment about how you might determine where a mitochondrial enzyme resides. Then read on, pausing after each step to see whether you can guess what comes next.

To begin with, you will need a centrifuge. As described at the end of Chapter 4, the centrifuge can be used to isolate a sample of mitochondria for study. After a relatively pure sample of mitochondria has been obtained, the matrix must be separated from the surrounding membranes. Think about osmosis (see Chapter 6). Think about lysis in a hypotonic solution (a membrane-bounded structure swells until it bursts). The next step is to transfer the mitochondria into a small volume of hypotonic solution (distilled water will do). Water will rush into the mitochondria, causing them to burst.

The result is a suspension consist-

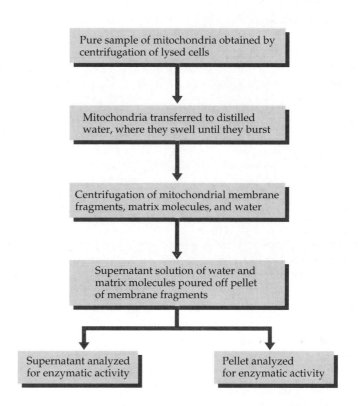

ing of the water, the mitochondrial matrix, and fragments of the membranes. What next? Think first. Centrifuge this suspension. The membranes sink to the bottom of the centrifuge tube, forming a pellet; the matrix remains in solution. Pour off the solution into a second tube; keep the membranous pellet in the first.

Now all that remains is to see which enzyme activities are in each

tube. If a particular enzyme (malate dehydrogenase from the citric acid cycle, for example) is found only in the tube containing the solution, we may reasonably conclude that it was originally present in the matrix. If another enzyme, such as succinate dehydrogenase, is found only in the first tube (which contains the pellet), it was most likely contained in the membrane.

---

inner mitochondrial membrane and is visible in electron micrographs, looks like a large knob.

The Chemiosmotic Mechanism of Mitochondria Summarized

To summarize the chemiosmotic mechanism: The flow of electrons through the respiratory chain transfers protons from the inside to the outside of the inner mitochondrial membrane, leading to an accumulation of protons on the outside. By the laws of diffusion (see Chapter 5), these excess protons tend to move spontaneously back into the matrix, which they can do only by passing through the channel-like ATP synthase molecules. This process provides the

conditions necessary for ATP production. Also, as protons diffuse away from an area of their high concentration, energy is released.

According to the chemiosmotic model, one would expect that the mitochondrion could be "fooled" into making more ATP by the following clever trick. A sample of isolated mitochondria is maintained in a solution at pH 8 (slightly basic) until it is fully adjusted; then suddenly the mitochondria are transferred into a second solution at pH 4 (fairly acidic) containing ADP and P_i. This transfer should lead to an excess of protons on the outside of the inner membrane, from where they should be able to proceed through the proposed ATP synthase channels, causing a burst of ATP production. This phenomenon

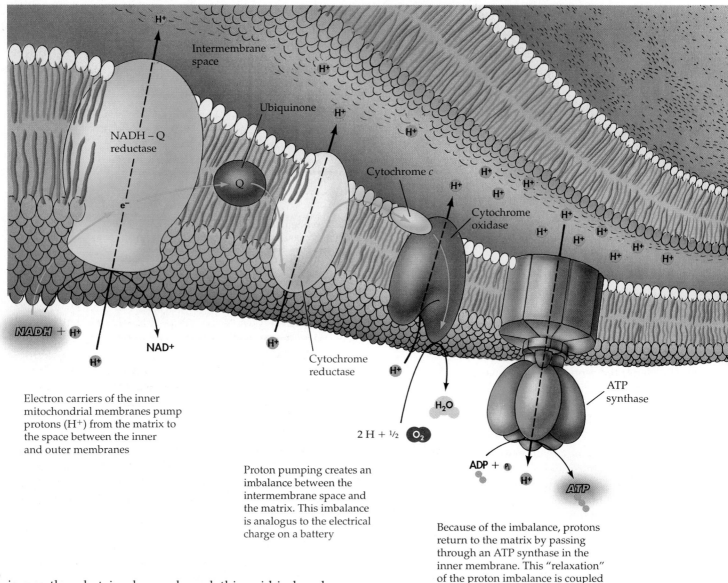

NADH – Q reductase

Intermembrane space

Ubiquinone

Cytochrome *c*

Cytochrome oxidase

e^-

NADH + H⁺

NAD⁺

Cytochrome reductase

H₂O

2 H + ½ O₂

ATP synthase

ADP + Pᵢ

ATP

Electron carriers of the inner mitochondrial membranes pump protons (H⁺) from the matrix to the space between the inner and outer membranes

Proton pumping creates an imbalance between the intermembrane space and the matrix. This imbalance is analogus to the electrical charge on a battery

Because of the imbalance, protons return to the matrix by passing through an ATP synthase in the inner membrane. This "relaxation" of the proton imbalance is coupled with the formation of ATP in the matrix

is exactly what is observed, and this acid-induced ATP production by isolated mitochondria stands as one of the stronger pieces of evidence favoring the chemiosmotic theory as the explanation for how oxidative phosphorylation proceeds in cells.

For the chemiosmotic mechanism to work, the inner membrane of the mitochondrion must be quite impermeable to protons; otherwise, the protons would leak back across the membrane as fast as they were pumped out by the respiratory chain. Cellular respiration would race along, and no ATP would form. The mechanism works correctly only because the respiratory chain and ATP formation are strictly coupled. Many years ago, biochemists discovered compounds that, when added to mitochondria or respiring tissue, uncouple respiratory metabolism from ATP formation. These **respiratory uncouplers** carry protons back across the membrane, discharging the proton-motive force. As a result, food is metabolized, but all the released energy is lost as heat, rather than being trapped in ATP. One naturally oc-

curring respiratory uncoupler plays an important role in regulating the temperature of some mammals: In "brown fat," uncoupling of respiration raises the body temperature (see Chapter 35).

The normal coupling of respiratory metabolism and ATP formation affords an important opportunity for the regulation of cellular respiration. Protons cannot return to the mitochondrial matrix unless ADP is available for conversion to ATP. When a cell's ATP level is high, there is little ADP, and this shortage stops the flow of protons. The excessive buildup of the proton-motive force stops the respiratory chain, and thus the citric acid cycle and glycolysis. On the other hand, if a cell is working hard and thus using lots of ATP, the ADP level rises, and respiratory metabolism speeds up. This phenomenon is called respiratory control.

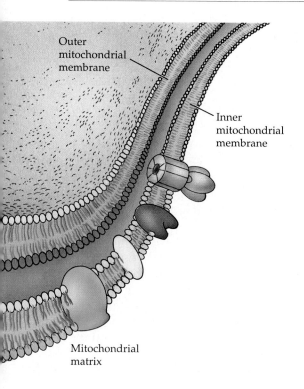

Outer
mitochondrial
membrane

Inner
mitochondrial
membrane

Mitochondrial
matrix

7.18 A Chemiosmotic Mechanism Produces ATP
As electrons pass along the series of carriers in the respiratory chain, protons are pumped from the mitochondrial matrix to the intermembrane space. As the protons return to the matrix through an ATP synthase, ATP forms, as shown in Figure 7.8.

FERMENTATION

Suppose that the supply of oxygen to a respiring cell is cut off, perhaps by drowning or by extreme exertion. As we can see in Figure 7.16, the first consequence of an insufficient supply of O_2 is that the cell cannot reoxidize cytochrome *c*, so all of that compound is soon in the reduced form. When this happens, there is no oxidizing agent to reoxidize QH_2, and soon all the Q is in the reduced form. So it goes, until the entire respiratory chain is reduced. By this point, there remains no NAD^+ and no oxidized FAD; therefore, the oxidative steps in glycolysis and the citric acid cycle also stop. If the cell has no other way to obtain energy from its food, it will die.

If, however, the cell is one—such as a muscle cell—that has the necessary enzymes, it will switch to fermentation. This process has two defining characteristics. First, a fermentative reaction uses NADH + H^+ to reduce pyruvate or one of its metabolites, with the important consequence that NADH is oxidized, regenerating NAD^+. Once the cell has some NAD^+, it can carry more glucose through glycolysis (that is, through the early steps of fermentation). The amount

of NAD^+ obtained from the fermentative step is just enough to take a comparable amount of glucose through glycolysis, with none left over to carry the pyruvate into the citric acid cycle. Instead, this newly produced pyruvate is also fermented, producing more NAD^+ to oxidize more glucose, and so forth. This illustrates the second characteristic of fermentation: By allowing glycolysis to continue, fermentation enables a sustained production of ATP—only as much as can be obtained from substrate-level phosphorylation in glycolysis (not the much greater yield obtainable with the citric acid cycle, the respiratory chain, and oxidative phosphorylation), but enough to keep the cell going.

In fact, when cells capable of fermentation become anaerobic, the rate of glycolysis speeds up tenfold or even more. Thus a substantial rate of ATP production is maintained, although the efficiency in terms of ATP molecules per glucose molecule is greatly reduced. Some bacteria of the genus *Clostridium*, while growing anaerobically in the presence of glucose, grow and multiply as rapidly as the fastest-growing aerobic bacteria. This rapid growth is made possible by the fact that the *Clostridium* bacteria are running the glycolytic reactions much more rapidly than the aerobes do.

Figure 7.19 shows the inputs and outputs of a particular type of fermentation, called lactic acid fermentation (because its end product is lactic acid, or lactate). Lactic acid fermentation takes place in many microorganisms and in our muscle cells. Unlike muscle cells, however, nerve cells (neurons) are incapable of fermentation because they lack the enzyme that reduces pyruvate to lactate. For this reason, in the absence of oxygen our nervous system (including the brain) is rapidly destroyed and is the first part of the body to die.

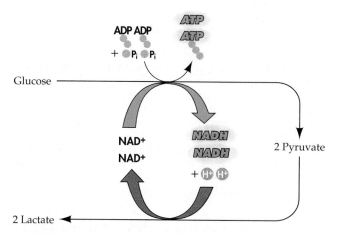

7.19 Lactic Acid Fermentation
Glycolysis produces pyruvate from glucose, as well as ATP and NADH + H^+. In lactic acid fermentation, pyruvate is then reduced to lactic acid (lactate) using NADH + H^+ as the reducing agent.

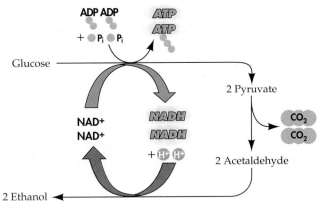

7.20 The Basis for the Brewing Industry
In alcoholic fermentation, pyruvate from glycolysis is converted to acetaldehyde, with the release of CO_2. The NADH + H$^+$ from glycolysis acts as a reducing agent, reducing acetaldehyde to ethanol.

Different forms of fermentation are observed in other kinds of organisms. Certain yeasts and some plant cells in anaerobic conditions carry on a process called **alcoholic fermentation** (Figure 7.20). Carbon dioxide is removed from pyruvate during alcoholic fermentation, leaving the compound acetaldehyde. This acetaldehyde is then reduced by NADH + H$^+$ to produce ethyl alcohol (ethanol). Remember that recycling NAD allows the fermenting cell to produce ATP by glycolysis. The brewing industry relies on alcoholic fermentation to produce wine and beer.

As noted earlier, some organisms carry on no energy metabolism other than fermentation. Some of these organisms are confined to totally anaerobic environments; others can carry on fermentation if they find themselves in the presence of oxygen. And several bacteria carry on cellular respiration—not fermentation—without using oxygen gas as an electron

7.21 Cellular Respiration Yields More Energy Than Glycolysis Does
Glycolysis yields two molecules of ATP for every glucose molecule entering the pathway. The ensuing citric acid cycle and respiratory chain produce an additional 34 ATP molecules for every glucose molecule. The source of most of these ATP molecules is the oxidation of reduced carriers (produced in glycolysis and the citric acid cycle) by the respiratory chain. We get three molecules of ATP for each NAD$^+$ regenerated by the respiratory chain and two molecules of ATP for each FAD. Thus the total gross yield of ATP from one molecule of glucose taken through glycolysis and respiration is 38. However, we must subtract two from that gross, for a net yield of 36 ATP. This is because the inner mitochondrial membrane is impermeable to NADH, and a "toll" of one ATP must be paid for each NADH (produced in glycolysis) that is shuttled into the mitochondrial matrix. The 36 molecules of ATP from the oxidation of glucose through glycolysis combined with cellular respiration still far exceeds the net of two molecules of ATP from fermentation.

acceptor. Instead, to oxidize their cytochromes these bacteria reduce nitrate ions (NO_3^-) to nitrite ions (NO_2^-).

COMPARATIVE ENERGY YIELDS

The total yield of stored energy from fermentation is two molecules of ATP per molecule of glucose oxidized. The maximum yield that can be obtained from glycolysis followed by complete aerobic respiration of a molecule of glucose is much greater—about 36 molecules of ATP (Figure 7.21). (Study Figures 7.11, 7.13, and 7.18 to review where the ATP molecules come from.) Why is so much more ATP produced by aerobic respiration? Because carriers (mostly NAD$^+$) are reduced in the citric acid cycle and then oxidized by the respiratory chain, with the accompanying production of ATP by the chemiosmotic mechanism. In an aerobic environment, a species capable of this type

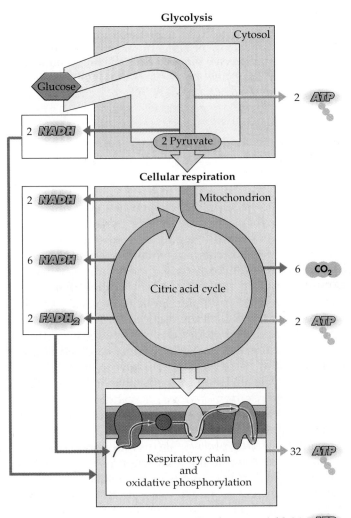

of metabolism will be at an advantage (in terms of energy availability per glucose molecule) over one that is limited to fermentation.

If glucose is simply burned, the reaction is

$$\text{Glucose} + 6\ O_2 \rightarrow$$
$$6\ CO_2 + 6\ H_2O - 686\ \text{kcal/mol}$$

with all of the 686 kcal of energy being released as heat and light. The complete biological "combustion"—respiration—of glucose could be represented by the same overall reaction, with the key difference that 36 molecules of ATP are formed for each molecule of glucose used. If each mole of ATP stores 12 kcal of energy (the actual amount varies as a function of the concentrations of ATP, ADP, and P_i in the cell), then 432 kcal (36 × 12 kcal) are stored to drive non-spontaneous reactions later, instead of being lost as heat.

CONNECTIONS WITH OTHER PATHWAYS

The respiratory pathways do not operate in isolation from the rest of metabolism. Rather, there is an interchange, with traffic flowing in both directions. For example, materials other than glucose can serve as the starting materials for respiratory ATP production. Other monosaccharides may be used, after being converted to glucose. Polymers such as starch and glycogen are **digested** (hydrolyzed) to glucose and subsequently metabolized to yield ATP. Fats are first digested to yield glycerol and fatty acids (see Chapter 3); the glycerol is then readily converted to glyceraldehyde 3-phosphate (an intermediate in glycolysis), and the fatty acids are broken down to form acetyl CoA (an early intermediate in the citric acid cycle).

Each of these reactions also operates in reverse. Thus, in the synthesis of fats, fatty acids form from acetyl CoA and glycerol forms from glyceraldehyde 3-phosphate. This occurs only when the cell has an adequate energy supply; otherwise the acetyl CoA would be needed strictly for the citric acid cycle and ATP formation. With an abundant supply of starting material (food, such as glucose), however, the cell can divert some acetyl CoA to fatty acid production and some glyceraldehyde 3-phosphate to glycerol formation. The fat that forms on our bodies, adding baggage, is a result of this diversion.

Some intermediates of the citric acid cycle are used in the synthesis of various important cellular constituents. Succinyl CoA (succinyl coenzyme A) is a starting point for chlorophyll synthesis, and α-ketoglutarate is a key starting material for amino acid (and hence protein) production. Other amino acids are formed from oxaloacetate. (Still other amino acids derive from pyruvate, which is not an intermediate in the cycle.) Acetyl CoA has many different fates:

In addition to its role in fatty acid production, it is a building block for various pigments, plant growth substances, rubber, and the steroid hormones of animals—among other functions.

The number of possible uses for acetyl CoA also presents a problem: If too many molecules of citric acid cycle intermediates are withdrawn from the cycle for use in other pathways, the oxaloacetate concentration could be lowered so much that there would no longer be enough to react with incoming acetyl CoA to keep the cycle going. This problem is avoided by a number of replenishing reactions, which bring in material from other parts of metabolism or which bypass some of the steps of the citric acid cycle, keeping atoms in the pathway. One such reaction bypasses the first step of the citric acid cycle (in which pyruvate is converted to acetyl CoA with the loss of a carbon atom as CO_2.) In the alternative reaction, pyruvate combines with a CO_2 molecule, forming oxaloacetate, the four-carbon substance at the end of the cycle. Thus the pool of citric acid cycle intermediates is increased, making up for materials that are lost from other parts of the cycle. We will refer to this combination again in the next section, which discusses the regulation of respiratory metabolism.

FEEDBACK REGULATION

Whereas fermentation produces only two molecules of ATP for every glucose molecule, passing the pyruvate from glycolysis to the citric acid cycle and respiratory chain yields 36 ATPs per glucose molecule. Thus an aerobically respiring organism obtains 18 times more energy per mole of glucose oxidized than one that is respiring anaerobically. In other words, when a yeast cell switches from aerobic respiration to anaerobic fermentation at low concentrations of oxygen, it must metabolize glucose 18 times faster to obtain the same amount of energy. But as soon as aerobic respiration begins again, glycolysis in the yeast cell slows down. The amount of glucose used is only as much as is needed for energy production under the existing conditions, anaerobic or aerobic. This phenomenon is called the **Pasteur effect**, after its discoverer, Louis Pasteur. What is the mechanism that slows down glycolysis when the respiratory chain begins to operate?

The mechanism by which glycolysis, the citric acid cycle, and the respiratory chain are regulated is allosteric control of the enzymes (see Chapter 6). Some products of later reactions, if they are in oversupply, can suppress the action of enzymes that catalyze earlier reactions. On the other hand, an excess of the products of one branch of a synthetic chain can speed up reactions in another branch, diverting raw materials away from their own synthesis (Figure 7.22).

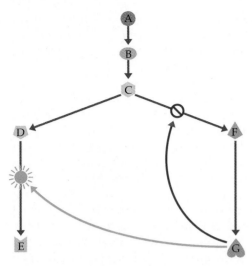

7.22 Allosteric Regulation
Compound G inhibits the enzyme for the conversion of C to F, blocking that reaction and ultimately its own synthesis, demonstrating negative feedback by allosteric regulation. Compound G also provides positive feedback to the enzyme catalyzing the step from D to E, changing that enzyme to a form that will catalyze the reaction.

These negative and positive feedback control mechanisms are used at many points in the energy-extracting processes and are summarized in Figure 7.23.

The main control point in glycolysis is the conversion of fructose 6-phosphate to fructose 1,6-bisphosphate by the enzyme phosphofructokinase. This enzyme is allosterically inhibited by ATP and activated by ADP or AMP. The enzyme is also inhibited by citrate (see Figure 7.23), for reasons that will become clear shortly. As long as fermentation proceeds, yielding a relatively small amount of ATP, phosphofructokinase operates at full efficiency. But when aerobic respiration begins producing ATP 18 times faster than before, the excess ATP allosterically inhibits the conversion of fructose 6-phosphate, and the rate of glucose utilization drops.

Pyruvate occupies a key position in the network diagrammed in Figure 7.23. In fermentation it is reduced to lactate, which can be either returned as pyruvate or used to resynthesize glucose for storage. Under aerobic conditions, pyruvate is converted to acetyl CoA, which enters the citric acid cycle by combining with oxaloacetate. Finally, as we noted in the last section, pyruvate can be used to produce more oxaloacetate by reaction with CO_2. The pathway pyruvate takes depends on conditions and needs in the cell. In the pyruvate-to-lactate conversion of fermentation, pyruvate is reduced by the $NADH + H^+$ produced in glycolysis. However, the affinity of the respiratory chain for NADH is much greater than that of the enzyme that forms lactate. (Recall that in the respiratory chain NADH-Q reductase oxidizes

$NADH + H^+$; see Figure 7.15.) Thus if the respiratory chain is operating, it steals all the available NADH, turning fermentation off.

At a second control point for pyruvate reactions, acetyl CoA regulates oxaloacetate production. If enough oxaloacetate is present to keep the citric acid cycle going as fast as acetyl CoA is produced, the concentration of acetyl CoA remains low. If too little oxaloacetate is available, acetyl CoA builds up, activating the enzyme that produces oxaloacetate and restoring the level of oxaloacetate needed for the operation of the citric acid cycle. Acetyl CoA is an **allosteric activator** for the reaction.

Thus the concentration of acetyl CoA determines the balance point between two competing reactions: one that uses oxaloacetate in turning the citric acid cycle and another that makes oxaloacetate if it is in short supply. ADP is an opposing **allosteric inhibitor** of this same oxaloacetate-producing enzyme. If the cell is low in ATP, it is high in ADP; by inhibiting the oxaloacetate-producing enzyme, ADP directs more pyruvate to become citrate, causing the citric acid cycle and respiratory chain to operate more rapidly and form more ATP.

The main control point for the citric acid cycle is the conversion of isocitrate to α-ketoglutarate. ATP and NADH are feedback inhibitors of this reaction; ADP and NAD^+ are activators. If too much ATP is accumulating, or if $NADH + H^+$ is being produced faster than it can be used by the respiratory chain, the isocitrate reaction is almost completely blocked and the citric acid cycle is essentially shut down. This would lead to a pileup of large amounts of isocitrate and citrate, except that the conversion of acetyl CoA to citrate is also slowed by ATP and $NADH + H^+$. The negative effects of halting the isocitrate reaction are thus spread backward through the chain of reactions. A certain excess of citrate does accumulate, however, and this excess acts as a negative feedback inhibitor to slow the fructose 6-phosphate reaction early in glycolysis. Consequently, if the citric acid cycle has been slowed down because of an excess of ATP (and not because of a lack of oxygen), glycolysis is shut down as well. Both processes resume when the ATP level falls and they are needed. Allosteric control keeps the process in balance.

Another control point in Figure 7.23 involves a method for storing excess acetyl CoA by using it to synthesize fatty acids. Excess citrate is an allosteric activator for one of the enzymes in the pathway for making fatty acids. If too much ATP is being made and the citric acid cycle is shut down, the accumulation of citrate switches acetyl CoA to the synthesis of fatty acids for storage. These may be metabolized later to produce more acetyl CoA.

Allosteric control of this sort is one of the most impressive examples of the tight organization that

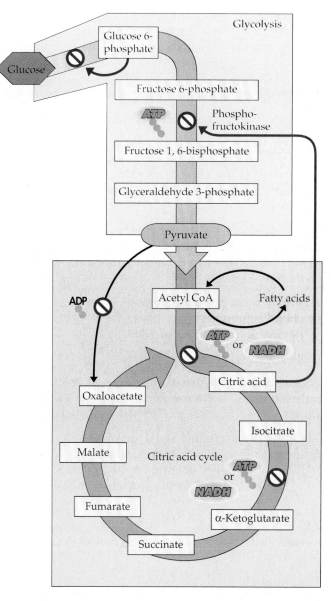

Positive regulation

Negative regulation

7.23 Feedback Regulation of Glycolysis and the Citric Acid Cycle

Positive and negative allosteric regulation control glycolysis and the citric acid cycle. Because it provides both negative feedback on the enzyme catalyzing the third step in glycolysis and positive feedback for the synthesis of fatty acids from acetyl CoA, citric acid is much like compound G in Figure 7.22. Note that feedback controls glycolysis and the citric acid cycle at crucial early steps in the pathways, increasing the efficiency of the pathways and preventing the excessive buildup of intermediates. Note also that the compounds inhibiting or activating enzymes are often the energy-carrying compounds themselves—ATP, ADP, NAD+, NADH, and so forth. If too much ATP accumulates, for example, it inhibits a key reaction and thus slows down ATP production. If much ATP has been consumed, resulting in the formation of ADP and P$_i$, the ADP activates enzymes in the pathway to stimulate the production of more ATP.

can evolve through natural selection, when selection pressure favors efficient operation in the competition among organisms for limited resources. Each of the feedback controls regulates a part or various parts of the energy-releasing pathways and keeps them operating in harmony and balance. It is unnecessary (and therefore inefficient and disadvantageous) to run the glycolytic mechanism too fast if it is supplemented by the more energy-efficient processes of the citric acid cycle and respiratory chain. In terms of energy production, it is wasteful to produce more acetyl CoA if there is insufficient oxaloacetate to handle it in the citric acid cycle. It is also senseless to shunt too much pyruvate into making oxaloacetate and to neglect production of the fuel acetyl CoA because, in the step that keeps the citric acid cycle turning, a two-carbon molecule of acetyl CoA must react with a four-carbon molecule of oxaloacetate to produce six-carbon citrate. Allosteric control maintains the proper balance among the uses of pyruvate by being sensitive to shortages or oversupplies of

acetyl CoA. All the other allosteric feedback controls help make the system more efficient and hence contribute to the success of the individual that carries them.

SUMMARY of Main Ideas about Energy Release in Cells

Energy for life comes from photosynthesis, glycolysis, cellular respiration, and fermentation.
Review Figures 7.1 and 7.9

Cells use ATP for short-term storing of energy, for transferring energy, and for releasing energy to do work.

Energy released by an exergonic reaction may be used to form ATP, and the splitting of ATP provides energy that can drive an endergonic reaction.
Review Figure 7.3

ATP may form by substrate-level phosphorylation or chemiosmotically.
Review Figures 7.7 and 7.18

Cells store and transfer electrons.

As a material is oxidized, the electrons it loses are transferred to another material, which is thereby reduced.

Many highly exergonic steps are oxidations that require specific oxidizing agents, notably, NAD^+; much of the energy liberated by the oxidation of a substrate is captured in the reduction of the oxidizing agent.
Review Figures 7.4 and 7.6

Glycolysis is a pathway of reactions catalyzed by enzymes in the cytosol.

The inputs to glycolysis are glucose, NAD^+, and $ADP + P_i$. Its outputs are pyruvate, $NADH + H^+$, and ATP.
Review Figure 7.11

Glycolysis provides starting materials for cellular respiration and fermentation; it also releases energy.
Review Figures 7.9 and 7.12

Cellular respiration proceeds through the citric acid cycle and the respiratory chain in the mitochondria of eukaryotes and in membrane systems in certain bacteria.
Review figure 7.17

The inputs to the citric acid cycle are pyruvate, NAD^+, FAD, and $ADP + P_i$. Its outputs are CO_2, $NADH + H^+$, $FADH_2$, and ATP.
Review Figure 7.13

Reduced electron carriers from glycolysis and the citric acid cycle are reoxidized by the respiratory chain.

The inputs to the respiratory chain are oxygen, $NADH + H^+$ (or $FADH_2$), and $ADP + P_i$. Its outputs are ATP, NAD^+ (or FAD), and water.
Review Figure 7.18

Many organisms derive their energy supply from fermentation, which proceeds in the absence of oxygen gas.

Fermentation oxidizes the $NADH + H^+$ produced in glycolysis, allowing glycolysis to continue.
Review Figures 7.19 and 7.20

For each molecule of glucose used, fermentation yields 2 molecules of ATP. In contrast, glycolysis combined with the citric acid cycle and the respiratory chain yields up to 36 molecules of ATP per molecule of glucose.
Review Figures 7.19, 7.20, and 7.21

Allosteric feedback controls regulate the web of reactions that constitute energy metabolism.
Review Figure 7.23

SELF-QUIZ

1. Which statement about ATP is *not* true?
   a. It is formed only under aerobic conditions.
   b. It is used as an energy currency by all cells.
   c. Its formation from ADP and phosphate is an endergonic reaction.
   d. It provides the energy for many different biochemical reactions.
   e. Some ATP is used to drive the synthesis of storage compounds.

2. Oxidation and reduction
   a. entail the gain or loss of proteins.
   b. are defined as the loss of electrons.
   c. are both endergonic reactions.
   d. always occur together.
   e. proceed only under aerobic conditions.

3. NAD$^+$
   a. is a type of organelle.
   b. is a protein.
   c. is an oxidizing agent.
   d. is a reducing agent.
   e. is formed only under aerobic conditions.

4. Glycolysis
   a. takes place in the mitochondrion.
   b. produces no ATP.
   c. has no connection with the respiratory chain.
   d. is the same thing as fermentation.
   e. reduces two molecules of NAD for every glucose molecule processed.

5. Fermentation
   a. takes place in the mitochondrion.
   b. takes place in all animal cells.
   c. does not require O_2.
   d. requires lactic acid.
   e. prevents glycolysis from taking place.

6. Which statement is *not* true of pyruvate?
   a. It is the end product of glycolysis.
   b. It gets reduced during fermentation.
   c. It feeds into the citric acid cycle.
   d. It is a protein.
   e. It contains three carbon atoms.

7. The citric acid cycle
   a. takes place in the mitochondrion.
   b. produces no ATP.
   c. has no connection with the respiratory chain.
   d. is the same as fermentation.
   e. reduces two molecules of NAD for every glucose molecule processed.

8. Which statement is *not* true of the respiratory chain?
   a. It takes place in the mitochondrion.
   b. It uses O_2 as an oxidizing agent.
   c. It leads to the production of ATP.
   d. It regenerates oxidizing agents for glycolysis and the citric acid cycle.
   e. It operates simultaneously with fermentation.

9. Which statement is *not* true of the chemiosmotic mechanism?
   a. Protons are pumped across a membrane.
   b. Protons return through the membrane by way of a channel protein.
   c. ATP is required for the protons to return.
   d. Proton pumping is associated with the respiratory chain.
   e. The membrane in question is the inner mitochondrial membrane.

10. Which statement is *not* true of oxidative phosphorylation?
   a. It is the formation of ATP during the operation of the respiratory chain.
   b. It is brought about by the chemiosmotic mechanism.
   c. It requires aerobic conditions.
   d. In eukaryotes, it takes place in mitochondria.
   e. Its functions can be served equally well by fermentation.

FOR STUDY

1. Trace the sequence of chemical changes that occurs in mammalian brain tissue when the oxygen supply is cut off. (The first change is that the cytochrome oxidase system becomes totally reduced, because electrons can still flow from cytochrome c but there is no oxygen to accept electrons from cytochrome oxidase. What are the remaining steps?)

2. Trace the sequence of chemical changes that occurs in mammalian muscle tissue when the oxygen supply is cut off. (The first change is exactly the same as that in Study Question 1.)

3. Some cells that use the citric acid cycle and the respiratory chain can also thrive by using fermentation under anaerobic conditions. Given the lower yield of ATP (per molecule of glucose) in fermentation, why can these cells function so efficiently under anaerobic conditions?

4. Describe the mechanisms by which the rates of glycolysis and of aerobic respiration are kept in balance with one another.

READINGS

Alberts, B., D. Bray, J. Lewis, M. Raff, K. Roberts and J. D. Watson. 1994. *Molecular Biology of the Cell*, 3rd Edition. Garland Publishing, New York. Chapter 7 develops the themes introduced in this chapter; Chapter 3 is also useful as an introduction.

Hinkle, P. C. and R. E. McCarty. 1978. "How Cells Make ATP." *Scientific American*, March. Discussion of the chemiosmotic mechanism, in which ATP is formed by protons passing back through a membrane after being pumped out by the respiratory chain.

Lodish, H., D. Baltimore, A. Berk, L. Zipursky, P. Matsudaira and J. Darnell. 1995. *Molecular Cell Biology*, 3rd Edition. Scientific American Books, New York. Chapter 17 gives an excellent more detailed treatment of the topics in this chapter.

Stryer, L. 1995. *Biochemistry*, 4th Edition. W. H. Freeman, New York. Although more advanced than this chapter, the section on glycolysis and respiration is straightforward and does not demand an advanced knowledge of chemistry.

Voet, D. and J. G. Voet. 1990. *Biochemistry*. John Wiley & Sons, New York. A general textbook with a full discussion of energy, enzymes, and catalysis. Outstanding illustrations.

Sunlight: The Source of Our Energy

8

Photosynthesis

We are the creatures of the sun. Its light is the source—direct or indirect—of the free energy that powers life on Earth. This is the important message about photosynthesis, which you may already know. As biologists, however, we might express a more refined appreciation of how we are able to use the sun's energy.

As biologists we can say, "We are creatures of the chloroplasts." This abundant organelle, which gives green plants their color, captures the sun's energy for us. Inside plants, within stacks of connected, inner membranes wrapped in two outer membranes, chemical reactions convert and store the energy that we draw upon. What are the reactants and what are the products? What sorts of molecules embedded in the inner membranes or floating in their vicinity participate in the reactions? What, exactly, happens inside this organelle that feeds the world? These are the central topics of this chapter, but before taking them up, let's consider photosynthesis in a more general way.

SUNLIGHT AND LIFE ON EARTH

Our own dependence on the sun, although absolute, is indirect. Like the other animals, the fungi, many protists, and most monerans, we depend upon a ready supply of partially reduced, carbon-containing compounds as a food source. From such compounds we get all the free energy that keeps us alive and functioning. From them, too, we obtain the carbon atoms used in every organic molecule in our bodies. In a word, we are **heterotrophs**: We need to feed upon something else. In a world populated exclusively by heterotrophs, all life would grind to an end as the food gradually disappeared.

Our world owes the continued existence of life to the presence of **autotrophs**—organisms that do not need previously formed organic substances from their environment. For autotrophs, an energy source (such as light) and an inorganic carbon source (such as carbon dioxide gas) suffice as a diet. From these simple ingredients, autotrophs make the reduced carbon compounds from which their bodies are built and their food needs met. By feeding on autotrophs, the heterotrophs of the world meet their needs for energy and matter. The principal autotrophs are photosynthetic organisms that use visible light as their energy source. From light, carbon dioxide, and water, they begin the chemistry that sustains almost the entire biosphere.

Photosynthesis is the conversion of light energy to chemical energy by living things. Organisms that conduct photosynthesis—plants and photosynthetic protists and monerans—stand at the gateway to the

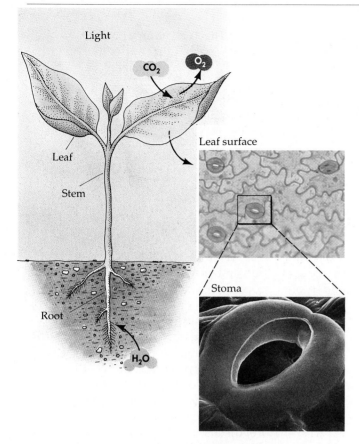

8.1 Ingredients for Photosynthesis
A typical terrestrial plant uses light from the sun, water from the soil, and carbon dioxide from the atmosphere to form organic compounds by photosynthesis. Carbon dioxide enters the leaves through openings called stomata. The top micrograph shows several stomata, magnified about 100 times; the stoma on the far right was closed when the photograph was taken. Below is a scanning electron micrograph of a single open stoma.

living world, at the interface where inorganic becomes organic, where nonlife becomes life. The worldwide extent of photosynthetic activity is stunning: Each year, tens of billions of tons of carbon atoms are taken from carbon dioxide and incorporated into molecules of sugars, amino acids, and other compounds.

EARLY STUDIES OF PHOTOSYNTHESIS

By the beginning of the nineteenth century, scientists understood the broad outlines of photosynthesis. Photosynthesis was known to use three principal ingredients—water, carbon dioxide (CO_2), and light—and to produce not only food but also oxygen gas (O_2). Scientists had learned that the water for photosynthesis comes primarily from the soil (for plants living on land) and must travel from the roots to the leaves; that carbon dioxide is taken in from the atmosphere through tiny openings, called **stomata**

(singular, *stoma*), in the leaves (Figure 8.1); and that light is absolutely necessary for the production of oxygen and food. The last of the important early discoveries, made during the first decade of the nineteenth century, was that carbon dioxide uptake and oxygen release are closely related and that both depend on light action. By 1804, scientists could summarize photosynthesis in plants by writing

$$CO_2 + H_2O + \text{light energy} \rightarrow \text{sugar} + O_2$$

but many details of the process remained hidden.

It was almost a century and a half before it was determined whether the oxygen released during photosynthesis comes from the carbon dioxide or from the water. The direct demonstration depended on one of the first uses of an isotopic tracer (see Chapter 2) in biological research. In the experiments, two groups of green plants were allowed to carry on photosynthesis. Plants in the first group were supplied with water containing the heavy-oxygen isotope ^{18}O and with CO_2 containing only the common isotope ^{16}O; plants in the second group were supplied with CO_2 labeled with ^{18}O and water containing only the common isotope. Oxygen gas was collected from each group of plants, and it was found that O_2 containing ^{18}O was produced in abundance by the plants given ^{18}O-labeled water but not by plants given labeled CO_2. From these results, scientists concluded that the oxygen gas produced during photosynthesis comes from water (Figure 8.2).

With the information about where the O_2 comes from in hand, and taking into account the number of

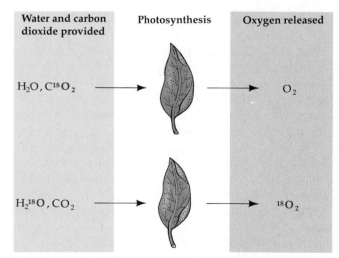

8.2 Water Produces the Oxygen Liberated during Photosynthesis
Experimenters gave some plants isotope-labeled carbon dioxide, $C^{18}O_2$, and unlabeled water (top); they gave other plants isotope-labeled water, $H_2^{18}O$ and unlabeled CO_2 (bottom). Because only plants in the bottom group released isotope-labeled oxygen gas, $^{18}O_2$, we know that water is the source of the oxygen.

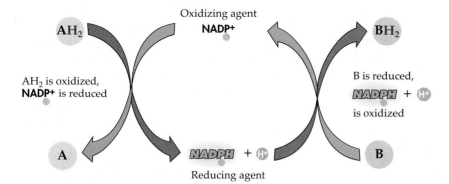

8.3 The Two Forms of NADP
When NADP$^+$, the oxidized form of nicotinamide adenine dinucleotide phosphate, reacts with a reduced molecule (AH$_2$), AH$_2$ becomes oxidized and NADP$^+$ becomes reduced to NADPH + H$^+$. In turn, NADPH + H$^+$, the reduced form, may react with an oxidized molecule such as B, becoming oxidized to NADP$^+$ and reducing B to BH$_2$. The red arrows trace the path of electrons in the reactions.

CO$_2$ molecules needed to form a simple sugar such as glucose, we may now rewrite the overall equation for photosynthesis as

$$6\ CO_2 + 12\ H_2O \rightarrow C_6H_{12}O_6 + 6\ O_2 + 6\ H_2O$$

Water appears on both sides of the equation because water is used as a reactant (the 12 molecules on the left) and released as a product (the 6 new ones on the right). Note that this equation is essentially the reverse of the overall equation for cellular respiration, given in Chapter 7.

Although the O$_2$ is in one sense a waste product, it is vital to all oxygen-requiring organisms. The photosynthetic production of oxygen by green plants is an important source of atmospheric oxygen, which most organisms—including the green plants themselves—require in order to complete their respiratory chains and thus obtain the energy to live.

THE PATHWAYS OF PHOTOSYNTHESIS

The overall photosynthetic reaction just shown cannot proceed in a single step. There is no precedent in all of chemistry for such a complex reaction being a single step. Rather, there must be a whole series of simpler steps. By the middle of the twentieth century, it was clear that photosynthesis comprises two pathways: one, driven by light, produces ATP; the other uses the ATP to produce sugar.

Just as NAD (nicotinamide adenine dinucleotide; see Chapter 7) bridges the two pathways of cellular respiration, a very similar compound bridges the two pathways of photosynthesis. This electron carrier is **NADP** (nicotinamide adenine dinucleotide phosphate). NADP is virtually identical to NAD, differing from it only in the possession of another phosphate group attached to the ribose portion of the molecule. Whereas NAD participates in metabolic breakdown reactions and energy transfers, NADP participates in synthetic reactions that require energy and reducing power. Like NAD, NADP exists in two forms. One (NADP$^+$) is an oxidizing agent, whereas the other (NADPH + H$^+$) is a reducing agent (Figure 8.3).

NADPH + H$^+$ is an intermediary for energy and reducing power. ATP and NADPH + H$^+$ are carriers of reducing power because reduction is always an endergonic process requiring both energy and electrons.

One photosynthetic pathway uses light to produce ATP. The reactions of this pathway are catalyzed by enzymes called **photosystems**, and the pathway itself is called **photophosphorylation**. During our discussion of this pathway, we refer to a version of photophosphorylation (*noncyclic* photophosphorylation), carried out by plants, that produces NADPH as well as ATP. Before discussing the second pathway of photosynthesis (in which ATP is used to produce sugar) in detail, we will briefly describe another version of photophosphorylation (*cyclic* photophosphorylation), which produces only ATP.

The NADPH + H$^+$ and ATP produced in the first pathway of photosynthesis are used in the second pathway, where reactions trap CO$_2$ and reduce the resulting acid to sugar. These sugar-producing reactions constitute the Calvin–Benson cycle (Figure 8.4), also known as the photosynthetic carbon reduction cycle, or simply the dark reactions (because none of them uses light, as does photophosphorylation). The reactions of both pathways proceed within the chloroplast, but, as we will see, they reside in different parts of the organelle. *Both* pathways stop in the dark because ATP synthesis and NADP$^+$ reduction require light. The rate of each set of reactions is dependent upon that of the other. They are tied together by the exchange of ATP and ADP and of NADP$^+$ and NADPH.

LIGHT AND PIGMENTS

The living world makes marvelous use of light. In photosynthesis, light is a source of *energy*; in most other light-related phenomena, it is involved in the transmission of *information*. Many of these phenomena will be described in later chapters. In them, we will find that light can be modulated in many ways to carry information: Its *brightness* may be varied, as

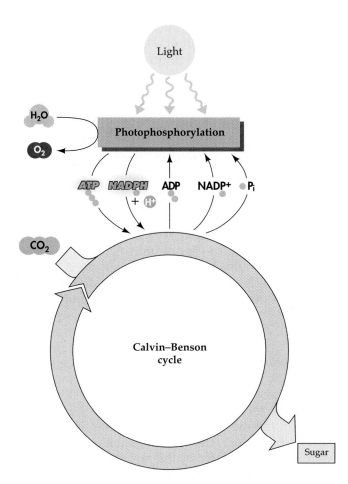

8.4   An Overview of Photosynthesis
Light energy and water are used in the first pathway, photophosphorylation, to produce ATP, NADPH + H⁺, and (in many organisms), O_2. Carbon dioxide and the ATP plus NADPH + H⁺ produced during photophosphorylation are used in the Calvin–Benson cycle (the second pathway) to produce sugars and other food molecules.

may its *color*, and it may be presented for various *durations*—continuously or in short, long, or variable periods. Some of the material to be covered will be more meaningful if we first learn to deal in a quantitative way with the brightness, color, and energy content of light.

Basic Physics of Light

Light is a form of radiant energy. It comes in discrete packets called **photons**. Light also behaves as if it were propagated in waves. The **wavelength** is the distance from the peak of one wave to the peak of the next (Figure 8.5). Different colors result from different wavelengths. Light and other forms of radiant energy—cosmic rays, gamma rays, X rays, ultraviolet radiation, infrared radiation, microwaves, and radio waves—are **electromagnetic radiation**. We have listed these forms of radiation here in order of in-

creasing wavelength and of decreasing energy per photon. Visible light fits into this **electromagnetic spectrum** between ultraviolet and infrared radiation (Figure 8.6). Although considerable attention has been devoted to the apparent paradox of light being simultaneously a wave phenomenon and a particle phenomenon, this is nothing to be concerned about in our study of photosynthesis.

The speed of light in a vacuum is one of the universal constants of nature. In a vacuum light travels at 3×10^{10} centimeters per second (186,000 miles per second), a value symbolized as c. In air, glass, water, and other media, light travels slightly more slowly. Let's consider light as a long train of waves moving in a straight line and see what the train would look like to a stationary observer. Successive peaks of the waves pass the observer with a uniform **frequency** (v) determined by the wavelength and the speed of light. The exact relationship is $v = c/\lambda$, where v (the Greek letter *nu*) is the frequency; c is the speed of light; and λ (Greek *lambda*) is the wavelength. Often v is expressed in hertz (Hz), c in centimeters per second (cm/sec), and λ in nanometers (nm). (One nanometer equals 10^{-9} meter or 10^{-7} centimeter; see the conversion table inside the back cover.)

Humans perceive light as having distinct colors (the reason for this will be explained in Chapter 39). The colors relate to the wavelengths of the light, as shown in Figure 8.6. Most of us can see electromagnetic radiation with wavelengths from 400 to 700 nm. At 400 nm we are at the blue end of the visible spectrum, whereas 700 nm is the red end. Wavelengths in the range from about 100 to 400 nm are ultraviolet radiation; those immediately above 700 are referred to as infrared.

The amount of energy, E, contained in a single photon is directly proportional to its frequency. The constant of proportionality that describes this relationship, h, is named Planck's constant after Max Planck, who first introduced the concept of the photon. With this information we can write the equation

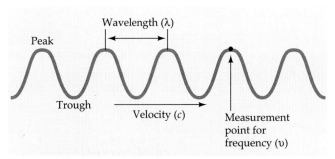

8.5   Light Has Wavelike Properties
The wavelength, λ, is the distance between the peaks of successive waves; and the frequency, v, is the number of peaks passing an observation point in a second. The velocity with which the train of waves moves is c.

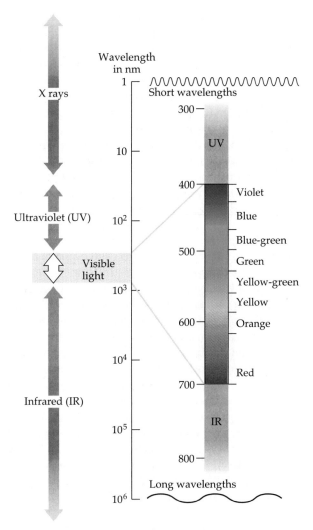

8.6 The Electromagnetic Spectrum
Wavelengths of electromagnetic radiation can be arranged on a scale called the electromagnetic spectrum. A portion of the spectrum in the vicinity of light visible to humans is represented here. Visible light comprises wavelengths between about 400 and 700 nm, although not everyone can see over this entire range. Ultraviolet radiation extends from the short-wavelength end of the visible spectrum, infrared radiation from the long-wavelength end.

$E = h\nu$, where ν is the frequency in Hz. Substituting c/λ for ν (from the equation relating λ, ν, and c), we see that $E = hc/\lambda$. Thus shorter wavelengths mean greater energies—that is, energy is inversely proportional to wavelength. A photon of red light of wavelength 660 nm has less energy than a photon of blue light at 430 nm; an ultraviolet photon of wavelength 284 nm is much more energetic than either of these. For any light-driven biological process—such as photosynthesis—a photon can be active only if it consists of enough energy to perform the work required.

The brightness, or **intensity**, of light at a given point is the amount of energy falling on a defined area—such as 1 cm^2—per second. Light intensity is usually expressed in energy units (such as calories) per square centimeter per second, but pure light of a single wavelength may also be expressed in terms of photons per square centimeter per second.

Pigments

When a photon meets a molecule, one of three things takes place. The photon may bounce off the molecule or it may pass through it. In other words, it may be reflected or transmitted. Neither of these causes any change in the molecule, and neither has any biological consequences. The third possibility is that the photon may be *absorbed* by the molecule. In this case, the photon simply disappears. Its energy, however, cannot disappear, because energy is neither created nor destroyed. The molecule acquires the energy of the absorbed photon and is thereby raised from a **ground state** (lower energy) to an **excited state** (higher energy). The difference in energy between this excited state and the ground state is precisely equal to the energy of the absorbed photon. The

8.7 Exciting a Molecule
(a) When a molecule, initially in the ground state, absorbs a photon, the molecule is raised to an excited state possessing more energy. (b) The absorption of the photon "boosts" one of the molecule's electrons to an orbital farther from the nucleus.

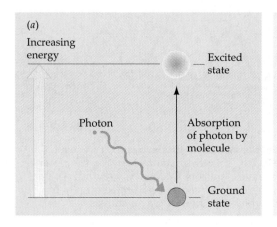

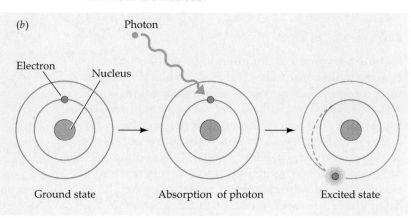

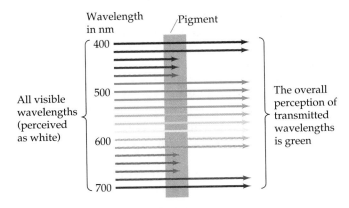

8.8 Why a Leaf Looks Green
The pigment chlorophyll (vertical bar), present in the leaves of all green plants, absorbs photons from specific wavelengths of visible light (short arrows). We see the combination of wavelengths that are not absorbed (long arrows) as the characteristic color of the pigment.

increase in energy boosts one of the electrons in the molecule into an orbital farther from the nucleus; in a sense, this electron is now held less firmly by the molecule (Figure 8.7). We will see the chemical consequence of this later in this chapter.

All molecules absorb electromagnetic radiation. The specific wavelengths absorbed by a particular molecule are characteristic of that type of molecule. Some molecules cannot absorb wavelengths in the *visible* region; those that can are called **pigments**.

When a beam of white light (light containing visible light of all wavelengths) falls on a pigment, certain wavelengths of the light are absorbed. The remaining wavelengths, which are reflected or transmitted, make the pigment appear to us to be colored. If, for example, a pigment absorbs both blue and red light, as does the pigment chlorophyll, what we see is the remaining light—primarily, green (Figure 8.8). The fact that chlorophyll absorbs light in both the blue and the red region of the spectrum indicates that it has two excited states of differing energy levels, both close enough to the ground state to be reached with the energy of photons of visible light.

Absorption Spectra and Action Spectra

A given type of molecule can absorb radiant energy of only certain wavelengths. If we plot a compound's absorption of light as a function of the wavelength of the light, the result is an **absorption spectrum** (Figure 8.9). Absorption spectra are good "fingerprints" of compounds; sometimes an absorption spectrum contains enough information to enable us to identify an unknown compound. The fact that the peaks in an absorption spectrum are smoothly rounded, rather than sharp spikes, tells us that a given excited state is not an extremely narrow range of energies. Rather, it consists of a substantial family of energy sublevels, differing by tiny increments of energy much smaller than those contained in a pho-

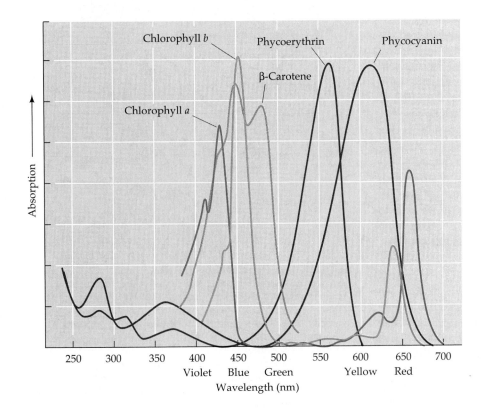

8.9 Photosynthetic Pigments Have Distinctive Absorption Spectra
Because these pigments, all of which participate in photosynthesis, absorb photons most strongly at different wavelengths, photosynthesis uses most of the visible spectrum. Notice how much of the visible spectrum would go to waste if chlorophyll *a* were the only pigment absorbing light for photosynthesis.

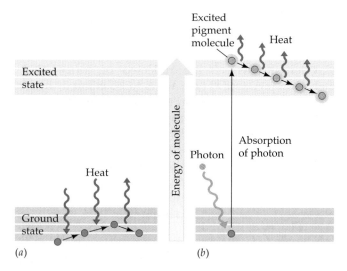

8.10 A Pigment's Energy Sublevels
(a) The ground state of a pigment consists of several sublevels of slightly different energy. When a molecule in the ground state rises from one sublevel to the next, it absorbs a tiny amount of heat; when the molecule falls to the next lower sublevel, it gives off heat. A molecule may be raised from any of these sublevels to an excited state. (b) All excited states also consist of energy sublevels. After reaching an excited state by absorbing a photon, a pigment molecule may give off minute amounts of heat as it drops from one sublevel to the next within the excited state.

ton of visible light. As molecules move from one of these sublevels to another they absorb or release small amounts of heat (Figure 8.10).

Light may be analyzed for its biological effectiveness, the magnitude of its effect on a particular activity such as photosynthesis. We may plot the effectiveness of light as a function of wavelength. The resulting graph is an **action spectrum.** Figure 8.11 shows the action spectrum for photosynthesis by a freshwater plant. As you can see, all wavelengths of visible light are at least somewhat effective in causing photosynthesis, although some are more effective than others. Because light must be absorbed in order to produce a chemical or biological effect, action spectra are helpful in determining what pigment or pigments are used in a particular photobiological process such as photosynthesis; that is, we should be able to find which pigment or pigments have absorption spectra that match the action spectrum of the process.

The Photosynthetic Pigments

Certain pigments are important in biological reactions, and we will discuss them as they appear in the book. Here we discuss pigments, found in leaves and in other parts of photosynthetic organisms, that play roles in photosynthesis. Of these, the most important are the **chlorophylls.** Chlorophylls occur universally in the plant kingdom, in photosynthetic protists, and

in virtually all photosynthetic bacteria (the exception being the halobacteria; see Box 8.A on page 174). A mutant individual lacking chlorophyll is unable to perform photosynthesis and will starve to death. In green plants, two chlorophylls predominate, **chlorophyll *a*** and **chlorophyll *b*;** they differ only slightly in structure. Both have a complex ring structure of a type referred to as a chlorin: a lengthy hydrocarbon "tail," and a central magnesium atom in the chlorin ring (Figure 8.12). (In Chapter 7 we learned about porphyrins, such as the heme found in hemoglobin and the cytochromes; porphyrins have structures very similar to those of chlorins.)

We saw in Figure 8.9 that the chlorophylls absorb blue and red wavelengths, which are near the two ends of the visible spectrum. Thus if *only* chlorophyll pigments were active in photosynthesis, much of the visible spectrum would go unused. However, all photosynthetic organisms possess **accessory pigments** that absorb photons intermediate in energy between the red and the blue wavelengths and then transfer a portion of the energy to chlorophyll to use in photosynthesis. Among these accessory pigments are **carotenoids** such as β-carotene (see Chapter 3); the carotenoids absorb photons in the blue and blue-green wavelengths and appear deep yellow. The

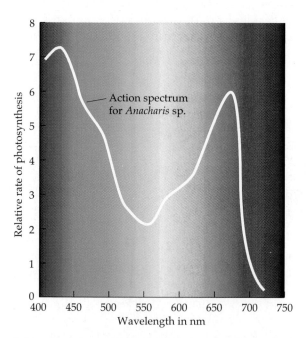

8.11 Action Spectrum of Photosynthesis
An action spectrum plots the biological effectiveness of wavelengths of radiation against the wavelength. Here the rate of photosynthesis in the freshwater plant *Anacharis* is plotted against wavelengths of visible light. You can see that wavelengths in the blue and orange-red regions of the visible spectrum cause the highest rates of photosynthesis. If we compare this action spectrum with the absorption spectra of specific pigments, such as those in Figure 8.9, we can identify which pigments are responsible for the process.

Chlorophyll *a*: R = — CH$_3$

Chlorophyll *b*: R = — C$\overset{H}{\underset{O}{\diagdown}}$

8.12 A Molecule of Chlorophyll
Chlorophyll consists of a chlorin ring with magnesium (shaded area), plus a hydrocarbon "tail." Chlorophyll *a* and chlorophyll *b* differ only in the groups attached to the position on the chlorin ring designated with an R (for "residue").

phycobilins (phycocyanin and phycoerythrin), which are found in red algae and in cyanobacteria (contributing to their respective colors), absorb various yellow-green, yellow, and orange wavelengths (see Figure 8.9). Such accessory pigments, in collaboration with the chlorophylls, constitute an energy-absorbing "antenna" covering much of the visible spectrum.

In the energy-absorbing antenna, any pigment molecule with a suitable absorption spectrum may absorb an incoming photon and become excited. The excitation passes from one pigment molecule to another in the antenna, moving to those pigments absorbing longer wavelengths (lower energies) of light. Thus the excitation must end up in the one pigment molecule in the antenna that absorbs the longest wavelength—the molecule that occupies the **reaction center** of the antenna. The reaction center is the part of the antenna that converts light absorption into chemical energy. The pigment molecule in the reaction center is always a molecule of chlorophyll *a*. There are many other chlorophyll *a* molecules in the antenna, but all of them absorb light at shorter wavelengths than does the molecule in the reaction center.

PHOTOPHOSPHORYLATION

Using the Excited Chlorophyll Molecule

A pigment molecule enters an excited state when it absorbs a photon (see Figure 8.7). The molecule usually does not stay in the excited state for very long. One means of returning to the ground state is by **fluorescence**. In this process, the boosted electron falls back from its higher orbital to the original, lower one. This process is accompanied by a loss of energy, which is given off as another photon (Figure 8.13*a*). The molecule absorbs one photon and within approximately 10^{-9} seconds emits another photon of longer wavelength than the one absorbed. The emitted light is fluorescence. If energy is simply absorbed and then rapidly returned as a photon of light, there can be no chemical or biological consequences.

For biological work to be done, something must happen to transfer energy in some other way during the billionth of a second before a photon is emitted as fluorescence. Photosynthesis conserves energy by using the excited chlorophyll molecule in the reaction center as a reducing agent (Figure 8.13*b*). Ground state chlorophyll (symbolized as Chl) is not much of a reducing agent, but excited chlorophyll (Chl*) is a good one. To understand the reducing capability of Chl*, recall that in an excited molecule, one of the electrons is zipping about in an orbital farther from its nucleus than it was before. Less tightly held, this electron can be passed on in a redox reaction to an oxidizing agent. Thus Chl* (but not Chl) can react with an oxidizing agent A in a reaction like this:

$$\text{Chl*} + \text{A} \rightarrow \text{Chl}^+ + \text{A}^-$$

This, then, is the first biochemical consequence of light absorption by chlorophyll in the chloroplast: The chlorophyll becomes a reducing agent and participates in a redox reaction that would not have occurred in the dark. As we are about to see, the further adventures of that electron (the one passed from chlorophyll to A) produce ATP and a stable reducing agent (NADPH), both of which are required in the Calvin–Benson cycle.

Noncyclic Photophosphorylation: Formation of ATP and NADPH

The NADPH-producing version of photophosphorylation we have been discussing is called **noncyclic photophosphorylation**. With its appearance, the evolution of life on Earth made a crucial advance because noncyclic photophosphorylation uses light energy not only to form ATP and NADPH + H$^+$, but also to release O$_2$.

In noncyclic photophosphorylation, electrons from water replenish chlorophyll molecules that have given up electrons. These electrons are transferred to oxidizing agents and ultimately to NADP$^+$, reducing

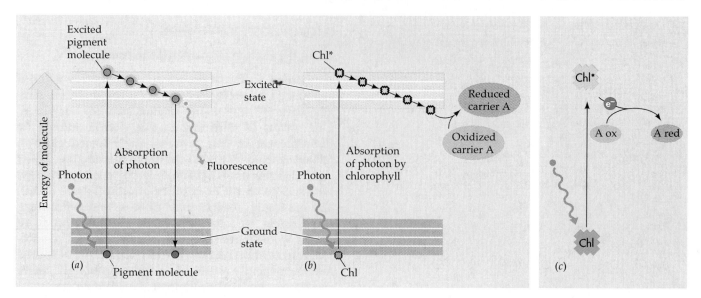

8.13 Excitation, Fluorescence, Redox

(a) When a pigment molecule absorbs a photon, boosting an electron to a higher orbital, the pigment moves to an excited state from its ground state. Although the molecule may then pass from one energy sublevel to the next, it spends very little time in the excited state. An excited molecule may return to the ground state by fluorescence, in which the electron falls back to its original lower orbital and a photon is emitted. Alternatively, an excited pigment molecule may pass the excitation to another pigment molecule, such as chlorophyll. (b) When a ground-state chlorophyll molecule (Chl) becomes excited (Chl*), it may become a reducing agent; the electron boosted to a higher orbital may pass to an oxidized electron carrier (A), reducing the carrier. Thus much of the energy of the excited state is preserved rather than being lost, as it is in fluorescence. (c) This diagram, which presents the same information as (b), illustrates the conventions to be used in Figures 8.14 and 8.15.

it to NADPH + H^+. The original source of the electrons is a plentiful one: water. As the electrons are passed from water to, ultimately, NADP, they go through a series of electron carriers. These spontaneous redox reactions are exergonic, and some of the free energy released is used ultimately to form ATP.

Noncyclic photophosphorylation requires the participation of two distinct molecules of chlorophyll—actually, two separate sets of chlorophyll molecules, in separate energy-absorbing antennas of pigment molecules. One of these sets, called **photosystem I**, is used to make a reducing agent strong enough to reduce $NADP^+$ to NADPH + H^+. The reaction center of the antenna for photosystem I contains a chlorophyll *a* molecule in a form called P_{700} because it can absorb light of wavelength 700 nm. **Photosystem II**, the other set of pigment molecules, takes electrons from water and passes them up to the series of redox carriers involved in the conversion of ADP + P_i to ATP. The reaction center of the antenna for photosystem II contains a chlorophyll *a* molecule in a form called P_{680} because it absorbs maximally at 680 nm. Thus photosystem II requires somewhat more energetic photons than does photosystem I. To keep noncyclic photophosphorylation going, both photosystems I and II must constantly be absorbing light, thereby boosting electrons to higher orbitals from which they may be captured by specific oxidizing agents.

We can follow the noncyclic pathway from water to NADP in Figure 8.14. Photosystem II (P_{680}) absorbs photons, sending electrons from P_{680} to an oxidizing agent (pheophytin-I) and causing P_{680} to become oxidized to P_{680}^+. Electrons from the oxidation of water, which forms H^+ ions and O_2, are passed to P_{680}^+ of photosystem II, reducing it once again to P_{680}, which can absorb photons, and so on. The electron donated by photosystem II to its oxidizing agent passes through a series of exergonic redox reactions, storing energy that is later used to form ATP. In photosystem I, P_{700} absorbs photons, becoming excited to P_{700}^*, which then reduces its own oxidizing agent (ferredoxin) while being oxidized to P_{700}^+. Then P_{700}^+ is returned to the ground state by accepting electrons passed through the chain from photosystem II. Now photosystem I is accounted for, and we must consider only the electrons from photosystem I and the protons from the original oxidation of water at the beginning of the scheme. These are used in the last step of noncyclic photophosphorylation, in which two electrons and two protons (from two operations of the noncyclic scheme) are used to reduce a molecule of $NADP^+$ to NADPH + H^+.

In sum, noncyclic photophosphorylation uses a molecule of water, four photons (two each absorbed by photosystems I and II), one molecule each of $NADP^+$ and ADP, and one P_i ion; from them it produces one molecule each of NADPH + H^+ and ATP,

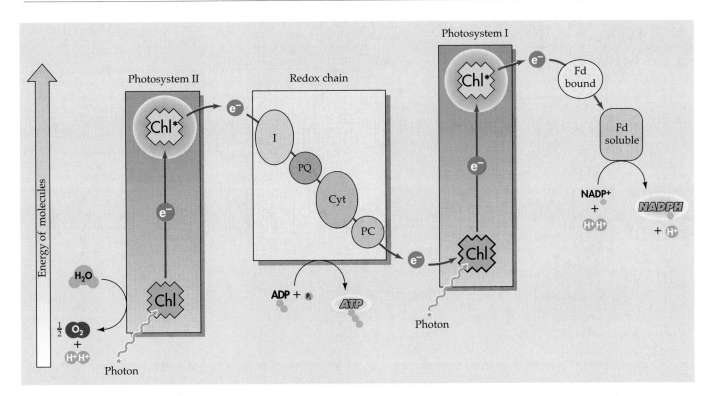

8.14 Noncyclic Photophosphorylation Uses Two Photosystems

Photosystems I and II—each containing chlorophyll—operate to keep noncyclic photophosphorylation going. ATP is produced by the redox chain between the photosystems, NADPH + H$^+$ is produced by the redox reactions that follow the electron's passing through photosystem I, and O$_2$ is produced as a by-product of the breakdown of water. Abbreviations: Cyt, cytochrome; Fd, ferredoxin; I, pheophytin-I; PC, plastocyanin; PQ, plastoquinone.

and one-half molecule of oxygen (review Figure 8.14). A substantial fraction of the light energy absorbed in noncyclic photophosphorylation is lost as heat, but another significant fraction is trapped in ATP and NADPH + H$^+$.

Cyclic Photophosphorylation: Formation of ATP but Not NADPH

Photophosphorylation that produces only ATP is called **cyclic photophosphorylation** because the electron passed from an excited chlorophyll molecule at the outset cycles back to the chlorophyll molecule at the end of the chain of reactions. Water, which supplies electrons to restore chlorophyll molecules to the ground state in noncyclic photophosphorylation, does not enter these reactions; thus no O$_2$ is produced from them.

Before cyclic photophosphorylation begins, P$_{700}$, the reaction center chlorophyll of photosystem I, is in the ground state. It absorbs a photon and becomes the reducing agent P$_{700}$*. The P$_{700}$* then reacts with oxidized ferredoxin (Fd$_{ox}$) to produce reduced ferredoxin (Fd$_{red}$). The reaction is spontaneous—that is, it is exergonic, releasing free energy.

In noncyclic photophosphorylation, this Fd$_{red}$ reduces NADP$^+$ to form NADPH + H; but Fd$_{red}$ is a good enough reducing agent to pass its added electron to *another* oxidizing agent, plastoquinone (PQ, a small organic molecule). This is what happens in cyclic photophosphorylation (Figure 8.15), which occurs in some organisms when the ratio of NADPH + H$^+$ to NADP$^+$ in the chloroplast is high. The Fd$_{red}$ reduces PQ (which is part of the electron transport chain that connects photosystems I and II; see Figure 8.14) in the reaction Fd$_{red}$ + PQ$_{ox}$ → Fd$_{ox}$ + PQ$_{red}$. Acting as if the electron passed to it came from pheophytin-I (as happens in noncyclic photophosphorylation), PQ$_{red}$ passes the electron to a cytochrome complex (Cyt). The electron continues down the chain until it completes its cycle by returning to P$_{700}$. This cycle is a series of redox reactions, each exergonic, and the released energy is stored in a form that ultimately can be used to produce ATP.

Remember that when P$_{700}$* passed its electron on to Fd, we were left with a molecule of positively charged P$_{700}$$^+$ (having lost an electron, the chlorophyll has one unbalanced positive charge). In due course, P$_{700}$$^+$ interacts with a reducing agent that donates an electron, converting it back to uncharged P$_{700}$. This reducing agent, plastocyanin (PC), is the last member of the electron transport chain in Figure

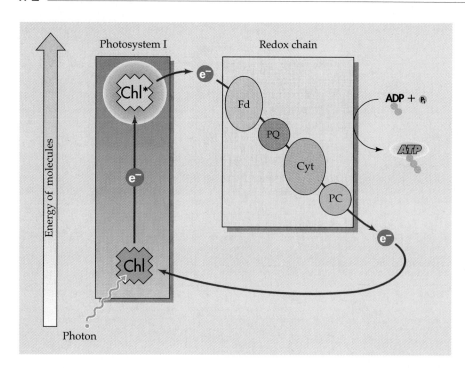

In cyclic photophosphorylation, excited chlorophylls pass electrons to an oxidizing agent, Fd, becoming positively charged and reducing Fd. Reduced Fd then reduces PQ, and so forth, in the cascade of redox reactions from Fd through PC. A chemiosmotic mechanism generates ATP from ADP + P_i. At the end of the redox chain, the last reduced electron carrier (PC_{red}) passes electrons to electron-deficient chlorophylls, returning them to a ground state ready to absorb another photon. Abbreviations: Cyt, cytochrome; Fd, ferredoxin; PC, plastocyanin; PQ, plastoquinone.

8.15. By the time the electron (passed from P_{700}^* and on through the redox chain) comes back to P_{700}^+ and reduces it, all the energy from the original photon has been released. In each of the redox reactions, some free energy is lost, until all of the original energy has been converted to heat *except* for that used to form ATP.

Comparing the cartoon in Figure 8.16 with Figure 8.15 may help make the concept of cyclic photophosphorylation clearer to you.

The Mechanism and Location of ATP Formation

How does electron transport in the photophosphorylation pathways form ATP? In Chapter 7 we considered the **chemiosmotic mechanism** for ATP formation in the mitochondrion. The chemiosmotic mechanism also operates in photophosphorylation. In chloroplasts, as in mitochondria, protons (H^+ ions) are transported across a membrane, resulting in a difference in pH and in electric charge across the membrane. In the mitochondrion, protons are pumped from the matrix, across the internal membrane, and into the space between the inner and outer mitochondrial membranes (see Figure 7.18). In the chloroplast, the electron carriers are located in the thylakoid membranes (see Figure 4.16). The electron carriers are oriented so that protons move into the interior of the thylakoid, so the inside becomes acidic with respect to the outside. This difference in pH leads to the passive movement of protons back out of the thylakoid, through protein channels in the membrane. These proteins are ATP synthases, enzymes that catalyze the formation of ATP; they are

activated by the movement of protons through the channels, just as in mitochondria (Figure 8.17).

The hypothesis that this chemiosmotic mechanism is responsible for the formation of ATP in chloroplasts was tested by Andre Jagendorf (of Cornell University) and Ernest Uribe (now at Washington State University) in the following way. Chloroplast thylakoids were isolated from spinach leaves and then kept in the dark, so that there would be no light energy to drive the production of ATP. The thylakoids were moved from a neutral solution to one with a low pH so that by diffusion the interiors of the thylakoids would become acidic. Then they were transferred to a solution that had a higher pH, so that the interiors of the thylakoids would be more acidic than the outsides—mimicking the situation created by light-driven pumping of protons into the interiors. This

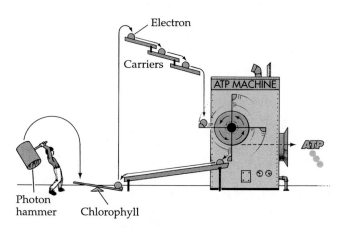

8.16 Cyclic Photophosphorylation Cycles Electrons

8.17 Chloroplasts Form ATP Chemiosmotically

Protons (H⁺) pumped across the thylakoid membrane from the stroma during noncyclic photophosphorylation make the interior of the thylakoid more acidic than the outside. Driven by this pH difference, the protons then return to the stroma through ATP synthase channels, activating the synthases to catalyze the formation of ATP from ADP + P_i. These reactions take place in several places on the thylakoid membrane simultaneously. Compare this chemiosmotic model with the one in Figure 7.18 that explains the activities of the inner mitochondrial membrane.

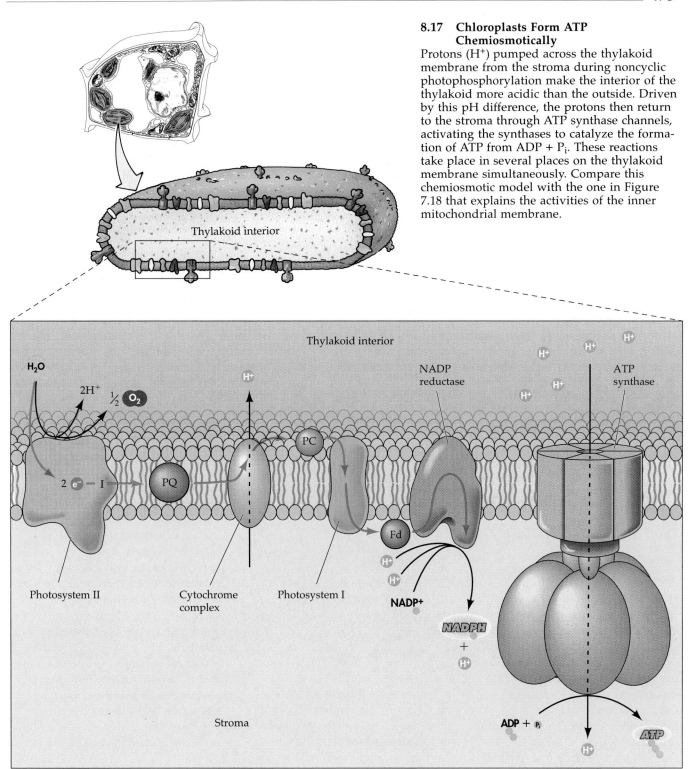

final step immediately resulted in the formation of ATP, even though no light was available to serve as the energy source (Figure 8.18). This is precisely the result predicted by the chemiosmotic model. (A very similar experiment, using mitochondria, pH changes, and ATP formation, was described in Chapter 7.)

Box 8.A describes an unusual use of chemiosmotic mechanisms to form ATP from sunlight.

Noncyclic and Cyclic Photophosphorylation Revisited

Photosystem I evolved before photosystem II; thus, cyclic photophosphorylation evolved before noncyclic photophosphorylation. Early in evolutionary history, photosynthetic bacteria used photosystem I and cyclic photophosphorylation to make ATP. This

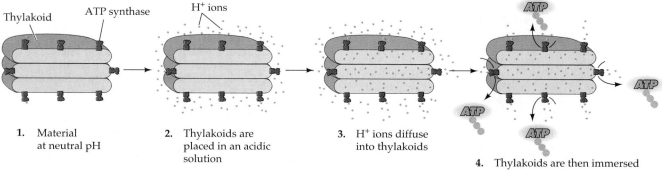

1. Material at neutral pH

2. Thylakoids are placed in an acidic solution

3. H^+ ions diffuse into thylakoids

4. Thylakoids are then immersed in a basic solution; movement of H^+ out of thylakoids is coupled with ATP formation

8.18   ATP Formation in the Dark
The experiment described in the text artificially induced ATP formation even without the presence of light energy.

form of photosynthesis evolved long before Earth's atmosphere contained significant quantities of oxygen gas. Nearly 3 billion years ago the cyanobacteria produced photosystem II, thus gaining the ability to perform noncyclic photophosphorylation—and to extract electrons from water and use them to reduce $NADP^+$. Over hundreds of millions of years, noncyclic photophosphorylation by cyanobacteria, algae, and plants poured enough oxygen gas into the atmosphere to make possible the evolution of cellular respiration. Today the evolutionarily ancient photosynthetic bacteria still have only cyclic photophos-

BOX 8.A

Photosynthesis in the Halobacteria

From time to time we discover that some group of organisms conducts its metabolic affairs in ways that previously were totally unexpected. In 1971, for example, biologists were surprised by discoveries about the metabolism of certain bacteria that live in salt ponds and salt lakes, places where the salt concentration is much higher than in the ocean. Because sunlight reaches to depths where these "salt-lovers" live, the surprise was not that these **halobacteria** trap sunlight energy. The surprise was that they do so without using chlorophyll.

These halobacteria species do not always use their unique light-trapping equipment. They are heterotrophs that get ATP by using oxygen

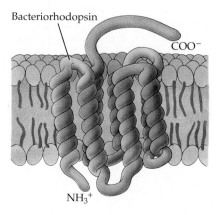

to metabolize substances they take in. But when the level of oxygen in the environment drops, the respiratory electron transport chains of the halobacteria shut down ATP production. Under these conditions, halobacteria turn on their light-trapping equipment.

The light-trapping equipment includes **retinal**, a carotenoid pigment. Retinal combines with a protein to form **bacteriorhodopsin**, which is incorporated into the plasma membrane. Each bacteriorhodopsin mole-

cule is organized in seven helical regions roughly perpendicular to the plane of the membrane. When one of these molecules absorbs light, protons are pumped through the membrane out of the cell. ATP forms by a chemiosmotic mechanism of the sort described in this chapter and in Chapter 7. This ATP is the immediate energy source for the halobacterium's metabolism.

Retinal is purple. Clusters of bacteriorhodopsin molecules form purple patches that cover as much as half of the cell surface. When a salt pond contains high densities of the tiny cells, the pond itself seems to be purple.

As you progress through this book, you will hear of retinal again. This pigment is also found in the vertebrate eye, where it plays a key role in vision. It is striking that retinal is used in two such different processes—vision and photophosphorylation—in organisms so widely separated in ancestry.

phorylation. Cyanobacteria, algae, and plants, which perform mostly noncyclic photophosphorylation, still perform cyclic photophosphorylation to produce ATP when the ratio of NADPH + H$^+$ to NADP$^+$ in their chloroplasts is high.

THE CALVIN–BENSON CYCLE

Real progress in understanding the dark reactions—the second main pathway of photosynthesis—came only after World War II. Satisfactory experimental techniques had not been developed before then. A group of scientists at the University of California, Berkeley, led by Melvin Calvin and including Andrew Benson and James Bassham, broke the problem wide open. The problem, as posed by the Berkeley group, was to learn the biochemical steps between the uptake of CO_2 and the appearance of the first complex carbohydrates in the chloroplast. Its solution depended on three advances in technique: (1) the discovery and availability of a radioactive carbon isotope, ^{14}C; (2) the development of a technique—paper partition chromatography—for the rapid separation of individual compounds from complex solutions; and (3) the development of autoradiography, a technique for locating colorless but radioactive compounds on a paper chromatogram. These advances are described in Box 8.B.

Armed with ^{14}C, paper partition chromatography,

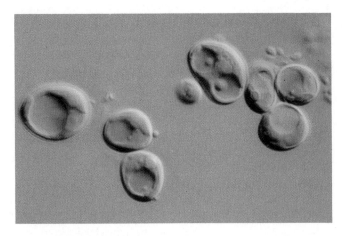

8.19 Algae Used in Experiments on Photosynthesis
The alga *Chlorella pyrenoidosa* was one of several used in studies by Calvin's group; this image is magnified about 1,600 times. Single-celled algae were chosen over leafy plants because their enzymes could be inactivated very quickly by ethanol, which penetrated the cell walls rapidly. This allowed the researchers to stop the reactions in the cells at a chosen time after treating the cells with radioactive $^{14}CO_2$ in dissolved form. Algae also could be grown continuously in cultures, providing a ready supply of research material with little variation.

and autoradiography, the Berkeley group set out to investigate how photosynthetic organisms metabolize CO_2. They worked mostly with unicellular aquatic algae, such as the green alga *Chlorella* (Figure 8.19). Algae were grown in dense suspensions in a flattened flask (called a "lollipop" because of its shape) between two bright lights, ensuring a rapid rate of photosynthesis (Figure 8.20a). To start an experiment, a solution containing dissolved $^{14}CO_2$ was suddenly squirted into the lollipop. At a carefully measured time after this squirt, a sample of the culture was rapidly drained into a container of boiling ethanol. The ethanol performed two functions: It killed the algae, stopping photosynthesis immediately, and it extracted the ^{14}C-containing intermediates of $^{14}CO_2$ metabolism from the algae (along with many other compounds). A sample of this ethanolic extract was spotted on filter paper for paper chromatography followed by autoradiography. Typical results of a 30-second exposure to $^{14}CO_2$ are shown in Figure 8.20b. As you can see, many biochemical reactions had taken place during that short interval.

The First Stable Product of Carbon Dioxide Fixation

During 30 seconds of continuous exposure to $^{14}CO_2$, many different products were formed in Calvin's lollipop. To determine which of them formed first, the experiment had to be repeated several times, using ever-shorter exposures. Even after exposures of less than two seconds, half a dozen or more labeled compounds appeared in the autoradiographs. One, however, was produced most rapidly and in greatest abundance: **3PG** (3-phosphoglycerate, also called 3-phosphoglyceric acid, or PGA), which we have already encountered as an intermediate in glycolysis. 3PG is the first stable product of CO_2 fixation. The Berkeley group isolated the individual carbon atoms from this 3PG and found that the carbon of the carboxyl group was much more intensely radioactive than the other two carbon atoms (Figure 8.21). (A single atom either is or is not radioactive. What we mean by "more intensely radioactive" is that in a *population* of molecules, the fraction of carboxyl carbons labeled with radioactivity was greater than the fractions of radioactive carbons in the other two carbon positions.)

The Berkeley group drew two important conclusions from finding the higher fraction of radioactive carboxyl carbons. First, the heavy labeling showed that the carboxyl carbon is obtained directly from CO_2. The existence of label in the other carbon atoms led to the second conclusion: Some kind of *cyclical* process is involved, a process by which 3PG is made by adding CO_2 to another compound that is itself produced from photosynthetic 3PG.

BOX 8.B

Tools that Cracked the Calvin–Benson Cycle

Melvin Calvin's group at Berkeley needed a way to keep track of the carbon atom from CO_2 taken up during photosynthesis in order to solve one of the mysteries of the dark reactions: What happens to the carbon after the CO_2 molecule reacts? It was the program to produce the first atomic bomb that indirectly provided the solution when the radioactive carbon isotope ^{14}C was made available to scientists. With this carbon isotope, samples of CO_2 could be prepared in which some of the carbon atoms were ^{14}C rather than the stable isotope ^{12}C found in nature.

Paper partition chromatography

Any compound incorporating the radioactive material would also be radioactive.

Next, Calvin's group needed an improved method of separating complicated mixtures into their individual components. Living things contain thousands of different chemical components. If we want to study just one of them, we must separate it from all the thousands of others. A powerful new separation tool, called **paper partition chromatography**, became available shortly before the Berkeley group began its work.

In paper chromatography, a drop of a solution containing the compounds to be separated is placed near one end of a strip of filter paper (as shown in part *a* of the figure below). The paper is then lowered into a container with a suitable mixture of organic solvents (liquids such as chloroform and alcohols) until the end of the paper is submerged. The solvent works its way up through the paper by capillary action. As the solvent climbs, so do the compounds to be separated. For now, let us say that

the compounds in the mixture are pigments, so that we can see what happens to them. By the time the solvent has moved several centimeters up the paper, one can usually see that the mixture of pigments has separated because the different pigments are visible as colored spots at various distances up the paper.

If repeated, such experiments show that a given compound always travels the same relative distance. If we take the distance (from the starting spot) moved by a particular compound and divide it by the distance (from the same point) moved by the solvent, that ratio is the "front ratio" (R_F) of that compound under those conditions of temperature and solvent composition. Part *b* of the figure shows how we determine the R_F. Once calculated, the R_F of a compound can be used to identify it.

Some compounds are difficult to separate by paper partition chromatography. In such cases, **two-dimensional chromatography** sometimes

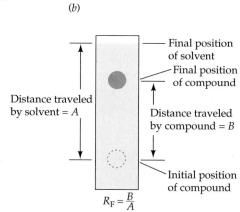

(a)

Paper strip

End of strip placed in solvent; solvent begins to rise

Solvent

Mixture of compounds to be separated

(b)

Distance traveled by solvent = A

Final position of solvent

Final position of compound

Distance traveled by compound = B

Initial position of compound

$$R_F = \frac{B}{A}$$

What Is the Carbon Dioxide Acceptor?

What is this compound, obtained from the metabolism of 3PG, that binds with CO_2 to make more 3PG? Given the structure of 3PG, it was reasonable to expect that the mysterious CO_2 acceptor would be a *two*-carbon compound that could react with CO_2 to become a *three*-carbon compound—3PG—with the CO_2 becoming a carboxyl group. If this idea were

correct, it should have been possible for the Berkeley group to find, on their chromatograms, a compound with only two carbon atoms, both of which were radioactive after a lollipop experiment. They did *not* find such a compound, and this made the problem more difficult. The "obvious" answer—that the CO_2 acceptor is a two-carbon compound—was wrong. Where would they go from here?

Two-dimensional chromatography

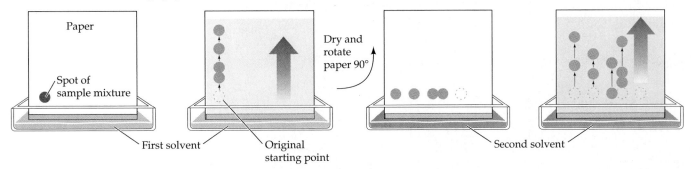

Autoradiography

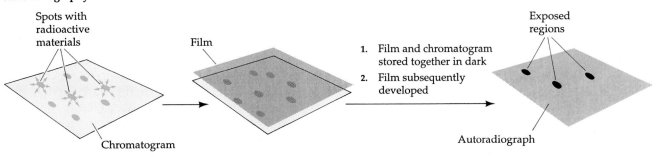

works. In this technique, the sample mixture is applied as a single spot near one corner of a square of filter paper. Paper chromatography is run as before, using one solvent and achieving a partial separation of the mixture along one edge of the paper. The paper is then dried and turned so that the row of partially separated compounds is now along the bottom edge. Chromatography is run again with a different solvent. If the properties of the second solvent are markedly different from those of the first, the second run will separate compounds that stayed together in the first run. The result looks like a scatter plot. Because each compound has a certain R_F in each of the solvents,

the precise pattern can be repeated again and again.

The third thing Calvin's group needed was a method for locating tiny amounts of radioactive material visually. Suppose now that you are doing paper chromatography runs not with a mixture of pigments, but with a mixture of radioactive and nonradioactive compounds. The radioactivity comes from ^{14}C incorporated into precursors as described at the beginning of this box. You could take advantage of the fact that radioactive emissions, like light, expose photographic film. The method that does this, **autoradiography**, works as follows: The paper chromatogram (the filter-paper sheet on which chro-

matography has been performed) has spots of ^{14}C-containing materials at unknown places. This chromatogram is taken into a darkroom, covered either with a type of photographic film or with a liquid photographic emulsion, and kept in the dark for a suitable length of time, during which the radioactive decay of ^{14}C releases particles that expose the film. Later the film is developed, revealing dark spots (composed of silver grains in the film; see Box 2.A) that correspond to the places on the chromatogram where there were accumulations of ^{14}C-containing materials. The R_F values of the radioactive compounds can thus be determined from the positions of the dark spots.

At this point in the investigation, concentrating on a photosynthetic *cycle* became useful. Consider a tentative cycle of the sort shown in Figure 8.22*a*: CO_2 reacts with X, the CO_2 acceptor, to produce 3PG. From the 3PG, photosynthetic organisms make products (things like glucose) and more X; this new X can react with another molecule of CO_2 and keep the cycle going. But what is X, and what are the other intermediates in the cycle?

It was observed that 3PG was the only acid phosphate produced in significant amount, whereas many kinds of *sugar* phosphates appeared on the chromatograms. On this basis, the Berkeley group guessed that the first thing to happen to 3PG is its conversion to a three-carbon sugar phosphate (glyceraldehyde 3-phosphate, which we will call G3P). Such a reaction is a reduction, and since reductions are highly endergonic, the Berkeley group proposed that this re-

(a)

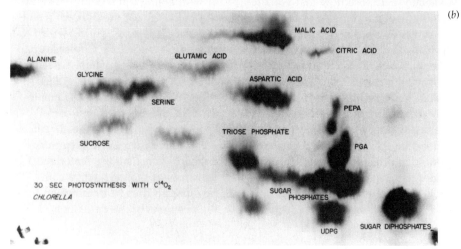

(b)

8.20 A Lollipop and Its Products

(a) The lollipop used in experiments on photosynthesis. The thin flask was filled with a suspension of algae and illuminated from both sides. After injection of $^{14}CO_2$, a sample was drained from the flask into boiling ethanol. (b) A chromatogram showing products of algal photosynthesis. The dark spots are compounds containing ^{14}C—all formed in the 30 seconds following injection of $^{14}CO_2$. The spot labeled PGA (an older notation) corresponds to the position of 3-phosphoglycerate (3PG), a compound we will discuss in the next section.

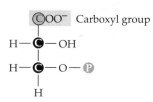

3-Phosphoglycerate

8.21 3PG Is the First Stable Product of CO_2 Fixation

In experiments in which an algal sample from a lollipop was killed with ethanol only a few seconds after the introduction of $^{14}CO_2$, it was found that most 3-phosphoglycerate molecules had a ^{14}C in the carboxyl group, as indicated by the red symbol; smaller numbers of 3PG molecules had the label in the other two carbon atoms. The heavy labeling in the carbon of the carboxyl group indicated that the carbon was obtained directly from $^{14}CO_2$.

action would require ATP (Figure 8.22b). They supported this proposal with two lollipop experiments, one in the light and one in the dark. When they supplied $^{14}CO_2$ to the algae in the lollipop, 3PG rapidly became radioactively labeled in both experiments, but G3P became radioactively labeled only in the light, showing that the reduction does require energy—presumably in the form of ATP generated in the light.

At this point an extremely clever suggestion was made: If we assume that Figure 8.22b accurately models what occurs in the chloroplast, then it should be possible to regulate this cycle in the laboratory by two simple means. First, turning off the light would specifically block the step from 3PG to G3P, the sugar phosphate, because the necessary ATP and NADPH + H$^+$ can only be produced with an input of energy; in photosynthesis, the energy source is light. There-

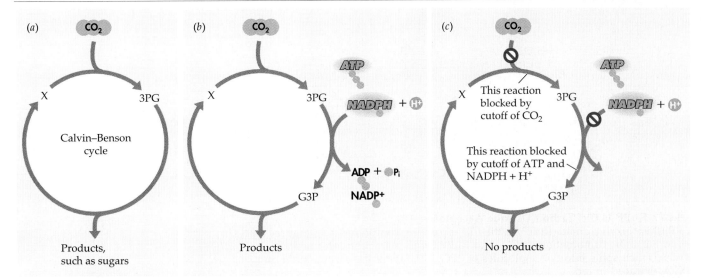

8.22 Manipulating the Dark Reactions of Photosynthesis

(a) After it became apparent that CO_2 combines with some molecule to form 3PG and that a cyclical process is involved, a pathway could be devised in which a molecule of compound X combines with CO_2 to form 3PG and in which further reactions regenerate X. (b) Ensuing speculation and experimentation led to the proposition that 3PG was reduced to a sugar phosphate, glyceraldehyde 3-phosphate (G3P); this endergonic reaction, a reduction, would require ATP, as shown here. (c) In the proposed model, a cutoff of CO_2 blocks the formation of 3PG from compound X and CO_2; a cutoff of ATP (by turning off the light) blocks the formation of G3P from 3PG.

are the changes we would expect to see *if* our model (Figure 8.22c) is correct. Similar reasoning should convince you that when the CO_2 supply is cut off (with the lights on), the concentration of X rises and that of 3PG falls. With no CO_2 available, X is no longer used up and 3PG is no longer formed, but 3PG can still be reduced to G3P.

Having devised this model, the Berkeley group proceeded to study the effects of changes in light intensity and CO_2 supply on the concentrations of all the major radioactive compounds found in the

fore ATP should be made in the chloroplast only when the light is on. Second, the reaction from X to 3PG could easily be blocked by cutting off the supply of CO_2 to the lollipop (Figure 8.22c).

Assume that photosynthesis is proceeding at a steady pace, so the concentrations of 3PG and the CO_2 acceptor X in the cells are constant. The lights are on, of course, and there is plenty of CO_2. Suddenly the investigator turns off the lights, thus blocking the cycle as proposed in Figure 8.22c: What change in 3PG concentration occurs immediately? What change in the concentration of X occurs immediately? *Stop.* Think about this before you read on.

When the light is turned off, no more ATP is made. Without ATP, the reaction from 3PG to the three-carbon sugar G3P cannot take place. Therefore, 3PG is no longer being used up. However, there is nothing to stop the formation of 3PG from incoming CO_2 and X, as long as there is any X around. Therefore, the immediate consequence of turning off the light is an *increase* in the level of 3PG. On the other hand, X continues to be used up, because its reaction with CO_2 does not depend on light or ATP; but the formation of X slows down because 3PG can no longer be reduced and ultimately form new X. Therefore, the concentration of X *decreases* (Figure 8.23). These

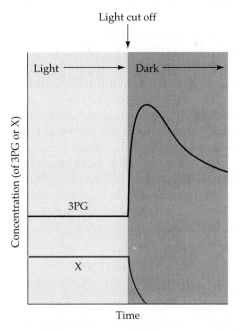

8.23 Changes in the Dark

When the light to a lollipop is turned off, 3PG accumulates as compound X is combined with CO_2. The concentration of X, on the other hand, falls when the light is turned off. X is not replenished because 3PG is not converted to G3P to continue the cycle (light is necessary to provide the ATP for the endergonic 3PG → G3P reaction).

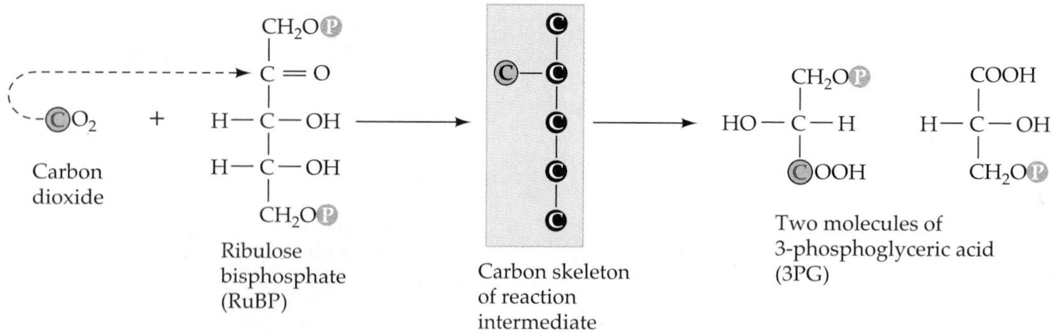

8.24 RuBP Is the Carbon Dioxide Acceptor
Ribulose bisphosphate (RuBP) is the CO_2-accepting compound X in Figures 8.22 and 8.23. The combination of CO_2 and RuBP forms a reaction intermediate, which then splits into two molecules of 3PG. The fate of the carbon atom in CO_2 is traced in red.

lollipop experiments. The first thing they noticed was that only one compound underwent the concentration changes proposed for 3PG—and that was 3PG itself. This result showed that their model was likely to be accurate. But would any compound behave in the way predicted for the mysterious compound X, the CO_2 acceptor? Yes—and only one: a five-carbon sugar phosphate called **RuBP** (ribulose bisphosphate). It seemed, then, that instead of the originally proposed reaction (in which CO_2 was thought to react with a two-carbon sugar to form 3PG), there must be a reaction in which CO_2, with its single carbon, combines with the five-carbon RuBP to give two molecules of the three-carbon 3PG (Figure 8.24).

The best way to prove that a proposed reaction does in fact take place is to find an *enzyme* that catalyzes it. In this case, such an enzyme was soon discovered. The enzyme, RuBP carboxylase, now commonly called **rubisco**, was found in the algae studied by the Berkeley group and also in the leaves of spinach. Best of all, studies of spinach revealed that rubisco is found in only one part of the cell: the chloroplast—exactly where one should find an enzyme concerned with photosynthesis. It was concluded, then, that RuBP is the CO_2 acceptor, the previously unknown compound X.

Filling the Gaps in the Calvin–Benson Cycle

Having discovered the first product of CO_2 fixation (3PG) and the CO_2 acceptor (RuBP), the Berkeley group proceeded to work out the remaining reactions of the cycle. They found some relatively complicated steps between G3P and RuBP; among the intermediates are sugar phosphates with four, five, six, and seven carbon atoms. All the proposed intermediates have been found in chloroplasts, as have all the necessary enzymes. It was also discovered that ATP is needed at one more point in the Calvin–Benson cycle:

in the step producing RuBP (ribulose *bis*phosphate) from RuMP (ribulose *mono*phosphate). This additional ATP requirement makes it even easier to understand why turning off the light drastically reduces the concentration of RuBP (X in Figure 8.23).

For now, you need learn only the material in Figure 8.25, which summarizes the key features of the Calvin–Benson cycle. In the chloroplast, most of the enzymes that catalyze the reactions of this pathway are dissolved in the stroma (see Figure 4.16), and this is where the reactions take place. RuBP reacts with CO_2 to form 3PG. 3PG is then reduced—in a reaction requiring ATP as well as hydrogens provided by NADPH + H^+—to a three-carbon sugar phosphate (G3P). What follows this step is a complex sequence of reactions with two principal outcomes: the formation of more RuBP, and the release of products such as glucose. The production of one molecule of glucose ($C_6H_{12}O_6$) requires the Calvin–Benson cycle to operate six times on successive CO_2 molecules. Just as the respiration of one mole of glucose *yields* 686 kcal of energy (see Chapter 7), it *requires* 686 kcal to make one mole of glucose from CO_2.

Figure 8.26 gives a general summary of photosynthesis. The glucose produced in photosynthesis is subsequently used to make other compounds besides sugars. The carbon of glucose is incorporated into amino acids, lipids, and the building blocks of the nucleic acids. The products of the Calvin–Benson cycle are of crucial importance to the entire biosphere, for they serve as the food for all of life. Their covalent bonds represent the total energy yield from the harvesting of light by plants. Most of this stored energy is released by the plants themselves in their own glycolysis and cellular respiration. However, much plant matter ends up being consumed by animals. Glycolysis and cellular respiration in the animals then releases free energy from the plant matter for use in the animal cells.

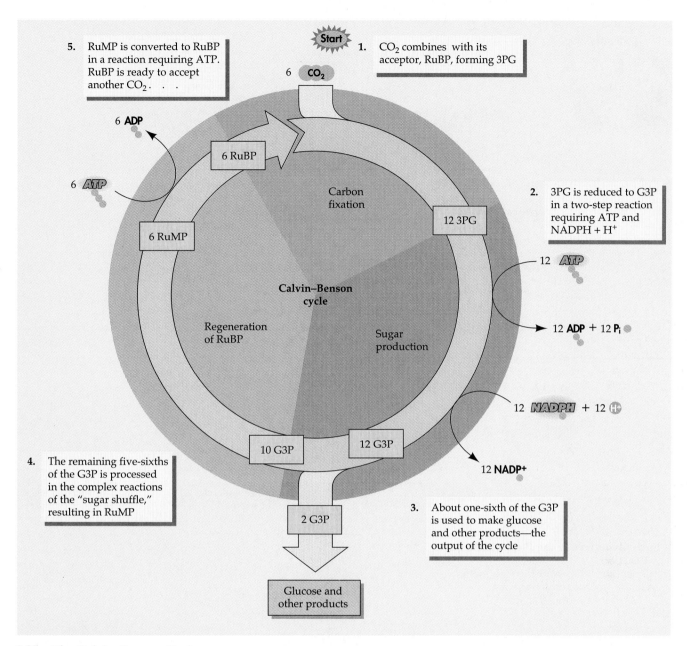

8.25 The Calvin–Benson Cycle
The Calvin-Benson cycle, sometimes called the dark reactions of photosynthesis, uses CO_2 to produce glucose and other organic molecules that contain the carbon and energy necessary for the many and varied processes of life. This diagram shows only the key steps; the values given are those necessary to make one molecule of glucose, which requires six "turns" of the cycle.

PHOTORESPIRATION

The substrate specificity of the enzyme rubisco is not limited to CO_2. Rubisco also catalyzes a reaction of RuBP with oxygen. (*Rubisco* stands for *ri*bulose *bis*phosphate *c*arboxylase/*o*xygenase.) The oxygenase reaction is favored when CO_2 levels are low or O_2 levels are high. One of the products of this reaction is glycolate, a two-carbon compound that leaves the chloroplast and diffuses into organelles called microbodies (see Chapter 4). In the microbody, glycolate is oxidized (in an oxygen-requiring reaction); later the product undergoes reactions in mitochondria leading to the release of CO_2. The rate of the overall process (from RuBP and O_2 to the release of CO_2) is roughly proportional to the light intensity. Because of its dependence on light and because it takes up oxygen and releases carbon dioxide, this process is called **photorespiration** (Figure 8.27).

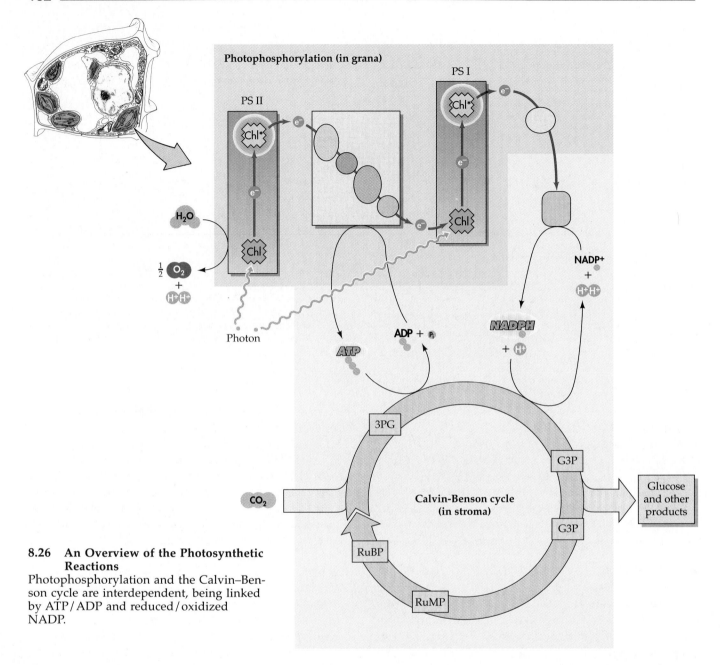

8.26   An Overview of the Photosynthetic Reactions
Photophosphorylation and the Calvin–Benson cycle are interdependent, being linked by ATP/ADP and reduced/oxidized NADP.

You can see that photorespiration interferes with photosynthesis; in fact, it apparently reverses it—but without resulting in ATP formation as does cellular respiration. The role of photorespiration in the life of the plant is unknown; it may simply be wasteful, with no positive role. With this in mind, many scientists are attempting to develop a gene that codes for a form of rubisco that recognizes only CO_2 as its substrate, and to insert that gene into crop plants. (See Chapter 14 for a discussion of this recombinant DNA technology.)

It seems odd, though, that rubisco, the most abundant single protein in the living world, should apparently function less than optimally. Most types of plants photorespire away a substantial fraction of the CO_2 initially fixed in photosynthesis. But, as we are

about to see, some plants have minimized photorespiration, thus maximizing the efficiency of their photosynthesis.

ALTERNATE MODES OF CARBON DIOXIDE FIXATION

The discoveries of the Berkeley group led to the expectation that the exposure of a plant to both light and $^{14}CO_2$ would always lead to the appearance of 3-phospho[^{14}C]glycerate as the first labeled product of CO_2 fixation. Thus scientists were surprised when they learned that such treatment of chloroplasts extracted from sugarcane leaves leads instead to the formation of four-carbon acids as the first ^{14}C-con-

(a)

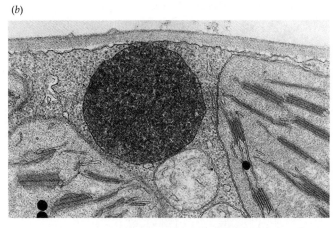

8.27 Chloroplasts, Microbodies, and Photorespiration *(a)* In the chloroplast, the enzyme rubisco may catalyze a reaction between RuBP and O_2; further reactions lead to the reduced two-carbon compound glycolate (CC_{red}). Glycolate leaves the chloroplast and diffuses into a microbody, where it is oxidized. Later reactions lead to the formation of CO_2, which diffuses out of the cell. With the uptake of O_2 and the release of CO_2, the overall result resembles respiration, and because the process occurs in rough proportion to light intensity, it has been dubbed "photorespiration." *(b)* The dark, round object is a microbody in a mesophyll cell of a tobacco leaf. Portions of adjacent chloroplasts are visible to the lower left and far right, a mitochondrion is to the immediate lower right, and the plasma membrane and cell wall are above the microbody in this view.

taining products. Subsequently it was shown that many plants follow this pattern in fixing CO_2. Known as **C₄ plants** because the first products of their CO_2 fixation are four-carbon compounds, these plants perform the normal Calvin–Benson cycle, but they have an additional early step that traps CO_2 without losing carbon to photorespiration, greatly increasing the overall photosynthetic yield. Because this step functions even at low levels of CO_2 within the leaf, C₄ plants very effectively optimize photosynthesis.

C₄ plants live in environments where water is not always available. To prevent excessive water loss, the leaves keep their stomata closed much of the time. This leads to a depletion, by photosynthesis, of CO_2 within the leaf. However, because their leaves contain the enzyme **PEP carboxylase** (phosphoenolpyruvate carboxylase), C₄ plants have a means of compensating for this depletion. PEP carboxylase catalyzes the reaction of PEP (phosphoenolpyruvate, a three-carbon acid) with CO_2 to yield the four-carbon compound oxaloacetate as the first product of CO_2 fixation. PEP carboxylase has a much greater affinity for CO_2 than does rubisco and, because it lacks an oxygenase function, PEP does not support photorespiration, so C₄ plants can trap CO_2 even when that gas is present in a much reduced concentration. (You may recall that PEP is a late intermediate in glycolysis—see Figure 7.11—and that oxaloacetate is the last intermediate in the citric acid cycle—see Figure 7.13. Evolution has led to the use of certain compounds in a number of different ways in living things.)

The leaf anatomy of C₄ plants differs from that of **C₃ plants** (plants that produce the three-carbon compound 3PG in the first step of the Calvin–Benson cycle). C₃ plants have only one type of cell capable of photosynthesis (Figure 8.28*a*), but the leaves of C₄ plants have two classes of photosynthetic cells, each with a distinctive type of chloroplast. The cells are arranged as shown in Figure 8.28*b*, with a photosynthetic **mesophyll** layer surrounding an inner layer of **bundle sheath cells**, which are also photosynthetic. Like those of C₃ plants, the mesophyll cells of C₄ plants contain chloroplasts filled with grana that trap carbon dioxide for photosynthesis. From this trapped CO_2, C₄ plants produce four-carbon compounds that diffuse into the bundle sheath cells. Once in the bundle sheath cells these compounds are decarboxylated (a carboxyl group is removed) to release CO_2, which is recaptured by rubisco and used in the Calvin–Benson cycle—the cycle that in C₃ plants takes place entirely within the mesophyll cells (Figure 8.29). The chloroplasts in the bundle sheath cells of C₄ plants lack well-developed grana but typically have substantial starch grains deposited in them because they, rather than the mesophyll chloroplasts, are the sites where sugars are finally formed and starches are stored.

What this system does is pump CO_2 from a region where its concentration is low (the intercellular spaces within the leaf) to one where it is relatively more abundant (the bundle sheath layer). At the same time, C₄ photosynthesis bypasses photorespiration, thus retaining more of the carbon that is fixed in photosynthesis. PEP carboxylase in the mesophyll chloroplasts can take up CO_2 when rubisco cannot, and, because it lacks the oxygenase activity of ru-

(a) Arrangement of cells in a C₃ leaf

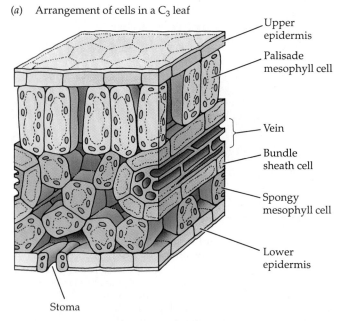

(b) Arrangement of cells in a C₄ leaf

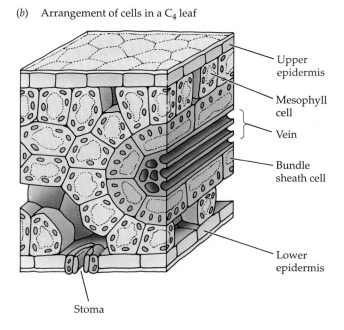

8.28 C₄ Plants Differ from C₃ Plants in Leaf Anatomy
(a) In the leaf of a C₃ plant, the bundle sheath cells surrounding the vascular elements of a vein are relatively small; the upper part of the leaf is filled with upright palisade mesophyll cells; and loosely arranged spongy mesophyll cells allow gases to circulate in the lower layers within the leaf. Both mesophyll layers carry on photosynthesis, but the bundle sheath cells do not. (b) In a C₄ leaf, the bundle sheath cells are usually larger and contain prominent chloroplasts toward their outer edges; uniform mesophyll cells surround the entire vascular bundle. This arrangement facilitates the incorporation of carbon from CO_2 into four-carbon compounds by the mesophyll cells and the passage of these carbon-containing compounds to the bundle sheath cells, where the reactions of the Calvin–Benson cycle take place. The mesophyll serves as a CO_2 pump.

bisco, does not cause photorespiration. The temporary products (four-carbon acids) are then loaded into the bundle sheath, allowing the release of sufficient CO_2 to keep the rubisco in the bundle sheath busy. The O_2 level in the bundle sheath is not so high, however, as to cause photorespiration. Table 8.1 compares C₃ and C₄ photosynthesis.

A related but distinguishable system called **Crassulacean acid metabolism** (CAM) functions in certain other plants that face frequent water shortages. Many of these are members of the family Crassulaceae, which includes the ice plants and some other succulent plants. Many cacti also perform CAM. Because of the way in which the stomata of CAM plants are regulated, these plants have access to atmospheric CO_2 only at night. By day their stomata are closed, preventing water loss, and no CO_2 can enter the leaf. Using PEP carboxylase to trap CO_2 at night allows these plants to store great quantities of CO_2 in the form of carboxyl groups of four-carbon acids. By day, behind closed stomata, the CO_2 is released within the leaves; it is recaptured by rubisco, and photosynthesis then proceeds by way of the Calvin–Benson cycle. The difference between this system and C₄ photosynthesis is that here the PEP comes from the respiratory breakdown of sugars at night; in C₄ photosynthesis, PEP is produced photosynthetically in a light-requiring reaction by day.

TABLE 8.1
Comparison of Photosynthesis in C₃ and C₄ Plants

VARIABLE	C₃ PLANTS	C₄ PLANTS
Photorespiration	Extensive	Minimal
Perform Calvin–Benson cycle	Yes	Yes
Primary CO_2 acceptor	RuBP	PEP
CO_2-fixing enzyme	Rubisco (RuBP carboxylase)	PEP carboxylase
First product of CO_2 fixation	3PG (3-carbon compound)	Oxaloacetate (4-carbon compound)
Affinity of carboxylase for CO_2	Moderate	High
Leaf anatomy: photosynthetic cells	Mesophyll	Mesophyll + bundle sheath
Classes of chloroplasts	One	Two

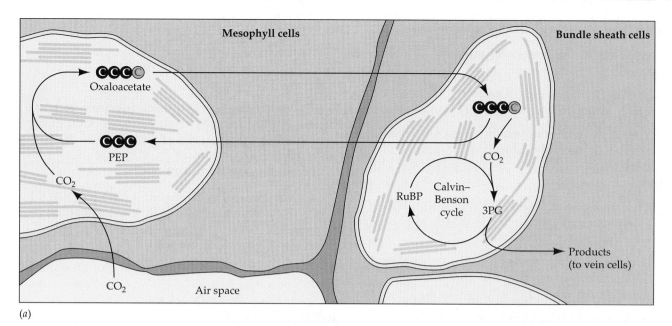

(a)

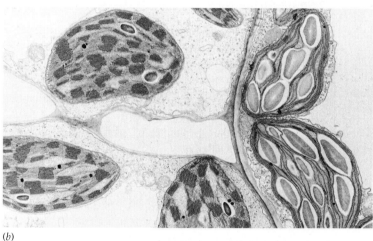

(b)

8.29 C₄ Photosynthesis

(a) In C$_4$ photosynthesis, mesophyll cells in the leaf take up CO_2 and incorporate the carbon atom into four-carbon compounds. The four-carbon compounds diffuse into adjacent bundle sheath cells, where they are decarboxylated, releasing CO_2. The enzyme rubisco picks up this CO_2, and the usual Calvin–Benson cycle of C$_3$ photosynthesis ensues. (b) Portions of two mesophyll cells (left) and two chloroplasts in a single bundle sheath cell (right) from the leaf of a C$_4$ plant. Note the numerous grana and few starch grains in the chloroplasts of the mesophyll cells; in the chloroplasts of the bundle sheath cell, where the Calvin–Benson cycle forms the products of photosynthesis, there are many large, oval starch granules but very few membranes organized into grana.

PHOTOSYNTHESIS AND CELLULAR RESPIRATION

In plants, cellular respiration takes place both in the light and in the dark, whereas photosynthesis takes place only in the light. The site of glycolysis is the cytosol, that of respiration is the mitochondria, and that of photosynthesis is the chloroplasts. Thus photosynthesis and respiration can proceed simultaneously but in different organelles.

For a plant to live, it must photosynthesize more than it respires, giving it a net gain of carbon dioxide and energy from the environment. Accordingly, the plant world, along with photosynthetic bacteria and protists, exports food—and oxygen—to the animal kingdom and to all other nonphotosynthetic organisms (with the exception of a few types of bacteria). Animals require both food and oxygen; they return carbon dioxide that plants may use in photosynthesis. Thus both carbon dioxide and oxygen have natural cycles.

Photosynthesis and respiration have important similarities. In eukaryotes, both processes reside in specialized organelles that have complex systems of internal membranes. ATP synthesis in both processes relies on the chemiosmotic mechanism, involving the pumping of protons through a membrane. Another key feature of both respiration and photosynthesis is electron transport, that is, the passing of electrons from carrier to carrier in a series of exergonic redox reactions. In respiration, the carriers receive electrons from high-energy food molecules and pass them ultimately to oxygen, forming water. On the other hand, photosynthesis requires an input of light energy to make chlorophyll, a reducing agent strong enough to initiate the transfer of electrons. In photosynthesis, water is the source of the electrons, and oxygen is released from water in a very early step. The electrons from water end up in NADPH and, finally, in food molecules.

SUMMARY of Main Ideas about Photosynthesis

The sun provides the energy for virtually all biological work.

Photosynthesis traps sunlight and uses the converted light energy to synthesize ATP and sugars.

The first pathway of photosynthesis is photophosphorylation; the second pathway is the Calvin–Benson cycle.
Review Figure 8.4

Photophosphorylation begins with the absorption of light by a pigment molecule.
Review Figure 8.8

Several chloroplast pigments help trap sunlight and pass the excitation to the pigment cholorophyll.
Review Figure 8.9

In photophosphorylation, an excited chlorophyll molecule passes an electron and energy to an oxidizing agent; the electron then passes through a series of carriers, and some of the energy thus released is used to form ATP chemiosmotically.
Review Figures 8.13 and 8.18

Noncyclic photophosphorylation produces NADPH + H$^+$ as well as ATP; it also produces O$_2$ by the breakdown of water.

Two chlorophyll-containing photosystems participate in noncyclic photophosphorylation.
Review Figure 8.14

ATP is the sole output of cyclic photophosphorylation, which uses only photosystem I.
Review Figure 8.15

The Calvin–Benson cycle uses carbon from carbon dioxide to produce organic compounds. CO$_2$ reacts with RuBP in the presence of rubisco, forming two molecules of 3PG.
Review Figures 8.22, 8.23, and 8.24

The Calvin–Benson cycle requires ATP and NADPH + H$^+$, the products of noncyclic photophosphorylation.
Review Figure 8.25 and 8.26

Because rubisco catalyzes the reaction of O$_2$ with RuBP, as well as that of CO$_2$, C$_3$ plants may photorespire when CO$_2$ levels are low or O$_2$ levels are high.
Review Figure 8.27

C$_4$ plants rarely photorespire because they augment the Calvin–Benson cycle with further reactions that permit photosynthesis even at low CO$_2$ levels.
Review Figure 8.28 and 8.29 and Table 8.1

C$_4$ photosynthesis and CAM use the same enzyme to trap CO$_2$, but the CO$_2$ acceptor is produced by energy-requiring light reactions in C$_4$ plants and by glycolysis in CAM plants.

SELF-QUIZ

1. Which statement about light is *not* true?
   a. Its velocity in a vacuum is constant.
   b. It is a form of energy.
   c. The energy of a photon is directly proportional to its wavelength.
   d. A photon of blue light has more energy than one of red light.
   e. Different colors correspond to different frequencies.

2. Which statement about light is true?
   a. An absorption spectrum is a plot of biological effectiveness versus wavelength.
   b. An absorption spectrum may be a good means of identifying a pigment.
   c. Light need not be absorbed to produce a biological effect.
   d. A given kind of molecule can occupy any energy level.

   e. A pigment loses energy as it absorbs a photon.

3. Which statement is *not* true of chlorophylls?
   a. They absorb light near both ends of the visible spectrum.
   b. They can accept energy from other pigments, such as carotenoids.
   c. Excited chlorophyll (Chl*) can either reduce another substance or fluoresce.
   d. Excited chlorophyll is an oxidizing agent.
   e. They contain magnesium.

4. In cyclic photophosphorylation
   a. oxygen gas is released.
   b. ATP is formed.
   c. water donates electrons and protons.
   d. NADPH + H$^+$ is formed.
   e. CO$_2$ reacts with RuBP.

5. Which does *not* happen in noncyclic photophosphorylation?
   a. Oxygen gas is released.
   b. ATP is formed.
   c. Water donates electrons and protons.
   d. NADPH + H$^+$ is formed.
   e. CO$_2$ reacts with RuBP.

6. In the chloroplast
   a. light leads to the pumping of protons out of the thylakoids.
   b. ATP is formed when protons are pumped into the thylakoids.
   c. light causes the stroma to become more acidic than the thylakoids.
   d. protons return passively to the stroma through protein channels.
   e. proton pumping requires ATP.

7. Which is *not* true of the Calvin–Benson cycle?
   a. CO$_2$ reacts with RuBP to form 3PG.

b. RuBP is formed by the metabolism of 3PG.
c. ATP and NADPH + H^+ are formed when 3PG is reduced.
d. The concentration of 3PG rises if the light is switched off.
e. Rubisco catalyzes the reaction of CO_2 and RuBP.

8. In C_4 photosynthesis
a. 3PG is the first product of CO_2 fixation.
b. rubisco catalyzes the first step in the pathway.
c. four-carbon acids are formed by PEP carboxylase in the bundle sheath.

d. photosynthesis continues at lower CO_2 levels than in C_3 plants.
e. CO_2 released from RuBP is transferred to PEP.

9. C_4 photosynthesis and the acid metabolism in Crassulaceae differ in that
a. only C_4 photosynthesis uses PEP carboxylase.
b. CO_2 is trapped by night in Crassulaceae and by day in C_4 plants.

c. four-carbon acids are formed only in C_4 photosynthesis.
d. only Crassulaceae commonly grow in dry or salty environments.
e. only C_4 photosynthesis helps conserve water.

10. Photorespiration
a. takes place only in C_4 plants.
b. includes reactions carried out in microbodies.
c. increases the yield of photosynthesis.
d. is catalyzed by PEP carboxylase.
e. is independent of light intensity.

FOR STUDY

1. Both photophosphorylation and the Calvin–Benson cycle stop when the light is turned off. Which specific reaction stops first? Which stops next? Continue answering the question "Which stops next?" until you have explained why both pathways have stopped.

2. In what principal ways are the reactions of photophosphorylation similar to the respiratory chain and oxidative phosphorylation discussed in Chapter 7? Differentiate between cyclic and noncyclic photophosphorylation in terms of (1) the *products* and (2) the *source* of electrons for reduction of oxidized chlorophyll.

3. The development of what three experimental techniques made it possible to elucidate the Calvin–Benson cycle? How were these techniques used in the investigation?

4. If water labeled with ^{18}O is added to a suspension of photosynthesizing chloroplasts, which of the following compounds will first become labeled with ^{18}O: ATP, NADPH, O_2, or 3PG? If water labeled with 3H is added to a suspension of photosynthesizing chloroplasts, which of those compounds will first become radioactive? If CO_2 labeled with ^{14}C is added to a suspension of photosynthesizing chloroplasts, which of those compounds will first become radioactive?

READINGS

Alberts, B., D. Bray, J. Lewis, M. Raff, K. Roberts and J. D. Watson. 1994. *Molecular Biology of the Cell*, 3rd Edition. Garland Publishing, New York. Chapter 14 on energy conversion contains a good discussion of photosynthesis.

Bjorkman, O. and J. Berry. 1973. "High-Efficiency Photosynthesis." *Scientific American*, October. A discussion of C_4 photosynthesis and what it means to the plants in which it is found.

Clayton, R. K. 1980. *Photosynthesis*. Cambridge University Press, New York. An advanced general treatment of photosynthesis by a prominent photobiologist.

Govindjee and W. J. Coleman. 1990. "How Plants Make Oxygen." *Scientific American*, February. A "clock" in photosystem II that splits water into oxygen gas, protons, and electrons.

Hall, D. O. and K. K. Rao. 1987. *Photosynthesis*, 4th Edition. Edward Arnold, New York. An intermediate-level treatment of all the major topics in photosynthesis and excellent bibliography, all in 100 pages.

Stryer, L. 1995. *Biochemistry*, 4th Edition. W. H. Freeman, New York. Chapter 22 gives an advanced but clear treatment of topics in photosynthesis.

Voet, D. and J. G. Voet. 1990. *Biochemistry*. John Wiley & Sons, New York. Chapter 22 discusses photosynthesis.

Weinberg, C. J. and R. H. Williams. 1990. "Energy from the Sun." *Scientific American*, September. Photosynthesis and biomass technology, along with other solar-derived technologies such as wind and solar-thermal, are considered as sources of energy for industrial and other uses.

Youvan, D. C. and B. L. Marrs. 1987. "Molecular Mechanisms of Photosynthesis." *Scientific American*, June. A difficult but interesting article on events in the first fraction of a millisecond of photosynthesis in a bacterium. Part of the article is better read after reading Part Two of this book.

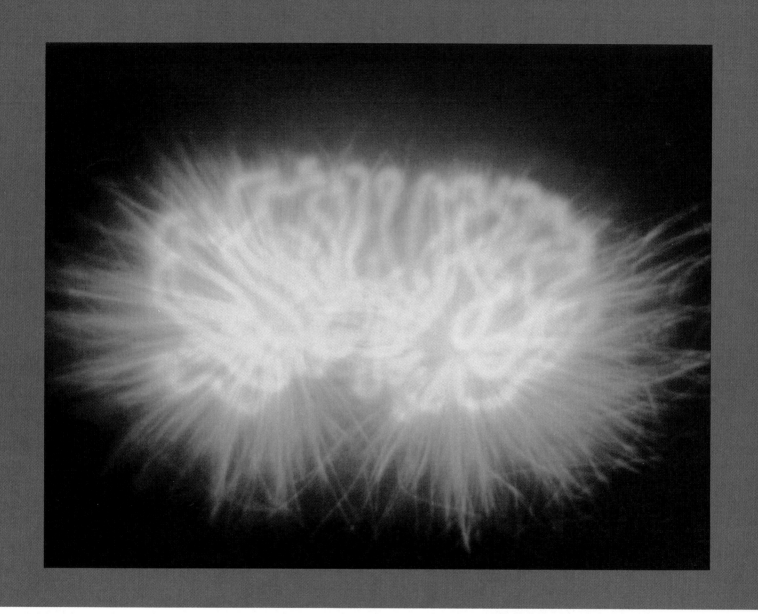

PART TWO
Information and Heredity

9 Chromosomes and Cell Division
10 Mendelian Genetics and Beyond
11 Nucleic Acids as the Genetic
    Material
12 Molecular Genetics of Prokaryotes
13 Molecular Genetics of Eukaryotes
14 Recombinant DNA Technology
15 Genetic Disease and Modern
    Medicine
16 Defenses against Disease
17 Animal Development

More than one hundred trillion (10^{14}) cells make up an adult human. As a fertilized human egg—a single cell—develops into a university student, its nucleus gives rise to over one hundred trillion nuclei, each containing basically the same genetic information as did the fertilized egg. An intricate mechanism first copies the genetic material in the nucleus and then partitions it into two daughter nuclei so that each gets one complete copy of the genetic information.

Part Two of this textbook deals with information and heredity. Multicellular organisms use their information to develop from a single cell, and each cell uses some of the information to build macromolecules and organelles. Each cell of an organism contains a full set of the information needed to build the whole organism, but a given cell acts on only the information that is relevant to it. The information is contained in the structures of nucleic acid molecules and is processed by enzymes. The structures of the enzymes, in turn, are encoded in the nucleic acids.

Heredity is the passing of information from one generation to the next. The orderly expression of hereditary information is required for development. This is especially evident in the development of the immune system, an important defense against disease. Recombinant DNA technology, an application of our knowledge about information and heredity, is providing us with new approaches to the treatment of many diseases.

Our early successes in understanding the molecular aspects of information and heredity came from studies of the genetics and molecular biology of prokaryotes. One entire chapter in Part Two (Chapter 12) is devoted to prokaryotes, and these fascinating organisms appear from time to time elsewhere in this part of the book. However, eukaryotic molecular biology is now making remarkable strides, and eukaryotes will occupy most of our attention. This chapter focuses almost exclusively on eukaryotic cells, although the last section addresses cell division in prokaryotes.

THE DIVISIONS OF EUKARYOTIC CELLS

The "intricate mechanism" that copies the genetic material for cell division is **mitosis**, which produces exact copies of a nucleus. This biological copying machine turns out the nuclei of the many cells of an organism's adult body.

A second mechanism for nuclear division is **meiosis**. This mechanism produces four daughter nuclei, each with only *half* the genetic information contained in the original cell, and each differing from the others in the exact information contained. When organisms reproduce sexually, pairs of such cells

9
Chromosomes and Cell Division

combine. Each sexual union of cells may produce new genetic combinations. Because of meiosis, offspring of the same parents differ.

The division of a eukaryotic cell consists of two steps: first the division of the nucleus (by mitosis or meiosis), then the division of the cytoplasm. Between divisions—that is, for most of its life—a eukaryotic cell is in a condition called **interphase**. What determines whether a cell in interphase will proceed to divide? How does mitosis lead to exact copies, and how does meiosis lead to diversity of products? Why do we need both exact copies *and* diverse products? Why do most organisms have sex in their life cycles? In this chapter we will learn the details of mitosis, meiosis, and interphase, as well as their biological consequences, which are of the utmost importance for heredity, development, and evolution.

EUKARYOTIC CHROMOSOMES AND CHROMATIN

All human cells, other than eggs and sperm, contain two full sets of genetic information, one from the mother and the other one from the father. Eggs and sperm, however, contain only a single set. Any particular egg or sperm in your body contains some information from your mother and some from your father. The genetic information consists of molecules of DNA packaged as **chromosomes** in the nucleus (see Chapter 4). Through a microscope, the nucleus appears relatively featureless (except for the nucleolus) during most of the life of a cell; the chromosomes cannot be seen (Figure 9.1).

The basic unit of the eukaryotic chromosome is a gigantic, linear, double-stranded molecule of DNA complexed with many proteins. During many stages of a eukaryotic cell's life cycle, each chromosome contains only one such DNA molecule. At other times, however, the DNA molecule doubles; the chromosome then comprises two joined **chromatids**, each made up of one DNA molecule complexed with proteins. At the particular times when chromosomes are visible in microscopes, the chromatids are joined in a specific, small region of the chromosome called the **centromere** (Figure 9.2). As we will see, centromeres direct the movement of chromosomes when a nucleus divides. A body that has a single centromere, whether it contains one or two DNA molecules, is properly called a *chromosome*.

The complex of DNA and proteins in a eukaryotic chromosome is referred to as **chromatin**. The DNA carries the genetic information, and the proteins organize the chromosome physically and regulate the activities of the DNA. Chromatin changes dramatically during mitosis and meiosis. During interphase, the chromatin of a chromosome is strung out so thinly that the chromosome cannot be seen clearly as a defined body under the light microscope. During most of mitosis and meiosis, however, the chromatin is coiled and compacted to a high degree, so that the chromosome appears as a bulky object (Figure 9.2*a*). This alternation of forms of chromatin relates to what the chromatin is doing during interphase and division. The genetic material is duplicated before each mitosis. Mitosis then separates this duplicated material into two new nuclei. This separation is accomplished more easily if the DNA is neatly arranged in compact units rather than being tangled up like a plate of spaghetti. During interphase, however, the DNA must direct the growth and other activities of

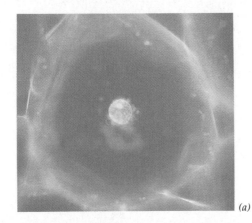

(a)

9.1 Nuclei
(a) The bright object in the middle of this spinach cell is a nucleus, as resolved through a light microscope. The smaller round spot inside the nucleus is a nucleolus, the site of ribosome assembly. (b) This view of the nucleus of an animal cell required an electron microscope. The two membranes of the nuclear envelope are distinct, as are a number of pores in the envelope. The nucleolus is at the lower right.

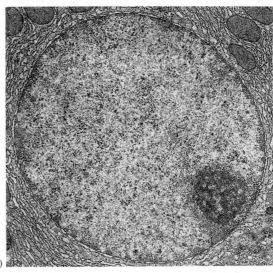

(b)

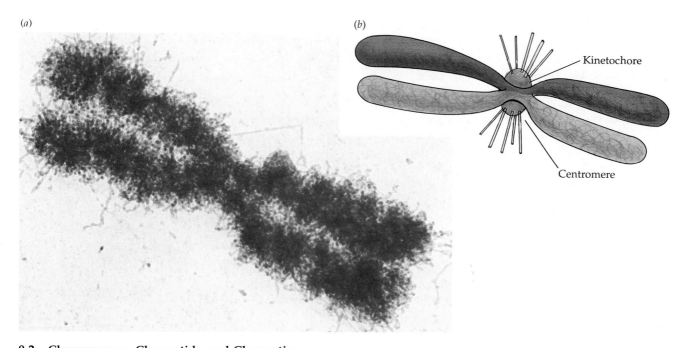

9.2 Chromosomes, Chromatids, and Chromatin

(a) A human chromosome in which the centromere is visible as a pinched-in region in the center. At the stage of the cell cycle captured here, the chromosome consists of two chromatids lying side by side. Individual fibers of chromatin are visible at the chromosome's edges. (b) A diagrammatic representation of the chromosome, with the two chromatids shown in different shades of blue. (Kinetochores will be described later in this chapter.)

the cell. As we will see in Chapters 11 and 13, DNA does this by interacting with enzymes while unwound and exposed.

Chromatin proteins associate closely with the DNA in chromosomes. Chromosomes contain large quantities of five classes of proteins, all of which are known as **histones**. Small for protein molecules, histones have a positive charge at pH levels found in the cell. The positive charge is a result of their par-

ticular amino acid compositions. Histone molecules join together to produce complexes around which the DNA is wound. Eight histone molecules, two of each of four of the histone classes, unite to form a core or spool shaped so that the DNA molecule fits snugly in a coil around it. The fifth class of histone (H1) appears to fit on the outside of the DNA, perhaps "clamping" it to the histone core. Strong evidence indicates that chromatin consists of a great number of these beadlike units, or **nucleosomes**, connected by a DNA thread (Figure 9.3).

A chromatid has a single DNA molecule running through vast numbers of nucleosomes. The many nucleosomes of a mitotic chromosome may pack together and coil as shown in Figure 9.4. During both

9.3 DNA Plus Histones Form Nucleosomes

(a) The DNA double helix coils around a central core of eight histone molecules to make a nucleosome. Another histone (H1) clamps the DNA to the core. (b) The "beads" of these chromatin fibers are nucleosomes. The "threads" connecting them are DNA.

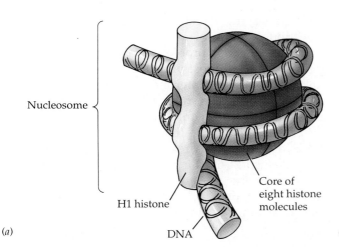

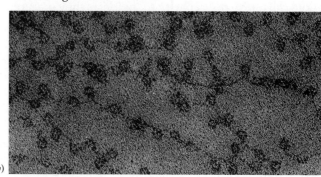

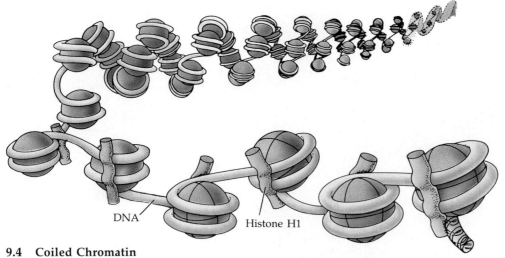

DNA

Histone H1

9.4   Coiled Chromatin
Nucleosomes are packed into a coil that twists into another, larger coil—and so forth—to produce condensed, supercoiled chromatin fibers such as those seen around the edges of the chromatids in Figure 9.2a. See also Figure 9.5.

mitosis and meiosis, the chromatin becomes ever more coiled and condensed, with further folding of the chromatin continuing up until the chromosomes separate (Figure 9.5). A diverse group of acidic proteins are also present in small quantities in chromosomes. The roles of these proteins will be considered in Chapter 13.

Although we know less about the organization of interphase chromatin than about that of mitotic chromatin, we do know that interphase chromatin has nucleosomes that are spaced at the same intervals as in supercoiled chromatin. During interphase, DNA thus remains associated with histone molecules while it replicates and directs synthesis of RNA. Current research is investigating the possibility that the structure of the nucleosomes changes as the cell proceeds with its interphase activities.

THE CELL CYCLE

A cell lives and functions until it divides or dies—or, if it is a sex cell, until it fuses with another sex cell. Some cells, such as red blood cells, muscle cells, and nerve cells, lose the capacity to divide as they mature; cells of certain other types rarely divide. And then there are cancerous cells, which, having escaped from the normal controls on division, divide rapidly and inappropriately. Most cells, however, have some probability of dividing, and some are specialized for rapid division. Thus for many kinds of cells we may speak of a **cell cycle** that has mitosis as one phase and interphase as the other (Figure 9.6). A given cell lives for one turn of the cycle and becomes two cells.

For life as a whole, the cycle repeats again and again as a constant source of renewal. The cell cycle, even for tissues engaged in rapid growth, consists mainly of interphase. Examining any collection of cells, such as a root tip or a slice of liver, reveals that most of the cells are in interphase most of the time. Only a small percentage of the cells are in mitosis at any given moment; this fact can be confirmed, in certain cultures of cells, by watching a single cell through its entire cell cycle.

The cell's DNA replicates during the **S phase** of interphase (the S stands for synthesis). The gap between the S phase and the onset of mitosis is referred to as Gap 2, or **G2**. Another gap phase—**G1**—separates the end of mitosis from the onset of the next S phase. If a cell is not going to divide, it may remain in G1 for weeks or even for many years until it dies—it seemingly will not waste effort replicating its genetic material. (There are some exceptions in which cells that will not divide do synthesize DNA and are thus stuck in G2, but continuation of the G1 phase is the rule in the vast majority of nondividing cells.)

Although some of a cell's biochemical activities change as the cell proceeds from one phase of its cycle to the next, most proteins are formed throughout all subphases of interphase. The histone proteins that we discussed in the previous section, however, are synthesized primarily during the S phase of the cell cycle, at the same time that DNA is being synthesized. While DNA replicates in the nucleus, histones are synthesized in the cytoplasm; the new histones enter the nucleus through the pores in the nuclear envelope (Figure 9.1b) and then combine with the DNA, forming nucleosomes.

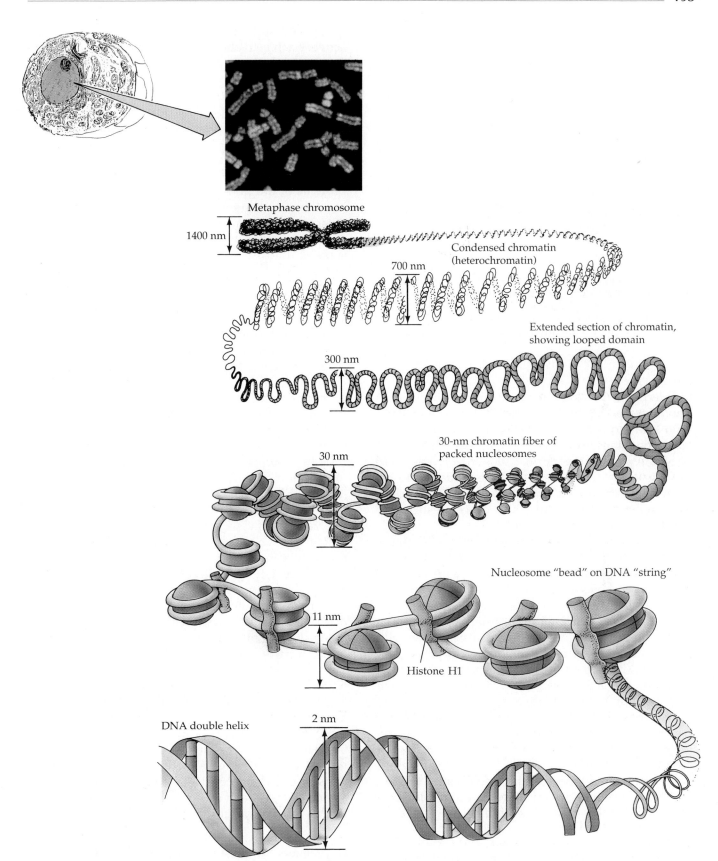

Metaphase chromosome

1400 nm

Condensed chromatin
(heterochromatin)

700 nm

Extended section of chromatin,
showing looped domain

300 nm

30-nm chromatin fiber of
packed nucleosomes

30 nm

Nucleosome "bead" on DNA "string"

11 nm

Histone H1

DNA double helix

2 nm

9.5 How DNA Packs into a Metaphase Chromosome

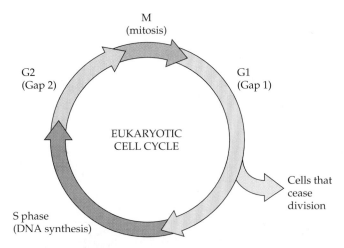

9.6 The Cell Cycle
A cell's life history is made up of a short mitosis and a longer interphase (green arrows). Interphase has three subphases in cells that divide. DNA is synthesized only during the short S phase between the G1 and G2 phases. Cells that do not divide are usually arrested in the G1 phase.

Interphase is a busy time—a time of "decisions." Should the nucleus replicate its DNA? Should it enter mitosis and divide? Some forms of cancer result from "bad" decisions. Some genes of a type called proto-oncogenes direct the normal sequence of events in interphase. If they mutate (become modified) to become what are called oncogenes, the normal decisions are no longer made, and cancer results (see Chapter 13).

How are appropriate decisions to enter the S phase or M phase (mitosis) made? These transitions—from G1 to S, and from G2 to M—require the activation of a protein complex called the maturation-promoting factor, MPF. The components of MPF are two proteins: one called cdc2 and the other belonging to a class called **cyclins**. When a cyclin combines with cdc2, enzymes act on the complex, converting it to an active MPF. One type of cyclin participates in the transition to the S phase; another type participates in the transition to M. In both cases, the active MPF not only brings about the appropriate phase of the cell cycle, but also activates other enzymes that degrade the cyclin, resulting eventually in the inactivation of the MPF itself! But then the *other* type of cyclin gradually increases in concentration sufficiently to bind with cdc2, leading once again to an active MPF that brings about the other transition in the cell cycle (Figure 9.7).

How does MPF produce its results? MPF has enzymatic activity that causes the addition of phosphate groups to other enzymes, making them active; the activity of these other enzymes is to add phosphate groups to still other enzymes, thus making *them* active. Each activation step multiplies the original effect, since each enzyme molecule catalyzes the activation of many molecules of the next kind of enzyme in this **cascade** of activation. (Other such cascades of phosphate transfer play roles in the action of many animal hormones, as explained in Chapter 36.) The products of the cascade from MPF are responsible for the observed effects at the cellular level. For example, one of the enzymes adds phosphate groups to the proteins of the nuclear lamina (see Chapter 4), causing the disintegration of the lamina and thus the detachment of chromatin from the nuclear envelope *and* the breakup of the nuclear envelope in an early stage of mitosis.

MITOSIS

In mitosis, a single nucleus gives rise to two nuclei that are genetically identical to each other and to the parent nucleus. Mitosis is a process of continuous change. Resist the temptation to think of it as a series of photographic slides in which one scene is replaced directly by another, distinctly different one. Rather, it is like a movie showing continuous action. For our discussion, however, it is convenient to look at mitosis as a series of important frames selected at intervals from the movie. Look at Figure 9.8 just one frame at a time, as we call for it in the text. After you have been through the entire story in the text, you can review mitosis by looking at Figure 9.8 as a whole; but you will find it easier to understand if you take it in smaller bites the first time through.

Let us begin with interphase, when the nucleus is between divisions. At the beginning of Figure 9.8 we see the nuclear envelope, the nucleoli, and a barely discernible tangle of chromatin. Immediately before mitosis (at the interphase–prophase transition), there is also a pair of **centrosomes** lying near the nucleus. The centrosomes are regions of the cell that help orchestrate chromosomal movement; these regions are not bounded by membranes and are not visible as discrete objects. In many organisms, each centrosome contains a pair of centrioles. The centrosomes of seed plants and some other organisms, however, do not have centrioles. Where present, each of these pairs of centrioles consists of one "parent" and one smaller "daughter" centriole at right angles to the parent centriole (Figure 9.9).

Development of the Chromosomes and Spindle

The appearance of the nucleus changes in the next frame of Figure 9.8, as the cell enters **prophase**, the

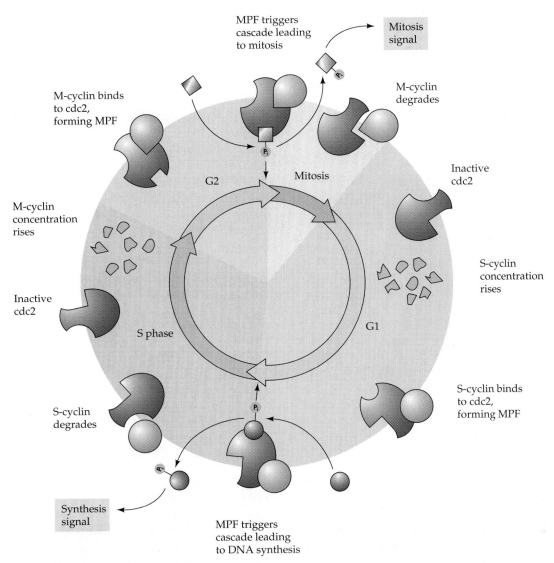

MPF triggers
cascade leading
to mitosis

Mitosis
signal

M-cyclin binds
to cdc2,
forming MPF

M-cyclin
degrades

Inactive
cdc2

M-cyclin
concentration
rises

S-cyclin
concentration
rises

Inactive
cdc2

S-cyclin binds
to cdc2,
forming MPF

S-cyclin
degrades

Synthesis
signal

MPF triggers
cascade leading
to DNA synthesis

G2

Mitosis

S phase

G1

9.7 Cyclins Trigger Decisions in the Cell Cycle
The signals for a cell to enter the M or S phase are triggered when the protein
cdc2 binds to a cyclin protein (M-cyclin or S-cyclin, respectively), producing a
protein complex called MPF (maturation-promoting factor). MPF enzyme activity in turn activates other enzymes in a cascade effect.

beginning of mitosis. The nucleolar material disperses. The centrosomes, with or without pairs of centrioles, move away from each other toward opposite ends of the cell. Each centrosome then serves as a **mitotic center** that organizes microtubules. In animal cells, some of the microtubules point away from the nuclear region and form starlike groupings called **asters**. Other microtubules, called **polar microtubules**, run between the mitotic centers and make up the developing **spindle** (Figure 9.10*a*). The spindle is actually two *half spindles*: Each polar microtubule runs from one mitotic center to the middle of the spindle, where it overlaps with polar microtubules of

the other half spindle (Figure 9.10*b*). Polar microtubules are unstable, constantly forming and falling apart until they contact polar microtubules from the other half spindle, at which point they become more stable.

The chromatin also changes during prophase. The extremely long, thin fibers take on a more orderly form as a result of coiling, supercoiling, and compacting (review Figure 9.5). At this level of magnification and at this stage of the nuclear cycle, each chromosome is seen to consist of two chromatids held tightly together over much of their length. The two chromatids of a single chromosome are identical

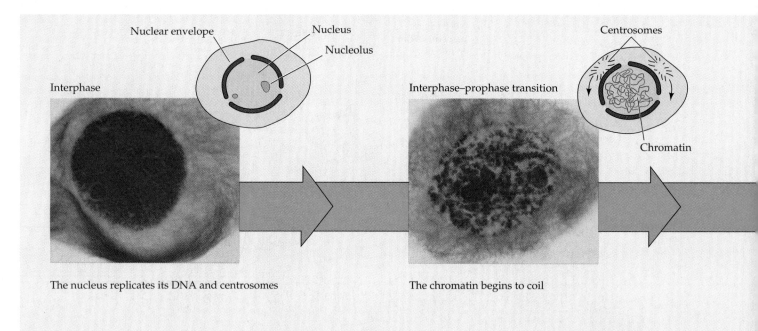

Interphase

The nucleus replicates its DNA and centrosomes

Interphase–prophase transition

The chromatin begins to coil

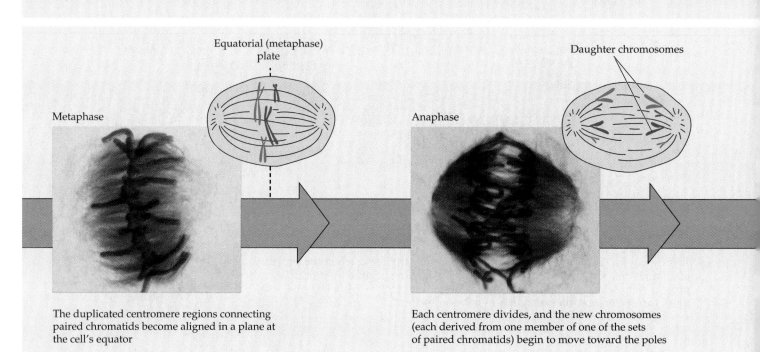

Metaphase

The duplicated centromere regions connecting paired chromatids become aligned in a plane at the cell's equator

Anaphase

Each centromere divides, and the new chromosomes (each derived from one member of one of the sets of paired chromatids) begin to move toward the poles

in structure, chemistry, and the hereditary information they carry because one chromatid, formed during the S phase of the previous interphase, is a replica of the other. Within the region of tight binding of the chromatids lies the centromere, which must be present in order for the chromatids to become associated with microtubules of the spindle. Very late in prophase, specialized three-layered structures called **kinetochores** develop in the centromere region, one on each chromatid (see Figure 9.2).

Dancing Chromosomes

The somewhat condensed chromosomes start to move at the end of prophase—the beginning of **prometaphase**, the next frame in Figure 9.8. The nuclear envelope suddenly disintegrates into membranous sacs (it takes only 20–30 seconds). This disintegration is caused by the activation of one of the MPFs, a process beginning with the accumulation of a cyclin during G2 of interphase. Groups of microtubules,

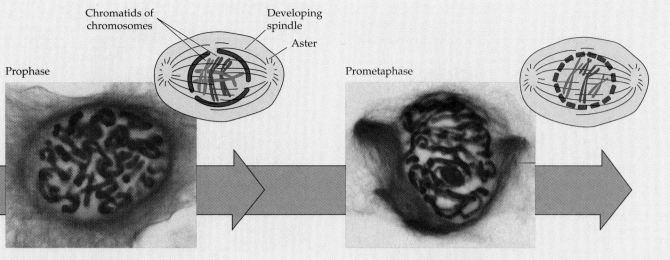

Prophase

Chromatids of
chromosomes

Developing
spindle

Aster

The chromatin continues to coil and supercoil,
making the chromatin more and more compact.
The chromosomes consist of identical,
paired chromatids

Prometaphase

The nuclear envelope breaks down. Kinetochore
microtubules appear and interact with the polar
microtubules of the spindle, resulting in
movement of the chromosomes

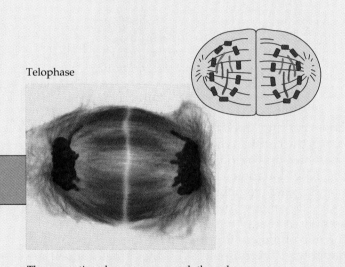

Telophase

The separating chromosomes reach the poles.
Telophase passes into the next interphase as the
nuclear envelopes and nucleoli re-form and the
chromatin becomes diffuse

9.8   Mitosis

Mitosis results in two nuclei, genetically identical to one
another and to the nucleus from which they formed.
These photomicrographs are of plant nuclei, which lack
centrioles and asters. The red dye stains microtubules and
thus the spindle; the blue dye stains the chromosomes. In
plants, the first steps toward division of the cell itself
cause changes in the telophase cell that disrupt staining,
causing the white line seen in the photomicrograph. The
diagrams are of corresponding phases in animal cells, in
order to introduce the structures not found in plants. In
the diagrams, the chromosomes are stylized to emphasize
the fates of the individual chromatids.

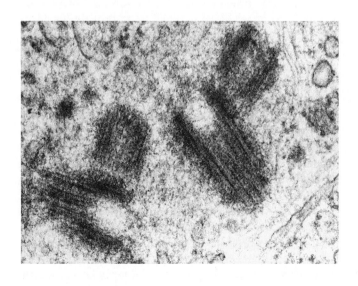

9.9   Centrioles

At a right angle to each parent centriole in this animal
cell is a daughter centriole that is about one-half as long.
Early in nuclear division one parent–daughter pair migrates to one side of the nucleus, the other to the opposite
side. Centrioles consist mostly of microtubules.

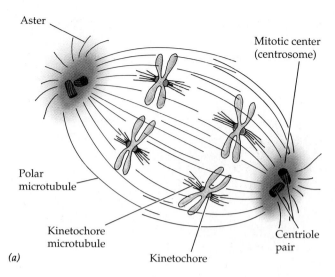

Aster

Mitotic center
(centrosome)

Polar
microtubule

Kinetochore
microtubule

Centriole
pair

Kinetochore

(a)

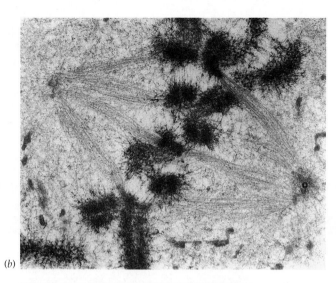

(b)

Kinetochore
microtubules

Kinetochore

(c)

9.10 The Mitotic Spindle Consists of Microtubules
(a) During prometaphase, polar microtubules extend from each pole of the spindle apparatus. Kinetochore microtubules attach to the kinetochores in the centromeres of the chromosomes and to polar microtubules. (b) Polar microtubules extending from the poles are visible in this electron micrograph of metaphase. The large dark objects in the middle are chromosomes. (c) Kinetochore microtubules extend down from the top of this electron micrograph to the kinetochore, which is seen as a dark, three-layered "plate."

called **kinetochore microtubules**, associate with the kinetochores (Figure 9.10c). Some of the polar microtubules attach to the kinetochore microtubules and become stable. The polar microtubules pull, so the kinetochore and its attached chromosome move toward one of the poles. The polar microtubules break down, and others from the same or the opposite pole attach to the kinetochore microtubules. Thus each chromosome may be pulled around seemingly aimlessly during prometaphase—until, randomly, the kinetochore of one chromatid is connected to microtubules from one pole while the kinetochore of the other chromatid is connected to microtubules from the other pole.

When this happens to a pair of kinetochores, the microtubules attached to them stop falling apart, perhaps because of the tension established by the opposing pulls from the two poles. The polar microtubules pull in such a way that the kinetochores approach a region halfway between the ends of the spindle. This region, which may be thought of as an invisible plane perpendicular to the long axis of the spindle, is called the **equatorial plate**, or metaphase plate.

The cell reaches **metaphase** when all the kinetochores arrive at the equatorial plate (see Figure 9.8). The condensation of chromatin that began with prophase continues until the end of metaphase. At this time, the centromeres divide. Metaphase is usually

brief, passing directly into **anaphase**, the phase in which the chromatids of each chromosome are pulled apart and drawn to opposite ends of the spindle. As the new **daughter chromosomes**—the former chromatids, each containing one double-stranded DNA molecule—move toward the opposite poles of the spindle, it is easy to see that the motion is caused by the microtubules "tugging" at the kinetochore of each daughter chromosome. As the kinetochores are pulled apart, the arms of the chromosomes drag along passively. The mechanism of the tugging is not fully understood. The microtubules are probably shortening, and the kinetochore microtubules may be sliding along the polar microtubules.

Also during anaphase, the poles of the spindle are often pushed farther apart by the action of some of the polar microtubules, thus contributing to the separation of the daughter chromosomes. The polar microtubules contain dynein, a protein also associated with the microtubules of cilia and flagella (Chapter 4). Presumably, then, the movement of the poles is produced in a manner similar to that in which eukaryotic flagella and cilia are made to beat.

Amazingly little energy is expended in moving a chromosome during anaphase. The hydrolysis of 20 ATP molecules is enough to move a chromosome from the equatorial plate to the pole.

The End of Mitosis

When the chromosomes stop moving at the end of anaphase, the cell enters **telophase** (the final frame in Figure 9.8). Two identical collections of chromosomes, which carry identical sets of hereditary instructions, are at the opposite ends of the spindle, which begins to break down (as do the asters, if present). A new nuclear envelope forms around each group of chromosomes. The chromosomes begin to uncoil, and continue uncoiling until they become the diffuse tangle of chromatin characteristic of interphase. The nucleolus or nucleoli reappear at specific sites on specific chromosomes. When these changes are complete, telophase—and mitosis—is at an end, and each of the daughter nuclei enters another interphase.

In interphase, the DNA duplicates and new chromatids form, so that each chromosome consists of two chromatids. The duplication of DNA is a major topic and is discussed in Chapter 11. Centrioles, if present, replicate during interphase: The two paired centrioles first separate, and then each acts as a parent for the formation of a new daughter centriole at right angles to it (see Figure 9.9).

Mitosis is beautifully precise. Its result is the formation of two nuclei *identical to each other* and to the parent nucleus in chromosomal makeup, and hence in genetic constitution.

Cytokinesis

Mitosis refers only to the division of the nucleus; it is not always immediately accompanied by the division of the rest of the cell, called **cytokinesis.** Generally, however, cytokinesis follows immediately upon mitosis. Animal cells, which lack cell walls, usually divide by a furrowing of the membrane, as if an invisible thread were tightening between the two parts (Figure 9.11a). The "invisible" thread consists of microfilaments of actin and myosin (Chapter 4), two proteins that interact here to produce a contraction, just as they do in muscles (Chapter 40).

Plant cells must divide differently because they have cell walls. As the spindle breaks down after mitosis, membranous vesicles derived from the Golgi apparatus appear in the equatorial region roughly midway between the two daughter nuclei of a dividing plant cell. With the help of microtubules, the vesicles begin to form a cell plate, that is, the beginning of a new cell wall (Figure 9.11b).

After cytokinesis, both daughter cells contain all the components of a complete cell. Mitosis ensures the precise distribution of chromosomes. Organelles such as ribosomes, mitochondria, and chloroplasts need not be distributed equally between daughter cells, as long as many of each are present in both cells; accordingly, there is no mechanism comparable to mitosis to provide for their equal allocation to daughter cells. Although centrioles, where present, were once thought to organize the mitotic spindle, scientists now speculate that the association of the

(a)

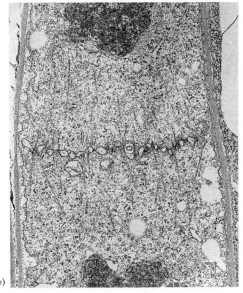

(b)

9.11 Cytokinesis
(a) A sea urchin egg has just completed cytokinesis at the end of the first division in its development into an embryo. The division furrow has completely separated the cytoplasm of one daughter cell from the other, although their surfaces remain in contact. Tiny, hairlike microvilli cover the surfaces of both cells. (b) The horizontal row of vesicles in this dividing plant cell in late telophase will join to form a cell plate between the cell above and the cell below. Microtubules, visible above and below in the cytoplasm, run between the vesicles.

centrioles with mitotic centers simply ensures that centrioles, like chromosomes, are distributed equally to the daughter cells.

SEX AND REPRODUCTION

As the cell cycle repeats itself, a single cell can give rise to a vast number of others. The cell could be a unicellular organism reproducing with each cycle, or a cell that divides to produce a multicellular organism. The multicellular organism, in turn, may be able to reproduce itself by releasing one or more of its cells, derived from mitosis and cytokinesis, as a spore *or* by having a multicellular piece break away and grow on its own (Figure 9.12). The multicellular organism reproducing by releasing cells and the unicellular organism are examples of **asexual reproduction**, sometimes called vegetative reproduction. Asexual reproduction is based on mitotic division of the nucleus and, accordingly, produces offspring that are genetically identical to the parent. It is a rapid and effective means of making new individuals and is widely practiced in nature.

A drawback of asexual reproduction is its very uniformity, which leads to the production of a **clone** of genetically identical progeny. Although the clone may be well adapted to its existing environment, it may be at great risk should conditions change. In contrast, organisms that produce genetically different offspring are more successful when the environment varies unpredictably in time and space, because at least *some* of their genetically diverse offspring may

be individuals able to meet the different challenges of a changing environment.

Diversity is fostered by **sexual reproduction**. Genetic information from two separate cells combines. In the reproduction of most animal species, the two cells are contributed by two separate parents. Each parent provides a sex cell, or **gamete**. Each gamete is **haploid**, meaning that it contains a single set of chromosomes; the number of chromosomes in such a single set is denoted by n. The two gametes—often identifiable as a female egg and a male sperm—fuse to produce a single cell, the **zygote** or fertilized egg. This fusion is called **fertilization**. Its consequence, the zygote, contains genetic information from both gametes and, hence, from both parents. The zygote also has *two* sets of chromosomes; it is said to be **diploid**, denoted by $2n$. In many species, including all animals, the zygote develops by mitotic divisions into a multicellular adult. Because the zygotic nucleus is diploid, all the body cells produced by mitosis are also diploid ($2n$). Sexual life cycles are summarized in Figure 9.13.

What happens when the adult from this zygote reproduces? If the gametes were produced by mitosis in this diploid parent, then they too would be diploid. Thus after fusion of two diploid ($2n$) gametes, the next-generation zygote would be $4n$. This is not a tenable situation because subsequent generations would contain more and more chromosomes. There must be a *reduction* step in the sexual life cycle, that is, a special type of nuclear division that reduces the chromosome number from diploid to haploid. This form of division in sexually reproducing organisms is meiosis.

Meiosis in animal cells directly produces haploid gametes. However, in plants and some fungi, meiosis gives rise to haploid **spores**, which undergo mitosis, producing multicellular haploid bodies (find the correct path in Figure 9.13). Particular cells in these haploid bodies ultimately give rise, by mitosis, to haploid gametes, and the life cycle continues. Details of some life cycles are considered in Chapters 23 to 27. The simplest possible sexual life cycle is one in which two haploid gametes fuse to give one diploid ($2n$) zygote, which immediately undergoes meiosis, yielding a new set of haploid (n) gametes. Embellishments on this scheme consist mainly of the addition of mitotic divisions leading to multicellularity in the haploid phase, the diploid phase, or both.

Keep in mind that sex and reproduction are *not* the same thing. Sex is the combining of genetic material from two cells. Reproduction is the formation of new individuals, whether of unicellular or multicellular organisms. Sex and reproduction can be widely separated in time in a unicellular species, a phenomenon illustrated in the life cycle of the protist *Paramecium* (see Chapter 23).

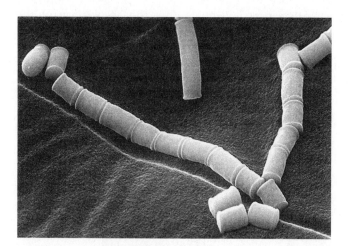

9.12 Asexual Reproduction
These spool-shaped cells are asexual spores formed by a fungus. Each spore contains a nucleus produced by a mitotic division. A spore and the fungal body that will grow from it following germination are the same genetically as the parent that fragmented to produce the spores. The general shape of the parent can be guessed from spores still in contact end-to-end.

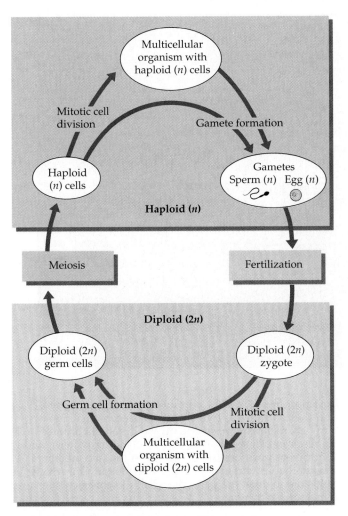

9.13 Fertilization and Meiosis Alternate in Sexual Reproduction
Haploid (*n*) cells or organisms alternate with diploid (2*n*) cells or organisms. A zygote, formed during fertilization, may differentiate into a germ cell, or it may form a multicellular organism that eventually produces germ cells. Whatever their origins, germ cells form haploid cells through meiosis. A haploid cell may differentiate into a gamete (sperm or egg), or it may form a multicellular organism that eventually produces gametes. Different organisms follow different paths around this cycle, and they may spend very different proportions of their life in different stages. Each variation is called a life cycle.

The essence of sexual reproduction is the selection of half of a parent's diploid chromosome set to make a haploid gamete, followed by the fusion of two such haploid gametes to produce a diploid cell containing genetic information from the two gametes. Both of these steps contribute to a shuffling of genetic information in the population, so that usually no two individuals have exactly the same genetic constitution. This result is the opposite of the situation with asexual reproduction. The diversity provided by sexual reproduction opened up enormous opportunities for evolution. Although both asexual and sexual modes of reproduction have existed for billions of years, there are many more species of sexually reproducing organisms than of asexually reproducing organisms.

THE KARYOTYPE

When nuclei are in metaphase of mitosis, the centromeres are spread out on the equatorial plate. At this time it is often possible to count and characterize the individual chromosomes. This process is relatively simple in some organisms, thanks to techniques that can capture cells in metaphase and spread out the chromosomes. A photograph of the entire set of chromosomes can then be made, and the images of the individual chromosomes can be cut out and pasted together in an orderly arrangement. Such a rearranged photograph reveals the number, forms, and types of chromosomes in a cell, all of which constitute its **karyotype** (Figure 9.14).

Individual chromosomes can be recognized by their length, the positions of their centromeres, and characteristic banding when they are stained and observed at high magnification. When the cell is diploid, the karyotype consists of pairs of chromosomes—23 pairs for a total of 46 chromosomes in our species, and more or fewer pairs in other diploid species (Table 9.1). In each recognizable pair of chro-

TABLE 9.1
Numbers of Pairs of Chromosomes in Different Species of Plants and Animals

COMMON NAME	SPECIES	NUMBER OF CHROMOSOME PAIRS
Mosquito	*Culex pipiens*	3
Housefly	*Musca domestica*	6
Garden onion	*Allium cepa*	8
Toad	*Bufo americanus*	11
Rice	*Oryza sativa*	12
Frog	*Rana pipiens*	13
Alligator	*Alligator mississipiensis*	16
Cat	*Felis domesticus*	19
House mouse	*Mus musculus*	20
Rhesus monkey	*Macaca mulatta*	21
Wheat	*Triticum aestivum*	21
Human	*Homo sapiens*	23
Potato	*Solanum tuberosum*	24
Cattle	*Bos taurus*	30
Donkey	*Equus asinus*	31
Horse	*Equus caballus*	32
Dog	*Canis familiaris*	39
Chicken	*Gallus domesticus*	≈39
Carp	*Cyprinus carpio*	52

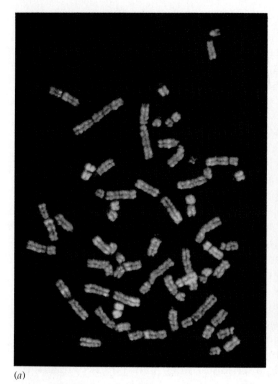

(a)

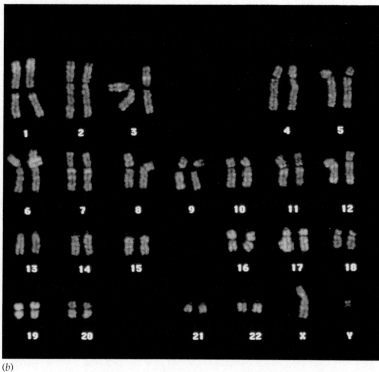

(b)

9.14 Human Cells Have 46 Chromosomes
Chromosomes of a human male. (a) The chromosomes are spread out because immersion in hypotonic solution bloated and then ruptured the cell, which was in metaphase of mitosis. You should be able to count 46 chromosomes. (b) A karyotype has been arranged from the metaphase spread of chromosomes. There are 23 pairs of homologous chromosomes, including a pair of sex chromosomes (XY). The chromosomes appear striped, and the centromeres (which appear as constrictions) occupy characteristic positions on the different chromosomes. You can see that the length, banding pattern, and centromere positions are the same on the two members of a homologous chromosome pair (except for XY), which helps distinguish the pair among all the chromosomes in a metaphase display.

mosomes, one chromosome comes from one parent and one from the other. The members of such a **homologous pair**, called homologs, are identical in size and appearance (with the exception of so-called sex chromosomes in some species; see Chapter 10), and they bear corresponding, though generally not identical, types of genetic information. Haploid cells contain only one of the homologs from each pair of chromosomes. Thus when haploid gametes fuse in fertilization, the resulting diploid zygote ends up with two homologs of each type.

MEIOSIS

Meiosis is the mechanism that reduces the diploid number of chromosomes to the haploid number for sexual reproduction. To understand the process and its specific details, it is useful to keep in mind the overall functions of meiosis: (1) to reduce the chromosome number from diploid to haploid, (2) to ensure that each of the four products has a complete set of chromosomes, and (3) to promote genetic diversity among the products. Pay particular attention to the fact that, although two divisions occur during meiosis, the DNA is replicated only once.

Two unique features characterize the first meiotic division, **meiosis I**. The first feature is that homologous chromosomes pair along their entire lengths, a process called **synapsis** that lasts from prophase to the end of metaphase of meiosis I. The second key feature is that homologous chromosomes separate during meiosis I. The individual chromosomes, each consisting of two joined chromatids, remain intact until the end of metaphase of **meiosis II**, the second meiotic division. In the discussion that follows, refer to Figure 9.17 (on pages 206 and 207) to help you visualize each step.

The First Meiotic Division

Meiosis I is preceded by an interphase during which each chromosome is replicated, so that each chromosome then consists of two sister chromatids. Meiosis I begins with a long **prophase I** (the first four frames of Figure 9.17), marked by a number of important changes. Very early in prophase I, the homologous chromosomes synapse; they are already

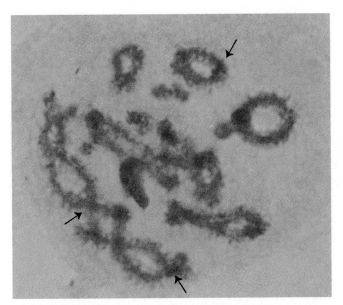

9.15 Chiasmata
Chiasmata—locations where segments of chromatids are being exchanged—are visible near the middles of some of the chromatids, and near the ends of others. Three of the many chiasmata in this micrograph are indicated with arrows.

tightly joined by the time they can be seen clearly under the light microscope. Throughout prophase I and metaphase I, the chromatin continues to coil and compact progressively, so that the chromosomes appear ever thicker and smoother.

Partway through prophase I, the homologous chromosomes seem to *repel* each other, especially near the centromeres, but they are held together by physical attachments (Figure 9.15). The regions having these attachments take on an X-shaped appearance and are called **chiasmata** (singular: *chiasma*, meaning "cross" in Greek). A chiasma reflects an exchange of material between chromatids on homologous chromosomes—what geneticists call crossing over (Figure 9.16). We will have a great deal to say about crossing over and its genetic consequences in coming chapters. Although the chromosomes exchange material shortly after synapsis begins, the chiasmata do not become visible until later, when the homologs are repelling each other.

Prophase I is followed by **prometaphase I** (not pictured in Figure 9.17), during which the nuclear envelope and the nucleoli disappear. A spindle forms, and microtubules become attached to the kinetochores of the chromosomes. In meiosis I, there is only one kinetochore per chromosome, not one per chromatid as in mitosis.

By **metaphase I**, the kinetochores have become connected to the poles, all the chromosomes have moved to the equatorial plate, and the homologous chromosomes are about to be pulled apart. Up to this point, they have been held together by chiasmata; it is this connection that provides the tension needed to stabilize the polar microtubules of the spindle. They separate in **anaphase I**, when individual chromosomes, each still consisting of *two* chromatids, are pulled to the poles, one homolog of a pair going to one pole and the other to the opposite pole (see Figure 9.17). (Note that this process differs from the separation of *chromatids* during mitotic anaphase.) Each of the two daughter nuclei from this division is haploid—that is, it contains only one set of chromosomes, compared to the two sets of chromosomes that were present in the original diploid nucleus. However, because they consist of two chromatids rather than just one, each of these chromosomes has twice the mass of a chromosome at the end of a mitotic division.

In some species, but not in others, there is a **telophase I**, with the reappearance of nuclear envelopes and so forth. When there is a telophase I, it is followed by an **interkinesis** phase similar to mitotic interphase. During interkinesis the chromatin is somewhat, but not completely, uncoiled. There is no replication of the genetic material because each chromosome already consists of two chromatids. In contrast to mitotic interphase, the sister chromatids are generally not genetically identical, because crossing over in prophase I has scrambled the original chromatids to some degree.

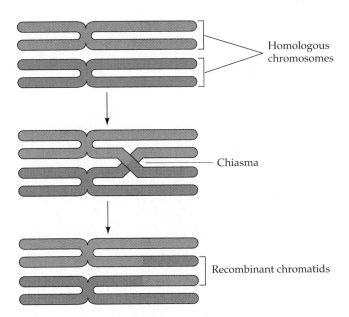

9.16 Crossing Over Forms Genetically Diverse Chromosomes
Early in prophase I, two chromatids of different homologs often cross over, break, and rejoin, so that each homolog has some DNA from the other. The products of crossing over are recombinant chromatids. This recombination can have important genetic and evolutionary consequences.

Meoisis I

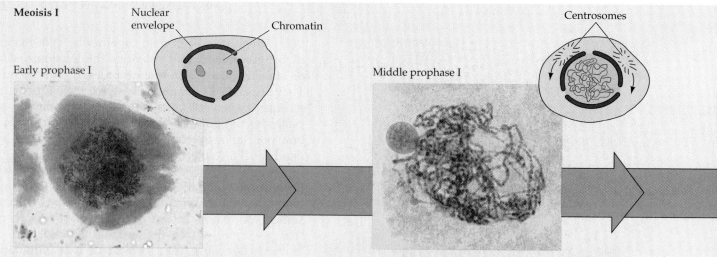

Early prophase I

The chromatin begins to condense following interphase

Middle prophase I

Synapsis aligns homologs, and chromosomes shorten

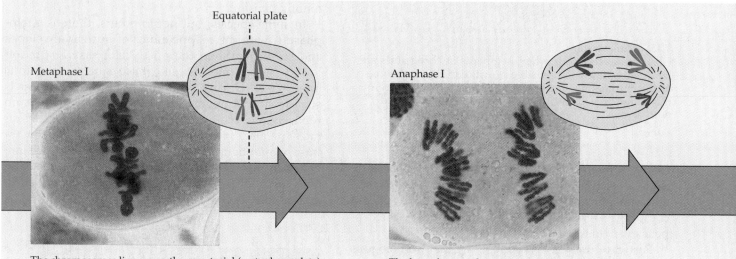

Metaphase I

The chromosomes line up on the equatorial (metaphase plate)

Anaphase I

The homologous chromosomes (each with two chromatids) move to opposite poles of the cell

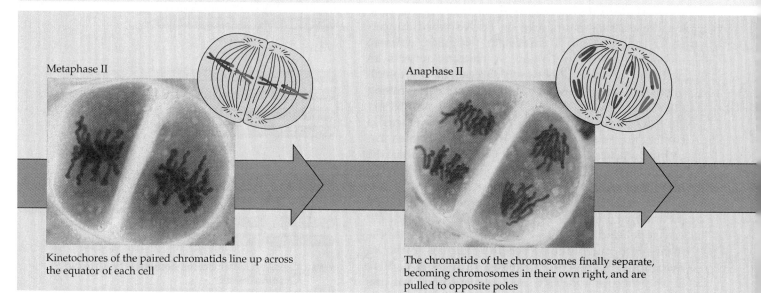

Metaphase II

Kinetochores of the paired chromatids line up across the equator of each cell

Anaphase II

The chromatids of the chromosomes finally separate, becoming chromosomes in their own right, and are pulled to opposite poles

9.17 Meiosis

In meiosis, two sets of chromosomes are divided among four cells, each of which then has half as many chromosomes as the original cell. This happens as a result of two successive nuclear divisions. The photomicrographs shown here are of meiosis in the male reproductive organ of a lily. As in Figure 9.8 (mitosis), the diagrams are of meiosis in an animal.

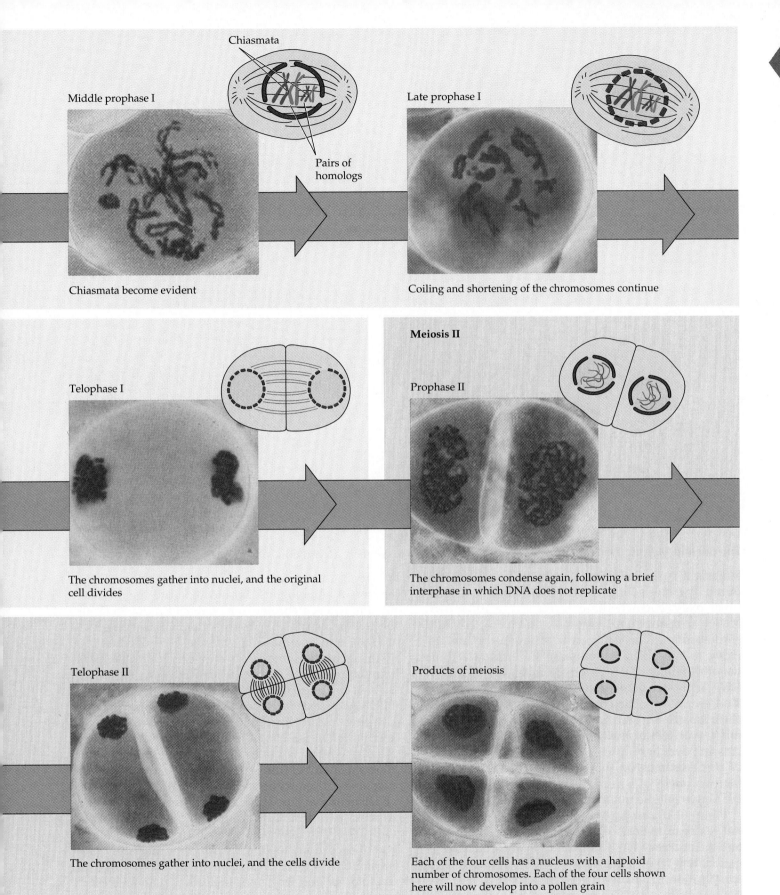

Middle prophase I

Chiasmata

Pairs of homologs

Chiasmata become evident

Late prophase I

Coiling and shortening of the chromosomes continue

Telophase I

The chromosomes gather into nuclei, and the original cell divides

Meiosis II

Prophase II

The chromosomes condense again, following a brief interphase in which DNA does not replicate

Telophase II

The chromosomes gather into nuclei, and the cells divide

Products of meiosis

Each of the four cells has a nucleus with a haploid number of chromosomes. Each of the four cells shown here will now develop into a pollen grain

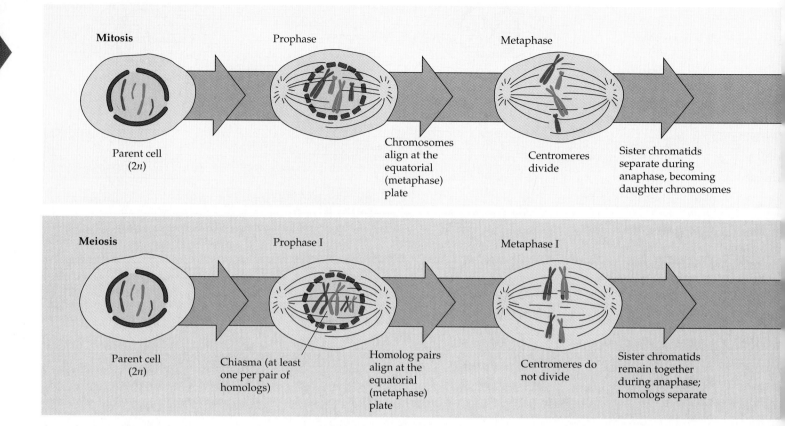

Mitosis

Parent cell (2*n*)

Prophase

Chromosomes align at the equatorial (metaphase) plate

Metaphase

Centromeres divide

Sister chromatids separate during anaphase, becoming daughter chromosomes

Meiosis

Parent cell (2*n*)

Prophase I

Chiasma (at least one per pair of homologs)

Homolog pairs align at the equatorial (metaphase) plate

Metaphase I

Centromeres do not divide

Sister chromatids remain together during anaphase; homologs separate

The Second Meiotic Division

Meiosis II is similar to mitosis. In each nucleus produced by meiosis I, the chromosomes line up at equatorial plates in metaphase II; the chromatids, each having a centromere, separate; and new daughter chromosomes (consisting now of single chromatids) move to the poles in anaphase II. There are three major differences between meiosis II and mitosis: (1) DNA replicates before mitosis but not before meiosis II; (2) in mitosis the chromatids making up a given chromosome are identical, whereas in meiosis II they differ over part of their length if they participated in crossing over in prophase of meiosis I; (3) the number of chromosomes on the equatorial plate of each of the two nuclei is *n* in meiosis II rather than 2*n* as in the single mitotic nucleus.

Figure 9.18 compares mitosis and meiosis. The final result of meiosis is four nuclei: Each nucleus is haploid and each has a single full set of chromosomes that differs from other such sets in its exact genetic composition. The differences, to repeat a very important point, result from crossing over during prophase I and from the separation of maternal and paternal chromosomes during anaphase I.

Synapsis, Reduction, and Diversity

What are the consequences of the synapsis and separation of homologous chromosomes during meiosis? In *mitosis*, each chromosome behaves independently of its homolog; its two chromatids are sent to opposite poles at anaphase. If we start a mitotic division with *x* chromosomes, we end up with *x* in each daughter nucleus (and each chromosome then consists of one chromatid, or double-stranded molecule of DNA). In *meiosis*, synapsis organizes things so that chromosomes of maternal origin pair with their paternal homologs. Then the separation during meiotic anaphase I ensures that each pole receives one chromosome member from each pair of homologous chromosomes. (Remember that each chromosome still consists of two chromatids.) For example, at the end of meiosis I in humans, each daughter nucleus contains 23 of the original 46 chromosomes—one member of each homologous pair. In this way, the chromosome number is decreased from diploid to haploid; in this way, too, meiosis I guarantees that each daughter nucleus gets a full set of chromosomes, for it must get one of each pair of homologous chromosomes.

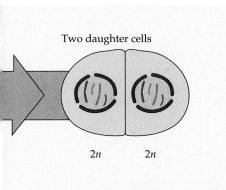

Two daughter cells

2n   2n

9.18 Mitosis and Meiosis Compared

Mitosis is a mechanism for constancy; the parent nucleus produces two identical daughter nuclei. Meiosis is a mechanism for diversity; the parent nucleus produces four daughter nuclei, each different from the parent nucleus and from its sister nuclei. The distinctive features of meiosis are synapsis and the failure of the centromeres to divide at the end of metaphase I.

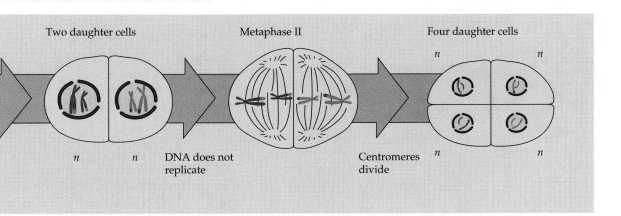

Two daughter cells   Metaphase II   Four daughter cells

n n DNA does not   Centromeres n n
replicate   divide

The products of meiosis I become genetically diverse for two reasons. First, synapsis during prophase I allows the maternal chromosome to interact with the paternal one; if there is crossing over, the recombinant chromatids contain some genetic material from each parent. Second, it is a matter of pure chance which member of a pair of chromosomes goes to which daughter cell at anaphase I. If there are two pairs of chromosomes in the diploid parent nucleus, a particular daughter nucleus could get paternal chromosome 1 and maternal chromosome 2, or paternal 2 and maternal 1, or both maternals, or both paternals. It all depends on the random way in which the chromosomes line up at metaphase I. Note that of the four possible chromosome combinations just described, two produce daughter nuclei that are essentially the same as one of the parental types (except for any material exchanged by crossing over). With three pairs, only two of the eight possible chromosome combinations are essentially the same as one of the parents. You can see that the probability of getting back the original parental combinations decreases rapidly as the number of chromosome pairs increases; most species of diploid organisms do, indeed, have more than two pairs (see Table 9.1).

Meiotic Errors and Their Consequences

A pair of homologous chromosomes may fail to separate during meiosis I, or sister chromatids may fail to separate during meiosis II or mitosis. This phenomenon, called **nondisjunction**, results in the production of aneuploid cells. **Aneuploidy** is a condition in which one or more chromosomes or pieces of chromosomes are lacking or are present in excess. If, for example, the chromosome-21 pair fails to separate during the formation of a human egg, allowing both members of the pair to go to one pole during anaphase I, then the resulting egg contains either two copies of chromosome 21 or none at all. If an egg with two of these chromosomes is fertilized by a normal sperm, the resulting zygote and infant has three copies of the chromosome: it is **trisomic** for chromosome 21 (Figure 9.19). As a result of carrying the extra chromosome 21, such a child demonstrates the symptoms of Down syndrome: impaired intelligence, characteristic abnormalities of the hands, tongue, and eyelids, and an increased susceptibility to diseases such as leukemia.

Other abnormal events also lead to aneuploidy. In a process called **translocation**, a piece of a chromo-

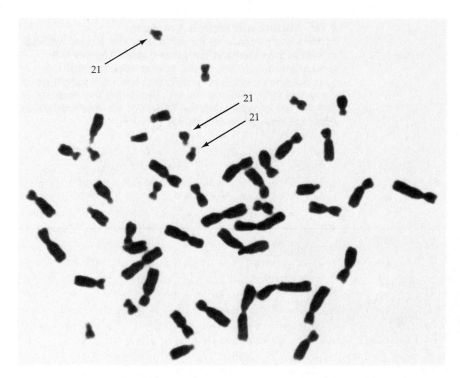

9.19 Chromosome Complement of Down Syndrome
The three copies of chromosome 21 in this spread of chromosomes from a cell in metaphase indicate that the person from whom the cell was taken has Down syndrome. Because of the extra chromosome 21, you should be able to count 47 chromosomes in the spread.

some may break away and become attached to another chromosome. For example, a particular large part of one chromosome 21 may be translocated to another chromosome. Individuals who inherit this translocated piece along with two normal chromosomes 21 also have Down syndrome.

Other human disorders result from particular chromosomal abnormalities. Sex chromosome aneuploidy causes such disorders as Turner syndrome and Klinefelter syndrome, which we discuss in Chapter 10 in connection with sex determination. Deletion of a portion of chromosome 5 results in cri-du-chat syndrome, so named because the afflicted infant's cry sounds like that of a cat. This syndrome includes severe mental retardation.

Trisomies (and the corresponding monosomies, where there is only one chromosome instead of two) are surprisingly common in human zygotes, but most of the embryos that develop from such zygotes do not survive to birth. Trisomies for chromosomes 13, 15, and 18 greatly reduce probabilities of surviving to birth and all lead to death before the age of one year; trisomies and monosomies for other chromosomes are lethal to the embryo. At least one-fifth of all recognized pregnancies self-terminate (miscarry) during the first two months, largely because of such trisomies and monosomies. (The actual fraction of self-terminated pregnancies is certainly much higher, because the earliest terminations usually go unrecognized.)

PLOIDY, MITOSIS, AND MEIOSIS

We have seen that both diploid and haploid nuclei divide by mitosis. Multicellular diploid and multicellular haploid individuals develop from single-celled beginnings by mitotic divisions. Mitosis may proceed in diploid organisms even when a chromosome from one of the haploid sets is missing or when there is an extra copy of one of the chromosomes (as in Down syndrome). There are also circumstances in which triploid ($3n$), tetraploid ($4n$), and higher-order polyploid nuclei are formed. Each of these **ploidy levels** represents an increase in the number of complete sets of chromosomes present. If by accident or (in some organisms) by design, the nucleus has one or more extra full sets of chromosomes (that is, if it is triploid, tetraploid, or of still higher ploidy), this condition in itself does not prevent mitosis. In mitosis, each chromosome behaves independently of the others.

In meiosis, by contrast, chromosomes synapse to begin division; if even one chromosome has no homolog, then anaphase I cannot send representatives of that chromosome to both poles. A diploid nucleus can undergo normal meiosis; a haploid one cannot. A tetraploid nucleus has an even number of each kind of chromosome, so it is possible for each chromosome to pair with its homolog, but a triploid nucleus cannot undergo normal meiosis because one-third of the chromosomes would lack partners. The requirement of an even number of chromosomes for

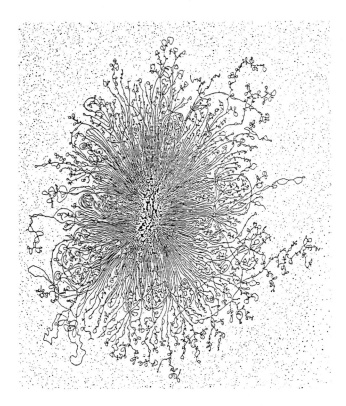

9.20   The Bacterial Chromosome Is a Circle
The long, looping fibers of DNA from this cell of the bacterium *Escherichia coli* are all part of one continuous circular chromosome.

longer than it is thick. The bacterium itself is about 1 µm (1,000 nm) in diameter and about 4 µm long. Thus the space into which the long thread of DNA is packed in the bacterial nucleoid is very small relative to the length of the DNA molecule. It is not surprising that the molecule usually appears in electron micrographs as a hopeless tangle of fibers.

When bacterial cells are gently lysed (broken open; see Chapter 5) to release their contents, the chromosome sometimes comes untangled and spreads out to its full length. Several techniques have shown that the bacterial chromosome is a closed circle rather than the linear structure found in eukaryotes. Circular chromosomes are probably to be found in all prokaryotes, as well as in some viruses and in the chloroplasts and mitochondria of eukaryotic cells. The bacterial chromosome is attached to the plasma membrane. When a new DNA molecule forms from the old one, it too attaches to the membrane. As the cell elongates during growth, the two attachment points separate so that when the new wall and membrane material form at fission, the two chromosomes are included in separate daughter cells (Figure 9.21).

meiosis has important consequences for the fertility of triploid, tetraploid, and other chromosomally unusual organisms that may be produced by intentional breeding or by natural accidents.

CELL DIVISION IN PROKARYOTES

In this chapter we have considered the structures and events of the division of the *eukaryotic* nucleus. Prokaryotic cells, by definition, lack nuclei and hence do not employ mitosis or meiosis in connection with their cell divisions. Still, prokaryotic division, called fission, must include an orderly distribution of genetic information to its daughter cells. Let's consider briefly how this is accomplished.

A prokaryote carries its genetic information on a chromosome that differs in composition and structure from the eukaryotic chromosome. The chromosome of a prokaryote is made of DNA, with protein components only temporarily bound to it and not part of the long-term structure. In the bacterium *Escherichia coli*, the main chromosome is a single circular molecule of DNA about 1.6 million nm (1.6 mm) long (Figure 9.20). The molecule is half a million times

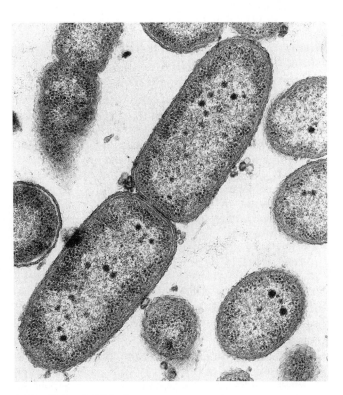

9.21   Bacterial Fission
These two cells of the bacterium *Pseudomonas aeruginosa* have almost completed fission. Plasma membranes have completely formed—separating the cytoplasm of one cell from that of the other—and only a small gap of cell wall remains to be completed. Each cell contains a complete chromosome in the light-toned nucleoid visible in the center of the cells.

SUMMARY of Main Ideas about Mitosis and Meiosis

Mitosis and meiosis are the processes by which the nuclei of eukaryotic cells divide.

Located within the nucleus, a eukaryotic chromosome carries genetic information in a continuous DNA molecule that wraps around aggregates of histone proteins to form nucleosomes.
Review Figures 9.3, 9.4, and 9.5

Microtubular spindles move chromosomes during nuclear divisions.
Review Figure 9.10

Mitotic division of a diploid nucleus produces two diploid daughter nuclei, each genetically and chromosomally identical to the parental nucleus.
Review Figure 9.8

In the cell cycle, mitosis alternates with interphase, which consists of a synthesis phase and two gap phases.
Review Figures 9.6 and 9.7

Meiotic division of a diploid nucleus produces four genetically diverse haploid daughter nuclei for sexual reproduction.
Review Figure 9.13 and 9.17

Meiosis is a pair of nuclear divisions with no intervening synthesis phase.

During the first meiotic division, homologous chromosomes, each consisting of two chromatids, synapse and then separate into daughter nuclei.
Review Figure 9.17 and 9.18

During the second meiotic division, chromatids separate into daughter nuclei.
Review Figures 9.17 and 9.18

Crossing over during meiosis forms recombinant chromatids.
Review Figure 9.16

After a nucleus divides by mitosis or meiosis, the rest of the cell may divide by cytokinesis.
Review Figure 9.11

The karyotype of an organism identifies its chromosomal makeup; the human karyotype consists of 23 pairs of homologous chromosomes.
Review Figure 9.14

Asexually reproduced offspring are genetically identical to the parent.

A prokaryotic cell, which has no nucleus, divides by fission after the DNA of its chromosome has replicated and separated to opposite areas of the plasma membrane.

SELF-QUIZ

1. Which of the following statements is *not* true of eukaryotic chromosomes?
   a. They sometimes consist of two chromatids.
   b. They sometimes consist of a single chromatid.
   c. They normally possess a single centromere.
   d. They consist of chromatin.
   e. They are always clearly visible as defined bodies under the light microscope.

2. Nucleosomes
   a. are made of chromosomes.
   b. consist entirely of DNA.
   c. consist of DNA wound around a histone core.
   d. are present only during mitosis.
   e. are present only during interphase.

3. Which of the following statements is *not* true of the cell cycle?

   a. It consists of mitosis and interphase.
   b. The cell's DNA replicates during G1.
   c. A cell can remain in G1 for weeks or much longer.
   d. Most proteins are formed throughout all subphases of interphase.
   e. Histones are synthesized primarily during the S phase.

4. Which of the following statements is *not* true of mitosis?
   a. A single nucleus gives rise to two identical daughter nuclei.
   b. The daughter nuclei are genetically identical to the parent nucleus.
   c. The centromeres divide at the onset of anaphase.
   d. Homologous chromosomes synapse in prophase.
   e. Mitotic centers organize the microtubules of the spindle fibers.

5. Which of the following statements is true of cytokinesis?
   a. A cell plate is formed in cytokinesis of animal cells.
   b. Furrowing of the membrane initiates cytokinesis in plant cells.
   c. Cytokinesis generally follows immediately upon mitosis.
   d. Actin and myosin are important in cytokinesis in plant cells.
   e. Cytokinesis is the division of the nucleus.

6. In sexual reproduction
   a. gametes are usually haploid.
   b. gametes are usually diploid.
   c. the zygote is usually haploid.
   d. the chromosome number is reduced during mitosis.
   e. spores are formed during fertilization.

7. In meiosis
   a. meiosis II reduces the chromosome number from diploid to haploid.
   b. DNA replicates between meiosis I and II.
   c. the chromatids making up a chromosome in meiosis II are identical.
   d. each chromosome in prophase I consists of four chromatids.
   e. homologous chromosomes are separated from one another in anaphase I.

8. In meiosis
   a. a single nucleus gives rise to two identical daughter nuclei.
   b. the daughter nuclei are genetically identical to the parent nucleus.
   c. the centromeres divide at the onset of anaphase I.
   d. homologous chromosomes synapse in prophase I.
   e. no spindle forms.

9. Which of the following statements is *not* true of aneuploidy?
   a. It results from chromosomal nondisjunction.
   b. It does not happen in humans.
   c. An individual with an extra chromosome is trisomic.
   d. Trisomies are common in human zygotes.
   e. A piece of one chromosome may translocate to another chromosome.

10. In prokaryotes
   a. there are no meiotic divisions.
   b. mitosis proceeds as in eukaryotes.
   c. the genetic information is not carried in chromosomes.
   d. the chromosomes are identical to those of eukaryotes.
   e. cell division follows division of the nucleus.

FOR STUDY

1. How does a nucleus in the G2 phase of the cell cycle differ from one in the G1 phase?

2. What is a chromatid? When does a chromatid become a chromosome?

3. Compare and contrast mitosis (and subsequent cytokinesis) in animals and plants.

4. Suggest two ways in which one might, with the help of a microscope, determine the relative durations of the various phases of mitosis.

5. Contrast mitotic prophase and meiotic prophase I. Contrast mitotic anaphase and meiotic anaphase I.

READINGS

All introductory genetics texts contain chapters on meiosis and how this process distributes genetic information.

Alberts, B., D. Bray, J. Lewis, M. Raff, K. Roberts and J. D. Watson. 1994. *Molecular Biology of the Cell*, 3rd Edition. Garland Publishing, New York. An outstanding book in which to pursue the topics of this chapter in greater detail. Chapters 8, 17, 18, and 20 have definitive modern treatments of the nucleus, mitosis, the cell cycle, meiosis, and more.

Glover, D. M., C. Gonzalez and J. W. Raff. 1993. "The Centrosome." *Scientific American*, June. New findings on the structure and function of the organelle that directs the assembly of the cytoskeleton and controls cell division—when it is present.

Mazia, D. 1961. "How Cells Divide." *Scientific American*, September. A classical description of mitosis by a leading researcher of cell division.

Mazia, D. 1974. "The Cell Cycle." *Scientific American*, January. This article discusses the four major stages of the cell cycle.

McIntosh, J. R. and K. L. McDonald. 1989. "The Mitotic Spindle." *Scientific American*, October. A fascinating description of the growth, disassembly, and interactions of the components of the spindle.

Mitchison, J. M. 1972. *The Biology of the Cell Cycle*. Cambridge University Press, New York. One of the few basic texts available on cell division and the cell cycle.

Murray, A. W. and M. W. Kirschner. 1991. "What Controls the Cell Cycle." *Scientific American*, March. A discussion of how cyclins, cdc2, and the maturation promoting factor (MPF) regulate mitosis and meiosis.

Sloboda, R. D. 1980. "The Role of Microtubules in Cell Structure and Cell Division." *American Scientist*, May/June. Includes a discussion of the spindle apparatus.

Mendel's Research Material
Studies of garden peas cracked the secrets of inheritance.

10

Mendelian Genetics and Beyond

You have seen how the nuclei of eukaryotic cells divide mitotically and meiotically. You know that by these processes cells pass copies of their genetic information to their descendants. By thinking about meiosis and sexual reproduction, you can account for the fact that offspring of the same parents may differ. But what if no one understood cell division—especially the complexities of meiosis? How could there be any understanding of how traits pass from parents to offspring without such background information? The answer is the subject of this chapter.

The term **Mendelian genetics** refers to certain basic inheritance patterns. The term honors the Austrian monk Gregor Johann Mendel (1822–1884), the person who first made rigorous, quantitative observations of the patterns of inheritance and proposed plausible mechanisms to explain them. In organisms that reproduce sexually and have more than one chromosome (and orderly meiosis), many traits pass from parent to offspring in accord with these patterns. When Mendel began his work with the garden pea in his monastery garden, little was known about the sex lives of plants or the consequences of sexual reproduction for the inheritance of traits. Chromosomes, mitosis, and meiosis were unknown.

MENDEL'S DISCOVERIES

Some observations that Mendel found useful in his studies had been made in the late eighteenth century by a German botanist, Josef Gottlieb Kölreuter. Kölreuter studied many plants by cross-pollinating them. He attempted many crosses between plants, produced many **hybrids** (the offspring of genetically different parents), and learned a great deal about pollination (see Chapter 33). In some instances he confirmed the common observation that hybrids are intermediate between their parents with respect to obvious traits such as size, color, and flower shape; more important, he found and emphasized that in some cases the hybrids are not intermediate but closely resemble just one of the parents.

Kölreuter also studied the offspring from **reciprocal crosses**. These are crosses made in both directions; that is, in one set of crosses, males with one form of a trait that we will call "A" are crossed with females having the "a" form of the same trait, while in a complementary set of crosses "a" males and "A" females are the parents. In an example of reciprocal crosses of plants, pollen (which carries the male sperm) from a plant with trait "A" is placed on the female organ—from which the sperm can travel to the eggs—of a plant with trait "a" in one set of crosses. In the reciprocal cross, pollen from "a" plants is placed on the female organs of "A" plants. (In many plant species the same individuals have both male and female reproductive organs; each plant may then reproduce as a male, as a female, or as both—which makes such plants excellent material for genetic studies.) In Kölreuter's experience, reciprocal crosses always gave identical results.

This was the state of knowledge in genetics when Mendel began his work. In one sense, the time was ripe for his discoveries, for it had recently been shown that one female gamete combines with one male gamete to bring about fertilization. On the other hand, the role of the chromosomes as bearers of genetic information was unknown, and mitosis and meiosis were yet to be discovered. Mendel himself was well qualified to make the big step forward. Although in 1850 he had failed an examination for a teaching certificate in natural science, he later undertook intensive studies in physics, chemistry, mathematics, and various aspects of biology at the University of Vienna. His work in physics and mathematics is probably what led to his applying experimental and quantitative methods to the study of heredity—and these were the key ingredients in his success.

Mendel worked out the basic principles of the heredity of plants and animals over a period of about nine years, the work culminating in a public lecture in 1865 and a detailed written account in 1866. However, his theory was not accepted. In fact, it was ignored. Perhaps the chief difficulty was that the physical basis of his theory was not understood until the discovery of meiosis, some years later. The most prominent biologists at the time Mendel published his results simply were not in the habit of thinking in mathematical terms, even the simple terms used by Mendel. Mendel's paper on plant hybridization appeared in a journal that was received by 120 libraries, and he sent reprinted copies (of which he had obtained 40) to several distinguished scholars. We know that at least one of these scholars died years later without even having opened the pages of the Mendel reprint. Whatever the reasons, Mendel's pioneering paper had no discernible influence on the scientific world for more than 30 years.

Then, in 1900, Mendel's discoveries burst into prominence as a result of independent experiments by the Dutchman Hugo de Vries, the German Karl Correns, and the Austrian Erich von Tschermak. Each of these scientists carried out crossing experiments and obtained quantitative data about the progeny; each published his principal findings in 1900; each cited Mendel's 1866 paper. At last the time was ripe for biologists to appreciate the significance of what these four geneticists had discovered—largely because meiosis had by then been described. That Mendel made his discoveries *prior* to the discovery of meiosis was due in part to the methods of experimentation he used.

Mendel's Strategy

Mendel chose the garden pea for his studies because of its ease of cultivation, the feasibility of controlled pollination, and the availability of varieties with differing traits. He controlled pollination by moving pollen from one plant to another; thus he knew the parentage of the offspring in his experiments. If untouched, the peas Mendel studied naturally self-pollinate—that is, the female organs of flowers receive pollen from the male organs of the same flowers—and he made use of this natural phenomenon in some of his experiments.

Mendel began by examining varieties of peas in a search for heritable traits (traits that can be passed from parent to offpsring) suitable for study. A suitable trait would be one that was "true-breeding." To be considered true-breeding, peas with white flowers, when crossed with one another, would have to give rise *only* to progeny with white flowers; tall plants bred to tall plants would have to produce only tall progeny. The suitable traits were also ones that had well-defined, contrasting alternatives that could be obtained in true-breeding form, such as purple flowers versus white flowers. For most of his work, Mendel concentrated on the seven pairs of contrasting traits shown in Figure 10.1. Before performing a given cross, he made sure that each potential parent was from a true-breeding strain; this was an essential point in his analysis of his experimental results.

Mendel then placed pollen he collected from one parental strain onto the stigma (female organ) of flowers of the other strain. The plants providing and receiving the pollen were the parental generation, designated **P**. In due course, seeds formed and were planted. The resulting new plants constituted the first filial generation, F_1. Mendel and his assistants examined each F_1 plant to see which traits it bore and then recorded the number of F_1 plants expressing each trait. In some experiments the F_1 plants were allowed to self-pollinate and produce a second filial generation, or F_2. Again, each F_2 plant was characterized and counted. Mendel performed other crosses in which the F_2 was produced by crossing F_1 hybrids with one of the true-breeding parental strains.

Always, each type of progeny was counted and recorded. This attention to quantitative detail was a unique advance in experimental biology; it allowed Mendel to make numerical comparisons and ultimately to develop a hypothesis, or model—a proposed explanation for the numbers he observed. In sum, Mendel devised a well-organized plan of research, pursued it faithfully and carefully, recorded great amounts of quantitative data, and analyzed the numbers he recorded to explain the relative proportions of the different kinds of progeny. His 1866 paper stands to this day as a model of clarity. His results

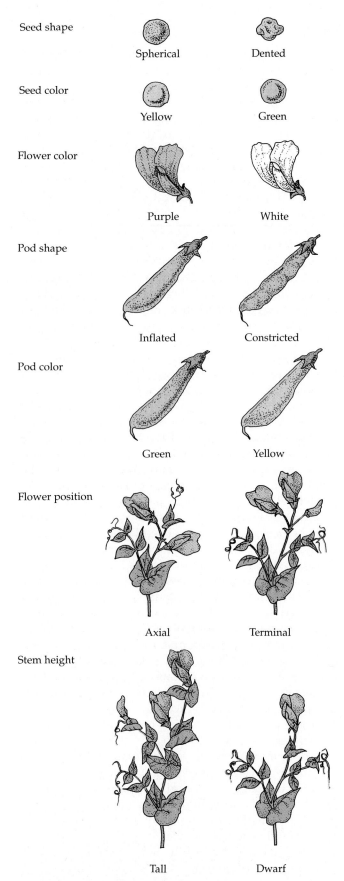

Seed shape

Spherical   Dented

Seed color

Yellow   Green

Flower color

Purple   White

Pod shape

Inflated   Constricted

Pod color

Green   Yellow

Flower position

Axial   Terminal

Stem height

Tall   Dwarf

10.1 Inherited Traits Studied by Mendel
Mendel's work on the genetics of peas focused on these seven pairs of traits. He isolated each as a true-breeding trait before he began his studies of various crosses.

and the conclusions to which they led are the subject of the next few sections.

Experiment 1

"Experiment 1" in Mendel's paper included a **monohybrid cross**, one in which the parents were both hybrids for a single trait. He took pollen from plants of a true-breeding strain with dented seeds and placed it on the stigmas of flowers of a true-breeding, spherical-seeded strain. He also performed the reciprocal cross, placing pollen from the spherical-seeded strain on the stigmas of flowers of the dented-seeded strain. In both cases, all the F_1 seeds that were produced were spherical—it was as if the dented trait had disappeared completely. The following spring Mendel grew 253 F_1 plants from these spherical seeds, each of which was allowed to self-pollinate—a monohybrid cross—to produce F_2 seeds. In all, there were 7,324 F_2 seeds, of which 5,474 were spherical and 1,850 dented (Figure 10.2).

Mendel observed that the spherical seed trait was **dominant** because it was expressed over the dented seed trait, which he called **recessive**. In each of the other six pairs of traits studied by Mendel, one proved to be dominant over the other. When he crossed plants differing in any of these traits, only one of each pair of traits was evident in the F_1 generation. However, the trait that was not seen in the F_1 *reappeared* in the F_2. Most important, the ratio of the two traits in the F_2 was always the same: approx-

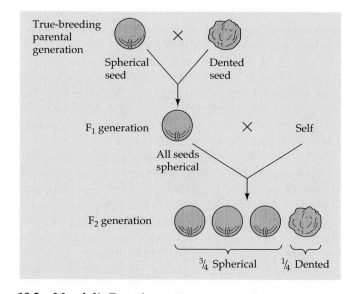

True-breeding parental generation

Spherical seed   Dented seed

F_1 generation

All seeds spherical   Self

F_2 generation

¾ Spherical   ¼ Dented

10.2 Mendel's Experiment 1
Mendel crossed plants of true-breeding varieties of peas with different seed shapes. Plants grown from the spherical F_1 seeds and then self-pollinated produced F_2 in which about ¼ of the seeds were dented and ¾ were spherical. The pattern was the same regardless of which variety contributed the pollen in the parental generation.

TABLE 10.1
Mendel's Results from Monohybrid Crosses

	P		F$_2$			
DOMINANT	×	RECESSIVE	DOMINANT	RECESSIVE	TOTAL	RATIO
Spherical	×	Dented seeds	5,474	1,850	7,324	2.96:1
Yellow	×	Green seeds	6,022	2,001	8,023	3.01:1
Purple	×	White flowers	705	224	929	3.15:1
Inflated	×	Constricted pods	882	299	1,181	2.95:1
Green	×	Yellow pods	428	152	580	2.82:1
Axial	×	Terminal flowers	651	207	858	3.14:1
Tall	×	Dwarf stems	787	277	1,064	2.84:1

imately 3:1—that is, three-fourths of the F$_2$ showed the dominant trait and one-fourth showed the recessive trait (Table 10.1). In his Experiment 1, the ratio was 5474:1850 = 2.96:1. Reciprocal crosses in the parental generation gave similar outcomes in the F$_2$.

Terminology for Mendelian Genetics

All by themselves the results from Experiment 1 disproved the widely believed theory that inheritance is a "blending" phenomenon. According to the blending theory, Mendel's F$_1$ seeds should have had an appearance intermediate between those of the two parents—they should have been slightly dented. Furthermore, the blending theory offered no explanation for the reappearance of the dented trait in the F$_2$ seeds after its apparent absence in the F$_1$ seeds. From his results Mendel proposed a **particulate theory**, in which the hereditary carriers are present as discrete units that retain their integrity in the presence of other units.

As he wrestled mathematically with his data, Mendel reached the conclusion that each pea has two such units for each character, one derived from each parent. Each gamete contains one unit, and the resulting zygote (and each cell of the adult that develops from it) contains two. This conclusion is the core of his model of inheritance. Mendel's "unit" is now called a **gene**.

Mendel reasoned that in Experiment 1, the spherical-seeded parent had a pair of genes of the same type, which we will call S, and the parent with dented seeds had two s genes. The SS parent produced gametes each containing a single S, and the ss parent produced gametes each with a single s. Each member of the F$_1$ generation had an S from one parent and an s from the other; an F$_1$ could thus be described as Ss. We say that S is dominant over s because s is not evident when both genes are present.

The physical appearance of a character is its **phenotype**. Mendel correctly supposed the phenotype to be the result of the **genotype**, or genetic constitution, of the organism showing the phenotype. In Experiment 1 we are dealing with two phenotypes (spherical seeds and dented seeds) and three genotypes: The dented-seed phenotype is produced only by the genotype ss, whereas the spherical-seed phenotype may be produced by the genotypes SS and Ss. The different forms of a gene (S and s in this case) are called **alleles**. Individuals that breed true for a character contain two copies of the same allele. For example, a strain of true-breeding peas with dented seeds must have the genotype ss—if S were present, the plants would produce spherical seeds. We say individuals that produce dented seeds are **homozygous** for the allele s, meaning that they have two copies of the same allele. Some peas with spherical seeds—the ones with the genotype SS—are also homozygous. However, other spherical-seeded plants are **heterozygous** because they have two different alleles of the gene in question; these plants have the genotype Ss. To illustrate these terms with a more complex example, one in which there are three gene pairs, an individual with the genotype AABbcc is homozygous for two genes—because it has two A alleles and two c alleles—but heterozygous for the gene with alleles B and b. An individual that is homozygous for a character is called a homozygote; a heterozygote is heterozygous for the character in question.

Segregation of Alleles

How does Mendel's model explain the composition of the F$_2$ generation in Experiment 1? Consider first the F$_1$, which has the spherical-seeded phenotype and the genotype Ss. According to the model, when any F$_1$ individual produces gametes, the alleles **segregate**, or separate, so that each gamete receives only *one* member of the pair of genes. Half the gametes contain the S allele and half the s allele. The random combination of these gametes produces the F$_2$ generation (Figure 10.3). Three different F$_2$ genotypes

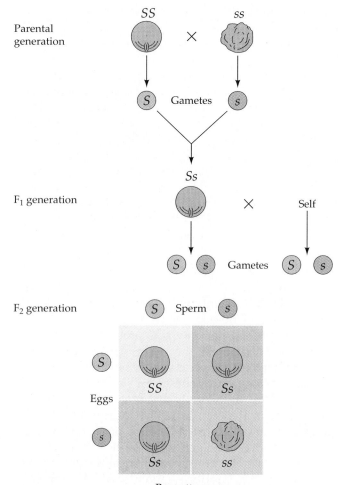

Parental generation

Gametes

F₁ generation

Self

Gametes

F₂ generation

Sperm

Eggs

Punnett square

10.3 Mendel's Explanation of Experiment 1
Mendel concluded that heredity depends on factors from each parent, and that these factors do not blend in the offspring. The drawing shows a modern version of Mendel's explanation of the experiment in Figure 10.2. A parent homozygous for the allele for spherical seeds is crossed with a parent homozygous for the allele for dented seeds. Each parent makes gametes of only one kind, either S or s, and these combine at fertilization to form plants that all have the genotype Ss and the spherical-seeded phenotype. When the F₁ plants self-pollinate they produce two kinds of eggs, S and s, and the same two types of male sex cells. These combine randomly in four different ways to form F₂ plants, as shown in the box at the bottom of the figure. Three of these combinations produce genotypes that determine the spherical-seeded phenotype and one produces the genotype for the dented-seed phenotype, resulting in the observed ratio of 3:1. The box in which the F₂ plants are displayed is called a Punnett square; it is a convenient device for keeping track of all the ways gametes can combine at fertilization.

are possible: SS, Ss (which is the same thing as sS), and ss. Our quantitative way of looking at things may lead us to wonder about what proportions of these genotypes we might expect to observe in the F₂ progeny. The expected frequencies of these three genotypes in our example may be determined in either of

two ways: by using the "Punnett square," devised in 1905 by the British geneticist Reginald Crundall Punnett, or by using simple probability calculations. The Punnett square is illustrated in Figure 10.3 and other figures in this chapter; probabilities are discussed in Box 10.A. By either method (they amount to the same thing), it becomes apparent that self-pollination of the F₁ genotype Ss will give the three F₂ genotypes in the expected ratio 1 SS:2 Ss:1 ss. Because S is dominant and s recessive, the ratio of *phenotypes* is 3 spherical (SS and Ss) to 1 dented (ss), just as Mendel observed.

Mendel did not live to see his theory placed on a sound physical footing based on chromosomes and DNA. Genes are now known to be portions of the DNA molecules in chromosomes. More specifically, a gene is a portion of the DNA that resides at a particular **locus**, or site, within the chromosome and that encodes a particular function. Remember that the cells in a multicellular organism have the same genotype because they are all derived by mitosis from a single cell, the zygote (Chapter 9). Each diploid cell has two homologous chromosomes of each type, and therefore has two alleles at each locus. Consistent with Mendel's model, if the two homologous chromosomes have copies of the same allele at a given locus, the cell and the organism are homozygous at that locus; if the homologs have differing alleles, the cell and the organism are heterozygous at that locus. Because meiosis reduces the number of chromosomes per cell, each gamete contains only one member of each homologous pair of chromosomes and, hence, only one allele at any given locus. If you visualize S and s as occupying specific, homologous sites on a pair of homologous chromosomes in an F₁ individual, you will see how they would be inherited through successive generations, by way of meiosis and the random fusion of gametes.

On the basis of monohybrid crosses such as that of Experiment 1, Mendel proposed his first law, called the **law of segregation**, which says that *alleles segregate from one another during the formation of gametes.* Mendel did not know about chromosomes or meiosis, but we do and we can picture the alleles segregating as chromosomes separate into gametes in meiosis (Figure 10.4).

The Test Cross

Mendel's theory adequately explains the ratios of phenotypes observed in F₁ and F₂ generations obtained from crosses of differing, true-breeding strains. To be regarded as fully satisfactory, however, the theory must also be able to predict—accurately—the outcome of other kinds of experiments. One such challenge was posed by Mendel himself. According to his theory, ⅔ of the F₂ spherical seeds from Ex-

BOX 10.A

Elements of Probability

Many people find it easiest to solve genetics problems using probability calculations, perhaps because the basic underlying considerations are a familiar part of daily life. When we flip a coin, for example, we expect it to have an equal probability of landing "heads" or "tails." When we roll a fair die, we expect to have equal chances of getting any of the numbers from one to six. We properly bet more money on a coin's giving at least one heads in a pair of tosses than we do on it coming up heads in a single toss—yet if we are even slightly sophisticated, we recognize that on a given toss, the probability of heads is independent of what happened in all the previous tosses. (For a fair coin, a run of 10 straight heads implies nothing about the next toss.

No "law of averages" increases the likelihood that the next toss will come up tails, and no "momentum" makes an eleventh occurrence of heads any more likely. On the eleventh toss, the odds are still 50:50.)

The basic conventions of probability are simple: If an event is absolutely certain to happen, its probability is 1. If it cannot happen, its probability is 0. Otherwise, its probability lies between 0 and 1. A coin toss results in heads half the time, and the probability of heads is ½—as is the probability of tails. If *two* coins (a penny and a dime, say) are tossed, each acts independently of the other. What, then, is the probability of both coins coming up heads? Half the time, the penny comes up heads; of that fraction, half the time the dime also comes up heads. Therefore, the joint probability of two heads is half of one-half, or $\frac{1}{2} \times \frac{1}{2} = \frac{1}{4}$. To find the joint probability of *independent* events, *multiply* the probabilities of the individual events.

To apply this to the crosses we have been discussing, we need only deal with gamete formation and random fertilization. A homozygote can produce only one type of gamete, so, for example, an SS individual has a probability equal to 1 of producing gametes with the genotype S. The heterozygote Ss produces S gametes with a probability of ½, and s gametes with a probability of ½ as well. Consider, now, the F_2 progeny of the cross of Figure 10.3. They are obtained by self-pollinating F_1 hybrids of genotype Ss. The probability that an F_2 plant is SS must be $\frac{1}{2} \times \frac{1}{2} = \frac{1}{4}$—there is a 50:50 chance of the sperm's being S, and this is independent of the 50:50 chance of the egg's being S. Similarly, the probability of ss offspring is $\frac{1}{2} \times \frac{1}{2} = \frac{1}{4}$. The probability of getting S from the sperm and s from the egg is also ¼, but the same genotype can also result from s in the sperm and S in the egg, with a probability of ¼. Thus the probability that an F_2 plant is a heterozygote is $\frac{1}{4} + \frac{1}{4} = \frac{1}{2}$. All three of the genotypes are expected in the ratio ¼ SS:½ Ss:¼ ss—hence the 1:2:1 ratio of genotypes and the 3:1 ratio of phenotypes seen in Figure 10.3.

periment 1 should be heterozygous, each carrying both S and s alleles (recall the genotype ratio 1:2:1). Therefore, if all the spherical seeds were allowed to grow into F_2 adults and self-pollinate, the ⅔ of the plants that were heterozygous would produce seeds of which about ¾ would be spherical and ¼ dented (recall the phenotype ratio 3:1). The other ⅓ of the F_2 plants, being SS homozygotes, would produce only spherical seeds (Figure 10.5). This result is what Mendel observed. It is impossible to know the genotype of a plant displaying a dominant phenotype simply by looking at it, but looking at the phenotypes of progeny obtained by self-fertilizing the plant gives us the answer.

The **test cross** is another way to test whether a given individual showing a dominant trait is homozygous or heterozygous. In a test cross, the individual in question is crossed with an individual known to be homozygous for the recessive trait—an easy individual to identify because its only phenotype is the recessive one. For the gene that we have been considering, the recessive homozygote for the test cross is ss. The individual being tested may be described initially as $S-$ because we do not yet know the identity of the second allele. If the individual being tested is homozygous dominant (here, SS), all offspring of the test cross will be Ss and show the dominant character (spherical seeds). If, however, the tested individual is heterozygous (Ss), then approximately ½ of the offspring of the test cross will show the dominant trait, but the other ½ will be homozygous recessive (Figure 10.6). These are exactly the results that are obtained; thus Mendel's model predicts accurately the results of such test crosses.

Independent Assortment of Alleles

What happens if a cross is made between two parents that differ at two or more loci? When a double heterozygote (for example, $AaBb$) makes gametes, do the alleles of maternal origin go together to one gamete

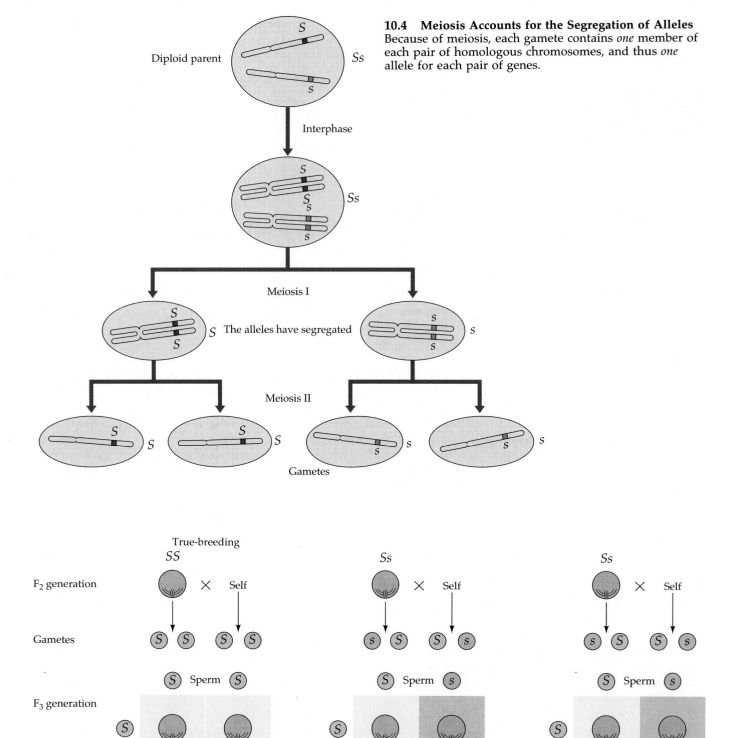

10.4 Meiosis Accounts for the Segregation of Alleles
Because of meiosis, each gamete contains *one* member of each pair of homologous chromosomes, and thus *one* allele for each pair of genes.

10.5 On to the F₃ Generation

According to Mendel's ideas, the spherical-seeded F₂ plants in Figure 10.3 are of two kinds. One-third of them (the *SS* homozygotes), upon self-pollination, will produce only spherical-seeded offspring. Two-thirds of them (the *Ss* heterozygotes) will, like the original F₁ plants in Figure 10.2, produce spherical- and dented-seeded plants in a 3:1 ratio.

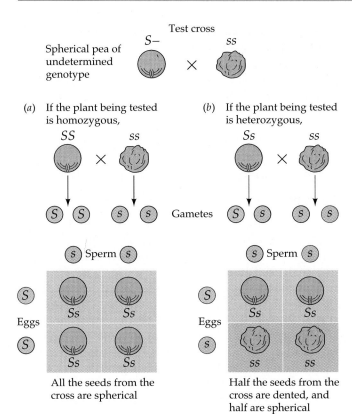

Test cross

Spherical pea of undetermined genotype $S-$ × ss

(a) If the plant being tested is homozygous,

SS × ss

↓ ↓ ↓ ↓ Gametes

S S s s

s Sperm s

Eggs S S

| | Ss | Ss |
| Ss | Ss |

All the seeds from the cross are spherical

(b) If the plant being tested is heterozygous,

Ss × ss

↓ ↓ ↓ ↓

S s S s

s Sperm s

Eggs S s

| Ss | Ss |
| ss | ss |

Half the seeds from the cross are dented, and half are spherical

10.6 Homozygous or Heterozygous? Try a Test Cross

A plant with a dominant phenotype may be homozygous or heterozygous. Its genotype can be deduced by observing the phenotypes of progeny produced by crossing it with a homozygous recessive plant—that is, by making a test cross. *(a)* If all progeny show the dominant phenotype, the plant being tested must have been homozygous for the dominant allele. *(b)* If half the progeny have the dominant phenotype and half have the recessive phenotype, the plant in question must have been heterozygous.

and those of paternal origin to another gamete? Or does a single gamete receive some maternal and some paternal alleles? To answer these questions Mendel performed a series of **dihybrid crosses**: crosses made between parents identically heterozygous at two loci.

In these experiments Mendel began with peas that differed for two characters of the seeds: seed shape and seed color. One true-breeding strain produced only spherical, yellow seeds and the other strain produced only dented, green ones. Plants of the first strain can be designated *SSYY*, indicating that they are homozygous both for the *S* allele at the seed-shape locus and for the *Y* allele at the seed-color

locus. The second doubly homozygous strain is *ssyy*. The doubly heterozygous F₁ offspring from a cross between these two strains are *SsYy*. Because the *S* and *Y* alleles are dominant, these F₁ seeds would all be yellow and spherical.

Now what about the dihybrid cross? There are two ways in which these doubly heterozygous plants might produce gametes, as Mendel saw it (remember that he had never heard of chromosomes, let alone of meiosis). First, if the alleles maintained the associations they had in the original parents, then only two types of gametes would be produced: *SY* and *sy*; and the F₂ progeny resulting from self-pollination of the F₁ plants would consist of three times as many plants bearing spherical, yellow seeds as ones with dented, green seeds. Were such results to be obtained, there would be no reason to suppose that seed shape and seed color were really regulated by two different genes, because spherical seeds would always be yellow, and dented ones green.

The second possibility is that the segregation of *S* from *s* is *independent* of the segregation of *Y* from *y* during the production of gametes. In this case, four kinds of gametes would be produced, and in equal numbers: *SY*, *Sy*, *sY*, and *sy*. When these gametes combined at random, they would produce an F₂ of nine different genotypes. The progeny can have any of three possible genotypes for shape (*SS*, *Ss*, or *ss*)

BOX 10.B

Probabilities in the Dihybrid Cross

If F₁ plants heterozygous for two independent traits self-pollinate, the resulting F₂ plants express four phe-

notypes. The proportions of these phenotypes are easily determined by probabilities. The probability of a seed's being yellow is ¾ (see Box 10.A); by the same reasoning, the probability of a seed's being spherical is also ¾. The two traits are determined by separate genes and are independent of one another, so the joint probability of a seed being both yellow and spherical is ¾ × ¾ = ⁹⁄₁₆. For the dented, yellow members of

the F₂, the probability of yellow is again ¾; the probability of dented seeds is ½ × ½ = ¼. The joint probability of a seed being both yellow and dented is, then, ¾ × ¼ = ³⁄₁₆. The same probability applies, for similar reasons, to the spherical, green F₂ seeds. Finally, the probability of F₂ seeds being both dented and green must be ¼ × ¼ = ¹⁄₁₆. Looking at all four phenotypes, we see they are expected in the ratio of 9:3:3:1.

and any of three for color (*YY*, *Yy*, or *yy*). These nine genotypes would produce just four phenotypes (spherical, yellow; spherical, green; dented, yellow; dented, green). By using either a Punnett square or simple probability calculations, we can show that these four phenotypes would be expected to occur in a ratio of 9:3:3:1.

Mendel's dihybrid crosses produced the results predicted by the second possibility. Four different phenotypes appeared in a ratio of about 9:3:3:1 in the F_2, rather than only the two parental types as predicted from the first possibility (Figure 10.7; Box 10.B). The parental traits appeared in new combinations in two of the phenotypic classes (spherical, green and dented, yellow). These are called **recombinant phenotypes**. These results led Mendel to the formulation of what is now known as Mendel's second law: *Alleles of different genes assort independently of one another during gamete formation.* This **law of independent assortment** is not as universal as the law of segregation because it applies only to genes that lie

on separate chromosomes, and not to those that lie on the same chromosome. It is, however, correct to say that *chromosomes* assort independently during the formation of gametes (Figure 10.8). It is interesting to note that all the genes Mendel studied were on different chromosomes; was that a matter of luck?

GENETICS AFTER MENDEL: ALLELES AND THEIR INTERACTIONS

Incomplete Dominance and Codominance

Some genes have alleles that are neither dominant nor recessive to each other. Instead, the heterozygotes show an intermediate phenotype superficially like that predicted by the old blending theory of inheritance. For example, if a true-breeding red snapdragon is crossed with a true-breeding white one, all the F_1 flowers are pink. That this phenomenon can still be explained in terms of Mendelian genetics rather than a blending theory is readily demonstrated by a further cross. If one of these pink F_1 snapdragons is crossed with a true-breeding white one, the blending theory predicts that all the offspring would be a still-lighter pink. In fact, approximately ½ of the offspring are white and ½ the same pink as the original F_1. Suppose now that the F_1 pink snapdragons are self-pollinated. The resulting F_2 plants are distributed in a ratio of 1 red : 2 pink : 1 white (Figure 10.9). Clearly the hereditary particles—the genes—have not blended, but they are readily sorted out in their original forms.

We can understand these results in terms of the Mendelian model. All we need to do in cases like this is recognize that the heterozygotes show a phenotype intermediate between those of the two homozygotes. Genes code for the production of specific proteins, many of which are enzymes. Different alleles at a locus code for alternative forms of a protein that differ in structure and, when the protein is an enzyme, often have different degrees of catalytic activity. In the snapdragon example, one allele codes for an enzyme that catalyzes a reaction leading to the formation of a red pigment in the flowers. The alternative allele codes for an altered protein lacking catalytic activity for pigment production. Plants homozygous for this alternative allele cannot synthesize red pigment, and their flowers are white. Heterozygous plants, with only one allele for the functional enzyme, produce just enough red pigment so that their flowers are pink. Homozygous plants with two alleles for the functional enzyme produce more red pigment, resulting in red flowers. When a heterozygous phenotype is intermediate, as in this example, the gene is said to be governed by **incomplete dominance**.

There are more examples of incomplete dominance

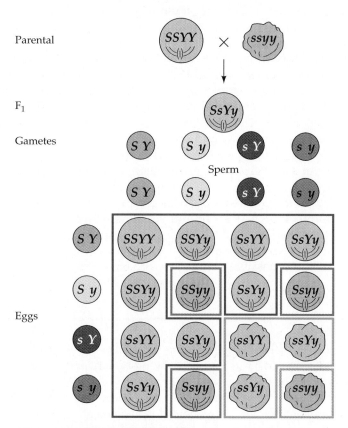

10.7 Independent Assortment
Plants heterozygous for two genes (*SsYy*) make four kinds of gametes in equal proportions. Random combination of these gametes produces equal numbers of the 4 × 4 = 16 combinations displayed in the boxes; these 16 combinations result in nine different genotypes. Because *S* and *Y* are dominant over *s* and *y*, respectively, the nine genotypes determine four phenotypes (indicated by four different outline colors in the Punnett square) in the ratio of 9:3:3:1.

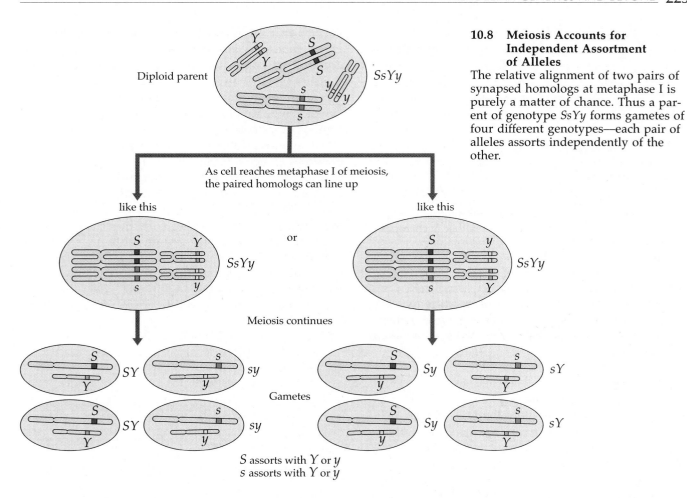

10.8 Meiosis Accounts for Independent Assortment of Alleles

The relative alignment of two pairs of synapsed homologs at metaphase I is purely a matter of chance. Thus a parent of genotype *SsYy* forms gametes of four different genotypes—each pair of alleles assorts independently of the other.

than of complete dominance. Thus an unusual feature of Mendel's report is that all seven of the examples he described (see Table 10.1) are characterized by complete dominance. For dominance to be complete, a single copy of the dominant allele must produce enough of its protein product to give the maximum phenotypic response. For example, just one copy of the dominant allele *T* at one of the loci studied by Mendel leads to the production of enough of a growth-promoting chemical so that the *Tt* heterozygotes are as tall as the homozygous dominant plants (*TT*)—the second copy of *T* causes no further growth of the stem. The homozygous recessive plants (*tt*) are much shorter because the allele *t* does not lead to the production of the growth promoter.

10.9 Incomplete Dominance Follows Mendel's Laws

Heterozygous snapdragons produce pink flowers because the allele for red flowers is incompletely dominant over the allele for white ones. When true-breeding red and white parents cross, all plants in the F₁ generation are pink. When these F₁ plants self-pollinate, they produce F₂ offspring that are white, pink, and red in a ratio of 1:2:1. A test cross, diagrammed at the right, confirms that pink snapdragons are heterozygous; see Figure 10.6 for the reasoning.

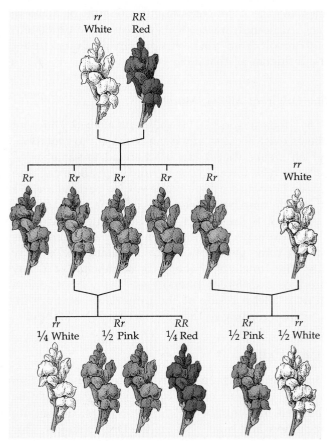

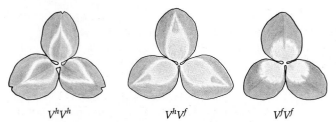

V^hV^h V^hV^f V^fV^f

10.10 Codominance in White Clover
White clover leaves have characteristic patterns of chevrons and colored areas, all genetically determined. The leaves on the left are homozygotes of genotype V^hV^h; those on the right are homozygotes of genotype V^fV^f. Since V^hV^f heterozygotes show both the chevron that is characteristic of V^h and the colored area that is characteristic of V^f, the V^h and V^f alleles are codominant.

Sometimes two alleles at a locus produce two different phenotypes, both of which appear in heterozygotes (Figure 10.10). This phenomenon is called **codominance**.

Pleiotropy

When a single allele has more than one distinguishable phenotypic effect, we say that the allele is **pleiotropic**. The most familiar example of pleiotropy is the allele responsible for the coloration pattern (light body, darker extremities) of Siamese cats, discussed later in this chapter. The same allele is also responsible for the characteristic crossed eyes of Siamese cats. Although these effects appear to be unrelated, both result from the same protein produced under the influence of that allele.

The Origin of Alleles: Mutation

Why does a gene have different alleles? Different alleles exist because any gene is subject to **mutation**, which means that it can be changed to some *stable, heritable* new form. In other words, an allele can mutate to become a different allele. One particular allele of a gene may be defined as **wild-type**, or standard, because it is present in most individuals and gives rise to an expected trait or phenotype. Other forms of that same gene, often called **mutant** alleles, may alter the function of the gene somewhat and may produce a different phenotype. The wild-type and mutant alleles reside at the same locus and are inherited according to the rules set forth by Mendel.

Multiple Alleles

Mutation, to be discussed in Chapter 11, is a random process; different copies of the same gene may be changed in a number of different ways, depending upon how and exactly where the DNA changes. This implies that there may be more than two alleles of a given gene in a group of individuals. (Any one individual has only two alleles, of course—one from the mother and one from the father.) In fact, there are many examples of such **multiple alleles**. Some clover leaves are plain green, while others have chevrons of other colors on their leaves. Seven alleles at a locus control the pattern of chevrons on the leaves of white clover (Figure 10.11). In the fruit fly *Drosophila melanogaster*, many alleles at one locus affect eye color by determining the amount of pigment produced (Table 10.2). The exact color of the fly's eyes depends on which two alleles are inherited.

The ABO blood group system in humans is determined by a set of three alleles (I^A, I^B, and i) at one locus. Different combinations of these alleles in different people produce four different blood types, or phenotypes: A, B, AB, and O (Table 10.3). Early at-

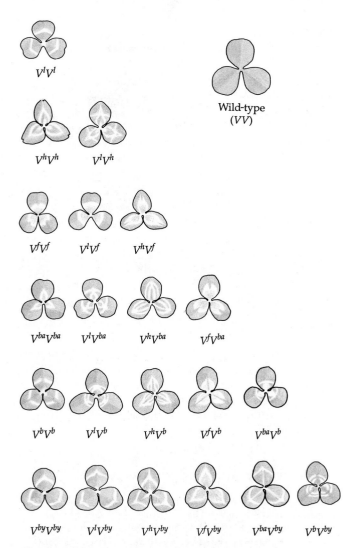

10.11 Multiple Alleles in White Clover
Seven alleles at the same locus determine the pattern of chevrons and colored areas on white clover leaves. Many of these alleles show codominance in heterozygotes.

TABLE 10.2
Multiple Alleles for Eye Color in *Drosophila melanogaster*

GENOTYPE	PHENOTYPE	DEGREE OF PIGMENTATION OF THE EYE
w^+w^+	wild-type (dull red)	0.6800
$w^{col}w^{col}$	colored	0.1636
$w^{sat}w^{sat}$	satsuma	0.1404
$w^{w}w^{col}$		0.1114
$w^{co}w^{co}$	coral	0.0798
$w^{w}w^{w}$	wine	0.0650
$w^{a3}w^{a3}$	apricot-3	0.0632
$w^{ch}w^{ch}$	cherry	0.0410
$w^{e}w^{e}$	eosin	0.0324
$w^{bl}w^{bl}$	blood	0.0310
$w^{a}w^{a}$	apricot	0.0197
$w^{t}w^{t}$	tinged	0.0062
ww	white	0.0044

The Rh factor, so named because it was first found in rhesus monkeys, is another substance on the surface of red blood cells. In most human populations, almost 100 percent of the individuals have the Rh factor, and their blood is said to be Rh$^+$ (Rh-positive). Among Caucasians, however, only 83 percent are Rh$^+$; the others lack the Rh factor, and their blood is called Rh$^-$ (Rh-negative). Like the ABO blood types, the Rh factor is genetically determined. A single locus with at least eight multiple alleles is responsible. Certain dominant alleles cause the production of the Rh factor; Rh$^-$ individuals are homozygous recessives.

Another system of multiple alleles is illustrated by the scallops in Study Question 1 at the end of the chapter. The question of how differing alleles may be maintained in a population through time will be examined in Chapter 19.

FOCUS ON CHROMOSOMES

Linkage

In the immediate aftermath of the rediscovery of Mendel's laws, the second law—independent assortment—was considered to be generally applicable. However, some investigators, including Punnett (the inventor of the square), began to observe strange deviations from the expected 9:3:3:1 ratio in some dihybrid crosses. In particular, they sometimes observed an apparent excess of parental phenotypes and a shortage of recombinant phenotypes among the F$_2$ progeny. Suppose that the original cross was between the genotypes *AABB* and *aabb*. If alleles at the *A* locus assorted independently of alleles at the *B* locus, the F$_2$ should consist of $\frac{9}{16}$ individuals with the dominant phenotypes for *A* and *B*, $\frac{3}{16}$ individuals dominant only for *A* (*A−bb*), $\frac{3}{16}$ individuals dominant only for *B* (*aaB−*), and $\frac{1}{16}$ double recessive homozygotes (*aabb*). What Punnett and others observed instead were large excesses of *aabb* over the $\frac{1}{16}$ expected.

These results become understandable when we assume that the two loci are on the *same chromosome*—

tempts at blood transfusion—made before these blood types were understood—often killed the patient. Around the turn of the century, however, the Austrian scientist Karl Landsteiner mixed blood cells and serum (blood from which cells have been removed) from different individuals. He found that only certain combinations of blood types are compatible. In other combinations, the red blood cells form clumps because of the presence in the serum of specific proteins, called antibodies (see Chapter 16), that react with foreign, or "nonself," cells and macromolecules (Figure 10.12). When transfusions are given, a perfect matchup of blood types between donor and patient is best. For example, if the patient has red blood cells of type A, then the blood donor should have type A cells as well. Certain combinations other than perfect matchups are also usually successful, as indicated in Table 10.3.

Blood type

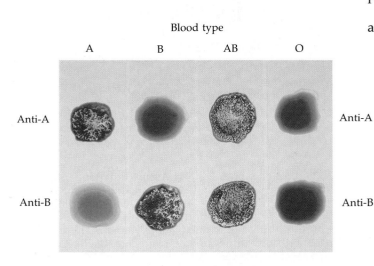

10.12 ABO Blood Reactions Are Important in Transfusions
Cells of blood types A, B, AB, and O were mixed with anti-A or anti-B antibodies. Red blood cells that react with antibodies clump together (speckled appearance in the photograph); red blood cells that do not react with antibody remain evenly dispersed. Note that anti-A reacts with A and AB cells but not with B or O. Which blood types do anti-B antibodies react with? As you look down the columns, note that each of the types, when mixed separately with anti-A and with anti-B, gives a unique pair of results; this is the basic method by which blood is typed.

TABLE 10.3
The ABO Blood System

BLOOD TYPE	GENOTYPE	REACTION WITH ANTI-A SERUM	REACTION WITH ANTI-B SERUM	TYPE OF DONOR BLOOD ACCEPTED
A	$I^A I^A$ or $I^A i$	Clumping of red blood cells	No clumping	A or O
B	$I^B I^B$ or $I^B i$	No clumping	Clumping of red blood cells	B or O
AB	$I^A I^B$	Clumping of red blood cells	Clumping of red blood cells	A, B, AB, or O
O	ii	No clumping	No clumping	O

that is, they are linked together. After all, since the number of genes in a cell far exceeds the number of chromosomes, each chromosome must contain many genes. (The human genome consists of perhaps 50,000 genes, distributed over 23 pairs of chromosomes.) Suppose, now, that the A and B loci are on the same chromosome. To remind ourselves that the genes are linked in this fashion, let us write the genotypes differently: One parent is $\overline{AB}\,\overline{AB}$ and the other $\overline{ab}\,\overline{ab}$. The former produces gametes that are of one type, $\overline{AB}$; the latter produces $\overline{ab}$ gametes. Thus the genotype of the F₁ is $\overline{AB}\,\overline{ab}$. Now, the key difference between dihybrid crosses with linkage and those without is in the formation of gametes by the F₁. Without linkage, as we have seen, four types of gametes are produced in equal frequency (AB, Ab, aB, and ab). *With* linkage, however, most of the gametes must be either $\overline{AB}$ or $\overline{ab}$ because the two loci are physically "tied together" on the same chromosome—they are part of the same DNA molecule. Instead of the 9:3:3:1 ratio of four F₂ phenotypes produced from a dihybrid cross without linkage (see Figure 10.7), a dihybrid cross with linkage yields only parental phenotypes in the F₂ (Figure 10.13). (As we will see, a few recombinant gametes will appear when genes are less closely linked than in our example. They result from crossing over between loci; see Chapter 9.)

The full set of loci on a given chromosome constitutes a **linkage group**. The number of linkage groups in a species, determined by experiments such as the dihybrid cross described here, should equal the number of homologous chromosome pairs, as determined by microscopic examination of nuclei undergoing meiosis or mitosis.

Sex Determination

In Kölreuter's experience, and later in Mendel's, reciprocal crosses apparently always gave identical results. This is because in diploid organisms, chromosomes come in pairs. One member of each chromosome pair derives from each parent; it does not matter, for example, whether a dominant allele was contributed by the mother or by the father. But this is not always the case; sometimes the parental origin of a chromosome does matter. To understand the types of inheritance in which parental origin is important, we must consider the ways in which sex is determined in different species.

In maize, a plant much studied by geneticists, every diploid adult has both male and female structures. These two types of tissue are genetically identical, just as roots and leaves are genetically identical.

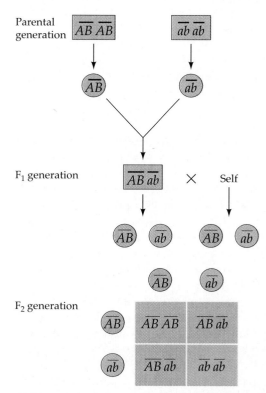

10.13 Linked Genes Do Not Assort Independently
When two genes are very tightly linked on the same chromosome (indicated by the line above the letters), all the F₂ offspring from a dihybrid cross have parental phenotypes. In the cross illustrated, genes A and B are so closely linked that they always segregate together, as if they were a single gene.

Plants such as maize, and animals such as earthworms, which produce both male and female gametes in the same organism, are said to be **monoecious** (from the Greek for "single house"). Some plants, such as date palms and oak trees, and most animals are **dioecious**, meaning some of the individuals can produce only male gametes and the others can produce only female gametes. In most dioecious organisms, sex is determined by differences in the chromosomes; but such determination operates in a bewildering variety of ways (Figure 10.14). The sex of a honeybee, for example, depends on whether it develops from a fertilized or an unfertilized egg. A fertilized egg is diploid and gives rise to a female bee—either a worker or a queen, depending on the diet during larval life. An unfertilized egg is haploid and gives rise to a male drone.

In many other animals, including ourselves, sex is determined by a single **sex chromosome** or by a pair of them. Both males and females have two copies of each of the rest of the chromosomes, which are called **autosomes**. The sex chromosome that is present in different numbers is the **X chromosome**. For example, female grasshoppers have two X chromosomes, whereas males have only one. These females form eggs containing one copy of each autosome and one X chromosome. The males form two types of sperm. Half contain an X chromosome and one copy of each autosome, and the other half contain only autosomes. This is a natural consequence of meiosis in the two sexes. In females, the two X chromosomes synapse in prophase I; one goes to each of the daughter nuclei from meiosis I. Males have but a single X chromosome, so there is no synapsis. Thus half the sperm end up containing an X chromosome, but the others get none. Female grasshoppers are described as being XX (ignoring the autosomes) and males as XO (pronounced "ex-oh"). When an X-bearing sperm fertilizes an egg, the zygote is XX and develops into a female. When a sperm without an X fertilizes an egg, the zygote is XO and develops into a male. This chromosomal mechanism ensures that the two sexes are produced in approximately equal numbers. No such mechanism for numerical equality of the sexes exists in the diploid–haploid system of bees.

Female mammals have two X chromosomes and males have one (Figure 10.14b). However, male mammals also have a sex chromosome that is not found in females: the **Y chromosome**. Females may be represented as XX and males as XY. The males produce two kinds of gametes, each with a complete set of autosomes but differing with respect to their sex chromosomes: Half the gametes carry an X chromosome and the rest carry a Y. When an X-bearing sperm fertilizes an egg, the resulting XX zygote is female; when a Y-bearing sperm fertilizes an egg, the XY zygote is male.

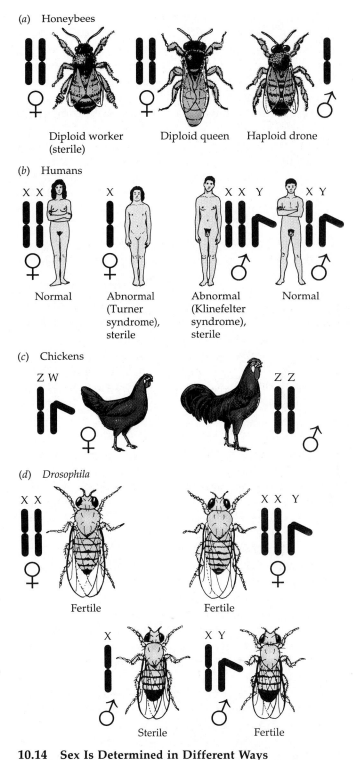

10.14   Sex Is Determined in Different Ways
(a) In honeybees, fertilized eggs develop into diploid females and unfertilized eggs develop into haploid males. In other animals, sex is determined by special sex chromosomes. *(b)* Normal human females (left) carry two X chromosomes; normal males (right) carry one X and one Y chromosome. Persons who have some other number of sex chromosomes may develop abnormally. *(c)* In birds, it is the males that carry two identical sex chromosomes (ZZ) and females that have differing ones (ZW). *(d)* *Drosophila* (fruit fly) females have two X chromosomes and may also have a Y chromosome; males have an X chromosome and, if they are fertile, a Y chromosome.

Some subtle but important differences show up clearly in mammals with abnormal sex chromosome constitutions. These conditions tell us something about the functions of the X and Y chromosomes. In both humans and mice, XO individuals sometimes appear. In humans, XO individuals are females who are moderately physically abnormal but mentally normal and are almost always sterile. The XO condition in humans is called Turner syndrome. In mice, XO individuals are fertile females that are virtually normal. XXY individuals also arise. XXY humans (a condition known as Klinefelter syndrome) are decidedly abnormal, always sterile, and always males. In brief, in humans the Y chromosome carries the genes that determine maleness. The absence of Y leads to femaleness, while the presence of Y has a definite masculinizing effect.

The Y chromosome functions differently in the fruit fly *Drosophila melanogaster* (Figure 10.14*d*). Superficially, *Drosophila* follows the same pattern as mammals: Females are XX and males are XY. However, XO individuals are males (rather than females as in mammals) and almost always are indistinguishable from normal XY males except that they are sterile. XXY *Drosophila* are normal, fertile females. In *Drosophila*, sex is determined strictly by the ratio of X chromosomes to autosome sets. If there is one X chromosome for each set of autosomes, the individual is a female; if there is only one X chromosome for the two sets of autosomes, the individual is a male. The Y chromosome plays no sex-determining role in *Drosophila*, but it is needed for male fertility.

In birds, moths, and butterflies, males are XX and females are XY. To avoid confusion, this is usually expressed as ZZ (male) and ZW (female) (Figure 10.14*c*). In these organisms, it is the female that produces two types of gametes. Thus the egg determines the sex of the offspring, rather than the sperm as in humans and fruit flies.

Sex Linkage

How does the existence of sex chromosomes affect patterns of inheritance? In *Drosophila* and in humans, the Y chromosome carries few known genes, whereas a substantial number of genes affecting a great variety of traits are carried on the X chromosome, which leads to an important deviation from the usual Mendelian ratios for the inheritance of genes located on the X chromosome. Any such gene is present in two copies in females, but in only one copy in males. Therefore, females may be heterozygous for genes that are on the X chromosome, but males will always be **hemizygous** for these genes—they will have only

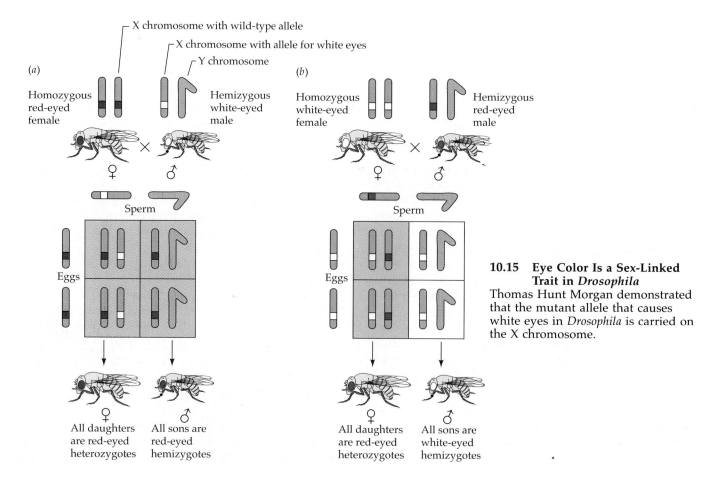

10.15 Eye Color Is a Sex-Linked Trait in *Drosophila*
Thomas Hunt Morgan demonstrated that the mutant allele that causes white eyes in *Drosophila* is carried on the X chromosome.

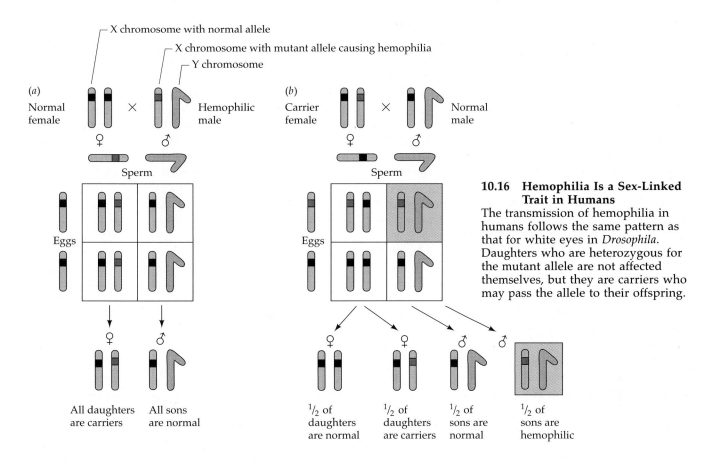

(a) Normal female × Hemophilic male

(b) Carrier female × Normal male

X chromosome with normal allele
X chromosome with mutant allele causing hemophilia
Y chromosome

Sperm

Eggs

All daughters are carriers

All sons are normal

Sperm

Eggs

½ of daughters are normal

½ of daughters are carriers

½ of sons are normal

½ of sons are hemophilic

10.16 Hemophilia Is a Sex-Linked Trait in Humans
The transmission of hemophilia in humans follows the same pattern as that for white eyes in *Drosophila*. Daughters who are heterozygous for the mutant allele are not affected themselves, but they are carriers who may pass the allele to their offspring.

one of each. It is useful here, as it is in many instances when studying genetics, to think of loci whose alleles govern easily observable phenotypes as **markers** of the chromosomes on which they are located. Reciprocal crosses of parents that have different markers on their sex chromosomes do not give identical results; this is a sharp deviation from the inheritance of markers on autosomes.

The first and still one of the best examples of **sex-linked inheritance**—inheritance of traits governed by loci on the sex chromosomes—is that of eye color in *Drosophila*. The wild-type eye color of these flies is red. In 1910, Thomas Hunt Morgan discovered a mutation that causes white eyes. He experimented by crossing flies of the wild-type and mutant phenotypes. His results demonstrated that the eye-color locus is on the X chromosome. When homozygous red-eyed females were crossed with (hemizygous) white-eyed males, all the sons and daughters had red eyes, because red is dominant over white and all the progeny had inherited a wild-type X chromosome from their mothers (Figure 10.15a). However, in the reciprocal cross, in which a white-eyed female was mated to a red-eyed male, all the sons were white-eyed and all the daughters red-eyed (Figure 10.15b). The sons from the reciprocal cross inherited their only X chromosome from their white-eyed mother; the Y chromosome they inherited from their father

does not carry the eye-color locus. The daughters, on the other hand, got an X chromosome with the white allele from their mother and an X chromosome bearing the red allele from their father; they were therefore red-eyed heterozygotes. When Morgan mated these same heterozygous females with red-eyed males, he observed that half their sons had white eyes, but all their daughters had red eyes.

The human X chromosome carries many loci. The alleles at these loci follow the same pattern of inheritance as those for white eyes in *Drosophila*. One X-chromosome locus, for example, has an allele that causes hemophilia, a hereditary disorder characterized by the failure of blood to clot properly; victims suffer from excessive and often fatal bleeding. Hemophilia appears in individuals who are homozygous for a mutant recessive allele. A hemophilic man married to a homozygous normal woman will not produce any hemophilic children. The sons inherit a single, normal X from their mother and will neither have the disease nor transmit it to their children. The daughters get an X chromosome bearing a normal allele from their mother and one bearing the allele for hemophilia from their father. Because hemophilia is recessive, however, the daughters will not be hemophilic (Figure 10.16a). They will, however, be heterozygous *carriers*. Such carriers of an X-linked trait will transmit the disease to half their sons and the carrier role to half their daughters (Figure 10.16b). What parental genotypes would produce a female hemophiliac? Her father would have to be a hemophiliac, and her mother a carrier. Because hemophilia is quite rare, two such people are unlikely to meet. Moreover, until recently hemophilic males rarely survived long enough to reproduce. One might thus expect hemophilic females to be extremely rare, and in fact very few have ever been found.

The small human Y chromosome carries very few loci. Among them are the maleness determiners, whose existence was suggested by the phenotypes of the XO and XXY individuals described earlier (see Figure 10.14). The pattern of inheritance of Y-linked alleles should be easy for you to work out. Give it a try now.

Mendelian Ratios Are Averages, Not Absolutes

You have now been introduced to the basic Mendelian ratios: 3:1, 1:1, 9:3:3:1; you will figure out others as you do homework or test problems. It is essential to remember, however, that these represent highest probabilities, not invariant rules. The X-Y system of sex determination in our species results in roughly equal numbers of males and females in a substantial population, but you know that a given family of four children may not consist of two girls and two boys. It is not unusual for four children to be of the same sex; in fact, one family in eight who have four children will have all boys or all girls. How do we know if this deviation is simply the result of chance?

When we are trying to understand the genetic basis for the results of a cross, it is important to know how much deviation from a predicted ratio we can reasonably expect due to normal chance. Statistical methods have been devised for determining whether the observed deviation from an expected ratio can be attributed to chance variation or whether the deviation is large enough to suggest that the observed ratio is caused by something specific. These methods take several factors into consideration, but one of the most important is **sample size**. If we expect to find a 3:1 ratio between two phenotypes and we look at a sample of only 8 progeny, we should not be surprised to find 7 individuals of one phenotype and 1 of the other (rather than the expected 6 and 2). It would be surprising, however, in a sample of 80 to find 70 of one phenotype and 10 of the other—you would question whether that really represented a 3:1 ratio. In experiments in genetics—and in quantitative biology in general—sample sizes should be large so that the data are easier to evaluate with confidence.

Special Organisms for Special Studies

Prokaryotic and eukaryotic organisms of many kinds have been used in genetic studies. A few species have been used many times because of one advantage or another. Gregor Mendel did his most famous work with the garden pea, but until recently the best-studied higher plant was maize, or corn (*Zea mays*). Originally, this emphasis on maize was due in part to its great agricultural importance. Maize has been examined so thoroughly that highly detailed **genetic maps** locating particular genes on each of the chromosomes are available.

The first animal to be studied in great detail was the fruit fly, *Drosophila melanogaster* (see Figure 10.15). Its small size, ease of cultivation, and short generation time made it an attractive experimental subject. Thomas Hunt Morgan and his students established *Drosophila* as a highly useful laboratory organism in Columbia University's famous "fly room," where such phenomena as sex linkage were discovered. *Drosophila* remains extremely important in studies of chromosome structure, population genetics, the genetics of development, and the genetics of behavior.

There was a period in genetics in which the focus was the common salmon-colored bread mold *Neurospora crassa*. It was used in a number of historic experiments. The products of meiosis in *Neurospora* are organized in an unusual way that makes it easy to visualize the results of crossing over, as you will see in the next section.

More recently, molecular geneticists and developmental biologists have directed heavy attention to two other organisms. We discuss research on the tiny worm *Caenorhabditis elegans* in Chapter 17 and the small plant *Arabidopsis thaliana*, considered by many to be a weed, in Chapters 33 and 34.

Meiosis in *Neurospora*

The life cycle of *Neurospora*, like those of other fungi, is complex. *Neurospora* grows from haploid spores that divide mitotically to produce a feltlike mat of long strands called hyphae. Eventually, as the end result of a sexual process, haploid nuclei from two individuals unite to produce zygotes in a specialized fruiting structure. As soon as a diploid nucleus is formed, it undergoes meiosis. Thus the zygote itself constitutes a greatly reduced diploid generation.

Meiosis in *Neurospora* is a tidy process that packages all of the nuclei produced by the divisions of a single zygote in a long, thin sac called an ascus (plural: asci). The four haploid nuclei then divide once again by mitosis. The eight nuclei produced by this sequence of events are incorporated into eight spores, all neatly lined up within the ascus (Figure 10.17). Because the ascus is so thin, the nuclei cannot pass one another as the divisions proceed, so the pairs of spores can easily be identified with the meiotic division that produced them. This makes *Neurospora* an especially useful organism in which to examine segregation, assortment, and recombination

10.17 *Neurospora* **Packs Its Haploid Spores into an Ascus**
A rosette of asci of *Neurospora crassa*, resulting from a cross of a spore-color mutant with a wild-type strain. Each ascus contains eight spores; their arrangements reflect different segregation patterns.

of genetic markers. Accordingly, we use it to illustrate the material in the next section.

Recombination in Eukaryotes

We have seen that each chromosome has many loci and that all the loci on one chromosome are linked to each other. If homologous chromosomes did not undergo crossing over when paired (see Chapter 9), all the markers on a given chromosome would all segregate together as a unit. A geneticist would have no way of knowing that linked loci are actually different loci. Mendel's second law (independent assortment of alleles of different loci) would apply only to loci on different chromosomes.

What actually happens is more complex and therefore more interesting. Markers located at different places on the same chromosome do sometimes separate from one another as the result of crossing over. The farther apart two markers are on a chromosome, the greater the likelihood that they will separate and recombine. Geneticists use **recombination frequencies** (the observed frequencies in the offspring of marker combinations different from those of the parents) to generate genetic maps that indicate the arrangement of markers along the chromosome (Box 10.C).

Genetic markers on the same chromosome pair recombine by **crossing over**, which results from the physical exchange of corresponding genetic segments between two homologous chromosomes during prophase I of meiosis. In other words, recombination occurs at the stage when homologous chromosomes are paired. Recall that the DNA has duplicated by this stage, and each chromosome consists of two chromatids. Thus crossing over occurs at the four-

strand stage. The exchange event at any point along the length of the chromosome involves only two of the four chromatids, one from each member of the chromosome pair (Figure 10.18). The lengths of chromosome are exchanged reciprocally, so both chromatids involved in crossing over are recombinant (that is, each chromatid contains genes from both parents); and no genes are created or destroyed. The points at which the chromatids break in the exchange seem to correspond perfectly (at the level of base pairs in the DNA), so that the amount of material donated by a chromatid exactly equals the amount it receives.

At any point along the paired chromosomes, only two of the four chromatids participate in crossing over; but other crossovers may occur at other points. These other crossovers may involve the same pair

Prophase I of meiosis

Paired homologous chromosomes

Nucleus

Crossover (chiasma)

Remainder of meiosis I

Recombinant chromosomes

Meiosis II

Mitosis

Ascus   Spore

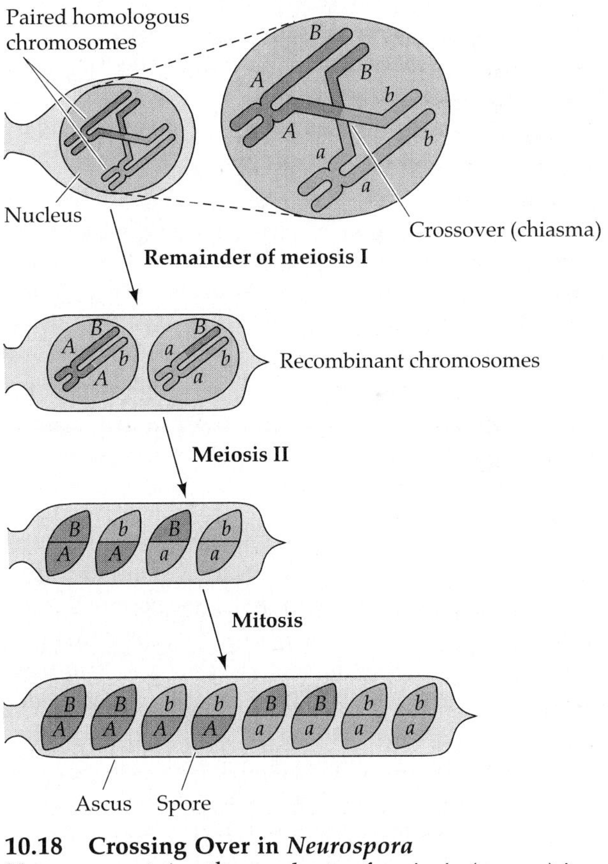

10.18 Crossing Over in *Neurospora*
Neurospora retains the products of meiosis (spores) in a single package (the ascus). This diagram shows only one of *Neurospora*'s several chromosome pairs, beginning with the diploid nucleus during prophase I of meiosis. A single crossover between genes *A* and *B* forms two recombinant chromatids. In this particular ascus, at the end of meiosis II the nuclei with recombinant chromosomes (carrying *Ab* and *aB*) lie in the middle of the ascus, while those with parental chromosomes (carrying *AB* and *ab*) lie at the ends. A mitotic division follows meiosis II, increasing the number of spores in the ascus to eight. Note that the recombinant chromosomes remain at the middle of the ascus.

BOX 10.C

Gene Mapping in Eukaryotes

Neurospora is an excellent organism for illustrating the principles of genetic mapping. In mapping experiments, we do not make use of the orderly packaging of spores in asci; rather, spores are collected at random and examined. Remember that this organism is haploid for most of its life cycle; haploid genotypes thus characterize strains. Suppose that we cross a strain of genotype *AB* with another strain of genotype *ab*. We then determine the **frequency of recombination** between the two markers as follows. Let us say that spores of genotype *AB* make up 40 percent of the total, *ab* 40 percent, *Ab* 10 percent, and *aB* 10 percent. Of all these spores, 40 + 40 = 80 percent are of the parental genotypes, 10 + 10 = 20 percent are recombinant. The frequency of recombination between the two markers is thus 20 percent.

To determine the linear sequence and spacing of markers on the chromosome, we can perform a three-factor cross. If *Neurospora* strains that are *ABC* and *abc* are crossed, the following classes of spore genotypes might be observed:

1. *ABC* 38.0 percent
2. *abc* 40.2 percent
3. *Abc* 7.2 percent
4. *aBC* 6.6 percent
5. *ABc* 3.1 percent
6. *abC* 3.7 percent
7. *AbC* 0.5 percent
8. *aBc* 0.7 percent

Classes 1 and 2 are parental types. Classes 3 and 4 are single recombinants between *A* and *B*. Classes 5 and 6 are single recombinants between *B* and *C*. Note that all four of the classes 3 through 6 are also recombinants between *A* and *C*. Classes 7 and 8 are recombinant between *A* and *B* and between *B* and *C*, but not between *A* and *C*; they represent double crossover types. Note that 7 and 8 are the least frequent classes, which is what we would expect of double crossover types. The two crossovers cancel each other in recombining *A* and *C*, leaving them in the parental arrangement relative to each other.

To compute map distances, we add the crossovers in all the classes that are recombinant between a given pair of markers. The "distance" between *A* and *B* is thus the sum of classes 3, 4, 7, and 8: 7.2 + 6.6 + 0.5 + 0.7 = 15 percent recombination. The distance between *B* and *C* (8 percent recombination) is obtained by summing classes 5, 6, 7, and 8. The distance between *A* and *C* is the sum of classes 3 through 6, plus two times class 7, plus two times class 8 (because each of these last two classes contains two crossovers between *A* and *C*): so the distance between *A* and *C* = 7.2 + 6.6 + 3.1 + 3.7 + 2(0.5) + 2(0.7) = 23 percent recombination. We can thus draw a map locating the three markers, showing their map "distances," as follows:

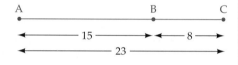

chromatids or any other possible pair that includes one member from each of the homologous chromosomes. The precise arrangement of spores in the ascus of *Neurospora* makes this an ideal organism in which to study the details of double and multiple crossovers (Figure 10.19). The probability that there will be more than one crossover in the segment between two particular markers depends on the distance between them—the greater the distance, the more likely crossover events will take place.

Cytogenetics

By making experimental crosses and calculating the recombination frequencies, geneticists can show that certain genes are associated in a linkage group, in a specific order (Box 10.C). Such a linkage group is logical, but to what extent does it actually correspond with the physical structure of a chromosome as seen under the microscope? To establish a relationship between a genetic linkage group and a chromosome, the cytogeneticist (a person who studies the microscopic appearance of chromosomes in relation to genetics) tries to find an individual in whom the normal linkage relationships are changed. The cytogeneticist then examines that individual's cells under the microscope, looking for a corresponding visible change in one or more chromosomes.

In the tissues of most species, the chromosomes are too small for an observer to see any except the most gross changes. One exception is the giant chromosomes in the salivary glands of the larvae of *Drosophila*. Called **polytene chromosomes**, they have replicated their DNA many times without cytokinesis, so that many copies of each DNA molecule lie side by side to form thick, snakelike structures that can be seen clearly even with a low-magnification lens (Figure 10.20). Condensed thickenings, or chro-

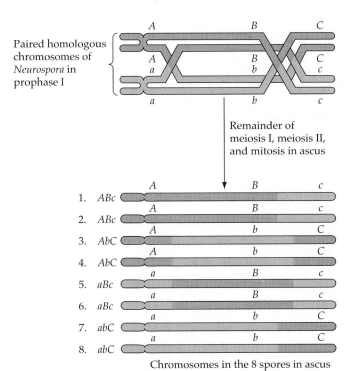

Paired homologous chromosomes of *Neurospora* in prophase I

Remainder of meiosis I, meiosis II, and mitosis in ascus

1. *ABc*
2. *ABc*
3. *AbC*
4. *AbC*
5. *aBc*
6. *aBc*
7. *abC*
8. *abC*

Chromosomes in the 8 spores in ascus

10.19 There Can Be Multiple Crossovers
A chromosome carrying alleles *A*, *B*, and *C* synapses with its homolog, which carries *a*, *b*, and *c*. The chromosomes cross over at three points. The crossovers are faithfully recorded in the order of spores, and all eight spores have recombinant chromosomes.

momeres, along these chromosomes are paired with sister chromomeres on the parallel strands of the polytene chromosome so the chromosomes appear to have a pattern of transverse bands. (These bands, visible in interphase in polytene chromosomes, are not the same as the bands seen by special staining of chromosomes in mitosis.) Each polytene chromosome has a characteristic pattern of bands. The two homologous polytene chromosomes are closely synapsed, causing the two chromosomes to look like a single structure. The thickness, spacing, and sharpness or diffuseness of the bands are so characteristic that an experienced cytogeneticist can often tell at a glance if the order or position of a group of bands has been changed.

One type of chromosomal change is an **inversion**. Inversions can be detected by genetic mapping as a change in the linkage relations of markers on a chromosome. If the normal order of the markers is *ABCDEFGHI*, an inversion may have the order *ABCDGFEHI*. When the polytene chromosomes of such individuals are examined, the order of a group of bands in one of the chromosomes also appears reversed. A linkage group with an inversion in it can be correlated with a chromosome whose bands are inverted, and the geneticist can infer that certain

genes are located in certain regions of the visibly altered chromosome.

Together with inversions, other chromosomal changes observable in polytene chromosomes, such as the **deletion**, or loss, of genetic material, can be useful in correlating the recombination map with the physical chromosome. Where there is genetic evidence of a deletion, a small group of bands (or even a single band) often is missing from one of the chromosomes. From the study of a great many chromosomal changes in *Drosophila*, it is well demonstrated that the order of genes deduced by the two methods agrees, but the distances do not. You can see this for yourself in Figure 10.21. Because of the exceedingly complex folding of DNA in a eukaryotic chromosome, one can never be sure that the microscopically observed length of a segment of chromosome is a good reflection of the length of DNA in that segment, nor is recombination-map distance necessarily a reliable reflection of the length of DNA. Despite these limitations, the combination of cytogenetics and recombination analysis has been successful in probing the composition of chromosomes of eukaryotes.

INTERACTIONS OF GENES WITH OTHER GENES AND WITH THE ENVIRONMENT

Thus far we have treated the phenotype of an organism, with respect to a given trait, as a simple result of its genotype, and we have implied that a single trait results from the alleles of a single locus. In fact, several loci may interact to determine a trait's phenotype. To complicate things further, the physical environment may interact with the genetic constitution of an individual in determining the phenotype.

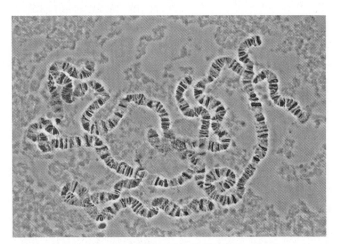

10.20 Polytene Chromosomes are Huge
A complete set of banded polytene chromosomes in a cell from the salivary gland of a *Drosophila* larva.

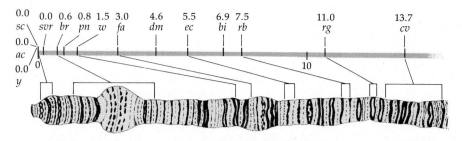

10.21   Map of One End of a Polytene X Chromosome
From the genetic map at the top you can read distances (in map units) between loci (italic letters) determined by recombination studies. Notice that these studies place the three loci at the left at the same point on the chromosome. The cytological map below gives, for 8 of the 14 loci, positions along the chromosome determined from studies of polytene chromosomes. Notice that the *order* of these 8 loci is the same on both maps. Notice also, by using the lines running between the maps, how different the *distances* on the two maps are.

Epistasis

When a particular trait is the result of a series of chemical reactions, each controlled by a different locus, the gene that acts at the earliest step in the series may, in one of its allelic forms, mask the expression of one or all of the other loci. This phenomenon of **epistasis**, in which one gene alters the effect of another, is illustrated by several loci that determine coat color in mice. The wild-type color is agouti, a grayish pattern resulting from bands on the individual hairs. The dominant allele *B* determines that the hairs will have bands and thus that the color will be agouti, whereas the homozygous recessive genotype *bb* results in unbanded hairs. Another, unlinked locus affects an early step in the formation of hair pigments. The dominant allele *A* at this locus allows normal color development, but *aa* blocks all pigment production and results in an all-white albino. As a result, *aa* is said to be epistatic over the *B* locus. Whether the genotype is *BB*, *Bb*, or *bb*, the result is an albino if the alleles of the other locus are both *a* (Figure 10.22). If a mouse with genotype *AABB* (and thus the agouti phenotype) is crossed with an albino of genotype *aabb*, the F₁ is *AaBb* and of the agouti phenotype. If the F₁ mice are crossed with each other to produce an F₂, the epistasis of *aa* will result in an expected phenotypic ratio of 9 agouti:3 black:4 albino. Can you show why this is so? The underlying ratio is the usual 9:3:3:1 for a dihybrid cross with unlinked genes, but be sure to look closely at each genotype and watch out for epistasis.

Epistasis can work in both directions between two genes, as first observed by William Bateson and Reginald Punnett. They performed a cross between two sweet pea plants (not the edible peas studied by Mendel). Each parent had white flowers. To the astonishment of Bateson and Punnett, the F₁ all had purple flowers! The F₂, obtained by self-pollinating the F₁, were in the ratio of 9 purple:7 white. This looks like a modification of the standard 9:3:3:1 ratio, with the last three groups lumped into one. Let's try, as did Bateson and Punnett, to figure it out. First, because this looks like a dihybrid cross ratio, we assume that two different loci are involved. We recall that both dominant alleles are present in 9/16 of the offspring of a dihybrid cross (see Figure 10.7), and we notice that this is the proportion of the F₂ that are purple. We therefore decide that each individual in this group has at least one copy of each dominant allele and write *A−B−* as their genotype. Because only 9/16 are purple, it must be that having a dominant allele for only one of the genes will *not* produce a color. Thus the genotypes *A−bb* and *aaB−* fail to give purple flowers.

Here's how we may represent the whole experiment: If the original parents were *AAbb* and *aaBB*, both would have been white, as observed; all the F₁ would have been *AaBb* and purple, also as observed. You should now work out the genotypes of the F₂ and then convert them to phenotypes, remembering that all the genotypes give white flowers *except* the ones that are *A−B−*. If you do this exercise carefully, you will obtain the observed 9:7 ratio. We may say that *aa* is epistatic to *B* and *bb* is epistatic to *A*, in that both of these doubly recessive genotypes alter the expression of the dominant allele at the other locus, thereby determining that the phenotype will be white. Another way to describe this kind of situation is to say that the two loci are complementary. Complementary loci are mutually dependent, the expression of each being dependent upon the alleles of the other.

The epistatic action of complementary genes may be explained as follows: The dominant alleles *A* and *B* in this example code for the production of enzymes that catalyze two separate reactions in the production

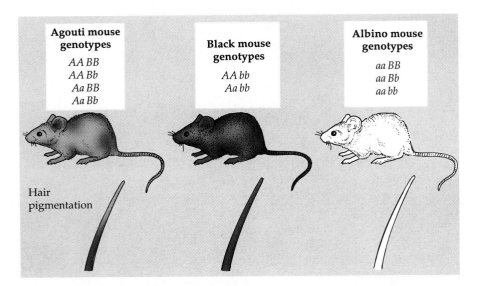

Agouti mouse genotypes

AA BB
AA Bb
Aa BB
Aa Bb

Black mouse genotypes

AA bb
Aa bb

Albino mouse genotypes

aa BB
aa Bb
aa bb

Hair pigmentation

10.22 Genes May Interact Epistatically
Mice that have at least one dominant allele at each locus are agouti. Mice with genotype *aa* are albino regardless of their genotype for the other locus, because the *aa* genotype blocks all pigment production. Mice with *bb* genotypes are black unless they also are *aa* (which makes them albino).

of a purple pigment. In order for the pigment to be produced, both reactions must take place. If a plant is homozygous for either *a* or *b*, the corresponding reaction will not occur, no purple pigment will form, and the flowers will be white.

Quantitative Inheritance and Environmental Effects

Individual heritable traits are often found to be controlled by many genes, each contributing to the final outcome. As a result, variation in such traits is **continuous** rather than, as in the examples we have been considering, **discontinuous**. In the experiment of Bateson and Punnett, for example, the sweet pea flowers were either white or purple; variation was discontinuous. But many traits that are under genetic control—such as height and other aspects of size, or skin color—vary continuously. We may think of these continuously varying traits as being controlled by multiple **polygenes**: loci whose alleles increase or decrease the observed character (Figure 10.23). Polygenes affecting a particular quantitative trait are common on many chromosomes. One of Mendel's wise decisions was to deal only with discontinuous variation, which is relatively simple. Had he, like Charles Darwin and others, concentrated on continuous variation, we might still not know the basic rules of heredity!

Humans differ with respect to the amount of a dark pigment, melanin, in the skin. There is great variation in the amount of melanin among different people, but much of this variation is determined by

alleles at just four (possibly three) loci. None of the alleles at these loci demonstrates dominance. Of course, skin color is not entirely determined by the genotype, since exposure to sunlight can cause the production of more melanin—that is, tanning.

Such environmental variables as light, temperature, and nutrition can sharply affect the translation of a genotype into a phenotype. A familiar example is the Siamese cat (Figure 10.24). This handsome animal normally has darker fur on its ears, nose, paws, and tail than on the rest of its body. These darkened parts are ones that have a somewhat lower temperature. A few simple experiments show that the Siamese cat has a genotype that results in dark fur, but only at temperatures somewhat below the general body temperature. If some dark fur is removed from the tail and the cat is kept at higher-than-usual temperatures, the new fur that grows in is light. Conversely, removal of light fur from the back, followed by local chilling of the area, causes the spot to fill with dark fur.

Genotype and environment interact to determine the phenotype of an organism. It is sometimes possible to determine the proportion of individuals in a group with a given genotype that actually show the expected phenotype. This proportion is called the **penetrance** of the genotype. The environment may also affect the **expressivity** of the genotype, that is, the degree to which it is expressed. For an example of environmental effects on expressivity, consider Siamese cats that are kept indoors and outdoors in different climates.

Uncertainty over how much of the observed vari-

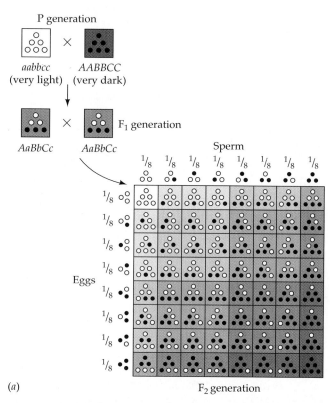

(a)

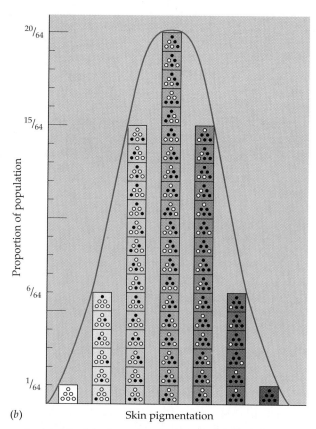

(b)

10.23 Polygenes Determine Human Skin Pigmentation
A model of polygenic inheritance based on three genes. The alleles *A*, *B*, and *C* contribute dark pigment to the skin, but the alleles *a*, *b*, and *c* do not. The more *A*, *B*, and *C* alleles an individual possesses, the darker that person's skin will be. The pattern of inheritance of the alleles is shown in (*a*). The black circles represent the alleles *A*, *B*, and *C*; the white circles represent the alleles *a*, *b*, and *c*. The frequencies of the phenotypes in (*a*) are graphed in (*b*). If both members of a couple have intermediate pigmentation (*AaBbCc*, for example), they are unlikely to have children with either very light or very dark skin.

ation is due to the environment and how much to the effects of the several polygenes complicates the analysis of quantitative inheritance. A useful approach that avoids this difficulty is to study identical twins. Since these individuals are genetically identical, any differences between such twins must be attributed to environmental effects.

The phenotype of an organism depends on its total genetic makeup and on its environment. Some of the interactive effects will become more obvious when we focus, in Chapter 12, on the regulation of gene expression.

NON-MENDELIAN INHERITANCE

You have studied the basic patterns of Mendelian inheritance in terms of chromosomal behavior. Does all inheritance in eukaryotes conform to the Mendelian pattern? Consider the four-o'clock plant shown in Figure 10.25. This particular four-o'clock shows three different patterns of chlorophyll distribution in three different parts of the shoot. One branch is all

10.24 The Environment Affects the Phenotype
This Siamese cat has dark fur on its extremities, where temperature is below the general body temperature.

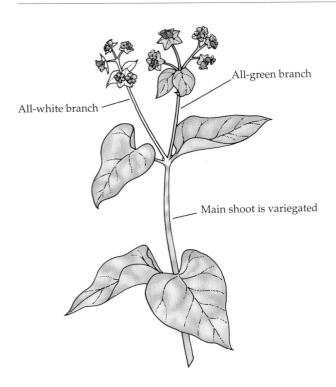

All-green branch

All-white branch

Main shoot is variegated

10.25 A Variegated Four-o'Clock with Green and White Branches
This is the plant used for the experiment summarized in Table 10.4. Pollen is transferred from a flower on one part of the plant to a flower on another part.

TABLE 10.4
Results of the Four-o'Clock Experiments

PHENOTYPE OF BRANCH WITH FEMALE PARENT	PHENOTYPE OF BRANCH WITH MALE PARENT	PHENOTYPE OF PROGENY
White	White	White
White	Green	White
White	Variegated	White
Green	White	Green
Green	Green	Green
Green	Variegated	Green
Variegated	White	Variegated, white, or green
Variegated	Green	Variegated, white, or green
Variegated	Variegated	Variegated, white, or green

white; it has no chlorophyll. Another branch is entirely green. The rest of the shoot is variegated—that is, there is chlorophyll in some patches of cells and none in others. Each part of the plant has flowers, so we can do the following experiment. Take pollen (which produces male gametes) from some flowers and transfer it to the female parts of flowers on another part of the plant, thus performing a cross. Table 10.4 shows the nine possible crosses and their outcomes. Study the table for a moment before reading further and try to discern the basic pattern of inheritance. Do you see the surprising feature? Look again. What should become obvious is that the phenotype of the "father" (the parent producing the pollen) is irrelevant to the outcome! Only the "mother" (the parent producing the egg) seems to play a role in determining the phenotype of the offspring.

This pattern of inheritance, in which the progeny's phenotype is unaffected by the father, is found in all sorts of eukaryotic organisms and is referred to as **maternal inheritance**. How does it work? The essence of Mendelian inheritance is that information carried on chromosomes is partitioned with great precision during meiosis, but eukaryotic cells have other self-reproducing entities besides the nuclear chromosomes. Chloroplasts and mitochondria carry some genetic information in small circular chromosomes (see Chapter 4). The DNA of these organelles is subject to mutation just as is the DNA in the chromosomes of the nucleus, so we may speak of alleles of non-nuclear genes. These genes are not inherited in the same way as nuclear chromosomal genes are because the eggs of most species contain large amounts of cytoplasm, but the sperm contain hardly any. Generally speaking, all the mitochondria in a zygote come from the cytoplasm of its mother's egg, even though half the zygote's nuclear chromosomes come from its father. In plant zygotes all the chloroplasts come from the maternal cytoplasm. Hence any particle that is inherited through the cytoplasm is said to be maternally inherited. In such cases, reciprocal crosses give quite different results, as we saw in Table 10.4.

SUMMARY of Main Ideas about Mendelian Genetics

By crossing different strains of pea plants and counting each type of progeny, Mendel discovered some characteristic ratios of inheritance.

F_1 progeny of a cross between a dominant homozygote and a recessive homozygote express only the dominant trait.

Both parental phenotypes reappear in the F_2 produced by a monohybrid cross, in the ratio 3:1.

The F_2 consists of three genotypes in the ratio 1:2:1.
Review Figures 10.2, 10.3, 10.4, and Table 10.1

A monohybrid cross with incomplete dominance results in a 1:2:1 phenotype ratio in the F_2.
Review Figure 10.10

A test cross shows whether an individual expressing a dominant phenotype is homozygous or heterozygous.
Review Figure 10.6

Heterozygous organisms display both traits when the two alleles are codominant.
Review Figure 10.10

Some loci are represented by multiple alleles; different alleles arise by mutation.
Review Figures 10.11, 10.12, and Tables 10.2 and 10.3

In a dihybrid cross with unlinked genes, alleles of the two loci assort independently at meiosis.
Review Figure 10.8

Dihybrid crosses with unlinked genes yield a 9:3:3:1 ratio of four phenotypes in the F_2.
Review Figure 10.7

If the markers in a dihybrid cross are linked, there is a much higher proportion of parental phenotypes in the F_2 and fewer—or no—recombinant individuals.
Review Figure 10.13

Crossing over between chromatids produces recombination.
Review Figures 10.18 and 10.19

From the frequency of recombination, taken as a measure of the distance between genes, mapping techniques reveal the linear order of genes on a chromosome.

The presence or absence of certain chromosomes determines sex in many organisms.

In humans and *Drosophila*, XX determines female, and XY male.
Review Figure 10.14

Traits governed by loci on the X or the Y chromosome are not inherited in the same ratios as those observed for autosomal markers.

Reciprocal crosses for sex-linked markers give nonidentical results.
Review Figures 10.15 and 10.16

Some genes modify the expression of other genes.
Review Figure 10.22

Interactions of multiple genes control many traits.
Review Figure 10.23

The environment may influence both the penetrance and the expressivity of a genotype.

Some traits are coded for by DNA contained in the mitochondria or chloroplasts, which are inherited maternally.

For traits that are inherited maternally, reciprocal crosses do not give identical results.
Review Figure 10.25 and Table 10.4

SELF-QUIZ

1. Which statement is *not* true for Mendel's cross of *TT* peas with *tt* peas?
   a. Each parent can produce only one type of gamete.
   b. F_1 individuals produce gametes of two types, each gamete being *T* or *t*.
   c. Three genotypes are observed in the F_2 generation.
   d. Three phenotypes are observed in the F_2 generation.
   e. This is an example of a monohybrid cross.

2. The phenotype of an individual
   a. depends at least in part on the genotype.
   b. is either homozygous or heterozygous.
   c. determines the genotype.
   d. is the genetic constitution of the organism.
   e. is either monohybrid or dihybrid.

3. Which of the following statements is *not* true of alleles?
   a. They are different forms of the same gene.
   b. There may be several at one locus.
   c. One may be dominant over another.
   d. They may show incomplete dominance.
   e. They occupy different loci on the same chromosome.

4. Which statement is *not* true of an individual that is homozygous for an allele?
   a. Each of its cells possesses two copies of that allele.
   b. Each of its gametes contains one copy of that allele.
   c. It is true-breeding with respect to that allele.
   d. Its parents were necessarily homozygous for that allele.
   e. It can pass that allele to its offspring.

5. Which of the following statements is *not* true of a test cross?
   a. It tests whether an unknown individual is homozygous or heterozygous.
   b. The test individual is crossed with a homozygous recessive individual.
   c. If the test individual is heterozygous, the progeny will have a 1:1 ratio.
   d. If the test individual is homozygous, the progeny will have a 3:1 ratio.
   e. Test cross results are consistent with Mendel's model of inheritance.

6. Linked genes
   a. must be immediately adjacent to one another on a chromosome.
   b. have alleles that assort independently of one another.
   c. never show crossing over.
   d. are on the same chromosome.
   e. always have multiple alleles.

7. In the F$_2$ generation of a dihybrid cross
   a. four phenotypes appear in the ratio 9:3:3:1 if the loci are linked.
   b. four phenotypes appear in the ratio 9:3:3:1 if the loci are unlinked.
   c. two phenotypes appear in the ratio 3:1 if the loci are unlinked.
   d. three phenotypes appear in the ratio 1:2:1 if the loci are unlinked.
   e. two phenotypes appear in the ratio 1:1 whether or not the loci are linked.

8. The sex of a honeybee is determined by
   a. ploidy, the male being haploid.
   b. X and Y chromosomes, the male being XY.
   c. X and Y chromosomes, the male being XX.
   d. the number of X chromosomes, the male being XO.
   e. Z and W chromosomes, the male being ZZ.

9. In epistasis
   a. nothing changes from generation to generation.
   b. one gene alters the effect of another.

   c. a portion of a chromosome is deleted.
   d. a portion of a chromosome is inverted.
   e. the behavior of two genes is entirely independent.

10. Individual heritable traits
    a. are always determined by dominant and recessive alleles.
    b. always vary discontinuously.
    c. can sometimes be controlled by many genes.
    d. were first studied in this century.
    e. do not exist outside the laboratory.

FOR STUDY

1. Utilizing the Punnett squares below, show that for typical dominant and recessive autosomal traits, it does not matter which parent contributes the dominant allele and which the recessive allele. Cross true-breeding tall plants (*TT*) with true-breeding dwarf plants (*tt*).

   Tall Female × Dwarf Male   Dwarf Female × Tall Male

	Male gametes	
Female gametes		

	Male gametes	
Female gametes		

2. Show diagrammatically what occurs when the F$_1$ offspring of the cross in Question 2 self-pollinate.

	Male gametes	
Female gametes		

3. A new student of genetics suspects that a particular recessive trait in fruit flies (dumpy wings) is sex-linked. A single mating between a fly having dumpy wings (*dp*; female) and a fly with wild-type wings (*Dp*; male) produces 3 dumpy-winged females and 2 wild-type males. On the basis of these data, is the trait sex-linked or autosomal? What were the genotypes of the parents? Explain how these conclusions can be reached on the basis of so few data.

4. The sex of fishes is determined by the same X–Y system as in humans and *Drosophila*. An allele of one locus on the Y chromosome of the fish *Lebistes* causes a pigmented spot to appear on the dorsal fin. A male fish with a spotted dorsal fin is mated with a female fish with an unspotted fin. Describe the phenotypes of the F$_1$ and the F$_2$ from this cross.

5. In *Drosophila melanogaster*, the recessive allele *p*, when homozygous, determines pink eyes. *Pp* or *PP* results in wild-type eye color. Another gene, on another chromosome, has a recessive allele, *sw*, that produces short wings when homozygous. Consider a cross between females of genotype *PPSwSw* and males of genotype *ppswsw*. Describe the phenotypes and genotypes of the F$_1$ generation and of the F$_2$ generation produced by allowing the F$_1$ to mate with one another.

6. On the same chromosome of *Drosophila melanogaster* that carries the *p* (pink eyes) locus, there is another locus that affects the wings. Homozygous recessives, *byby*, have blistery wings, while the dominant allele *By* produces wild-type wings. The *p* and *by* loci are very close together on the chromosome; that is, the two loci are tightly linked. In answering these questions, assume that no crossing over occurs.
   a. For the cross *PPByBy* × *ppbyby*, give the phenotypes and genotypes of the F$_1$ and of the F$_2$ produced by F$_1$ interbreeding.

   b. For the cross *PPbyby* × *ppByBy*, give the phenotypes and genotypes of the F$_1$ and of the F$_2$.
   c. For the cross of Question 6b, what further phenotype(s) would appear in the F$_2$ generation if crossing over occurred?
   d. Draw a nucleus undergoing meiosis, at the stage in which the crossing over (Question 6c) occurred. In which generation (P, F$_1$, or F$_2$) did this crossing over take place?

7. Consider the following cross of *Drosophila melanogaster* with alleles as described in Question 6. Males with genotype *Ppswsw* are crossed with females of genotype *ppSwsw*. Describe the phenotypes and genotypes of the F$_1$ generation.

8. In the Blue Andalusian fowl, a single pair of alleles controls the color of the feathers. Three colors are observed: blue, black, and splashed white. Crosses among these three types yield the following results:

Parents	Progeny
Black × blue	Blue and black (1:1)
Black × splashed white	Blue
Blue × splashed white	Blue and splashed white (1:1)
Black × black	Black
Splashed white × splashed white	Splashed white

   a. What progeny would result from the cross blue × blue?
   b. If you wanted to sell eggs, all of which would yield blue fowl, how should you proceed?

9. In *Drosophila melanogaster*, white (*w*), eosin (*w^e*), and wild-type red (*w^+*) are multiple alleles of a single locus for eye color. This locus is on the X chromosome. An eosin-eyed female is crossed with a male with wild-type eyes. All the female progeny are red-eyed; half the male offspring have eosin (pale orange) eyes, and half have white eyes.

    a. What is the order of dominance of these alleles?

    b. What are the genotypes of the parents and progeny?

10. Color blindness is a recessive trait. Two people with normal vision have two sons, one color-blind and one with normal vision. If the couple also has daughters, what proportion of them will have normal vision? Explain.

11. A mouse with an agouti coat is mated with an albino mouse of genotype *aabb*. Half the offspring are albino, one-quarter are black, and one-quarter are agouti. What are the genotypes of the agouti parents and of the various kinds of offspring? (Hint: see the section "Epistasis.")

12. Sweet peas (genotype *aaBB*) with white flowers are crossed with sweet peas with purple flowers. Of the progeny, half have purple flowers and half have white flowers. What can you say about the genotype of the purple-flowered parent? (Hint: This is another problem dealing with epistasis.)

13. The photograph shows the shells of 15 bay scallops, *Argopecten irradians*. These scallops are hermaphroditic—that is, a single individual can reproduce sexually, as did the pea plants of the F_1 generation in Mendel's experiments. Three color schemes are evident: yellow, orange, and black and white. The color-determining locus has three alleles. The top row shows a yellow scallop and a representative sample of its offspring, the middle row shows a black-and-white scallop and its offspring, and the bottom row shows an orange scallop and its offspring. Assign a suitable symbol to each of the three alleles participating in color control; then determine the genotype of each of the three parent individuals and tell what you can about the genotypes of the different offspring. Explain your results carefully.

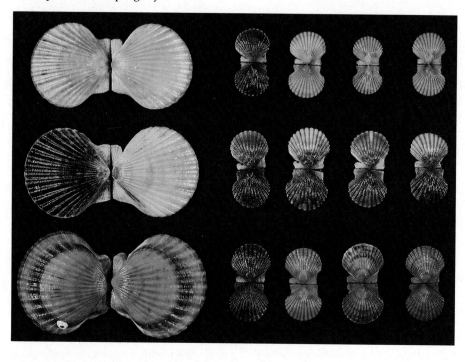

READINGS

Cooper, N. G. 1995. *The Human Genome Project: Deciphering the Blueprint of Heredity.* University Science Books, Mill Valley, CA. This unique book follows an introduction to the ideas of classical and molecular genetics with a complete discussion of the purpose, approach, technology, pitfalls, and implications of the Human Genome Project.

Griffiths, A. J. F., J. H. Miller, D. T. Suzuki, R. C. Lewontin and W. M. Gelbart. 1993. *An Introduction to Genetic Analysis*, 5th Edition. W. H. Freeman, New York. An excellent textbook of modern genetics. Chapters 2 and 3 are particularly relevant to this chapter; Chapters 4 and 5 are also useful.

Mange, E. J. and A. P. Mange. 1994. *Basic Human Genetics*. Sinauer Associates, Sunderland, MA. Genetics, especially chromosomal inheritance, can be studied using humans as examples; this book does so at an introductory level.

Russell, P. J. and J. M. Nickerson. 1992. *Genetics*, 3rd Edition. Harper/Collins, New York. A well-balanced treatment of a broad range of topics in genetics. Highly recommended.

Sapienza, C. 1990. "Parental Imprinting of Genes." *Scientific American*, October. When reciprocal crosses aren't equivalent.

Stern, C. and E. R. Sherwood (Eds.). 1966. *The Origin of Genetics: A Mendel Source Book.* W. H. Freeman, New York. A collection of the writings of researchers at the dawn of the science of genetics, including translations of Mendel's papers and letters. The last two articles discuss the likelihood that Mendel fudged his data.

Sturtevant, A. H. and G. W. Beadle. 1962. *An Introduction to Genetics*. Dover, New York. First published by W. B. Saunders in 1939. Though old, this text holds up as a fine introduction to formal chromosome genetics.

Gregor Mendel described the basic patterns of inheritance in plants and animals and devised a powerful explanation for the mechanisms underlying these patterns. The second of these accomplishments is most impressive because Mendel had no way of knowing the physical basis for his proposed mechanisms. He never knew what a gene is, in chemical terms, nor did he know about the behavior of chromosomes. Hence he could not have known how genes are copied between generations or how new alleles arise. He could not have had the slightest inkling of how a gene works—that is, how the genotype produces a phenotype.

During the same years Mendel was analyzing how the characteristics of pea plants were inherited, a Swiss chemist, Friedrich Miescher, was at work trying to identify the chemical composition of cell nuclei. Miescher focused on a material that has a high ratio of nuclear to cytoplasmic volume: pus cells from bandages discarded from the wounds of soldiers. Miescher found that the nuclei contained large amounts of protein and of a previously undescribed compound that we now call DNA.

Neither Mendel's nor Miescher's work was understood or appreciated for the remainder of the nineteenth century, and genetics and DNA did not meet until 1944. During the half-century since then, geneticists, biochemists, biophysicists, and molecular biologists have developed a detailed picture of the chemistry and functioning of the genetic material. In this and the next two chapters, we will try to show you how the main questions of molecular genetics have been studied.

WHAT IS THE GENE?

During the first half of the twentieth century, the hereditary material was generally assumed to be protein. The impressive chemical diversity of proteins made this assumption seem reasonable. Also, some proteins—notably enzymes and antibodies—show great specificity. By contrast, nucleic acids were known to have only a few components and seemed too simple to carry the complex information expected in the genetic material. The recognition that the gene is not a protein, but rather deoxyribonucleic acid, or DNA, was slow in coming and depended on the interaction of several types of research.

The Transforming Principle

The history of biology is filled with incidents in which research on some specific topic has—with or without answering the question originally under investigation—contributed richly to another, apparently un-

The Double Helix of DNA

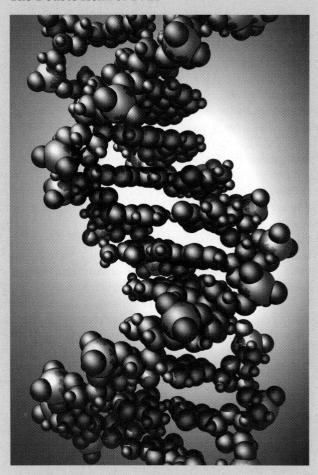

11

Nucleic Acids as the Genetic Material

related area. Such a case is the work of Frederick Griffith, an English physician. In the 1920s Griffith was studying the disease-causing behavior of the bacterium *Streptococcus pneumoniae*, or pneumococcus, one of the agents that produce pneumonia in humans. He identified two strains of pneumococcus, designated S and R because the former produces shiny, smooth (S) colonies when grown in the laboratory, whereas the colonies of the latter are rough (R) in appearance. When the S strain was injected into mice, they died within a day, and the hearts of the dead mice were found to be teeming with the deadly bacteria. When the R strain was injected instead, the mice did not become diseased. In other words, the S strain is virulent (disease-causing) and the R strain is nonvirulent. This difference was eventually shown to be due to a difference in the chemical makeup of the bacterial surface: The S strain has a polysaccharide capsule that protects the bacterium from the defense mechanisms of the host (see Chapter 16). The R strain lacks this capsule and can be inactivated by a mouse's defenses.

In hopes of developing a vaccine against pneumonia, Griffith inoculated other mice with heat-killed S pneumococci. Neither heat-killed S nor living R pneumococci produced infection. Griffith inoculated mice with a mixture of living R bacteria and heat-killed S bacteria. To his astonishment, all these mice died of pneumonia. When he examined blood from the hearts of these mice, he found it full of living bacteria, many of them belonging to the virulent S strain! He concluded that, in the presence of the dead S pneumococci, some of the inoculated R pneumococci had been transformed into virulent organisms (Figure 11.1).

Did transformation of the bacteria depend upon something the mouse did? No. It was soon shown that the same transformation occurs when living R and heat-killed S bacteria are simply incubated together in a test tube. Next it was discovered that a cell-free extract of heat-killed S cells also transforms R cells. (A cell-free extract contains all the contents of ruptured cells, but no intact cells.) This result demonstrated that some substance—called at the time a chemical **transforming principle**—from the dead S pneumococci can cause a permanent change in the affected R cells. (Remember that great numbers of *living* bacteria of the S type were always found in the mice that died as a result of being inoculated with

the mixture of heat-killed S and living R bacteria.) From these observations scientists concluded that the transforming principle carried heritable information; thus it could be thought of as genetic material. We now know that any genetic trait in pneumococci or in several other types of bacteria can be passed, by way of transforming principles, from one bacterium to another.

The Transforming Principle is DNA

A crucial step in the history of biology was the identification of the transforming principle, accomplished over a period of several years by Oswald T. Avery and his colleagues at what is now Rockefeller University. They treated samples of the transforming principle in a variety of ways to destroy different types of substances—proteins, nucleic acids, carbohydrates, lipids—and then tested the treated samples to see if they had retained transforming activity. The answer was always the same: If the DNA in the

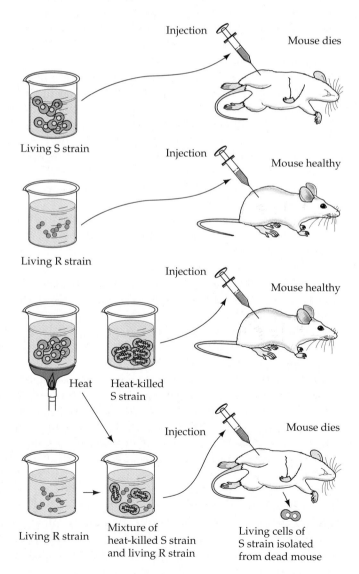

11.1   Genetic Transformation of Nonvirulent R Pneumococci
Griffith's experiments demonstrated that some factor in the virulent S strain could transform nonvirulent R-strain bacteria into a lethal form, even when the S-strain bacteria had been killed by high temperatures.

sample was destroyed, transforming activity was lost; everything else was dispensable. As a final step, Avery, with Colin MacLeod and MacLyn McCarty, isolated virtually pure DNA from a sample of pneumococcal transforming principle and showed that it was highly active in causing bacterial transformation. Their work, published in 1944, was a milestone in establishing that DNA is the genetic material in cells; at the time, however, it did not receive the attention it deserved, and scientists in other labs continued their attempts to identify the hereditary material.

The Genetic Material of a Virus

A report published in 1952 by Alfred D. Hershey and Martha Chase of the Carnegie Laboratory of Genetics had a much greater immediate impact than did Avery's 1944 paper. The Hershey–Chase experiment was carried out with a virus that infects bacteria. This virus, called T2 bacteriophage, consists of a DNA core packed within a protein coat (Figure 11.2a). The virus is thus made of the two materials that were, at the time, the leading candidates for the genetic material.

When a T2 bacteriophage attacks a bacterial cell, part—but not all—of the virus enters the cell; about 20 minutes later the bacterial cell lyses (breaks apart), releasing 200 to 1,000 new T2s. Hershey and Chase set out to determine which part of the virus—protein or DNA—is the hereditary material that gets inside the bacterial cell. The idea was to trace these two components during the life cycle of the virus, so Hershey and Chase labeled each with a specific radioactive tracer. All proteins contain some sulfur (in the amino acids cysteine and methionine), an element that is not present in DNA, and sulfur has a radioactive isotope, ^{35}S. The deoxyribose–phosphate "backbone" of DNA, on the other hand, is rich in phosphorus (see Chapter 3), an element not present in proteins—and phosphorus also has a radioactive isotope, ^{32}P. Thus Hershey and Chase grew one batch of T2 in a bacterial culture in the presence of ^{32}P, so that all the viral DNA was labeled with ^{32}P. Similarly, the proteins of another batch of T2 were labeled with ^{35}S (Figure 11.2b).

In separate experimental runs, Hershey and Chase combined radioactive viruses containing either ^{32}P or ^{35}S with bacteria (Figure 11.2c). After a few minutes, the mixtures were swirled vigorously in a kitchen blender, which (without bursting the bacteria) stripped away the parts of the virus coats that had not penetrated the bacteria. The bacteria were then separated from the rest of the material in a centrifuge. It was found that more than three-fourths of the ^{35}S (and thus the protein) had separated from the bacteria, and that two-thirds or more of the ^{32}P (and thus the DNA) had settled to the bottom of the centrifuge

tube along with the bacteria. Although the numbers were not as clear-cut as one might like, these results suggested that the DNA was transferred to the bacteria while the protein remained outside.

Confirmation of this tentative conclusion came when other batches of bacteria and labeled T2 were incubated together for longer periods, allowing a progeny generation of viruses to be collected. When this was done, Hershey and Chase found that the resulting T2 progeny contained less than 1 percent of the original ^{35}S but about one-third of the original ^{32}P—and thus, presumably, one-third of the DNA. This result showed that T2 injects the DNA from its head into a bacterium while the external protein structures remain outside the bacterial cell. Because DNA was carried over in the virus from generation to generation, whereas protein was not, a logical conclusion was that the hereditary information of the viruses is contained in the DNA. The Hershey–Chase experiment convinced most scientists that DNA is indeed the carrier of hereditary information.

NUCLEIC ACID STRUCTURE

Once scientists agreed that the genetic material is DNA, they wanted to learn just what the DNA molecule looks like. In its structure they hoped to find clues to two questions: how the molecule is replicated between nuclear divisions, and how it causes the synthesis of specific proteins. Both expectations were fulfilled.

Evidence from X-Ray Crystallography and Biochemistry

The structure of DNA was deciphered only after many types of experimental evidence and theoretical considerations were put together. The most crucial "hard" evidence was obtained by X-ray crystallography. The positions of atoms in a crystalline substance can be inferred by the pattern of diffraction of X rays passed through the crystal, but even today this is not an easy task when the substance is of enormous molecular weight. In the early 1950s, even a highly talented X-ray crystallographer could (and did) look at the best available images from DNA preparations and fail to see what they meant. Nonetheless, the attempt to characterize DNA would have been impossible without the crystallographs prepared by the English chemist Rosalind Franklin. Franklin's work, in turn, depended upon the success of the English biophysicist Maurice Wilkins in preparing very uniformly oriented DNA fibers, which made far more manageable samples for crystallography than had previous samples.

(a) **The virus:** T2 bacteriophage

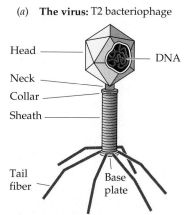

Head — DNA

Neck

Collar

Sheath

Tail
fiber

Base
plate

11.2   T2 and the Hershey-Chase Experiment

(a) The external structures of the bacteriophage T2 consist entirely of protein. This cutaway view shows a strand of DNA within the head. (b) What made the Hershey–Chase experiment possible was the production of two sets of uniquely tagged T2 bacteriophage—one for the DNA and one for the protein coat. (c) Because progeny generations of viruses incorporated the radioactively tagged DNA but not the radioactively tagged proteins, the experiment demonstrated that DNA, not protein, is the hereditary material.

(b) **Preparation of specifically labeled viruses**

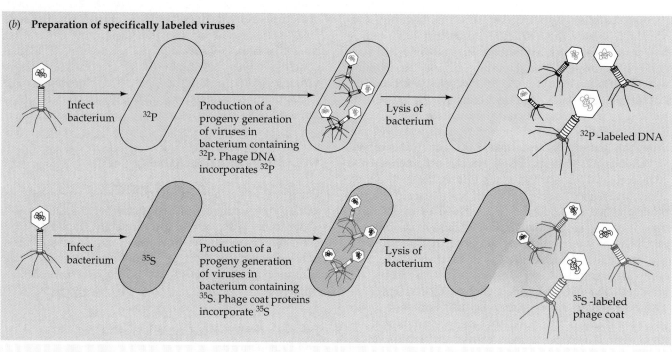

Infect bacterium → ^{32}P → Production of a progeny generation of viruses in bacterium containing ^{32}P. Phage DNA incorporates ^{32}P → Lysis of bacterium → ^{32}P-labeled DNA

Infect bacterium → ^{35}S → Production of a progeny generation of viruses in bacterium containing ^{35}S. Phage coat proteins incorporate ^{35}S → Lysis of bacterium → ^{35}S-labeled phage coat

(c)   **The experiment**

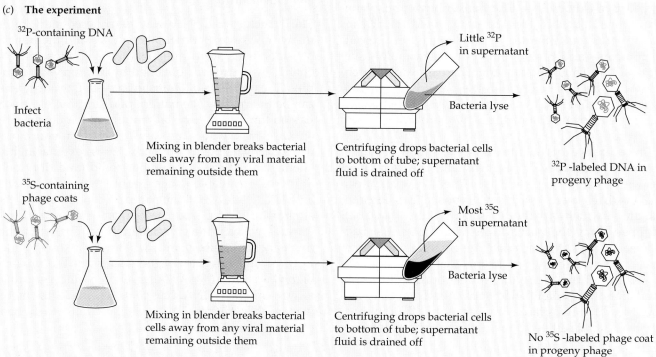

^{32}P-containing DNA

Infect bacteria

Mixing in blender breaks bacterial cells away from any viral material remaining outside them

Centrifuging drops bacterial cells to bottom of tube; supernatant fluid is drained off

Little ^{32}P in supernatant

Bacteria lyse

^{32}P-labeled DNA in progeny phage

^{35}S-containing phage coats

Mixing in blender breaks bacterial cells away from any viral material remaining outside them

Centrifuging drops bacterial cells to bottom of tube; supernatant fluid is drained off

Most ^{35}S in supernatant

Bacteria lyse

No ^{35}S-labeled phage coat in progeny phage

The chemical composition of DNA also provided important clues about its structure. Biochemists knew the chemical structures of the four monomers, or nucleotides, of DNA (see Chapter 3), as well as how one nucleotide is joined to another to form a polynucleotide chain. Recall that a nucleotide of DNA consists of a molecule of the sugar deoxyribose, a phosphate group, and a nitrogen-containing base (see Figures 3.22 and 3.23). The only differences between the four nucleotides of DNA are in their nitrogenous bases: the purines **adenine** and **guanine** and the pyrimidines **cytosine** and **thymine**. In 1950 Erwin Chargaff at Columbia University reported observations of major importance. He and his colleagues had found that DNA from many different species—and from different sources within a single organism—exhibits certain regularities. In any DNA the following rules hold: The amount of adenine equals the amount of thymine, and the amount of guanine equals the amount of cytosine. As a result, the total amount of purines equals the total amount of pyrimidines. The structure of DNA could scarcely have been worked out without this information, yet its significance was overlooked for at least three years.

Watson, Crick, and the Double Helix

Solving the puzzle was accelerated by the technique of model building: assembling three-dimensional representations of possible molecular structures. This technique, originally exploited in structural studies by the American chemist Linus Pauling, was used by the English physicist Francis Crick and the American geneticist James D. Watson, then both at the Cavendish Laboratory of Cambridge University. Watson and Crick attempted to combine all that had been learned so far about DNA structure into a single, coherent model. The crystallographers' results convinced Watson and Crick that the DNA molecule is **helical** (cylindrically spiral) and provided the values of certain distances within the helix. The results of density measurements and model building suggested that there are two polynucleotide chains in the molecule. The modeling studies also led to the conclusion that the two chains in DNA run in opposite directions, that is, that they are **antiparallel**.

Crick and Watson attempted several models. Late in February of 1953, they built the one that established the general structure of DNA. There have been minor amendments to their first published structure, but the principal features remain unchanged.

Key Elements of DNA Structure

Four features summarize the molecular architecture of DNA: *The molecule is (1) a double-stranded helix, (2) of uniform diameter, (3) twisting to the right (that is,*

twisting in the same direction as the threads on most screws), with (4) the two strands running in opposite directions. As you know, the two strands are polynucleotide chains. The sugar–phosphate backbones of the chains coil around the outside of the helix, with the nitrogenous bases pointing toward the center. The two chains are held together by hydrogen bonding between specifically paired bases. As implied by Chargaff's studies, adenine (A) pairs with thymine (T) by forming two hydrogen bonds, and guanine (G) pairs with cytosine (C) by forming three hydrogen bonds. Because the A–T and G–C pairs, like rungs of a ladder, are of equal length and fit identically into the double helix, the diameter of the helix is uniform (Figure 11.3a). A pair of purines would cause a swelling in the molecule and a pair of pyrimidines would cause a constriction, but DNA never includes such pairs. Every base pair consists of one purine (A or G) and one pyrimidine (C or T). Two grooves, one broad (the major groove) and one narrow (the minor groove), spiral around the outside of the DNA molecule (Figure 11.3b).

We say the two DNA strands run in opposite directions, but what does this mean? The direction of a polynucleotide can be defined by looking at the linkages between adjacent nucleotides. (These linkages are called phosphodiester bonds.) To avoid confusion with the carbon and nitrogen atoms in the ring structure of the bases (which are numbered 1, 2, 3, . . .), a prime symbol is placed after a number that refers to a carbon atom of a sugar: 1′, 2′, 3′, and so on. In the sugar–phosphate backbone of DNA, the phosphate groups connect to the 3′ carbon of one deoxyribose molecule and the 5′ carbon of the next, linking successive sugars together (Figure 11.4). Thus the two ends of a polynucleotide differ. Polynucleotides have a free (not connecting to another nucleotide) 5′ phosphate ($-OPO_3^{3-}$) group at one end—the **5′ end**—and a free 3′ hydroxyl ($-OH$) group at the other—the **3′ end**—just as polypeptides have a free amino group at one end and a free carboxyl group at the other end. The 5′ end of one strand in a DNA double helix is paired with the 3′ end of the other strand, and vice versa; that is, the strands run in opposite directions (see Figure 11.3a).

Alternative Structures for DNA

We've been talking about DNA as if its shape never varied—that is, as if it were always a right-handed double helix, with two grooves of unequal width spiraling up its side. This is the form of most DNA. Over the past 40 years, however, some minor variations on this theme have been discovered, and in 1979, a strikingly different structure was observed in some samples of DNA. These molecules twist to the *left* rather than to the right, and they have only one

(a)

(b)

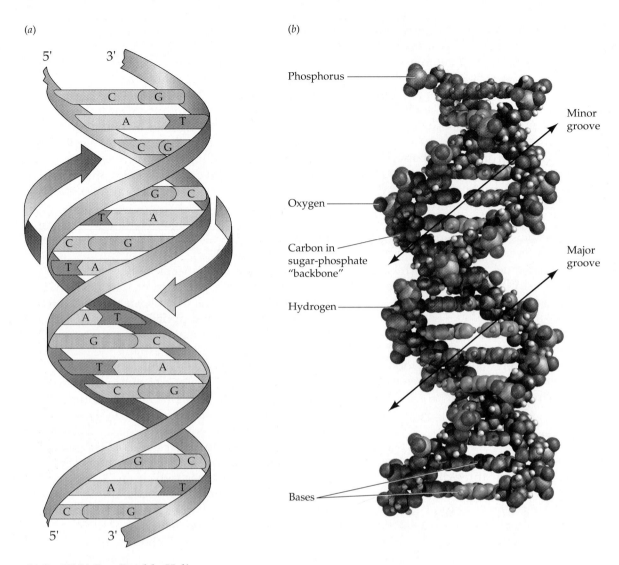

11.3 DNA Is a Double Helix
(a) Watson and Crick proposed that DNA is a double-helical molecule. The brown bands represent the two sugar–phosphate chains, with pairs of bases forming horizontal connections between the chains. The two chains run in opposite directions. (b) Biochemists can now pinpoint the position of every atom in a DNA macromolecule. To see that the essential features of the original Watson–Crick model have been verified, follow with your eyes the double-helical ribbons of sugar–phosphate groups and note the horizontal rungs of the bases.

groove. This form of DNA is referred to as **Z-DNA** because its sugar–phosphate backbones follow a zig-zag course rather than the smooth spiral of normal DNA backbones (Figure 11.5). The structure of Z-DNA is a left-handed double helix. It was first observed when scientists synthesized short DNA molecules that had alternating purine and pyrimidine bases on each strand. Then researchers discovered that short stretches of Z-DNA appear naturally in the DNA of living organisms, for example, in the chromosomes of the fruit fly *Drosophila melanogaster*.

Whether Z-DNA has a role distinct from that of normal DNA is not yet known. One possibility, which is mere speculation at this time, is that the

sharply different shape of Z-DNA makes it recognizable to one or more proteins that play regulatory roles.

Structure of RNA

To understand how DNA functions, you need to know about RNA. RNA (ribonucleic acid) is a polynucleotide similar to DNA (see Figure 3.25) but different in three ways: (1) RNA generally consists of only one polynucleotide strand (thus Chargaff's equalities, G = C and A = T, are true only for DNA and not for RNA); (2) the sugar molecule found in ribonucleotides is ribose, rather than deoxyribose as

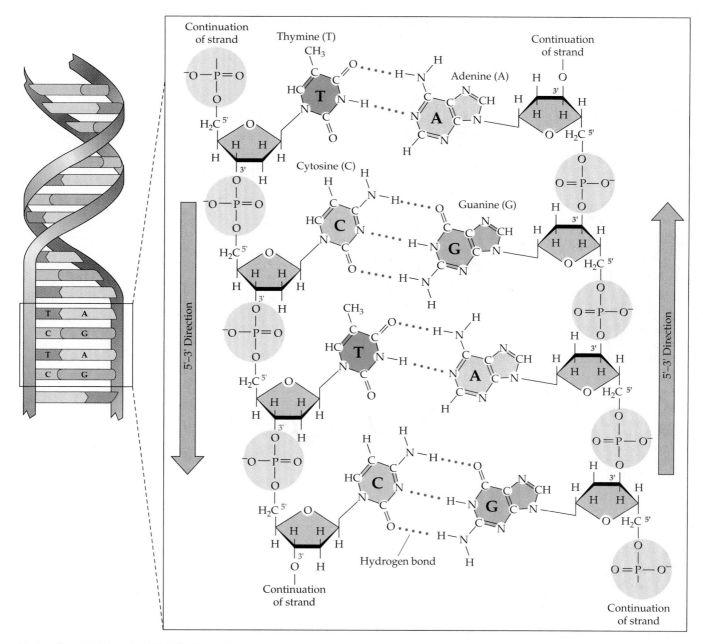

11.4 Base Pairing in DNA Is Complementary

In the sugar–phosphate backbone of DNA, each phosphate group links the 3′ carbon of one sugar to the 5′ carbon of the next sugar along the backbone. This asymmetry gives each DNA strand a 5′ "head" and a 3′ "tail." Complementary strands line up head-to-tail. Pairs of complementary bases form hydrogen bonds that hold the two strands of a DNA double helix together. T–A pairs form two hydrogen bonds; G–C pairs form three hydrogen bonds.

strand of RNA may pair between complementary bases. As we will see in the discussion of transfer RNA later in this chapter, when RNA bases do pair, uracil is like thymine in that it is complementary to adenine.

Implications of the Double-Helical Structure of DNA

Watson and Crick's double-helical model gave hints about how DNA carries out its biological role. First, the molecule is, in a sense, boring—it runs on and on, nucleotide pair after nucleotide pair, with no kinks and no bulges. Such a molecule can carry and convey information in only one way: The information

in DNA; and (3) although three of the nitrogenous bases (adenine, guanine, and cytosine) found in ribonucleotides are identical with those bases found in deoxyribonucleotides, the fourth base in RNA is uracil (U), which is similar to thymine but lacks the methyl (—CH₃) group. Different stretches of a single

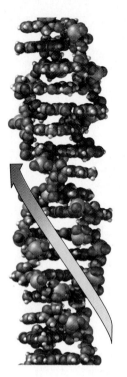
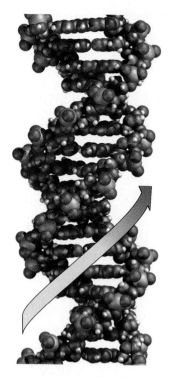

Z-DNA   Normal DNA

11.5   One Form of DNA Has a Reverse Twist
Z-DNA twists to the left rather than to the right as does normal DNA. The difference between the two forms is evident if you compare the phosphate backbones (yellow atoms) on the two models; the zigzag (Z) pattern of the line connecting the atoms in Z-DNA is clearly different from the smooth spiral in normal DNA.

must lie in the linear sequence of the nitrogenous bases.

An implication of **complementary base pairing**, A with T and G with C, was pointed out by Crick and Watson in the original publication of their findings in the journal *Nature* (1953): "It has not escaped our notice that the specific pairing we have postulated immediately suggests a possible copying mechanism for the genetic material." Each strand of the double helix is complementary to the other—that is, at each point, the base on one strand is complementary with the base on the other strand. If the spirals unwound, each strand could serve as a guide for the synthesis of a new one; thus each single strand of the parent double helix could produce a new double-stranded molecule identical to the original. All the information needed to construct new DNA molecules is present in a single strand of DNA, because that information is the sequence of bases. The double-helical structure of DNA also suggested an explanation for some mutations: They might simply be changes in the linear sequence of nucleotide pairs.

In sum, *gene function, gene replication, and gene mutation could all be accounted for by the double-helical structure of DNA.*

REPLICATION OF THE DNA MOLECULE

Just three years after Watson and Crick published their paper in *Nature*, their prediction that the DNA molecule contains all the information needed for its own replication was proved by the work of Arthur Kornberg at Washington University in St. Louis. Kornberg showed that DNA can replicate in a test tube with no cells present. All that is required is a mixture containing DNA, a specific enzyme (which he called DNA polymerase), and a mixture of the four precursors: the nucleoside triphosphates deoxy-ATP, deoxy-CTP, deoxy-GTP, and deoxy-TTP (Figure 11.6). If any one of the four nucleoside triphosphates is omitted from the reaction mixture, DNA does not replicate itself. The intact DNA serves as a template for the reaction—a guide to the exact placement of nucleotides in the new strand. Where there is a T in the template, there must be an A in the new strand, and so forth.

Two other models of how the double helix might replicate were suggested after the original paper by Watson and Crick. In the model of conservative replication (Figure 11.7a), the original double helix would serve somehow as a template but would either

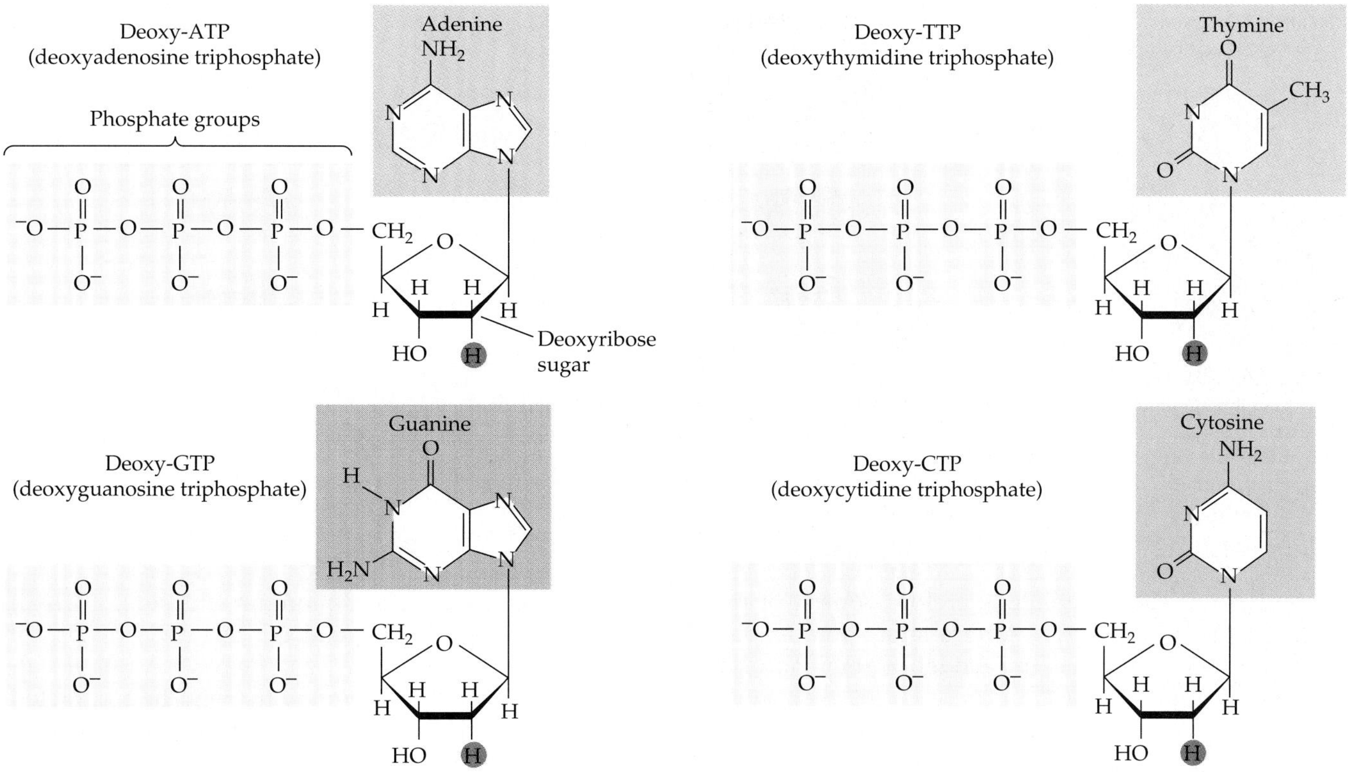

Deoxy-ATP
(deoxyadenosine triphosphate)

Adenine
NH_2

Phosphate groups

Deoxyribose
sugar

Deoxy-TTP
(deoxythymidine triphosphate)

Thymine
CH_3

Deoxy-GTP
(deoxyguanosine triphosphate)

Guanine

Deoxy-CTP
(deoxycytidine triphosphate)

Cytosine
NH_2

11.6 Building Blocks of DNA
The four deoxyribonucleoside triphosphates that form DNA differ only in their nitrogenous bases.

11.7 Three Models for DNA Replication All Obeyed Base Pairing Rules

In each model, the stretches of original DNA are blue and newly synthesized DNA is orange. (a) Conservative replication would preserve the original molecule and generate an entirely new molecule. (b) Dispersive replication would produce two molecules with old and new DNA interspersed along each strand. (c) Semiconservative replication would also produce molecules with both old and new DNA, but each molecule would contain one old strand and one new one.

be reconstituted or, perhaps, would never unwind at all. Thus the new molecule would contain none of the atoms of the original. According to the model of dispersive replication (Figure 11.7b), fragments of the original molecule would serve as templates, assembling two molecules, each containing old and new parts, perhaps at random. In **semiconservative replication** (the model proposed by Watson and Crick; Figure 11.7c), the original two strands would separate, and each would function as the template for a new partner. Each molecule produced by semicon-

Original DNA
double helix

DNA molecules
after one
round of
replication

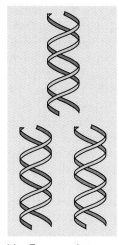

(a) Conservative
replication

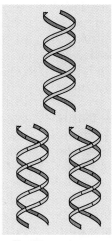

(b) Dispersive
replication

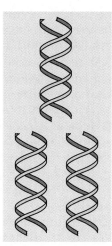

(c) Semiconservative
replication

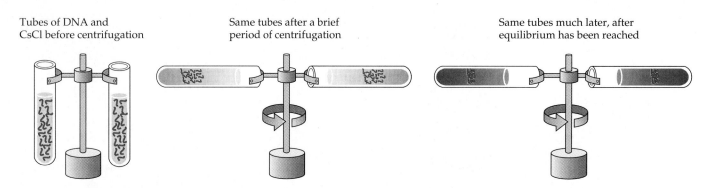

Tubes of DNA and CsCl before centrifugation

Same tubes after a brief period of centrifugation

Same tubes much later, after equilibrium has been reached

11.8 Density Gradient Centrifugation
When a solution of cesium chloride is centrifuged at extremely high speed, the cesium ions tend to sink slightly, forming a density gradient along the tube. Another substance in the tube will float at the point where its density matches that of the gradient. In this illustration, the red DNA molecules aggregate in a single band.

servative replication would therefore consist of one old and one new strand. After a short time, experimental work confirmed the model of semiconservative replaction.

Demonstration of Semiconservative Replication

A clever experiment by Matthew Meselson and Franklin Stahl convinced the scientific community that semiconservative replication is the correct model. Working at the California Institute of Technology in 1957, they devised a simple way to distinguish old strands of DNA from new ones. The key was to use a "heavy" isotope of nitrogen. Heavy nitrogen (^{15}N) is a rare, nonradioactive isotope that makes molecules more dense than chemically identical molecules containing the common isotope ^{14}N. To study DNA of different densities (that is, DNA containing ^{15}N versus DNA containing ^{14}N), Meselson, Stahl, and Jerry Vinograd invented a new centrifugation procedure that allowed them to determine the density of DNA from a specific sample. At a certain molarity, a solution of cesium chloride (CsCl) has a density very close to that of DNA; at high gravitational forces produced in an ultracentrifuge, cesium ions sediment to some extent, thus establishing a density gradient. When a DNA sample is dissolved in CsCl and centrifuged at about 100,000 times the force of gravity, the DNA gathers in a layer in the centrifuge tube at a position where the density of the CsCl solution equals that of the DNA (Figure 11.8). To cluster in this way, DNA that is initially lower in the tube, where the density is greater than its own, must rise; DNA that is in a region of lower density must sink.

After developing this method of measuring DNA densities, Meselson and Stahl could begin experimenting. They grew a culture of the bacterium *Escherichia coli* for 17 generations on a medium in which the nitrogen source (ammonium chloride, NH$_4$Cl) was made with ^{15}N instead of ^{14}N. As a result, all the DNA in the bacteria was "heavy." Another culture was grown on medium with ^{14}N. They extracted DNA from both cultures. When these extracts were combined and centrifuged with CsCl, two separate DNA bands formed, showing that this method would work for separating DNA samples of slightly different densities.

Meselson and Stahl then conducted their main experiment. They grew another culture on ^{15}N medium and *transferred* it to normal ^{14}N medium. *E. coli* reproduces every 20 minutes, and Meselson and Stahl collected some of the bacteria from each generation after the transfer. They extracted DNA from the samples. After DNA was duplicated and the cells divided to produce each new generation, the DNA banding in the density gradient was different from the original banding. Initially, the DNA was uniformly labeled with ^{15}N and hence was relatively dense. After one generation, when the DNA had been duplicated once, all the DNA was of an intermediate density. After two generations, there were two equally large DNA bands: one of low density and one of intermediate density. In samples from subsequent generations, the proportion of low-density DNA increased steadily.

These data can be explained by the semiconservative model of DNA replication. The high-density DNA had two ^{15}N strands, the intermediate-density DNA had one ^{15}N and one ^{14}N strand, and the low-density DNA had two ^{14}N strands. In the first round of DNA replication, the strands of the double helix, both heavy with ^{15}N, separated; during the process of separation, each acted as the template for a second strand, which contained only ^{14}N and hence was less dense. Each double helix then consisted of one ^{15}N and one ^{14}N strand and was of intermediate density. In the second replication, the ^{14}N-containing strands

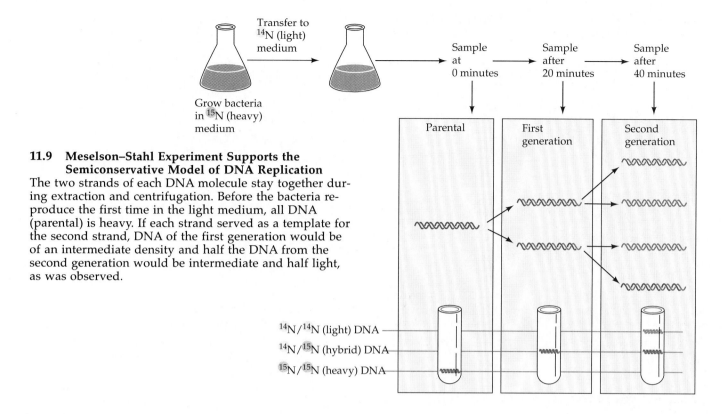

11.9 Meselson–Stahl Experiment Supports the Semiconservative Model of DNA Replication
The two strands of each DNA molecule stay together during extraction and centrifugation. Before the bacteria reproduce the first time in the light medium, all DNA (parental) is heavy. If each strand served as a template for the second strand, DNA of the first generation would be of an intermediate density and half the DNA from the second generation would be intermediate and half light, as was observed.

directed the synthesis of partners with ^{14}N, creating low-density DNA, and the ^{15}N strands got new ^{14}N partners (Figure 11.9).

If the DNA had replicated in accord with either of the other models, the results would have been quite different. Under the conservative model, after one generation there would have been two bands—one for heavy DNA (^{15}N–^{15}N) and the other for light (^{14}N–^{14}N); no DNA of intermediate density would have formed at any time. If dispersive replication had taken place, the first round of replication would have produced DNA of intermediate density, but the density of all the DNA would have decreased with each subsequent replication. The crucial observation proving the semiconservative model was that intermediate-density DNA (^{15}N–^{14}N) appeared in the first generation and continued to appear in subsequent generations.

Replicating an Antiparallel Double Helix

How is semiconservative DNA replication accomplished? That it depends upon enzymes should not surprise you. We now know that there are different types of DNA polymerases, with different functions, and that the replication process is intricate. Kornberg's basic observations, including the need for a DNA template and for a mixture of nucleoside triphosphates, still hold. He also showed that nucleotides are always added to the growing chain at the same end: the 3′ end, the end at which the DNA

strand has a free hydroxyl group on the 3′ carbon of its terminal deoxyribose (Figure 11.10; see also Figure 11.4). This hydroxyl group reacts with a phosphate group on the 5′ carbon of the deoxyribose of a deoxyribonucleoside triphosphate (see Figure 11.6), and thus the chain grows. Bonds linking the phosphate groups of the deoxyribonucleoside triphosphate break and thereby release the energy for this reaction. Two of the phosphate groups diffuse away, and one becomes part of the sugar–phosphate backbone of the growing DNA molecule.

For double-stranded DNA to replicate, the strands must unwind and separate from each other. Only then can they function as templates for the synthesis of new, complementary strands. The unwinding results in a **replication fork**—a moving, Y-shaped structure that is the region where new DNA strands are being synthesized. Recall that the two original strands are antiparallel, with the 3′ end of one strand paired with the 5′ end of the other. As the replication fork moves along the parent DNA molecule, an enzyme, DNA polymerase III, catalyzes the replication of both strands. How can this be accomplished, given that new nucleotides are added only at the 3′ end of a polynucleotide chain?

One parent strand is being exposed beginning at its 3′ end, which presents no problem—its complementary strand is synthesized continuously as the replication fork proceeds. This daughter strand is called the **leading strand**. The other daughter strand, the **lagging strand**, is produced in discontinuous

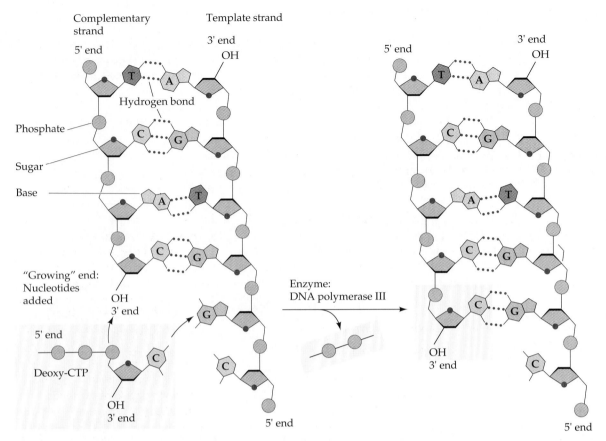

11.10   A Replicating DNA Strand Grows from 5′ to 3′
A DNA strand, with its 3′ end at the top and its 5′ end at the bottom, is the template for the synthesis of the complementary strand at the far left. The new strand has its 5′ end at the top and its 3′ end at the bottom—that is, it is antiparallel to the template strand. DNA polymerase III adds the next deoxyribonucleotide, with the base C, at the free —OH group at the 3′ end of the growing chain. At the right are the same template and the growing strand, now one base longer.

spurts (100 to 200 nucleotides at a time in eukaryotes; 1,000 to 2,000 at a time in prokaryotes). The discontinuous stretches are synthesized just as the leading strand is, by adding the 5′ end of a nucleotide to the 3′ end of the daughter strand, but the stretches are synthesized in the opposite direction with respect to the replication fork. These stretches of new DNA for the lagging strand are called **Okazaki fragments** after their discoverer, the Japanese biochemist Reiji Okazaki. While the leading strand grows continuously "forward," the lagging strand grows in shorter, "backward" stretches with gaps between them (Figure 11.11). The gaps between the Okazaki fragments are then filled in by DNA polymerase I, and another enzyme, **DNA ligase**, links the fragments.

Working together, two DNA polymerases, DNA ligase, and several other proteins (Box 11.A) do the complex job of DNA synthesis with a speed and accuracy that are almost unimaginable. In *E. coli*, the complex makes new DNA at a rate in excess of 1,000 base pairs per second and makes mistakes in fewer than one base in 10^8–10^{12}.

On a bacterial chromosome, DNA synthesis begins at just one point, the **origin of replication**. Each chromosome of a eukaryote, on the other hand, has many origins of replication. DNA synthesis may thus proceed simultaneously in many areas of a single eukaryotic chromosome. Synthesis proceeds in both directions from an origin of replication as two replication forks move away from it (Figure 11.12).

PROOFREADING AND DNA REPAIR

DNA must be faithfully replicated and maintained. The price of failure is great—it may even be death.

BOX 11.A

Collaboration of Proteins at the Replication Fork

The replication of a DNA molecule is an amazingly complex process. We have already described the general problem of synthesizing both a leading and a lagging strand at the same time—but this is only the most obvious difficulty. DNA replication includes all the following steps: (1) unwinding the two parent strands, (2) providing a primer for the synthesis of a new strand, (3) elongating each of the daughter strands, (4) filling in the gaps between the Okazaki fragments of the lagging strand, (5) connecting the completed Okazaki fragments, and (6) editing the newly synthesized strands for accuracy of replication. Each of these steps requires one or more specific proteins, many of which are enzymes.

The unwinding of the double helix is mediated by two related enzymes called **helicases**, one of which attaches to each of the parental DNA strands. Energy to separate the strands comes from the hydrolysis of ATP. The separated strands would tend to interact with each other because of their complementarity, but this is prevented by the attachment of **single-stranded DNA-binding proteins** to each of the separated strands. These binding proteins hold the single strands of DNA in a configuration that binds readily to DNA polymerase III. A pair of DNA polymerase III molecules at the replication fork catalyzes elongation of the leading and lagging strands. The discontinuous production of the lagging strand, however, results in a repeated need for a new primer to start the synthesis of the next Okazaki fragment. The primer is a short single strand of RNA rather than of DNA; it is formed, complementary to the template DNA strand, by an enzyme called a **primase**, which is one of several polypeptides bound together in an aggregate called a **primosome**. DNA polymerase III ex-

tends the primer. DNA polymerase I later replaces the RNA primer segments with DNA segments. Finally, each newly completed Okazaki fragment is linked to the completed portion of the lagging strand in a reaction catalyzed by DNA ligase.

Besides catalyzing elongation of the leading and lagging strands, DNA polymerase III plays another crucial role in DNA replication: It checks the accuracy of its own work. After adding a monomer to a strand, DNA polymerase III tests the new base pair to see that it is complementary to the nucleotide in the template strand. If an incorrect nucleotide has been inserted, the DNA polymerase excises the erroneously selected nucleotide and tries again. As a result, DNA replication is startlingly faithful, with an error rate on the order of one wrong nucleotide per billion—even though the error rate before the proofreading process is on the order of one wrong nucleotide per 10,000.

Even this description of DNA replication is simplified. In *E. coli*, for example, more than 30 polypeptides participate. The largest of the proteins, DNA polymerase III, has a molecular weight of 760,000 and consists of seven or eight polypeptide subunits.

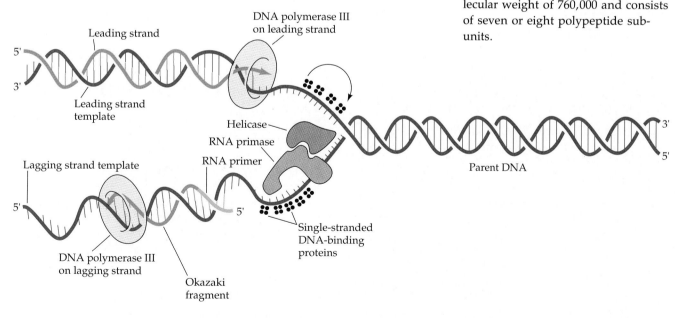

Leading strand

5'

3'

Leading strand template

DNA polymerase III on leading strand

Helicase

RNA primase

RNA primer

Lagging strand template

5'

5'

DNA polymerase III on lagging strand

Okazaki fragment

Single-stranded DNA-binding proteins

Parent DNA

3'

5'

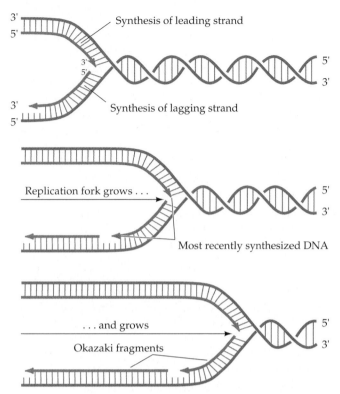

11.11 DNA Is Synthesized at a Replication Fork
As the original DNA strands unwind and separate, both daughter strands are synthesized in the 5'-to-3' direction, although their template strands are antiparallel. The leading strand is synthesized continuously, but the lagging strand is synthesized in Okazaki fragments that are joined later. Abbreviated here, eukaryotic Okazaki fragments are hundreds and prokaryotic fragments thousands of nucleotides long.

Heredity itself is at stake, as is the functioning of a cell or multicellular organism. Yet the replication of DNA is *not* perfectly accurate, and the DNA of nondividing cells is subject to damage by environmental agents. In the face of these threats, why has life gone on so long?

What saved us are DNA repair mechanisms. We have just asserted that replication proceeds with mistakes in fewer than one base in 10^8–10^{12}. In fact, DNA polymerases initially make a significant number of mistakes in assembling polynucleotide strands; in *E. coli*, for example, the error rate would result in flaws in approximately one out of every ten genes each time the cell divides. In humans, about 1,000 genes in every cell would be affected each time the cell divided. After introducing a new nucleotide into a growing polynucleotide strand, however, the DNA polymerases carry out a "proofreading" function. When a DNA polymerase "notices" a mispairing of bases, it removes the improperly introduced nucleotide and tries again. Odds are that the polymerase will be successful in inserting the correct monomer

the second time because the error rate is about 1 in 10^4 base pairs. This repair mechanism greatly lowers the overall error rate for replication.

What if a DNA molecule becomes damaged during the often long life of a cell? Some cells live and play important roles for many years, even though their DNA is constantly at risk from such hazards as high-energy radiation, chemicals that induce mutations, and random spontaneous chemical reactions. Cells owe their lives to the many DNA repair mechanisms that have evolved. An important class of repair mechanisms that includes the proofreading function just described is called excision repair. Certain enzymes "inspect" the cell's DNA. When they find mispaired bases, chemically modified bases, or spots in which one strand has more bases than the other (with the result that one or more bases of one strand form an unpaired loop), these enzymes make a cut in the defective strand. Another enzyme cuts away the bases adjacent to and including the offending base, and DNA polymerase and DNA ligase synthesize a new, usually correct piece to replace the excised one. Our dependence on such repair mechanisms is underscored by our susceptibility to a number of diseases that arise from DNA-repair defects. One example is the skin disease xeroderma pigmentosum. People with this disease lack a mechanism that normally repairs damage caused by the ultraviolet radiation in sunlight. Without this mechanism, a person exposed to sunlight invariably develops skin cancer.

Normal, undamaged DNA is vital to life. But what exactly is DNA for? What do genes do? What do they control?

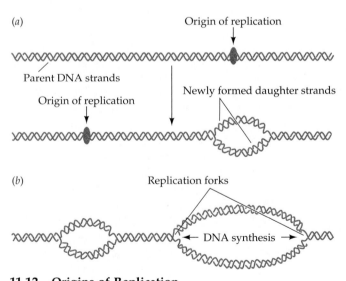

11.12 Origins of Replication
DNA synthesis can begin at many origins in a eukaryotic chromosome. (*a*) The parent strands separate at an origin of replication. (*b*) Synthesis proceeds in both directions from an origin of replication.

WHAT DO GENES CONTROL?

There are many steps between genotype and phenotype. Genes cannot, all by themselves, directly produce a phenotypic result such as the color of an eye or a flower, the shape of a seed, or a cleft chin, any more than a thermostat without a furnace can heat a house. What are the steps between the genes on the chromosomes and the phenotype of the organism? The first hints came early in this century from the work of the English physician Archibald Garrod. Alkaptonuria is a hereditary disease in which the patient's urine turns black when exposed to air. Garrod recognized that this symptom showed that the biochemistry of the affected individual was different from that of other persons. He suggested in 1908 that alkaptonuria and some other hereditary diseases are consequences of "inborn errors of metabolism." He proposed that what makes the urine dark is a defect in an enzyme. Garrod studied the pattern of inheritance of alkaptonuria and reasoned that it affects individuals who are homozygous for the recessive allele of a particular gene (see Chapter 10), which in normal individuals codes for an enzyme needed for metabolism of the amino acid tyrosine. His proposals were the first plausible approach for explaining how genes are expressed. However, like Mendel's explanation of inheritance in the garden pea, Garrod's proposals were too advanced for their time and sat almost unnoticed for over 30 years.

A series of experiments performed by George W. Beadle and Edward L. Tatum at the California Institute of Technology in the 1940s confirmed and extended Garrod's ideas. Beadle and Tatum experimented with the bread mold *Neurospora crassa*. *Neurospora* can be grown on a simple, completely defined medium (that is, one in which all the ingredients are known) containing inorganic ions, a simple source of nitrogen (such as ammonium chloride), an organic source of energy and carbon (such as glucose), and a single vitamin (biotin; see Chapter 43). From this minimal medium, the enzymes of wild-type *Neurospora* can catalyze the metabolic reactions needed to make all the chemical constituents of its cells. Beadle and Tatum reasoned that mutations might alter the enzymes so that they could no longer do their jobs. In that case, mutants of *Neurospora* might be found that could not make certain compounds they needed; such mutants would grow only on media to which those compounds were added. Mutants of this type have since been named **auxotrophs** ("increased eaters"), in contrast to the wild-type **prototrophs** ("original eaters") that constituted the original *Neurospora* population. Prototrophs grow on minimal medium, whereas auxotrophs require specific additional nutrients.

Beadle and Tatum irradiated cells of *Neurospora* with X rays to increase the frequency of mutations and then isolated some nutritional mutants. These auxotrophs did not grow on the minimal medium that supported the growth of the wild-type strain, but they did grow on a complex medium enriched with amino acids (the monomers of proteins), purines and pyrimidines (the nitrogenous bases of nucleic acids), and vitamins. Beadle and Tatum tested these mutants to determine the simplest nutritional supplements that would support their growth (Figure 11.13). Among the collection of mutants were individual strains that required a specific amino acid or vitamin. In almost every case, the nutritional requirement was simple: Only a single compound had to be added to the minimal medium to support the growth of any given mutant. This result supported the idea that mutations have simple effects—and, perhaps, that each mutation causes a defect in only one enzyme in the metabolic pathway leading to the synthesis of the required nutrient.

The auxotrophs identified in this way could be divided into classes on the basis of the nutritional supplements that would support their growth. For example, all mutants that did not grow on minimal medium but grew on minimal medium supplemented with the amino acid arginine were classified as *arg* mutants. Other sets of mutants were found that required adenine, or proline, or vitamin B_1, and so forth. Within a group of mutants with the same nutritional requirement, mapping studies established that some of the individual mutations were at different loci on a chromosome or were on different chromosomes, indicating that different genes can govern a common biosynthetic pathway. For example, Beadle and Tatum found no fewer than 15 different *arg* mutants. These mutants were then grown in the presence of various suspected intermediates in the synthetic metabolic pathway for arginine. Different mutants were able to grow on different intermediates, as well as on arginine-supplemented medium (Figure 11.14).

Growing mutants on different intermediates helped to determine the biochemical steps by which various amino acids and other compounds are synthesized in *Neurospora*. Much more important, however, this work led to the formulation of the one-gene, one-enzyme theory. According to this theory, the function of a gene is to control the production of a single, specific enzyme. This proposal strongly influenced the subsequent development of the sciences of genetics and molecular biology. Garrod had pointed in the same direction more than three decades earlier, but only now were other scientists prepared to act on the suggestion.

The one-gene, one-enzyme hypothesis was re-

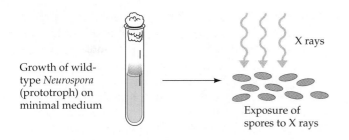

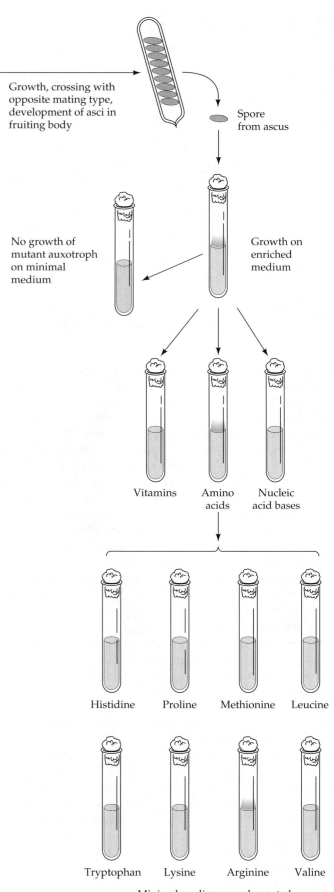

11.13   Identifying Nutritional Mutations of *Neurospora*
Offspring (spores) of cultures that have been irradiated in order to increase their mutation rate are individually placed on an enriched medium. Each spore forms a fungal mat that contains only one kind of haploid nucleus. Part of each mat is placed on minimal medium to test whether it is a mutant auxotroph. The auxotroph's nutritional requirement (in this instance, arginine) is determined by finding which substance supports its growth (indicated here by an orange "glow"). Because each spore is haploid, by the end of the procedure the phenotype reveals the genotype—an important reason for using *Neurospora*.

fined in the mid-1950s as a result of work on sickle-cell anemia. This serious disease is the consequence of a recessive allele that, when homozygous, results in defective red blood cells. Where oxygen is abundant, as in the lungs, the cells are normal in structure and function. But at the low oxygen levels characteristic of working muscles, the red blood cells collapse into the shape of a sickle (Figure 11.15). Linus Pauling, at the California Institute of Technology, speculated that the disease results from a defect in hemoglobin, a protein that fills red blood cells and carries oxygen. Human hemoglobin is a tetramer, consisting of two each of two different polypeptide chains. After Pauling's suggestion, it was shown that one of the two kinds of polypeptides differs by one amino acid between normal and sickle-cell hemoglobin. This result suggested the more satisfying one-gene, one-polypeptide theory: The function of a gene is to control the production of a single, specific polypeptide. Much later, it was discovered that some genes code for forms of RNA that do not get translated into polypeptides.

FROM DNA TO PROTEIN

How does DNA function? We have learned about its structure and how it replicates. Next we need to know what it does. Beadle and Tatum demonstrated that genes are responsible for the production of proteins. But how does a gene specify the synthesis of a protein?

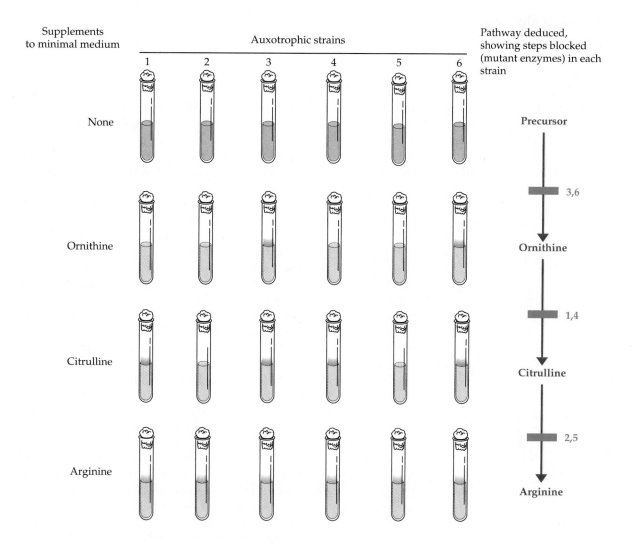

| Supplements to minimal medium | Auxotrophic strains | | | | | | Pathway deduced, showing steps blocked (mutant enzymes) in each strain |

None — Precursor

3,6

Ornithine → Ornithine

1,4

Citrulline → Citrulline

2,5

Arginine → Arginine

11.14 Dissecting a Biochemical Pathway

You have just isolated six arginine-requiring auxotrophic mutants of *Neurospora*. You know that ornithine and citrulline are chemically related to arginine, and you think that they may be intermediates in its synthesis. Let's use these six mutants to deduce the biosynthetic pathway to arginine. All six strains grow when supplied with arginine (fourth row of cultures) but not on unsupplemented minimal media (top row). Strains 2 and 5 grow only on media supplemented with arginine. What does this suggest? Because these strains cannot grow on either ornithine or citrulline, their mutations must interfere with the synthesis of arginine itself—the final step. Strains 1 and 4 grow when supplied with citrulline but not with ornithine. What does this suggest? Their mutations must block the synthesis of citrulline but permit the conversion of citrulline to arginine. Thus part of the pathway must be citrulline → arginine. Note that strains 1 and 4 do not grow when supplemented with ornithine, suggesting that if ornithine is an intermediate, then it must occur before this genetic block. What can you infer from the behavior of strains 3 and 6? We leave that for you to decide. The simplest pathway that is consistent with all these observations is shown at the right.

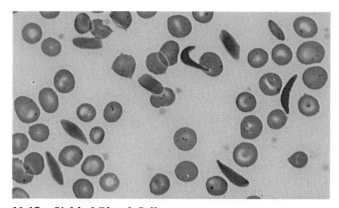

11.15 Sickled Blood Cells

Most of these human red blood cells are normal: They are flattened and roughly circular, with concave centers. Some of the cells, recognizable by their shape, are sickled. This change in shape results from a single amino acid substitution in one of the polypeptides of the protein hemoglobin. The sickled cells are fragile and are eliminated in the spleen, with anemia the result.

The Central Dogma of Molecular Biology

The **central dogma** of molecular biology is one of the most important concepts to have emerged in the attempt to explain how genes make polypeptide chains. The central dogma is, simply, that DNA codes for the production of RNA (transcription), RNA codes for the production of protein (translation), and protein does *not* code for the production of protein, RNA, or DNA (Figure 11.16). In Crick's words, "once 'information' has passed into protein *it cannot get out again.*"

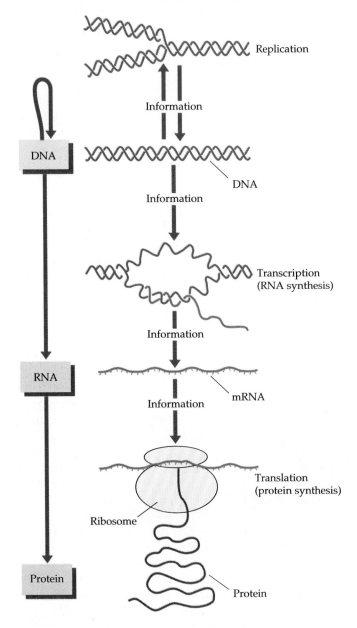

11.16 The Central Dogma
Information coded in the sequence of base pairs in DNA is replicated in new molecules of DNA and passed to molecules of RNA. Information in RNA is passed to proteins.

Crick contributed two key ideas to the development of the central dogma. The first solved a difficult problem: How could one explain the relationship between a specific nucleotide sequence (in DNA) and a specific amino acid sequence (in protein), since nucleotides do not attach to amino acids. Crick made a clever suggestion: He proposed that there is an adaptor molecule that carries a specific amino acid at one end and that recognizes a sequence of nucleotides with its other end. In due course, other molecular biologists found and characterized these adaptor molecules. They are small RNAs called transfer RNAs, or **tRNAs**. Because they recognize the genetic message *and* simultaneously carry specific amino acids, tRNAs can translate the language of DNA into the language of proteins.

Crick's second major contribution to the central dogma addressed another problem: How does the genetic information get from the nucleus to the cytoplasm? (Most of the DNA of a eukaryotic cell is confined to the nucleus, but proteins are synthesized in the cytoplasm.) Crick, together with the South African geneticist Sydney Brenner and the French molecular biologist François Jacob, developed the "messenger hypothesis" in response to this question. According to this hypothesis, a specific type of RNA molecule forms as a complementary copy of one strand of the gene. Because the RNA molecule contains the information from that gene, there should be as many different messengers as there are different genes. This messenger RNA, or **mRNA**, then travels from the nucleus to the cytoplasm. In the cytoplasm, mRNA serves as a template on which the tRNA adaptors line up to bring amino acids in the proper order into a growing polypeptide chain in the process called **translation** (see Figure 11.16).

Summarizing the main features of the central dogma, the messenger hypothesis, and the adaptor hypothesis, we may say that *a given gene is transcribed to produce a messenger RNA complementary to one of the DNA strands*, and that *transfer RNA molecules translate the sequence of bases in the mRNA into the appropriate sequence of amino acids.*

RNA Viruses and the Central Dogma

According to the central dogma of molecular biology, DNA codes for RNA and RNA codes for protein. All cellular organisms have DNA as their hereditary material. Only among viruses (which are not cellular) is a variation on the central dogma found. Many viruses, such as the tobacco mosaic virus, have RNA rather than DNA as their nucleic acid. Heinz Fraenkel-Conrat of the University of California at Berkeley separated the protein and RNA fractions of the tobacco mosaic virus and then recombined them to obtain active virus particles. When he took RNA from one mutant strain of this virus and combined it with protein from another, the resulting viruses replicated to produce more virus particles like the first (the RNA-donating) strain. Thus he showed that RNA is the genetic material of the tobacco mosaic virus. RNA itself is the template for the synthesis of the next generation of viral RNA and viral proteins. In this virus, DNA is left out of the information flow.

Rous sarcoma virus is an RNA virus that causes a cancer in chickens. The virus enters a chicken cell and subsequently causes the cell to make a DNA "transcript" of the viral RNA, the reverse of the usual process. The afflicted cell does not burst, but it changes permanently in shape, metabolism, and growth habit. The new DNA becomes part of the hereditary apparatus of the infected chicken cell. In 1964, Howard Temin of the University of Wisconsin discovered that the virus carries an enzyme for the manufacture of DNA, using viral RNA as the informational template. The enzyme was named **reverse transcriptase** because it transcribes DNA from RNA rather than RNA from DNA. Viruses that employ reverse transcriptase are known as **retroviruses**; one of the most studied retroviruses is the human immunodeficiency virus (HIV), which causes AIDS (see Chapter 16). The central dogma requires slight modification to account for the flow of information in retroviruses and their hosts. However, it is still true that information does not flow from protein back to the nucleic acids.

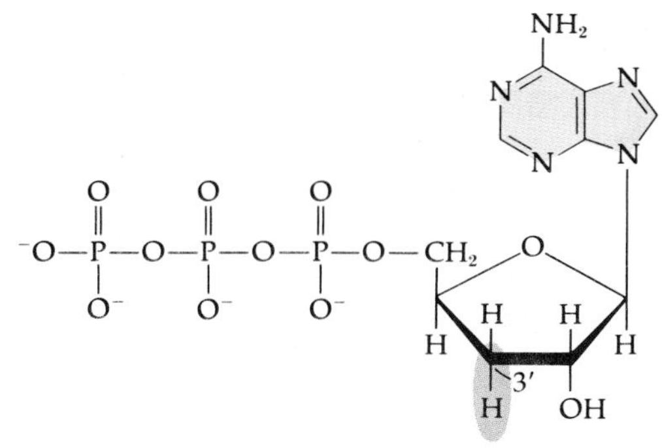

Transcription

The formation of a specific RNA under the control of a specific DNA is called **transcription**. Transcription requires the enzyme **RNA polymerase**, the appropriate ribonucleoside triphosphates (ATP, GTP, CTP, and UTP), and the DNA template. Only *one* of the DNA strands—the **template strand**—is transcribed. The other, complementary DNA strand remains untranscribed. The DNA molecule must partially unwind, as in DNA replication, to expose the bases on the template strand that will be transcribed.

RNA polymerase catalyzes the continuous transcription of DNA in only one direction. This fact was shown in an ingenious experiment using an artificial nucleoside triphosphate (cordycepin triphosphate) that has an unusual nucleoside portion (3'-deoxyadenosine, or cordycepin; Figure 11.17). Cordycepin triphosphate lacks a hydroxyl group at the 3' position of its sugar; thus only its 5' end (and not its 3' end) can be attached in an RNA strand. The nucleotides in a strand of RNA, as in DNA, are covalently bonded such that the 3' end of one attaches to the 5' end of the next, and so on. If mRNA grew by adding nucleotides to its 5' end, cordycepin triphosphate could not form a bond with it. Thus, the presence of this

Cordycepin triphosphate

11.17 Cordycepin Triphosphate
This molecule is very similar to adenosine triphosphate (ATP; see Figure 7.2) but it lacks an —OH group at its 3' carbon end (blue oval). The incorporation of cordycepin triphosphate into mRNA blocks elongation of the strand because it has no attachment site for the next nucleotide.

compound among the available nucleoside triphosphates would have no effect on transcription. When added to a transcription reaction mixture, however, cordycepin triphosphate actually *does* strongly inhibit mRNA formation. Since this result makes sense only if the 3' position of the RNA molecule is the growing point, this experiment clearly shows that mRNA grows from the 5' end to the 3' end.

Formation of mRNA is inhibited because RNA polymerase mistakes cordycepin triphosphate for normal adenosine triphosphate (ATP), and attaches the 5' carbon of cordycepin triphosphate to the 3' end of the growing RNA chain. After cordycepin triphosphate is attached, the RNA chain cannot react with another nucleoside triphosphate because cordycepin triphosphate has no hydroxyl group on its 3' end. Once cordycepin triphosphate has joined a chain, then, no other nucleotides can be added, mRNA elongation ceases, and only short strands of mRNA with terminal cordycepin molecules can be recovered.

When DNA polymerase catalyzes the replication of DNA, the two strands of the parent molecule are unwound, and each strand becomes paired with a new strand. In transcription, DNA is unwound, but it must then be *rewound*. The DNA strand that is to be transcribed must be partly separated from its partner so that it may serve as a template for mRNA synthesis, but the RNA transcript peels away as it is formed, allowing the DNA that has already been transcribed to be rewound (Figure 11.18).

Transcription of genes begins at regions called **initiation sites** that tell the RNA polymerase where to attach and which strand to copy. Similarly, particular base sequences in the DNA specify the termination of transcription. The transcription of DNA is under precise control: Particular genes are transcribed in

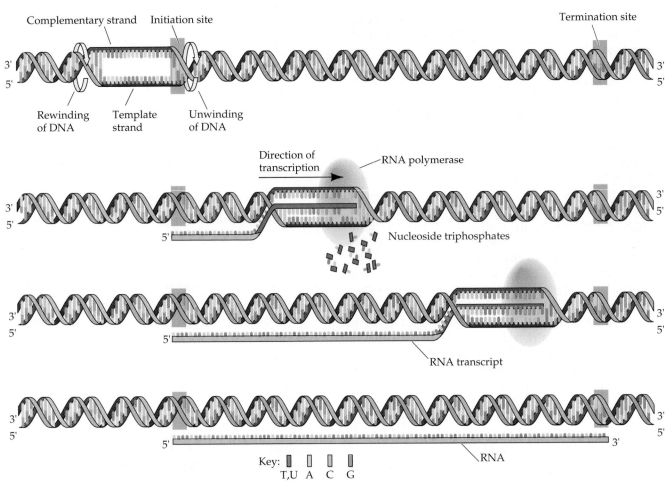

Complementary strand   Initiation site   Termination site

3′
5′

Rewinding   Template   Unwinding
of DNA   strand   of DNA

Direction of   RNA polymerase
transcription

3′
5′

5′
                                          Nucleoside triphosphates

3′
5′

5′
                                          RNA transcript

3′
5′

5′   3′

Key:
T,U A   C   G   RNA

11.18   DNA Is Transcribed into RNA
The DNA double helix unwinds to give RNA polymerase, moving in the 5′-to-3′ direction along the template strand, access to the nucleotide sequence. As the growing RNA transcript is released from the template, the two DNA strands rewind. RNA transcripts are made from only one strand of a DNA double helix. The base-pairing rules are similar to those for DNA: adenine with uracil and guanine with cytosine. The RNA polymerase, which is much larger than shown here, actually covers about 50 base pairs.

certain cells at specific times, whereas other genes are transcribed at other times. This intriguing regulatory process is discussed in Chapter 12.

It is not just mRNA that is produced by transcription. The same process is used in the synthesis of tRNA and of ribosomal RNA, or **rRNA**, which constitutes a major fraction of the ribosome. These other forms of ribonucleic acid are coded for by specific regions of the DNA. In prokaryotes, most of the DNA acts as a template for the production of mRNA, tRNA, or rRNA. The situation in eukaryotes is more complicated, as will be explained later in this chapter and in Chapter 13.

Transfer RNA

You can think of the genetic information transcribed in an mRNA molecule as a series of three-letter "words." Each sequence of three nucleotides (the "letters") along the chain specifies a particular amino acid. The three-letter "word" is called a **codon**. Let's see how a codon is related to the amino acid for which it codes. As predicted by Crick, the codon and the amino acid are related by way of an adaptor—a specific type of tRNA. For each of the 20 amino acids, there is at least one specific tRNA molecule.

A tRNA molecule is small, consisting of only about 75 to 80 nucleotides. Robert Holley of Cornell University was the first to work out the complete nucleotide sequence of a particular tRNA species. He noticed that several regions, apparently separate when the molecule was viewed stretched out, had complementary base sequences. It seemed that these regions could come together by folding and then could stabilize the fold with complementary base pairing. Some years later it became possible to deter-

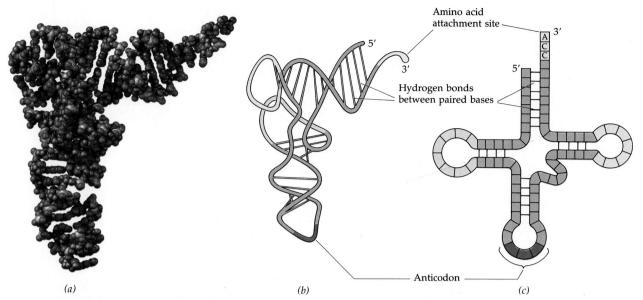

(a)   *(b)*   *(c)*

11.19   Transfer RNA: Crick's "Adaptors"
Molecules of tRNA "read" the genetic code. *(a)* This computer-generated space-filling representation shows the three-dimensional structure of a tRNA, which is then drawn in *(b)*. This three-dimensional shape, brought about by the four regions of base pairing, is diagrammed for clarity in *(c)*. Notice that the region of tRNA that binds to the amino acid is far from the anticodon that interacts with the mRNA.

mine the three-dimensional structures of tRNAs, and it was found that such pairing does occur, giving all tRNAs certain shapes in common. At one end of every tRNA molecule is a site to which the amino acid attaches. At the opposite end is a group of three bases, called the **anticodon**, that constitutes the point of contact with mRNA (Figure 11.19). Each tRNA species has a unique anticodon, allowing it to unite by complementary base pairing with only one codon. This is the key to the specificity of translation.

The three-dimensional shape of tRNAs allows them to combine specifically with the binding sites on ribosomes. The structure of tRNA molecules relates clearly to their functions: They carry amino acids, associate with mRNA molecules, and interact with ribosomes.

How does a tRNA molecule combine with the correct amino acid? A family of **activating enzymes**, known more formally as aminoacyl-tRNA synthases, accomplishes this task. Each activating enzyme is specific for one amino acid. It must also find its appropriate tRNA, which it does by recognizing short sequences of bases on the tRNA, away from the anticodon. The enzyme reacts first with a molecule of amino acid and a molecule of ATP, producing a high-energy AMP-amino acid that remains bound to the enzyme. The high energy results from the breaking of the bonds in the ATP. The enzyme then catalyzes a shifting of the amino acid from the AMP (adenosine

monophosphate) to the 3'-terminal nucleotide of the tRNA, where it is held by a relatively high-energy bond. The activating enzyme finally releases this **charged tRNA** (tRNA with its attached amino acid), which can then charge another tRNA molecule (Figure 11.20). The high-energy bond in the charged tRNA provides the energy for the synthesis of a peptide bond joining adjacent amino acids.

The Ribosome

Ribosomes are required for translation of the genetic information into a polypeptide chain. Each ribosome consists of two subunits, a heavy (large) one and a light (small) one (Figure 11.21*a*). In eukaryotes, the heavy subunit consists of three different molecules of rRNA (ribosomal RNA) and about 45 different protein molecules, arranged in a precise pattern. The light subunit in eukaryotes consists of one rRNA molecule and 33 different protein molecules. The ribosomes of prokaryotes are somewhat smaller than those of eukaryotes. Mitochondria and chloroplasts also contain ribosomes, some of which are even smaller than those of prokaryotes. When not active in the translation of mRNA, the ribosomes exist as separated subunits.

Each ribosome has two tRNA-binding sites that participate in translation. The ribosome also binds to the mRNA that it is translating.

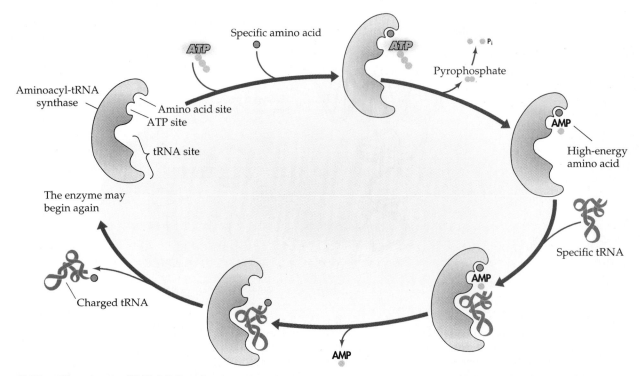

11.20 Charging a tRNA Molecule

An activating enzyme, aminoacyl-tRNA synthase, has a three-part active site that recognizes three smaller molecules: a specific amino acid, ATP, and a specific tRNA. The enzyme activates the amino acid, catalyzing a reaction with ATP in which high-energy AMP amino acid and a pyrophosphate ion are formed. The enzyme then catalyzes a reaction of the activated amino acid with the correct tRNA, producing a charged tRNA molecule with a high-energy bond that provides energy for synthesizing a peptide bond. The enzyme is free to charge another tRNA.

11.21 Translation of Genetic Information

(a) Each ribosome consists of a light and a heavy subunit, which separate when they are not in use. (b) An initiation complex. After the initiation complex forms, the heavy subunit joins it, completing the ribosome. (c) The polypeptide chain elongates as the mRNA is translated.

(d) Translation terminates when the ribosome encounters a stop signal on the mRNA. The completed polypeptide, final tRNA, and ribosomal subunits all dissociate from the mRNA. Circled three-letter abbreviations represent amino acids (see Table 3.1).

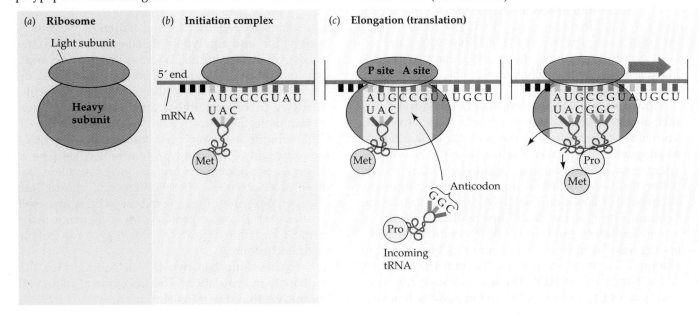

Translation

We have been working our way through the steps by which the sequence of bases in the template strand of a DNA molecule specifies the sequence of amino acids in a protein. We are now at the last step: translation, the RNA-directed assembly of a protein. The translation of mRNA begins with the formation of an initiation complex, which consists of a charged tRNA bearing the first amino acid and a light ribosomal subunit, both bound to the starting point on the mRNA chain. The anticodon of the charged tRNA binds to the appropriate point on the mRNA by complementary base pairing with the first codon. Hydrogen bonds form, linking the codon and the anticodon. The light ribosomal subunit binds to the same point (Figure 11.21*b*). The heavy subunit of the ribosome then joins the complex (Figure 11.21*c*).

There are two RNA-binding sites on the ribosome: the A site (which accepts a tRNA molecule bearing one amino acid) and the P site (which will carry a tRNA molecule bearing a growing polypeptide chain). The first charged tRNA now lies in the P site of the ribosome, and the A site is over the second codon. A charged tRNA whose anticodon is complementary to the second codon enters the open A site. The first amino acid joins the amino acid on the tRNA in the A site, with the peptide linkage forming in such a way that the first amino acid is the N-terminus of the new protein (see Chapter 3), while the second amino acid remains attached to its tRNA by its carboxyl group (—COOH). The first tRNA, having released its amino acid, dissociates from the complex, returning to the cytosol to become charged with another amino acid of the same kind. The second tRNA, now bearing a dipeptide, shifts to the P site of the ribosome, which moves along the mRNA by another codon. The process continues—the latest charged tRNA enters the open A site, picks up the growing polypeptide chain from the one in the P site, and then moves to the newly vacated P site—until a stop codon enters the A site and terminates translation (Figure 11.21*d*). (Stop codons are described later in this chapter, in the section on the genetic code.) The newly synthesized protein separates from the ribosome. The N-terminus of the new protein is the amino acid corresponding to the first codon on the mRNA; the C-terminus is the last amino acid to join the chain.

Several ribosomes can work simultaneously at translating a single mRNA molecule to produce a number of molecules of the protein at the same time. As soon as the first ribosome has moved far enough from the initiation point, a second initiation complex can form, then a third, and so on. The first ribosome to initiate translation is the first to finish translating the message and be released. The assemblage of a thread of mRNA with its beadlike ribosomes and their growing polypeptide chains is called a polyribosome, or **polysome**. Cells that are actively synthesizing proteins contain large numbers of polysomes and fewer free ribosomes or ribosomal subunits. Figure 11.22 shows a polysome in action.

A given ribosome is not specifically adapted to produce just one kind of protein. It was once thought that there were specific ribosomes for each type of protein produced in a cell, but that is not the case. A ribosome can combine with any mRNA and any tRNAs and thus can be used to make different polypeptide products. The mRNA contains the informa-

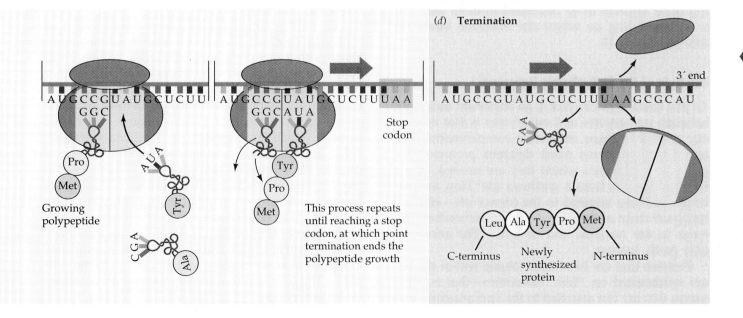

(*d*) **Termination**

Growing polypeptide

This process repeats until reaching a stop codon, at which point termination ends the polypeptide growth

Stop codon

3′ end

C-terminus

Newly synthesized protein

N-terminus

(a)

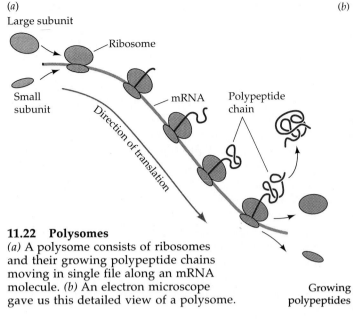

Large subunit

Ribosome

Small subunit

Direction of translation

mRNA

Polypeptide chain

11.22 Polysomes

(a) A polysome consists of ribosomes and their growing polypeptide chains moving in single file along an mRNA molecule. *(b)* An electron microscope gave us this detailed view of a polysome.

(b)

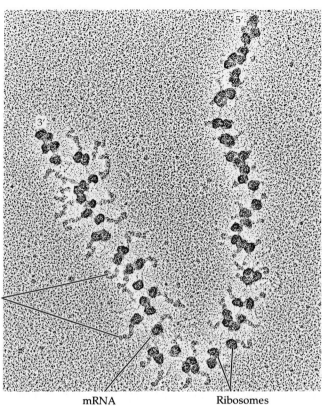

5′

3′

Growing polypeptides

mRNA

Ribosomes

tion that specifies the polypeptide sequence. The ribosome is simply the molecular machine that accomplishes the task. Its structure allows it to hold the mRNA and tRNAs in the right positions, thus allowing the growing polypeptide to be assembled efficiently.

We have simplified this account of translation in at least two ways. First, we have not mentioned the specific proteins and small molecules that play roles in the polypeptide elongation and release processes. Second, we have described protein synthesis as it occurs in prokaryotes. In eukaryotes, things are more complex (see Chapter 13); in particular, several steps come between transcription and translation.

In at least one virus certain genes overlap, allowing some mRNAs to be translated in more than one way, depending on where the ribosome binds the mRNA (Box 11.B).

The Role of the Endoplasmic Reticulum

As you learned in Chapter 4, an important difference between prokaryotes and eukaryotes is that eukaryotic cells have many individual compartments. Different compartments need different proteins. Are proteins synthesized where they are needed, or are they transported from a synthesis site? How are particular proteins targeted to the correct site—electron transport chain components to the mitochondria, histones to the nucleus, and so forth? The answer is only partly known.

Proteins that are to remain soluble within the cell are synthesized on "free" ribosomes—that is, ribosomes that are not attached to the endoplasmic retic-

ulum (ER). On the other hand, proteins that are to become parts of membranes, or are to be exported from the cell, or are to end up in lysosomes or peroxisomes, are generally synthesized on the ribosomes of the rough ER. All protein synthesis, however, *begins* on free ribosomes. The first few amino acids of a polypeptide chain determine whether production of the protein will be completed on the rough ER or on free ribosomes.

If a specific sequence of amino acids, the **signal sequence**, is present at the beginning of the chain, the finished product will be a membrane protein or a protein destined for export. The signal sequence attaches to a signal recognition particle composed of protein and RNA. This attachment blocks further protein synthesis until the ribosome can become attached to a specific receptor protein in the membrane of the ER. The receptor protein becomes a channel through which the growing polypeptide is extruded, either into the membrane itself or into the interior of the ER, as synthesis continues. An enzyme within the ER interior then removes the signal sequence from the new protein, which ends up either built into the membrane or retained within the ER rather than in the cytosol (Figure 11.23). From the ER the newly formed protein can be transported to its appropriate location—to other cellular membranes or to the outside of the cell—without mixing with other molecules in the cytoplasm.

BOX 11.B

Making the Most of Your DNA

A virus named φX174, one of the smallest bacteriophages, is made up of a few types of protein molecules plus a small, circular molecule of DNA. This viral DNA must code for nine types of proteins, including some that are not part of the mature phage but that are needed during viral replication or release. We know the lower limit for the length of a DNA base sequence that can code for a protein. Scientists puzzled over φX174 because its total DNA content (about 5,400 nucleotides) appears to be too low by 10–15 percent to code for all nine proteins. This problem was resolved in a most unexpected way in 1977, when the English

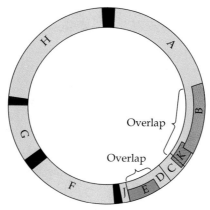

The circle represents the chromosome of the φX174 virus; the letters indicate the proteins coded for by the genes in the colored regions. Note that some regions coding for one protein also include the information for other proteins; for example, the region coding for protein A also codes for protein B and part of protein K. The few short stretches of the chromosome that do not code for proteins are shown in black.

biochemist Frederick Sanger and his colleagues reported the complete structure, nucleotide by nucleotide, of the DNA molecule of φX174. With the nucleotide map spread out before

them, and with their knowledge of the primary structures of the proteins, these scientists were able to discover that the genes coding for two of the proteins are embedded within two of the other genes, as indicated in the figure. Putting it another way, the same stretch of DNA can participate in coding for two entirely different proteins. The mRNA transcribed from such a shared stretch of DNA contains codes for both proteins. Which protein is produced depends on where translation begins—that is, on which of two initiation sites becomes bound to a ribosome. When you read about frameshift mutations in this chapter, you might recall the story of φX174 and marvel that such an improbable thing as shared DNA should ever have come about. The phenomenon of overlapping genes does not appear to be common; it may have evolved in φX174 because of the limitation on the size of the DNA molecule imposed by the small protein coat.

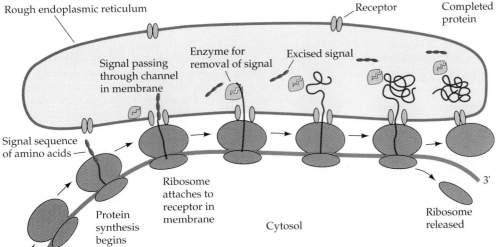

11.23 A Signal Sequence Moves Polypeptide Elongation into the ER
Synthesis of a protein to be exported from the cell begins free in the cytoplasm, as ribosomal subunits and mRNA come together and the first few amino acids, including a signal sequence, are linked. Synthesis continues only after a signal–receptor interaction creates a channel through which the polypeptide will elongate into the interior of the endoplasmic reticulum. When the protein is complete, it is released to the ER interior, and the ribosomal subunits separate from the membrane and from one another.

The Genetic Code

Which mRNA codons translate into which amino acids? Molecular biologists broke the code in which genetic information is stored in the early 1960s. The problem seemed overwhelming at the outset: How could more than 20 "code words" be written with an "alphabet" consisting of only four "letters"? How, in other words, could four bases code for 20 or so different amino acids? It was not yet possible to determine the base sequence in a nucleic acid, so scientists could not simply compare the primary structure (the amino acid sequence) of a protein with the base sequence of the appropriate DNA or mRNA molecule. However, there were ways of getting partial information about the code, even though nucleic acid chemistry was not yet very far advanced.

Marshall W. Nirenberg and J. H. Matthaei, at the National Institutes of Health, made the first breakthrough when they realized that they could use a very simple artificial polymer instead of a complex, natural mRNA as a messenger. They could then see what the simple, artificial messenger coded for. In 1961, Nirenberg and Matthaei published the first of their papers on polypeptide synthesis directed by artificial mRNAs. Nirenberg had prepared an artificial mRNA in which all the bases were uracil; the molecule was called poly U. When poly U was added to a reaction mixture containing ribosomes, amino acids, activating enzymes, tRNAs, and other factors, a polypeptide formed. This polypeptide contained only one kind of amino acid: phenylalanine (Phe). Poly U coded for poly Phe! Accordingly, it appeared that UUU was the mRNA code word—the codon—for phenylalanine (Figure 11.24a). Following up on this success, Nirenberg and Matthaei easily showed that CCC codes for proline and AAA for lysine. (Poly G presented some chemical problems and was not tested initially.) UUU, CCC, and AAA were three of the easiest codons; different approaches were required to work out the rest.

Har Gobind Khorana, at the University of Wisconsin, painstakingly synthesized artificial mRNAs such as poly CA (CACACA. . .) and poly CAA (CAA-CAACAA. . .). Khorana found that poly CA codes for a polypeptide consisting of threonine (Thr) and histidine (His), in alternation (His–Thr–His–Thr. . .; Figure 11.24b). There are thus two possible codons in poly CA, CAC and ACA. One of these must code for His and the other for Thr—but which is which? The answer came from the results with poly CAA, which produces three different polypeptides: poly Thr; poly Gln (polyglutamine); and poly Asn (polyasparagine). To understand this, we must know that an artificial messenger can be read beginning at any point in the chain; there is no specific initiator region. Thus poly CAA can be read as a polymer of CAA, of

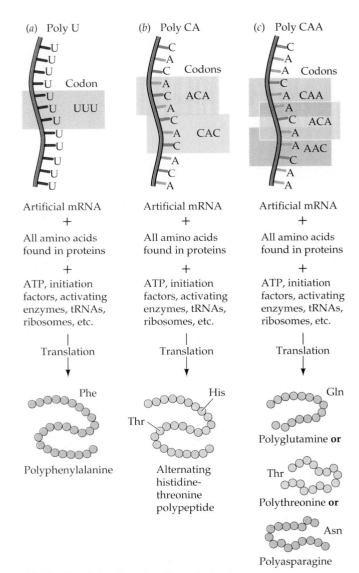

11.24 Deciphering the Genetic Code
The experiments summarized here helped reveal how information is coded in mRNA. (a) The translation of an artificial mRNA containing only uracil (poly U) into a polypeptide containing only phenylalanine suggested that the RNA code for phenylalanine contains only uracil; we now know it is UUU. (b) The translation of poly CA into a polypeptide in which the amino acids threonine and histidine alternate suggested that ACA and CAC are the codons. (c) Poly CAA can be read in three different ways: in units of CAA, ACA, or AAC. Each unit is translated into a different polypeptide containing only one type of amino acid.

ACA, or of AAC (Figure 11.24c). Comparing the results of the poly CA and poly CAA experiments, you should be able to figure out which code word is for Thr and which is for His.

Nirenberg discovered something that led to the decoding of the remaining words in the code book in 1964 and 1965. He found that simple "mRNAs," only three monomers long and each amounting to a

First letter	Second letter: U	C	A	G	Third letter
U	UUU Phenylalanine	UCU Serine	UAU Tyrosine	UGU Cysteine	U
U	UUC Phenylalanine	UCC Serine	UAC Tyrosine	UGC Cysteine	C
U	UUA Leucine	UCA Serine	UAA Stop codon	UGA Stop codon	A
U	UUG Leucine	UCG Serine	UAG Stop codon	UGG Tryptophan	G
C	CUU Leucine	CCU Proline	CAU Histidine	CGU Arginine	U
C	CUC Leucine	CCC Proline	CAC Histidine	CGC Arginine	C
C	CUA Leucine	CCA Proline	CAA Glutamine	CGA Arginine	A
C	CUG Leucine	CCG Proline	CAG Glutamine	CGG Arginine	G
A	AUU Isoleucine	ACU Threonine	AAU Asparagine	AGU Serine	U
A	AUC Isoleucine	ACC Threonine	AAC Asparagine	AGC Serine	C
A	AUA Isoleucine	ACA Threonine	AAA Lysine	AGA Arginine	A
A	AUG Methionine; initiation codon	ACG Threonine	AAG Lysine	AGG Arginine	G
G	GUU Valine	GCU Alanine	GAU Aspartic acid	GGU Glycine	U
G	GUC Valine	GCC Alanine	GAC Aspartic acid	GGC Glycine	C
G	GUA Valine	GCA Alanine	GAA Glutamic acid	GGA Glycine	A
G	GUG Valine	GCG Alanine	GAG Glutamic acid	GGG Glycine	G

11.25 The Universal Genetic Code
Genetic information is encoded in mRNA in three-letter units—codons—made up of the bases uracil (U), cytosine (C), adenine (A), and guanine (G). To decode a codon, find its first letter in the left column, then read across the top to its second letter, then read down the right column to its third letter. The amino acid the codon specifies is in the corresponding row. For example, AUG codes for methionine, and GUA codes for valine.

codon, can bind to ribosomes and that the resulting complex can then cause the binding of the corresponding charged tRNA. Thus, for example, simple UUU causes phenylalanyl-tRNA charged with phenylalanine to bind to the ribosome. After this discovery, complete deciphering of the code book was relatively simple. To find the "translation" of a codon, Nirenberg could use a sample of that codon as an artificial mRNA and see which amino acid became bound.

The complete genetic code is shown in Figure 11.25. Notice that there are many more RNA codons than there are different amino acids in proteins. Combinations of the four "letters" (the bases) give 64 different three-letter codons, yet these determine only 20 amino acids. Three of the codons (UAA, UAG, UGA) are **stop codons**, or chain terminators; when the translation machinery reaches one of these codons, translation stops and the polypeptide is released from the ribosome–mRNA–tRNA complex. AUG, which codes for methionine, is also the **start codon**, which acts as the initiation signal for translation. Still, 61 codons are far more than enough to code for 20 amino acids—and indeed there are repeats. Thus we say that the code is **degenerate**, which means that an amino acid may be represented by more than one codon. The degeneracy is not evenly divided among the amino acids. For example, methionine and tryptophan are represented by only one codon each, whereas leucine is represented by six different codons.

The term *degeneracy* should not be confused with *ambiguity*. To say that the code was ambiguous would mean that a single codon could specify either of two (or more) different amino acids. There would be doubt whether to put in, say, leucine or something else. The genetic code is not ambiguous. Degeneracy in the code means that there is more than one unequivocal way to say, "Put leucine here." In other words, a given amino acid may be coded for by more than one codon, but a codon can code for only one amino acid.

The code appears to be universal, applying to all the species on our planet. The code must be an ancient one that has been maintained intact throughout the evolution of living things. Only one exception is known: Within mitochondria the code differs slightly but detectably from that in prokaryotes and elsewhere in eukaryotic cells. The code is not even quite the same in the mitochondria of different eukaryotes. The significance of this difference is not yet clear.

You should remember that the codons in Figure 11.25 are mRNA codons. The master words on the DNA strand that was transcribed to produce the mRNA are complementary to these codons, so, for example, AAA in the template DNA strand corresponds to phenylalanine (which is coded for by the mRNA codon UUU). Does this code really work? Final proof was obtained by synthesizing artificial DNA of known sequence, introducing it into bacteria, and inducing the bacteria to produce the specific protein coded for by that DNA. We can now program bacteria to synthesize proteins no organism ever made before.

MUTATIONS

Accurate DNA replication, transcription, and translation all depend upon the reliable pairing of complementary bases. Errors occur, though infrequently, in all three. In particular, errors in the replication of DNA during the production of the gametes are crucial to evolution. If there were no mutations—heritable changes in the genetic information—there would be no evolution. Minute changes in the genetic material often lead to easily observable changes in outward form and function. The detection of a mutation depends on our ability to observe these phenotypic effects. Some effects of mutation in humans are obvious—dwarfism, for instance, or the presence of more than five fingers on each hand. A mutant genotype in a microorganism may be obvious if, for example, it results in a change in color or in nutritional requirements, as we have discussed for *Neurospora*. Other mutations, however, may be virtually unobservable. In humans, for example, there is a mutation that drastically lowers the level of an enzyme called glucose 6-phosphate dehydrogenase that is present in many tissues, including red blood cells. The red blood cells of a person carrying the mutant gene are abnormally sensitive to an antimalarial drug called primaquine; when such people are treated with this drug, their red blood cells rupture, causing serious medical problems. People with the normal allele have no such problem. Before the drug came into use, no one was aware that such a mutation existed. Similarly, distinguishing a mutant bacterium from a normal one may be a very subtle matter, dependent on what tools are available.

Some mutations cause their phenotypes only under certain restrictive conditions and are not detectable under other conditions. We call organisms that carry such mutations conditional mutants. Many conditional mutants are temperature-sensitive, unable to grow at some restrictive temperature, such as 37°C, but able to grow normally at a lower, permissive temperature, such as 30°C. The mutant allele in such an organism may code for an enzyme with an unstable tertiary structure that is altered at the restrictive temperature.

All mutations are alterations in the nucleotide sequence in DNA. We divide mutations into two categories based on the extent of the alteration. **Point mutations** are mutations of single genes. One allele becomes another because of small alterations in the sequence or number of nucleotides—even as small as the substitution of one nucleotide for another. **Chromosomal mutations**, introduced in Chapter 10, are more extensive alterations. Chromosomal mutations may change the position or direction of a DNA segment without actually removing any genetic information, or they may cause a segment of DNA to be irretrievably lost. Both point mutations and chromosomal mutations are heritable.

Point Mutations

Many point mutations consist of the substitution of one base for another in the DNA and hence in the mRNA. Often base-substitution mutations change the genetic message so that one amino acid substitutes for another in the protein. Figure 11.15 shows the sickled blood cells that result from one such mutation. A **missense mutation** such as this may sometimes cause the protein not to function, but often the effect is only to reduce the functional efficiency of the protein. Individuals carrying missense mutations may survive even though the affected protein is essential to life. In the course of evolution, some missense mutations even improve functional efficiency.

Nonsense mutations, another type of base-substitution mutation, are more often disruptive than are missense mutations. A nonsense mutation is one in which the base substitution results in the formation of a chain-terminator codon, such as UAG, in the mRNA product. A nonsense mutation results in a shortened protein product, since translation does not proceed beyond the point where the mutation occurred.

Not all point mutations are base substitutions. Single base pairs may be inserted into or deleted from DNA. Such mutations are known as **frame-shift mutations** because they interfere with the decoding of the genetic message by throwing it out of register. Think again of codons as three-letter words, each corresponding to a particular amino acid. Translation proceeds codon by codon; if a base is added to the message or subtracted from it, translation proceeds perfectly until it comes to the one-base insertion or deletion. From that point on, the three-letter words in the message are one letter out of register. In other words, such mutations shift the "reading frame" of the genetic message. Frame-shift mutations almost always lead to the production of completely non-functional proteins.

Consider the following example to see how a frame-shift mutation works. If a template strand of DNA reads from 3′ to 5′ as follows:

3′—AGATACGTGCTGCAT—5′

it is transcribed to yield an mRNA with the following sequence from 5′ to 3′:

5′—UCUAUGCACGACGUA—3′

which can be divided up into the following codons:

. . . UCU AUG CAC GAC GUA . . .

which, as you can determine from Figure 11.25, translates to the following amino acid sequence (from N-terminus to C-terminus):

. . . serine methionine histidine aspartic acid valine . . .

Now suppose that, through a deletion, the DNA strand loses the fifth base in the sequence above, an A, so it reads as follows:

3′—AGATCGTGCTGCAT—5′

Transcription yields the following mRNA sequence:

5′—UCUAGCACGACGUA—3′

or

. . . UCU AGC ACG ACG UA . . .

which, in turn, is translated to the amino-acid sequence

. . . serine serine threonine threonine . . .

No wonder frame-shift mutations are so disruptive. An organism carrying such a mutant gene can survive only if the affected gene product is not essential to the cellular machinery, or if the organism carries another copy of the gene in its normal form.

Certain chemicals, called **mutagens**, can induce mutations. Among these are base analogs: purines or pyrimidines not found in natural DNA but enough like the natural bases that they can be incorporated into DNA. Base analogs are mutagenic presumably because they are more likely than natural DNA bases to mispair. For example, 5-bromouracil is very similar to thymine and it is easily incorporated into DNA in place of thymine (Figure 11.26). But 5-bromouracil is much more likely than thymine to engage in an abnormal pairing with guanine, and it therefore induces

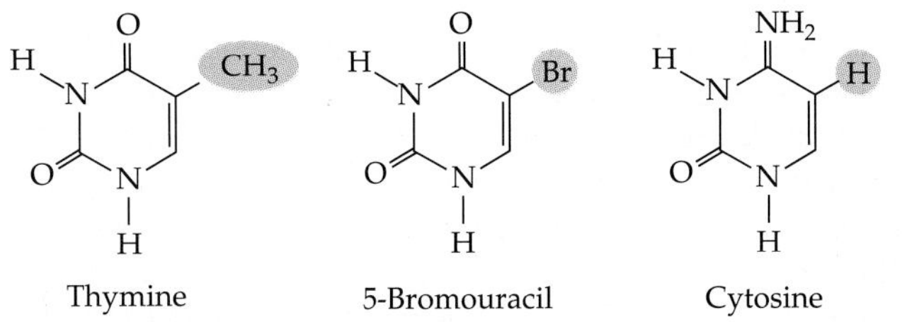

Thymine   5-Bromouracil   Cytosine

11.26 Base Analogs Are Similar to Bases

Enzymes that replicate DNA cannot distinguish between the base analog 5-bromouracil and the base thymine because the two molecules are so similar in shape. Once incorporated in a DNA strand, 5-bromouracil may rearrange itself to a form that resembles cytosine and then pair with guanine.

mutations of A–T to G–C and G–C to A–T. Thus 5-bromouracil is a potent chemical mutagen.

Chromosomal Mutations

Genetic strands can break and rejoin, grossly disrupting the sequence of genetic information. There are four types of such chromosomal mutations: deletions, duplications, inversions, and reciprocal translations (Figure 11.27).

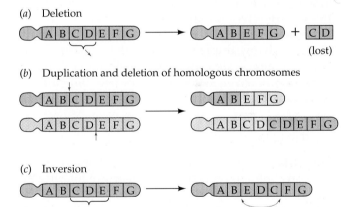

(a) Deletion

(b) Duplication and deletion of homologous chromosomes

(c) Inversion

(d) Reciprocal translocation between nonhomologous chromosomes

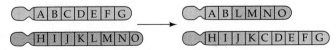

11.27 Chromosomal Mutations

Chromosomes may break during replication and parts of chromosomes may rejoin. Letters on the colored chromosomes represent segments along the length of the chromosome. (a) A deletion results when a chromosome breaks in two places and a segment is lost when the parts rejoin. (b) A duplication and a deletion result when homologous chromosomes break at different points and swap segments. (c) An inversion results when a broken segment is reinserted in reverse order. (d) A reciprocal translocation results when two nonhomologous chromosomes exchange segments.

Deletions remove part of the genetic material. Like frame-shift point mutations, they cause death unless they affect unnecessary genes or are masked by the presence, in the same cell, of normal copies of the deleted genes. It is easy to imagine one mechanism that could produce deletions: A DNA molecule might break at two points and the two end pieces might rejoin, leaving out the DNA between the breaks.

Another mechanism by which deletion mutations might arise would lead simultaneously to the production of a second kind of chromosomal mutation: a **duplication**. Duplication would come about if homologous chromosomes broke at different positions and then reconnected to the wrong partners. One of the two molecules produced by this mechanism would lack a segment of DNA, and the other would have two tandem copies of the information that was deleted from the first.

Breaking and rejoining can also lead to **inversion**—the removal of a segment of DNA and its reinsertion into the same location, but "flipped" end for end so that it runs in the opposite direction. If an inversion includes part of a segment of DNA that codes for a protein, the resulting protein will be drastically altered and almost certainly nonfunctional.

The fourth type of chromosomal mutation, called **translocation**, results when a segment of DNA breaks, moves from a chromosome, and is inserted into a different chromosome. Translocations may be reciprocal, as in Figure 11.27d; notice that the mutation involving duplication and deletion in Figure 11.27b is also a *non*reciprocal translocation. Translocations can make synapsis in meiosis difficult and thus sometimes lead to aneuploidy (see Chapter 9).

The Frequency of Mutations

All mutations are rare events, but mutation frequencies vary from organism to organism and for different genes within a given organism. Usually the frequency of mutation is much lower than one mutation per 10^4 genes per DNA duplication, and sometimes the frequency is as low as one mutation per 10^9 genes per duplication. Most mutations are point mutations in which one nucleotide is substituted for another during the synthesis of a new DNA strand.

THE ORIGIN OF NEW GENES

Most mutations harm the organism that carries them. Some mutations are neutral (they have no effect on the organism's ability to survive or produce offspring). Once in a while, however, a mutation improves an organism's adaptation to its environment or becomes favorable when environmental variables change. Duplication mutations may be the source of "extra" genes. Most of the more complex creatures living on Earth have more DNA and therefore more genes than the simpler creatures do. Humans, for example, have 1,000 times more genetic material than bacteria have.

How do new genes arise? If whole genes were sometimes duplicated by the mechanism described in the previous section, the bearer of the duplication would have a surplus of genetic information that might be turned to good use. Subsequent mutations in one of the two copies of the gene might not have an adverse effect on survival because the other copy of the gene would continue to produce functional protein. The extra gene might mutate over and over again without ill effect because its function would be fulfilled by the original copy. If the random accumulation of mutations in the extra gene led to the production of some useful protein (for example, an enzyme with an altered specificity for the substrates it binds, allowing it to catalyze different—but related—reactions), natural selection would tend to perpetuate the existence of this new gene. New copies of genes also arise through the activity of transposable elements, which are discussed in Chapters 12 and 13.

SUMMARY of Main Ideas about Nucleic Acids as the Genetic Material

DNA is the genetic material of all cellular organisms and many viruses.

Genetic information can be transferred from dead bacteria to genetically different live bacteria by transformation. The transforming principle is DNA.
   Review Figure 11.1

The Hershey–Chase experiment gave compelling evidence that DNA is the genetic material.
   Review Figure 11.2

DNA consists of two polynucleotide chains forming a double helix.

DNA chains are held together by hydrogen bonding between their nitrogenous bases. Base pairing is complementary: A–T, T–A, G–C, C–G.

Normal DNA has a right-handed helical structure with two grooves.
   Review Figures 11.3 and 11.4

RNA is usually a single polynucleotide chain.

DNA replication, catalyzed by DNA polymerases, is semiconservative.
Review Figures 11.7 and 11.9

Because the two antiparallel DNA strands are synthesized in the 5'-to-3' direction, the lagging strand must be synthesized discontinuously.
Review Figures 11.10 and 11.11

DNA polymerases introduce occasional mistakes, but the proofreading function of these enzymes corrects most of the mistakes.

Study of blocks in biochemical pathways showed that many genes encode single polypeptides.
Review Figures 11.13 and 11.14

DNA codes for RNA and RNA codes for protein.
Review Figure 11.6

In retroviruses, RNA serves as a template for DNA synthesis in the host cell as carried out by reverse transcriptase.

Transcription forms an RNA molecule with a base sequence complementary to that of the template strand of the DNA.
Review Figure 11.18

In translation, the information transcribed into mRNA is used to make a polypeptide from amino acids.

tRNAs bind amino acids.
Review Figures 11.19 and 11.20

A ribosome binds to mRNA, and binds charged tRNAs at two other sites.
Review Figure 11.21

Several ribosomes can simultaneously translate the same mRNA.
Review Figure 11.22

Most amino acids are specified by more than one codon. There are also a start codon and three stop codons.
Review Figure 11.25

Mutations, which are rare, change the genetic information.

Point mutations change single nucleotide pairs.

Chromosomal mutations affect larger stretches of a DNA molecule.
Review Figure 11.27

SELF-QUIZ

1. Griffith's studies of *Streptococcus pneumoniae*
   a. proved that DNA is the genetic material of bacteria.
   b. proved that DNA is the genetic material of bacteriophages.
   c. demonstrated the phenomenon of bacterial transformation.
   d. proved that bacteria reproduce sexually.
   e. proved that protein is not the genetic material.

2. In the Hershey–Chase experiment
   a. DNA from parent bacteriophage appeared in progeny bacteriophages.
   b. most of the phage DNA never entered the bacteria.
   c. more than three-fourths of the phage protein appeared in progeny phages.
   d. DNA was labeled with radioactive sulfur.
   e. DNA formed the coat of the bacteriophage.

3. Which statement about complementary base pairing is *not* true?
   a. It plays a role in DNA replication.
   b. In DNA, T pairs with A.
   c. Purines pair with purines, and pyrimidines pair with pyrimidines.
   d. In DNA, C pairs with G.
   e. The base pairs are of equal length.

4. In semiconservative replication of DNA
   a. the original double helix remains intact and a new double helix forms.
   b. the strands of the double helix separate and act as templates for new strands.
   c. polymerization is catalyzed by RNA polymerase.
   d. polymerization is catalyzed by a double-helical enzyme.
   e. DNA is synthesized from amino acids.

5. Which of the following does not occur during DNA replication?
   a. Unwinding of the parent double helix
   b. Formation of short pieces that are united by DNA ligase
   c. Complementary base pairing
   d. Use of a primer
   e. Polymerization in the 3'-to-5' direction

6. Transcription
   a. produces only mRNA.
   b. requires ribosomes.
   c. requires tRNAs.
   d. produces RNA growing from the 5' end to the 3' end.
   e. takes place only in eukaryotes.

7. Which statement is *not* true of translation?
   a. It is RNA-directed polypeptide synthesis.
   b. An mRNA molecule can be translated by only one ribosome at a time.
   c. The same genetic code is in effect in all organisms.
   d. Any ribosome can be used in the translation of any mRNA.
   e. There are both start and stop codons.

8. Which statement is *false*?
   a. Transfer RNA functions in translation.
   b. Ribosomal RNA functions in translation.
   c. RNAs are produced in transcription.
   d. Messenger RNAs are produced on ribosomes.
   e. DNA codes for mRNA, tRNA, and rRNA.

9. The genetic code
   a. is different for prokaryotes and eukaryotes.
   b. has changed during the course of recent evolution.
   c. has 64 codons that code for amino acids.
   d. is degenerate.
   e. is ambiguous.

10. A mutation that results in the codon UAG where there had been UGG
    a. is a nonsense mutation.
    b. is a missense mutation.
    c. is a frame-shift mutation.
    d. is a large-scale mutation.
    e. is unlikely to have a significant effect.

FOR STUDY

1. The genetic code is described as degenerate. What does this mean? How is it possible that a point mutation, consisting of the replacement of a single nitrogenous base in DNA by a different base, might not result in an error in protein production?

2. Suppose that Meselson and Stahl had continued their experiment on DNA replication for another 10 bacterial generations. Would there still have been any ^{14}N–^{15}N DNA present? Would it still have appeared in the centrifuge tube? Explain.

3. Look back at Khorana's experiment with poly CA and poly CAA, in which the codons for histidine and threonine were determined. Using the genetic code (Figure 11.25) as a guide, deduce what results Khorana would have obtained had he used poly UG and poly UGG as artificial messengers. (In fact, very few such artificial messengers would have given useful results.) For an example of what could happen, consider poly CG and poly CGG. Using poly C as the messenger, a mixed polypeptide of arginine and alanine (−Arg−Ala−Arg−Ala−) would be obtained; poly CGG would give three polypeptides: polyarginine, polyalanine, and polyglycine. Can any codons be determined from only these data? Explain.

4. What causes transcription to start? to stop? What causes translation to start? to stop?

READINGS

Felsenfeld, G. 1985. "DNA." *Scientific American*, October. A well-illustrated description of DNA structure and function.

Griffiths, A. J. F., J. H. Miller, D. T. Suzuki, R. C. Lewontin and W. M. Gelbart. 1993. *An Introduction to Genetic Analysis*, 5th Edition. W. H. Freeman, New York. An excellent textbook of modern genetics. Chapters 11, 12 and 13 are particularly relevant.

Judson, H. F. 1979. *The Eighth Day of Creation*. Simon and Schuster, New York. A sparkling history of molecular biology, with the best available description of the events surrounding the discovery of the structure of DNA.

Radman, M. and R. Wagner. 1988. "The High Fidelity of DNA Duplication." *Scientific American*, August. How error avoidance and error correction work. Why don't they work even better?

Smith, M. 1979. "The First Complete Nucleotide Sequencing of an Organism's DNA." *American Scientist*, vol. 67, pages 57–67. How nucleotide sequences in DNA are worked out and the interesting discovery of overlapping genes in the bacteriophage φX174.

Stent, G. S. and R. Calendar. 1978. *Molecular Genetics*, 2nd Edition. W. H. Freeman, New York. A brilliant technical and historical introduction to molecular genetics and the role of DNA.

Upton, A. C. 1982. "The Biological Effects of Low-Level Ionizing Radiation." *Scientific American*, February. How radiation leads to mutations.

Watson, J. D. 1968. *The Double Helix*. Atheneum, New York. A captivating and, to some, infuriating book in which Watson describes the events leading to the discovery of DNA structure.

Watson, J. D., N. H. Hopkins, J. W. Roberts, J. A. Steitz and A. M. Weiner. 1987. *Molecular Biology of the Gene*, 4th Edition. Two volumes. Benjamin/Cummings, Menlo Park, CA. See especially Chapters 3, 9, 10, and 12 to 15 of Volume I.

You are looking at *Escherichia coli*, the best-understood prokaryote that ever lived. We know its molecular biology more intimately than that of any other organism. *E. coli* was the "host" species in 1973, when biologists first succeeded in "transplanting" a gene from one species into another species and in seeing that gene expressed—transcribed and translated—in its new host. Since then *E. coli* has hosted many kinds of foreign or synthetic genes of interest to biomedical scientists. For even longer, however, biologists have poked and probed *E. coli*, seeking to understand basic genetic and molecular biological principles.

Prokaryotes such as *E. coli* and bacteriophages (bacterial viruses) have often been easier subjects than eukaryotes for experimental study. Molecular biologists working with bacteria and viruses in the 1950s and 1960s discovered most of the principles described in Chapter 11. Perhaps these principles would still be hidden from us if work had been limited to garden peas, *Neurospora*, corn, and fruit flies. What are the advantages of working with bacteria and viruses? First, it is easier to work with small amounts of DNA. A typical bacterium contains about $\frac{1}{1000}$ as much DNA as a single human cell, and a typical bacteriophage contains about $\frac{1}{100}$ as much DNA as a bacterium. Second, data on large numbers of organisms can be obtained easily from prokaryotes, but not from most eukaryotes. A single milliliter of medium can contain more than 10^9 *E. coli* cells or 10^{11} bacteriophages and cost less than a penny. In addition, a culture of *E. coli* can be grown under conditions that allow it to double every 20 minutes. By contrast, 10^9 mice would cost more than 10^9 dollars and would require a cage that would cover about 3 square miles. Growth of a generation of mice takes about 3 months instead of 20 minutes.

To be of use to geneticists and molecular biologists, bacteria and viruses must be genetically variable and, preferably, have some form of sex life. That bacteria mutate was demonstrated by Salvador Luria and Max Delbrück in 1943. (Remember that mutation is the source of genetic variation.) Three years later, Joshua and Esther Lederberg and Edward Tatum proved that some bacteria can engage in a form of sexual reproduction. The only similarity between this "sexual reproduction" and sexual reproduction in eukaryotes is that genetic information is transferred from one individual to another.

The ease of growing and handling bacteria and their viruses permitted the explosion of molecular biology that came shortly thereafter (you read about some of these discoveries in Chapter 11). The relative biological simplicity of bacteria and bacteriophages contributed immeasurably to the discoveries about the genetic material, the replication of DNA, and the mechanisms of gene expression. Later these bacteria

The Prettiest *E. coli* You Ever Saw
This transmission electron micrograph of *Escherichia coli* was artificially colored to illuminate the bacteria's interior components.

12

Molecular Genetics of Prokaryotes

and bacteriophages were the first subjects of recombinant DNA technology (see Chapter 14). Questions of interest to all biologists continue to be studied in prokaryotes, and prokaryotes continue to be important tools for biotechnology and for research on eukaryotes.

MUTATIONS IN BACTERIA AND BACTERIOPHAGES

E. coli—or any other bacterial species—reproduces by the division of single cells into two identical offspring. A single cell gives rise to a clone—a population of genetically identical individuals. As long as conditions remain favorable, a population of *E. coli* can double every 20 minutes. Researchers grow *E. coli* on the surface of a solid medium containing a sugar, minerals, a nitrogen source such as ammonium chloride, and a solidifying agent such as agar. Aseptic conditions are maintained so that there are no other microorganisms to compete with the bacterial cells on the medium. Each bacterium gives rise to a small, rapidly growing colony. If enough cells (10^7–10^8) are used, the resulting 10^7–10^8 colonies grow until they merge, forming what is called a bacterial lawn. If the culture begins with a pure type, such as the strain called *E. coli* K, the entire lawn will be of the same type.

We can do an experiment to demonstrate that bacterial mutants arise. First we mix a large sample of *E. coli* K with a suspension of a bacteriophage, such as the one called T4, and pour the mixture over growth medium in a petri plate. Wherever the virus finds a bacterial cell, it attaches to it, infects it, and eventually causes it to burst, killing the bacterium and releasing many new viruses. These viruses, in turn, attack neighboring cells. Soon visible **plaques**, or circular clearings, begin to appear in the lawn wherever the viruses have killed bacteria (Figure 12.1). The plaques, which are caused by the virus-induced bursting, or lysis, of bacteria, grow and grow. As you scan several such plates, however, here and there you will find a bacterial colony growing in the midst of a plaque, in spite of the surrounding hordes of viruses. Each of these colonies has arisen from a mutant bacterium that is resistant to the virus. We call these bacteria *E. coli* K/4 (pronounced "K-bar-4") because they are resistant to phage T4. We know that resistance is a heritable trait because the mutant bacteria give rise to colonies of cells that are similarly resistant to T4.

The phages also mutate. If you prepare plates inoculated simultaneously with *E. coli* K/4 and phage T4, you expect to see no plaques because K/4 is resistant to T4. Occasional plaques *are* found, however. These plaques must arise from mutations, but what

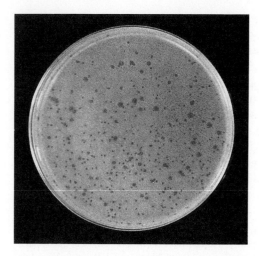

12.1 Bacteriophages Clear Bacterial Lawns
The dark circles are clear plaques in an opaque lawn of *E. coli*. Plaques form where bacteriophages have lysed bacterial cells.

have mutated—the bacteria or the viruses? Give this question a moment's thought. This time it must be the T4 that have mutated. A mutation of one *E. coli* K/4 back to the wild-type K would not result in plaque formation because only that single cell would be infected by the phage and burst; the neighboring bacteria would still be resistant and would still form an even lawn. However, a mutant phage T4 can infect a K/4 cell and, through its progeny phages, lead to the formation of a plaque. Such a phage is designated T4*h* because it has changed with regard to its host—the type of bacterial cell it can infect. The mutation of the phage, like that of the bacterium, is heritable, as the ability of the progeny phages to lyse the K/4 and form a growing plaque shows.

We have just seen how bacteria and bacteriophages are grown in the laboratory and what the consequences of certain mutations are. What we have seen, by the way, is an example of evolution on a small scale. *E. coli* K mutate to K/4 at a low rate all the time in nature, and bacteriophage T4 mutate to T4*h*. The mutants normally do not take over the entire population but exist in low frequency as members of the bacterial or phage population. However, when the environment favors one genotype in a population over others, the proportions of the different genotypes in the population change. Here, for example, T4 kills *E. coli* K, but not K/4, so K/4 soon becomes predominant.

BACTERIAL CONJUGATION

The existence and heritability of mutations in bacteria and their viruses made these microbes attractive subjects for genetic investigations. But if their reproduction were solely asexual, bacteria and their phages would not be useful for genetic analysis. Some form of exchange of genetic information between individuals is necessary. Luckily, such exchange does occur. Genetic recombination—a sexual phenomenon—is a rare event in *E. coli*; it was demonstrated by Joshua and Esther Lederberg and Edward Tatum in 1946.

The Lederbergs and Tatum used two nutrient-requiring (auxotrophic) strains of *E. coli* K12 as parents. Because strain I required the amino acid methionine and the vitamin biotin for growth, its genotype is symbolized as $met^- bio^-$. Strain II requires neither of these substances but cannot grow without the amino acids threonine and leucine. Considering all four factors, we note that strain I is $met^- bio^- thr^+ leu^+$ and strain II $met^+ bio^+ thr^- leu^-$. These two mutant strains were mixed and cultured together for several hours on a medium supplemented with methionine, biotin, threonine, and leucine, so that both could grow. The bacteria were then removed from the medium by centrifugation, washed, and transferred to minimal medium. Neither parental strain could grow on this medium because of their nutritional requirements. However, a few colonies *did* appear on the plates. Because they grew in the minimal medium, these colonies must have consisted of bacteria that were $met^+ bio^+ thr^+ leu^+$—that is, they were prototrophic

(Figure 12.2). These colonies appeared at a rate of approximately 1 for every 10 million cells put on the plates.

From what you have learned thus far, try to formulate at least three possible explanations for the appearance of these prototrophic colonies. One possibility is mutation, but this hypothesis can be rejected for the following reasons: The observed colonies would have had to arise from *double* mutations because each parental strain started with two defective alleles. Given the range of single-mutation frequencies, one would expect such double mutations to occur in, at most, 1 out of every 10^{12} cells—100,000 times less frequently than was actually observed. Neither parental strain, when grown alone under the same conditions, was ever observed to give rise spontaneously to prototrophic colonies.

A second possibility is transformation, the incorporation of DNA from dead bacteria into live ones, as first described by Griffith (see Figure 11.1). This explanation was ruled out by Bernard D. Davis of the U.S. Public Health Service. Davis conducted his experiment with a U-shaped tube, the two arms of which were separated by a very fine filter. The pores in the filter were large enough for DNA to pass through, but small enough to prevent the passage of bacteria. He placed a culture of strain I *E. coli* K12 in one arm and a culture of strain II in the other. Then he applied alternating pressure and suction so that the growth medium was flushed back and forth. The flushing mixed the solutions from the two arms without mixing the bacteria. DNA from dead bacteria *would* have been able to pass through the filter, but

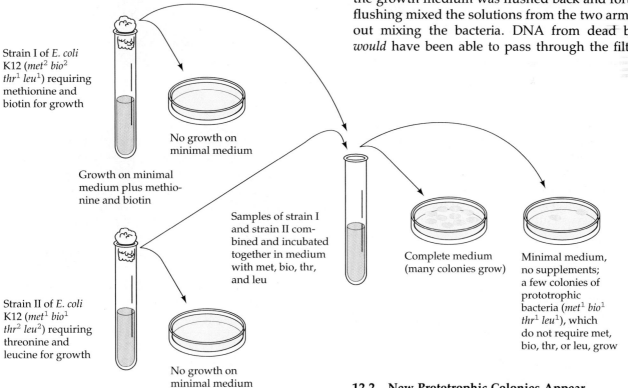

Strain I of *E. coli* K12 ($met^2 bio^2 thr^1 leu^1$) requiring methionine and biotin for growth

No growth on minimal medium

Growth on minimal medium plus methionine and biotin

Samples of strain I and strain II combined and incubated together in medium with met, bio, thr, and leu

Strain II of *E. coli* K12 ($met^1 bio^1 thr^2 leu^2$) requiring threonine and leucine for growth

No growth on minimal medium

Growth on minimal medium plus threonine and leucine

Complete medium (many colonies grow)

Minimal medium, no supplements; a few colonies of prototrophic bacteria ($met^1 bio^1 thr^1 leu^1$), which do not require met, bio, thr, or leu, grow

12.2 New Prototrophic Colonies Appear
After growing together, a mixture of complementary auxotrophic strains contains a few cells that can give rise to new prototrophic colonies.

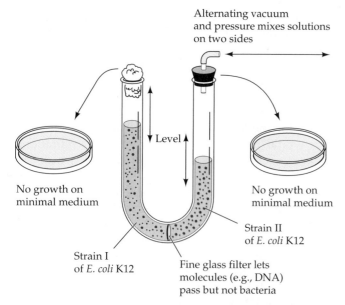

Alternating vacuum and pressure mixes solutions on two sides

↕ Level ↕

No growth on minimal medium

No growth on minimal medium

Strain I of *E. coli* K12

Strain II of *E. coli* K12

Fine glass filter lets molecules (e.g., DNA) pass but not bacteria

12.3 The Davis U-Tube Experiment
Because no prototrophic bacterial cell was recovered from either side of the U-tube, transformation was ruled out as a possible explanation for the appearance of prototrophic colonies in the Lederberg–Tatum experiment (Figure 12.2). Davis's results suggested that the appearance of prototrophic cells requires physical contact between cells of the parental strains.

Davis found *no* wild-type bacteria on either side of the filter (Figure 12.3). These results showed that the phenomenon observed by the Lederbergs and Tatum requires physical contact between cells of the two strains, not just incorporation of DNA.

The third possibility is that the bacteria had **conjugated** in pairs, allowing their genetic material to mix and recombine to produce prototrophic colonies from cells containing *met*⁺ and *bio*⁺ alleles from strain II and *thr*⁺ and *leu*⁺ alleles from strain I. This explanation was confirmed by other experiments, in which two cells of differing genotype mated, and one cell—the recipient—received DNA that included the two wild-type alleles for the loci in the recipient. Recombination then created a genotype with four wild-type alleles. The physical contact required for conjugation was later observed under the electron microscope (Figure 12.4).

What sort of a process brings about the recombination of genes after bacteria conjugate? We will learn about this shortly.

Isolating Specific Bacterial Mutants

Throughout this chapter we will consider experiments that use bacteria and phages with various specific genotypes, as we have just done in the conjugation experiment. How can one obtain a strain with a particular genotype, such as *met*⁻*bio*⁻*thr*⁺*leu*⁺?

To isolate bacteria carrying a particular mutation—

say *met*⁻—we start with a strain carrying the wild-type allele for which mutations are desired. We then subject these bacteria to procedures that increase the mutation rate, such as irradiation with ultraviolet or X rays, or the addition of a chemical mutagen. Now the search begins. First, we let the bacteria in the culture grow and increase their numbers by keeping them in a medium that includes the compound that will be needed by the desired mutant strain (in our example, the medium must include methionine). The overwhelming majority of the cells in the culture are unchanged; these wild-type cells must be eliminated.

There is more than one way to eliminate the wild-type cells, but we will describe one invented by Bernard Davis. He knew that the antibiotic penicillin kills only growing bacteria. Therefore, his method was to put a mixed culture of many wild-type and a few mutant bacteria into a medium *lacking* the nutrient for which the desired mutants were auxotrophic (again, methionine in our example), and add penicillin. In this experiment, the cells that do not need methionine grow rapidly—and commit "penicillin suicide." Because they grow, they die. The desired mutants, on the other hand, fail to grow (because the needed nutrient is unavailable), so they avoid damage by the penicillin. These mutants are then trans-

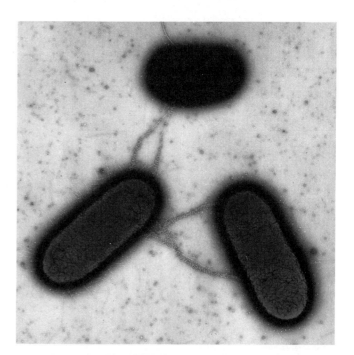

12.4 Bacteria Conjugate
In this electron micrograph of *E. coli*, the "male" cell on the left is connected to two "female" cells by thin tubes called F-pili. In this instance, two F-pili are in contact with each female cell. The tiny "beads" on the pili are bacteriophages that attach specifically to F-pili, making the pili more visible. After cells are joined by F-pili, they are drawn into closer contact, and DNA is transferred from one cell to the other.

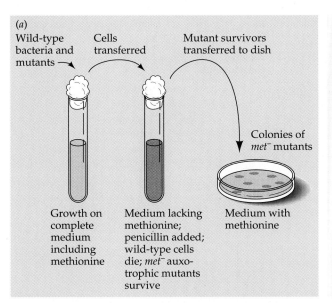

(a)

Wild-type bacteria and mutants → Cells transferred → Mutant survivors transferred to dish

Colonies of *met⁻* mutants

Growth on complete medium including methionine

Medium lacking methionine; penicillin added; wild-type cells die; *met⁻* auxotrophic mutants survive

Medium with methionine

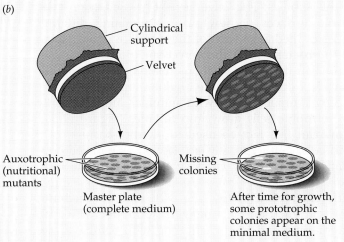

(b)

Cylindrical support

Velvet

Auxotrophic (nutritional) mutants

Missing colonies

Master plate (complete medium)

After time for growth, some prototrophic colonies appear on the minimal medium.

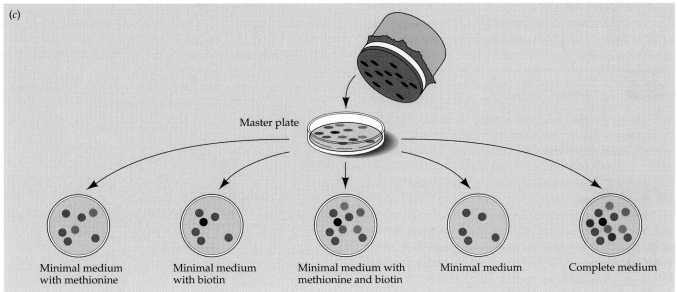

(c)

Master plate

Minimal medium with methionine

Minimal medium with biotin

Minimal medium with methionine and biotin

Minimal medium

Complete medium

12.5 Isolating and Identifying Auxotrophic Mutants
(a) Penicillin kills growing wild-type cells; the nongrowing, methionine-requiring auxotrophic mutants survive. *(b)* In the replica plating process, the velvet replicates the spatial pattern, permitting identification of auxotrophic colonies present on the master plate but missing from the replica plate with minimal medium. *(c)* The colonies shown in green on the master plate grow only on the plate containing both methionine and biotin (middle) and are therefore *met⁻bio⁻*. Can you identify the nutritional requirements of each remaining colony on the master plate?

ferred to medium that has no penicillin but does contain methionine, so that they may grow and form colonies (Figure 12.5*a*).

A particular methionine mutant can then be chosen and used as a parental strain for selecting a *met⁻bio⁻* (*thr⁺leu⁺*) double mutant by the same general approach as was used to select the single (*met⁻*) mutant. In this case, of course, methionine must be present in the medium at all times, and the absence of biotin is used to protect the *met⁻bio⁻* double mutants from penicillin-induced death. By selecting for the second mutation, one obtains the desired double mutant (*met⁻bio⁻*).

How does one identify the progeny of various recombinant types after performing a cross between two strains of bacteria? One method is **replica plating**, a technique invented by Joshua and Esther Lederberg. A small sample (about 0.1 milliliter) of a mixed suspension, presumably including the desired genotype as well as others, is spread on a plate with complete medium and allowed to produce colonies. A sterilized piece of velvet, mounted on a cylindrical support that fits easily into a petri plate, is now pressed gently against the medium. Its fuzzy surface picks up substantial numbers of bacteria from each of the colonies. The velvet is next pressed against the sterile surfaces of new plates containing different types of media. In each of these replica plates, some of the bacteria from the velvet stick to the agar medium—in the same positions relative to one another that they occupied on the original, "master" plate.

On a replica plate with minimal medium, only prototrophic cells grow into colonies (Figure 12.5b). How can this fact be used to identify bacteria of a particular genotype, say *met⁻bio⁻*, from a mixed population obtained by crossing two strains (one *met⁻bio⁺*, the other *met⁺bio⁻*)? Assume that we make five different replica plates: one with minimal medium, one with methionine, one with biotin, one with both methionine and biotin, and one with complete medium. Wild-type colonies from the master plate would grow on each of these replica plates. What about single mutants (*met⁻bio⁺* and *met⁺bio⁻*)? And what about the desired double mutants? Think about these before studying Figure 12.5c. As you can see, replica plating is a powerful means for characterizing mutant colonies.

The Bacterial Fertility Factor

Transfer of genetic material during conjugation in *E. coli* is a one-way process from a donor to a recipient. English microbiologist William Hayes characterized many strains and found that each strain is either recipient or donor, which can also be called female and male, respectively.

The female bacterium becomes male after conjugation—in bacteria, maleness is an infectious venereal disease! Hayes also found that a particular strain of male bacteria gives rise to occasional mutants that no longer function as males—but can now act as females. Hayes rationalized these observations by proposing that maleness in bacteria is due to the presence of a fertility factor, called **F**. Males possess the factor and are F^+; females, lacking the factor, are F^-. In a cross of $F^+ \times F^-$, a copy of the F factor is transferred to the female, thus rendering it F^+, while the original male remains F^+ (Figure 12.6).

The F factor is an extra piece of DNA that can replicate itself and persist in the cell population as if it were a second chromosome independent of the normal bacterial chromosome. Males can change into females simply by losing the F factor through mutation. Genes on the F factor direct a number of processes, among which is the formation on the surface of the male bacterium of long, thin, hairlike projections called **F-pili** (singular: *pilus* = hair). These are tubes with ends that attach to the surface of female cells (see Figure 12.4). Initial contact is made by an F-pilus, and subsequently a mating contact is made that allows DNA to be transferred.

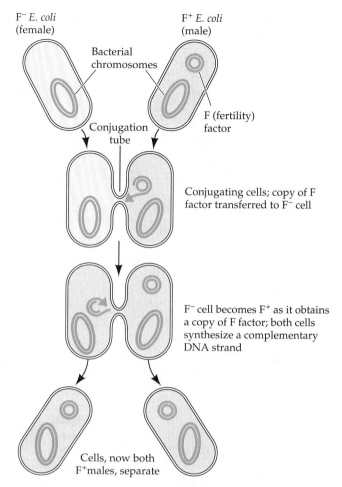

12.6 Infectious Fertility in *E. coli*
During conjugation, the F^- recipient cell receives a copy of the F factor—an extra piece of DNA—from the donor cell by way of a connecting tube (the conjugation tube) and becomes F^+.

Transfer of Male Genetic Elements

The discovery that genetic recombination could follow conjugation opened the possibility of mapping the genetic material of bacteria. However, early attempts at mapping were complicated by the fact that very few recombinant offspring arose from $F^+ \times F^-$ crosses, thus making it difficult to obtain reliable quantitative data. The situation changed with the discovery of certain mutant male strains. Recombinant offspring *were* obtained when these males were used in crosses. These strains were called **Hfr** mutants, for *High frequency of recombination*. Hfr males, unlike ordinary F^+ males, do not generally transfer their F factor to the female. Also, they transfer only certain marker genes with high frequency, transferring other markers no more frequently than ordinary F^+ males do. We know now that in Hfr strains the F factor is actually incorporated into the bacterium's chromosome. Work in 1955 by the French biologists Elie Wollman and François Jacob explained these observations.

Jacob and Wollman showed that the markers from an Hfr male enter the female one at a time. In their most dramatic experiments, they used the technique of **interrupted mating**. They mixed Hfr and F⁻ bacteria at high concentration to initiate conjugation; at various times thereafter they diluted samples of the mixture and agitated them in a kitchen blender for two minutes. Such agitation separates conjugating bacteria but does not damage them. The number of Hfr markers passed to the females depended upon the length of time allowed before conjugation was interrupted—the longer the conjugation, the more markers were transferred (Figure 12.7a). The markers always entered in a particular order from any particular Hfr strain. The Hfr mutant almost never transferred the F factor.

Jacob and Wollman recognized immediately that this interrupted mating technique provided a simple way to map the chromosome. They prepared different mutant strains and crossed pairs of strains; then they interrupted successive samples from the crosses. Because the markers are transferred in a particular sequence, the length of mating time required before a particular marker is transferred and thus available to appear in recombinant progeny is a measure of its location on the chromosome (Figure 12.7b).

Different Hfr mutants have different genetic maps (Table 12.1). If you examine the table, however, you may be able to spot a regularity in the different maps. Jacob and Wollman noticed that although different markers are transferred first in different Hfr strains, the maps are always consistent in that a marker *B* that lies between markers *A* and *C* always does so in every Hfr strain. That is, the starting points may vary, but the *order* of genes remains constant. Even when genes are in a reversed order, *B* is still between *A* and *C*. The simplest conclusion is that the bacterial chromosome is *circular* (Figure 12.8). If you break the circle in different places and convert the results into linear form, you can see how the maps are generated.

From these and other experiments, Jacob and Wollman concluded that (1) the *E. coli* chromosome is circular, (2) Hfr males have the F factor incorporated into their chromosome, (3) the location where the F factor is inserted varies, giving rise to different Hfr strains, (4) the inserted F factor marks the point at which the chromosome "opens" as conjugation begins, and (5) one end—always the same one—of the opened chromosome leads the way into the female. The piece of chromosome continues to move through the conjugation tube until mating is inter-

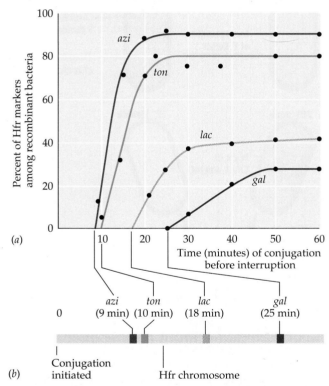

12.7 Chromosome Maps from Interrupted Matings
Chromosome maps of *E. coli* were constructed by interrupting conjugating cells at various times and counting recombinants from the matings. (a) The map in (b) was constructed from data plotted on this graph. In the particular Hfr strain used, recombinants carrying the *azi* allele were first detected from matings interrupted 10 minutes after mixing, as were recombinants carrying the *ton* allele. By extrapolation from other points on the graph, it was determined that the *azi* gene was transferred at 9 minutes, the *ton* gene at 10 minutes. Timings for the *lac* and *gal* genes were established in the same way. (b) The units on this chromosome map are minutes rather than recombination frequencies.

TABLE 12.1
Sequences of Markers Transferred by Various Hfr Strains

ORDER OF ENTRY	Hfr STRAIN				
	H	1	2	3	4
1	T	L	pro	ade	B₁
2	L	T	T₁	lac	ilu
3	azi	B₁	azi	pro	mal
4	T₁	ilu	L	T₁	trp
5	pro	mal	T	azi	gal
6	lac	trp	B₁	L	ade
7	ade	gal	ilu	T	lac
8	gal	ade	mal	B₁	pro
9	trp	lac	trp	ilu	T₁
10	mal	pro	gal	mal	azi
11	ilu	T₁	ade	trp	L
12	B₁	azi	lac	gal	T

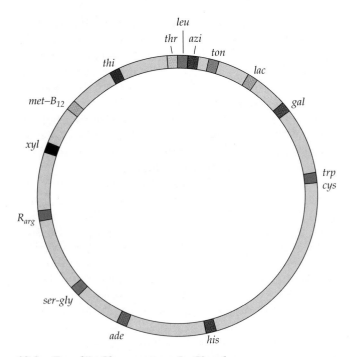

12.8   *E. coli*'s Chromosome Is Circular
This map of the *E. coli* chromosome summarizes data from interrupted conjugation experiments with many Hfr strains (see Figure 12.7 and Table 12.1). The three-letter abbreviation for each gene is derived from its phenotype: *leu*cine-requiring, *azi*de-resistant, *xyl*ose-utilizing, and so on. This early version of the map shows only a few genes. By now, well over half of the estimated 3,000 *E. coli* genes have been mapped.

rupted naturally or otherwise. At the very end of the opened chromosome lies the portion of the F factor that determines maleness.

What moves from the Hfr to the F⁻ is just one strand of the double-stranded Hfr chromosome. Transfer is initiated when one strand within the F factor is nicked (Figure 12.9). The 5′ end of the nicked strand begins to unravel from the chromosome and moves to the F⁻. Meanwhile, the transferred strand is replaced in the Hfr by DNA synthesis at the 3′ end of the nick, using the intact circular strand as the template. Thus the male still contains a double-stranded set of DNA sequences even after donating a fair amount of DNA to the F⁻. As it enters the F⁻, the nicked DNA strand replicates, becoming double-stranded. Markers on this piece of DNA will give rise to recombinant bacteria only if they become incorporated into the F⁻ chromosome by crossing over (Figure 12.10). About half of the transferred Hfr markers get incorporated in this way; the others are lost as the cell divides.

Sexduction

Sometimes the F factor of an Hfr male separates from the chromosome. In the separation process, the F

factor may carry with it a bit of the chromosome. Any genes thus captured by the F factor are transferred to the F⁻ recipient when conjugation occurs. The process in which genes are carried by the autonomous F factor into the F⁻ is called **sexduction**. The modified F factor is called an F′ (F-prime) factor (Figure 12.11).

An F or F′ factor, like the bacterial chromosome, is a circular DNA molecule. Genes carried by the F′ factor are allelic to genes on the main chromosome in the recipient cell. Thus cells harboring an F′ factor may contain more than one allele of a particular gene. Such cells may be used to study whether there is dominance among the alleles of a gene present on the F′ factor—a question that usually cannot be investigated in these normally haploid bacteria.

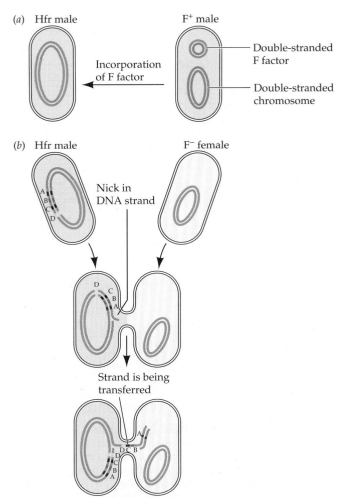

12.9   Origin and Behavior of Hfr Strains
(*a*) A cell becomes an Hfr male when the F factor of an F⁺ cell is incorporated into its chromosome. (*b*) During conjugation, the Hfr male's chromosome opens within the inserted F factor and one strand of the DNA double helix is transferred to the recipient cell. Because most of the F factor is the last DNA to be transferred, the recipient cell usually does not become a male, since the complete chromosome is rarely transferred.

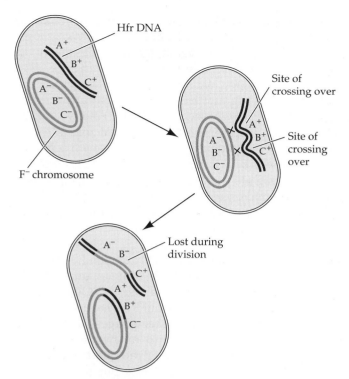

12.10   Recombination Following Conjugation
DNA from an Hfr donor may become incorporated into the recipient cell's chromosome through crossing over. Only part of the donor chromosome here was incorporated—the part containing A^+ and B^+. The $A^+B^+C^-$ sequence becomes a permanent part of the recipient genotype, and the reciprocal $A^-B^-C^+$ segment is lost.

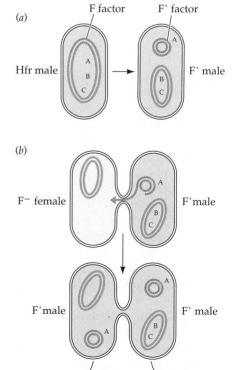

12.11   Sexduction
(a) An F factor in an Hfr cell may carry some genes with it as it leaves the cell's chromosome, changing the cell into an F′ cell. (b) A recipient that conjugates with an F′ cell receives any genes carried by the F′ factor by sexduction and becomes an F′ cell.

BACTERIOPHAGES

Recombination in Phages

Genetic recombination in phages was demonstrated in 1946, the same year the Lederbergs and Tatum revealed the sex life of bacteria. Here too, mutant phenotypes were needed as markers. The alleles of some phage genes affect the appearance of plaques (see Figure 12.1), and we have already noted that mutations can change the ability of a phage to infect certain hosts. To understand the basis for genetic recombination in phages, recall that phages reproduce by injecting their genetic material into a host bacterium, which then supports the synthesis of a large number of progeny phages.

Alfred Hershey and Raquel Rotman, at Washington University in St. Louis, performed a series of experiments in which *E. coli* were simultaneously infected by *two* different mutant strains of the bacteriophage T2. In their first experiment, one of the phage strains was genotypically h^+r and the other was hr^+. (We need not worry here about what the phenotypes were; just note that h^+ and h are alleles at one locus and r^+ and r are alleles at another locus.) We would expect the addition of these phages to a culture of *E. coli* to produce substantial numbers of phages of both parental types. Hershey and Rotman found not only the parental types, however, but also many phages of genotypes h^+r^+ and hr—that is, recombinant phages (Figure 12.12). As more markers in such phage crosses were studied, a map began to take form. In due course, it was learned that the phage has a single, circular chromosome.

Lysogeny and the Disappearing Phages

Up to now we have treated bacteriophages as if their life cycle was always **lytic**: a cycle in which a phage infects a bacterial cell, the phage replicates, the cell lyses, and many new phages are released to renew the cycle. With some bacteria and some phages, however, infection does not always result in lysis of the bacteria. The phages seem to disappear from the culture, leaving the bacteria immune to further attack by the same strain of phage. In such cultures, however, a few free phages are sometimes detected. Bacteria harboring phages that are not lytic are called **lysogenic**. When lysogenic bacteria are combined with other bacteria that are sensitive to the phages, they cause the sensitive cells to lyse.

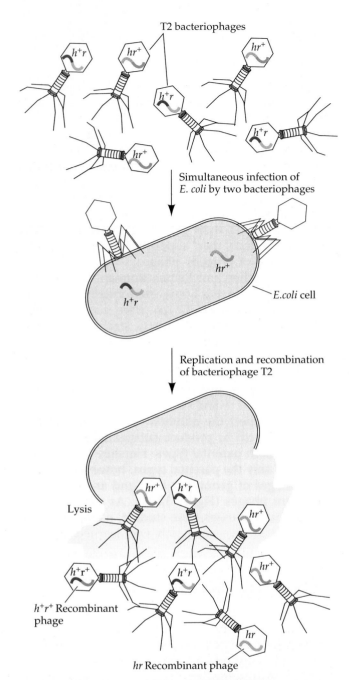

T2 bacteriophages

Simultaneous infection of
E. coli by two bacteriophages

E.coli cell

Replication and recombination
of bacteriophage T2

Lysis

h^+r^+ Recombinant
phage

hr Recombinant phage

12.12 Bacteriophage Genetic Recombination
Bacteriophage strains can be crossed by infecting a culture of susceptible bacteria with both strains simultaneously.

To see where the free phages come from in a culture of lysogenic bacteria, the French microbiologist André Lwoff performed the following delicate experiment with the large bacterium species *Bacillus megaterium*. From a lysogenic culture of *B. megaterium*, Lwoff isolated a single bacterial cell, mounting it in a drop of medium on a microscope slide. He watched patiently until the cell divided and then removed one of the daughter cells with a micropipette. This cell was transferred to an agar medium to see whether

its offspring would be lysogenic. Meanwhile Lwoff watched the daughter cell still under the microscope. When it divided, he again farmed out one of its daughters to solid medium while retaining the other on the microscope slide. He repeated this procedure 19 times, finding that each of the cells transferred to solid medium gave rise to a lysogenic colony. At each transfer, he also sampled the original microdrop to see whether it contained any free phages. It did not. In repetitions of this experiment, however, he sometimes would see the *B. megaterium* burst while it was under the microscope. Whenever this happened, the drop of medium was found to be teeming with free phages.

These experiments revealed that the lysogenic bacteria contain a noninfective entity, which Lwoff called a **prophage**. Prophages could remain quiet within bacteria through many cell divisions. Occasionally, however, a lysogenic cell would be induced to lyse, releasing a large number of free phages, which could then infect other bacteria and renew the life cycle. Lwoff learned that ultraviolet radiation is a potent inducer of the production of free phages. Work by many investigators established finally that the prophage is a molecule of phage DNA that has been incorporated into the bacterial chromosome. Notice the similarity between prophage behavior and that of the F factor that inserts into the chromosome to give rise to an Hfr male. Just as the F factor sometimes leaves the chromosome, so may the prophage—whereupon the phage DNA is activated to multiply rapidly, to make many new phages, and to lyse the bacterium. The lytic and lysogenic cycles are contrasted in Figure 12.13.

Transduction

The prophage can escape from the chromosome. As you might expect, bacterial genes are occasionally taken along by the departing phage DNA. The resulting phages can then introduce these markers into other bacteria that they infect, and the markers may be incorporated into the chromosomes of the new hosts (Figure 12.14*a*). This phenomenon, discovered in 1956 by Joshua Lederberg, is called **restricted transduction** (here *transduce* means "to transfer"). Transducing phages, which carry bacterial markers, cause newly infected bacteria to become lysogenic. In restricted transduction, only the chromosomal genes that are adjacent to the site of attachment of the prophage may be taken along with the phage DNA.

A related phenomenon, **general transduction**, results from the incorporation of part of the *bacterial* chromosome, *without* the prophage, into a phage coat (Figure 12.14*b*). The resulting particle, even though it lacks any phage genes, can infect another bacter-

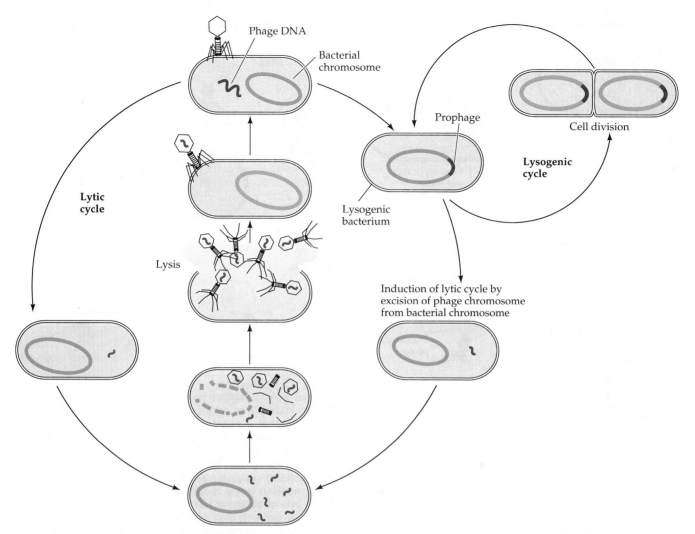

12.13 Lytic and Lysogenic Phage Cycles
Phage DNA injected into a host bacterium may be incorporated into the host's chromosome, becoming a noninfective prophage (lysogenic cycle), or it may remain free (lytic cycle). The lytic cycle, which produces new phages and lyses the host cell, may be repeated until all host cells are lysed. In the lysogenic cycle, the prophage is replicated as part of the host's chromosome; but if a prophage separates from the host's chromosome, it becomes a lytic phage.

ium, injecting the piece of DNA from its former host. A bacterium thus infected with a piece of foreign bacterial DNA does *not* become lysogenic, nor does it form new phages and burst as in the lytic cycle—it has not really been infected by a phage. It simply contains extrachromosomal bacterial DNA, as if it had conjugated with an Hfr cell. If crossing over takes place between the host chromosome and the transduced DNA, the transfer of markers is completed. In contrast with restricted transduction, general transduction can move any part of the bacterial chromosome. There is no limitation on what chromosomal markers might become enclosed in a phage coat. The phage coat is big enough to house several adjacent bacterial genes. General transduction therefore is another powerful tool for mapping the bacterial chromosome—with viral assistance.

EPISOMES AND PLASMIDS

The F factor and viral prophages are examples of **episomes**. Episomes are nonessential genetic elements that can exist in either of two states: independently replicating within a cell, or integrated into the main chromosome. Episomes cannot arise by mutation; they must be obtained by infection from outside the bacterium. The infection can come from a virus or from another bacterium. As we have seen, episomes may be used as vehicles for transferring genetic markers from one bacterium to another. Other nonessential genetic elements, which exist only as free, independently replicating circles of DNA that cannot be incorporated into the bacterial chromosome, are called **plasmids**. (An episome is simply a plasmid that has the possibility of becoming part of the chromosome.)

(a) **Restricted transduction**

Prophage

Separate prophage with bacterial marker

A^+

C^+ B^+

A^+ B^+ C^+

Viral replication and lysis

A^+ A^+

A^+

A^+ A^+

Prophage plus A^+ marker

A^+

C^+ B^+ A^-

Infection of bacteria

Transducing phage

A^- cell becomes A^+ with incorporation of marker carried by transducing phage

(b) **General transduction**

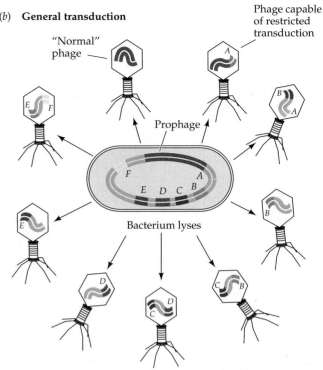

"Normal" phage

Phage capable of restricted transduction

A

E F

B A

Prophage

F A

E D C B

Bacterium lyses

E

B

D C

D C

C B

12.14 Transduction

Phages may transduce bacterial DNA from cell to cell. *(a)* In this example of restricted transduction, an A^- bacterial cell—one that is not producing the gene product associated with marker A—becomes A^+ when a transducing phage introduces the marker into the chromosome of the recipient bacterium. *(b)* In general transduction, parts of the bacterial chromosome are incorporated into phage coats without being accompanied by phage DNA. Each of these bits of bacterial DNA may be injected into a new bacterial cell and may become part of the bacterium's DNA by recombination.

Resistance factors, or **R factors**, are important plasmids. R factors first came to the attention of biologists in 1957 during a dysentery epidemic in Japan, when it was discovered that some strains of the dysentery bacterium *Shigella* were resistant to several antibiotics. Researchers found that resistance to the entire spectrum of antibiotics could be transferred by conjugation even when no markers on the main chromosome were transferred. Also, F⁻ cells could serve as donors, indicating that the genes for antibiotic resistance were not carried by the F factor. Eventually it was shown that the genes were carried on plasmids. Each of these plasmids (the R factors) carries one or more genes conferring resistance to particular antibiotics. As far as biologists can determine, R factors appeared long before antibiotics were discovered and used, but they seem to have become more abundant in modern times. Can you propose a hypothesis to explain why R factors might be more widespread now than they were in the past?

Bacteria do not require plasmids to live. Thus in order for a particular type of plasmid to be maintained within a population of bacteria, it must have an origin of replication (see Chapter 11). That is, the plasmid must be a **replicon**, capable of independent replication, so that it divides at roughly the same rate as the bacterium. Otherwise, it is simply diluted out of the population.

TRANSPOSABLE ELEMENTS

As we have seen, plasmids, episomes, and even phage coats can transport genes from one bacterial cell to another. Another type of "gene transport" within the individual cell relies on segments of chromosomal or plasmid DNA called **transposable elements**. Copies of transposable elements can be inserted at other points in the same or other DNA molecules, often producing multiple physiological effects resulting from the disruption of the genes into which the transposable elements are inserted (Figure 12.15*a*).

The first transposable elements to be discovered in prokaryotes were large pieces of DNA, typically 1,000 to 2,000 base pairs long, found in many places in the *E. coli* chromosome. The sequence of a transposable element can replicate independently of the rest of the chromosome. The copy then inserts itself at other, seemingly random places in the chromosome. The genes encoding the enzymes necessary for this insertion are found within the transposable

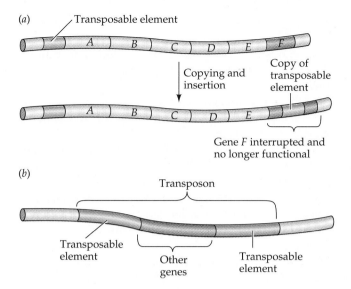

(a) Transposable element

Copying and insertion

Copy of transposable element

Gene *F* interrupted and no longer functional

(b)

Transposon

Transposable element

Other genes

Transposable element

12.15   Transposable Elements
(a) If a transposable element appears in the middle of another gene, that gene can no longer be transcribed to yield an appropriate mRNA; thus the interrupted gene cannot function. *(b)* A transposon consists of two transposable elements flanking another gene or genes; the entire transposon is copied and inserted as a unit.

element itself. Many transposable elements discovered later were longer (about 5,000 base pairs) and carried one or more additional genes. These longer elements with additional genes are called **transposons** (Figure 12.15*b*).

Initially, transposable elements were called jumping genes, but the frequency of their insertion is usually very low. Transposition is closely regulated—random insertion would often lead to the inactivation of an essential gene and the death of the cell. The process by which transposable elements move is complex and incompletely understood. In one step, the DNA at the new site is cut by an enzyme, transposase, that is encoded by the transposable element. The enzyme makes a staggered cut, and the copy of the transposable element inserts itself between the ends. When the gaps are repaired by DNA polymerase, a short sequence of chromosomal DNA at one end of the inserted sequence appears again, in duplicate, at the other end; the duplication results from the original staggered cut.

Transposable elements have contributed to the evolution of plasmids. The plasmids called R factors originally gained their genes for antibiotic resistance through the activity of transposable elements; one piece of evidence for this conclusion is that each resistance gene in an R factor is part of a transposon. Transposons on the F factor and on the bacterial chromosome interact to direct the insertion of the F factor into the chromosome in the development of an Hfr male.

CONTROL OF TRANSCRIPTION IN PROKARYOTES

Let us now consider how the activities of prokaryotic genes are *regulated*. As a normal inhabitant of the human gut, *Escherichia coli* has to adjust to sudden changes in its chemical environment. Its host may present it with one foodstuff one hour and another the next. For example, the bacteria may suddenly be deluged with milk, the main carbohydrate of which is lactose. This sugar is a β-galactoside—a disaccharide containing galactose β-linked to glucose (see Chapter 3). Before lactose can be of any use to the bacteria, it must first be taken into their cells by an enzyme called β-galactoside permease. Then it must be hydrolyzed to glucose and galactose by another enzyme, β-galactosidase. A third enzyme, thiogalactoside transacetylase, is also required for lactose metabolism. When *E. coli* is grown in a medium that does not contain lactose or other β-galactosides, the levels of all three of these enzymes within the bacterial cell are extremely low—the cell does not waste energy and material making the unneeded enzymes. If, however, the environment changes so that lactose is the predominant sugar and very little glucose is present, the synthesis of all three of these enzymes begins promptly and their levels may rise more than a thousandfold. Regulation of enzyme synthesis by the genes that code for them thus promotes efficiency in the cell.

Compounds that evoke the synthesis of an enzyme (as does lactose in this example) are called **inducers**. The enzymes that are evoked are called **inducible enzymes**, whereas enzymes that are made all the time at a constant rate are called **constitutive enzymes**. If lactose is removed from *E. coli*'s environment, synthesis of the three enzymes stops almost immediately. The enzyme molecules that have already been formed do not disappear; they are merely diluted during subsequent growth and reproduction until their concentration falls to the original low level within each bacterium.

The blueprints for the synthesis of these three enzymes are called **structural genes**, indicating that they specify the primary structure (that is, the amino acid sequence) of a protein molecule. When Jacob, Wollman, and Monod mapped the particular structural loci coding for enzymes that metabolize lactose, they discovered that all three lie close together in a region that covers only about 1 percent of the *E. coli* chromosome.

It is no coincidence that these three genes lie next to one another. The information from them is transcribed into a single, continuous molecule of mRNA. A molecule that contains transcripts of more than one gene is called a **polycistronic messenger**. Because this particular polycistronic messenger governs the synthesis of all three lactose-metabolizing enzymes, either all or none of the enzymes are made, depending on whether their common message—their mRNA—is present in the cell.

Processing a Polycistronic Messenger

How can a single mRNA molecule make three different polypeptides? The answer is that the polycistronic mRNA contains punctuation marks to specify the end of one polypeptide chain and the start of the next. A molecule of tRNA is always attached to a growing polypeptide chain, but a finished molecule of protein does not contain any tRNA. This indicates that the last step in prokaryotic protein synthesis must involve not only the termination of the polypeptide chain but also the removal of the terminal tRNA. The termination signal is encoded in the mRNA. Three codons of the genetic code (UAA, UAG, and UGA) mean "terminate translation," and one of them must be present at the end of each structural gene.

It is easy to see how a polycistronic messenger can give rise to one polypeptide for each structural gene. A ribosome begins at one end of the message, translates until it comes to the termination signal of the first structural gene transcript, and then releases the first polypeptide. The ribosome itself may remain bound and start translating the second structural gene at the next initiation site and, when it finishes, release the second polypeptide, and so on. When the ribosome has translated all the structural gene transcripts on the mRNA, it is released (Figure 12.16).

Promoters

Some genes are transcribed more often than others. In Chapter 11 we said that RNA polymerase attaches to DNA and starts transcribing, but we did not mention where it attaches. The polymerase does not attach itself randomly; special regions for attachment, called **promoters**, are built into the DNA molecule. There is one promoter for each structural gene or set of structural genes to be transcribed into mRNA. Promoters serve as a punctuation, telling the RNA polymerase where to start and which strand of DNA to read. A promoter and one or more structural genes are enough to specify the synthesis of an mRNA molecule.

Not all promoters are identical. One promoter may bind RNA polymerase very effectively and therefore trigger frequent transcription of its structural genes; in other words, it competes effectively for the available RNA polymerase. Another promoter may bind the polymerase poorly, and its structural genes are rarely transcribed. The efficiency of the promoter sets a limit on how often each structural gene can be transcribed. An enzyme that is needed in large amounts is encoded by a structural gene whose promoter is efficient, but the synthesis of an enzyme that is needed only in tiny amounts is controlled by an inefficient promoter.

What about enzymes, such as those that metabolize lactose, that bacteria need in large amounts at some times but not at all at others? The genes coding for these enzymes must contain a very efficient promoter so that the maximum rate of mRNA synthesis is high, but there must also be a way to stop mRNA synthesis when the enzymes are not needed.

Operons

Prokaryotes have evolved a mechanism that meets these two needs: they place an obstacle between the

12.16 Translation of Polycistronic mRNA
A polycistronic mRNA codes for several polypeptides. Each structural gene along the mRNA codes for a different polypeptide, and these genes are "punctuated" by termination signals that signal the ribosome to release the completed polypeptide. The ribosome is released when it has translated all the structural gene transcripts.

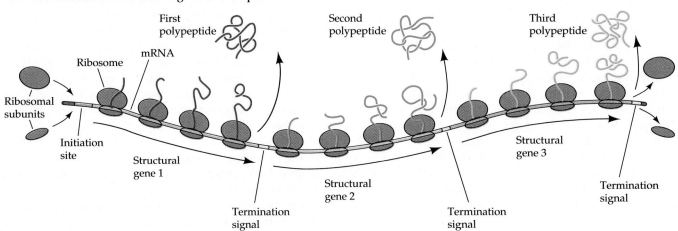

promoter and its structural genes. A short stretch of DNA called the **operator** can bind a special type of protein molecule, the **repressor**, creating the obstacle. When the repressor (the protein) is bound to the operator (the DNA), it blocks the transcription of mRNA (Figure 12.17). When the repressor is not attached to the operator, messenger synthesis proceeds rapidly. The whole unit, consisting of closely linked structural genes and the stretches of DNA that control their transcription, is called an **operon**. An operon always consists of two binding sites on the DNA molecule—a promoter and an operator—and one or more structural genes.

E. coli has three different ways of controlling transcription of operons. Two depend on interactions with the operator, and the third depends on interaction with the promoter. The two operator controls differ: one induces transcription, and the other represses it. We will now consider each of the three control systems in turn.

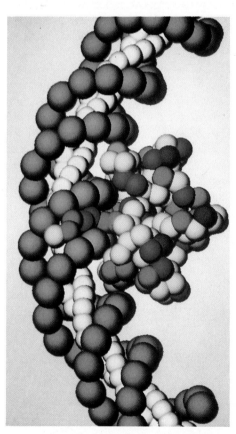

12.17 Repressor Bound to Operator Blocks Transcription
The yellow and orange spheres denote parts of the DNA molecule, which contains the operator. A portion of the repressor has already bound to the minor groove of the operator region of the DNA, and the remaining portion is about to bind to the major groove. The amino acids of the repressor are colored according to the following conventions: pale blue, hydrophobic; green, hydrophilic; red, positive charge; blue, negative charge.

Operator–Repressor Control That Induces Transcription: The *lac* Operon

The operon that controls and contains the genes for the three lactose-metabolizing enzymes is called the *lac* operon. As just explained, RNA polymerase binds the promoter, and the repressor binds to the operator. How is the operon controlled? The key lies in the repressor and its binding to the operator. The repressor is able to bind not only to its specific operator but also to inducers. Inducers of the *lac* operon are molecules of lactose and certain other β-galactosides. Binding of the inducer changes the shape of the repressor (by allosteric modification; see Chapter 6), and the change in shape makes the repressor fall off the operator.

For example, when lactose (an inducer) is added to a culture of *E. coli*, it enters the cell and promptly combines with the *lac* operon's repressor, changing the repressor's shape and causing it to detach from the operator. RNA polymerase can then bind to the promoter and start transcribing the structural genes of the *lac* operon. The mRNA transcribed from these genes is translated by ribosomes, which synthesize the enzyme products of the operon.

What happens if the concentration of lactose drops? As lactose concentration decreases, the inducer (lactose) molecules separate from the repressor, the repressor quickly becomes bound to the operator, and transcription of the *lac* operon stops. Translation stops soon thereafter, because the mRNA that is already present quickly degrades. The inducer, which is the target of the enzyme products of the operon, regulates the binding of the repressor to the operator.

Repressor proteins are coded for by **regulatory genes**—genes that control the activity of structural genes. The regulatory gene that codes for the repressor of the *lac* operon is called the *i* gene (for "inducibility"). The *i* gene happens to lie close to the operon that it controls, but many other regulatory genes are distant from their operons. Like all genes, the *i* gene itself has a promoter, which can be designated p_i. This promoter is very inefficient, allowing the production of just enough mRNA to synthesize about ten molecules of repressor per cell per generation. There is no operator between p_i and the *i* gene. Therefore, the repressor of the *lac* operon is constitutive, that is, it is made at a constant rate not subject to environmental control. Figure 12.18 shows the sequence of the regulatory gene and the *lac* operon, and Figure 12.19 outlines how the *lac* operon is regulated.

Let us review the important features of inducible systems such as the *lac* operon. The unregulated condition of the *lac* operon is one of being turned *on*. Control is exerted by a regulatory protein—the repressor—that turns the operon *off*. Some genes, such

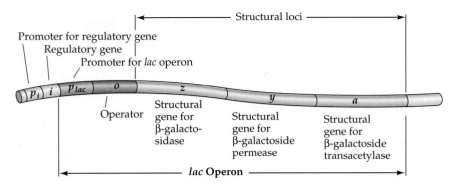

12.18   The *lac* Operon of *E. coli* and Its Regulator
An operon is a segment of DNA that includes structural genes along with sequences that regulate the transcription of those genes. The *lac* operon includes a promoter, an operator, and three structural genes. The regulatory gene (*i*) encodes a repressor protein that controls the operon.

as *i*, produce proteins whose sole function is to regulate the expression of other genes, and certain other DNA sequences (namely, operators and promoters) do not code for any proteins. Promoters are not even transcribed.

Operator–Repressor Control That Represses Transcription: The Tryptophan Operon

We saw that *E. coli* benefits from having an inducible system for lactose metabolism. Only when lactose is present does the system switch on. Equally valuable to a bacterium is the ability to switch off the synthesis of certain enzymes in response to something in the environment. For example, if the amino acid tryptophan, an essential nutrient, is present in ample concentration, it is advantageous to stop making the enzymes for tryptophan synthesis. When the formation of an enzyme is turned off in response to such a biochemical cue, the enzyme is said to be **repressible**.

Monod realized that repressible systems, such as the one for tryptophan synthesis, could work by mechanisms similar to those of inducible systems, such as the *lac* operon. In repressible systems, the repressor cannot shut off its operon unless it first unites with a **corepressor**, which may be either the nutrient itself (tryptophan in this case) or an analog of it. If the nutrient is absent, the operon is transcribed at a maximum rate. If the nutrient is present, the operon is turned off (Figure 12.20).

The difference between inducible and repressible systems is small but significant. In inducible systems, a substance in the environment (the inducer) interacts with the regulatory-gene product (the repressor), rendering it *incapable* of binding to the operator and thus incapable of blocking transcription. In repressible systems, a substance in the environment (the

corepressor) interacts with the regulatory-gene product to make it *capable* of binding to the operator and blocking transcription. Although the effects of the substances are exactly opposite, the systems as a whole are strikingly similar.

In both the inducible lactose system and the repressible tryptophan system, the regulatory molecule functions by binding the operator. Let us next consider an example of control by binding the *promoter*.

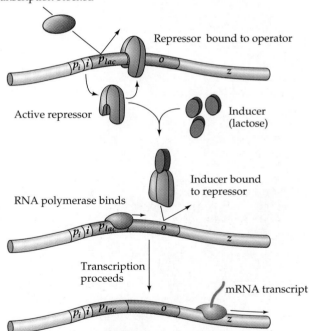

12.19   The *lac* Operon: Transcription Induced by Removal of Repressor
In an *E. coli* cell growing in the absence of lactose, the repressor protein coded for by gene *i* prevents transcription by binding to the operator. Lactose induces transcription by binding to the repressor, which cannot then bind to the operator. As long as the operator remains free of repressor, RNA polymerase that recognizes the promoter can transcribe the operon. Refer to Figure 12.18 for an explanation of the colors in the gene sequence.

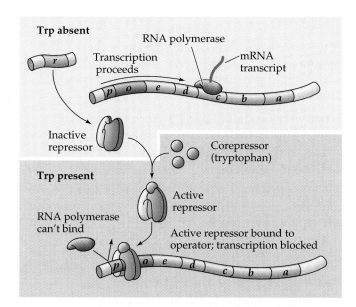

Trp absent

Transcription proceeds — RNA polymerase — mRNA transcript

Inactive repressor

Corepressor (tryptophan)

Trp present

Active repressor

RNA polymerase can't bind

Active repressor bound to operator; transcription blocked

12.20 The Tryptophan Operon: Transcription Repressed by Binding of Repressor

In an *E. coli* cell growing in the absence of tryptophan (Trp), regulatory gene *r* produces an inactive repressor that does not bind to the operator of the tryptophan operon. RNA polymerase can thus transcribe the operon's structural genes into mRNAs that are translated into enzymes of the tryptophan pathway. When tryptophan is present, it converts the inactive repressor into an active one, which does bind to the operator, thus blocking RNA polymerase from transcribing the structural genes and preventing synthesis of the enzymes of the tryptophan pathway. Because tryptophan activates an otherwise inactive repressor, it is called a corepressor.

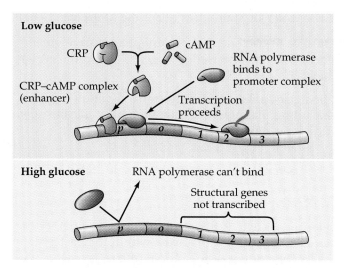

Low glucose

CRP — cAMP

CRP–cAMP complex (enhancer)

RNA polymerase binds to promoter complex

Transcription proceeds

High glucose

RNA polymerase can't bind

Structural genes not transcribed

12.21 Transcription Enhanced at the Promoter Site

This operon's structural genes encode enzymes that break down a food source other than glucose. When supplies of glucose are low, a receptor protein (CRP) and cAMP form a complex that binds to the promoter and activates it, allowing transcription of structural genes that encode enzymes for catabolizing the alternative energy source. A cell that contains ample glucose and does not require energy from other sources contains little cAMP and little CRP–cAMP; in such a cell, the structural genes are not transcribed and the catabolic enzymes are not formed.

Control by Increasing Promoter Efficiency

A bacterial cell has the means to increase the transcription of certain relevant genes when it needs a new energy source. *E. coli* prefers to get its energy from glucose in the environment. When glucose is unavailable, *E. coli* must get energy from another source, such as lactose or certain other sugars or even amino acids. The alternative energy source must be catabolized (degraded) by reactions requiring catabolic enzymes. There are regulatory molecules that enhance the transcription of operons containing the genes for these enzymes. The mechanism of this effect is entirely different from the two operator–repressor mechanisms just discussed, which turn the operator on or off. This third type of mechanism makes the promoter function more efficiently.

Suppose that a bacterial cell lacks a glucose supply but has access to another food that can be catabolized to yield energy. In operons containing genes for catabolic enzymes, the promoters bind RNA polymerase in a series of steps (Figure 12.21). First, a special protein (abbreviated **CRP**, for *cAMP receptor protein*) binds the low-molecular-weight compound adeno-sine 3′,5′-cyclic monophosphate, better known as cAMP. Next, the CRP–cAMP complex binds close to the binding site of the RNA polymerase and enhances the binding of the polymerase 50-fold. The promoters of the *lac* operon and many other genes responsible for sugar metabolism are activated in this way.

When glucose becomes abundant in the medium, breakdown of the alternative food molecules is not needed, so the cell diminishes or abolishes the synthesis of the corresponding catabolic enzymes. Glucose lowers the concentration of cAMP by a mechanism that is not yet understood but is sometimes called catabolite repression.

Preventing versus Enhancing Transcription

The inducible and repressible systems—the two operator–repressor systems—are examples of *negative* control of transcription because the regulatory molecule (the repressor) in each case *prevents* transcription. The promoter system is an example of *positive* control of transcription because the regulatory molecule (the CRP–cAMP complex) *enhances* transcription.

As we will see in Chapter 13, the regulation of gene expression in eukaryotes is more intricate than in prokaryotes. Transcription in eukaryotes is subject primarily to positive control.

SUMMARY of Main Ideas about the Molecular Genetics of Prokaryotes

Studies of bacteria and bacteriophages have greatly increased our knowledge of genetics and molecular biology.

Biologists isolate mutant bacteria in various ways, including plating on selective media, penicillin suicide, and replica plating.
Review Figure 12.5

The bacterial chromosome is a single, circular DNA molecule.
Review Figure 12.8

Plasmids, such as R factors, are nonessential, circular DNA molecules that exist within the bacterial cell independent of the chromosome.

Episomes, such as the F factor and prophages, are plasmids that can also be incorporated into the chromosome.

Genetic recombination in bacteria may follow conjugation, transduction, or transformation.

In conjugation, DNA is transferred from a male bacterium to a female bacterium while the two cells are physically attached.
Review Figure 12.6

Some strains of male bacteria donate DNA much more frequently than others.
Review Figure 12.9

In transduction, DNA is transferred from one bacterium to another by a bacteriophage.
Review Figure 12.14

In transformation, living cells take up DNA from their environment.

All known bacteriophages have lytic life cycles; some also have lysogenic cycles.

During a lysogenic cycle, the phage DNA is maintained as a prophage incorporated into the bacterial chromosome.
Review Figure 12.13

A lytic cycle may follow a lysogenic cycle, and the prophage may carry bacterial genes when it separates from the chromosome, leading to restricted transduction.

When genetically different bacteriophages simultaneously infect the same bacterial cell, recombination of the phage genetic material may take place.
Review Figure 12.12

Transposable elements are segments of DNA that can cause copies of themselves to be inserted elsewhere in the chromosome.
Review Figure 12.15

Transposable elements can inactivate genes by inserting into them.

Structural genes are expressed by being transcribed into an mRNA and translated into a protein; regulatory genes control the expression of other genes.

Much of the genetic material of bacteria is organized into operons, each consisting of one or more structural genes, an operator, and a promoter.
Review Figure 12.18

Interactions with the operator control some operons.

Transcription of structural genes is prevented when the operator is bound by a repressor.

Operator–repressor interactions underlie repressible systems, such as the tryptophan operon, as well as inducible ones, such as the lactose operon.
Review Figures 12.19 and 12.20

Interactions with the promoter underlie control of some catabolic enzymes.

Binding of the CRP–cAMP complex to the promoter greatly enhances binding of RNA polymerase. Glucose availability regulates the concentration of cAMP.
Review Figure 12.21

SELF-QUIZ

1. In bacterial conjugation
   a. each cell donates DNA to the other.
   b. a bacteriophage carries DNA between bacterial cells.
   c. one partner possesses a fertility factor.
   d. the two parent bacteria merge like sperm and egg.
   e. all the progeny are recombinant.

2. Which statement is *not* true of the bacterial fertility factor?
   a. It is a plasmid.
   b. It confers "maleness" on the cell in which it resides.
   c. It can be transferred to a female cell, making it male.
   d. It has thin projections called F-pili.
   e. It can become part of the bacterial chromosome.

3. Hfr mutants
   a. are female bacteria that are highly efficient recipients of genes.
   b. rarely transfer all the markers on the chromosome.
   c. keep their F factor separate from the chromosome at all times.
   d. are unable to conjugate with other bacteria.
   e. transfer markers in random order.

4. Lysogenic bacteria
   a. lack a prophage.
   b. are accompanied by free phages when growing in culture.
   c. lyse immediately.
   d. cannot release their phages.
   e. are susceptible to further attack by the same strain of phage.

5. Which statement is *not* true of transduction?
   a. The viral DNA is an episome.
   b. Transduction is a useful tool for mapping a bacterial chromosome.
   c. In restricted transduction, the newly infected cell becomes lysogenic.
   d. Transduction results in genetic recombination.
   e. To carry bacterial markers, the viral coat must contain viral DNA.

6. Plasmids
   a. are circular protein molecules.
   b. are required by bacteria.
   c. are tiny bacteria.
   d. may confer resistance to antibiotics.
   e. are a form of transposable element.

7. Which statement is *not* true of a transposable element?
   a. It can be copied to another DNA molecule.
   b. It can be copied to the same DNA molecule.
   c. It is typically 100 to 500 base pairs long.
   d. It may be part of a plasmid.
   e. It encodes the enzyme transposase.

8. In an inducible operon
   a. an outside agent switches on enzyme synthesis.
   b. a corepressor unites with the repressor.
   c. an inducer affects the rate at which repressor is made.
   d. the regulatory gene lacks a promoter.
   e. the control mechanism is positive.

9. The promoter
   a. is the region that binds the repressor.
   b. is the region that binds RNA polymerase.
   c. is the gene that codes for the repressor.
   d. is a structural gene.
   e. is an operon.

10. The CRP–cAMP system
    a. produces many catabolites.
    b. requires ribosomes.
    c. operates by an operator–repressor mechanism.
    d. is a form of positive control of transcription.
    e. relies on operators.

FOR STUDY

1. Viruses sometimes carry DNA from one cell to another by transduction. Sometimes a segment of bacterial DNA is incorporated into a phage protein coat without any phage DNA. These particles can infect a new host. Would the new host become lysogenic if the phage originally came from a lysogenic host? Why or why not?

2. For studies of metabolism in a particular species of bacteria, you need to isolate the following mutant strains: a histidine auxotroph (a strain, *his⁻*, that cannot synthesize the amino acid histidine) and a tryptophan auxotroph (*trp⁻*, unable to synthesize the amino acid tryptophan). After irradiating a culture of the bacteria with ultraviolet light to increase the mutation rate, you expect to find some *his⁻* and *trp⁻* auxotrophs in the culture. Describe all the steps you would take in order to increase the percentages of *his⁻* and *trp⁻* auxotrophs, using the "penicillin suicide" technique.

3. You are provided with two strains of *Escherichia coli*. One, an Hfr strain that is sensitive to streptomycin, carries the markers A^+, B^+, and C^+. The other is an F^- strain that is resistant to streptomycin and carries the markers A^-, B^-, and C^-. You mix the two cultures. After 20, 30, and 40 minutes you take samples of the mixed culture and swirl them vigorously in a blender. Next you add streptomycin to the swirled cultures. You examine surviving bacteria by replica plating. Some of the bacteria from the 20-minute sample are B^+; in the 30-minute sample there are both B^+ and C^+ bacteria; but A^+ bacteria are found only in the 40-minute sample. What can you say about the arrangement of the A, B, and C loci on the bacterial chromosome? Explain your answer fully.

4. You have isolated three strains of *E. coli*, which you name I, II, and III. You attempt to cross these strains, and you find that recombinant progeny are obtained when I and II are mixed or when II and III are mixed, but not when I and III are mixed. By diluting a suspension of II and plating it out on solid medium, you isolate a number of separate clones. You find that almost all these clones can conjugate with strain I to produce recombinant offspring. One of the clones derived from strain II, however, lacks the ability to conjugate with strain I. Characterize strains I, II, and III and the nonconjugating clone of strain II in terms of the fertility (F) factor.

5. In the lactose operon of *E. coli*, repressor molecules are coded for by the regulatory gene. The repressor molecules are made in very small quantities and at a constant rate per cell. Would you surmise that the promoter for these repressor molecules is efficient or inefficient? Is synthesis of the repressor constitutive, or is it under environmental control?

6. A key characteristic of a repressible enzyme system is that the repressor molecule must react with a corepressor (typically, the end product of a pathway) before it can combine with the operator of an operon to shut the operon off. How is this different from an inducible enzyme system?

READINGS

Darnell, J. E., Jr. 1985. "RNA." *Scientific American*, October. A discussion of aspects of transcription and of gene regulation in prokaryotes and eukaryotes.

Griffiths, A. J. F., J. H. Miller, D. T. Suzuki, R. C. Lewontin and W. M. Gelbart. 1993. *An Introduction to Genetic Analysis*, 5th Edition. W. H. Freeman, New York. An up-to-date revision of one of the field's classic textbooks.

Judson, H. F. 1979. *The Eighth Day of Creation*. Simon and Schuster, New York. A constantly fascinating history of molecular biology, with much attention to the regulation of gene expression. For a lay audience.

Nomura, M. 1984. "The Control of Ribosome Synthesis." *Scientific American*, February. A discussion of how ribosomes are assembled and the roles of operons in regulating ribosome production in bacteria.

Stent, G. S. and R. Calendar. 1978. *Molecular Genetics*, 2nd Edition. W. H. Freeman, New York. Technical and historical information charmingly presented.

Your life depends on the continual delivery of oxygen to your living tissues. This job is accomplished by your red blood cells with their cargo of hemoglobin molecules. Each red blood cell functions for about four months and then is destroyed, to be replaced by a new one. The replacement rate to keep you going is 100 billion red blood cells per day! DNA deep within your bones directs these specialized cells to form and mature, by which time they are quite unlike other cells. Red blood cells are simple, membranous bags that lack compartments and are full of hemoglobin. They are simple, but each one must be exactly right. The plasma membrane must have the right proteins, in the right proportions, to govern the movement of oxygen and other substances into and out of the cell.

The DNA deep within your bones must direct not only the perfect assembling of each red blood cell, but also the production of hundreds of thousands of perfect hemoglobin molecules for each one of today's hundred billion new red blood cells. Each hemoglobin molecule requires four perfect polypeptides, coded for by two different genes, plus many enzymes, coded for by other genes, to do the assembling.

Why are these genes expressed only in the special places where their products are needed? What keeps other genes in the DNA deep within your bones switched off, so that their products never appear there? In this chapter you will learn about the structure of the eukaryotic gene and see how that structure permits the control of gene expression.

EUKARYOTES AND EUKARYOTIC CELLS

Most eukaryotic cells are much larger and more internally complex than prokaryotic cells. In particular, eukaryotic cells typically contain an array of membrane-bounded organelles specialized for various functions. Eukaryotic cells contain much more DNA than do prokaryotic cells. For example, most mammals have about 1,000 times as much DNA per cell as does the bacterium *Escherichia coli*. DNA is packaged differently in eukaryotic cells than it is in prokaryotic cells: The eukaryotic chromosome is organized into nucleosomes (see Chapter 9). For these and other reasons, gene expression and the patterns of inheritance in eukaryotes differ from the same phenomena in prokaryotes.

Eukaryotes evolved as unicellular organisms, and only unicellular eukaryotes existed for a very long time before multicellularity evolved. Single-celled eukaryotes had some real advantages in a world full of prokaryotes. Their internal organelles allowed them to gain efficiency by separating various activities into special compartments. The eukaryotes were the first

DNA in Your Body Directs Production of a Hundred Billion New Red Blood Cells—Every Day

13
Molecular Genetics of Eukaryotes

organisms with sex as we usually understand it—that is, with equal samples from the genomes of two individuals making up the genome of the offspring, and with sexual processes typically taking place in the life of each individual. In prokaryotes, sexual processes occur very rarely, and the "male" usually contributes only a small fraction of its genome to the offspring. Sexual reproduction afforded eukaryotes a greater ability to produce offspring that could be successful in heterogeneous (mixed) and changing environments.

Many eukaryotes today are multicellular. There is usually a division of labor among the cells; therefore a multicellular organism has various types of cells that contain different proteins and are capable of performing different specialized functions. The human body has at least 200 different cell types, differing in a few major proteins and many minor proteins. During the development of the multicellular body, different genes are expressed at different times, or in different specific tissues (see Chapter 17). Such differential gene expression is important even in the development of the tiny but complex bodies of unicellular eukaryotes. How is gene expression managed? Or, in other words, how are eukaryotic genes turned on and off?

Before we can address this question, we must examine the structure of the eukaryotic gene itself, as well as the complex series of steps from gene to protein product in eukaryotes. Eukaryotic genes tend to differ from those of most prokaryotes in that they have stretches of DNA that are not expressed in polypeptide products. That is, most eukaryotic genes are "split" by the presence of noncoding DNA in the midst of the coding sequences. As we shall see, this characteristic requires processing at the molecular level that is not found in prokaryotes.

What is known about eukaryotic gene structure and expression has been revealed by application of the techniques of molecular biology. Because understanding one of these techniques in particular—nucleic acid hybridization—makes it easy to learn about some of the important discoveries, we will begin by considering this technique.

HYBRIDIZATION OF NUCLEIC ACIDS

Nucleic acid hybridization depends on the association, through complementary base pairing, of single-stranded nucleic acids. If we carefully heat a sample of DNA, the hydrogen bonds forming the base pairs are destroyed and the two strands of each double helix separate—we say that the DNA **denatures**. If we then lower the temperature slowly, the complementary strands join again, or **reanneal**, to form double-stranded DNA, with each base pair obeying the A–T, G–C pairing rules (Figure 13.1a,b). To make this procedure work efficiently, one must enzymatically or mechanically "cut" the DNA into short segments a few hundred bases long and carefully regulate the temperature and salt concentration in the test solution.

Suppose, now, that we denature a sample of DNA with heat and then combine it with a sample of RNA that has been transcribed from part of that DNA. Because the RNA is complementary to the one strand of DNA that coded for it (the template strand), it may anneal with that DNA strand to form a hybrid (Figure 13.1c). As the temperature is lowered, the RNA transcript and the other DNA strand compete to anneal with the template strand. The reannealing of the DNA strands can be prevented by immobilizing the denatured DNA on a nitrocellulose filter before it cools, which keeps the separated DNA strands from coming together, thereby favoring RNA binding. (The immobilized DNA is still accessible for hybridization with nucleic acids in solution.) The tendency for DNA strands to reanneal can also be reduced by outnumbering the DNA strands with RNA strands.

EUKARYOTIC GENE STRUCTURE

Now that we understand how nucleic acid hybrids are formed, we are ready to consider the eukaryotic gene and how it differs from the prokaryotic gene. The structure of the eukaryotic gene has been studied by comparing the DNA with its RNA transcripts. One way of making this comparison is as follows: We denature a sample of DNA and then add corresponding mRNA, such as would be found in the cytoplasm. From the resulting hybridization we obtain uniform, double-stranded DNA–mRNA hybrid structures associated with *single-stranded*, looped structures. The loops are displaced, noncoding DNA strands. This method revealed in 1977 that double-stranded hybrid regions are studded with both single- and double-stranded loops. What could this mean?

The mRNA is a faithful transcript of the information required for protein synthesis. After hybridization, all the mRNA is bound, through complementary base pairing, with the appropriate region of the single-stranded DNA. However, there is some DNA *in the middle of the gene* that is not represented in the mRNA. That is, some DNA sequences are not represented in the information that ends up in the mRNA that encodes the protein product. In fact, most (but not all) vertebrate genes as well as many other eukaryotic genes contain such intervening sequences, called **introns**: segments of DNA that do not encode any part of the polypeptide product of the gene. Later in this chapter we will learn that not all the transcribed RNA gets into the cytoplasm to be

(a) Upon being carefully heated, the two polynucleotide
strands of a DNA molecule denature (separate)

DNA

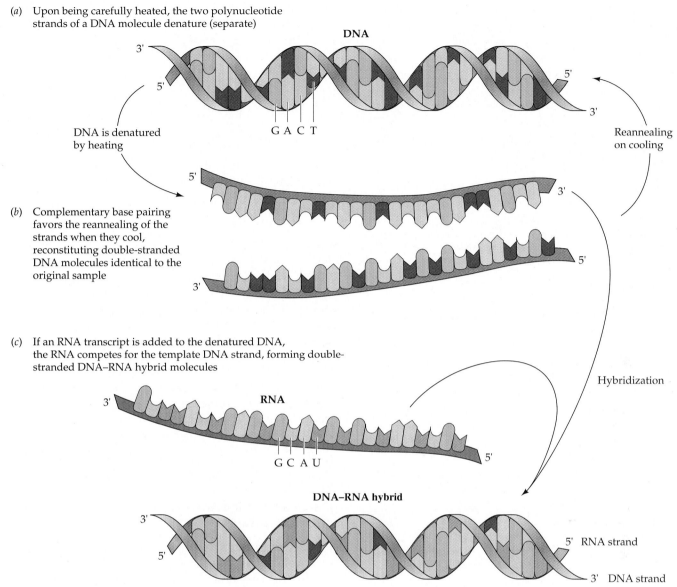

DNA is denatured
by heating

Reannealing
on cooling

(b) Complementary base pairing
favors the reannealing of the
strands when they cool,
reconstituting double-stranded
DNA molecules identical to the
original sample

(c) If an RNA transcript is added to the denatured DNA,
the RNA competes for the template DNA strand, forming double-
stranded DNA–RNA hybrid molecules

Hybridization

RNA

DNA–RNA hybrid

5' RNA strand

3' DNA strand

13.1 Nucleic Acid Hybridization
Denatured strands of DNA anneal with complementary
sequences of either DNA or RNA.

mRNA. The parts of the gene that *are* represented in
the mRNA product are called **exons** because they are
*ex*pressed regions of the gene (Figures 13.2 and 13.3).

Introns do not scramble the sequence that codes
for a polypeptide; they just reside in the middle of
it. The base sequence of the exons, taken in order, is
exactly complementary with that of the mature
mRNA product. The introns simply separate the coding sequence for the protein into parts. Exons and
introns are found in all groups of eukaryotes and
even in a few prokaryotes. We are still seeking to
understand the significance of introns to the organisms that possess them. Later in this chapter we will
describe the posttranscriptional events that remove
the transcripts of introns.

REPETITIVE DNA IN EUKARYOTES

What else can we learn by using nucleic acid hybridization? If we denature eukaryotic DNA and then let
the complementary strands reanneal, we can collect
data on how long it takes to renature. From these
data we observe that some parts of the genome anneal only very slowly, whereas other segments
quickly find partners. Why should one DNA sequence anneal quickly and another slowly? The answer is that there are multiple copies of some, but
not all, stretches of DNA. If a particular single strand
of DNA has, say, a few hundred complementary
segments with which it can anneal, it will be able to
find a partner much more rapidly than one for which
only a single acceptable partner exists.

(a)

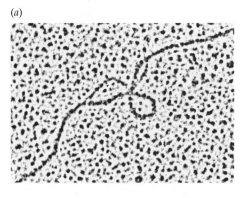

(b)

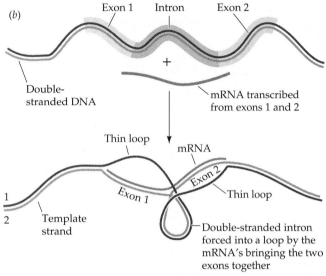

Exon 1   Intron   Exon 2

Double-
stranded DNA

mRNA transcribed
from exons 1 and 2

Thin loop

mRNA

1

2   Template
strand

Exon 1

Exon 2

Thin loop

Double-stranded intron
forced into a loop by the
mRNA's bringing the two
exons together

(c)   With no introns:

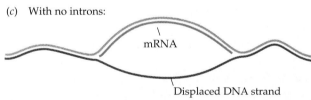

mRNA

Displaced DNA strand

13.2   Evidence for Extra DNA in the Eukaryotic Gene
Mouse DNA was partially denatured and mixed with
mRNA transcribed from one of the genes in the DNA.
(a) Examination of the resulting mixture by electron microscopy revealed thick nucleic acid, bearing thick loops
(here one points downward) and thinner loops. (b) A diagrammatic interpretation of (a). The mRNA hybridized
with the template DNA strand, forming double strands
and thin loops; these thin loops are the unpaired, complementary strand of DNA. The double-stranded DNA from
the intron reannealed because it had no counterpart in
the mRNA. The mRNA joined the two exons, forcing the
intron DNA into a thick loop. (c) Hybridization pattern
observed when no introns are present, as in prokaryotic
DNA.

Researchers applying this reannealing technique,
called liquid hybridization, discovered that there are
three classes of eukaryotic DNA. The class that reanneals the slowest consists of **single-copy sequences**—
genes that, like prokaryotic genes, have only one

copy in each **genome**, or haploid set of chromosomes.
Single-copy sequences code for most of the enzymes
and structural proteins in eukaryotes. Some single-
copy sequences form long spacers between successive genes.

The class of DNA sequences that reanneals the
fastest consists of **highly repetitive DNA**. This fraction varies widely from species to species and may
make up a third or more of the genome. Although
there are half a million copies per genome of some
of these segments, their function is not understood.
Much of the highly repetitive DNA is located near
the centromeres (see Chapter 9) and may help maintain the integrity of chromosomes during mitosis and
meiosis. The large number of identical DNA sequences at a centromere may be related to the attachment of multiple spindle fibers to each chromosome. DNA of this class is usually not transcribed.

The class of eukaryotic DNA that reanneals at rates
between the two extremes is **moderately repetitive
DNA**, which is present in a few hundred to 10,000
copies per genome. Some of these moderately repetitive genes may be important in the regulation of
development, and some are duplicate genes, such as
those for rRNA. The ends of eukaryotic chromosomes, called **telomeres**, consist of moderately repetitive DNA. Transposable elements are moderately
repetitive genes with special properties. The next two
sections discuss telomeres and transposable elements
in detail.

Disadvantages and Advantages
of Ends: Telomeres

Is it better for the DNA in a chromosome to be linear
or circular? This question cannot really be answered,
but the possession of ends is a real problem for chromosomes when it comes to DNA replication. The
problem is that the monomers of DNA cannot simply
string together from scratch; they can be added only
to the 3' end of a previously existing nucleic acid

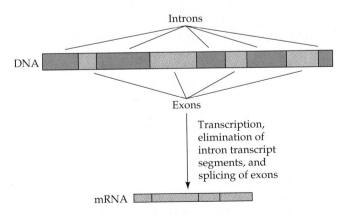

Introns

DNA

Exons

Transcription,
elimination of
intron transcript
segments, and
splicing of exons

mRNA

13.3   mRNA Is Encoded Only in Exons

strand. That is, they require a **primer**. In both prokaryotes and eukaryotes primers consist of RNA laid down using the other strand of the existing DNA as a template. At a later stage in DNA replication, a growing strand of DNA approaches the other (5′) end of the primer and eventually replaces the primer, so the entire new molecule consists of DNA, with no inserted stretches of RNA (see Box 11.A).

Recall from Chapter 11 that replication proceeds differently on the two strands of a DNA molecule. Both new strands form in the 5′-to-3′ direction, but one (the leading strand) grows continuously from one end to the other, while the other (the lagging strand) grows as a series of Okazaki fragments. With the circular prokaryotic chromosome, as both DNA strands extend continuously around the whole circular chromosome, production of a complete series of Okazaki fragments is not a problem—there is always some DNA at the 5′ end of a primer, ready to replace it.

Now think about replication at an end of a linear eukaryotic chromosome. The leading strand can grow without incident to the very end. But how does the last Okazaki fragment for the end of the lagging strand form? Replication must begin with an RNA primer at the 5′ end of the forming strand, but there is nothing beyond the primer in the 5′ direction to replace the RNA. Thus the new daughter chromosome in a eukaryote lacks a bit of double-stranded DNA, a bit of genetic information, at each end. The ends of the chromosome have been clipped off. This clipping problem threatens to shorten all eukaryotic chromosomes; why don't they all disappear over the course of many replications? The answer is that a repeated DNA sequence forms a telomere at each end of the chromosome. This sequence of moderately repetitive DNA can double back on itself and thus serve as the primer for completing the lagging strand as the chromosome is replicated.

This activity is essential for the preservation of eukaryotic chromosomes, as the requirements for making a viable artificial chromosome illustrate. We can make artificial chromosomes that, when inserted into yeast cells, behave like normal chromosomes—they are replicated, they are distributed normally in mitosis, and their genes are transcribed and translated. For this process to work, however, the artificial chromosome must contain three features: a centromere (to allow attachment to mitotic and meiotic spindles); an origin of replication (to allow replication); and a telomere at each end (to protect the chromosome from shortening). The rest of the artificial chromosome may consist of any DNA sequences we wish to put there.

The telomeres may serve another important function in the behavior of eukaryotic chromosomes. By attaching to the nuclear envelope, telomeres may help initiate the process of synapsis early in meiosis (see Chapter 9).

Transposable Elements in Eukaryotes

Eukaryotes have **transposable elements** that are moderately repetitive segments of DNA. Because these transposable elements insert themselves into different parts of chromosomes, they have been called "jumping genes" (see Chapter 12). The first evidence for transposable elements came from studies on maize, a eukaryote, conducted by Barbara McClintock at Cold Spring Harbor Laboratory.

The method by which transposable elements are copied in eukaryotes differs from that in prokaryotes. The copying mechanism in eukaryotes requires an RNA intermediate. Recall that in prokaryotes the DNA of the transposable elements can simply copy itself. The transposable elements of both prokaryotes and eukaryotes are always parts of chromosomes; unlike plasmids, they do not function as independent pieces of DNA.

Thus far, scientists have learned very little about the roles of transposable elements in the life of a cell. Although the protein products of some eukaryotic transposable elements have been identified, the cellular functions of these products have not been determined. Transposable elements may, in effect, be parasites that simply replicate themselves; on the other hand, they may play important roles in the survival of the organisms in whose chromosomes they reside.

Transposable elements can act as mutators—that is, by jumping into genes, they can eliminate or change the functions of those genes (see Chapter 12). Transposable elements can bring about deletions, insertions, transpositions, and inversions; they also may be a source of duplications, in which multiple copies of genes are created. Some genes may be inactivated by the insertion of transposable elements; other genes may be placed in new relative positions, affecting their transcription. For some genes, transposable element insertions constitute more than 99 percent of all mutations.

Recall from Chapter 4 the endosymbiotic theory of the origin of chloroplasts and mitochondria, which proposes that chloroplasts and mitochondria are the descendants of once free-living prokaryotes. Transposable elements seem to have played a part in this process. Chloroplasts and mitochondria possess DNA, and some parts of these organelles are encoded by genes on this extranuclear DNA. Other parts of the organelles are coded for by nuclear genes, a finding that might appear to weaken the endosymbiotic theory. It has recently been shown, however, that in the course of evolution genes of some organisms have been transposed to the nuclei from both chloroplasts and mitochondria. The insertion of transposable elements, and the subsequent loss of the originals of these genes from the chloroplasts and mitochondria, can therefore be used to counter this argument against the endosymbiotic theory. Thus, because of our relatively new knowledge about transposable elements, a finding once considered to be evidence *against* the endosymbiotic theory can now be used as evidence *for* the theory.

How did transposable elements arise? Their source is still unknown, but there are interesting hints at a relationship between retroviruses (tumor viruses that use reverse transcriptase to transcribe their RNA to DNA) and the transposable elements of eukaryotes, as we will see when we discuss cancer-causing genes in Chapter 15. It is possible that some transposable elements arose from retroviruses.

GENE DUPLICATION AND GENE FAMILIES

Some genes have just two or only a few copies. These copies arise, evolutionarily, in various ways. In the previous section we noted that transposable elements are a source of gene duplication. Another source is unequal crossing over (see Chapter 10), in which mispaired chromosomes cross over in such a way as to put both copies of a gene on the same chromosome. A set of duplicated genes is called a **gene family**. Members of a gene family may reside on different chromosomes, or they may be bunched tightly on a single chromosome.

Once more than one copy of a gene exists, the copies may evolve differently. One copy must retain the original function, or the organism may not survive. As long as one copy does this, however, the others may change slightly, extensively, or not at all.

An evolutionarily ancient gene family found in vertebrates codes for the globin proteins. Globins are required for the binding and transport of oxygen; some of them are components of hemoglobin, the oxygen-carrying pigment of red blood cells, and one is found in myoglobin, a related protein that binds

and stores oxygen within muscle fibers (see Chapter 41). Each molecule of hemoglobin is a tetramer. The four globin polypeptides that make up the tetramer in adult humans are two of one type (α) and two of another (β). The hemoglobin tetramers of human fetuses also contain two types of globin polypeptides, both of which differ from those found in adult hemoglobin. Still other α-like and β-like polypeptides are found in the earliest embryonic stages.

All these globin polypeptides are coded for by globin genes descended from a single ancestral globin gene. This fact has been ascertained from similarities in the amino acid sequences of the polypeptides and from the locations of introns within the genes. Like the members of other gene families, the globin genes differ from one another more in their introns than in their exons; the exons are probably conserved with little variation because their gene products perform essential functions, whereas the introns have no products.

The genes for human α-like globins lie in a tight cluster on one of our chromosomes, and the genes for β-like globins lie in a cluster on a different chromosome. In both clusters there are additional stretches of DNA, closely similar in base sequence to the globin genes, that are not expressed. Such apparently nonfunctional genes are called **pseudogenes** (Figure 13.4). How were pseudogenes discovered, if they have no function? Nucleic acid hybridization revealed the existence of pseudogenes because they hybridize to a significant extent with adult globin mRNA (as do all other members of the globin gene family).

Most pseudogenes are probably duplicate genes that changed so much during evolution that they no longer function. The changes may have inactivated promoters, caused nonsense mutations in exons, or eliminated the clipping out of an intron. Some pseudogenes, however, did not arise by gene duplication; they were derived by reverse transcription of the mRNA. Such "processed pseudogenes" lack introns and are found away from the rest of a gene cluster. Whatever functions the pseudogenes had in the past

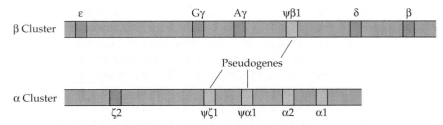

13.4 Gene Families May Include Pseudogenes
The human α-like and β-like globin gene families are organized into clusters. Each gene cluster includes both functional genes and pseudogenes; pseudogenes are prefixed by the Greek letter psi (ψ).

(when they were not pseudogenes) are now performed by other genes in the family. In many gene families, pseudogenes outnumber functional genes, often by several fold. As we mentioned already, different active members of the human globin gene family serve at different times in development. There are other gene families in which different active members function in different tissues or under different environmental conditions instead of at different times. The gene families for rRNA, tRNAs, and histones are examples in which duplication is used to meet the demand for large amounts of the gene product.

RNA PROCESSING IN EUKARYOTES

The RNA transcript of a eukaryotic gene contains both introns and exons. How do we get from this product to an mRNA, and from there to the protein encoded by the gene? The original product of transcription of a eukaryotic gene is a heterogeneous nuclear RNA, or **hnRNA**, so called because of the great range of sizes of RNAs of this class. As we are about to see, a great deal more must be done to the hnRNA to produce a mature mRNA that is ready to be translated.

Capping and Tailing RNA

An early step in the processing of an hnRNA molecule is the addition of a **cap**—a modified molecule of guanosine triphosphate (GTP)—at the 5′ end of the hnRNA. This modified G cap is retained during the processing of mRNA and facilitates the binding of the mRNA to a ribosome for translation (see Chapter 11); it may also help protect the mRNA from degradation.

At the other end of some hnRNA molecules, the 3′ end, a string of 100 to 200 adenine nucleotides, called a **poly A tail**, is added by a process called polyadenylation. (Poly A tails constitute 5 to 20 percent of the length of mature mRNA molecules produced from hnRNA.) Neither the modified G cap nor the poly A tail is coded for in the DNA; both are added as part of the early processing of hnRNA.

The poly A tail is thought to protect the modified hnRNA and the mRNA from degradation; some evidence suggests that the tail is needed for translation as well. Somewhat less than one-third of the hnRNA molecules in mammalian cells have poly A tails, and about 70 percent of the resulting mRNA molecules have the tails. We do not yet know why some RNAs get poly A tails and others do not. For those hnRNAs that do get poly A tails, however, the poly A is essential in order for the hnRNA to mature into mRNA.

Splicing RNA

The next step in the processing of eukaryotic RNA is the removal of the regions coded for by introns in the DNA. If these RNA regions, which are also called introns, were not removed, a nonfunctional mRNA or an improper protein would be produced. A process called **RNA splicing** removes introns and splices exons together. Figure 13.5 illustrates how splicing of an hnRNA transcript 7,700 nucleotides long results in a mature mRNA of only 1,872 nucleotides. Capping, tailing (polyadenylation), the removal of seven introns, and the splicing of eight exons do the job.

In the mRNA splicing reaction, a loop forms, extruding the intron and bringing the adjacent exons together. How do the exons become linked? At the boundaries between introns and exons there are **consensus sequences**—short stretches of DNA that appear, with little variation, in many different genes. Eukaryotic nuclei contain molecules of small nuclear RNA, or **snRNA**, that contain regions complementary to the consensus sequences. To accomplish the splicing, an snRNA combines with a set of proteins to produce a small nuclear ribonucleoprotein particle, or **snRNP**. One of the snRNPs recognizes and binds the consensus sequence at one end of the intron; a second, different snRNP recognizes and binds the consensus sequence at the other end of the intron; other snRNPs recognize and bind a sequence in the intron itself. Together, six different snRNPs constitute a "splicing machine" called a spliceosome. The spliceosome joins the exons and releases the introns (Figure 13.6).

Splicing mechanisms differ among RNA classes: mRNAs, rRNAs, and tRNAs are all spliced in different ways. Molecular biologists were startled to learn in 1981 that in the protist *Tetrahymena thermophila*, the RNA precursor of rRNA can catalyze the splicing of its own intron. That is, the RNA, in the absence of any protein, is catalytic. Another case of a catalytic RNA (not involving RNA splicing) has been discovered, this time in *E. coli*. The existence of RNAs with catalytic powers may help explain evolution at the dawn of life (see Chapter 18).

The Stability of mRNAs

In eukaryotes, RNA exits the nucleus through pores in the nuclear envelope (see Figure 4.10). The transport of RNA from the nucleus to the cytoplasm is mediated by carrier proteins, but the details of this mechanism are not yet known. Mature mRNA in the cytoplasm may be relatively stable, lasting for hours or days. In prokaryotes, however, an mRNA molecule usually lasts for only a few minutes following transcription—its life is so short that translation be-

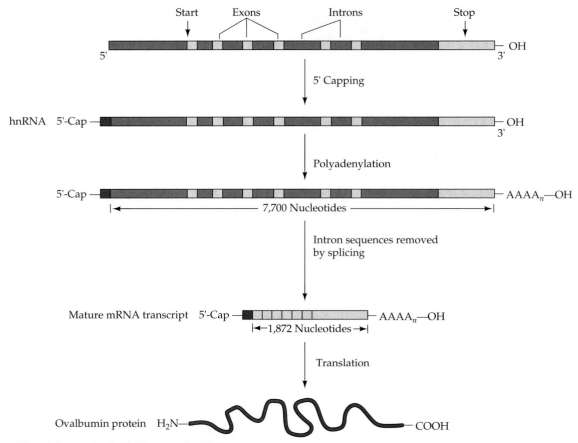

13.5 Eukaryotic Cells Process hnRNA
The heterogenous nuclear RNA (hnRNA) of the gene that codes for ovalbumin, the major protein in egg whites, is capped at the 5' end and has a poly A tail added at the 3' end. It is 7,700 nucleotides long and contains seven introns and eight exons. Splicing removes over three-quarters of the nucleotides and results in a mature mRNA that can be translated to yield ovalbumin.

gins before the mRNA is completely formed. As we will see, different eukaryotic RNAs may differ from one another in their stability. The stability of mRNAs plays an important role in the development of animals and plants.

CONTROL OF GENE EXPRESSION IN EUKARYOTES

For development to proceed normally, the right genes must be expressed at just the right times and in just the right cells. Indeed, the expression of eukaryotic genes is precisely regulated. The modes of this regulation are many. In a few cases, gene regulation depends on *changes in the DNA itself*—genes are actually rearranged on the chromosomes (an example is given at the end of this chapter). *Transcription* in eukaryotes is subject to complex mechanisms of regulation, which have a surprising variety. *Translation* may be regulated by a variety of means. Even the polypeptide products of translation may require

further processing before they become biologically active, so we can also speak of *posttranslational control* of the expression of some genes.

We begin with some cases in which gene expression is under *transcriptional* control. There are at least three ways in which transcription may be controlled: genes may be inactivated, specific genes may be amplified, or—most frequently—specific genes may be transcribed selectively.

Transcriptional Control: Gene Inactivation

As mitosis or meiosis concludes, chromosomes uncoil, but not completely (see Chapter 9). During interphase, one portion of the chromatin—the **euchromatin**—is diffuse and thus does not stain. The **heterochromatin**, however, retains its coiling and continues to be stainable by the dyes that stain mitotic chromosomes. Heterochromatin generally is not transcribed, and any genes that it contains are thus inactivated.

How such inactivation controls gene expression is

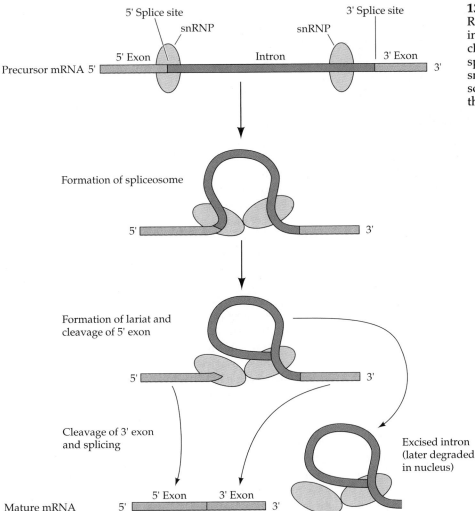

5' Splice site

snRNP

3' Splice site

snRNP

5' Exon

Intron

3' Exon

Precursor mRNA 5'   3'

Formation of spliceosome

5'   3'

Formation of lariat and
cleavage of 5' exon

5'   3'

Cleavage of 3' exon
and splicing

Excised intron
(later degraded
in nucleus)

Mature mRNA   5'   5' Exon   3' Exon   3'

13.6   snRNPs Splice RNA
RNA splicing depends on the binding of several small nuclear ribonucleoprotein particles to form the spliceosome. (Only two of the snRNPs are shown here.) The transcript of the intron forms a "lariat" that is cleaved away.

easy to understand if you think about the X and Y chromosomes of mammals. The normal female mammal has two X chromosomes; the normal male has one X and one Y. The Y chromosome has few, if any, genes that are also present on the X chromosome, and the Y appears to be transcriptionally inactive in most cells. Hence there is a 100 percent difference between females and males in the dosage of X-chromosome genes. Why is this not a case of aneuploidy involving a rather large chromosome? Aneuploidy for an autosome of comparable size is invariably lethal. Why then is not one sex or the other grossly deformed or completely inviable?

The answer was found in 1961 by Mary Lyon and Liane Russell, working independently. Lyon suggested that one of the X chromosomes in each cell of a normal female mammal is inactivated early in embryonic life and remains inactive ever after. The choice of which X in any pair of X chromosomes remains active is random. Because many cells are ultimately produced from each cell in which the choice is made, female mammals contain patches of

tissue in which one or the other X is active. This interpretation is supported by genetic, biochemical, and cytological evidence. Interphase cells of normal female mammals have a single, stainable nuclear body called a **Barr body** (after its discoverer, Murray Llewellyn Barr) that is not present in males. The Barr body is the inactive X chromosome, condensed into heterochromatin (Figure 13.7). The cells of women who have only one X chromosome, like those of normal men, contain *no* Barr bodies. Other women, who have a chromosomal constitution of XXX, have cells with *two* Barr bodies; there are even XXXXY males who have three Barr bodies in each cell. We may thus infer that interphase cells of each individual, male or female, contain a *single* active X chromosome, making the dosage of *expressed* X-chromosome genes constant and the same in both sexes.

In individual chromosomes, the presence of limited regions of transcriptionally active euchromatin may sometimes be observed. This activity is most obvious in polytene chromosomes (see Chapter 10), the giant chromosomes found in insect salivary

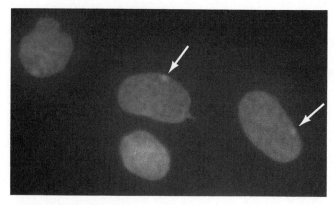

13.7   The Barr Body
The spots marked with arrows in these nuclei of human female cells are Barr bodies. A Barr body is the condensed, inactive member of the pair of X chromosomes in the cell.

glands. In some preparations of polytene chromosomes, puffs are visible. These **chromosome puffs** are regions of maximally extended chromatin, whose DNA is being transcribed (Figure 13.8).

Forming part of heterochromatin is not the only way that genes are inactivated. Another mechanism of gene inactivation is a chemical modification, called **DNA methylation**, that adds methyl groups to some cytosine residues in certain genes. The presence of the methyl groups prevents transcription of the genes. In humans and chickens, the DNA coding for globin synthesis is unmethylated in developing red blood cells. In cells that do not need to produce globin, however, the cytosine residues of the globin genes are highly methylated, so no globin is produced.

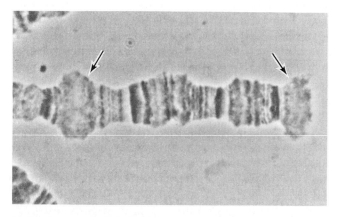

13.8   Chromosome Puffs
Arrows point to two puffs in this chromosome from a salivary gland cell of an insect larva. Puffs reveal regions where DNA is being transcribed to RNA. As the development of a larva proceeds, different regions of a chromosome puff when different gene products are needed.

Transcriptional Control: Gene Amplification

One obvious way for a cell to make more of one enzyme or RNA product than another cell does is to have more copies of the appropriate gene and to transcribe them all. The process of creating more genes of one kind in order to enhance transcription is called **gene amplification**. Such gene duplication results in more DNA per cell than there would be if there were only one copy of the gene. As we saw earlier in this chapter, one class of eukaryotic DNA—moderately repetitive DNA—is composed of such multiple gene copies.

The genes that code for histones (the proteins that interact with DNA to form nucleosomes; see Chapter 9) are present in great numbers of copies—perhaps 500 per cell in sea urchins and tens to hundreds per cell in mammals and the fruit fly *Drosophila*. It is not surprising, then, that each human cell contains over 2,000 genes coding for tRNA synthesis. Because 61 types of specific tRNA are required in protein synthesis, there are many different tRNA genes, but even so there are many copies of each. In addition, every eukaryotic cell contains in the nucleolar region of the nucleus many copies of the DNA that codes for rRNA. The multiple copies are arranged in a series, one after another in what is called a tandemly repetitive region.

In most cells, the tandemly repetitive region contains enough gene copies to produce rRNA as fast as it is needed. In some cells, however, even these multiple genes are apparently insufficient. The germ cells in amphibians that are destined to become eggs, for example, store large amounts of ribosomes for the early development of the embryo. During egg formation, each of these cells, called oocytes, multiplies the number of copies of rRNA genes until there are about a thousand nucleoli, containing more than a million rRNA genes in all, floating free in its nucleoplasm. Each of these nucleoli consists of DNA with repeating segments coding for rRNA, and each is transcribed to furnish the tremendous amount of rRNA that is stored in a mature oocyte for use by the embryo. This amplification ensures that the oocyte cytoplasm has enough rRNA to sustain rapid development during the entire period from fertilization to the formation of the gastrula (see Chapter 17), a structure consisting of hundreds of thousands of cells.

In frog oocytes, each haploid set of chromosomes contains about 500 copies of the rRNA genes before gene amplification; afterward, there are nearly a million. Evidently no genes in these cells except those responsible for rRNA synthesis are amplified. It is thought that genes are amplified in the oocytes of many animals and, under certain conditions, in vascular plants.

Figure 13.9 shows active rRNA genes from an amphibian nucleolus. The axial strands, or connecting threads, are nuclear DNA that codes for rRNA. The fuzzy-looking "triangles" attached to the DNA are many strands of rRNA in the making. Each partial rRNA molecule is attached to the DNA by an RNA polymerase molecule. Many polymerases can transcribe the DNA simultaneously. The apex of each triangle is the point at which RNA synthesis starts, so the RNA strands protruding from this region are very short. As transcription proceeds along the DNA, the RNA strands become longer and longer as more and more of the DNA coding for rRNA is read. Many of these fuzzy triangles are repeated along the DNA of the nucleolar organizer, showing us that the rRNA-coding DNA sequence itself is repetitive. Notice, too, that between the triangles there is quite a bit of silent "spacer," DNA that does not seem to be transcribed. The function of these spacers is unknown, but there is evidence suggesting that they serve as a "loading zone" for the RNA-polymerase proteins that transcribe the genes.

Transcriptional Control: Selective Gene Transcription

Very few genes become amplified. A much more common type of control simply switches the transcription of individual single-copy genes on or off. In some cases **selective transcription** of genes is mediated by eukaryotic steroid hormones (see Chapter 36). Insects with polytene chromosomes are good subjects for experimental studies of such hormonal control because transcription switched on by the hormones is accompanied by the formation of easily observable chromosome puffs (see Figure 13.8). The insect hormone ecdysone has three different types of specific effects on particular genes. Transcription of some genes, as indicated by puff formation, begins within four hours after treatment with ecdysone. Other genes form puffs several hours later—the later puffs seem to depend on both ecdysone and the protein products of earlier puffs. In addition, certain other genes stop producing puffs when ecdysone is present. Thus a complex repertoire of transcriptional events is under hormonal control.

Recall that transcriptional control in prokaryotes relies on operons, which are subject to either negative or positive control. Eukaryotic genes do not have operons, and the control of their transcription is almost always positive. In the negative control systems of prokaryotes, genes are expressed unless they are turned *off* by regulatory proteins. The positively regulated eukaryotic gene is turned *on* by proteins. Eukaryotes have much more DNA than do prokaryotes, so the danger that regulatory proteins will bind to inappropriate sites is much greater. This danger is reduced, however, by requiring the protein to bind at *multiple* sites in order to initiate the transcription of a single gene. As a result, the **promoter** for a eukaryotic gene is more complex than a prokaryotic promoter (see Chapter 12). In both eukaryotes and prokaryotes, the promoter is the stretch of DNA to which RNA polymerase binds to initiate transcription.

Unlike prokaryotes, which have only one type of RNA polymerase, eukaryotes have *three* RNA polymerases. RNA polymerase I transcribes the DNA that encodes rRNA; not surprisingly, this is the most abundant RNA polymerase in a eukaryotic cell. RNA polymerase II transcribes the structural genes that encode mRNAs and thus has the greatest diversity of products. RNA polymerase III transcribes the DNA that encodes tRNAs and some other small RNA species. We focus on RNA polymerase II for the rest of this discussion.

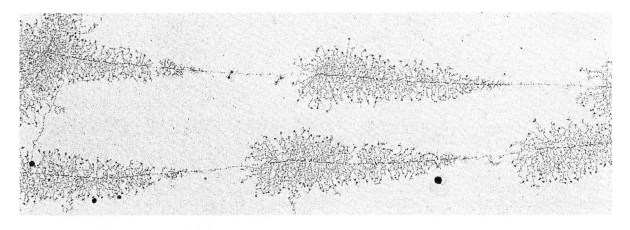

13.9 Transcription in the Nucleolus
Elongating strands of rRNA transcripts form arrowhead-shaped regions, each centered on a strand of DNA that codes for the rRNA.

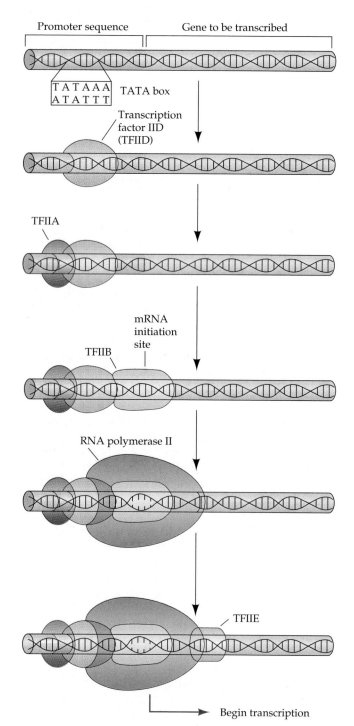

Promoter sequence   Gene to be transcribed

T A T A A A
A T A T T T   TATA box

Transcription
factor IID
(TFIID)

TFIIA

mRNA
initiation
site

TFIIB

RNA polymerase II

TFIIE

Begin transcription

13.10 Forming a Transcription Complex
Interactions among the TATA box, four transcription factors, and RNA polymerase II lead to the formation of the transcription complex.

RNA polymerase II by itself cannot simply bind to the chromosome and initiate transcription. Rather, it can bind and act only after various regulatory proteins, or **transcription factors**, have assembled on the chromosome. Each transcription factor recognizes and binds to a particular sequence of the DNA. One

such sequence is the TATA box, an eight-base-pair sequence consisting only of T–A pairs. The TATA box is found in many eukaryotic promoters about 25 base pairs before the starting point for transcription. One of the transcription factors binds to the TATA box. Next, other transcription factors bind to other sequences on the promoter. Then RNA polymerase II joins the growing transcription complex, which is completed by the addition of yet another transcription factor (Figure 13.10). Still other DNA sequences, each requiring a different transcription factor, precede the TATA box on the chromosome (Figure 13.11). Transcription begins only after all the sequences have bound their transcription factors. These DNA sequences and their transcription factors have two principal roles in the regulation of transcription: The TATA box helps pinpoint the exact starting point for transcription, and the other sequences determine the efficiency of the promoter.

Some sequences, such as the TATA box, are common to the promoters of many genes and are recognized by transcription factors found in all the cells of an organism. Other sequences are specific to only a few genes and are recognized by transcription factors found only in certain tissues; these play important parts in differentiation.

Another important type of regulation is brought about by **enhancers**, DNA sequences that bind transcription factors and stimulate specific promoters, thus enhancing the transcription of specific genes. Enhancers can act at greater distances from the regulated genes than do promoters. In fact, enhancers may lie far away, in either direction along the sequence, from the genes they regulate. Many enhancers are specific to particular cell types and are inactive in others.

How do transcription factors and other DNA-binding proteins recognize and interact with specific sequences of bases in DNA? Such proteins have do-

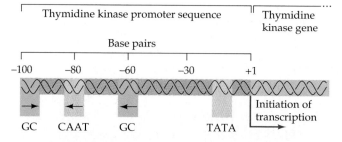

Thymidine kinase promoter sequence   Thymidine
kinase gene

Base pairs

−100   −80   −60   −30   +1

Initiation of
transcription

GC   CAAT   GC   TATA

13.11 DNA Modules in a Eukaryotic Promoter
The promoter for the gene that encodes the enzyme thymidine kinase is typical of eukaryotic promoters. It contains the TATA box and three other DNA sequences, two of them identical (GC) but oriented in opposite directions. Transcription factors bind to the sequences to initiate transcription.

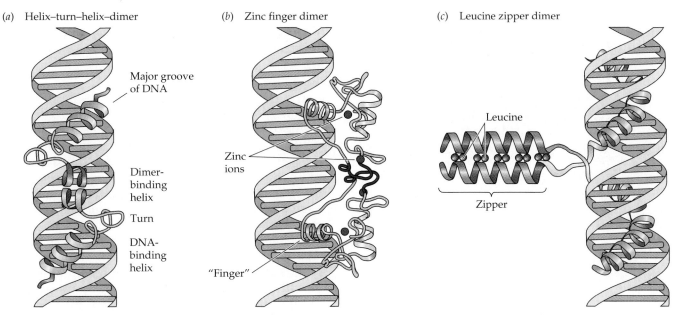

(a) Helix–turn–helix–dimer   (b) Zinc finger dimer   (c) Leucine zipper dimer

Major groove of DNA

Dimer-binding helix

Turn

DNA-binding helix

Zinc ions

"Finger"

Leucine

Zipper

13.12 Transcription Factors Bind DNA

The structures of transcription factor proteins favor DNA binding; structural motifs include these examples. (a) Helix-turn-helix proteins are dimers with two pairs of α-helical regions that fit in the major groove of DNA and bind specific base sequences. (b) Many transcription factors have "zinc fingers"—regions, held in shape by zinc ions, that protrude precisely into the major groove. (c) Some protein dimers that bind DNA are held together by "leucine zippers" in which multiple copies of the amino acid leucine in the two polypeptide strands attract one another strongly by hydrophobic bonding. In these renditions, the red areas indicate the portion of the molecule that joins the monomers together in a dimer; green regions are where the transcription factor dimers bind to the DNA; and yellow indicates the linker regions that hold the red and green regions in the correct relative positions.

mains that fit into the major groove of a DNA molecule and other parts that bind these domains together in a way that holds the protein tightly to the DNA (Figure 13.12).

Translational Control

We might guess that eukaryotes do *not* control the expression of their genes at the translational level because each of their mRNAs codes for only one polypeptide (suggesting that the production of a given polypeptide could be adequately regulated by mechanisms acting on transcription). Our guess would be wrong, however; translation of some genes *is* regulated. This level of control is a very rapid one and acts closest to the formation of the polypeptide product. Let's consider some examples of translational control of differentiation.

Several different mechanisms provide translational control. One involves a hormone acting on mRNA. As mammals prepare to lactate—to produce milk—a hormone, prolactin, acts on the mammary gland as the final trigger for milk production. The primary effect of this hormone is a dramatic increase in the translation of mRNA for casein, a major milk protein. Prolactin increases the longevity of casein mRNA, allowing it to be translated 25 more times than it is in the absence of the hormone.

A second mechanism of translational control—the capping of mRNA—is evident in the oocytes of sea urchins and certain moths. As already noted, most mRNAs become capped with a modified G unit during RNA processing. Uncapped messages are not translated. Stored mRNA in a tobacco hornworm oocyte, for example, has the G portion of the cap,

but the G has not been modified; hence these uncapped messages cannot be translated. When a tobacco hornworm egg is fertilized, the uncapped message is modified to complete the cap, and the mRNAs can then be translated.

In another example, an elegant set of controls ensures that hemoglobin is synthesized efficiently and in appropriate quantities in developing red blood cells. Hemoglobin is a moderately complex molecule that consists of two α-globin chains, two β-globin chains (both types of globins are proteins), and four smaller heme molecules, one for each globin chain. Thus the complete hemoglobin molecule consists of three distinct types of components, in a ratio of 2:2:4 (Figure 13.13). The components are usually synthesized in just this ratio; if their production gets out of balance, severe illnesses beset the organism. Three separate mechanisms maintain the ratio. First, any excess of heme results in feedback inhibition (see Chapter 6) of heme synthesis, thus reducing the imbalance. Second, excess heme increases the transla-

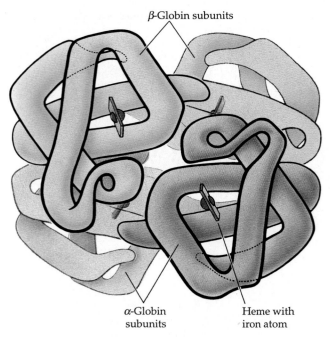

β-Globin subunits

α-Globin subunits

Heme with iron atom

13.13 Hemoglobin Consists of Three Types of Molecules
The hemoglobin molecule is made of two α-globin (purple) and two β-globin (green) polypeptides, and each of these globins contains a heme molecule (red).

tion of globin messengers. Third, an appropriate ratio of α-globin to β-globin chains is brought about through control of the translation of the two globin mRNAs.

Posttranslational Control

We have considered how gene expression may be regulated by the control of transcription, of RNA processing, and of translation. The story does not end here, however, because the expression of some genes may be modified even *after* translation has taken place. Here is a brief summary of four types of **posttranslational control**.

(1) Some proteins are specifically *inactivated* by specific degradation shortly after their formation. (2) Others, such as insulin and certain other hormones, are not produced in an active form by translation, but are made active by later chemical modification (Figure 13.14). (3) Some proteins have to be inserted into particular compartments of the cell, such as mitochondria, and others must be directed through the endoplasmic reticulum and inserted into the plasma membrane before they become active. Proteins destined to associate with or pass through particular membranes contain **leader sequences** of amino acids at their N-terminal (NH₂) ends. A given leader se-

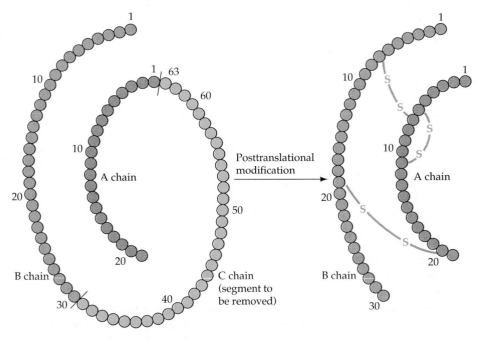

Posttranslational modification

A chain

B chain

C chain (segment to be removed)

Proinsulin, the product of translation

A chain

B chain

Insulin, the active hormone

13.14 Posttranslational Events in Insulin Synthesis
The translation product of the mRNA that codes for the hormone insulin is a larger molecule, proinsulin, that does not act as a hormone. *After* translation, part of proinsulin is removed, leaving the active polypeptide hormone, insulin.

13.15 Some Proteins Are Inactive Until They Reach Their Destination

After being synthesized in the cytoplasm, the inactive forms of certain proteins move to their destinations, where, with final modifications, they become active molecules. The genes are finally expressed only where their products are needed. Leader sequences of amino acids guide the proteins across the membranes to their destinations. This example shows two proteins, one destined for the mitochondrial matrix and the other for the intermembrane space.

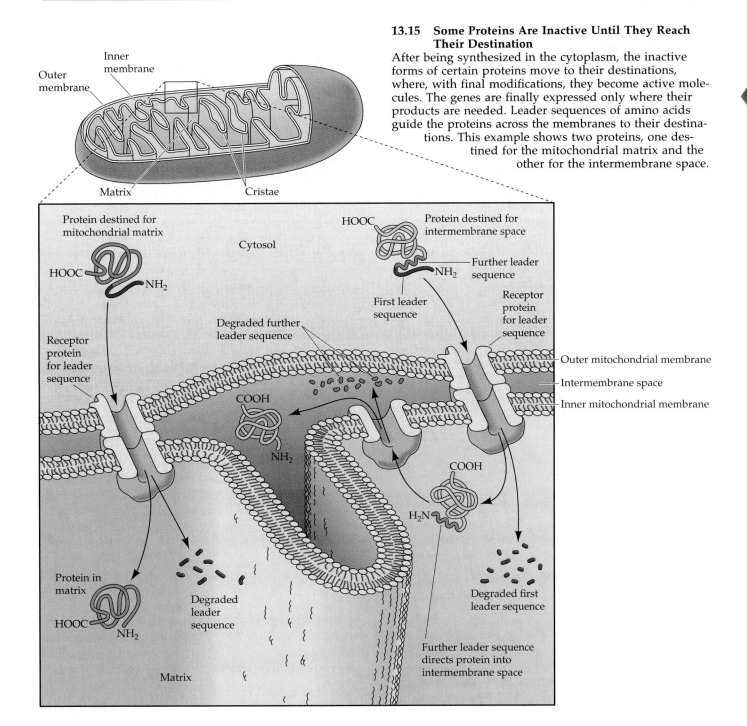

quence is recognized by a specific recognition protein in the appropriate membrane, allowing transit through that membrane. Proteases then remove the leader sequence, leaving the final, active protein as a product (Figure 13.15). (4) Finally, some proteins are inactive until they are incorporated into larger, compound structures. The proteins tubulin and actin, which form microtubules and microfilaments, respectively (see Chapter 4), do not become active until these structures have formed.

This brief overview of posttranslational control completes our consideration of molecular events in gene expression. In looking at the many mechanisms for controlling gene expression in eukaryotes, we gave no details about the very first mechanism we mentioned: rearrangement of genes on chromosomes. Box 13.A describes a dramatic example of such gene swapping. In Chapter 17 we will consider specific examples of the control of gene expression during animal development.

BOX 13.A

Cassettes and the Mating Type of Yeasts

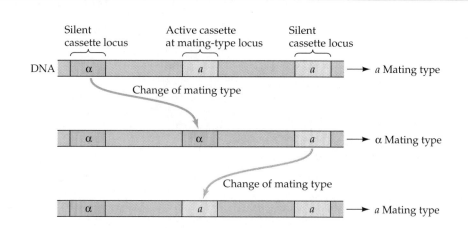

Silent cassette locus   Active cassette at mating-type locus   Silent cassette locus

DNA   α   a   a   → a Mating type

Change of mating type

α   α   a   → α Mating type

Change of mating type

α   a   a   → a Mating type

Yeasts have two mating types, *a* and α. The mating type is determined by a locus—the mating-type locus—on one of the chromosomes. A cell with the *a* allele at the mating-type locus is of mating type *a*; a cell with the α allele at the locus is of mating type α. In some yeasts, the mating type may change with almost every generation of cells. How does the mating type change so rapidly?

The mechanism has been likened to a cassette recorder, with the mating-type locus being the tape deck into which a cassette is inserted and "played." The yeast cell keeps two cassettes—unexpressed copies of the *a* and α alleles—at other loci on the chromosome that bears the mating-type locus, as shown here. From time to time the allele in the mating-type locus is removed, and one of the two unexpressed loci is copied to the mating-type locus. If this newly inserted cassette differs from the previous occupant of the mating-type locus, the mating type changes, because yeasts are unicellular and are haploid throughout most of their life cycle.

SUMMARY of Main Ideas about Gene Expression in Eukaryotes

Different genes are expressed at different times or in different tissues.

Eukaryotic DNA and its RNA transcripts are made up of introns that are removed and exons that code for the polypeptide product.
    Review Figures 13.2 and 13.3

The three classes of eukaryotic DNA are single-copy DNA, which codes for proteins; moderately repetitive DNA, which codes for telomeres and transposable elements; and highly repetitive DNA, whose function is unknown.

The hnRNA produced by the transcription of a gene is capped and tailed, and introns are spliced from it; by these modifications it becomes mRNA.
    Review Figures 13.5 and 13.6

The expression of structural genes is controlled in many different ways and at different points in the sequence from DNA to RNA to protein.

Euchromatin can be transcribed; heterochromatin is generally not transcribed.

DNA methylation inactivates genes.

Certain genes whose products are needed in enormous quantities are amplified before the time for transcription.

The transcription of many single-copy genes is switched on and off according to the need for their products.
    Review Figures 13.10 and 13.12

After some mRNAs are produced, translational controls block synthesis of their proteins until the proteins are needed.

Posttranslational control results in a final polypeptide that differs from the polypeptide product of translation.
    Review Figures 13.14 and 13.15

Transposable elements modify genes and chromosomes.

Duplication of genes in the course of evolution has given rise to families of genes.
    Review Figure 13.4

SELF-QUIZ

1. Which statement is *not* true of nucleic acid hybridization?
   a. It depends upon complementary base pairing.
   b. A DNA strand can hybridize with another DNA strand.
   c. An RNA strand can hybridize with a DNA strand.
   d. A polypeptide can hybridize with a DNA strand.
   e. Double-stranded DNA denatures at high temperatures.

2. Which statement is *not* true of introns?
   a. Their name is short for "intervening sequence."
   b. They do not encode any part of the polypeptide product of the gene.
   c. They are found in all vertebrate genes.
   d. They are transcribed.
   e. They interrupt, but do not scramble, the coding sequence for a polypeptide.

3. In regard to repetitive DNA
   a. much highly repetitive DNA lies near the centromeres.
   b. highly repetitive DNA reanneals most slowly of the three classes of DNA.
   c. highly repetitive DNA is transcribed often and rapidly.
   d. single-copy DNA is rare in eukaryotes.
   e. transposable elements are single-copy genes.

4. Capping of hnRNA
   a. takes place at its 3′ end.
   b. facilitates binding of the mRNA to a ribosome.
   c. is coded for in the DNA.
   d. prevents its translation.
   e. consists of the addition of a poly A tail.

5. Which statement is *not* true of RNA splicing?
   a. It removes introns.
   b. It is performed by small nuclear ribonucleoprotein particles (snRNPs).
   c. There are different splicing mechanisms for mRNAs, rRNAs, and tRNAs.
   d. It is directed by consensus sequences.
   e. It lengthens the RNA molecule.

6. Which genes are *not* commonly present in many copies in eukaryotes?
   a. Those that encode histones
   b. Those that encode mRNAs
   c. Those that encode tRNAs
   d. Those that encode rRNA
   e. Those that are present in a tandemly repetitive region

7. Which statement is *not* true of selective transcription in eukaryotes?
   a. Different classes of RNA polymerase transcribe different parts of the genome.

   b. Transcription requires transcription factors.
   c. Genes are transcribed in groups called operons.
   d. Control is almost always by positive regulation.
   e. The promoter is more complex than in prokaryotes.

8. Transcription factors in eukaryotes
   a. consist of DNA.
   b. consist of RNA.
   c. include such sequences as the TATA box.
   d. allow the binding of RNA polymerase to the promoter.
   e. cause operons to be transcribed.

9. Translational control
   a. is not observed in eukaryotes.
   b. is a slower form of regulation than is transcriptional control.
   c. occurs by only one mechanism.
   d. requires that mRNAs be uncapped.
   e. ensures that hemoglobin is synthesized in appropriate quantity.

10. Which statement is *not* true of telomeres?
   a. They contain repetitive DNA.
   b. They appear at the ends of eukaryotic chromosomes.
   c. They are the sites at which spindle fibers attach.
   d. They protect chromosomes from shortening during replication.
   e. They may play a role in synapsis during meiosis.

FOR STUDY

1. In rats a gene 1,440 base pairs long codes for an enzyme made up of 192 amino acid units. Discuss this apparent discrepancy.

2. Describe the steps in the production of mature, translatable mRNA from a eukaryotic gene that contains introns.

3. How can nucleic acid hybridization techniques be used to determine whether a gene possesses introns?

4. Describe the origin and development of gene families such as the one containing genes that code for globins.

5. Prepare a list of the possible ways in which transcription and translation may be regulated in eukaryotes. Contrast this with the situation in prokaryotes.

READINGS

Cech, T. R. 1986. "RNA as an Enzyme." *Scientific American*, November. A description of exciting discovery of the catalytic activity of certain RNAs and its roles in the molecular biology of eukaryotes. These findings help us to understand the origin of life.

Chambon, P. 1981. "Split Genes." *Scientific American*, May. Introns and exons—their origin and how they are handled.

Donelson, J. E. and M. J. Turner. 1985. "How the Trypanosome Changes Its Coat." *Scientific American*, February. Trypanosomes evade the host's immune system by constantly switching on new genes that code for different surface antigens.

Griffiths, A. J. F., J. H. Miller, D. T. Suzuki, R. C. Lewontin and W. M. Gelbart. 1993. An Introduction to Genetic
Analysis, 5th edition. W. H. Freeman,
New York. An up-to-date revision of
one of the field's classic textbooks. See
especially Chapter 17.

Grunstein, M. 1992. "Histones as Regulators of Genes." *Scientific American*, October. Histones not only organize the
nucleosomes, but they regulate transcription.

McKnight, S. L. 1991. "Molecular Zippers
in Gene Regulation." *Scientific American*, April. A description of an important class of DNA-binding proteins,
some of which are transcription factors.

Ptashne, M. 1989. "How gene activators
work." *Scientific American*, January.
How genes are turned on and off in
eukaryotic cells.

Rennie, J. 1993. "DNA's new twists." *Scientific American*, March. Transposable
elements and other features.

Rhodes, D. and A. Klug. 1993. "Zinc Fingers." *Scientific American*, February.
How zinc fingers help some proteins
bind to DNA and regulate transcription.

Steitz, J. A. 1988. "Snurps." *Scientific
American*, June. How spliceosomes remove intron transcripts.

Varmus, H. 1987. "Reverse Transcription." *Scientific American*, September.
Reverse transcription is not confined
to the retroviruses. This article describes reverse transcription in eukaryotes, stressing its likely relevance to
the emergence of DNA as the genetic
material.

S uppose you could "teach" bacteria or other unicellular organisms to produce important chemicals normally produced only by humans. Might that be medically useful? Suppose that you could teach bacteria to clean up oil spills in the ocean. Might that be a powerful tool for environmental protection? Suppose that you could teach important crop plants to make their own fertilizer. Wouldn't *that* be useful as we attempt to feed the ever-expanding human population?

The first item on this "wish list" is already a reality, the second has been achieved to a very limited extent, and the third is being vigorously pursued. In each case, the "teaching" process consists of providing an organism with genes from another organism that is capable of doing something the organism receiving the genes can't do.

This process brings together techniques from molecular biology, microbial genetics, and biochemistry. It has become known as **recombinant DNA technology**, popularized as "genetic engineering" and "cloning." By recombinant DNA, we mean DNA made up of connected segments from mixed sources—perhaps from two different species, or perhaps a combination of natural and synthetic DNA. An example of combining DNA from different species is the insertion of a gene from a human into the DNA of a yeast. This feat can convert the yeast cells into "factories" producing the protein product of the human gene. An early example of combining natural and synthetic DNA was the use of bacteria to produce the human hormone somatostatin, a 14-amino acid polypeptide whose sequence was known. Biologists at the City of Hope Medical Center and at the University of California, San Francisco, synthesized a completely artificial stretch of DNA, part of which was designed to code for the amino acid sequence of somatostatin. This DNA was first inserted into a plasmid that had been isolated from a bacterial cell. The scientists then introduced the recombinant plasmid into a culture of *Escherichia coli*, where the plasmid replicated, producing multiple copies of itself in each bacterial cell. These *E. coli* could then be induced to synthesize human somatostatin. Somatostatin may be useful in treating pancreatitis (a disease of the pancreas), acromegaly (abnormal enlargement of bones in the hands, feet, and face), and insulin-dependent diabetes. A short time ago somatostatin could not be considered for medical use because almost none of it was available from natural sources. Recombinant DNA technology, however, has made possible the large-scale production of somatostatin, opening vistas for treatment of these diseases.

As a tool, recombinant DNA technology has revolutionized much of experimental biology. Most recent advances in understanding how the genes of eukaryotes are regulated and organized have come

Brewing Medicine
Hepatitis vaccine produced by recombinant DNA technology "brews" in an 800-liter fermenter.

14

Recombinant DNA Technology

through the application of recombinant DNA technology. Recombinant DNA technology is also revolutionizing agriculture, medicine, and other areas of the chemical industry, as well as forensics and the battle against environmental pollution. In this chapter we will consider the basic techniques of this technology and some of its applications in the laboratory and beyond.

THE PILLARS OF RECOMBINANT DNA TECHNOLOGY

Scientists realized that chemical reactions used in living cells for one purpose may be applied in the laboratory for other, novel purposes. Recombinant DNA technology is based on this realization, and on the recognition of the properties of certain enzymes and of DNA itself. Naturally occurring enzymes that cleave DNA, help it grow, and repair it are numerous and diverse, and many of them are now used in the laboratory to manipulate and combine DNA molecules from different sources.

The nucleic acid base-pairing rules underlie many of the fundamental processes of molecular biology. The mechanisms of DNA replication, transcription, and translation all rely on complementary base pairing. Similarly, all the key techniques of recombinant DNA technology—sequencing, splicing, locating, and identifying DNA fragments—make use of the complementary pairing of A with T (or U) and of G with C.

CLEAVING AND SPLICING DNA

The basic operations of cleaving and splicing DNA are good examples of how scientists use enzymes and complementary base pairing creatively. The enzymes are used in the laboratory to achieve different overall purposes than they would in the living cell.

Restriction Endonucleases

All organisms must have mechanisms to deal with their enemies. As we saw in Chapter 12, bacteria are attacked by bacteriophages that inject their genetic material into their hosts. Eventually the phage genetic material may be replicated by the enzyme systems of the host. Some bacteria defend themselves against such invasions by producing enzymes called **restriction endonucleases** that can cleave double-stranded DNA molecules—such as those injected by many phages—into smaller, noninfectious fragments (Figure 14.1). There are many such enzymes, each of which cleaves DNA at a specific site defined by a *sequence of bases* and called a **recognition site**. The

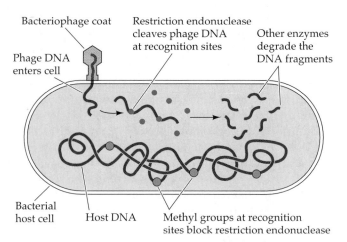

14.1 Bacteria Fight Phages with Restriction Endonucleases
Bacteria produce restriction endonucleases that break up phage DNA. Other enzymes protect the bacteria's own DNA from being cleaved; they do this by methylating the host DNA.

DNA of the bacterial host is not cleaved by its own restriction endonucleases because of the activity of specific methylases, enzymes that add methyl ($—CH_3$) groups to certain of the bases within the recognition sites. The methylation of the bases makes the recognition sites unrecognizable to the restriction endonucleases, thus preventing cleavage of the host DNA, but the unmethylated phage DNA is efficiently degraded.

A specific sequence of bases defines each recognition site. For example, the restriction endonuclease *Eco*RI (named after its source, *E. coli*) cuts DNA only where it encounters these paired sequences in the double helix:

5'. . .GAATTC. . .3'

3'. . .CTTAAG. . .5'

Other restriction endonucleases recognize different base sequences. The sequence recognized by *Eco*RI occurs on the average about once in 4,000 base pairs—about once per four prokaryote genes. This restriction endonuclease can chop a typical big piece of DNA into smaller pieces containing, on the average, just a few genes. Remember that "on the average" does not mean that it cuts at regular intervals along all stretches of DNA. The *Eco*RI recognition sequence does not occur even once in the 40,000-base-pair sequence of the DNA of phage T7—a characteristic that is crucial for the survival of T7, because *E. coli* is its host. Fortunately for *E. coli*, the DNA of other phages does contain this recognition sequence, which prevents *E. coli* from being overrun by phages.

Different restriction enzymes that recognize different recognition sites may cut the same sample of

DNA. Several hundred restriction endonucleases have been extracted from various bacteria, and many are available for recombinant DNA research. Thus cutting a sample of DNA in many different, specific places is an easy task in the laboratory. We can use restriction endonucleases as "knives" for genetic "surgery."

Sticky Ends and DNA Splicing

An important property of some restriction endonucleases is that they make staggered cuts in the DNA rather than cutting both strands at the same point. For example, *Eco*RI cuts DNA as shown at the top of Figure 14.2*a* (note that the cut is within the recognition sequence given in the previous section). After the two cuts are made, the two strands are held together by only four base pairs. The hydrogen bonds of those base pairs are too weak to persist at warm temperatures (room temperature or above), so the pieces separate. As a result, there are single-stranded tails at the site of each cut. These tails are called **sticky ends** because they have a specific base sequence that is capable of binding (by complementary base pairing, at low temperature) with complementary sticky ends.

After a piece of DNA has been cut by a restriction endonuclease, it is possible for the complementary sticky ends to rejoin. The original ends can join, or an end may pair with another fragment. If more than one recognition site for a given restriction endonuclease is present in a sample, the enzyme can make a number of cut pieces, all with the *same* sequences in their sticky ends. When the temperature is lowered, the pieces reassociate at random. At the lower temperature, base pairs are more stable and four base pairs may hold the two pieces of DNA together. The new associations are unstable, however, because they are maintained by only a few hydrogen bonds.

The joined sticky ends can be permanently united by a second enzyme, DNA ligase, which makes the joining very stable. The usual function of DNA ligase in the cell, as mentioned briefly in Chapter 11, is to unite the Okazaki fragments of the lagging strand during DNA replication (see Box 11.A). DNA ligase also mends breaks in polynucleotide chains, thus helping in DNA repair.

A piece of DNA can be inserted into a plasmid as shown in Figure 14.3, as long as both the plasmid and the source for the DNA piece contain recognition sites for the same restriction endonuclease. The DNA to be inserted is cleaved from within its molecule by cutting both ends with a particular restriction endonuclease. The circular plasmid is cleaved with the same endonuclease, transforming it into a linear molecule with sticky ends. The sticky ends of the piece of DNA join the sticky ends of the cleaved plasmid, and DNA ligase seals the joining, regenerating a circular plasmid. The plasmid now contains the inserted

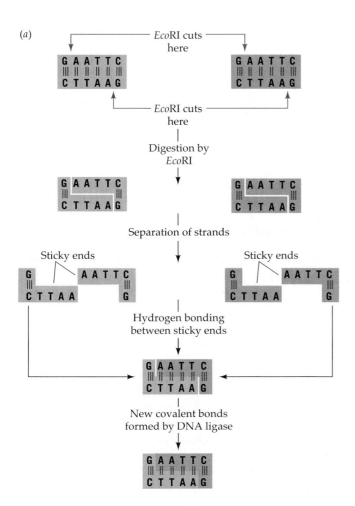

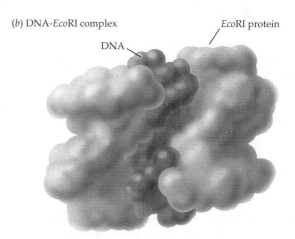

14.2 Cutting and Splicing DNA
(*a*) Some restriction endonucleases sever one strand of the double helix at one point and the other strand at a different point. The separated pieces have single-stranded sticky ends capable of combining with the sticky ends of complementary single strands. Newly joined pieces are stabilized by the action of DNA ligase. (*b*) The binding of the restriction endonuclease *Eco*RI to its recognition site on a DNA molecule.

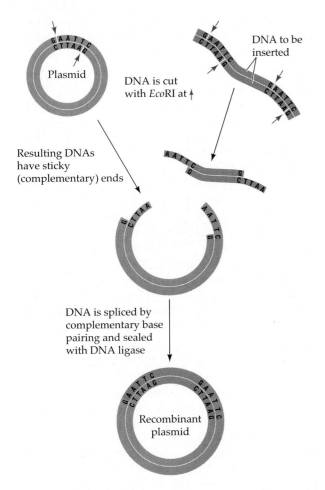

14.3 Insertion of a DNA Sample into a Plasmid
DNA from an outside source is inserted into a plasmid in the middle of a recognition site (red arrows) for the restriction endonuclease.

DNA—in the middle of what used to be a recognition site for the restriction endonuclease. Notice that the recombinant plasmid has two recognition sites, whereas the original plasmid had only one.

CLONING GENES

A typical aim of recombinant DNA work is to obtain many copies of a particular gene. To do so, we make **transgenic** bacterial or yeast cells containing the desired gene, then allow the transgenic cells to multiply. A transgenic cell or organism is one that contains foreign DNA integrated into its own genetic material. To ensure that the foreign gene, natural or synthetic, gets integrated, we must first insert it into a suitable **cloning vector**—such as a virus or a plasmid—before introducing it into the bacterial or yeast host. A stretch of DNA that is introduced by itself into a cell does not get replicated and will eventually be degraded; a cloning vector (and any gene contained in it), however, *does* get replicated.

Host Cells

The early successes of recombinant DNA technology were achieved with bacteria as hosts. However, bacteria are not ideal organisms for processing and studying eukaryotic genes. By now, cells of some eukaryotes have also been used as hosts.

Yeasts are now the most commonly used host cells for recombinant DNA studies of eukaryotic genes. Because it is widely employed in such studies, *Saccharomyces cerevisiae* (baker's or brewer's yeast), is rapidly becoming the eukaryote best understood at the molecular level, just as *Escherichia coli* is the best-understood prokaryote. Unlike most other eukaryotes, yeasts are unicellular; however, yeasts are typical eukaryotes in most respects, including their genetic mechanisms. The DNA of *S. cerevisiae* is organized into 17 chromosomes (in the haploid phase) that can be separated in the laboratory using a new technique that will be described later in this chapter.

Yeasts share three of the greatest advantages of *E. coli* for work in molecular biology. Because they are tiny and unicellular, they are easy to grow in vast numbers in small volumes of medium. They multiply almost as rapidly as *E. coli*. (The division time of a typical culture of *E. coli* is in the range of 20 to 60 minutes and that of *S. cerevisiae* is 2 to 8 hours.) Finally, although a mammalian cell has about 1,000 times more DNA than does *E. coli*, a haploid cell of *S. cerevisiae* is much more manageable—the yeast has only three and a half times more DNA than an *E. coli* cell has.

Foreign genes can be carried into a yeast cell in three different ways: (1) The recombinant DNA is present in one or a few copies of a plasmid into which a centromere (cloned from a yeast chromosome) and a replication site have been inserted. The centromere makes the plasmids stable, so the mitotic spindle can deliver them to the daughter nuclei of the yeast. (2) The recombinant DNA is present in many copies of a plasmid without a centromere. These plasmids are unstable and are passed on to daughter yeast cells at random. (3) The recombinant DNA is incorporated directly into a yeast chromosome.

Vectors

Plasmids—extrachromosomal circles of DNA—were the first vectors to be used in cloning genes. Many different plasmids have been isolated from bacterial and eukaryotic sources. For a given experiment, we select a plasmid with certain specific characteristics. First, the plasmid must be a **replicon**—that is, it must have an origin of replication (see Chapter 11); otherwise it will not be replicated when the host cell divides. Second, the plasmid should carry one or more genes conferring particular properties—such as resistance to specific antibiotics—that may be used for selection purposes. Third, the plasmid should have only one recognition site where the restriction endonuclease will cut it. If it has no such site, it cannot be opened for the insertion of the new genes; if it has two or more, the restriction endonuclease and DNA ligase may form many diverse products rather than just one. With a single site, the probability of achieving the desired insertion is higher but still not great (Figure 14.4).

Plasmids are efficient vectors for cloning small DNA fragments, but they multiply more slowly—and are less stable—when they contain large DNA fragments. Nowadays much recombinant DNA research uses viruses, in part because they are stable for cloning larger DNA fragments. Genes from other sources are inserted into the DNA or RNA of the viruses used as vectors. Viral vectors are commonly used for studies of eukaryotic DNA. The recombinant viral chromosome produced for such a study must be able to fit into the viral coat. Thus a nonessential piece of the chromosome may be "edited out" before the foreign gene is inserted so that the new recombinant will be small enough to be packaged into the viral coat for delivery to the cell.

Inserting Foreign DNA into Host Cells

Although some viral cloning vectors can infect host cells directly, most vectors require assistance in inserting their DNA into host cells.

As you will recall from Chapters 11 and 12, bacteria are capable of **transformation**—they can contact and take up isolated DNA from other bacteria. This ability provided researchers with the method for introducing plasmid vectors into bacteria. Bacteria are mixed with a solution containing plasmids and, under the appropriate conditions, some of the bacteria take up the plasmid DNA.

A similar process in eukaryotic cells, **transfection**—the uptake, incorporation, and expression of foreign DNA—often includes other steps. The cell walls must be removed before DNA can enter plant or fungal cells. The walls can be digested by appropriate enzymes, such as an enzyme found in snail guts that the snail uses to digest its meals. The resulting plant or fungal cell, without a wall, is called a **protoplast**. Plant and yeast protoplasts can be transfected with plasmids.

DNA may be inserted into eukaryotic cells in other, more drastic ways. In one method, **micropro-**

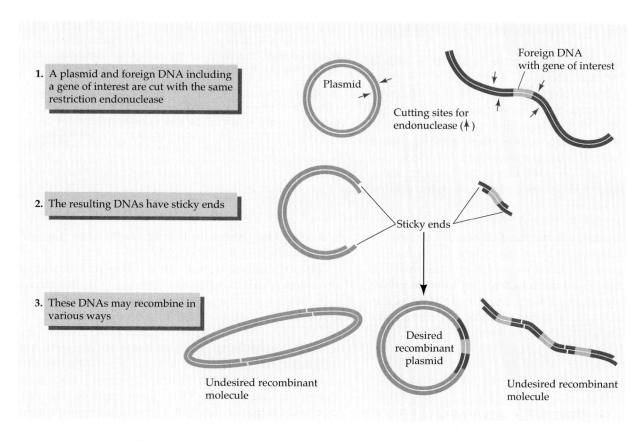

1. A plasmid and foreign DNA including a gene of interest are cut with the same restriction endonuclease

2. The resulting DNAs have sticky ends

3. These DNAs may recombine in various ways

Plasmid

Foreign DNA with gene of interest

Cutting sites for endonuclease (↟)

Sticky ends

Undesired recombinant molecule

Desired recombinant plasmid

Undesired recombinant molecule

14.4 Insertion into Plasmids Gives Multiple Products
These are only three of the many ways in which the plasmid and foreign DNA can combine. If the plasmid had more than one cutting site for the restriction endonucle-ase, the range of combinations would be much greater. As we will soon see, there is an elegant way to isolate the desired recombinant plasmid.

(a)

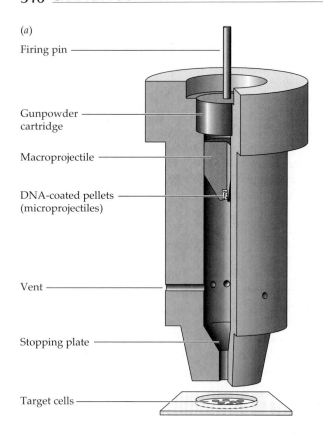

Firing pin

Gunpowder cartridge

Macroprojectile

DNA-coated pellets (microprojectiles)

Vent

Stopping plate

Target cells

(b)

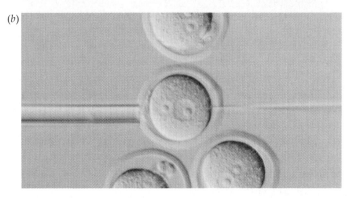

14.5 Inserting DNA into Cells

(a) This "DNA particle gun"—diagrammed on the left and photographed in use on the right—shoots DNA into eukaryotic cells. Gunpowder blasts the macroprojectile against a stopping plate, bringing it to an abrupt halt in a dead end. The resulting momentum releases the microprojectiles of DNA-coated tungsten and carries them through cell walls and into the target cells. *(b)* And then there is the labor-intensive approach. The micropipette at the right of the photograph injects foreign DNA into a fertilized mouse egg (the small nucleus is from the egg and the larger one is from the sperm). The injected DNA is entering the sperm nucleus. The pipette on the left holds the cell steady during the "operation."

jectiles—tiny, high-velocity particles of tungsten, coated with the DNA to be used for transfection—are shot into plant cells (Figure 14.5*a*). In a second method, **electroporation**, cells are exposed to rapid pulses of high-voltage current, which temporarily renders the plasma membrane permeable to macromolecules in the medium. For recombinant DNA studies of mammalian cells a third method, direct injection of DNA with a micropipette, is used. This method is more labor-intensive than others, since each recipient cell is treated individually (Figure 14.5*b*).

DNA may also be coated in various ways to allow it to pass through plasma membranes. It can be complexed with lipids, or it can be enclosed in natural or artificial plasma membranes, which are then fused to the host cells. The nonbiological methods described here are reasonably efficient, resulting in transfection of 1 percent or more of the potential recipient cells. Higher yields are obtained with some of the biological methods.

Selecting Transgenic Cells

Following interaction with a preparation of plasmid vectors, a population of host yeast or bacterial cells is heterogeneous, since only a small percentage of the cells have taken up the plasmid. Also, only a few of the plasmids that have moved into host cells contain the DNA sequence we wish to clone. How can

we select only those cells that contain the plasmid *with* the desired foreign DNA?

The experiment we are about to describe illustrates an elegant, commonly used approach to this problem. In this example, we will use bacteria as hosts, but a similar method is used in studies of yeast. We select, as our vector, a plasmid that contains two key genes. One of these genes provides resistance to an antibiotic; the other is the *E. coli z* gene—the gene in the *lac* operon that encodes the enzyme β-galactosidase (see Chapter 12). This *z* gene contains the recognition sequence for the restriction endonuclease we will use. The following description is diagrammed in Figure 14.6.

When we add the endonuclease to the plasmid preparation, it cuts the plasmid within the *z* gene, leaving sticky ends. We then add the foreign DNA,

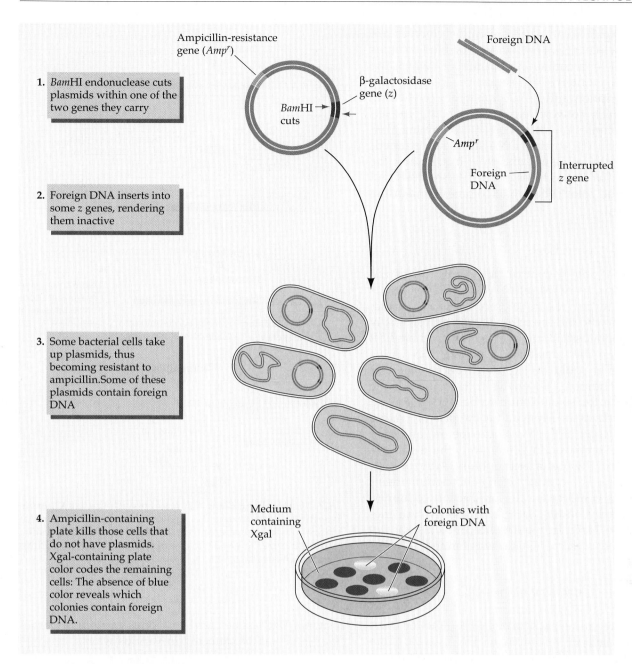

1. *Bam*HI endonuclease cuts plasmids within one of the two genes they carry

2. Foreign DNA inserts into some *z* genes, rendering them inactive

3. Some bacterial cells take up plasmids, thus becoming resistant to ampicillin. Some of these plasmids contain foreign DNA

4. Ampicillin-containing plate kills those cells that do not have plasmids. Xgal-containing plate color codes the remaining cells: The absence of blue color reveals which colonies contain foreign DNA.

Ampicillin-resistance gene (*Amp^r*)

Foreign DNA

β-galactosidase gene (*z*)

*Bam*HI cuts

Amp^r

Foreign DNA

Interrupted *z* gene

Medium containing Xgal

Colonies with foreign DNA

14.6 Color Coding the Colonies We Want

Inserting genes that confer antibiotic resistance into plasmids allows us to use ampicillin to eliminate those bacterial cells that do not contain the plasmid. To discover which plasmid-containing cells also took up the foreign DNA, we place the cells in Xgal. Those cells with no foreign DNA produce β-galactosidase, which turns the Xgal blue. In colonies of cells that took up the foreign DNA, however, β-galactosidase transcription is interrupted and no blue color is seen.

which has been cut with the same endonuclease, and allow the sticky ends of this DNA and of the plasmids to recombine. This process gives us a heterogeneous population of plasmids, some containing the foreign DNA. Combination of the plasmid population with host cells leads to transformation, giving a hetero-

geneous population of bacteria, some containing a plasmid.

Which cells contain plasmids? Which plasmids contain the foreign DNA? We add the antibiotic to which the gene in the plasmid confers resistance. This treatment kills all the bacteria that do not contain the plasmid. But which of the surviving bacteria contain the plasmids with the foreign DNA? β-Galactosidase, the product of the plasmid's *z* gene, catalyzes the hydrolysis of a number of β-galactosides, including a colorless synthetic compound called Xgal (pronounced "ex-gal"). The products of hydrolysis of Xgal include an intensely blue compound. Therefore, if we grow the antibiotic-resistant cells on a medium containing Xgal, some of the resulting colonies are deep blue, while others are white. The blue colonies consist of bacteria containing plasmids in which the z gene is an uninterrupted sequence. In these bacteria, transcription and translation produce β-galactosidase, which converts Xgal to the blue product. The white colonies consist of bacteria containing plasmids in which the foreign DNA has been inserted, interrupting the sequence and thus inactivating the z gene, so that it produces no β-galactosidase and hence no blue color! The white colonies contain the cells we want.

Controlling a Cloned Gene

What if we want to control the expression of the foreign gene in its new home? We may have reasons for wanting to turn the gene on or off. For example, the scientists who did the original work on bacterially produced somatostatin, mentioned earlier, thought it safest to keep the somatostatin gene under wraps except when it could be expressed under carefully controlled conditions. To enable transcription of the inserted gene for somatostatin to be turned on and off, the investigators arranged for the plasmid vector to include a copy of the E. coli lac operon, and they inserted the somatostatin gene into z, the β-galactosidase structural gene of that operon. After putting the modified plasmid into bacteria and cloning them, the scientists could cause the production of human somatostatin by adding lactose to the growth medium (Figure 14.7). In the absence of lactose or another inducer, no somatostatin was formed.

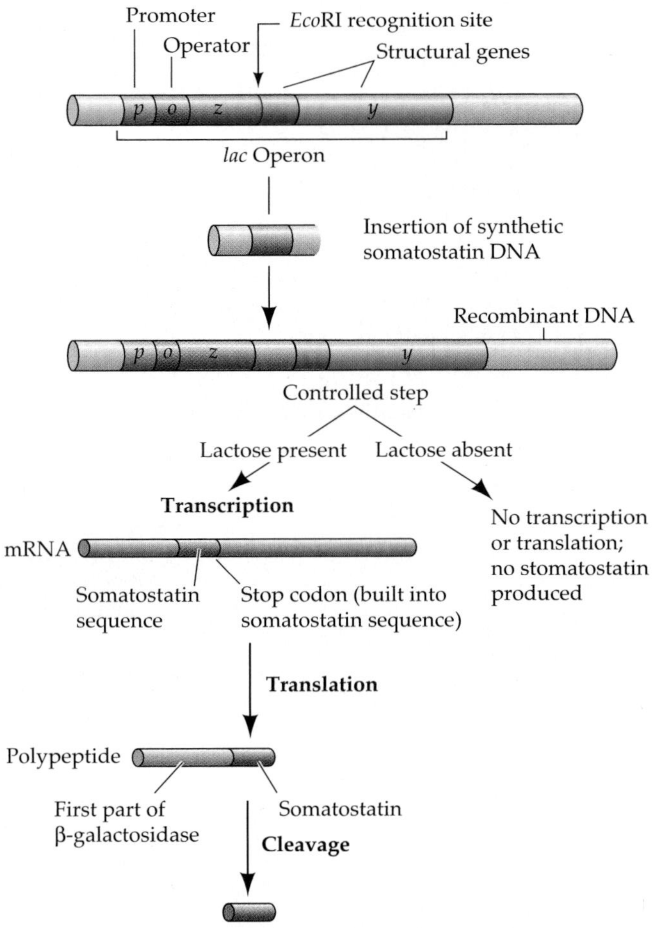

SOURCES OF GENES FOR CLONING

There are three principal sources of the genes or DNA fragments used in recombinant DNA work. One source is pieces of chromosomes inserted into vectors; these DNA-vector units are maintained as gene libraries. A second source is complementary DNA, obtained by reverse transcription from specific RNAs. The third source is laboratory synthesis of specific polynucleotide sequences.

Gene Libraries and Shotgunning

DNA from a desired source can be isolated and broken into small fragments, usually by restriction endonucleases. These fragments can then be incorporated into bacteriophage DNA or plasmids that are, in turn, cloned in bacteria. This technique, called **shotgunning** because the DNA is fragmented at random rather than into specific desired pieces, produces a collection of clones called a **gene library** (Figure 14.8). Each clone carries a fragment of the original DNA, and the library as a whole carries all of the

14.7 Controlling Gene Expression with the lac Operon Inserting a foreign gene (in this case, the gene for somatostatin) into the lac operon is a "trick" that allows investigators to turn the gene's transcription on or off by adding or withholding lactose.

original DNA. The clone or clones carrying a particular gene can be detected in various ways.

Complementary DNA

If a specific RNA, such as a particular mRNA, is available, one can make a complementary DNA, or **cDNA**, by using the RNA as a template. Although the starting amount of specific mRNA is usually small, the cDNA can be cloned.

The steps that produce cDNA are illustrated in Figure 14.9. Recall that most eukaryotic mRNAs have a poly A tail—a string of adenine, or A, residues at their 3' end. The first step in cDNA production is to allow an mRNA to hybridize with a molecule (called oligo-dT) consisting of a string of T residues. After the hybrid forms, the oligo-dT can serve as a primer and the mRNA can serve as a template for the enzyme **reverse transcriptase**—the enzyme that retroviruses use to synthesize DNA from RNA templates in host cells. If the primer and reverse transcriptase

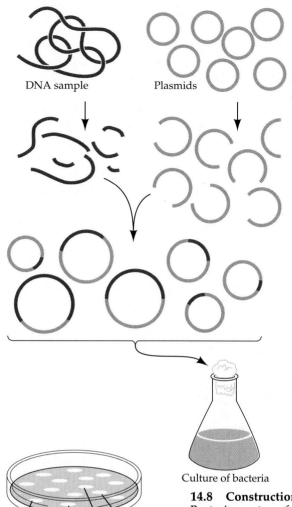

1. A DNA sample and plasmids are cleaved with the same restriction endonuclease

2. DNA fragments and opened plasmids are mixed and spliced with DNA ligase

3. A mixture of different plasmids results

4. Plasmids are mixed with bacteria and placed on a nutrient medium

5. Colonies contain clones of different fragments of original DNA

6. Individual recombinant colonies are isolated and each is maintained as a pure culture; each such culture is a "volume" in the gene library

DNA sample

Plasmids

Culture of bacteria

Individual recombinant cultures

14.8 Construction of a Gene Library
Bacteria are transformed by DNA inserted into plasmids or other vectors. Each recombinant bacterium gives rise to a colony containing part of the original DNA sample. Cultures of these colonies can be analyzed to determine which genes they contain.

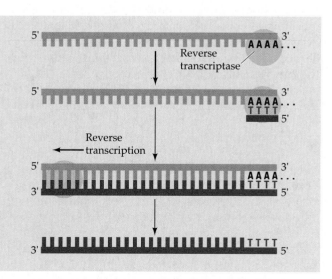

1. mRNA template with 3' poly A tail is combined with reverse transcriptase enzyme

2. A short oligo-dT primer is added and allowed to hybridize with the poly A tail

3. Reverse transcriptase synthesizes cDNA using the mRNA template and deoxyribonucleoside triphosphate substrates, creating a DNA–RNA hybrid

4. When synthesis is completed, the mRNA is removed, leaving single-stranded cDNA

Reverse transcriptase

Reverse transcription

14.9 Synthesis of Complementary DNA
Synthesis of single-stranded cDNA requires the enzyme reverse transcriptase, an mRNA template with a poly A tail on the 3' end, a short primer chain of oligo-dT, and the four deoxyribonucleoside triphosphates.

are given a source of DNA precursors (the four deoxyribonucleoside triphosphates; see Figure 11.6), they will synthesize a strand of cDNA complementary to the mRNA. After the cDNA strand is removed from the mRNA (by increasing the pH of the solution, thus denaturing the cDNA-mRNA hybrids and degrading the mRNA), it can be used for cloning, hybridization, or other experiments.

If the next step in a research project is to clone the cDNA by inserting it into a DNA vector, the single-stranded cDNA must first be converted into double-stranded DNA, using DNA polymerase. A collection of complementary DNAs incorporated into bacteriophage vectors is called a **cDNA library**. Such cDNA libraries contain only *expressed* DNA, that is, DNA that is represented by mRNAs. In comparing cDNA libraries from different tissues or different stages of development, one can gain insight into which genes are being turned on or off. By contrast, the advantage of a *gene* library is that is *does* include DNA that is not transcribed, as well as DNA that is processed out of the mRNAs (including introns, promoters, and regulatory signals such as the TATA box). "Chromosome walking," which we discuss later in this chapter, can be done only with gene libraries, not with cDNA libraries.

Synthetic DNA

When we know the amino acid sequence of the desired polypeptide product, we can get the DNA we need for cloning by synthesizing it directly. This process works well for small DNA molecules. This approach was used by the group that first cloned a DNA sequence coding for somatostatin. Commercially made instruments for automated DNA synthesis are now available.

How do we design a small, synthetic gene? Using the genetic "code book" (see Figure 11.25), we can figure out the appropriate base sequence for the synthetic gene. Using this sequence as the starting point, we can add other sequences to serve specific purposes. What else might we add? For instance, how can we ensure that translation begins and ends at the right places? We can add codons for initiation and termination of translation. How can we prepare the synthetic gene for insertion into a cloning vector? We can add to the ends of our synthetic gene appropriate recognition sequences for the restriction endonuclease that we will use to splice the synthetic DNA into the cloning vector. Other refinements are also possible.

Besides being used to synthesize small polypeptides such as somatostatin, synthetic DNA has been put to two other principal uses. One use, to be discussed in the next section, is as "probes" for detecting specific DNA fragments. The other use is in "directed" mutagenesis—the production of specific alterations in DNA regions with known sequences.

EXPLORING DNA ORGANIZATION

Much current research uses recombinant DNA techniques to study the organization of DNA, including whole chromosomes. These studies may be conducted at many levels, from examining the sequence of genes on a chromosome down to examining the sequence of bases in a DNA fragment.

Separation of Intact Chromosomes

How can we obtain the largest DNA molecules—chromosomes—in pure form? Human chromosomes can be separated by an elegant automated technique. A suspension of chromosomes (chromosomes mixed with but not dissolved in a liquid) is stained with two fluorescent dyes. Each chromosome takes up the two dyes in a particular ratio. The suspension is then passed at high speed through a fine tube, where it is exposed to laser beams of two different wavelengths, each of which is absorbed by one of the dyes. The resulting fluorescence from the dyes in the chromosomes is analyzed. Each chromosome exhibits a distinct pattern of fluorescence. When the desired chromosome (recognized by its fluorescence pattern) passes the observation point, the drop in which it is contained is given a tiny electric charge. As the drop falls away from the nozzle of the apparatus, it is attracted by a charged plate and falls into a tube, while the other, uncharged drops fall into another tube (Figure 14.10). In this way, one type of chromosome can be isolated from all others in the suspension. This system can be adjusted so that any one of the 22 autosomes, or the X or Y chromosome, is collected—at rates in the hundreds per second.

Separation and Purification of DNA Fragments

DNA fragments differing in length can be separated by **gel electrophoresis** (Figure 14.11a). A mixture of negatively charged DNA fragments is placed in a porous gel, and an electric field with a positive charge is applied across the gel. The negative charge on each DNA fragment is proportional to its length because each phosphate group in the chain is negatively charged. As the fragments move through the field, attracted by the positively charged electrode, the gel resists their movement with different intensity, depending on their length. Longer fragments are retarded more strongly than small fragments. The smallest fragments move the fastest and therefore travel the farthest across the gel. Each DNA fragment thus stops at a different point on the gel, and the

14.10 Automated Separation of Chromosomes

This apparatus rapidly separates specific human chromosomes based on how each chromosome stains with two fluorescent dyes. Droplets containing the desired chromosome are identified by their stain, electrically charged, and deflected into a collecting tube.

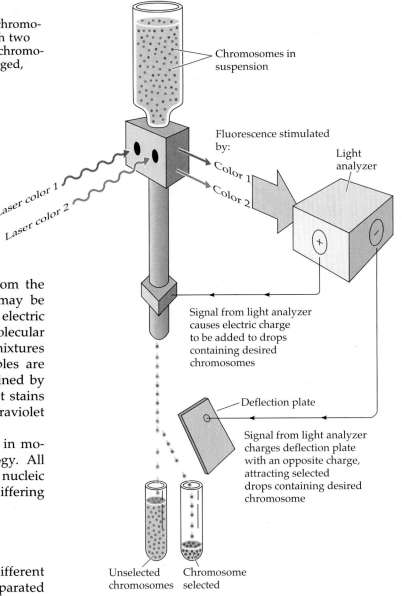

Chromosomes in suspension

Fluorescence stimulated by:

Color 1

Color 2

Laser color 1

Laser color 2

Light analyzer

Signal from light analyzer causes electric charge to be added to drops containing desired chromosomes

Deflection plate

Signal from light analyzer charges deflection plate with an opposite charge, attracting selected drops containing desired chromosome

Unselected chromosomes

Chromosome selected

separated materials can then be removed from the gel in their pure form. Different samples may be "run" side by side in different "lanes" in the electric field. DNA fragments of known length or molecular weight are often run next to experimental mixtures to provide a size reference. After the samples are run, the DNA on the gel plate can be examined by covering the plate with a fluorescent dye that stains the DNA and then placing the plate under ultraviolet light (Figure 14.11*b*).

Electrophoresis is a very useful technique in molecular biology, biochemistry, and cell biology. All sorts of molecules, particularly proteins and nucleic acids, can be separated by virtue of their differing sizes and electric charges.

Detection of Specific DNA Fragments

Electrophoresis separates DNA fragments of different sizes but does not itself show which of the separated fragments contains a particular piece of DNA. Detection of a particular piece depends on complementary base pairing with a suitable **probe**—a strand of DNA or RNA known to have the base sequence complementary to the piece of DNA sought. For example, a specific mRNA can be used as a probe to locate the gene from which it was transcribed. Such hybridization experiments cannot be done in the gel, however; first the DNA must be transferred to a nitrocellulose filter that binds and immobilizes single-stranded DNA.

A technique called **Southern blotting** is often used in such a search for a specific DNA fragment. A mixture of DNA fragments including the one of interest is separated by electrophoresis. The electrophoresis gel is then soaked in an alkaline solution, which breaks the hydrogen bonds of the DNA fragments, separating the strands. The gel is then "blotted" with a sheet of nitrocellulose, in a setup similar to that shown in Figure 14.12, to transfer some of the

DNA to the sheet. The sheet is removed and heated, which "fixes" the DNA so that it is immobilized on the nitrocellulose in its original position. Then the nitrocellulose filter is soaked in a solution containing the probe (mRNA or cDNA that has been labeled, radioactively or otherwise). When probe molecules meet DNA strands to which they are complementary, they are trapped by complementary base pairing, forming double-stranded nucleic acid molecules. These trapped molecules remain attached to the sheet, while the other probe molecules are washed away. The positions of the desired fragments on the sheet may then be detected by finding the bound, labeled probe.

Southern blotting was named after its inventor, Scottish molecular biologist E. M. Southern. Related techniques have been named Northern blotting (RNA blotting) and Western blotting (protein blotting).

(a)

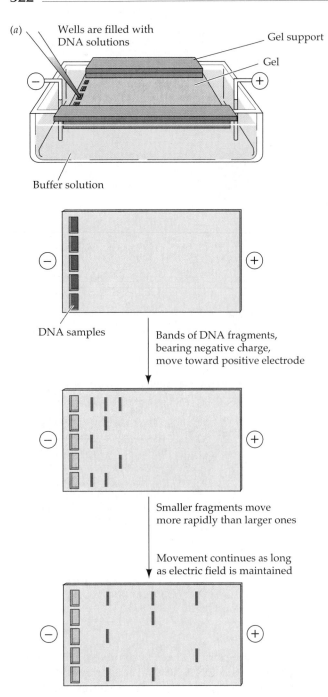

Wells are filled with DNA solutions

Gel support

Gel

Buffer solution

DNA samples

Bands of DNA fragments, bearing negative charge, move toward positive electrode

Smaller fragments move more rapidly than larger ones

Movement continues as long as electric field is maintained

(b)

14.11 Separating Macromolecules by Gel Electrophoresis
(a) One version of a setup for separating nucleic acid fragments or other macromolecules by gel electrophoresis. Samples are injected by pipette into wells in a horizontal gel slab, and an electric field is applied. DNA fragments move at different rates determined by their sizes. (b) Examining a stained slab under ultraviolet light. The pink bands contain the separated DNA fragments.

14.12 Southern Blotting
A buffer solution moves through a wick, an electrophoresis gel, a nitrocellulose filter, and into a stack of absorbent paper that acts as a blotter. As the buffer moves upward, it transfers DNA strands from the gel to the nitrocellulose filter, which immobilizes the DNA strands.

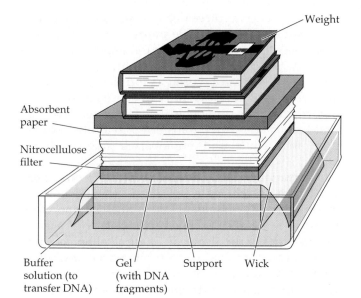

Weight

Absorbent paper

Nitrocellulose filter

Buffer solution (to transfer DNA)

Gel (with DNA fragments)

Support

Wick

Autoradiography

Radioactive materials release energy or subatomic particles that can darken a photographic emulsion. This phenomenon is the basis of autoradiography, a technique illustrated in Box 2.A and described in Box 8.B. Autoradiography can be combined with some of the techniques we have just described to help scientists locate particular DNA fragments. It plays an important part in the technique we describe next.

Localization of Genes on Chromosomes

Radioactive probes (RNA or cDNA) can be used to determine on which chromosome a gene or other specific DNA segment lies and even where it lies on the chromosome. A microscope slide is prepared with cells in metaphase of nuclear division (Chapter 9). The slide is treated with weak base to separate the DNA strands slightly—to break the hydrogen bonds and expose single-stranded DNA for hydrogen bonding. The probe is then poured onto the slide. The probe binds only to the DNA sequence to which it is complementary, and the unbound probe is washed away. The slide is stained to reveal the bands on the chromosomes, and autoradiography shows where the probe has bound in relation to the banding pattern. This technique, summarized in Figure 14.13, is called **in situ hybridization** (*in situ* means "in place").

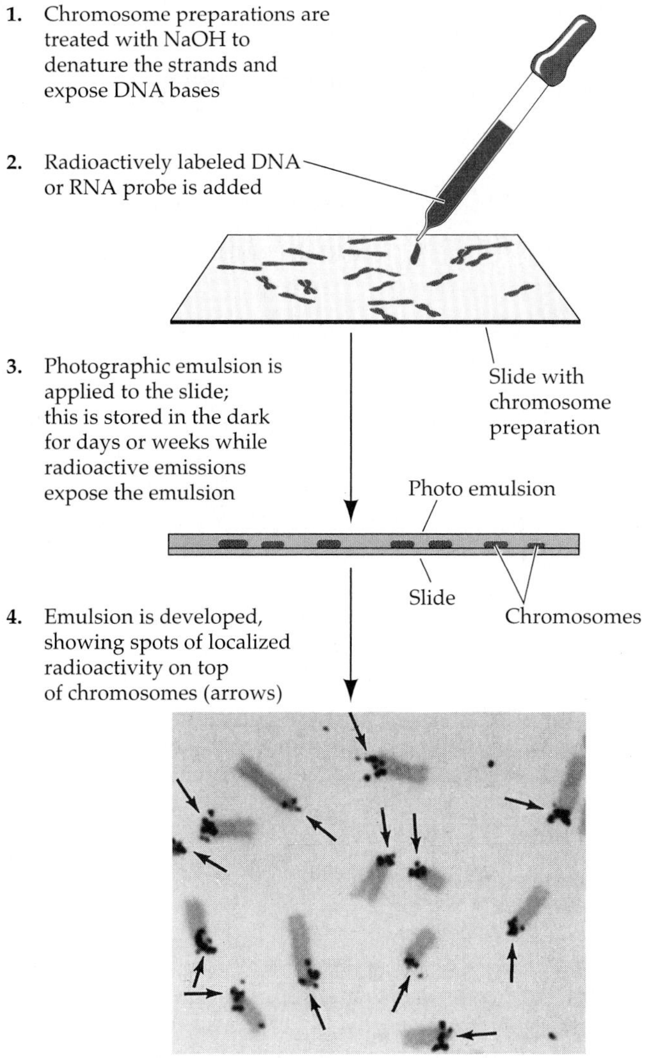

Restriction Mapping

For some research purposes we may want to determine the actual sequence of bases in samples of DNA (gene fragments, genes, or chromosomes). Before we can begin that project, however, we must obtain a preliminary, less detailed map of a DNA molecule. Restriction endonucleases, with their highly specific recognition sequences, can be used to subdivide a DNA molecule into fragments that can then be ordered to form a **restriction map** (Figure 14.14). Sometimes the base sequences of the individual fragments themselves can be determined, a step on the way to sequencing the entire molecule.

Chromosome Walking

Chromosomes are too large for direct DNA sequencing, and they are too large to clone using viruses or plasmids as vectors. To study a chromosome, then, we must first cleave it into a large number of smaller fragments. The chromosome can be treated with one or more restriction endonucleases in such a way that cleavage is incomplete, so individual molecules in the sample are cleaved in different places and the differ-

1. Chromosome preparations are treated with NaOH to denature the strands and expose DNA bases

2. Radioactively labeled DNA or RNA probe is added

Slide with chromosome preparation

3. Photographic emulsion is applied to the slide; this is stored in the dark for days or weeks while radioactive emissions expose the emulsion

Photo emulsion

Slide

Chromosomes

4. Emulsion is developed, showing spots of localized radioactivity on top of chromosomes (arrows)

14.13 In Situ Hybridization
A radioactive DNA or RNA probe locates specific genes on specific chromosomes. Autoradiography reveals where the probe has bound in relation to the banding pattern of the chromosome.

ent fragments overlap one another. After we determine the base sequence in each fragment, we need to know the order of the fragments themselves if we are to know the base sequence of the whole chromosome. How do we determine the order of these smaller fragments in the original chromosome?

An easy and elegant approach, called **chromosome walking**, is illustrated in Figure 14.15. First two samples of the original DNA are cleaved with different restriction endonucleases. The fragments from each sample are then cloned, creating two gene libraries. A clone from the first library is made single-stranded and used as a probe of the second gene library; the probe hybridizes with a fragment only if the fragment contains a base sequence complementary to the probe

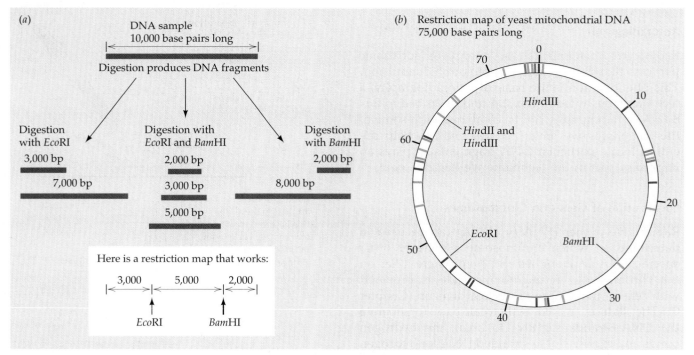

14.14 Restriction Mapping
(a) In three experiments, restriction endonucleases cut samples of the same DNA molecule. The red bars summarize the data obtained. By piecing together the fragments, we can determine the positions of the endonuclease recognition sites in the original DNA. (b) The restriction map of DNA isolated from yeast mitochondria. Cutting sites for four restriction endonucleases are shown. Try to reconstruct the experimental data that gave rise to this map. (Numbers represent thousands of base pairs.)

on one of its strands. If hybridization occurs, then the other strand of the hybrid fragment must contain the same sequence as the probe—that fragment overlaps the probe fragment. The cloned fragment from the second library is then used as a probe of the *first* library to identify a *third* overlapping fragment, the third fragment is used as a probe of the *second* library, and so forth. By identifying successive overlapping fragments, chromosome walking reveals the order of the fragments in the original DNA.

Sequencing DNA

The discovery of restriction endonucleases opened possibilities for determining the base sequence of a fragment of DNA. By using various restriction enzymes, one could cut a sample of DNA into multiple fragments. Hybridization techniques enabled biologists to determine which genes are associated with which fragments and to construct a partial map of the DNA sample. But there was still the problem of determining the base sequences within these smaller pieces. In the mid-1970s, the British biochemist Frederick Sanger (who in 1953 had determined the first protein primary structure), Allan Maxam and Walter Gilbert (both of Harvard University), and others de-

vised techniques for the rapid sequencing of DNA. A number of methods are in current use; one is described in Box 14.A.

Why should we want to determine the base sequence of a DNA sample? One reason is that this knowledge, along with our knowledge of the genetic code, enables us to determine the primary structure of the gene's protein product. Although we could analyze the protein directly, it turns out to be easier to determine the primary structure of a protein by analyzing the corresponding DNA.

Most important, knowing base sequences may help us determine how regulatory sites (such as promoters) function. Also, closely related genes may have quite different functions, and sequencing helps us understand how they differ. In addition, we can use DNA sequencing to identify precise molecular defects, as in genetic diseases. Gene cloning coupled with DNA sequencing has provided important information about the insertion of transposable elements in eukaryotes as well as in prokaryotes. DNA sequencing can provide information of interest to evolutionary biology as well. By comparing the base sequences of homologous genes from different organisms, we can extend our knowledge of the evolutionary relationships among species.

1. Two DNA samples are treated separately with two restriction endonucleases, generating two gene libraries

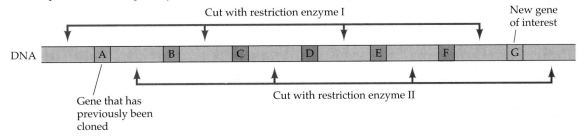

2. Gene libraries from the two enzymes contain overlapping fragments

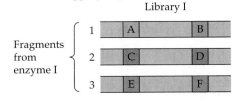

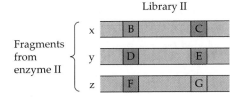

3. The libraries are probed, first with A and then with newly identified probes, "walking" down the DNA

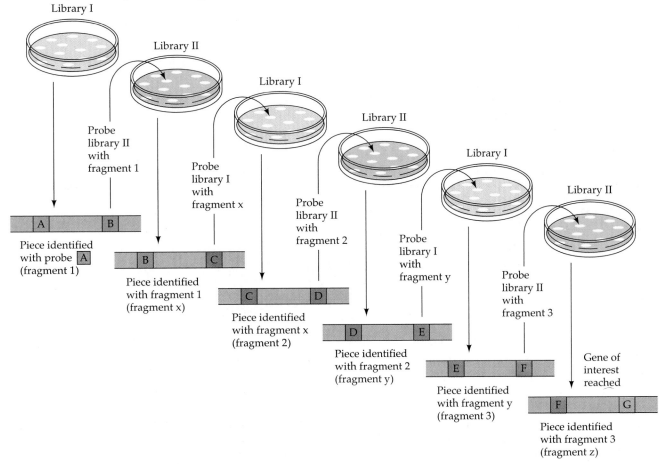

4. DNA is mapped from identified fragments

14.15 Chromosome Walking
The "volumes" in a gene library (see Figure 14.8) can be put in the proper order—their order in the original DNA molecule—by chromosome walking.

BOX 14.A

Determining the Base Sequence of DNA

One of the simplest, fastest, and most accurate methods for sequencing DNA samples was developed by Leroy Hood and his colleagues. This method has similarities to the experiment described in Chapter 11 (in the section entitled "Transcription") in which the compound cordycepin triphosphate is used to interrupt RNA synthesis. This experiment demonstrated that RNA synthesis, like DNA synthesis, proceeds from the 5' to the 3' end of the molecule.

The nucleoside triphosphates normally used as substrates for DNA synthesis are the 2'-deoxyribonucleoside triphosphates (dNTPs). The Hood technique also makes use of 2',3'-dideoxyribonucleoside triphosphates, or ddNTPs (shown in figure (a)). Suppose that a DNA strand is being synthesized, by the addition of one dNTP after another. If a ddNTP is picked up instead, it joins the growing chain; but, since it lacks a free hydroxyl group at C3', it cannot accept the next dNTP. Thus synthesis stops at the point where the ddNTP is inserted—like cordycepin

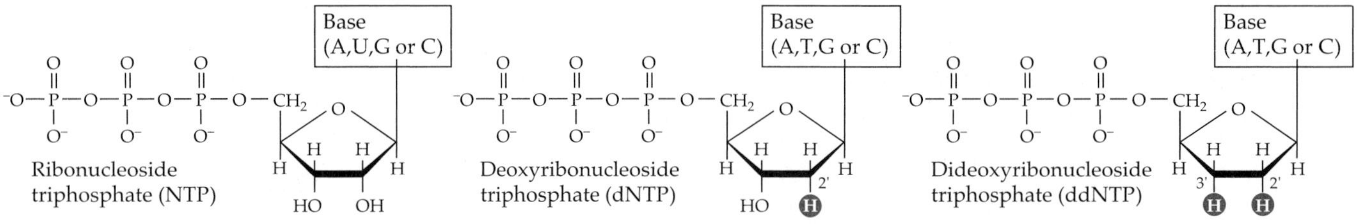

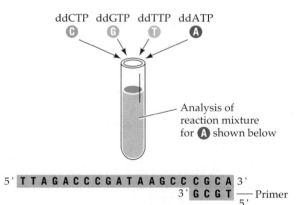

1. Single-stranded DNA sequence to be determined

5' TTAGACCCGATAAGCCCGCA 3'

2. A sample of this unknown DNA is combined with primer, DNA polymerase, 4 dNTPs, and 4 ddNTPs each bound to a fluorescent dye, and synthesis begins. The results are illustrated here by what binds to a T in the unknown strand. If dATP is picked up from the mixture, synthesis continues. If ddATP is picked up, synthesis stops. A series of fragments of different lengths is made, each ending with a ddNTP

ddCTP ddGTP ddTTP ddATP

Analysis of reaction mixture for A shown below

5' TTAGACCCGATAAGCCCGCA 3'
                    3' GCGT — Primer
                          5'

GENE COPIES BY THE BILLION

Biologists often want to obtain particular pieces of DNA—a particular gene, for example—in quantity sufficient for biochemical studies. A powerful technique, the **polymerase chain reaction**, has made this process of DNA amplification relatively easy. First described in 1984, the polymerase chain reaction has become one of the most widely used techniques in molecular biology. It can produce billions of copies of a single piece of a DNA molecule—the target se-

quence—in a few hours. By contrast, the conventional techniques of cloning recombinant DNA require days to weeks.

The polymerase chain reaction is a cyclic process in which the following sequence of steps is repeated over and over again. Double-stranded DNA is heat-denatured into single strands, primers for DNA synthesis are added to the 3' ends of the target DNA sequence on the separated strands, and DNA polymerase catalyzes the production of new complementary strands. A single cycle doubles the amount of triphosphate, the ddNTP terminates the chain.

In this technique for sequencing DNA, single-stranded DNA to be sequenced is combined with DNA polymerase (to synthesize the complementary strand), a primer, the four dNTPs (dATP, dGTP, dCTP, and dTTP)—and small amounts of each of the four ddNTPs (ddATP, ddGTP, ddCTP, or ddTTP), each attached to a fluorescent molecule emitting a different color of light (see part (b) of the figure). The reaction mixture soon contains a DNA mixture made up of the unknown single strand and shorter complementary strands. The complementary strands, each terminating in a ddNTP, are of various lengths.

For example, each time a T is reached on the template strand, the growing complementary strand adds, at random, either dATP or ddATP. If ddATP is added, chain growth terminates at that point. By chance, some of the replicating strands grow to greater lengths than others before coming to a ddATP stopping point. The other ddNTPs also produce strands of various lengths. The strands can then be displayed by gel electrophoresis, resulting in a series of colored bands. The color of each band tells us which ddNTP terminated the strands in that band. Because we thus know which ddNTP gave rise to each strand, we also know which base is last in the strands of each set. Given that these strands are complementary to the original sample, we can then determine the exact sequence of bases in the original DNA sample, as shown in the figure.

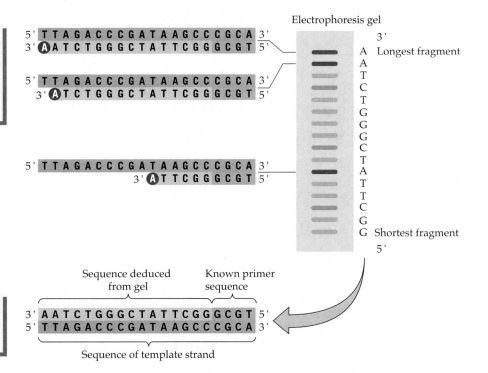

3. The resulting strands of various lengths are separated by gel electrophoresis, which can detect length differences as short as one base. Each strand fluoresces a color that identifies the ddNTP that terminated the strand

4. The sequence of the newly synthesized strand of DNA can now be deduced and converted to the sequence of the template strand

the DNA sequence in the reaction mixture and leaves the new DNA in the double-stranded state (Figure 14.16). Each cycle takes only one to a few minutes.

To use the polymerase chain reaction, a scientist must know a sequence of about 20 bases at the 3' end of the target sequence on each DNA strand. Knowing that sequence, the scientist can create the two oligonucleotides that are complementary to the sequence to use as primers for DNA synthesis. The primers are crucial: Whether the starting DNA sample is ample or limited, pure or crude, long or short, the primers indicate where DNA synthesis is to take place. The primers give the polymerase chain reaction its versatility.

Separating the strands of a DNA molecule requires a temperature of about 94°C. Therefore, the reaction mixture must be heated to that temperature in each cycle of the polymerase chain reaction. At first this was a problem, because the heating destroyed the DNA polymerase that catalyzed the polymerase chain reaction. Now, however, the polymerase chain reaction is run with a temperature-resistant DNA

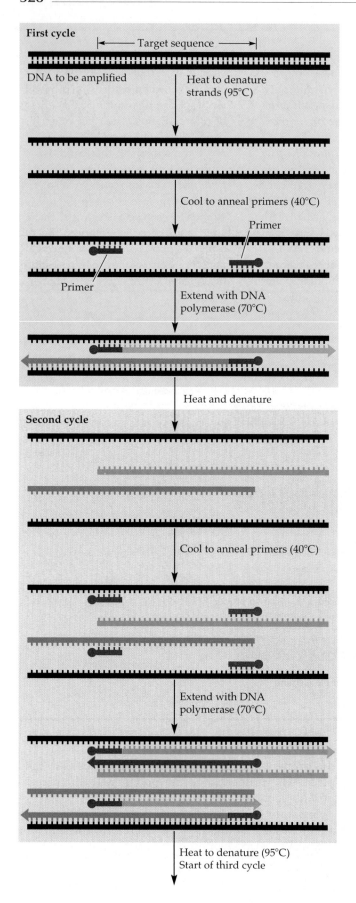

First cycle

Target sequence

DNA to be amplified

Heat to denature strands (95°C)

Cool to anneal primers (40°C)

Primer

Primer

Extend with DNA polymerase (70°C)

Heat and denature

Second cycle

Cool to anneal primers (40°C)

Extend with DNA polymerase (70°C)

Heat to denature (95°C)
Start of third cycle

14.16 The Polymerase Chain Reaction

Because it is used so often and in so many laboratories, this technique was named "molecule of the year" by the journal Science in 1990. In the presence of appropriate primers, a heat-resistant DNA polymerase copies both strands of the desired DNA sequence; by alternating heating (to separate DNA strands) and synthesis phases, a PCR machine can rapidly make enormous numbers of copies.

polymerase from the bacterium *Thermus aquaticus*, which lives in hot springs. This polymerase survives the high temperatures used to separate the DNA strands. The whole process has been automated, and the compact "PCR machine" (polymerase chain reaction machine) is found in thousands of laboratories all over the world.

The amplification of DNA samples by the polymerase chain reaction has found many uses, ranging from the diagnosis of hereditary diseases in human fetuses to the identification of semen samples in criminological investigations and the study of ancient DNA in frozen samples from the last Ice Age.

PROSPECTS

The techniques of recombinant DNA technology have become standard tools for investigators of the molecular biology of both prokaryotes and eukaryotes and in the pharmaceutical industry as well as the clinical laboratory. These techniques are being used to turn selected strains of bacterial, yeast, plant, and animal cells into factories for the production of important polypeptides. Thus we can supply patients with gene products they cannot make for themselves; surely this technology will be a major tool of medicine well into the twenty-first century. We will consider some of the medical applications of recombinant DNA technology in Chapter 15.

Concerns about safety were expressed in the early days of recombinant DNA research, and at one point the research was called to a temporary halt. The worst of the fears have abated, however, and the technology is now widely agreed to be safe as long as certain reasonable precautions are observed. These precautions, set forth as guidelines by the U.S. National Institutes of Health, include precisely defined containment provisions (methods for keeping experimental organisms from escaping from the laboratory) and rules for handling different biological materials. Most experiments are thought to pose no hazard. Some are considered risky and accordingly are conducted under more restrictive conditions. When first issued, the guidelines specifically restricted recombinant DNA work with the genes of cancer viruses. But gradually, as various concerns have been put to rest, the guidelines have been relaxed. Relaxation of the guidelines has led to major advances in our understanding of the cancer-producing oncogenes (see Chapter 15).

Plant Agricultural Biotechnology

Earth's human population is increasing by more than 100 million individuals each year. In less than four decades the population will reach nine billion. To feed all those people we will have to triple world food production. Most humans depend on only a few crop plants for essential calories—rice, wheat, maize, sorghum, and several others (such as potatoes) that dominate in certain local agricultures. Because only a few genetic strains of each of these crops are planted in a given year, the food supply of hundreds of millions of people is in constant peril from mutations in disease-causing organisms that could overcome the plants' resistance and destroy entire harvests. Equally important threats are changes in weather patterns, or political and economic disruptions that affect supplies of essential agricultural chemicals, fertilizers, or fossil fuels.

Other than the overriding need to contain population growth, the greatest potential for improving human welfare with modern biological technology lies in the search for economically feasible crop plants that are higher-yielding; more nutritious; disease-resistant; drought-, salt-, and pollution-tolerant; and otherwise able to meet the challenges of our overextended planetary resources.

Plant breeding is a form of "genetic engineering" that has been used for a long time; it has been practiced intensively in the twentieth century. By the judicious crossing of existing strains of plants, we have been able to increase crop yields about 1 percent per year for the past century. Among the greatest triumphs of the plant breeders were the hybridization of corn (in the 1930s) and the "Green Revolution" (in the 1950s and 1960s, during which improved strains of wheat and rice were used to increase food production in many parts of the world). Although plant breeding will continue to play a key role in the development of agriculture, it has three significant limitations. (1) Plant breeding is a slow process, and it requires many acres of land. (2) It is nonspecific in that the entire genomes of the parents participate in a cross (the genome of an organism is all the genetic information the organism contains); thus unwanted genes may appear in the progeny along with the desirable ones for which the breeder is selecting. (3) The breeder can work only with strains (and, more rarely, species) that can interbreed with one another.

The addition of recombinant DNA technology to the tools of the breeder addresses each of those three limitations. First, recombinant DNA techniques are often applied to many millions of independent cells at a time, all within a single flask, thus eliminating the space problem; and the cells, unlike whole plants, multiply many times a day. Second, individual genes can be transferred from one plant to another without dragging along other, undesired, genes. Third, gene transfer is not restricted to closely related plants; in fact, a plant may be given genes from *any* organism. When we know the specific biochemical basis for a change that we want to make in a plant, recombinant DNA technology is extraordinarily useful.

For both conventional plant breeding and recombinant DNA work, there is a continuing and growing need for a source of suitable genes, sometimes referred to as **germ plasm**, to introduce into existing species. One goal of conservationists and biologists is to maintain an adequate global supply of germ plasm, as seeds in repositories and as plants in nature, to ensure genetic diversity in nature and a continuing supply of tools for the breeder and the biologist. A lost species is a lost treasury of germ plasm. Recombinant DNA, perhaps as gene libraries, could supplement a gene repository for germ plasm.

The cloning vector commonly used in recombinant DNA work with plants is a plasmid found in *Agrobacterium tumefaciens*, a bacterium. *A. tumefaciens* is a pathogen that causes the plant disease crown gall, which is characterized by large tumors (Figure 14.17*a*). The bacterium contains a large plasmid, called Ti (for *T*umor-*i*nducing). Part of the Ti plasmid is T-DNA, a transposon (see Chapter 12) that produces copies of itself in the chromosomes of infected plant cells. This transposon is the key to gene cloning in plants. The gene to be cloned is inserted into the transposon in a Ti plasmid, the plasmid is inserted (by transformation) into *A. tumefaciens*, the bacterium infects the plant, and the gene is copied into the plant's chromosomes along with the rest of the transposon (Figure 14.17*b,c*). Since the recombinant DNA is found only in tumor cells, this might seem like a nonheritable change in the plant's genome. However, tumor cells can be isolated and grown in culture, eventually giving rise to a complete, new, normal plant (Figure 14.18). If a plant grows from a tumor cell containing recombinant DNA, each of its cells is transgenic, as explained earlier in this chapter.

Scientists have produced transgenic crop plants that are resistant to herbicides or to insects and others that are resistant to viral diseases (Figure 14.19). Still other transgenic plants produce an enzyme that breaks down a plant growth substance that normally causes senescence (aging; see Chapter 34) and thus spoilage of harvested fruit.

Transgenic bacteria will also play growing roles in agriculture. For example, nitrogen fixation—the conversion of atmospheric nitrogen gas to ammonium ions usable as a nitrogen source for plants—is crucial

(a)

14.17 Crown Gall and *Agrobacterium tumefaciens*

(a) *A. tumefaciens* causes tumors—crown gall—on infected plants, such as this geranium. New shoots are forming within the gall, which lies at the base of the geranium stem. (b) *A. tumefaciens* contains the Ti plasmid which, in turn, contains a transposon called T-DNA. Copying the transposon to a chromosome of an infected plant is a key step in the development of the disease. (c) A scientist can clone a desired gene by inserting it into the transposon of a modified Ti plasmid, transforming *A. tumefaciens* with the plasmid, and infecting plants with the bacteria.

(b)

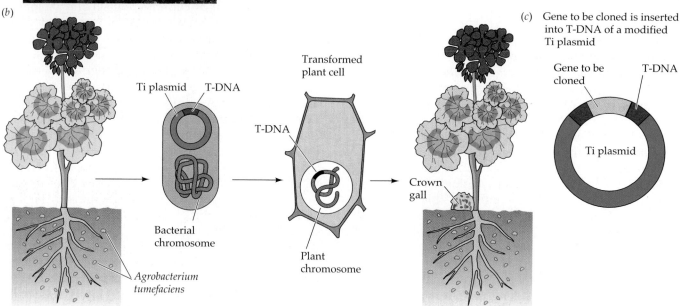

(c)

14.18 Multiplying a Clone

This rosette of transgenic cotton plantlets is developing from a culture of crown gall tumor tissue. The plantlets contain a gene they picked up from the *Agrobacterium tumefaciens* that caused the tumor.

to agriculture and to life on Earth (see Chapter 32). Much current research is focusing on modifying both the nitrogen-fixing bacteria and the plants that harbor them in order to improve the efficiency of nitrogen fixation. In another example, genetically altered bacteria have already been produced that, when present on plant surfaces, prevent frost formation (Figure 14.20).

Recombinant DNA Technology and the Environment

With disturbing frequency, great quantities of oil have been accidentally released from tankers into surrounding waters, causing severe damage to sensitive environments (Figure 14.21). Genetically engineered bacteria, equipped with DNA that codes for enzymes that cleave hydrocarbons and other constituents of oil, are being developed to combat such

14.19 Genetically Engineered Resistance to a Virus
When tested in the field, transgenic tomato plants (right) treated to resist infection by the cucumber mosaic virus produced marketable fruit. The control plant (left) was stunted by viral infection and produced no tomatoes.

spills. Several biotechnology companies are producing these and other bacterial strains modified to deal with other types of organic pollutants, such as sewage and dioxins.

At the other end of the petroleum industry, genetically engineered bacteria will be useful in the production of chemicals to enhance yields from oil drilling. Oil yield in the United States continues to decline, and products of recombinant DNA technology represent an approach to increasing yields from existing and newly discovered reservoirs.

14.20 Fighting Frost with Bacteria
Strawberry blossoms and *Pseudomonas syringae* bacteria in water at –2.8°C. The tube on the left contains normal *P. syringae* and is frozen solid. The *P. syringae* on the right, however, lack a protein on their surface that provides a "template" for water molecules to align as ice—the gene that codes for the protein has been deleted. Ice does not form readily in the presence of these transgenic bacteria. Spraying such bacteria on crops may provide protection from freezing.

Yeasts and other microorganisms can concentrate metals such as nickel, gold, and plutonium from dilute solutions. Thus they can be used both in combating pollution and in increasing the recovery of important metals from waste. Genetic manipulation of bacteria and yeast for metal recovery and for the separation of metals from their ores is being actively pursued.

Release of transgenic organisms into the environment in large numbers must be done with care. Generally, bacteria prepared for release into the soil, lakes, or oceans will be designed to do their work effectively but not to persist long in the environment. In spite of such care, debates on the safety and merits of this kind of work should continue. We must always consider the environmental impact and health concerns as well as the technological and economic advantages.

Genome Projects and Modern Medicine

There are hundreds of human hereditary diseases, and a new one is discovered every two or three days.

14.21 An Environmental Disaster
This oil slick resulted from a spill in the Delaware River near Philadelphia. Biotechnologists are working on methods to deal biologically with such spills.

It has been estimated that each person is heterozygous for no fewer than 30 recessive genetic diseases. Recombinant DNA technology affords new methods—often the first methods available—for combating these diseases. Several **genome projects** are already under way. These are major efforts to *map* and *sequence* the human genome and, in part or in whole, the genomes of certain other species. Mapping precedes sequencing and in itself yields much valuable information. The development of effective gene therapies (see Chapter 15) will depend on the mapping and characterization of the genes responsible for the diseases, and this provides major motivation for supporting human genome projects. Genes responsible for cystic fibrosis and other diseases have already been mapped and cloned and can now be studied to learn about the gene products, the diseases themselves, and their possible treatment.

Genome projects require the gathering of vast amounts of data, including the locations of genes, their functions, their base sequences, and more. Processing—and simply storing—so much information requires massive computing resources, and the analysis of it all will depend heavily on advances in computing theory. Biology and computer science will be full partners in this venture. While computer science develops new algorithms and parallel processing techniques, biology and biotechnology will develop new physical and biochemical techniques for DNA manipulation and analysis.

The return on this investment will be magnificent, including a revolution in the diagnosis and treatment of hereditary diseases. Even if we sequence every base in the human genome, however, we still will not know what all the DNA sequences *do*—or how their products do what they do. An understanding of genetic regulation in eukaryotes requires knowledge of much more than base sequences. But without the mapping and sequencing data from genome projects, these greater problems will remain unsolved.

SUMMARY of Main Ideas about Recombinant DNA Technology

Recombinant DNA technology serves research in most areas of present-day biology; in addition, its applications are important in agriculture, medicine, forensics, environmental protection, and the chemical industry.

Foreign DNA can be inserted into a vector (a virus or plasmid) with the assistance of restriction endonucleases and other enzymes.
Review Figures 14.2, 14.3, and 14.4

The modified vectors are inserted into suitable hosts for cloning.
Review Figures 14.5 and 14.6

Bacteria are widely used as hosts, but eukaryotic hosts, particularly yeasts, are increasingly used for transfection with eukaryotic genes.

Gene libraries, cDNA production, and direct chemical synthesis are sources of DNA for cloning.
Review Figures 14.8 and 14.9

DNA fragments are commonly separated by gel electrophoresis and then identified by Southern blotting.
Review Figures 14.11 and 14.12

Specific DNA fragments can be amplified rapidly by means of the polymerase chain reaction.
Review Figure 14.16

DNA fragments and chromosomes may be analyzed by restriction mapping, chromosome walking, and DNA sequencing.
Review Figures 14.14 and 14.15

SELF-QUIZ

1. Restriction endonucleases
   a. play no role in bacteria.
   b. cleave single-stranded DNA molecules.
   c. cleave DNA at highly specific recognition sites.
   d. are inserted into bacteria by bacteriophages.
   e. add methyl groups to specific DNA base sequences.

2. "Sticky ends"
   a. are double-stranded ends of DNA fragments.
   b. are complementary to other specific sticky ends.
   c. rejoin best at elevated temperatures.
   d. are removed by restriction endonucleases.
   e. are identical for all restriction endonucleases.

3. Which statement is *not* true of DNA ligase?
   a. It is an enzyme.
   b. It is a normal constituent of cells.
   c. It can unite sticky ends in recombinant DNA work.
   d. It functions in the normal replication of DNA.
   e. It mends breaks in polypeptide chains.

4. Which feature is undesirable in a plasmid for cloning a gene?
   a. Possession of an origin of replication
   b. Possession of genes conferring resistance to antibiotics
   c. Possession of recognition sites for multiple restriction endonucleases
   d. Possession of multiple recognition sites for the endonuclease to be used
   e. Possession of genes other than the one to be cloned.

5. Transfection can be accomplished by
   a. using high-velocity particles of tungsten coated with DNA.
   b. Southern blotting.
   c. gel electrophoresis.
   d. the polymerase chain reaction.
   e. heating the material to denature it.

6. Complementary DNA
   a. is produced from ribonucleoside triphosphates.
   b. is produced using oligo-dU.
   c. is produced by reverse transcription.
   d. requires no primer.
   e. requires no template.

7. Southern blotting
   a. is used to detect a specific DNA fragment.
   b. is used to detect a specific RNA fragment.
   c. is used to detect a specific polypeptide fragment.
   d. is a technique for separating nucleic acid fragments.
   e. is used to separate chromosomes.

8. Restriction mapping
   a. is a useful tool for separating DNA fragments.
   b. is a useful tool for subdividing a DNA molecule into manageable fragments.
   c. cannot be used on prokaryotic DNA.

   d. can be used to produce large quantities of specific DNA.
   e. is an expensive and controversial procedure.

9. The polymerase chain reaction
   a. is a method for sequencing DNA.
   b. is used to detect a specific DNA fragment.
   c. is used to produce large quantities of specific DNA.
   d. is used to map genes.
   e. is used to transcribe specific genes.

10. Genome projects
   a. are a matter for the distant future.
   b. all focus on the human genome.
   c. will tell us what all our genes do.
   d. have already yielded medically useful results.
   e. will probably all be carried out in one carefully chosen university.

FOR STUDY

1. Using examples from this chapter, describe how molecular biologists have found new uses in recombinant DNA technology for enzymes produced by bacteria.

2. Make a thorough list of the phenomena and techniques discussed in this chapter that depend on complementary base pairing.

3. You have attempted to insert a copy of a particular gene into a plasmid, specifically placing your gene in the middle of a gene conferring resistance to the antibiotic streptomycin. The plasmid also has a gene conferring resistance to the antibiotic aureomycin. You have transformed bacteria (sensitive to both antibiotics) with your plasmid suspension. Describe the procedures you would use to select the bacteria that have taken up the plasmid. What additional steps would be required to select the bacteria that have taken up copies of the plasmid containing your gene?

4. Discuss (a) what you see as important positive features of a human genome project and (b) what you consider to be negative features of such a project.

READINGS

Gasser, C. S. and R. T. Fraley. 1992. "Transgenic Crops." *Scientific American,* June. A brief overview of some major projects for the improvement of crops through biotechnology.

Gilbert, W. 1991. "Toward a Paradigm Shift in Biology." *Nature,* January 10. A discussion of the impact of recombinant DNA technology on pure research in biology.

Griffiths, A. J. F., J. H. Miller, D. T. Suzuki, R. C. Lewontin and W. M. Gelbart. 1993. *An Introduction to Genetic Analysis,* 5th Edition. W. H. Freeman, New York. An outstanding genetics textbook, thoroughly up to date.

Murray, A. W. and J. W. Szostak. 1987. "Artificial Chromosomes." *Scientific American,* November. Tools for cloning human genes in yeast and for the investigation of chromosomal behavior during mitosis and meiosis.

Neufeld, P. J. and N. Colman. 1990. "When Science Takes the Witness Stand." *Scientific American,* May. Ethical issues in the use of DNA evidence.

Watson, J. D., J. Tooze and D. T. Kurtz. 1989. *Recombinant DNA: A Short Course,* 2nd Edition. Scientific American Books, New York. Begins at the elementary level but ends up presenting a great deal of molecular biology. A short, intense book, but readable.

Weinberg, R. A. 1985. "The Molecules of Life." *Scientific American,* October. Introductory chapter to a special issue on the molecules of life; gives a good overview of the role of recombinant DNA technology in various aspects of molecular biology.

Weintraub, H. M. 1990. "Antisense RNA and DNA." *Scientific American,* January. Deactivation of specific genes—a powerful research tool, perhaps someday a medical tool as well.

Searching for Genetic Causes—and Cures
Andrew Slay (seated) suffers from spinal muscular atrophy, a severe neuromuscular disease caused by a defect in a single gene. He is with Dr. T. Conrad Gilliam, a geneticist who has conducted a long search to find the gene responsible. Once the gene is located, modern genetic techniques may be able to correct the defect and cure the disease.

15

Genetic Disease and Modern Medicine

What happens when a gene doesn't work? If a person carries alleles that fail to code correctly for a particular protein, the effect may be medically insignificant—blue eyes instead of brown eyes, for example. If the missing protein catalyzes a step in a pathway such as cellular respiration, however, the zygote itself is not viable, and there will *be* no person. Between these extremes lie many inherited genetic diseases, such as sickle-cell anemia and cystic fibrosis. Our species is subject to more than 4,000 known heritable metabolic defects, most of which block production of necessary enzymes or other proteins. Many of these defects are recessive traits, whose alleles are unable to code for the normal proteins. It has been estimated that each of us is heterozygous—a carrier—for 30 or more of these recessive disorders.

Does all genetic ill health result from having defective alleles? The answer is no. We discussed another type of genetic ill health, chromosomal aberrations, in Chapters 9 and 10. Conditions such as Down syndrome result from either inappropriate numbers of chromosomes or inappropriate chromosomal structures. These conditions are inherited.

Is all genetic ill health inherited? Again, the answer is no. Cancer is not usually inherited, but it *is* a genetic disease because it results from effects on certain important genes that we will discuss later in this chapter. The change may occur in any part of the body, and in some cases, it may be expressed in many parts of the body.

Is there any hope that cancer and other genetic diseases can be defeated? Modern biomedical science has made considerable strides since gaining the ability to locate and clone genes whose defective alleles are responsible for genetic diseases. In the next few years we should see major advances in our ability to treat and prevent some of these diseases. Let's begin by considering a few of the better-known human genetic disorders.

SOME INHERITED DISEASES

Human populations were isolated from one another for long periods of evolutionary time. As will be explained in Chapter 19, an allele that remains rare in one isolated population may, with time, become much less rare in another population. Some defective alleles responsible for inherited diseases became less rare in certain human groups. The probability of your carrying a defective allele for certain diseases thus depends on your ancestry.

Cystic fibrosis, caused by an autosomal recessive allele (see Chapter 10), is a major killer of young people; most of its victims die during their early to mid twenties. Among Caucasians, cystic fibrosis is

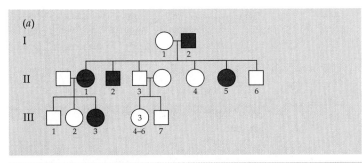

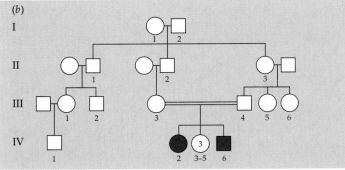

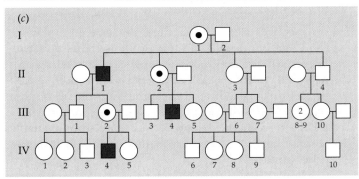

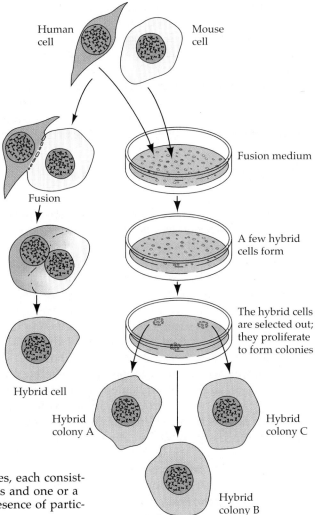

15.4 Patterns of Inheritance

In these pedigrees, roman numerals identify generations; arabic numerals identify individuals within each generation. The symbols are explained in Figure 15.3; numbers inside of symbols mean multiple individuals of that discription (e.g., three sisters in generation IV). (a) Inheritance of an autosomal dominant trait. Individual I-2 is a male heterozygous for the trait; if he were homozygous, all of the progeny in generation II would show the trait. None of the affected individuals in generations II or III can be homozygotes, because each has received a normal recessive allele from one parent. This is the pattern of inheritance for Huntington's disease. (b) Inheritance of an autosomal recessive trait. If the trait were dominant rather than recessive, either III-3 or III-4 would have to show the trait in order for IV-2 and IV-6 to have inherited it. III-3 and III-4 are first cousins—their mating is represented by the double horizontal line. There is no indication of sex linkage. This is the pattern of inheritance for cystic fibrosis. (c) Inheritance of a sex-linked recessive trait. The affected individuals—II-1, III-4, and IV-4—are all males. The carriers are all females. This is the pattern of inheritance for hemophilia.

availability of data from pedigree analysis. Once a gene has been localized on a particular chromosome by somatic-cell genetic techniques, we know that any other genes linked to that gene (as previously discovered by pedigree analysis) must reside on the same chromosome.

Another technique of somatic-cell genetics is chromosome-mediated gene transfer. With this method, instead of combining cells from two species, the researcher combines somatic cells of a recipient species (such as a mouse) with *chromosomes* isolated from metaphase cells of a donor species (such as *Homo sapiens*). When recipient cells are incubated with isolated chromosomes, only some parts of some chromosomes are taken up by the cells, and only a small fraction of what is taken up becomes associated with

15.5 Mouse and Human Cells Can Fuse

The final result of this hybridization process is a series of colonies, each consisting of many cells with a full complement of mouse chromosomes and one or a few human chromosomes. Each colony may be tested for the presence of particular genes and chromosomes.

(a)

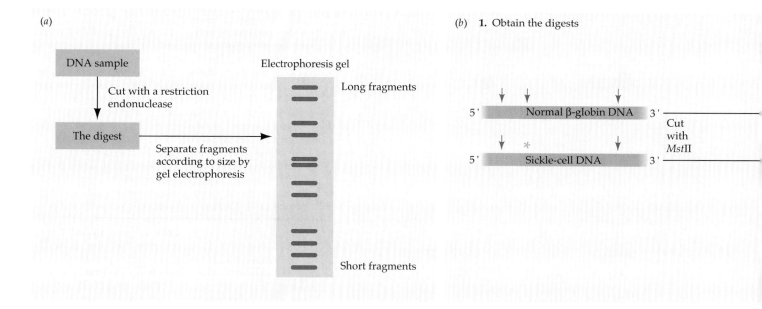

1. Obtain the digests

the host chromosomes of the mouse and is thus sta- bilized. The chances that two particular genes of the donor will both be stabilized are extremely slight unless the genes are closely linked. Thus by observ- ing the frequency of two loci that are transferred together, biologists can determine the map distance between them.

Somatic-cell genetics provides useful, but not very delicate, tools for mapping genes. In the next section we will see how restriction endonucleases help iden- tify more specifically the location of a particular gene.

Restriction Fragment Length Polymorphisms

Restriction endonucleases cut DNA molecules at spe- cific locations; each restriction endonuclease recog- nizes a unique sequence of base pairs and cuts DNA only at that sequence (see Chapter 14). If we treat a sample of DNA with a restriction enzyme and then subject the fragments (referred to as the digest) to gel electrophoresis, the fragments separate according to their lengths into groups, which appear as bands on the gel (Figure 15.6*a*). A given single-stranded DNA probe may bind to one or more of the bands, as revealed by Southern blotting (see Chapter 14). The bands to which the probe binds are pieces of DNA that are, in part, complementary to the probe. If we apply the same probe to Southern blots of restriction digests from different individuals, we may find differences in the binding patterns (Figure 15.6*b*).

What do such differences mean? The differences must result from mutations that eliminate or add restriction endonuclease cutting sites in the region corresponding to the probe. A single point mutation

can easily cause such a change. The existence of more than one band pattern for a probe is called a restric- tion fragment length polymorphism, or **RFLP** (pro- nounced "rifflip"). A RFLP band pattern is inherited in Mendelian fashion and can be followed through a pedigree—and it can serve as a genetic marker. Many RFLPs have been discovered and are now available as landmarks that can be related to the positions of other genes.

How can we use RFLPs to help determine the location of a gene? We can study pedigrees for evi- dence that the gene of interest and certain RFLP band patterns are inherited together, implying that the gene and the markers lie in the same part of a chro- mosome (Figure 15.7). RFLP mapping enabled sci- entists to determine that the cystic fibrosis gene is located on a portion of the long arm of human chro- mosome 7. But that portion was still huge—about 1.5 million base pairs long. How can we find the gene in a "haystack" that big?

Narrowing the Search

Chromosome walking (see Chapter 14) is a suitable technique for finding a gene in a relatively small stretch of DNA. For larger stretches—such as the 1.5- million-base-pair stretch containing the cystic fibrosis gene—an analogous procedure called chromosome jumping was developed. Using this technique, sci- entists established a series of starting points for chromosome walks and narrowed the location of the cystic fibrosis gene to a much smaller part of chro- mosome 7.

Additional experiments narrowed the search even further. Eventually the start and stop signals were

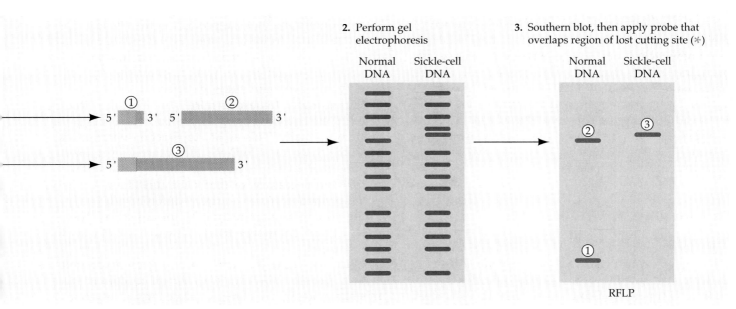

2. Perform gel electrophoresis

3. Southern blot, then apply probe that overlaps region of lost cutting site (✻)

15.6 Probing Restriction Digests

(a) DNA fragments cut by a restriction endonuclease can be put in order by size using gel electrophoresis. (b) Probing restriction digests may reveal that DNA from two individuals differs, as in this comparison of DNAs from two homozygous humans, one having the allele that codes for normal β-globin polypeptide (a subunit of the hemoglobin molecule) and the other having the sickle-cell allele. The restriction endonuclease *Mst*II has three cutting sites (red arrows) in or near the β-globin gene, but one of these sites is eliminated (green asterisk) by the mutation that forms the sickle-cell allele. A probe overlapping that site reveals the RFLP in this region.

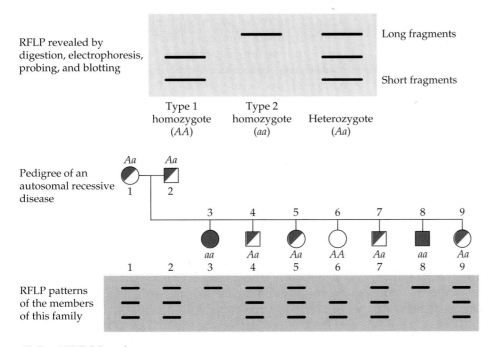

15.7 RFLP Mapping

In the hypothetical pedigree shown here, the inheritance of the RFLP shown at the top center and that of an autosomal recessive disease are closely linked. Affected individuals have the genotype *aa*. Note the fragment pattern from child number 6, which suggests that he has genotype *AA*. This RFLP can serve as a marker in mapping the locus of the disease gene.

located, and ultimately the nucleotide sequence of the gene was determined. Thanks to these sophisticated new techniques, the gene for cystic fibrosis has been identified, mapped, and sequenced. As we noted earlier, a few other disease genes have been located by similar means, and the number is increasing steadily. Locating the gene that causes a particular disease is the first step to developing an effective treatment. What do we do after we have found a gene related to a disease?

DEALING WITH A DEFECTIVE GENE

After locating a gene related to a disease, the next thing we want to know is what the gene *does*—that is, what the function of the protein encoded by the gene is. Ultimately, of course, we want to learn what to do about the defective allele—how to cure the disease. Recent advances are helping us toward both of these goals. We can study specific gene function in laboratory animals called "knockout mice" that are genetically engineered to contain the defective allele. We can also try to get around the effects of a defective allele by inserting functioning alleles into the cells of persons with the disease.

Knockout Mice

Since it is usually impossible to use humans as the laboratory subjects when studying the functions of defective alleles, what researchers needed was an **animal model** for the specific disease. The ideal animal would show the symptoms of the disease, would be useful in biochemical and molecular biological studies, and could be the test organism for possible therapies. Unfortunately for us, nonhuman animals do not usually get the same genetic diseases humans do. Intensive laboratory work, however, resulted recently in the development of a variety of transgenic rodents. Particularly useful in studies of human diseases are the so-called **knockout mice**— mice in which defective alleles have been inserted to replace the functioning alleles, which were "knocked out." Many of the genes studied in this fashion are associated with specific human diseases. Knockout mice are now available as models for cystic fibrosis, sickle-cell anemia, atherosclerosis, and many other human diseases.

With knockout mice as their subjects, scientists are now able to study human genetic diseases in ways that would be difficult or impossible in work with humans. Using the techniques of biochemistry, molecular biology, physiology, and pharmacology, we can explore how the defective alleles lead to the expression of a disease and we can test potential drugs and other treatments. Work with knockout mice

helped determine that the product of the cystic fibrosis gene is a chloride channel, as we saw earlier in this chapter. Knockout mice are useful in many other types of investigations as well; in principle, we can modify any gene and study its effect.

Surprisingly, knockout mice that lack the functional allele encoding a crucial protein sometimes still behave normally. This ability to function without an essential gene may indicate that otherwise unused members of a gene family (see Chapter 13) can take over when necessary, or that some functions may be accomplished through entirely different pathways. To be an effective animal model for a disease, however, a knockout mouse must show the symptoms of the disease.

Once we know the function of a gene associated with a disease, how can we attempt to correct the mistake made by the defective allele? As we are about to see, there is a promising strategy for dealing with this problem.

Gene Therapy

Perhaps the most obvious thing to do when a person lacks a functional allele is to provide one. If a person cannot make an essential protein because he or she is homozygous for a defective recessive allele, can we simply give the person a supply of cells with correct genetic instructions for making the protein? Such **gene therapy** is under intensive investigation. The first federally approved gene therapy in the United States began in September 1990 on a patient suffering from immune deficiency because the enzyme adenosine deaminase (ADA) her body produced was defective. Biomedical scientists obtained some of her white blood cells, transferred normal ADA genes into them, and returned the genetically modified cells to the patient's body (Figure 15.8). As we write this, the patient has been symptom-free for more than three and a half years. Gene therapy is being performed on other patients with ADA deficiency, as well as with other genetic diseases.

Like many other types of cells in vertebrates, white blood cells cannot divide. **Stem cells** of different types, however, give rise to differentiated cell types such as white blood cells. Stem cells that produce white blood cells have been the target cells for most of the early attempts at gene therapy. Certain other white blood cells have also been used, as in the case of the early trials on ADA deficiency. Because the cells used in these gene therapies cannot divide, patients require repeated treatments, including isolation of their cells, genetic engineering of the cells, and return of the cells to the body. Stem cells inserted with the functional allele offer the possibility of one-step gene therapy because they should continue to generate new cells with the functional allele.

Sick patient

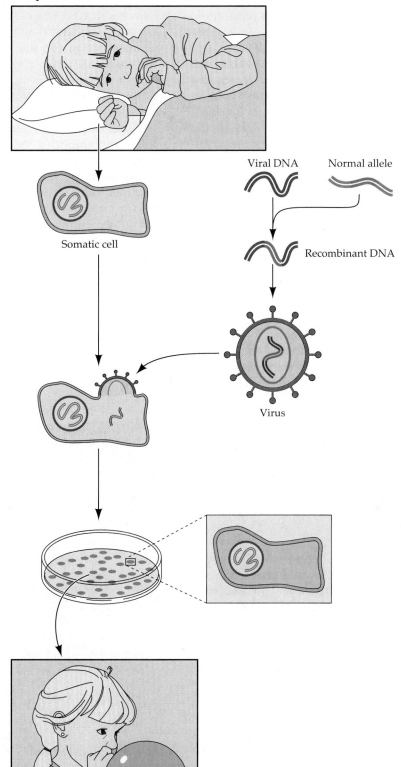

1. Isolate somatic cells from patient homozygous for defective allele

Somatic cell

Viral DNA   Normal allele

2. Insert copy of normal allele into viral DNA

Recombinant DNA

3. Infect isolated somatic cells with virus containing the recombinant DNA

Virus

4. Viral DNA carrying the normal allele inserts into the patient's somatic cell chromosome

5. Culture somatic cells containing the normal allele

6. Inject cultured cells into patient

7. Symptoms relieved by expression of normal allele

Well patient

15.8   Gene Therapy
Recombinant DNA technology provides techniques for inserting a functional gene into a patient's somatic cells, rendering the patient capable of producing the protein encoded in that allele.

How likely are we to practice this specific form of gene therapy—isolating, modifying, and returning a patient's cells—on a large scale? Unless the intermediate recombinant DNA technological steps can be automated, this approach to treating genetic disease appears to be too labor-intensive, too costly, and too dependent on sophisticated laboratory equipment and techniques for large-scale use. It has been suggested that, if these problems are resolved, the techniques are more likely to be applied to the treatment of cancer or infectious diseases than to single-gene genetic defects.

In any case, there are many ethical concerns associated with gene therapy. For example, should such therapy be limited to the modification of somatic cells, such as stem cells and white blood cells, or should it be extended to germ cells (gametes) in order to ensure that future generations do not inherit a genetic disease? The general opinion at this time inclines to the view that we should *not* genetically modify germ cells. The argument is that if we do this, we will be tempted to make other changes—inserting genes for increased height, for example—that are less nobly motivated and too strongly resemble playing God. Similar concerns about inserting genes for enhanced intelligence, on the other hand, can be dismissed as far beyond the realm of possibility for a long time to come, since no such genes have been identified. Some of these ethical considerations apply not only to gene therapy but also to the widespread practice of screening fetuses and newborn babies for genetic defects.

Screening for Harmful Alleles

During pregnancy, cells from the growing fetus may be sampled by methods such as amniocentesis or chorionic villi sampling (see Chapter 37) and then tested for harmful alleles. There are many tests, some focusing on chromosomes, others on proteins, still others on biochemical reactions; most recently it has become possible to test the DNA directly. RFLPs tightly linked to known disease genes are a powerful tool for such screening, as is the polymerase chain reaction (see Chapter 14). The goal of such screening, of course, is to prevent human suffering. The ill effects of phenylketonuria, as mentioned earlier in this chapter, can be prevented by an appropriate diet if the presence of the defective allele responsible for the condition is detected in the infant. In the case of some other diseases, abortion may be the only active step that can be undertaken in response to the detection of a disease-causing allele. This solution, although acceptable to many, is abominable to others. This is not the only ethical issue to arise from genetic screening.

At least as challenging is the issue of privacy. If genetic screening is conducted as a matter of course, who should be party to the information—only the affected individual? the mother? the father? the insurance company? the employer? the government? We must all think about this issue; we may find ourselves voting on it. *None* of these questions is simple, not even the one about the affected individual.

Other issues must also be addressed. For example, does the emphasis on genetic screening demean those who have the conditions for which we screen? What are the public and private costs of providing treatment or care for these people? Should we even be concerned with eliminating genetic defects, for are we not all humans and of equal value (Figure 15.9)?

In our discussion of gene defects and their cure, you may have found yourself thinking entirely in terms of inherited diseases. Recall, however, that alterations in genes in an individual's own somatic cells—*uninherited* alterations—can result in genetic disease. The best-studied example of such a disease is cancer. The term *cancer* refers not to one disease but to many diseases. More than 200 forms have been described.

15.9 Life with a Genetic Defect
Many children are born each year with Down syndrome, a genetic defect that results in mental retardation and other symptoms. Despite their handicap, such individuals enjoy the activities of life and often become productive members of society.

CANCER

Each year about 6.5 million new cases of cancer are diagnosed worldwide. What do these cases have in common? All forms of **cancer** differ from other diseases in two ways. First, a cancerous cell loses control over its own division. Cells that have been transformed to the cancerous state divide rapidly and inappropriately, ultimately forming tumors (large masses of cells). **Benign tumors**, which are not cancers, remain localized where they form, but **malignant tumors**—cancers—invade surrounding tissues and spread to other parts of the body. This spreading, called **metastasis**, is the second defining characteristic of cancer.

Metastasis proceeds in two stages: First the cancer cells extend into surrounding tissues; then they enter either the bloodstream or the lymphatic system (Figure 15.10). Cancer cells metastasizing by way of the lymph are slowed by lymph nodes, where they must pause before proceeding to the next node. The removal of a series of lymph nodes and the ducts between them can often end the disease in such cases. For example, a mastectomy (removal of breast tissue) to treat breast cancer often includes the removal of lymph nodes. Metastasis through the bloodstream is another matter. Less common than metastasis through the lymphatic system, it is rapid and often fatal.

Different forms of cancer affect different parts of the body. **Carcinomas**—cancers that arise in surface tissues such as the skin and the lining of the gut—account for more than 90 percent of all human cancers. Lung cancer, breast cancer, colon cancer, and liver cancer are all carcinomas. (Among these, the first three account for more than half of all cancer deaths in Europe and North America; liver cancer is the most common fatal tumor in Africa and Asia.) **Sarcomas** are cancers of tissues such as bone, blood vessels, and muscle. **Lymphomas** and **leukemias** affect the cells that give rise to the white and red blood cells, respectively. (White blood cells of several types participate in the body's immune system, and red blood cells carry oxygen throughout the body.)

Where does a cancer begin? That is, what triggers the *transformation of a normal cell to a cancer cell?*

Genes, Viruses, and Cancer

The first cancer-causing agent to be identified was a virus that produces sarcomas in chickens. Since that discovery, many other viruses have been demonstrated to cause cancers in animals. The ability of certain viruses to cause cancer in humans became apparent only many years later, and it appears likely that viruses cause only a small fraction of human cancer.

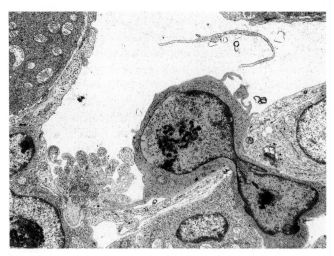

15.10 The Spread of Cancer
A cancer cell (shaded in color) squeezes into a small blood vessel through a cell in the vessel's wall. Cancer spreads through animal bodies as the malignant cells invade tissues and enter the bloodstream or lymphatic system. The blood or lymph then transports the cancer cells throughout the body, starting new tumors in various organs.

Because many cancers are diseases of old age, it was proposed that cancers result from an accumulation of mutations in a single cell. What could account for these mutations? Chemical **carcinogens** in the diet or in the immediate environment are substances that can cause cancer. Polluted air and tobacco smoke contain carcinogens. All known carcinogens react directly or indirectly with DNA to cause mutations. Ultraviolet radiation, X rays, and other forms of high-energy radiation also cause mutations and thus cancer. Excessive exposure to sunlight, with its substantial content of ultraviolet rays, can lead to cancers of the skin. Skin cancers are particularly common among fair-skinned persons and albinos, as well as persons whose enzymes for repairing damage to DNA are defective.

Chemical carcinogens, radiation, and tumor-inducing viruses all cause mutations or otherwise modify the DNA content of a nucleus. What, then, is the nature of these changes? Since cancerous transformation results from changes in genes required for normal growth, we we must first understand what normal growth is.

Growth Factors and Cancer-Related Genes

An animal consists of masses of cells of many types. The cells divide at different times and at different rates; some do not divide at all after a certain developmental stage. Cell division is tightly orchestrated in a healthy animal.

How is cell division regulated in the various parts of the animal body? Part of the answer lies in a group of proteins called **growth factors**, which circulate in the blood and trigger the normal division of cells. A growth factor acts only on certain cells—its target cells. What keeps it from triggering division in any cell it happens to meet? Each growth factor can be bound only by a unique receptor protein, and the receptor proteins are embedded in the plasma membranes of the target cells. After binding its growth factor, a receptor protein becomes active as an enzyme, catalyzing reactions that participate in cell division. These reactions are not triggered unless the growth factor is present.

Growth factors and receptor proteins are coded for by genes that may change and cause cancer. Cancers originate in the activities of cancer-producing alleles, or **oncogenes**: dominant alleles that *arise from the normal recessive alleles that encode, among other things, a growth factor and certain receptor proteins.* The normal alleles are called **proto-oncogenes**. Proto-oncogenes are absolutely essential to the normal development of the cell. It is the *mutant* form, the oncogene, that causes cancer. The question of how oncogenes produce cancer is an area of intense study. There is no single, simple answer. Ultimately the oncogene leads to unregulated cell multiplication. The regulation of cell division, so important to normal cells, is absent in cancerous cells.

Proto-oncogenes can lead to cancers by various mechanisms, three of which we describe here. One mechanism is simple mutation. A single point mutation can cause certain proto-oncogenes to become oncogenes and induce tumor formation. A second mechanism is overproduction. Multiple copies of the proto-oncogene may form through gene amplification (see Chapter 13) or because normal mechanisms of gene regulation fail. In a third mechanism the proto-oncogene moves (by transposition or by chromosomal translocation; see Chapter 11) to a new chromosomal site near the promoter of a very active gene, resulting in continual transcription of the proto-oncogene.

A different type of mutation—a recessive mutation of a dominant **tumor-suppressor gene**—initiates most human cancers. Tumor-suppressor genes encode proteins that inhibit cell division, thus suppressing the formation of tumors. The best-studied tumor-suppressor gene, *p53*, encodes a protein that halts the cell cycle at a point before division, allowing a cell to repair any damaged DNA before it divides. Mutations of *p53* are evident in tumors in half of all cancer cases. The first clear example of a cancer associated with mutation of a tumor-suppressor gene (in this case not *p53*) was retinoblastoma, a tumor of the eye most often observed in children.

Transformation to the cancerous state requires multiple mutations in a single cell. Different types of cancer result from different sequences, but usually more than one tumor-suppressor gene must mutate, and often a proto-oncogene mutates to an oncogene (Figure 15.11). The requirement for multiple mutations, which generally take a long time, is consistent with the observation that many cancers are diseases of old age. Not all the changes need arise by mutation, however; as explained in the previous section, viruses are a source of genetic change, and of cancer.

Many cancers in mammals and birds are triggered by retroviruses (see Chapter 11). Recall that these viruses have RNA, not DNA, as their genetic material. To replicate within cells of their hosts, retroviruses must have their RNA copied to DNA by reverse transcription (Figure 15.12). Reverse transcription is accomplished by an enzyme, reverse transcriptase, that is coded for by a viral gene. The DNA reverse transcript of viral RNA is inserted into a host chromosome, where it may trigger the transformation of the host cell into a cancerous cell.

Some retroviruses that lack oncogenes can cause tumors in animals, but this process typically requires many months. It has been discovered in such cases that the viral DNA was inserted at a locus very close to a proto-oncogene in the host chromosome. Under these circumstances cancer formation by oncogenes depends on viral sequences to induce gene activity in the host.

Retroviruses that are highly cancer-producing (oncogenic) contain, in addition to the three genes necessary for their own reproduction, an oncogene that causes cancerous growth and behavior in the infected cell. The promoters of viral oncogenes are strong and cause frequent transcription. How did oncogenic viruses get their oncogenes?

Origins of Oncogenic Viruses and Transposable Elements

The oncogenes of retroviruses most likely arose from animal proto-oncogenes by the incorporation of an RNA transcript of a proto-oncogene into a viral coat along with part of the viral RNA genome (a process similar to transduction, which was described in Chapter 12). This theory of the origin of retroviral oncogenes is supported by the observation that a retroviral oncogene is similar in base sequence to the suspected "parental" proto-oncogene, except that the proto-oncogene has introns, whereas the retroviral oncogene is uninterrupted. The oncogene differs further from the proto-oncogene in that it has been shortened at either the N-terminal or C-terminal end and has experienced one or more point mutations.

It has also been suggested that some (but not all) transposable elements may have arisen from retroviruses that became immobilized. This suggestion is

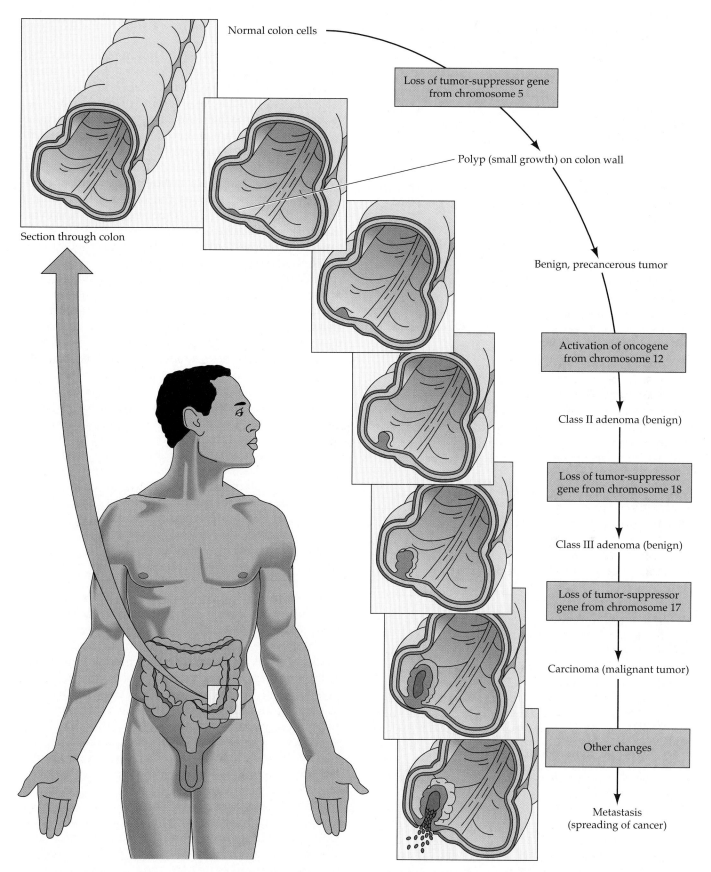

Normal colon cells

Loss of tumor-suppressor gene from chromosome 5

Polyp (small growth) on colon wall

Section through colon

Benign, precancerous tumor

Activation of oncogene from chromosome 12

Class II adenoma (benign)

Loss of tumor-suppressor gene from chromosome 18

Class III adenoma (benign)

Loss of tumor-suppressor gene from chromosome 17

Carcinoma (malignant tumor)

Other changes

Metastasis (spreading of cancer)

15.11 Multiple Steps May Transform a Normal Cell into a Cancerous Cell
These stages in the development of normal tissue into a cancer of the colon (the large intestine) reflect a series of mutations required to transform a normal cell into a cancerous one.

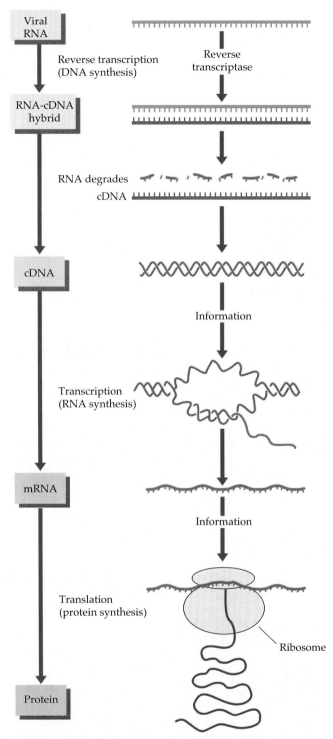

15.12 Reverse Transcription
Retroviruses have no DNA; when they reproduce, the enzyme reverse transcriptase produces DNA from an RNA template.

based on DNA-sequencing studies, which have revealed strong similarities between the base sequences of the ends of certain retroviruses and those of certain transposable elements, indicating a possible common origin.

Treating Cancer

Our bodies usually possess natural mechanisms that detect and eliminate cancerous cells as they form. These activities of the cellular immune system and of "killer cells" will be considered in Chapter 16. Obviously these mechanisms are imperfect, because cancers often do form and metastasize. When cancers arise and spread, what can we do?

There are currently three principal approaches to treating cancer: surgery, radiotherapy, and chemotherapy. Surgery removes the affected tissues and organs. For this approach to be successful, however, the surgeon must know the exact locations of all cancerous tissues. Also, the tissue to be removed must obviously not be irreplaceable.

Radiotherapy—exposure to massive doses of X rays or gamma rays—is a possible treatment for cancerous tissues that cannot be replaced. For example, treating cancer of the larynx (voice box) by radiotherapy may leave the vocal cords intact, whereas surgery would remove them altogether, leaving the patient without normal speech. Tissues exposed to massive doses of radiation suffer extensive chromosomal breakage; some cells not scheduled to divide again remain alive after the chromosomes are broken, but dividing cells such as cancerous cells and the cells of the immune system die when they attempt to undergo mitosis. Because radiation is also harmful to normal cells, it must be restricted to very specific areas of the body; thus the radiologist, like the surgeon, must know the extent and location of the cancer. Another danger of radiotherapy is that it can seriously damage the immune system, leaving the patient defenseless against bacterial, viral, and other infections.

The third major cancer treatment is **chemotherapy**, treatment with drugs that have their greatest effect on rapidly dividing cells. These drugs generally act by interfering with the metabolism of the building blocks of DNA. In chemotherapy, precise knowledge of the locations of cancerous tissues is not necessary; however, normal cells—especially those of the immune system—are also affected. The side effects of chemotherapy are often severe.

By using one or more of these approaches, we are able to achieve many cures. Can we use molecular biological techniques to open new avenues of treatment? One new approach takes advantage of the principle of complementary base pairing. **Antisense nucleic acids** are single-stranded stretches of RNA or DNA targeted against the mRNAs transcribed from harmful genes such as oncogenes. An antisense nucleic acid is complementary at each nucleotide to its target mRNA and therefore forms hydrogen bonds that join the two molecules (Figure 15.13). In this "duplex" condition the mRNA is inactive; it cannot

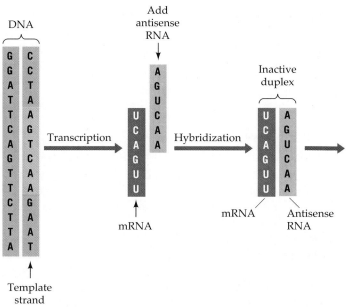

DNA

Template strand

Add antisense RNA

Transcription

mRNA

Hybridization

Inactive duplex

mRNA   Antisense RNA

Not translated; may be degraded or modified by enzymes

15.13 Antisense RNA Inactivates the Corresponding mRNA

In one type of cancer therapy, targeted single-stranded stretches of RNA are injected into cells, inactivating the transcription of oncogene mRNAs.

be translated by ribosomes, and it may be degraded or modified by enzymes.

Antisense RNA injected into a cell has succeeded in inactivating a previously injected mRNA, but wouldn't it be more effective to "teach" a cell how to make its own antisense RNA so that repeated injections would not be necessary? We could, for example, insert a gene encoding a particular antisense RNA into a suitable RNA virus and then use the virus to insert its RNA into the patient's cells. This approach has worked, both against RNAs that had themselves been inserted by viral vectors and against genes normally present in the cell. There is hope that antisense nucleic acids may eventually be used to combat viral infections, including those caused by oncogenic viruses. Another, longer-term possibility is the clinical use of antisense nucleic acids to inactivate cellular oncogenes and thus bring cancers under control. One of the most difficult aspects of such a therapy will be to design an antisense nucleic acid sufficiently specific to inactivate the oncogene without impairing the action of the normal allele—the proto-oncogene that is so important in the normal growth of a cell.

There is hope, too, that immunological techniques, such as vaccination against the few known oncogenic viruses, may reduce the incidence of cancer. What other preventive measures might we take to reduce the risk of developing cancer in the first place rather than having to fight a cancer after it is started?

Preventing Cancer

The single most effective way to minimize the risk of cancer is never to smoke—or, if you smoke, to stop immediately and forever. Avoiding second-hand smoke—smoke exhaled by smokers in your vicinity—is also highly advisable. The relationship between smoking and cancer has been established beyond question.

Other cancer-preventive measures are less firmly established. However, minimizing exposure to carcinogens—both chemical carcinogens and mutagenic forms of irradiation, such as ultraviolet radiation and X rays—makes obvious sense. Although a deep tan was once a seemingly important trophy of summer, we now know that there is no such thing as a "healthy" tan. Various commercially available "sunblocks" prevent ultraviolet radiation from reaching the body surface. Some of these compounds may themselves prove to be carcinogenic, however; therefore, the safest approach to sunlight is avoidance—wearing protective clothing and minimizing direct exposure, especially at midday. Avoidance is also the best approach to chemical carcinogens in the diet, in the home, and in the workplace. But how do we know which compounds are carcinogenic?

The Ames Test

The most widely employed test for chemical carcinogens is based on the observation that every known carcinogen either is itself a mutagen or is converted to a mutagen by the liver. In the **Ames test**, named for its inventor Bruce Ames, the suspected carcinogenic compound is combined with a suspension of ground-up liver cells and mutant bacterial cells that cannot grow in the absence of, say, a particular amino acid. We then look for the appearance of bacteria that can grow in the absence of that amino acid (Figure 15.14). Such bacteria must result from an alteration that reverses the original mutation; thus their presence indicates that the compound added to the suspension is carcinogenic.

In contrast to other tests, such as direct observation of a compound's ability to cause tumors when applied to mice or rats, the Ames test is simple, fast,

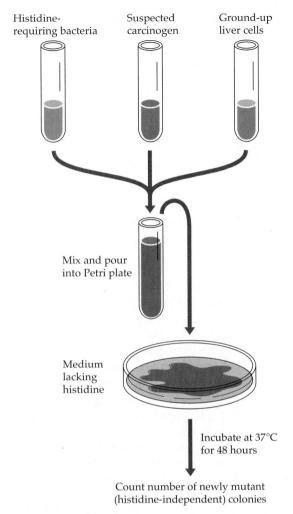

Histidine-requiring bacteria | Suspected carcinogen | Ground-up liver cells

Mix and pour into Petri plate

Medium lacking histidine

Incubate at 37°C for 48 hours

Count number of newly mutant (histidine-independent) colonies

15.14 The Ames Test
The Ames test for a suspected carcinogen's ability to cause mutations works because liver cells convert carcinogens to mutagens. In looking for "back mutations" in an already-mutant strain of bacteria, we recognize mutagenized cells by their ability to produce colonies on a medium on which the original strain could not grow.

and inexpensive. There is a strong correlation between activity in the Ames test and tumor-forming activity. Thus we can test many compounds for carcinogenic activity without unnecessarily sacrificing the lives of large numbers of rodents and other test animals.

Can We Cheat Death?

Why do people die? Violent death, whether by accident, war, or other directed violence, has always been with us. Ill health has always been a cause of death, but the patterns of death due to disease have changed during the twentieth century. Throughout recorded history, the leading killers were infectious diseases caused by pathogens. Typhus was such a potent killer that it has affected history directly, wiping out the army with which Napoleon invaded Russia. Plague took tremendous tolls in Europe, especially during the ninth and fourteenth centuries. In the twentieth century, advances in medicine and, even more important, in public health measures led to a startling reduction in infectious disease. By the 1960s the leading killers were *noninfectious* diseases, notably heart disease and cancer. Can we combat heart disease and cancer as effectively as we have dealt with infectious diseases? That remains to be seen. Success against noninfectious diseases, however, would still not mean that we have conquered death. There will always be a "number-one killer." Only its identity will change.

Consider a specific example: A new disease has killed almost two million people within the decade since its discovery. That disease is AIDS—acquired immunodeficiency syndrome. AIDS kills indirectly, by destroying the body's immune system—a major protection against other diseases. We will consider the immune system and AIDS in the next chapter.

SUMMARY of Main Ideas about Genetic Disease and Modern Medicine

Chromosomal aberrations cause one type of genetic ill health.
    Review Figures 9.19 and 10.16

Defective alleles that are recessive cause many genetic diseases; dominant defective alleles cause only a few.

Failure to produce a particular protein is the defect responsible for many genetic diseases.

The number of triplet repeats in some disease-causing alleles is much greater than the number in the corresponding normal alleles.
    Review Figure 15.2

Current efforts to map human genes will help in the diagnosis and treatment of genetic diseases.

Pedigree analysis reveals whether the inheritance pattern of a disease is dominant or recessive and whether the defective allele is carried on an autosome or on a sex chromosome.
    Review Figures 15.3 and 15.4

To determine which chromosome carries a certain gene, researchers use somatic-cell hybridization.
    Review Figure 15.5

Mutations in recognition sites for restriction endonucleases change the band pattern obtained by gel electrophoresis of restriction digests, giving rise to restriction fragment length polymorphisms (RFLPs); RFLPs are convenient guideposts for locating genes on chromosomes.

Review Figures 15.6 and 15.7

After a gene has been isolated, its functions can be studied in appropriately designed knockout mice.

Gene therapy is one way to deal with a defective gene, but it is unlikely to be used on a large scale.

Review Figure 15.8

Although most cancers are not inherited, cancer is a genetic disease because it appears when a cell's genetic control of its own division fails.

Cancer metastasizes in the bloodstream or lymphatic system.

Cancers begin when the DNA of genes required for normal growth changes, transforming normal cells into cancer cells.

Transformation to the cancerous state requires multiple mutations that may include the conversion of proto-oncogenes to oncogenes and the inactivation of tumor-suppressor genes.

Review Figure 15.11

About half of all cancers contain cells with mutations in the tumor-suppressor gene *p53*.

The principal approaches to treating cancer are surgery, radiotherapy, and chemotherapy.

The most effective way to prevent cancer is to avoid exposure to carcinogens.

SELF-QUIZ

1. Major symptoms of cystic fibrosis are caused by
   a. fibrous materials that collect in cysts.
   b. cysts that clog up the fibrosis organ.
   c. defective transport of calcium across the plasma membrane.
   d. defective transport of chloride across the plasma membrane.
   e. defective transport of sodium across the plasma membrane.

2. Phenylketonuria can be treated by
   a. dietary restrictions.
   b. tissue transplantation.
   c. depletion of an excessive substance.
   d. surgical repair of tissue.
   e. replacment of a missing gene product.

3. Fragile-X syndrome is an X-linked disorder
   a. that occurs with the same frequency in males and females.
   b. that occurs much more frequently in females than in males.
   c. that is extremely rare in all populations.
   d. whose expression within a family varies between generations.
   e. with a uniform phenotype.

4. The type of inheritance in which most affected persons (about ½ males and ½ females) have normal parents (and sometimes the parents are related to each other) is

   a. Y-linked.
   b. X-linked dominant.
   c. X-linked recessive.
   d. autosomal dominant.
   e. autosomal recessive.

5. The assignment of human autosomal genes to specific chromosomes
   a. was initially less difficult than assigning X-linked genes.
   b. was initially just as difficult as assigning X-linked genes.
   c. was initially more difficult than assigning X-linked genes.
   d. has been possible since 1900.
   e. will probably be entirely completed by 1999.

6. Researchers in somatic cell genetics fuse cells from different species for the purpose of
   a. amplifying certain genes by the polymerase chain reaction.
   b. producing transgenic animals.
   c. assigning genes to specific chromosomes.
   d. gene cloning.
   e. alleviating genetic deficiencies.

7. Restriction fragment length polymorphisms (RFLPs) are variations in the lengths of
   a. proteins produced by restrictions of mutation.
   b. time needed to produce certain types of mutations.
   c. restriction enzymes, due to new mutations.
   d. RNA segments produced by a restriction enzyme.

   e. DNA segments produced by a restriction enzyme.

8. Knockout mice
   a. readily contract communicable diseases.
   b. are exceptionally healthy.
   c. do not contract the genetic diseases humans do.
   d. have had defective alleles inserted in place of functional alleles.
   e. have not yet found uses in biomedical research.

9. Gene therapy
   a. has not yet been tested on human patients.
   b. can include isolating, modifying, and returning a patient's own cells.
   c. is likely to be widely used—and soon—in treating genetic disorders.
   d. could not be used to cure a genetic disease.
   e. is likely to focus on the germ cells rather than somatic cells.

10. A proto-oncogene is
    a. a transcription factor produced by an oncogene.
    b. an abnormal gene that prevents cells from dividing.
    c. a normal gene that regulates cell division.
    d. one of the few genes normally found in red blood cells.
    e. one of the few genes normally found in a retrovirus.

FOR STUDY

1. Genetic defects may be produced by either dominant or recessive alleles. Of the known autosomal dominant disorders, those whose symptoms appear at an early age are relatively benign. Why should this be so?

2. It seems possible that genes associated with disorders may often contain regions of triple repeats. Suggest how you might make use of this idea to search for new disease genes. (Hint: think in terms of RFLP mapping.)

3. Mouse–human cell hybridization gave rise to five hybrid colonies. The table at the right shows the human genes and human chromosomes that were present (plus sign) in each colony. From these data, what can you say as to which genes are located on which chromosomes?

4. Many cancers are diseases of old age. Account for this on the basis of what you have learned about the transformation of a healthy cells to the cancerous state.

5. If someone has a genetic disease, who has the right to know? The person's spouse or fiance? The person's parents or children? The person's employer? The company that insures the person? List arguments *for* and *against* informing each of these people or groups.

		HYBRID CELL LINES				
		A	B	C	D	E
Human genes	1	+	−	−	+	−
	2	−	+	−	+	−
	3	+	−	−	+	−
	4	+	+	+	−	−
Human chromosomes	1	−	+	−	+	−
	2	+	−	−	+	−
	3	−	−	−	+	+

From Griffiths et al., *An Introduction to Genetic Analysis*, 5th Edition, p. 168.

READINGS

Capecchi, M. R. 1994. "Targeted gene replacement." *Scientific American,* March. The story of "knockout mice."

Croce, C. M. and G. Klein. 1985. "Chromosome Translocations and Human Cancer." *Scientific American*, March. A clear treatment of oncogenes and their activation.

Culotta, E. and D. E. Koshland, Jr. 1993. "Molecule of the Year: p53 Sweeps Through Cancer Research." *Science*, vol. 262, pages 1958–1961. A brief look at the best-studied tumor-suppressor gene.

Golde, D. W. 1991. "The Stem Cell." *Scientific American*, December. The cell that gives rise to blood cells and immune systems. Applications in treatment of cancer and immune defects.

Griffiths, A. J. F., J. H. Miller, D. T. Suzuki, R. C. Lewontin and W. M. Gelbart. 1993. *An Introduction to Genetic Analysis*, 5th Edition. W. H. Freeman, New York. Chapters 6 and 15 are particularly relevant; Chapter 22 is also useful.

Hunter, T. 1984. "The Proteins of Oncogenes." *Scientific American*, August. A detailed treatment of several specific oncogenes and their products.

Lawn, R. M. and G. A. Vehar. 1986. "The Molecular Genetics of Hemophilia." *Scientific American*, March. The use of recombinant DNA technology and bacteria to produce a blood-clotting protein that may save the lives of hemophiliacs.

Liotta, L. A. 1992. "Cancer Cell Invasion and Metastasis." *Scientific American*, February. How cancer spreads in the body, and prospects for treatment.

Mange, E. J. and A. P. Mange. 1993. *Basic Human Genetics*. Sinauer Associates, Sunderland, MA. Genetics, especially chromosomal inheritance, can be studied using humans as examples; this book does so at an introductory level.

Miller, J. A. 1990. "Genes That Protect against Cancer." *BioScience*, vol. 40, pages 563–566. The roles of anti-oncogenes, also known as tumor-suppressor genes.

Patterson, D. 1987. "The Causes of Down Syndrome." *Scientific American*, August. Identification and mapping of genes responsible for the most common cause of mental retardation.

Rennie, John. 1994. "Grading the Gene Tests." *Scientific American*, June. A good start for a discussion of ethical issues in screening for hereditary diseases.

Verma, I. M. 1990. "Gene Therapy." *Scientific American*, November. A discussion of how healthy alleles are introduced to correct heritable disorders.

Weinberg, R. A. 1988. "Finding the Anti-Oncogene." *Scientific American*, September. Isolation of a gene that prevents a cell from proliferating out of control.

Weintraub, H. M. 1990. "Antisense RNA and DNA." *Scientific American*, January. Antisense nucleic acids as tools for basic research and as tools against viruses and, maybe, cancer.

White, R. and J.-M. Lalouel. 1988. "Chromosome Mapping with DNA Markers." *Scientific American*, February. A description of the powerful tool known as restriction-fragment length polymorphism.

By the year 2000 as many as 40 million people, worldwide, will have become infected with HIV, the human immunodeficiency virus, which causes AIDS (*acquired immune deficiency syndrome*). That number about equals the combined total populations of the states of New York, Michigan, and Pennsylvania. The number is an estimate by the World Health Organization, made in mid-1993, by which time over 13 million people had already become infected. Of these 13 million, over 2 million had developed AIDS, and of those 2 million, most had died. Who are these millions of people? They are women, men, children, and newborn infants. About half of those infected with HIV contracted the infection between the ages of 15 and 24. Most of them became infected through heterosexual intercourse.

AIDS does not kill people directly. The death of an AIDS patient is usually the result of infection by a disease-producing organism that is normally present in all of us, but that the AIDS victim cannot tolerate. Why are AIDS patients at extreme risk from infections that rarely trouble most other people? You may have read or heard that the development of AIDS is strongly associated with a reduction in numbers of certain types of blood cells. What does this have to do with the disastrous loss of ability to deal with infection?

In this chapter we focus on the mechanisms by which organisms combat infection and disease. The environment is alive with microorganisms capable of producing disease in other organisms. All organisms have defenses that help them cope with microorganisms and foreign macromolecules. After considering what these mechanisms are and how they work, we conclude with an examination of how they fail in persons infected with AIDS.

NONSPECIFIC DEFENSES AGAINST PATHOGENS

Consider the challenges faced by a potential pathogen (a disease-causing organism such as a bacterium, virus, fungus, protist, or animal parasite) as it approaches the body of an animal. There are hurdles that the pathogen must overcome: It must arrive at the body surface of a potential host, enter the host, multiply inside the host, and finally, prepare to infect the next host. Failure to overcome any one of these hurdles ends the reproductive career of a pathogenic organism. Animals have defenses that stop many different pathogens from invading their bodies. Because these initial defenses give general protection against different pathogens, they are called **nonspecific defenses**.

An Artist's Response
The work of American artist Keith Haring is dedicated to the victims of AIDS.

16

Defenses against Disease

Pathogens arrive at a potential host by way of airborne droplets (as from a sneeze), food or drink, an animal that bites, direct contact with an infected individual, or contact with some pathogen-carrying object in the environment. Many of the most massive improvements in public health have come with the control of sewage or with campaigns against insects, ticks, and other animals that carry pathogens; these measures prevent pathogens from reaching us. "Entering the host" means different things to different pathogens. Some pathogens simply multiply on the surface of a mucous membrane that lines, for example, the throat or intestine of the host. Other pathogens penetrate and multiply within these surface cells, and still others pass into deeper tissues of the body or into the bloodstream.

Skin is a primary nonspecific defense against invasion. Bacteria and viruses rarely penetrate healthy, unbroken skin. Damaged skin or other surface tissue, however, is another matter. The sensitivity of surface tissue accounts in part for the greatly increased risk of infection by HIV in a person who already has a sexually transmitted disease. It may also partially explain why a woman can become infected by HIV during heterosexual intercourse much more easily than a man can, since the vaginal lining is more frequently damaged by intercourse than is the skin of the penis.

In addition to skin, we have other defenses against invasion by pathogens. One of these is our **normal flora**: the bacteria and fungi that live and reproduce in great numbers on our surfaces without causing disease (Figure 16.1). These natural occupants compete with pathogens for locations and nutrients, and some of them produce inhibitor compounds that are toxic to potential pathogens.

The one type of healthy surface that *is* penetrable by bacteria is the mucous membrane, a type of mucus-secreting tissue found in parts of the visual, respiratory, digestive, and urogenital systems. However, these areas of the body have other defense mechanisms to discourage penetration by pathogens. Secretions such as tears, nasal drips, and saliva possess an enzyme called **lysozyme** that attacks the cell walls of many bacteria. Mucus in our noses traps most of the microorganisms in the air we breathe, and most of those that get past this filter end up trapped in mucus deeper in the respiratory tract. They are removed from the respiratory tract by the beating of cilia in the respiratory passageway, which moves a sheet of mucus and the debris it contains up toward the nose and mouth. Another effective means of removing microorganisms from the respiratory tract is the sneeze.

Pathogens that travel as far into a person as the digestive tract (stomach, small intestine, and large intestine) are met by other defenses. The stomach is

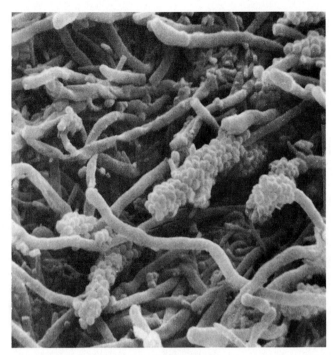

16.1 Normal Flora Gone Rampant
The human mouth harbors a wide variety of microorganisms, most of which cause no damage under normal conditions. When bacteria accumulate on the surfaces of teeth, the result is called plaque. The presence of plaque contributes to tooth decay. This electron micrograph shows plaque on a tooth three days after the person stopped brushing.

an unfriendly environment for most bacteria because of the hydrochloric acid that is secreted into it. The lining of the small intestine cannot be penetrated by bacteria, and some pathogens are killed by bile salts secreted into this part of the tract. The large intestine harbors many bacteria, which multiply freely; however, these are usually removed quickly with the feces. (The digestive system is described fully in Chapter 43.)

If Pathogens Evade the Blocks to Entry

Pathogens that manage to penetrate the surface cells of an animal's body encounter still other defenses. These defenses fall into two categories: nonspecific and specific (the specific defenses are the immune system). One of the nonspecific mechanisms is a "battle" for iron that takes place between some pathogens and the host. Both require iron for their metabolism, but they utilize it in different forms. Pathogen and host produce different substances to trap the iron, and competition may be intense. If iron is in limited supply, the pathogen seldom wins the battle.

The nonspecific defenses of the animal host also include antimicrobial proteins in its tissues and body fluids. Two important types of antimicrobial proteins are complement proteins and interferons, which we will discuss in more detail later in the chapter.

In your body, and in the bodies of all other vertebrates (animals with backbones), blood circulates through a system of vessels, pumped by a heart. Blood contains a mixture of different types of cells. As we will see, the circulating behavior and other activities of certain types of blood cells give vertebrate animals powerful defenses against pathogens. (The circulatory system is described fully in Chapter 42.)

White blood cells called **phagocytes** provide an extremely important nonspecific defense against pathogens that penetrate the surface of the host. Some phagocytes adhere to certain tissues; others travel freely in the circulatory system. Pathogens become attached to the membrane of a phagocyte (Figure 16.2). The phagocyte ingests the pathogens by endocytosis. Once inside an endocytic vesicle in a phagocyte, pathogens are destroyed by enzymes from lysosomes that fuse with the vesicle (see Figure 4.22). A single phagocyte can ingest 5 to 25 bacteria before it dies from the accumulation of toxic breakdown products. Even when phagocytes do not destroy all the invaders, they usually reduce the number of pathogens to the point where other defenses can finish the job. So important is the role of the phagocytes that if their functioning is impaired by disease, the animal usually soon dies of infection.

A class of small white blood cells, known as **natural killer cells**, can initiate the lysis of some tumor cells and some normal cells that are infected by a virus. It is possible that the natural killer cells seek out cancer cells that appear in the body.

Another important nonspecific defense is **inflammation**. The body employs this characteristic, highly generalized response in dealing with infections, mechanical injuries, and burns. The damaged cells themselves cause the inflammation by releasing various substances. You have experienced the symptoms of inflammation: redness and swelling, with heat and pain. The redness and heat result from dilation of blood vessels in the infected or injured area. The blood capillaries (the smallest vessels) become leaky, allowing some blood plasma (described later in this chapter) and phagocytes to escape into the tissue, causing swelling. The pain results from increased pressure (from the leakage) and from the action of some leaked enzymes. Certain of the plasma proteins and the phagocytes are responsible for most of the healing aspects of inflammation. The heat may also play a healing role if it raises the temperature beyond that at which the pathogen that triggered the inflammation can multiply effectively.

Viral Diseases and Interferon

When you have a viral disease, such as influenza, you are unlikely to develop another viral disease at the same time. An apparent explanation for this phenomenon was provided in 1957 by Alick Isaacs and Jean Lindemann of the National Institute for Medical Research in London. They found that inoculating the cells of a developing chick with influenza virus causes the cells to produce small amounts of an antimicrobial protein called **interferon** that increases the resistance of neighboring cells to infection by influenza or *other* viruses. Interferons have been found in many vertebrates and are one of the body's lines of nonspecific defense against viral infection.

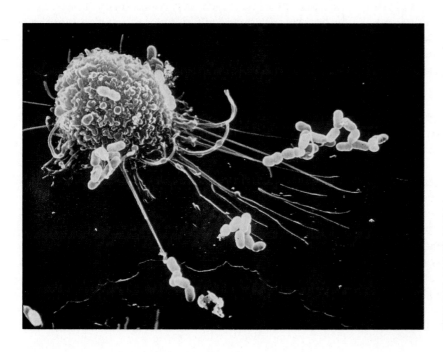

16.2 A Phagocyte and Its Bacterial Prey
Some bacteria (appearing yellow in this artificially colored scanning electron micrograph) have become attached to the surface of a phagocyte in the human bloodstream. Many of these bacteria will be taken into the phagocyte and destroyed before they can multiply and damage the human host. The long protuberances of the phagocyte probably help it move and adhere to other cells.

Interferons differ from species to species, and each vertebrate species produces at least three different interferons. All interferons are glycoproteins (proteins with a carbohydrate component) consisting of about 160 amino acid units. By binding to receptors in the membranes of cells, interferons inhibit the ability of the viruses to replicate. Interferons have been the subject of intensive research because of their possible applications in medicine—for example, the treatment of influenza and the common cold.

Nonspecific Defenses of Plants

Plants also have a variety of mechanisms, both mechanical and chemical, by which they resist or even actively oppose infection by pathogens. The outer surfaces of plants are protected by tissues such as the epidermis or cork. If pathogens get past these barriers, the differences between the defense systems of plants and animals become apparent. Animals generally *repair* tissues that have been infected—they heal, through appropriate developmental pathways. Plants, on the other hand, do not make repairs. Instead, they develop in ways that seal off the damaged tissue so that the rest of the plant does not become infected. Trees seal off damaged tissue by producing new wood that differs in orientation and chemical composition from the previously deposited wood. Some of the new cells also contain substances that resist the growth of microorganisms and hence tend to protect the rest of the plant.

The healing mechanism just described is primarily mechanical. Many plants have chemical defenses as well. For example, infection of one of these plants by certain fungi stimulates the plant to produce substances that are toxic to the fungi. Some of these toxic substances can act against some bacteria as well. Their antifungal activity is nonspecific—that is, the substances can destroy many species of fungi in addition to the one that originally triggered their production. Physical injuries, viral infections, and even certain chemical compounds can also induce the production of these substances.

SPECIFIC DEFENSES: THE IMMUNE SYSTEM

Our nonspecific defenses are numerous and effective, but some invaders nevertheless elude the nonspecific defenses and must be dealt with by defenses targeted against specific threats. The destruction of specific pathogens is an important function of an animal's **immune system**. The immune system recognizes and attacks specific invaders, such as bacteria and viruses. After responding to a particular type of pathogen once, the immune system can usually respond more rapidly and powerfully to the same threat in the future. Thus the functions of an immune system are to *recognize, selectively eliminate,* and *remember* foreign invaders.

An animal with a defective immune system can die from infection by even "harmless" bacteria. Some microorganisms routinely carried in or on an animal's body without harm are potentially pathogenic and will cause disease if the host's immune system is stressed in some way.

The immune system is made up of cells that travel in the body's fluids. Blood is a fluid tissue. About 55 to 65 percent of a human's blood is the yellowish liquid matrix called **plasma**; the remainder consists of red blood cells, white blood cells, and platelets. Plasma is mostly water, but contains many other important substances.

As we will see in Chapter 42, some of the circulating components of blood are returned to the heart not by veins but by the lymphatic system (Figure 16.3). **Lymph**, a blood filtrate that accumulates in the spaces outside the blood capillaries, contains water, solutes, and white blood cells that have left the capillaries, but no red blood cells. The lymph is collected in vessels called lymph ducts and routed back into blood vessels near the heart.

White blood cells are larger and far less numerous in the blood than red blood cells (Figure 16.4). White blood cells have nuclei and are colorless. Like the cancer cell in Figure 15.10, they can move through tissues by squeezing through junctions between the cells that make up the walls of blood capillaries. In response to invading pathogens, the number of white blood cells in the blood and lymph may rise sharply, providing medical professionals with a useful clue for detecting an infection.

The most abundant types of white blood cells are the phagocytes (which, as you already know, are important as nonspecific defenses) and the **lymphocytes**. Two groups of lymphocytes, the **B cells** and **T cells**, together with specialized cells that arise from them, are the important cells of the immune system. Both B cells and T cells originate from cells in the bone marrow. The precursors of T cells migrate to an organ called the thymus and develop their unique properties there, becoming mature T cells. The B cells migrate from the bone marrow to the outer regions of the body, circulate in the blood and lymph vessels, and pass through the lymph nodes and spleen. B and T cells look the same under the light microscope, but they have quite different functions in immune responses.

RESPONSES OF THE IMMUNE SYSTEM

Foreign organisms and substances that invade the animal body and escape the internal, nonspecific de-

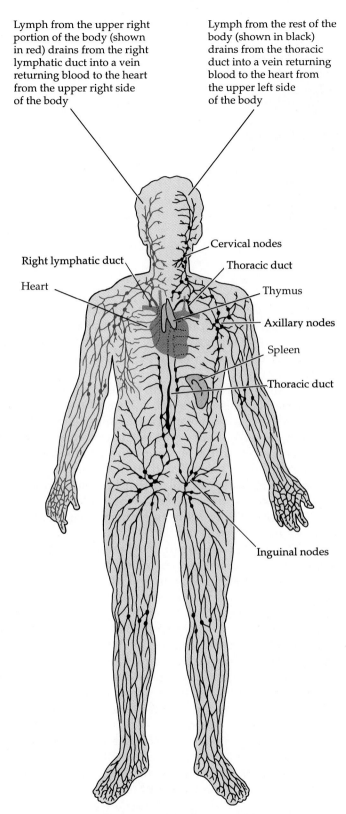

Lymph from the upper right portion of the body (shown in red) drains from the right lymphatic duct into a vein returning blood to the heart from the upper right side of the body

Lymph from the rest of the body (shown in black) drains from the thoracic duct into a vein returning blood to the heart from the upper left side of the body

Right lymphatic duct

Heart

Cervical nodes

Thoracic duct

Thymus

Axillary nodes

Spleen

Thoracic duct

Inguinal nodes

16.3   The Human Lymphatic System
A network of ducts collects lymph from the body's tissues and carries it toward the heart, where it mixes with blood to be pumped back to the tissues. There are major lymph nodes in the neck, armpits, and groin. What we call "swollen glands" in the neck are cervical lymph nodes in which invading bacteria are trapped.

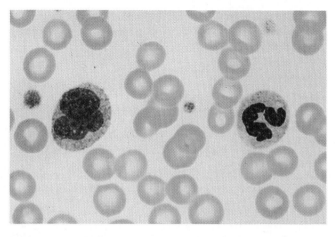

16.4   Blood Cells
The large, stained cell to the left is a white blood cell called a monocyte. The other stained cell is a white blood cell called a neutrophil. Notice that the numerous red blood cells in this blood smear are unstained because red blood cells do not have nuclei.

fenses come up against the immune system (Table 16.1). The immune system has two responses against invaders. Protein molecules produced by the immune system attack bacteria and viruses that are in fluids of the animal body but outside the cells. This is the **humoral immune response** (from the Latin *humor*, "fluid"). Certain cells of the immune system mount an attack called the **cellular immune response** against viruses that have become established within cells of the animal body and against fungi and microscopic animals.

The two responses operate in concert—simultaneously, cooperatively, and sharing mechanisms. Both responses aim at the same specific areas on the surfaces of the foreign bodies that invade an animal. We call the foreign bodies (organisms or molecules) **antigens**; the specific sites on antigens that the immune system attacks are **antigenic determinants**.

Highly specific protein molecules called **antibodies** carry out the humoral immune response against invaders in the fluids. An animal produces a vast diversity of antibodies that, among them, can react with virtually any conceivable antigen in the bloodstream or lymph. Some antibodies travel free in the blood and lymph; others exist as integral membrane proteins on B cells. Where do antibodies come from? Some of an animal's B cells differentiate to become **plasma cells**. Plasma cells then produce antibodies.

Each antibody recognizes and binds to a particular site—an antigenic determinant—on one or more antigens that invade an animal's body (Figure 16.5). An antigenic determinant is a specific chemical grouping that may be present on many different molecules. A large antigen such as a whole cell may have many

TABLE 16.1
Animal defenses against disease

NONSPECIFIC DEFENSES		SPECIFIC DEFENSES (IMMUNE SYSTEM)
BLOCK INVADERS	ATTACK ALL SUCCESSFUL INVADERS	ATTACK SPECIFIC SUCCESSFUL INVADERS
Skin	Competition for iron	Humoral immune
Normal flora	Antimicrobial proteins	response
Mucous membranes	(complement and	Cellular immune response
Protective secretions	interferons)	
	Phagocytes	
	Natural killer (NK) cells	
	Inflammation	

different antigenic determinants on its surface, each capable of binding a specific antibody.

The cellular immune response is directed against a multicellular antigen or an antigen that has become established within a cell of the animal. The animal cell is able to display the invader's antigenic determinants on its surface; we will explain how this happens later in the chapter. The cellular immune response, in contrast to the humoral response, does not use antibodies. Instead, it is carried out by T cells that roam through the bloodstream and lymph. The T cells have **T-cell receptors**: surface glycoproteins that recognize and bind to antigenic determinants.

Like antibodies, T-cell receptors have specific molecular configurations that bind to specific antigenic determinants. Once bound to a determinant, each particular type of T cell initiates characteristic activity. Some T cells recognize and help destroy foreign cells

Antigenic determinants

Antibodies (immunoglobulins)

(a)   *(b)*   *(c)*

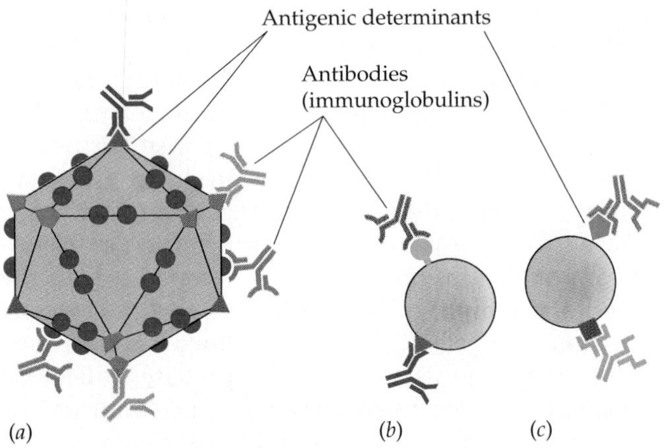

16.5 Each Antibody Matches an Antigenic Determinant
Three antigens are depicted in tan: a virus (*a*) and two distinct globular proteins (*b* and *c*). Each has on its surface antigenic determinants that are recognized by specific antibodies. An antibody recognizes and binds to its antigenic determinant wherever it is; for example, the antibody depicted in purple locates its unique determinant both on the virus (*a*) and on protein (*b*).

or any of the body's own cells that have been altered by viral infections. Because T cells recognize and mobilize attacks on foreign material, they are responsible for the rejection of certain types of organ or tissue transplants.

The immune responses act in concert not only with each other but also with the nonspecific defenses. We have seen that an animal's white blood cells (phagocytes and lymphocytes) defend it against disease. The important difference between phagocytes and lymphocytes (both T and B) is that a particular T or B cell reacts specifically—that is, with only one specific antigenic determinant—whereas phagocytes react nonspecifically with any foreign matter they encounter.

An animal does not require a previous encounter with a particular antigen to mount an immune response. An invading antigen will cause an immune response even if it is the body's first contact with that antigen. This observation prompted some biologists and chemists to propose that a specific antibody must be formed *after* the body encounters the corresponding antigen, perhaps by using the antigenic determinant as a template for the final three-dimensional folding of the antibody molecule. This idea had to be abandoned in the face of experimental results, however. We now know that a person can produce *millions* of distinct antibodies without foreign templates—even though that person may never have encountered the corresponding antigenic determinants. The problem of accounting for the origin of such a tremendous diversity of specific proteins will be considered later in this chapter.

Immunological Memory and Immunization

The effectiveness of the immune system is enhanced by **immunological memory**. The first time a vertebrate animal is exposed to a particular antigen, there is a time lag (usually several days) before the number of antibody molecules and the number of T cells circulating in the bloodstream catch up with the number of invaders. But for years afterward, sometimes for life, the immune system "remembers" that particular antigen and remains capable of responding more quickly than it did on the first encounter. Whereas the first exposure to the antigen results in some response, a second exposure causes a much greater, longer-sustained, and more rapid production of antibodies and T cells (Figure 16.6).

The ability of the human body to remember a specific antigen explains why **immunization** has virtually wiped out such deadly diseases as smallpox, diphtheria, and polio in medically sophisticated countries. (Smallpox, in fact, has been eliminated worldwide from the spectrum of infectious diseases affecting humans, thanks to a concentrated international effort by the World Health Organization. As far as we know, the only remaining smallpox viruses on Earth are those kept in some laboratories.) **Vaccination** means injecting a small amount of virus or bacteria or their proteins (usually treated to make them harmless) into the body. Later, if the same or very similar disease organisms attack, the body's cells are prepared because they have already been exposed to the antigen. They recognize the antigen and quickly overwhelm the invaders with a massive production of lymphocytes and antibodies.

Clonal Selection and Its Consequences

Each person possesses an enormous number of different B cells and T cells, apparently capable of dealing with practically any antigen ever likely to be encountered. How does this diversity arise? You may also wonder why some of our antibodies and T cells do not attack and destroy the components of our own bodies. An individual can mount an immune response against another person's proteins, yet it rarely mounts one against its own. The immune system can distinguish **self** (one's own antigens) from **nonself** (those from outside the body). The versatility of immune responses, immunological memory, and the recognition of self can all be explained satisfactorily by a particular theory of the origin of specific antibodies.

In 1954 the Danish immunologist Niels K. Jerne proposed a new view of the relationship between antigen and antibody. His idea was that the antigen does not itself specify the structures of the antibodies formed against it; instead, in Jerne's view, those an-

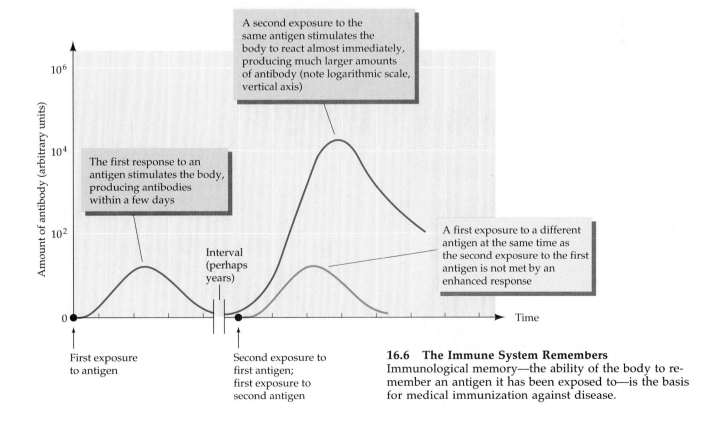

16.6 The Immune System Remembers
Immunological memory—the ability of the body to remember an antigen it has been exposed to—is the basis for medical immunization against disease.

tibodies are already being produced in small quantity, and the antigen specifically stimulates the cells that are making those particular antibodies to increase production. (Recall that antibodies are produced by plasma cells, which are differentiated B cells). Jerne hypothesized that there must be a population of different B cells corresponding to each of the antigenic determinants to which the organism can respond. Jerne's model was improved and extended by the Australian immunologist MacFarlane Burnet, who named it the **clonal selection theory**. According to the theory, the individual animal contains an enormous variety of different B cells, each type able to produce only one kind of antibody. Recent molecular work has shown that the DNA sequences encoding antibodies are arranged in slightly different ways in different B cells, so different B cells in a single animal have slightly different genotypes. After an antigen enters the body, it encounters B cells that can recognize its antigenic determinants. As a consequence of this meeting, each of these B cells begins to multiply, giving rise to a large clone of plasma cells, each of which secretes antibody of the same specificity (Figure 16.7).

The clonal selection theory accounts nicely for the body's ability to respond rapidly to any of a vast number of different antigens. In the extreme case, even a single B cell might be sufficient for an immunological response by the body, provided it encounters the antigen and then proliferates into a large enough clone rapidly enough to deal with the invasion. Clonal selection accounts for the proliferation of both B and T cells.

The clonal selection theory also explains two other phenomena: recognition of self (discussed in the next section), and immunological memory, evidence of which was illustrated in Figure 16.6. According to the clonal selection theory, an activated lymphocyte produces two types of daughter cells. The ones that carry out the attack on the antigen are **effector cells**—either plasma cells that produce antibodies, or T cells that bind antigenic determinants. The others, called **memory cells**, are long-lived cells that retain the ability to start dividing on short notice to produce more effector and more memory cells. Effector cells live only a few days, but memory cells may survive for decades. When the body first encounters a particular antigen, one or more types of lymphocytes become activated and divide to produce clones of effector and memory cells. (Why more than one type? Because the antigen may possess more than one antigenic determinant.) The effector cells destroy the invaders at hand and then die, but one or more clones of different memory cells have now been added to the immune system. Thus if the animal encounters the same antigen a second time, it can respond more rapidly and more massively, as Figure 16.6 shows.

Self, Nonself, and Tolerance

In addition to explaining immunological memory, the clonal selection theory may explain the recognition of self. Given the great array of different lymphocytes directed against particular antigens, how is it that a healthy animal apparently does not produce self-destructive immune responses? There appear to be

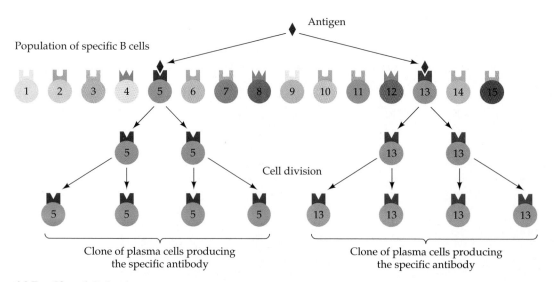

16.7 Clonal Selection
An animal produces many kinds of B cells, each genetically unique and specific for a particular antigen. Of the B cells shown here, only two—5 and 13 (red)—recognize the antigen. The antigen stimulates these B cells to proliferate, forming clones of plasma cells that produce the particular antibody specified by the ancestral cell's genotype.

two mechanisms of self-tolerance: clonal deletion and suppressor T cells.

Lymphocytes that have not yet fully differentiated are not capable of attacking antigens. When "antiself" lymphocytes in this undifferentiated state encounter corresponding self antigens, the antiself lymphocytes are either inactivated or eliminated; thus, no clones of antiself lymphocytes normally appear in the bloodstream. This phenomenon is referred to as **clonal deletion**.

The second mechanism that may account for self-tolerance depends upon a class of lymphocytes known as **suppressor T cells**. These antigen-specific cells inhibit the activities of effector T and B cells. At least some aspects of self-tolerance probably are the result of the inactivation of anti-self lymphocytes by specific suppressor T cells.

In 1945 Ray D. Owen at the California Institute of Technology observed that some *nonidentical* twin cattle contained some of each other's red blood cells, even though their blood cells were of differing types and might have been expected to cause immune responses resulting in their elimination. Four years later Burnet suggested that the blood cells had passed between the animals before the differentiation of immune specificities was complete and thus were regarded as "self" when the recognition of self developed. Burnet further proposed that, if this were true, one should be able to inject foreign antigen into an animal early in its development and cause that animal henceforth to recognize that antigen as "self." Inducing **immunological tolerance** in this way was demonstrated in 1953 by the English immunologist Peter B. Medawar, who used two strains of mice, each so highly inbred that they were almost a clone. Medawar injected cells from adult mice of one strain into newborn mice of another strain. Other newborn mice of the second strain served as uninjected controls. Eight to ten weeks later, he tested for tolerance in the treated and untreated mice of the second strain by grafting skin from the first strain onto them. The untreated mice rejected the grafts but the treated mice accepted them (Figure 16.8). Medawar thus discovered that immunological tolerance to an antigen can be induced by exposure to the antigen early in development.

The establishment of tolerance, whether by the production of appropriate suppressor T cells or by clonal deletion, must be repeated throughout the life of the animal because lymphocytes are produced constantly. Continued exposure to self-antigen helps to maintain tolerance. If for some reason an animal stops producing a given protein—that is, a self-antigen—a clone of lymphocytes directed against that protein may become established. Then if the protein is synthesized again later, it may cause a full-fledged immune response, resulting in an **autoimmune disease** (such as rheumatoid arthritis, in which an immune system attacks part of the body in which it resides).

Development of Plasma Cells

Let's consider in more detail the *mechanisms* of the immune responses. When a B or T cell is activated by an antigen, it proliferates, with both effector and memory cells being produced. The activation of a B cell begins with the binding of particular antigens to the antibodies carried on the B-cell surface. A particular type of T cell, called a helper T cell, must be present; the helper T cell also binds to the antigen. Then, cellular division and differentiation lead to the formation of plasma cells (the effector cells) and memory cells. As plasma cells develop, the number of ribosomes and the amount of endoplasmic reticulum in their cytoplasm increase greatly. These increases prepare the cells for synthesizing large amounts of antibodies for secretion (Figure 16.9). All the plasma cells arising from a given B cell produce antibodies of specificity identical to that of the receptors that bound antigen to the parent B cell.

For many years, scientists wanted to obtain cultures of plasma cells all producing the same antibody so that they could prepare large quantities of pure antibody. Methods were finally discovered for producing antibodies from single clones of cells (Box 16.A), but what *are* the molecules we call antibodies?

IMMUNOGLOBULINS: AGENTS OF THE HUMORAL RESPONSE

Antibodies are also called **immunoglobulins**. We generally use the latter term when we turn from the larger picture of how the humoral immune response attacks antigens in the body fluids to look more closely at the molecules that do the attacking. The chemical structure of the most common form of immunoglobulins was worked out by Gerald M. Edelman (Rockefeller University) and Rodney M. Porter (Oxford University), who found that the basic immunoglobulin molecule is a tetramer consisting of four polypeptides. The polypeptides are two identical "light" chains and two identical "heavy" chains. Disulfide bonds hold the chains together. Each chain consists of a **constant region** and a **variable region** (Figure 16.10). The constant regions of both light and heavy chains are similar from one immunoglobulin to another. On the other hand, the amino acid sequence of the variable region (considering the variable ends of the heavy and light chains together) is unique in each of the millions of different types of immunoglobulins that can mount a humoral immune response. Thus the variable regions of a light and a

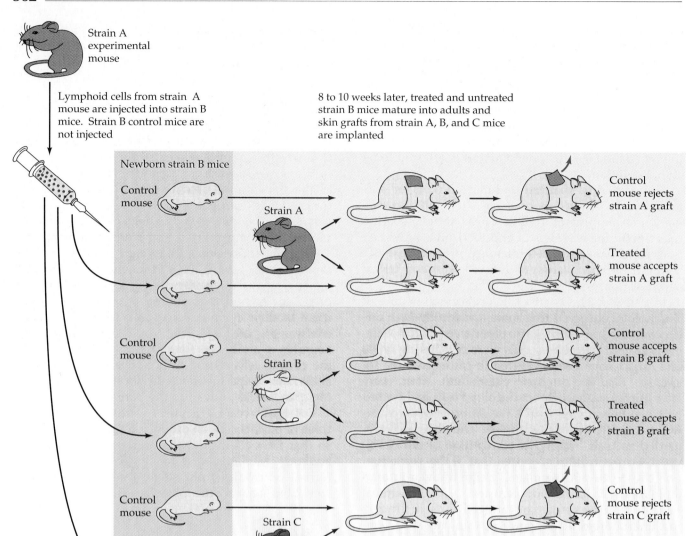

Strain A experimental mouse

Lymphoid cells from strain A mouse are injected into strain B mice. Strain B control mice are not injected

8 to 10 weeks later, treated and untreated strain B mice mature into adults and skin grafts from strain A, B, and C mice are implanted

Newborn strain B mice

Control mouse

Strain A

Control mouse rejects strain A graft

Treated mouse accepts strain A graft

Control mouse

Strain B

Control mouse accepts strain B graft

Treated mouse accepts strain B graft

Control mouse

Strain C

Control mouse rejects strain C graft

Treated mouse rejects strain C graft

16.8 Making Nonself Seem Like Self
The ability of adult mice to recognize and reject grafts of foreign skin can be overcome. Adult mice of strain B that were injected shortly after birth with lymphoid cells from strain A tolerate grafts from strain A or strain B, but reject grafts from other strains, such as strain C. The control mice—adults of strain B raised from uninjected newborn mice—accept grafts only from other strain B mice, rejecting skin from strain A as well as from strain C. What is recognized as "self" and "nonself" thus depends partly on when it is first encountered.

heavy chain combine to form, on each of the immunoglobulin's "arms," a highly specific, three-dimensional structure similar to the active site of an enzyme. This characteristic part of a particular immunoglobulin molecule is what binds with a particular, unique antigenic determinant.

Although the variable regions are responsible for the *specificity* of an immunoglobulin, the constant regions are equally important, for it is the constant regions that determine the type of action to be taken

in eliminating the antigen, as we will see in the next section. In particular, the constant regions of the heavy chains determine whether the antibody remains part of the cell's plasma membrane or is secreted into the bloodstream. The two halves of an antibody, each consisting of one light and one heavy chain, are identical, so each of the two arms can combine with an identical antigen, leading sometimes to the formation of a large complex of antigen and antibody molecules (Figure 16.11).

16.9   A Plasma Cell
Note the prominent nucleus (recognizable by the double membrane), the cytoplasm crowded with rough endoplasmic reticulum, and an extensive Golgi complex—all structural features of a cell actively synthesizing and exporting proteins.

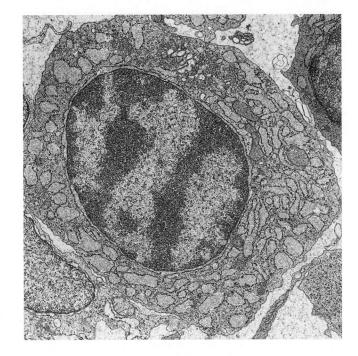

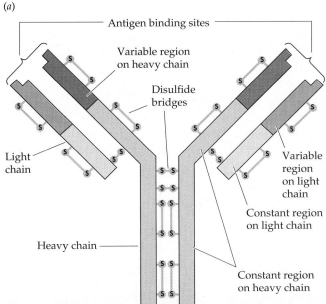

(a)

Antigen binding sites

Variable region on heavy chain

Disulfide bridges

Light chain

Variable region on light chain

Constant region on light chain

Heavy chain

Constant region on heavy chain

(b)

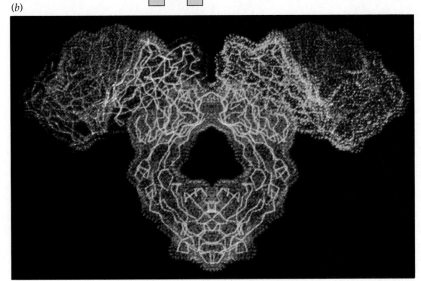

16.10   Structure of Immunoglobulins
(a) Disulfide bridges hold the four polypeptide subunits of an immunoglobulin together. The variable regions recognize and bind antigens. (b) An immunoglobulin molecule in roughly the same orientation as (a), drawn by a computer. Green indicates the light chains, mixed with red for the variable regions and yellow for the constant regions. Dark blue indicates the heavy chains, mixed with red for the variable regions and light blue for the constant regions.

BOX 16.A

Monoclonal Antibodies

Because most antigens carry many different antigenic determinants, animals usually produce a complex mixture of antibodies. Scientists found it virtually impossible to separate the individual antibody types for chemical study. In the mid-1970s, however, Cesar Milstein (an Argentine biochemist living in Cambridge, England) and a colleague from Switzerland, Georges Köhler, made an important breakthrough. They knew that a single lymphocyte produces only a single species of antibody. In principle, all one needed to do was to cause a single lymphocyte to multiply in pure culture to get a large population of cells, all dedicated to the production of the same antibody. However, the antibody-producing cells cannot be cultured. On the other hand, cancerous tumors of plasma cells, called myelomas, grow rapidly in culture. The cells of a given tumor all produce the same antibody, but they produce far too little of the antibody to be useful sources. Milstein and Köhler made use of both cell types—normal lymphocytes and myeloma cells—to produce hybrid cells (**hybridomas**) that made specific normal antibodies in quantity and that, like the myeloma cells, could proliferate rapidly and indefinitely in culture.

Clones of hybrid cells are made as follows. An animal is inoculated with an antigen to trigger specific lymphocyte proliferation. Later, the spleen is dissected out and lymphocytes are collected from it. (The spleen, like the lymph nodes and certain other lymphoid tissues associated with the gut, is a site of lymphocyte accumulation and maturation.) These lym-

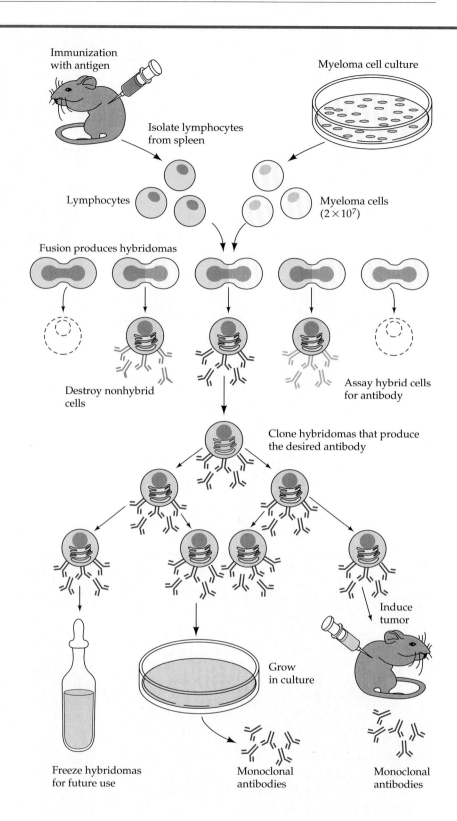

Immunization with antigen

Myeloma cell culture

Isolate lymphocytes from spleen

Lymphocytes

Myeloma cells (2×10^7)

Fusion produces hybridomas

Destroy nonhybrid cells

Assay hybrid cells for antibody

Clone hybridomas that produce the desired antibody

Induce tumor

Grow in culture

Freeze hybridomas for future use

Monoclonal antibodies

Monoclonal antibodies

phocytes are combined under appropriate conditions with myeloma cells from a single tumor. Some lymphocytes fuse with myeloma cells, giving rise to hybridomas. The cell mixture is then treated to destroy all non-hybrid cells. The hybridomas are grown in a suitable medium so that each one forms a clone. Individual clones are tested, and the ones that produce the desired antibodies—specific for one antigenic determinant—are selected. These clones produce **monoclonal antibodies** (uniform antibodies from a single clone of cells) in large quantities, either from a mass culture or following transfer into an animal where they can grow as a tumor. The hybridomas may also be frozen for storage.

Monoclonal antibodies are ideal for the study of specific antibody chemistry, and they have been used to further our knowledge of cell membranes as well as for specialized laboratory procedures such as tissue typing for grafts and transplants. Monoclonal antibodies have many practical applications. One possibility is passive immunization—inoculation with specific antibody rather than with an antigen that causes the patient to develop his or her own antibody (as most vaccines are designed to do). Monoclonal antibodies can also be used for detecting specific cancers; diagnostic kits using monoclonal antibodies for colon cancer are already available. Antitumor monoclonal antibodies have been successfully employed in immunotherapy for tumors. In yet another use, poisons directed against tumors can be attached to specific monoclonal antibodies as a form of cancer treatment. Developmental biologists (Chapter 17) have used monoclonal antibodies to show that the cell membranes of different kinds of cells in an animal contain different molecules and that the molecules change as the animal develops.

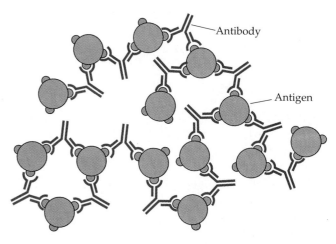

16.11 Antibody–Antigen Complex
An antibody has two sites that can bind to different molecules of antigen, and more than one antibody may bind to the same antigen molecule. Large antibody–antigen complexes may precipitate from the blood; they play a role in many types of clinical laboratory tests.

There are five immunoglobulin classes based on differences in the constant regions of the heavy chains (Figure 16.12). One, called immunoglobulin M, or IgM, is always the first antibody product of a plasma cell. The cell may later switch over to the production of other classes of immunoglobulins—but with the same antibody specificity. The four other classes—IgA, IgD, IgE, and IgG—play different roles in the immune system.

IgG molecules, which have the γ heavy-chain constant region, make up about 85 percent of the total immunoglobulin content of the bloodstream. They consist of a single immunoglobulin unit (two identical heavy chains and two identical light chains) and are produced in greatest quantity during a secondary immune response (see Figure 16.6). IgG defends the body in several ways. For example, some IgG molecules that have bound antigens become attached by their heavy chains to a type of phagocyte called macrophages. This IgG–macrophage union makes it easier for the phagocytes to ingest the antigens (Figure 16.13). Another major function of IgG is to activate antimicrobial proteins, a potent set of nonspecific defenses collectively called the complement system (discussed in the next section); this also enhances phagocytosis.

The bulk of the antibodies produced at the beginning of a primary immune response are IgM molecules. They differ from IgG in being composed of five immunoglobulin units (see Figure 16.12). Because they have more binding sites, IgM molecules are more active than IgG molecules in activating the complement system and promoting the phagocytosis of antibody-coated cells.

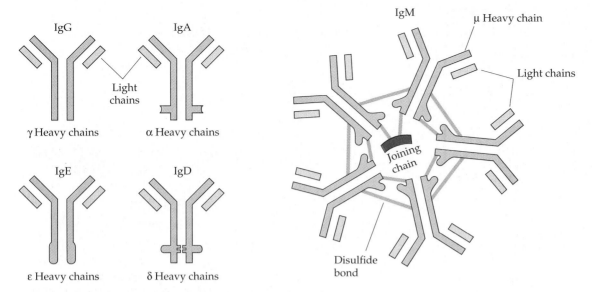

16.12 Classes of Immunoglobulins

Each class of immunoglobulin has its own type of heavy chain, identified by a Greek letter. IgM, unlike other antibodies, is made up of five immunoglobulin subunits. Two of the five units are bonded to a single joining chain.

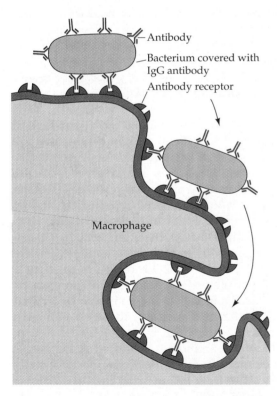

16.13 IgG Antibodies Promote Phagocytosis

Phagocytes called macrophages have receptors on their cell surfaces for part of the IgG molecule. Thus when a bacterium has reacted with IgG, the resulting complex binds readily to macrophages, activating phagocytosis (ingestion of the antigen by the macrophage).

Only small amounts of IgD antibody travel free in the bloodstream. The major role of IgD is to serve, as IgM does, as membrane receptors on B cells.

IgE antibodies take part in inflammation and allergic reactions. IgE helps kill worm pathogens such as those that cause the disease schistosomiasis, which affects some 200 million people in Africa and South America. Where inflammation occurs, IgE may participate in bringing white blood cells, components of the complement system, and other factors into the inflamed region. For most of us, the effect of IgE is most apparent when we suffer allergies. IgE molecules bind to antigenic determinants on the substances that provoke the allergy (the allergens), and they also bind to receptor sites on the surfaces of cells called mast cells. The mast cell–IgE–allergen complex stimulates the release of histamine and other compounds, leading in turn to inflammation. Hives, hay fever, eczema, and asthma are all common allergic reactions (Figure 16.14).

Body secretions such as saliva, tears, milk, and gastric fluids all contain immunoglobulins, specifically IgA. IgA molecules are transported across epithelial cells to join the secreted fluids. IgA exists as both monomers and dimers; Figure 16.12 shows only the monomeric unit.

ANTIBODIES AND NONSPECIFIC DEFENSES WORKING TOGETHER

Recall that antimicrobial proteins were listed at the beginning of the chapter among the nonspecific defenses. Vertebrate blood contains about 20 different antimicrobial proteins that make up the **complement system**. These proteins, in different combinations, provide three types of defenses. Their most impres-

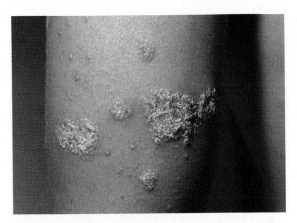

16.14   An Allergic Reaction
Eczema is a common, noncontagious allergic reaction characterized by itching, redness, and the appearance of crusted lesions.

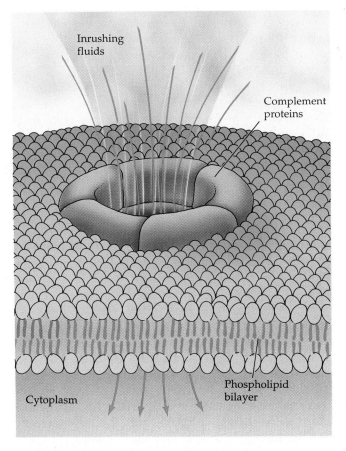

16.15   A Pore in a Lytic Complex Destroys a Foreign Cell
The end result of a cascade of reactions is a precisely arranged group of proteins extending through the phospholipid bilayer of the plasma membrane of a foreign cell. A pore in the complex makes the affected cell leaky; fluids rushing into the cell then cause it to burst.

sive defense, carried out with the help of antibodies, is to lyse (burst) foreign cells—bacteria, for example. When IgG antibodies bind to antigenic determinants on the surface of a foreign cell, this binding may bring about the binding of the first of the complement proteins to the cell surface. What follows is a cascade of reactions, with different complement proteins acting upon one another in succession. The final product of the complement cascade is a lytic complex—a doughnut-shaped structure in the membrane of the foreign cell that renders the membrane leaky, allowing fluids to enter the cell rapidly, causing lysis (bursting) of the foreign cell (Figure 16.15).

In the second type of defense, complement proteins help phagocytes destroy foreign microorganisms. The phagocytes can recognize foreign cells more easily after complement proteins attach to the foreign cells. The third defensive activity of the complement system is to attract phagocytes to sites of infection.

The complement system is *nonspecific* in its action. The specific recognition that is the first step toward forming a lytic complex is accomplished by the antibodies that bind to cell-surface antigens of the invading cells, not to the complement proteins themselves. Hereditary deficiencies in one or another of the complement proteins can cause characteristic diseases, mostly infections and hypersensitivity diseases.

THE ORIGIN OF ANTIBODY DIVERSITY

A newborn mammal possesses a full set of genetic information for immunoglobulin synthesis. At each of the loci coding for the heavy and light chains it has an allele from the mother and one from the father. Throughout the animal's life, each of its cells begins with the same full set. However, the genomes of B cells become modified during development in such a way that each cell eventually can produce one—and only one—type of antibody. Different B cells develop different antibody specificities. A question that vexed immunologists and geneticists for decades was how a single organism could produce so many different specific immunoglobulins—perhaps a million, or even a billion, antibody specificities. Research in recent years has effectively answered this question.

The most surprising part of the answer is that functional immunoglobulin genes are assembled from DNA segments that initially are spatially separate. Every cell has hundreds of DNA segments potentially capable of participating in the synthesis of the variable regions, the parts of the antibody molecule conferring immunological specificity. B-cell precursors differ from all other cells in that these DNA segments are *rearranged* during B-cell development. Pieces of the DNA are deleted, and DNA segments formerly distant from one another are spliced together; thus a gene is assembled from randomly selected pieces of DNA. Each B-cell precursor in the animal assembles its own unique set of immunoglobulin genes. This remarkable process generates many diverse antibodies from the same starting genome. The assembly of immunoglobulin genes from spatially separate DNA segments was first demonstrated by Susumu Tonegawa, in Switzerland, in 1976.

In both humans and mice, the DNA segments coding for immunoglobulin heavy chains are on one chromosome and those for light chains are on others. The variable region of the light chain is coded for by two families of DNA segments, and the variable region of the heavy chain is coded for by three families. (We discussed gene families in Chapter 13; see Figure 13.4.) Gene families are assembled randomly; thus the diversity afforded by several hundred DNA segments is multiplied by the combination of different families. Furthermore, since light and heavy chains are synthesized independently of one another, the combination of light and heavy chains introduces more diversity. If, say, 1,000 different light-chain variable regions and perhaps 10,000 different heavy-chain variable regions could be produced, this would allow some $1,000 \times 10,000 = 10,000,000$ different immunoglobulins to be produced.

Are we forgetting another contribution to antibody diversity? B-cells are diploid, so each contains *two* chromosomes with the heavy-chain genes and two of each of the others with light-chain genes. So shouldn't each B-cell produce two types of heavy chains and two types of light chains—and thus as many as four different types of antibodies? If this were true, the clonal selection theory, which depends upon a given cell producing a single antibody species, would be in shambles. The theory holds, however, because only one homologous chromosome from each relevant pair produces an mRNA that is translated into the corresponding antibody chain. (The underlying mechanism is not yet well understood.)

Although the diploidy of B cells does not contribute to antibody diversity, there is one more factor that does: *mutation*. Mutation occurs frequently in the genes of the variable region during B-cell development. It has been estimated that such mutations increase the total number of different antibody specificities by 10 to 100 times or more.

How a B Cell Produces a Particular Heavy Chain

To see how DNA rearrangement generates antibody diversity, let's consider how the heavy chain of IgM is produced. B cells produce this antibody, which then inserts itself into the plasma membrane of the B cells.

The locus governing heavy-chain synthesis is on chromosome 12 of mice and on chromosome 14 of humans. In mice, the locus is arranged as shown in Figure 16.16, with a long stretch of DNA occupied by a family of 100 or more V (variable) segments. Humans have about 300 V segments. In a given B cell, only *one* of these segments is used to produce part of the variable region of the heavy chain; the remaining V segments are discarded or rendered inactive. At a distance of many nucleotides from the V segments is a family of 10 or more D (diversity) segments. Again, only one of these is used to produce part of the variable region of the heavy chain of a given B cell, as is only one J (joining) segment from the family of four such segments (in mice) lying yet farther along the chromosome. This combination of one each of V, D, and J segments forms a complete variable region for a functional gene. Still farther along the chromosome, and separated from the suite of J segments, is a family of eight segments, one of which codes for the constant region of the heavy chain. Light chains are produced from similar families of DNA segments, but without D segments.

How does order emerge from this seeming chaos of DNA segments? Two important steps impose order. First, substantial chunks of DNA are deleted

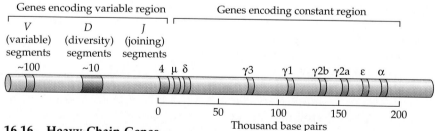

16.16 Heavy-Chain Genes
Immunoglobulin heavy chains are encoded by a gene that has many segments. The variable region for the heavy chain of a particular antibody is encoded by one V segment, one D segment, and one J segment. Each of these segments is taken from a pool of like segments. The constant region is selected from another pool of segments.

from the chromosome during the rearrangement of the segments. As a result of these deletions, a particular *D* segment is joined directly to a particular *J* segment, and then the *D* segment is joined to one of the *V* segments; thus a single "new" sequence, consisting of one *V*, one *D*, and one *J* segment, can now code for the variable region of the heavy chain. All the progeny of this cell constitute a clone having this same sequence for the variable region. Different B cells result from DNA being deleted in different places on the chromosome, leading to different variable-region sequences. In mice—which have about 100 *V* segments, 10 or more *D* segments, and 4 *J* segments to choose from—about $100 \times 10 \times 4 = 4,000$ different heavy-chain variable regions are possible. This estimate increases when we take into account the variation introduced by mutation.

The second step in organizing the synthesis of an immunoglobulin chain follows transcription. Splicing of the RNA transcript (see Chapter 13) removes the product of an intron that includes any *J* segments lying between the selected *J* segment and the first constant-region segment. Splicing also removes the products of introns contained in both the *V* segment and the *C* (constant-region) segment (Figure 16.17). The result is an mRNA that can be translated, directly yielding the heavy chain of the cell's specific antibody.

Note that two distinct types of nucleic acid splicing contribute to the formation of an antibody. *DNA* splicing, before transcription, joins the *V*, *D*, and *J* segments. *RNA* splicing, after transcription, joins the *J* segment to the constant region.

The Constant Region and Class Switching

Early in its life a plasma cell produces IgM molecules that are responsible for the specific recognition of a particular antigenic determinant. At this time, the constant region of the antibody's heavy chain is encoded by the first constant-region segment, the μ segment (see Figure 16.16). During an immunological response—later in the life of the plasma cell—another deletion may occur in the plasma cell's DNA, positioning the heavy-chain variable-region gene (consisting of the same *V*, *D*, and *J* segments) next to a constant-region segment farther down the original DNA, such as the γ, ε, or α constant region (Figure 16.18).

Such a deletion—called **class switching**—results in the production of an antibody with a different *function* but the same *antigen specificity*. The new antibody has the same variable regions of the light and heavy chains but a different constant region of the heavy chain. This antibody falls into one of the four other immunoglobulin classes (IgA, IgD, IgE, or IgG; see Figure 16.12), depending on which of the constant-region segments is adjacent to the variable-region gene. Once class switching has occurred, the plasma cell cannot go back to making the previous immu-

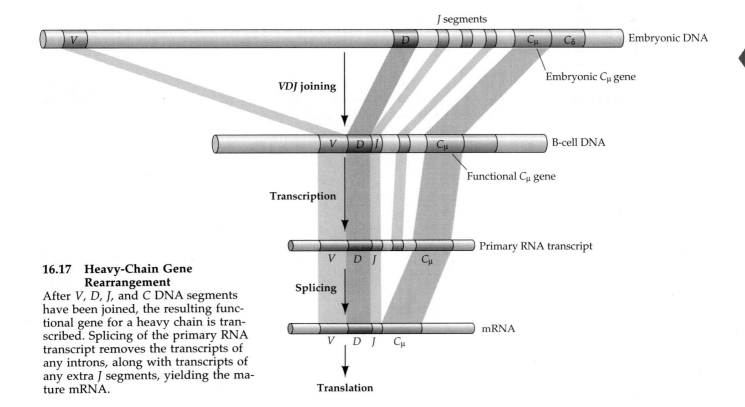

16.17 Heavy-Chain Gene Rearrangement
After *V*, *D*, *J*, and *C* DNA segments have been joined, the resulting functional gene for a heavy chain is transcribed. Splicing of the primary RNA transcript removes the transcripts of any introns, along with transcripts of any extra *J* segments, yielding the mature mRNA.

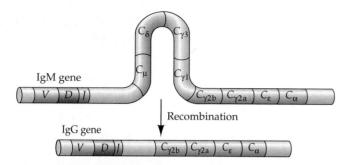

16.18 Class Switching
The functional gene produced by joining *V*, *D*, *J*, and *C* segments may later be modified, causing a different *C* segment to be transcribed. This modification, known as class switching, is accomplished by deleting part of the constant region. Shown here is class switching from an IgM gene to an IgG gene.

noglobulin class—that part of the DNA has been deleted. On the other hand, if additional constant-region segments are still present, another class switch may occur.

T CELLS: AGENTS OF BOTH RESPONSES

Thus far we have been concerned primarily with the humoral immune response, whose effector molecules are the antibodies secreted by plasma cells that develop from activated B cells. T cells are the effectors of the cellular immune response, which is directed against multicellular invaders and against antigens that have become established *within* the animal's cells. If the antigens are inside host cells, how do T cells encounter the antigenic determinants? We will address that key question after a brief introduction to the T cells themselves.

T cells participate in *both* the humoral and cellular immune responses. These responses are highly specific, and we will consider how T cells contribute to the specificity of both responses. Like B cells, T cells possess specific receptors.

T-cell receptors are glycoproteins with molecular weights about half that of an IgG. They are made up of two polypeptide chains, each encoded by a separate gene (Figure 16.19). The genes that code for T-cell receptors are similar to those for immunoglobulins, suggesting that both are derived from a single, evolutionarily more ancient group of genes. Like the immunoglobulins, T-cell receptors include both variable and constant regions and are assembled by *V–D–J* joining. Once formed, the receptors are bound to the plasma membrane of the T cell that produces them. In the next sections, which introduce the major histocompatibility complex, we discuss how T-cell receptors bind antigens.

When T cells are activated by contact with a spe-

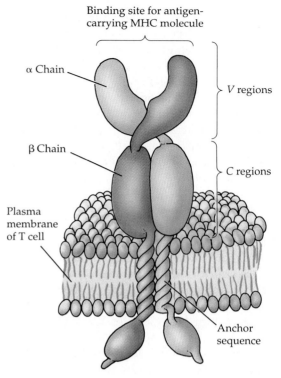

16.19 A T-Cell Receptor
A T-cell receptor consists of two polypeptide chains—an α chain and a β chain. Each polypeptide chain possesses a hydrophobic region that anchors the chain in the phospholipid bilayer of the T-cell plasma membrane.

cific antigenic determinant, they develop and give rise to several distinct types of effector cells. Cytotoxic T cells, or **T_C cells**, are one type of effector cell. T_C cells recognize virus-infected cells and kill them by causing them to lyse (Figure 16.20). Helper T cells,

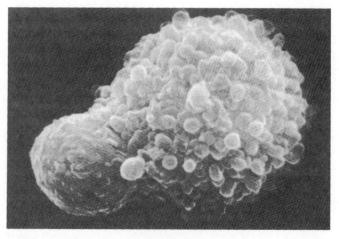

16.20 A Virus-Infected Cell Bites the Dust
A cytotoxic T cell (smaller sphere at lower left) has contacted a virus-infected cell, causing the infected cell to lyse, as indicated by the blisters all over its surface. The T_C cell induced the lysis by releasing a protein, aptly named perforin, that perforated the plasma membrane of the infected cell.

or **T$_H$ cells**, assist both the cellular and humoral immune systems. As mentioned already, a T$_H$ cell of appropriate specificity must bind an antigen before a B cell can be activated by it. Still other T cells differentiate into suppressor T cells, or **T$_S$ cells**. These regulatory cells inhibit the responses of both B and T cells to antigens. They probably play an important role in immune tolerance and in the acceptance of self antigens. Now that we are familiar with the major types of T cells, we can return to the seemingly difficult question of how T cells meet the antigenic determinants if the antigens themselves are inside host cells.

The Major Histocompatibility Complex

The key to the interactions of T cells and antigenic determinants, and to the interactions of B cells and the various classes of T cells, lies in a tight cluster of loci called the major histocompatibility complex, or **MHC**. These loci code for specific proteins on the surfaces of cells. Because of the number of MHC genes and the number of their alleles, different animals of the same species are highly likely to have different MHC genotypes—and that difference is what leads to the rejection of organ transplants.

Similarities in structure and base sequences between MHC genes and the genes coding for antibodies suggest that the MHC genes may be descended from the same ancestral genes as are those for antibodies and T-cell receptors. Major aspects of the immune systems may thus be woven together by a common evolutionary thread.

There are three classes of MHC loci. Class I MHC loci code for proteins (antigens) that are present on the surface of every cell in the animal. These proteins function in antiviral T-cell immunity. The products of class II loci are found only on the surfaces of B cells, T cells, and certain phagocytes called macrophages (Figure 16.21). It is this class of MHC products that is primarily responsible for the interaction of T$_H$ cells, macrophages, and B cells in antibody responses. Class III MHC loci code for some of the proteins of the complement system that interact with antigen–antibody complexes and result in the lysis of foreign cells (see Figure 16.15). Now let's see how class I and class II MHC products help T cells interact with antigenic determinants.

T Cells and the Humoral Immune Response

A macrophage ingests an antigen and breaks it into antigen fragments, each with one or more antigenic determinants. The fragments are called **processed antigen**. The products of class II MHC loci bind processed antigen, carry it to the surface of the cell, and display it on the outside of the cell where it is available to T$_H$ cells (Figure 16.22). Because the MHC products *present* antigen, the macrophages are thus referred to as **antigen-presenting cells**.

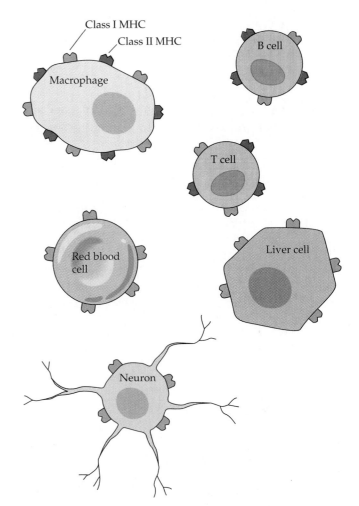

16.21 MHC Products in Plasma Membranes
Class I MHC proteins can be found on every cell in an animal; class II MHC proteins are found only on B cells, T cells, and macrophages.

A T$_H$ cell can bind to processed antigen only if the T-cell receptors correspond to the displayed antigenic determinant—and only if the processed antigen is carried by an MHC protein. The T-cell receptor binds both antigenic determinant and MHC protein. When binding is complete, the macrophage releases substances that activate the T$_H$ cell, causing it to produce a clone of differentiated cells capable of interacting with B cells. The steps to this point constitute the activation phase of the response. Next comes the effector phase, in which B cells are activated.

B cells, too, are antigen-presenting cells. B cells take up antigen bound to their receptors, process it, and display it on class II MHC products. An activated T$_H$ cell binds only if it recognizes both the displayed antigenic determinant and the MHC protein. The bound T$_H$ cell releases helping signals that cause the

16.22   MHC Products and Antigen-Presenting Cells
The products of class II MHC loci present the processed antigen to the T cells.

B cell to produce a clone of plasma cells. Finally, the plasma cells secrete antibody, completing the effector phase of the humoral immune response.

T Cells and the Cellular Immune Response

MHC proteins, this time the products of class I MHC loci, play a similar role in the cellular immune response. Here the virus-infected cells themselves are the antigen-presenting cells. Class I MHC proteins in the membrane of the infected cell display processed antigen (virus fragments), making it available to T$_C$ cells. When T$_C$ cells bind the complex of processed

16.23   The MHC Mediates Both Immune Responses
This diagram summarizes the activation and effector phases of the humoral immune response and the cellular immune response.

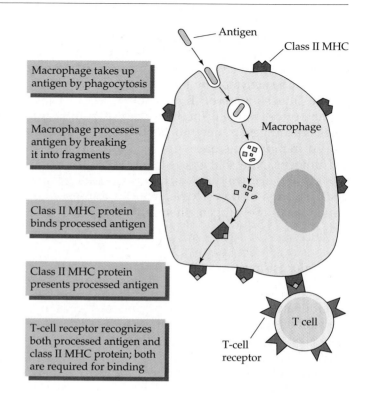

Macrophage takes up antigen by phagocytosis

Macrophage processes antigen by breaking it into fragments

Class II MHC protein binds processed antigen

Class II MHC protein presents processed antigen

T-cell receptor recognizes both processed antigen and class II MHC protein; both are required for binding

(a)   **Humoral response**

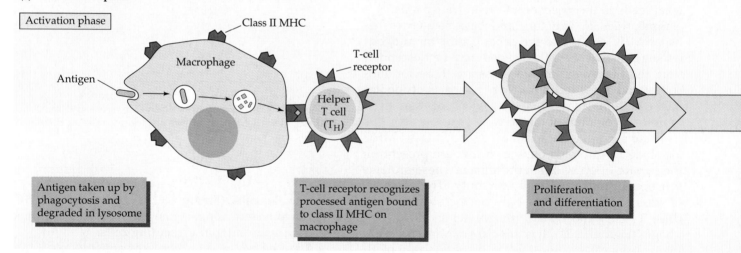

Activation phase

Antigen taken up by phagocytosis and degraded in lysosome

T-cell receptor recognizes processed antigen bound to class II MHC on macrophage

Proliferation and differentiation

(b)   **Cellular response**

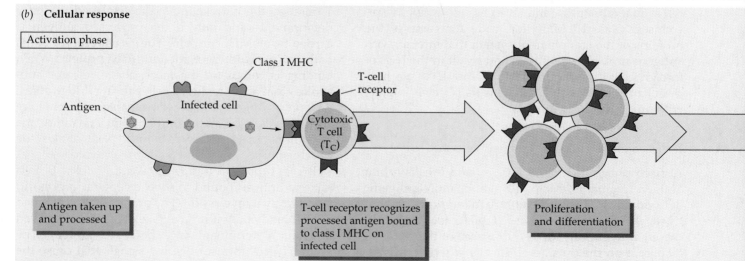

Activation phase

Antigen taken up and processed

T-cell receptor recognizes processed antigen bound to class I MHC on infected cell

Proliferation and differentiation

antigen and class I MHC protein, they become activated. In the effector phase of the cellular immune response, T-cell receptors on the activated T_C cells recognize processed viral antigen displayed by class I MHC proteins on the surface of virus-infected cells. The T_C cells then produce lytic signals (perforin; see Figure 16.20), causing the infected cell to lyse. Because T-cell receptors recognize self MHC products, they rid the body of its own virus-infected cells.

The roles of T cells in the humoral and cellular immune systems are similar (Figure 16.23). Both use MHC proteins on the surfaces of antigen-presenting cells, and both have well defined activation and effector phases. Next we consider another crucial interaction between T cells and the MHC.

The MHC and Tolerance of Self

The MHC plays a key role in establishing tolerance to self, without which an animal would be destroyed by its own immune system. Developing T cells undergo "testing" in the thymus. One test asks, Can this cell recognize the body's MHC proteins? A T cell unable to recognize self MHC would be useless to the animal because it could not participate in any immune reactions. Such a cell fails the test and dies within about three days. The second and more crucial test asks, Does this cell bind to self MHC protein *and* to one of the body's own antigens? A T cell that satisfied both of these criteria would be harmful or lethal to the animal; it fails the test and is destroyed immediately.

The T cells that survive this pair of tests are those that recognize the animal's own MHC proteins and that ignore the animal's own antigens. Such T cells are the ones that can do the work of the cellular immune system. They get their diplomas and mature into either T_C cells, T_H cells, or T_S cells.

Transplants

A major side effect of the MHC antigens became important with the development of organ transplant surgery, sometimes with devastating results. Because

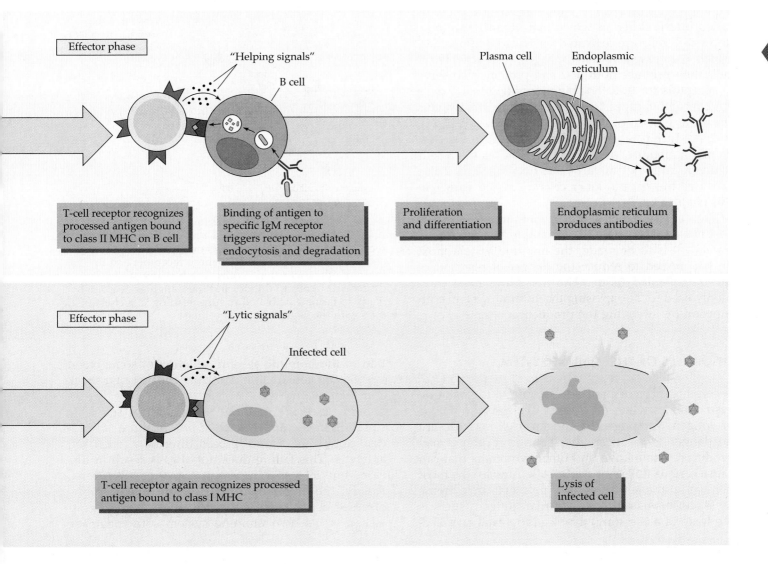

Effector phase

"Helping signals"

B cell

Plasma cell

Endoplasmic reticulum

T-cell receptor recognizes processed antigen bound to class II MHC on B cell

Binding of antigen to specific IgM receptor triggers receptor-mediated endocytosis and degradation

Proliferation and differentiation

Endoplasmic reticulum produces antibodies

Effector phase

"Lytic signals"

Infected cell

T-cell receptor again recognizes processed antigen bound to class I MHC

Lysis of infected cell

the proteins produced by the MHC are specific to each individual, they act as antigens if transplanted into another individual. An organ or a piece of skin transplanted from one person to another is recognized as nonself and soon provokes an immune response; the tissue is then killed, or "rejected," by the cellular immune system. But if the transplant is performed immediately after birth or if it comes from a genetically identical person (an identical twin), the material is recognized as self and is not rejected.

Physicians can overcome the rejection problem for a while by treating the patient with drugs, such as cyclosporin, that suppress the immune system. This technique, however, compromises the ability of patients to defend themselves against bacteria and viruses. Cyclosporin and some other immunosuppressants interfere with communication between cells of the immune system. How do the cells of the immune system normally communicate with one another?

Interleukins

Communication in the immune system consists of signals that pass from macrophages to T cells, and from T_H cells to B cells. These signals are proteins called **interleukins**, of which more than 10 are now known. T cells are activated by IL-1 (interleukin-1), which is released by macrophages. The activated T cells then produce both IL-2 and proteins that serve as receptors for IL-2. The binding of IL-2 to a T cell's IL-2 receptors causes the cell to divide. The result is the rapid growth of a clone of T cells (Figure 16.24). IL-2 produced by T_H cells helps B cells to start secreting antibodies, and it probably causes the B cells to divide after they are activated by antigen. IL-2 also activates the natural killer cells discussed earlier in this chapter.

The central role of IL-2 in the immune response is best illustrated by a medical example. When a tissue transplant is to be made, the immune system must be suppressed to reduce the danger of rejection of the transplanted tissue. The two drugs most commonly used for suppressing the immune system both function by inhibiting the production of IL-2.

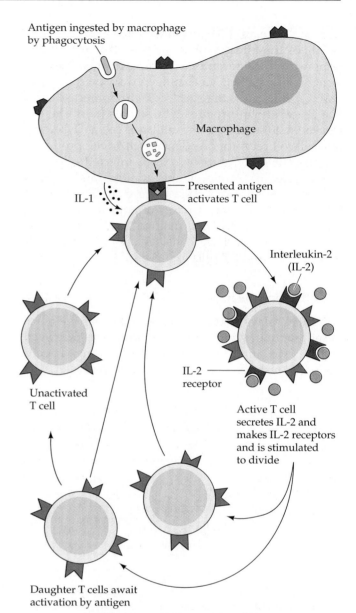

16.24 The Action of Interleukin-2
A T cell, activated by IL-1 from a macrophage, makes IL-2 receptors that become part of its own plasma membrane and secretes IL-2 that binds to these receptors and to those of other activated T cells. The binding of the IL-2 causes the T cells to divide, and the daughter cells are ready to become activated in turn, leading to a clone of T cells with the same antigen specificity.

DISORDERS OF THE IMMUNE SYSTEM

A common type of condition relating to the immune system arises when the system overreacts to a dose of antigen. Although the antigen itself may present no danger to the host, the immune response may produce inflammation and other symptoms that can cause serious illness or death. **Allergies** are the most familiar examples of such a problem. In extreme cases, an exposure to a particular antigen (such as the toxin of a bee sting) may lead to a fatal overreaction of the immune system—dilation of some blood vessels and constriction of others, causing first shock and then death.

Sometimes the immune recognition of self fails, resulting in the appearance of one or more "forbidden clones" of B and T cells directed against self antigens. This failure does not always result in disease, but in some instances it can be disastrous. Among the **autoimmune diseases** of our species—diseases in which components of the body are attacked by its own immune system—are rheumatic

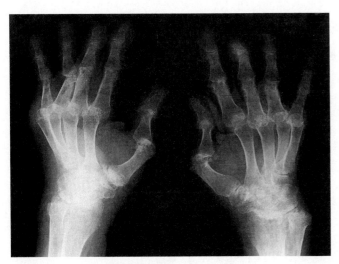

16.25 Rheumatoid Arthritis
This X ray shows the bones in the hands of a 50-year-old man with a 20-year history of rheumatoid arthritis. This autoimmune disease affects the joints, destroying the cartilage with an inflammatory reaction and causing pain and the severe deformities seen here.

fever, rheumatoid arthritis (Figure 16.25), ulcerative colitis, myasthenia gravis (muscle weakness), and several others. Many medical scientists believe that multiple sclerosis results from an abnormality in a type of T cell. The abnormal T cells mount an immune attack on the myelin sheath, an insulating material that surrounds many nerve cells (see Chapter 38). When the myelin sheath is damaged, the result is a severely debilitating loss of nerve function, including blindness and loss of motor control.

Immune Deficiency Disorders

There are various immune deficiency disorders, such as those in which B cells are never formed and others in which B cells lose the ability to give rise to plasma cells. In either case, the affected individual is unable to produce antibodies and thus lacks a major line of defense against microbial pathogens.

Because of its essential roles in both antibody responses and cellular immune responses, the T_H cell is perhaps the most central of all the components of the immune system—the worst one to lose to an immune deficiency disorder. The disease responsible for the epidemic discussed at the beginning of this chapter, acquired immune deficiency syndrome, or **AIDS**, homes in on the T_H cells. At the onset of the infection, the number of HIV (*h*uman *i*mmunodeficiency *v*irus) particles in the blood soars until the immune system mounts a response that clears most—but, unfortunately, not all—of the HIV particles. Some of the particles infect and remain latent in the T_H cells, often for many years. As more cells become infected, the immune system is weakened.

Eventually the HIV count in the blood soars again and the immune system fails completely.

HIV is a retrovirus—a virus with RNA as its genetic material, capable of inserting its own genome into the genome of its animal host. The structure of HIV is shown in Figure 16.26. A central core, with a protein coat (p24 capsule protein), contains two identical molecules of RNA, as well as certain enzymes. An envelope, derived from the plasma membrane of the cell in which the virus was formed, surrounds the core. The envelope is studded with an envelope protein (gp120, where gp stands for glycoprotein) that enables the virus to infect its target T_H cell.

HIV attacks host cells at a membrane protein called CD4, which is found primarily on T_H cells and macrophages. CD4 acts as the receptor for the viral envelope protein gp120. The binding of gp120 to CD4 starts a complex series of events (Figure 16.27). When HIV infects a cell, the viral core is admitted into the cell; then the core "uncoats" itself, releasing its contents. Among the enzymes in the core is **reverse transcriptase**, which catalyzes the formation of a double-stranded DNA molecule encoding the same information as the viral RNA (see Chapters 13 and 14). Another enzyme catalyzes the destruction of RNA molecules transcribed from the host cell's own DNA. The DNA transcript enters the nucleus of the host cell and is spliced into a chromosome, much as bacteriophage DNA may become incorporated into a bacterial chromosome. Another HIV enzyme, called **integrase**, catalyzes this splicing. The DNA transcript of the HIV RNA thus becomes a permanent part of the chromosome, replicating with it at each division of a host cell.

A DNA transcript of the HIV RNA, once incorporated into the genome of a T_H cell, may remain there latent for days, or even for a decade or more. The latent period ends if the HIV-infected T_H cell becomes activated. Then the viral DNA is transcribed, yielding many molecules of viral RNA, some of which are translated, forming the enzymes and structural proteins of a new generation of viruses. Other viral RNA molecules are incorporated directly into the new viruses as their genetic material. Formation of new viruses may be slow; the viruses bud from the infected cell, surrounding themselves with modified plasma membrane from the host. More-rapid virus production leads to lysis of the host cell. Several viral genes control the rate of production of individual proteins and of whole viruses. One viral gene, responsible for the antigenic properties of the envelope protein, has a high mutation rate, causing the antigen to change, making HIV a moving target for what is left of the host's immune system—and complicating efforts to develop a vaccine against AIDS.

We do not yet know the mechanisms that lead to

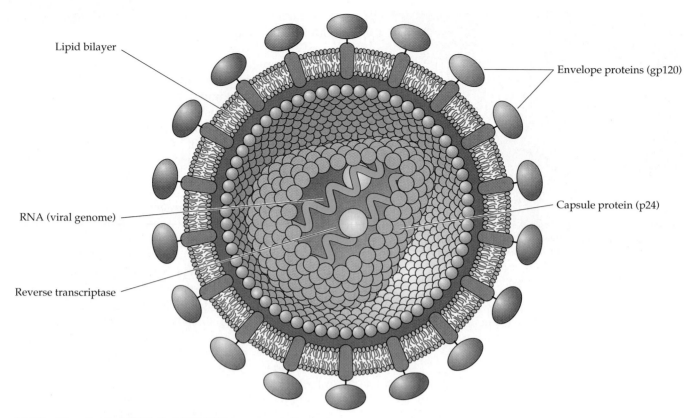

16.26   Structure of HIV, the AIDS Virus
HIV has a core containing RNA and various proteins, including the enzyme reverse transcriptase. Another protein (shown in blue) surrounds the core. The surface of the virus is complex: A phospholipid bilayer, studded with knobs of envelope protein, covers the protein layer surrounding the core.

Labels on figure: Lipid bilayer; Envelope proteins (gp120); RNA (viral genome); Capsule protein (p24); Reverse transcriptase

gradual, selective depletion of the T_H cells. As a consequence of this depletion, however, the host's immune system becomes unable to function. AIDS patients usually die of "opportunistic" infections, diseases caused by bacteria and fungi that are almost always eliminated by the immune systems of uninfected individuals. **AIDS-related complex**, a less severe form of the disease, has milder symptoms but appears to develop into full-blown AIDS in most patients. Thus far, HIV infection appears to lead to death in all cases.

In the United States, the highest incidence of AIDS occurs among drug addicts (from the use of shared, contaminated needles) and male homosexuals. AIDS can be transmitted by blood transfusions, by sexual activity (either homosexual or heterosexual), and from mother to fetus. HIV is not transmitted by mosquitoes or other insect vectors, or by kissing or casual contact. On a worldwide basis, the most common mode of transmission is heterosexual intercourse. At this writing, most of the HIV-infected people in the world live in Africa (Figure 16.28), but the virus is spreading fastest among people living in south and

southeast Asia. By the turn of the century, residents of developing countries will bear 90 percent of new HIV infections as the AIDS epidemic rages on throughout the world. There is not likely to be an effective vaccine or cure by that time. Millions of people will have died before the AIDS epidemic ends. In the meantime, medical and biological research on the subject is proceeding with great intensity.

Prospects for an AIDS Cure— or an AIDS Vaccine

Prospects for a *cure* for AIDS are exceedingly dim at this time. What would an AIDS cure entail? It would include the detection and elimination of every HIV-infected cell in the body. Given the long latent period of infection, this task appears overwhelming. If a cure seems far off, what then?

Many investigators are seeking a vaccine to forestall HIV infection. You have probably seen newspaper or television news reports of some of these studies, and you will certainly see many more. Here again, however, there are major stumbling blocks. One of the greatest problems is the exceptional genetic variation of HIV. New strains are constantly appearing, and a promising new vaccine against one strain may afford no protection whatsoever against new or even other old strains. What next?

There is substantial agreement that we can learn

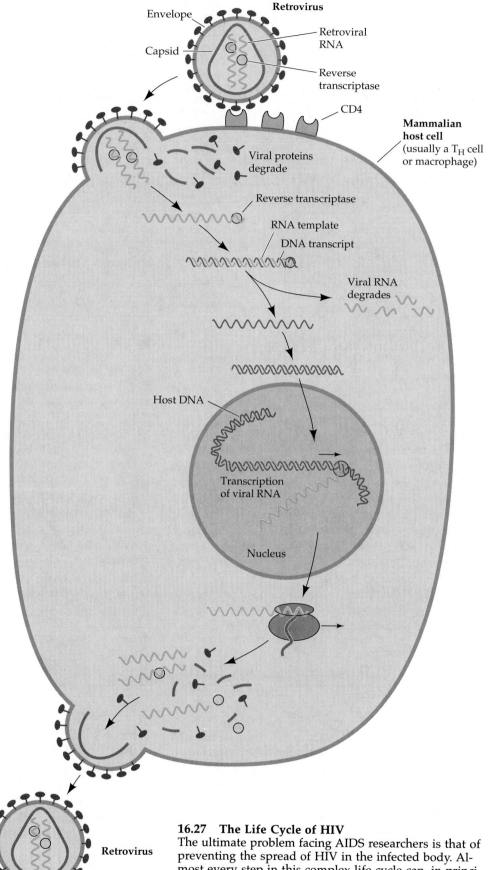

1. Retrovirus attaches to host cell at membrane protein CD4

2. The viral core is uncoated as it enters the host cell by endocytosis

3. Viral RNA uses reverse transcriptase to make complementary DNA

4. Viral RNA degrades

5. Single-stranded reverse transcript synthesizes second complementary DNA strand

6. DNA enters cell nucleus and is integrated into host chromosome, forming a provirus

7. Proviral DNA transcribes viral RNA, which is exported to cytoplasm

8. Viral RNA is translated

9. Viral proteins, new capsids, and envelopes are assembled

10. Assembled virus buds from the cell membrane and releases virus particles

11. Virus matures

16.27 The Life Cycle of HIV
The ultimate problem facing AIDS researchers is that of preventing the spread of HIV in the infected body. Almost every step in this complex life cycle can, in principle, be attacked by one or more potentially therapeutic agents. Which step will we manage to interrupt?

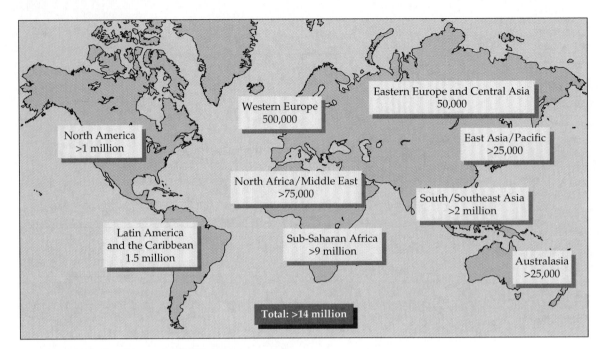

16.28   HIV Infections as of Mid-1994
The numbers are based on World Health Organization estimates. HIV is currently spreading most rapidly in south and southeast Asia.

ways to *control the replication* of HIV within the infected body and thus delay the onset of AIDS symptoms. There are at least 16 steps in the HIV life cycle. If we can block just *one* of them completely, we break the cycle and hold the infection in check. Potential therapeutic agents are being tested against almost all of the steps shown in Figure 16.27. It is of course crucial to work on steps that are unique to the virus so that we can block the step without killing the patient by blocking a corresponding step in the patient's own metabolism. Drugs currently in use to treat AIDS in the United States are AZT, ddI, and ddC—all directed at the reverse transcription step.

None of these is a "magic bullet" against the disease. In the short run, we hope to delay HIV replication by a combination of partially effective treatments.

What can be done until biomedical science provides the tools to bring the worldwide AIDS epidemic to an end? Above all, people must recognize that they are in danger whenever they have sex with a partner whose total sexual history is not known. The danger rises as the number of sex partners rises, and the danger is much greater if sexual intercourse is not protected by a latex condom. The danger of heterosexual intercourse transmitting HIV rises ten- to a hundredfold if either partner has another sexually transmitted disease. Face it: *Every* penetrative sex act with a partner of uncertain HIV status carries a real threat of HIV infection, and *no cure is in sight.*

SUMMARY of Main Ideas about Defenses Against Disease

All organisms have defenses that help them cope with pathogens.

Nonspecific defenses block attempts of pathogens to invade our bodies and destroy pathogens that do get in.
   Review Figure 16.2 and Table 16.1

The immune system recognizes, selectively eliminates, and remembers specific invaders, including nonself cells and macromolecules.

Antigens (the invaders) have one or more antigenic determinants.

Components of the immune system recognize specific antigenic determinants.
   Review Figure 16.5

An immune response includes the formation of effector and memory cells; memory cells allow a larger and more rapid response to a second exposure to the antigen.
   Review Figure 16.6

The humoral immune response is directed against antigens outside the host cells.

The humoral immune response relies on antibodies (immunoglobulins), each of which recognizes and binds one specific antigenic determinant.

B cells give rise to plasma cells, the effector cells that produce antibodies.

The cellular immune response is directed against multicellular pathogens and against viruses that have established themselves in host cells.

The cellular immune response relies on T-cell receptors, each of which recognizes and binds one specific antigenic determinant.
   Review Figure 16.19

Effector T cells include cytotoxic (T_C), helper (T_H), and suppressor (T_S) T cells.

Specific B and T cells become activated when exposed to specific antigenic determinants; they multiply and differentiate to give rise to clones of cells of identical specificity.
   Review Figure 16.7

Continued exposure to self antigens beginning at or before birth results in the deletion or inactivation (by suppressor T cells) of cells that would otherwise produce antiself antibodies.

The immunoglobulin molecule is a tetramer of two identical light-chain polypeptides and two identical heavy-chain polypeptides.
   Review Figure 16.10

Both light and heavy chains contain constant and variable regions.

Immunoglobulin specificity is determined by the variable regions.

Different heavy chains determine the type of effects produced by different antibody classes.

Antibody diversity results from the chance selection of different regions of a chromosome in constructing a single gene coding for the immunoglobulin molecule, as well as from frequent mutation of some of the genes.
   Review Figures 16.16 and 16.17

Genes of the major histocompatibility complex (MHC) encode cell-surface proteins that present antigen to helper T cells and effector cells of the cellular immune system.
   Review Figures 16.21, 16.22, and 16.23

Histocompatibility antigens also are the basis for the rejection of nonself tissue transplants.

Cells of the immune system communicate by releasing interleukins.
   Review Figure 16.24

HIV, a retrovirus that attacks helper T cells, causes AIDS.
   Review Figure 16.26

Reverse transcriptase, carried by the HIV virus, catalyzes the formation of a DNA transcript of the viral RNA.

The DNA transcript of HIV RNA is spliced into the host genome and may remain latent for years before being expressed and producing new viruses, eventually crippling the host's immune system.
   Review Figure 16.27

There is no immediate prospect of either a cure or a vaccine for AIDS.

SELF-QUIZ

1. Which statement is *not* true of phagocytes?
   a. Some travel in the circulatory system.
   b. They ingest microorganisms by endocytosis.
   c. A single one can ingest 5 to 25 bacteria before it dies.
   d. Although they are important, an animal can do perfectly well without them.
   e. Lysosomes play an important role in their function.

2. Immunoglobulins
   a. help antibodies do their job.
   b. recognize and bind antigenic determinants.
   c. are among the most important genes in an animal.
   d. are the chief participants in nonspecific defense mechanisms.
   e. are a specialized class of white blood cells.

3. Which statement is *not* true of an antigenic determinant?
   a. It is a specific chemical grouping.
   b. It may be part of many different molecules.
   c. It is the part of an antigen to which an antibody binds.
   d. It may be part of a cell.
   e. A single antigen has only one antigenic determinant on its surface.

4. T-cell receptors
   a. are the primary receptors for the humoral immune system.
   b. are carbohydrates.
   c. cannot function unless the animal has previously encountered the antigen.
   d. are produced by plasma cells.
   e. are important in combatting viral infections.

5. According to the clonal selection theory
   a. an antibody changes its shape according to the antigen it meets.
   b. an individual animal contains only one type of B cell.

c. the animal contains many types of B cells, each producing one kind of antibody.

d. each B cell produces many types of antibodies.

e. no clones of antiself lymphocytes appear in the bloodstream.

6. Immunological tolerance
   a. depends on repeated exposure throughout the life of the animal.
   b. develops late in life and is usually life-threatening.
   c. disappears at birth.
   d. results from the activities of the complement system.
   e. results from DNA splicing.

7. The extraordinary diversity of antibodies results in part from
   a. the action of monoclonal antibodies.
   b. the splicing of RNA molecules.
   c. the action of suppressor T cells.
   d. the splicing of gene segments.
   e. their remarkable nonspecificity.

8. Which of the following play no role in the antibody response?
   a. Helper T cells
   b. Interleukins
   c. Macrophages
   d. Reverse transcriptase
   e. Products of class II MHC loci

9. The major histocompatibility complex
   a. codes for specific proteins found on the surfaces of cells.
   b. plays no role in T-cell immunity.
   c. plays no role in antibody responses.
   d. plays no role in skin graft rejection.
   e. is coded for by a single locus with multiple alleles.

10. Which of the following plays no role in AIDS?
    a. Integrase
    b. Reverse transcriptase
    c. Transcription
    d. Translation
    e. Transfection

FOR STUDY

1. Describe the part of an antibody molecule that interacts with an antigenic determinant. How is it similar to the active site of an enzyme? How does it differ from the active site of an enzyme?

2. Contrast immunoglobulins and T-cell receptors, with respect to structure and function.

3. Discuss the diversity of antibody specificities in an individual in relation to the diversity of enzymes. Does every cell in an animal contain genetic information for all the organism's enzymes? Does every cell contain genetic information for all the organism's immunoglobulins?

4. Describe and contrast two ways in which DNA splicing plays roles in the immune response.

5. Discuss the roles of monoclonal antibodies in medicine and in biological research.

READINGS

Ada, G. L. and G. Nossal. 1987. "The Clonal-Selection Theory." *Scientific American*, August. A fascinating historical account of the development of the central concept of immunology.

Atkinson, M. A. and N. K. Maclaren. 1990. "What Causes Diabetes?" *Scientific American*, July. An explanation of the origin of diabetes, in terms of an autoimmune response, that may lead to the development of preventive therapies for insulin-dependent diabetes.

Boon, T. 1993. "Teaching the Immune System to Fight Cancer." *Scientific American*, March. Therapies designed to cause a patient's own T-cells to attack tumors.

Cohen, I. R. 1988. "The Self, the World, and Autoimmunity." *Scientific American*, April. The nature of autoimmune diseases, and an approach to their prevention.

Edelman, G. M. 1970. "The Structure and Function of Antibodies." *Scientific American*, August. The amino acid sequence of an antibody dictates its unique characteristics and determines its ability to interact with an antigen.

Engelhard, V. H. 1994. "How Cells Process Antigens." *Scientific American*, August. Clearly explains the roles of the MHC molecules and other key players.

Golde, D. W. 1991. "The Stem Cell." *Scientific American*, December. The cell that gives rise to blood cells and immune systems—applications in cancer treatment and immune defects.

Golub, E. S. and D. R. Green. 1991. Immunology: A Synthesis, 2nd Edition. Sinauer Associates, Sunderland, MA. An outstanding textbook of immunology, with special attention to the experimental basis for what we know.

Haynes, B. F. 1993. "Scientific and Social Issues of Human Immunodeficiency Virus Vaccine Development." *Science*, vol. 260, pages 1279–1286. The technical—as well as the social and ethical—difficulties in conquering AIDS are enormous.

Lerner, R. A. and A. Tramontano. 1988. "Catalytic Antibodies." *Scientific American*, March. A powerful new tool for biotechnology, combining the talents of enzymes and antibodies, which may also be useful in augmenting the capabilities of the immune system.

Marrach, P. and J. Kappler. 1986. "The T Cell and Its Receptor." *Scientific American*, February. A detailed consideration of the key actors in the cellular immune system.

Smith, K. A. 1990. "Interleukin-2." *Scientific American*, March. A clear description of the role of interleukin-2 in the expansion of a clone of T cells.

Tonegawa, S. 1985. "The Molecules of the Immune System." *Scientific American*, October. A beautifully illustrated account of the structures of antibodies and T-cell receptors and of how they are formed.

von Boehmer, H. and P. Kisielow. 1991. "How the Immune System Learns about Self." *Scientific American*, October. Experiments with transgenic mice established the clonal deletion theory.

The baby octopus in the photograph, like its still-unhatched sisters and brothers, has an intricate body. If it is lucky enough to survive, it will grow a hundredfold or more in size, and its adult form will differ from the hatchling you see here. The differences, however, will not be nearly as great as the differences between this newborn animal and the single cell it developed from.

A fertilized egg, or zygote, is a single cell and often very tiny; but within the miniscule structure is the astounding potential to create an entire organism. The human zygote, for example, is only about one-tenth of a millimeter in diameter, or one-fifth the diameter of the period at the end of this sentence. Small as it is, the zygote gives rise to a precisely shaped, extremely complex adult. What happens between the single-celled and adult states? For one thing, an amazing number of new cells forms—more than a hundred trillion (10^{14}) of them. Are these cells all alike? Scarcely. The human body consists of more than 200 types of cells, each with one or more important roles. The body changes form significantly as various tissues take up new positions. Myriad intricate steps make up the development of the body.

THE STUDY OF ANIMAL DEVELOPMENT

Development is a process of progressive change during which an organism successively takes on the forms of the several stages of its life cycle. In its early stages of development an animal (or plant) is called an **embryo**. Sometimes the embryo is contained within a protective structure such as an eggshell or a uterus. An embryo does not actively feed because it obtains its food directly or indirectly (by way of the egg, for example) from its mother. Although we focus on the embryo in this chapter, development is a process that continues through all stages, ceasing only with the death of the animal.

The goal of the developmental biologist is to understand the steps from zygote to adult and the molecular mechanisms that underlie them. The task is daunting—just look at the incredible diversity of animal body types on Earth and imagine the different paths that must lead to that diversity! Early twentieth-century work on development offered hope as important generalizations emerged from the study of the first several hours of development in organisms as diverse as sea urchins, frogs, and flies. But after more studies, the patterns of development of some animals appeared to differ from those of others, and the apparent lack of "rules" of development was discouraging.

Today, however, we find ourselves in a true "golden age" of developmental biology. Some of the pieces are coming together. A few of the gaps be-

Life Emerges from an Octopus Egg

17

Animal Development

17.1 The Cast of Characters
Some of the species scientists use extensively in research on animal development include sea urchins, frogs, and chickens. Here the contrast between adult organisms and their embryo or larval stages is clear.

tween different patterns of development have been bridged, and, as is true throughout biology, discoveries in one field have unexpected application in other fields. The work of geneticists, developmentalists, endocrinologists (those who study hormone action), biochemists, neurobiologists, and others has come to a common focus. Seemingly unrelated studies of roundworms, fruit fly larvae, and mice contribute to one another and to the understanding of human development. Current research has uncovered a stunning unity in certain developmental processes at the cellular and molecular levels, in spite of the equally stunning diversity at the level of the entire organism. Our ignorance still exceeds our knowledge, however.

In our study of animal development, we begin with an an overview of the *unity* that exists across animal groups in the earliest stages of development. We will see that, by cell division and cell movement, a zygote first becomes a solid sphere of cells. The

sphere then develops into a hollow ball of cells. Groups of cells begin to infold, forming a gut. Organs begin to form. Once we have the sequence—the unity—of these earliest stages, we will consider the important developmental processes of determination, differentiation, and pattern formation as we take a closer look at specific points in the development of several animal species.

CLEAVAGE

Some species have been particularly useful subjects in studies of animal development. Foremost among them are frogs, chickens, sea urchins, fruit flies, and a tiny roundworm (Figure 17.1). A sea urchin, a frog, a bird, and a roundworm all start life in the same way: as dividing cells. As we explore the unity in the development of these animals, however, our attention is drawn, even at the first cell divisions, to differences.

Becoming Multicellular

When an egg is fertilized, the resulting zygote nucleus is activated, and DNA replication and mitosis commence. The activation of the nucleus begins the process of cleavage, in which the zygote divides mitotically and gives rise, over a period of hours, to hundreds of cells, and eventually to thousands. In most embryos, initially all the cells divide at the same time in each round of division: First there is a single cell, then there are 2, then 4, 8, 16, and so forth.

From species to species, differences in cleavage, including the arrangement of the daughter cells, or **blastomeres**, depend especially on the amount and distribution of **yolk**—nutrients stored in the egg. The sea urchin, for example, has a small egg (0.15 mm in diameter) with only a small amount of yolk that is distributed uniformly. In such an egg, the blastomeres separate completely from one another as they are formed; division proceeds nearly simultaneously from both ends, or poles, of the cell (Figure 17.2a).

By comparison, a frog egg is rather larger (0.5–1 mm), with the yolk concentrated in one half (the vegetal hemisphere) and pigment concentrated in the other (the animal hemisphere). After fertilization, there is a substantial redistribution of the cytoplasmic contents of the zygote, including the movement of some pigmented material. As a result, a **gray crescent** forms on one side of the zygote, opposite the point of sperm entry and near the boundary between the animal and vegetal hemispheres. The gray crescent contains less pigment than does the rest of the animal hemisphere. As we will see, the gray crescent is of great significance in later development. The blastomeres of a frog zygote divide completely, but the division begins at the point farthest from the vegetal hemisphere. This point is called the **animal pole**; the point farthest from the animal pole is called the **vegetal pole**. The plane of the first cleavage passes through the animal pole, through the middle of the gray crescent on one side of the zygote, through the site of sperm entry on the other, and eventually through the vegetal pole (Figure 17.2b).

A bird egg, which is larger than that of a frog and

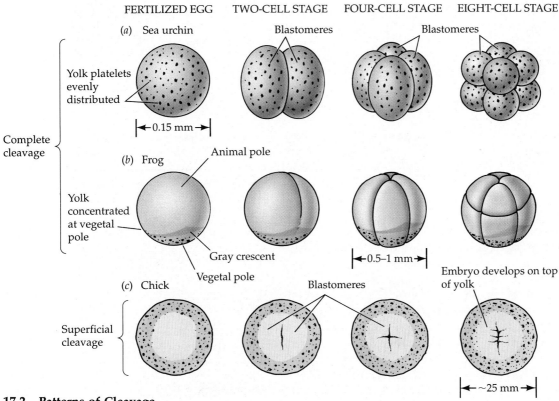

17.2 Patterns of Cleavage
Patterns of early embryonic development reflect differences in the way egg cytoplasm is organized. (a) A sea urchin egg is small, with little yolk (black dots). All of the first three cell divisions are complete, cutting through the entire egg. Both of the first two divisions divide the cell lengthwise; the third cuts across the middle of the cells in a plane perpendicular to that of the first two cleavages. After three divisions, then, the sea urchin embryo consists of eight approximately equal-size cells arranged in two layers. (b) A frog egg is larger than a sea urchin's, with yolk concentrated toward the vegetal pole. All cell divisions are complete: the first two begin near the animal pole; the third is transverse, cutting across the cells near the animal pole. As a result, the cells of the eight-cell stage differ in size, in the amount of yolk they contain, and in their proximity to the gray crescent. (c) A chicken egg is mainly yolk with only a small mass of cytoplasm. The first few divisions of the cytoplasm (shown here from above) do not extend through the yolk; they yield a thin, flat embryo on the surface.

contains much more yolk, is markedly different. Yolk occupies the bulk of the cell, with the yolk-poor cytoplasm confined to the surface or to one end of the cell. Cell division in such zygotes is incomplete: Following each mitosis, the nuclei are separated completely, but initially the daughter cells are not separated completely by plasma membranes. As cleavage continues, however, more and more cells with complete plasma membranes are formed. The embryo develops initially as a disc-shaped mass on top of the yolk (Figure 17.2c).

In some animals, such as roundworms and clams, sister blastomeres become committed to different developmental roles already when there are eight or fewer cells. The parts of the embryo that will be derived from each blastomere are fixed after the second or third division—in a few species, even after the first division. If a blastomere is lost from an embryo of this type, the corresponding portion of the animal is not produced. This developmental pattern is called **mosaic development**, with each blastomere contributing a specific set of "tiles" to the final "mosaic" of the organism. As a consequence, in such embryos the future fates of cells produced at each early division are predictable. Mosaic development is dramatically illustrated in a tiny roundworm, *Caenorhabditis elegans* (Figure 17.3).

By contrast to *C. elegans*, sea urchin and vertebrate embryos are characterized by **regulative development**, in which the loss of some cells during cleavage does not affect development because the remaining cells compensate for the loss. In humans, for example, the separation of one embryo into two masses at the 64-cell stage or beyond, and the subsequent de-

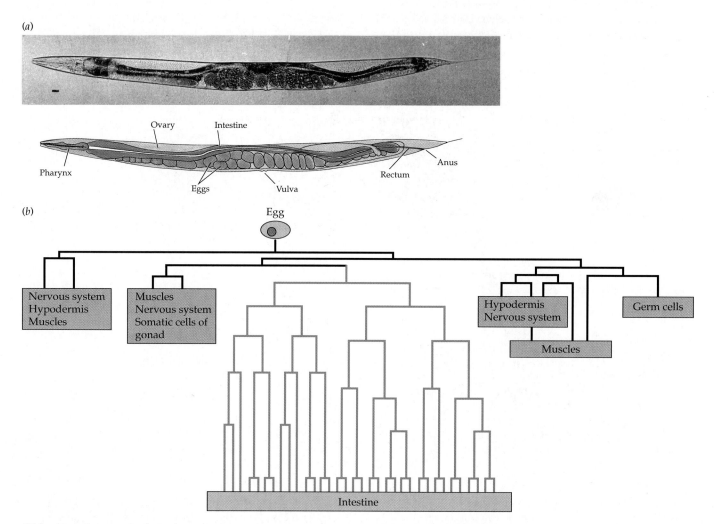

17.3 Development of *C. elegans*
Caenorhabditis elegans is a 1-millimeter-long nematode, or roundworm. There are two sexes: male and the hermaphrodite shown here, which has both male and female reproductive structures. The hermaphrodite reaches its adult stage just three and a half days after fertilization of the egg. (a) Because its internal structures are visible and its anatomy is relatively simple, the transparent, colorless worm is a useful organism for tracking cellular development. (b) It has been possible to trace all divisions from a single cell to the 959 cells found in the fully developed adult. Each fork in this tree represents the mitotic division of a cell. Here we focus on those cells that give rise to the intestine. At the eight-cell embryonic stage, a single cell is can be identified as the source of all future intestinal cells.

velopment of each, produces identical twins. Even in regulative development, however, there is a point beyond which a loss of cells does affect the outcome. By the same token, the development of primarily mosaic embryos includes some steps in which substitutions are possible.

In spite of the differences just discussed, a number of features of cleavage—the series of mitotic divisions that divides the zygote cytoplasm into many smaller cells—are common to most animals. First, mitosis is usually quite rapid during this stage. For a given organism, the blastomeres proliferate more rapidly than do any other cells. Second, there is little or no increase in overall volume during cleavage, so the blastomeres become smaller and smaller with each division; the entire ball of cells remains approximately the same size as the fertilized egg. Third, the ratio of nuclear volume to cytoplasmic volume in the embryo increases steadily throughout cleavage because the original supply of cytoplasm (from the egg) is being shared by ever more nuclei. Finally, during cleavage massive synthesis of DNA and associated chromosomal proteins takes place.

Formation of the Blastula

Throughout cleavage, the embryo retains roughly the same external spherical form. As cleavage proceeds in sea urchins, frogs, birds, and many other animals, the sphere becomes a hollow structure, the **blastula**. Its cavity, the **blastocoel**, forms because the blastomeres selectively detach from one another on their innermost surfaces during early cleavage, leaving a fluid-filled space in the center. The blastomeres themselves constitute a sheet of cells, the **blastoderm**. In sea urchins the blastoderm is only one cell thick; in

other animals, such as frogs, it may be several cells thick (Figure 17.4).

The developing blastula is a dynamic structure; each division of blastomeres results in changed contacts among the cells. Within the blastula, however, the contents of the embryo are distributed much as they were in the original zygote; in the frog embryo, for example, the yolk is still more concentrated near the vegetal pole. Only in the subsequent gastrula stage of development do massive rearrangements of cells and materials begin within the embryo. How does the embryo proceed next to lay the groundwork for the production of specific structures such as a digestive tube (a gut)?

GASTRULATION

In all animals, cells from the surface of the blastula move to the interior to form layers. Cells redistribute themselves to transform the blastula into a **gastrula**, an embryo with a gut connected to the outside world by either an anus or a mouth. During gastrula formation, called **gastrulation**, some surface cells move into the interior, resulting in a two- or three-layered embryo. These layers are called germ layers because they give rise to distinct tissues and organs (Table 17.1). The blastoderm **invaginates** (bends inward) to form a pocket of an inner germ layer, the **endoderm**, leaving an outer germ layer, the **ectoderm**. In a few animals such as jellyfish, a two-layered embryo consisting only of ectoderm and endoderm develops into a two-layered adult with little cellular diversity. Adults of most species, however, develop from a three-layered embryo: The third germ layer, the **mesoderm**, forms between ectoderm and endoderm. In

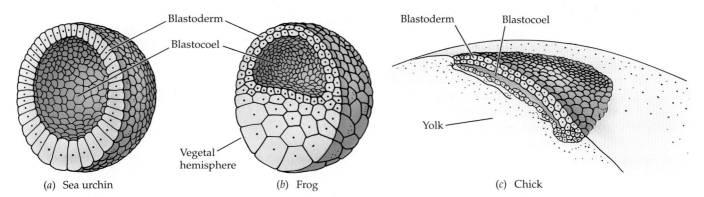

17.4 Blastulas Reflect Patterns of Cleavage
From the beginnings shown in Figure 17.2, continued cell divisions produce hollow blastulas, cut open here to show the blastocoel. *(a)* In sea urchins and organisms with similar eggs, the blastoderm is a single sheet of cells, and the blastocoel they enclose is spherical. *(b)* In frogs, the blastoderm is many cells thick in the vegetal hemisphere and the blastocoel is flattened on one side. *(c)* The blastocoel of birds is a lens-shaped cavity defined by a thin blastoderm layer below and a slightly thicker one above.

TABLE 17.1
Fates of Embryonic Germ Layers in Vertebrates[a]

GERM LAYER	FATE
Ectoderm	Brain and nervous system; lens of eye; inner ear; lining of mouth and of nasal canal; epidermis of skin; hair and nails; sweat glands, oil glands, milk secretory glands
Mesoderm	Skeletal system: bones, cartilage, notochord; gonads; muscle; outer coverings of internal organs; dermis of skin; circulatory system: heart, blood vessels, blood cells; kidneys
Endoderm	Inner linings of: gut, respiratory tract (including lungs), liver, pancreas, thyroid, and urinary bladder

[a]The final structures are complex, containing cells from more than one germ layer. Interactions among tissues are usually important in determining the composition and structure of an organ.

the illustrations throughout this chapter, endoderm cells are shown in yellow, mesoderm cells in pink, and ectoderm cells in blue.

Despite many variations in detail, some general features are common to the gastrulation of all animals. The rate of mitosis during gastrulation is much slower than it was during cleavage. There are massive movements of cells, giving rise to adjacent external and internal tissues: ectoderm, mesoderm, and endoderm. As development proceeds, different genes are activated in cells from the different germ layers, so that different gene products are formed.

Gastrulation in sea urchins results from several types of cell movements. Certain cells of a sea urchin blastula first become columnar, flattening one end of the embryo slightly. Then, as shown at the top of Figure 17.5, the central columnar cells bulge into the blastocoel, break free and actively crawl into the cavity in a process called **ingression**. These ingressing cells are **primary mesenchyme** cells. (*Mesenchyme* means a loosely organized array of cells, as distinguished from an *epithelium*, in which cells are packed tightly in sheets.) The primary mesenchyme cells are the beginning of the mesoderm. Next, the flattened end of the embryo invaginates as the columnar cells become wedge-shaped and buckle inward to produce the endoderm pocket. The ectoderm remains on the outside.

Invagination during gastrulation forms a new cavity, the **archenteron** or "primitive intestine," that opens to the exterior through the **blastopore**. As the archenteron lengthens, more cells move into the blastocoel from the archenteron's tip in a second round of ingression. These **secondary mesenchyme** cells

first form fine extensions, called filopodia, that extend through the blastocoel and attach to the inner surface of the ectoderm opposite the blastopore. Contraction of these filopodia lengthens the archenteron further, until the tip of the archenteron approaches the area of the future mouth. At this stage, the secondary mesenchyme cells detach from the endoderm. Migrating groups of mesenchyme cells eventually form a continuous mesoderm layer. The migration of the primary and secondary mesenchyme

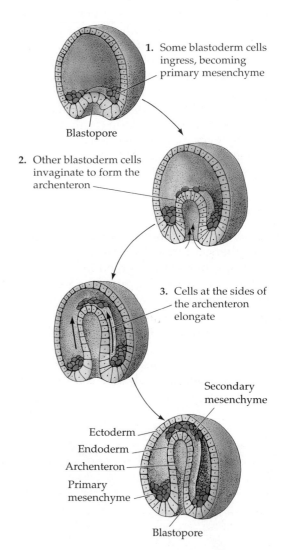

1. Some blastoderm cells ingress, becoming primary mesenchyme

Blastopore

2. Other blastoderm cells invaginate to form the archenteron

3. Cells at the sides of the archenteron elongate

Secondary mesenchyme

Ectoderm
Endoderm
Archenteron
Primary mesenchyme

Blastopore

17.5 Gastrulation in the Sea Urchin
During gastrulation, cells move to new positions and form the three germ layers from which all adult tissues develop. First, certain cells (at the bottom of the blastula in these drawings) leave the surface layer and crawl into the blastocoel, forming the primary mesenchyme (pink). Other surface cells buckle inward, and a new cavity—the archenteron—develops. Cells that line the archenteron become the endoderm layer (yellow), while cells that remain outside the blastopore become the ectoderm layer (blue). Groups of secondary mesenchyme cells (pink) ingress from the tip of the archenteron, eventually joining other mesenchyme cells to form the mesoderm.

cells is guided in part by fibers of an extracellular protein, fibronectin, laid down by ectodermal cells. The mesenchyme cells selectively attach to the fibronectin fibers and crawl along them.

The archenteron becomes the digestive cavity of the embryo, and its opening to the outside—the blastopore—becomes the anus of the sea urchin. The mouth develops later, from a perforation that forms at the other end. Animals such as sea urchins and vertebrates, in which the mouth develops from a second opening distant from the blastopore, are called deuterostomes (from the Greek for "mouth second"). Other animals, such as earthworms and insects, in which the mouth develops from the blastopore and the *anus* is formed by a second opening, are called protostomes ("mouth first"). The deuterostome–protostome distinction is one of the bases we use in classifying animals (see Chapter 26).

Gastrulation in Embryos with Much Yolk

In frogs, substantial quantities of yolk influence gastrulation and the formation of the three germ layers. Cells at one side of a frog embryo change shape and initiate invagination. Movement of cells starts just below the center of the gray crescent. Recall that the gray crescent is a product of cytoplasmic reorganization in the frog zygote. The first site of invagination marks the **dorsal lip** of the blastopore. Then two things happen. First, the cells at the animal end of the embryo begin to increase their surface area and to expand, a process that continues throughout gastrulation. As this expansion proceeds, the cells then turn inward at the dorsal lip of the blastopore and "flow" into the interior of the embryo (Figure 17.6). This inward turning of an expanding sheet of cells is called **involution**. The archenteron forms as involution proceeds. The dorsal lip of the blastopore extends, eventually forming a complete circle, with cells involuting at its dorsal (upper), lateral (side), and ventral (lower) lips. Within the circular blastopore, a "yolk plug" consisting of food-laden cells from the vegetal hemisphere of the embryo can be seen. The cells involuting over the lip of the blastopore give rise to both endoderm and mesoderm layers.

The formation of the mesoderm is complex, involving several types of cell movement. Some mesodermal precursors crawl along fibronectin fibers, as in sea urchin embryos. The major force moving future mesoderm to the interior and away from the blastopore, however, is convergent extension, a pattern of movement in which cells become narrower, elongate, and move between one another like cars rearranging themselves in heavy traffic. During convergent extension a strip of mesoderm differentiates to form the **notochord**, a stiff, supportive rod of cartilage. Later in the frog's development, the supportive function of the notochord is taken over by the vertebrae.

Gastrulation in Birds and Mammals

Because of their massive yolk content, bird eggs exhibit a pattern of gastrulation that differs from both sea urchin and frog eggs. Starting with the disc-shaped blastula (Figure 17.2c), surface cells lying on either side of the embryo migrate toward the center line and then forward, forming a depression called the **primitive streak** that is analogous to the blastopores of sea urchins and frogs. As migrating cells reach the primitive streak, they separate from one another and ingress through it (Figure 17.7). Once inside, they come together again, forming internal sheets that will become the mesoderm and endoderm. In this manner, the gut-forming cells of the endoderm are brought inside the embryo, the skin- and nerve-forming cells of the ectoderm remain external, and the mesoderm, which forms most of the organs, is brought between them. Interactions among cells of these germ layers then begin to determine the fates of specific cells in different regions of the embryo (see Table 17.1).

Gastrulation in mammals is somewhat similar to that in birds—and to that in reptiles, from which both birds and mammals evolved. Were it not for the shared evolutionary origins of birds and mammals, the primitive streak and other birdlike features of mammalian gastrulation would be surprising, because mammalian eggs are small and lacking in yolk in comparison with bird eggs. Other aspects of early mammalian development differ from the corresponding stages in birds and, in fact, in all other animals.

One striking feature of mammalian development is its slow pace. By contrast, the worm *Caenorhabditis elegans* reaches its adult form in three and a half days after fertilization; after two days a frog embryo has become clearly recognizable as a tadpole. But by two days a human embryo has reached only the 2-cell stage, and by three days only the 8-cell stage. At the 8-cell stage the mammalian embryo becomes more compact, and tight junctions (see Chapter 5) form between the outer cells of the mass, isolating the inside of the mass from the external environment. (Recall that tight junctions allow no gaps between adjacent cells.) Following additional cleavages, some cells form an inner cell mass enclosed by a surrounding layer called the trophoblast. Thus, the mammalian blastula, or **blastocyst,** differs in several respects from the blastulas of other animals (Figure 17.8). For example, in mammals the embryo develops within and is protected and nourished by the body of the mother. The blastocyst becomes implanted in the maternal uterus, where the trophoblast gives rise to the placenta (see Chapter 37), while certain cells of the

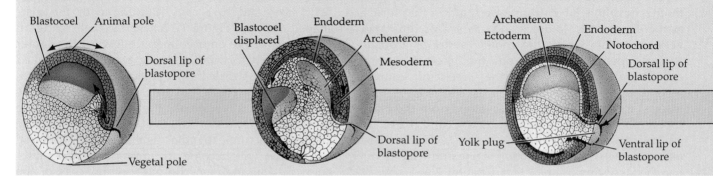

1. Gastrulation begins when cells just below the center of the gray crescent invaginate to form the dorsal lip of the future blastopore

2. Cells at the animal pole spread out, pushing surface cells below them toward and across the dorsal lip. Those cells involute into the interior of the embryo where they form the endoderm and mesoderm

3. This involution creates the archenteron and destroys the blastocoel. The dorsal lip forms a circle, with cells on both its dorsal and ventral surfaces; the yolk plug is visible through the blastopore

17.6 Gastrulation in the Frog

The developmental sequence begun in Figures 17.2*b* and 17.4*b* continues with gastrulation. The color conventions for the endoderm, ectoderm, and mesoderm remain as in previous figures; green is added, identifying cells that will form the nervous system.

inner cell mass give rise to protective membranes that are not part of the embryo. The disc-shaped remainder of the inner cell mass gives rise to the embryo itself by steps including a gastrulation process somewhat reminiscent of that of birds.

As we have seen, the patterns of gastrulation differ in detail from group to group of animals. However, the result in all cases is an embryo with a gut and two or three germ layers. The next step is the development of a nervous system.

NEURULATION

As an animal develops from gastrula to adult, many specialized organs and organ systems are formed, so it is not surprising that developmental patterns differ from one organ to another. One organ system is the central nervous system—the brain and spinal cord—which develops as a tube from a sheet of cells. At the beginning, this sheet is an external layer of the embryo; at the end, the tube is internal. The formation of such a tube in the development of the central nervous system is referred to as **neurulation**. We will consider neurulation in the frog embryo.

Recall that the late gastrula of a frog already has three tissue layers: ectoderm, endoderm, and mesoderm. At this stage, the embryo has anterior (front) and posterior (rear) ends, and the blastopore is positioned at the extreme rear. On the dorsal side of the embryo, the ectoderm begins to thicken, forming a flattened **neural plate** (Figure 17.9). The edges of the neural plate thicken and move upward to form neural folds. In the center of the plate, a neural groove forms and deepens as the folds begin to roll toward each other. As this is going on, the embryo as a whole elongates its anterior-to-posterior axis. The neural folds continue to roll toward one another

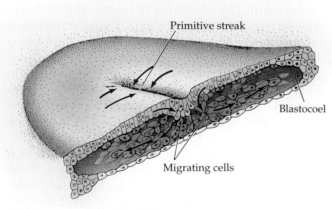

17.7 Gastrulation in Birds

During gastrulation in a bird embryo (shown from above), cells from the blastula's surface migrate toward the primitive streak and then separate from the surface layer and crawl into the blastocoel. Once in the blastocoel, the migrating cells come together in two sheets to form mesoderm (pink) and endoderm (yellow), while the cells that remain on the surface form ectoderm (blue).

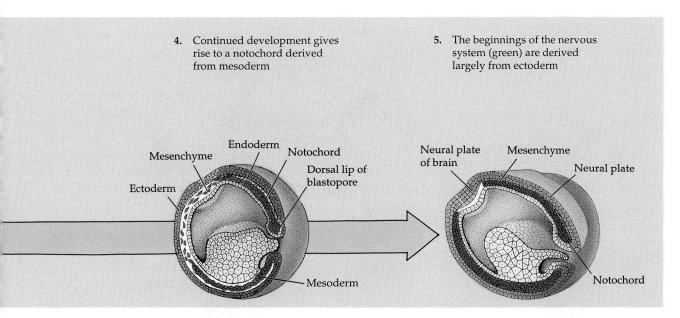

4. Continued development gives rise to a notochord derived from mesoderm

5. The beginnings of the nervous system (green) are derived largely from ectoderm

and finally, late in neurulation, fuse to form a narrow, hollow cylinder. This cylinder, the **neural tube**, becomes detached from the overlying ectoderm of the embryonic surface. Thus cells that once were part of a surface sheet are now incorporated into an internal tube (see the cross sections in Figure 17.9).

Many important events take place during neurulation, but the defining event is the formation of the neural tube. This type of process, in which a sheet of ectoderm from the surface becomes a tube inside the embryo, is repeated in many developmental contexts.

The entire nervous system derives from ectoderm,

and the anterior end of the neural tube ultimately becomes the brain. While the ectoderm is undergoing these dramatic changes, what is happening to the embryonic frog's mesoderm?

Late in gastrulation, as you saw in Figure 17.6, the notochord forms from mesoderm, providing support for the developing embryo. This cartilaginous rod continues to develop during neurulation, as shown in the longitudinal sections in Figure 17.9. After the notochord forms, other portions of the mesoderm to either side of the notochord and neural tube condense into blocks of cells called **somites** (Figure 17.10). Some somite cells give rise to muscles, others give rise to bones—the vertebrae. The somites develop soonest in the anterior end of the embryo and then sequentially toward the posterior end.

LATER STAGES OF DEVELOPMENT

Although we have seen differences in the ways frog, sea urchin, and chick embryos pass through their earliest stages of development, our main purpose so far in this chapter has been to show that various animals have much in common during the stages of cleavage, gastrulation, and neurulation. After neurulation, cells continue to divide, change shape, and move as embryos progress toward adulthood, but the later stages of development differ profoundly among animal groups.

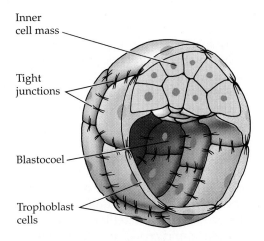

17.8 The Distinctive Blastocyst of a Mammal
The mammalian blastula, called a blastocyst, features an inner cell mass, some of whose cells give rise to the embryo after the blastocyst implants in the wall of the mother's uterus. Gastrulation follows; it resembles gastrulation in birds, despite the absence of a large mass of yolk.

Growth

Growth—irreversible increase in size—is often extensive. This increase in size results from cell multipli-

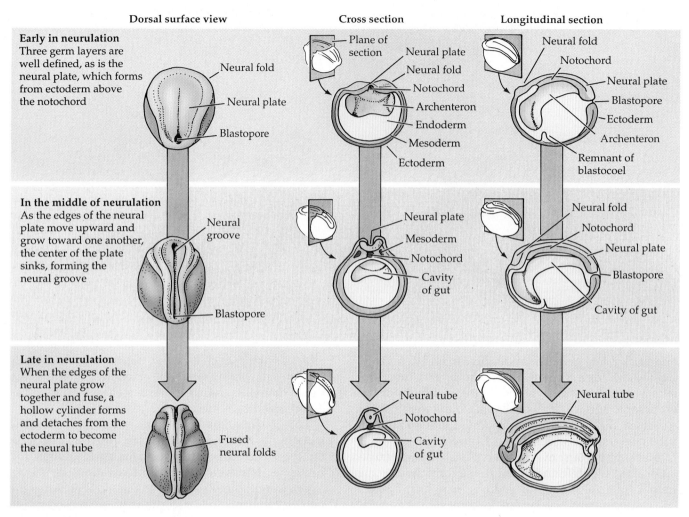

Dorsal surface view | **Cross section** | **Longitudinal section**

Early in neurulation
Three germ layers are well defined, as is the neural plate, which forms from ectoderm above the notochord

Labels: Neural fold, Neural plate, Blastopore; Plane of section, Neural plate, Neural fold, Notochord, Archenteron, Endoderm, Mesoderm, Ectoderm; Neural fold, Notochord, Neural plate, Blastopore, Ectoderm, Archenteron, Remnant of blastocoel

In the middle of neurulation
As the edges of the neural plate move upward and grow toward one another, the center of the plate sinks, forming the neural groove

Labels: Neural groove, Blastopore; Neural plate, Mesoderm, Notochord, Cavity of gut; Neural fold, Notochord, Neural plate, Blastopore, Cavity of gut

Late in neurulation
When the edges of the neural plate grow together and fuse, a hollow cylinder forms and detaches from the ectoderm to become the neural tube

Labels: Fused neural folds; Neural tube, Notochord, Cavity of gut; Neural tube

17.9 Neurulation in the Frog
Continuing from Figure 17.6, these drawings outline the development of the frog's neural tube. The three views across the top show a late gastrula. The longitudinal sections (right column) show the development of the notochord and its position relative to the neural tube.

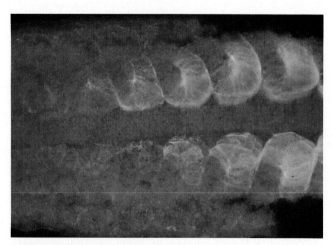

17.10 Somites Give Rise to Muscle and Bone
Somites in this amphibian embryo look like green blocks because of a fluorescent stain.

cation and cell expansion. In most animals, growth throughout the life of the individual follows what may be described as an S-shaped curve (Figure 17.11*a*). An initial period of slow growth is typically followed by a long phase of rapid growth, with growth slowing markedly at some stage. Details vary considerably among animal groups, however. In many groups—lobsters, for example—growth continues until the organism dies. In humans, overall growth ceases sometime after puberty, and we stay at a more or less constant size throughout most of our adult life (Figure 17.11*c*). Nonetheless, cell division continues at a rapid pace in some tissues throughout our lives, replacing cells such as those sloughed off by our skin (more than 1 gram a day) and intestinal lining, as well as the millions of blood cells turned over each minute (see Chapter 16).

Returning to the lobster, we see another departure from the pattern of Figure 17.11*a*, because the growth

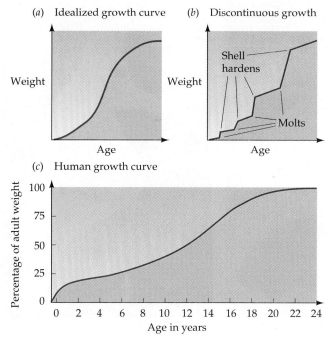

(a) Idealized growth curve

(b) Discontinuous growth

Shell hardens

Molts

(c) Human growth curve

17.11 Growth Patterns in Animals
All types of animals increase in weight as they grow older, but their growth rates differ, as these growth curves show. The steeper the curve, the faster the growth rate. *(a)* Idealized S-shaped growth curve characteristic of many species. There is some growth at every age, but the maximum growth rate is during the middle of the life cycle. *(b)* Discontinuous growth characteristic of hard-shelled animals that must molt to grow. A growth spurt follows each molt, before the new shell hardens. *(c)* A variant of the S-shaped curve shows the growth pattern for humans. A decade of gradual, preadolescent growth precedes a growth spurt during puberty; within a few years, the final adult weight is approached.

of a lobster is *discontinuous*. Because of its rigid external skeleton, the individual must molt (shed its skeleton) in order to grow. Accordingly, a lobster grows in spurts following successive molts (Figure 17.11*b*).

Larval Development and Metamorphosis

Many animals go through a larval stage in development. A **larva** is an immature form of an animal, differing in appearance from the adult. Examples of larvae include the caterpillar stage of a butterfly and the tadpole of a frog. A tadpole must change dramatically to become a four-legged adult frog, a radical rearrangement of structures called **metamorphosis**. Most larvae differ strikingly from adults of the same species.

During metamorphosis, new adult structures must be formed and old larval structures lost. Many changes must happen between the tadpole and the frog. The gut is shortened, corresponding to the transition from a vegetarian tadpole to a carnivorous

adult. The brain is remodeled to allow binocular (two-eyed) vision. Limbs appear and the tail disappears, corresponding to the transition from a swimming tadpole to a jumping frog. The loss of old tissues and old parts is accomplished by programmed cell death; that is, certain cells, such as those constituting the tadpole's tail, are purposely killed in developing organisms. Cell death plays a part not only in metamorphosis, but also in the embryonic development of most species, thus contributing in a general way to the development of form in animals.

The overall pattern of development in butterflies, moths, and many other insects is familiar to you (Figure 17.12). From a fertilized egg there develops a creeping larva that feeds voraciously, growing through a series of molts (the larval stages between molts are called **instars**). Some specialized cells, arranged in clusters called **imaginal discs**, remain undifferentiated throughout larval growth but later give rise to the tissues of the adult. The final instar stops feeding and then surrounds itself with a cocoon and transforms into a **pupa**. In the pupa, tremendous changes take place (Figure 17.13). Some larval cells die, others are reprogrammed to make different products characteristic of the adult, and the imaginal discs differentiate into new adult structures. Such a major revision between larva and adult is referred to as **complete metamorphosis**. In sharp contrast is the **gradual metamorphosis** characteristic of other insects, including grasshoppers and cockroaches. In gradual metamorphosis, the instars between molts are known as nymphs (or, when aquatic, naiads) and resemble miniature adults in many physical features.

LOOKING CLOSER AT DEVELOPMENT

With the understanding that animals develop from zygotes through cleavage, gastrulation, and neurulation stages, we can now examine three processes that reveal a great deal about how cells, and the animals themselves, change during development. These processes are determination, differentiation, and pattern formation.

Table 17.1 showed that cells of the three germ layers have fates. Some ectoderm cells, for example, will become the outermost layers of the animal's skin (the epidermis). How did developmental biologists learn what is summarized in Table 17.1? What is the experimental evidence?

Staining specific cells of the early embryo and observing which cells of older embryos contained the stain enabled biologists to determine which adult structures develop from certain parts of the blastula and early gastrula. For instance, the shaded area of the frog blastoderm shown in Figure 17.14 has the fate of becoming (that is, it is destined to become)

(a)   (b)   (c)   (d)   (e)

17.12   Complete Metamorphosis in a Moth
The comet-tail moth from Madagascar undergoes complete metamorphosis. (a) Eggs and first instar larvae (one in the process of hatching). (b) Third instar. (c) Fifth instar. (d) Pupa (removed from cocoon). The sweeping dis- posal of old tissues and the development of adult structures from imaginal discs in the pupa leads to the designation "complete." (e) Adult male moth, approximately 30 minutes after emergence from the pupa. Its long tails are not yet fully expanded.

part of the skin of the tadpole larva if left in place. If we cut out a piece from this region and transplant it to another place on an early gastrula, however, the type of tissue it becomes is determined by its *new* location, as Figure 17.14 shows. The developmental potential of blastoderm cells—that is, their range of possible development—is thus greater than their fate, which is limited to what normally develops.

Does developing embryonic tissue retain its broad developmental potential? Generally speaking, no. The developmental potential of cells becomes restricted fairly early. If taken from a region fated to develop into brain, for example, late gastrula tissue becomes brain tissue even if transplanted to parts of an early gastrula destined to become other structures. The tissue of the late gastrula is thus said to be *determined*: Its fate has been sealed, regardless of its surroundings. By contrast, the younger experimental tissue in Figure 17.14 has not yet become determined.

Determination is not something that is visible under the microscope—cells do not change appearance when they become determined. Changes in biochemistry, structure, and function constitute the *differentiation* of cells. Determination precedes differentiation. **Determination** means commitment; the final realization of this commitment is differentiation which we discuss next. We will then examine two mechanisms by which cells become determined.

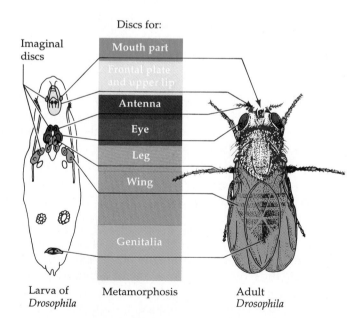

Discs for:

Imaginal discs

Mouth part

Frontal plate and upper lip

Antenna

Eye

Leg

Wing

Genitalia

Larva of *Drosophila*   Metamorphosis   Adult *Drosophila*

17.13   Complete Metamorphosis in Fruit Flies
In insects that undergo complete metamorphosis, such as the fruit fly, the embryo hatches from its egg into a soft-bodied larva that feeds for some time before entering the pupal stage. In the pupa, most of the larval tissues die and are reabsorbed, providing building blocks for subsequent development. The remaining larval tissues are specialized imaginal discs, which differentiate to form the organs of the adult insect.

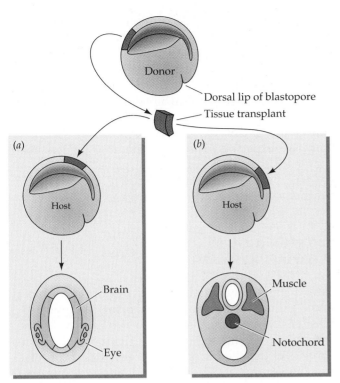

17.14 Developmental Potency in Early Gastrulas
Tissues in early gastrulas retain the capacity to develop in alternate ways, as demonstrated in transplantation experiments with frogs. Tissue destined to become part of a tadpole's skin is cut from an early gastrula and transplanted to another gastrula (the host). (a) When transplanted to a region of the host destined to become brain tissue, the donor tissue also develops into brain tissue. (b) When transplanted to a mesodermal region of the host, transplanted tissue becomes mesoderm and ultimately muscle and notochord. The wide developmental potential of early gastrulas is lost by the late gastrula stage.

DIFFERENTIATION

The process by which a cell achieves its fate—by which it acquires its final structure and physiological function—is **differentiation**. Because the cells of a multicellular organism arise by mitotic divisions of a single-celled zygote, they are for the most part genetically identical. In the absence of all mutation, all the cells in an organism have the same hereditary makeup; yet the adult organism is composed of many distinct types of cells. This apparent contradiciton results from the close regulation of the expression of various parts of the genome. Most of what we know about differentiation comes in large part from recombinant DNA technology (see Chapter 14).

The zygote is **totipotent**: It has the ability to give rise to every type of cell in the adult body. Its genetic "library" is complete, containing instructions for all the structures and functions that will arise throughout the life cycle. Later in the development of animals (and probably to a lesser extent in plants), the cellular descendants of the zygote lose their totipotency and become determined—that is, committed to form only certain parts of the embryo. When a cell achieves its determined fate, it is said to have differentiated. The mechanisms of differentiation relate primarily to changes in the transcription and translation of genetic information (see Chapter 13).

Is Differentiation Irreversible?

An early explanation of the mechanisms of differentiation was that the cell nucleus undergoes irreversible genetic changes in the course of development. It was suggested that chromosomal material is lost, or that some of it is irreversibly inactivated.

Differentiation is clearly irreversible in certain types of cells. The mammalian red blood cell, which loses its nucleus during development, is an example. Another is the tracheid, a water-conducting cell in vascular plants. The development of a tracheid culminates in the death of the cell, leaving only the pitted cell walls that were formed while the cell was alive (see Chapter 29). In these two extreme cases, the irreversibility of differentiation can be explained by the absence of a nucleus. Generalizing about mature cells that retain functional nuclei is more difficult. Most biologists tend to think of plant differentiation as reversible and of animal differentiation as irreversible, but this is not a hard-and-fast rule. A lobster can regenerate a missing claw, but a cat cannot regenerate a missing paw. Why is differentiation reversible in some cells but not in others? At some stage of development do changes within the nucleus permanently commit a cell to specialization?

At the Institute of Cancer Research in Philadelphia in the 1950s, Robert Briggs and Thomas J. King performed experiments to see whether genetic material was preserved or was permanently inactivated or lost during normal development. To find out whether the nuclei of frog blastulas had lost the ability to do what the zygote nucleus could do, they carried out a series of meticulous transplants. First they removed the nucleus from an unfertilized egg (thus forming what is called an enucleated egg). Then, with a very fine glass tube, they punctured a cell of a blastula and drew up part of its contents, including the nucleus, which they then injected into the enucleated egg, and the egg was activated. More than 80 percent of these operations resulted in the formation, from the egg and its new nucleus, of a normal blastula; of these blastulas, more than half developed into normal tadpoles and, ultimately, adult frogs. These experiments showed that no information has been lost from the nucleus by the time the blastula has formed. On the other hand, Briggs and King found that when the nuclei were derived from older embryonic stages, fewer larvae developed (Figure 17.15).

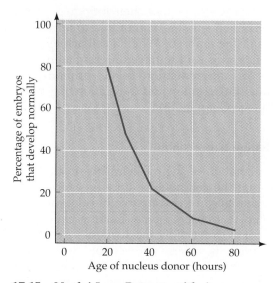

17.15 Nuclei Lose Potency with Age
Nuclei from older frog embryos transplanted into enucleated eggs are less likely to direct successful development of an embryo than are younger nuclei.

This work was carried further by John B. Gurdon and his associates at Oxford University, who performed similar transplants using nuclei from gastrulas, swimming tadpoles, and even adult frogs. Nuclei from differentiated adult cells were transplanted into enucleated eggs, the eggs were raised to the blastula stage, nuclei were isolated in turn from these blastulas, and these nuclei were used for further transplants. When placed in an enucleated egg, a nucleus obtained by such serial transplants occasionally was able to direct development to a tadpole stage, complete with brain, gut, blood, heart, and other parts. Work of this sort convinced many developmental biologists that the loss of particular genes is not the cause of differentiation, and that genes no longer expressed in certain cells are still present and can be expressed if they are placed in a "younger" environment such as an enucleated egg. In the next section we consider another example of development that does not result from the loss of genes.

Transdetermination of Imaginal Discs

The fates of the imaginal discs of insects are determined long before metamorphosis. If transplanted from one larva to another, an imaginal disc still develops into the same type of organ (a wing, for example, or an antenna) that it would have if left undisturbed, and that organ is formed wherever the imaginal disc is placed in the host body.

If transplanted into an *adult* insect, an imaginal disc remains undifferentiated (because the hormonal signal for its development is lacking), but the cells of the disc continue to divide within the new host. Later these transplanted disc cells may be transplanted to other adult insects or to larvae. If returned to a larva, the disc cells *almost* always develop into the adult organ for which they were originally determined. Occasionally, however, an imaginal disc will **transdetermine** in the course of a series of transplants; that is, it will develop into an organ other than that normally expected. Transdetermination shows that imaginal discs have not lost the genes they do not normally express, since a disc *may* express these genes and produce a different organ.

DETERMINATION BY CYTOPLASMIC SEGREGATION

How does determination come about? Even within a single animal, a number of mechanisms are involved. Most of the mechanisms fall into two categories, the first based on the segregation of cytoplasmic components of the egg into separate cells or parts of cells, and the second based on the influences of one part of the embryo on another part. We will consider cytoplasmic segregation first, beginning with its role in distinguishing one end of an animal from the other.

Polarity in the Egg and Zygote

Polarity, the difference of one end from the other, is obvious in development. Our heads are distinct from our feet, and the distal ends of our arms (wrists and fingers) differ from the proximal ends (shoulders) of our arms. An animal's polarity develops early, even in the egg itself. Yolk may be distributed asymmetrically in the egg and the embryo, and other chemical substances may be confined to specific parts of the cell or may be more concentrated at one pole than at the other. In some animals, the original polar distribution of materials in the egg's cytoplasm changes as a result of fertilization, yielding a new polar distribution in the zygote. As cleavage proceeds, the resulting blastomeres contain unequal amounts of the materials that were not distributed uniformly in the zygote. As we learned from the work of Briggs and King and of Gurdon, cell nuclei do not always undergo irreversible changes during early development; thus we can explain some embryological events on the basis of the *cytoplasmic* differences in blastomeres.

Even as apparently simple a structure as a sea urchin egg has polarity. As the gastrula forms (see Figure 17.5), one can see a slight difference in blastomere size; the cells in the vegetal half—the ones that ingress—are somewhat larger. A striking differ-

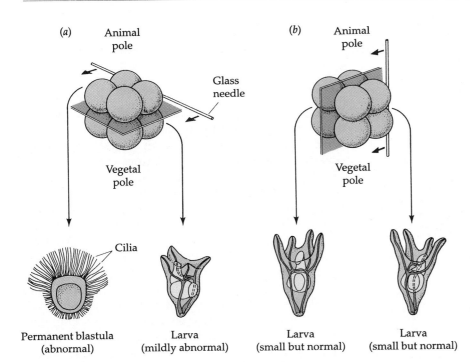

(a) Animal pole

Glass needle

Vegetal pole

Cilia

Permanent blastula (abnormal)

Larva (mildly abnormal)

(b) Animal pole

Vegetal pole

Larva (small but normal)

Larva (small but normal)

17.16 Early Asymmetry in the Embryo
The animal and vegetal halves of very young sea urchin embryos differ in their developmental potential. The vegetal half alone cannot direct development at all, and completely normal development requires cells from both halves.

ence between blastomeres can be demonstrated well before that, however. The Swedish biologist Sven Hörstadius showed in the 1930s that the development of sea urchin embryos that have been divided in half at the eight-cell stage depends on how the separation is performed. If the embryo is split into "left" and "right" halves, with each half containing cells from both the animal and the vegetal pole, normal-shaped but dwarfed larvae develop from the halves (Figure 17.16). If, however, the cut separates the four cells at the animal pole from the four at the vegetal pole, the result is different. The animal half develops into an abnormal blastula with large cilia at one end and cannot form a larva, whereas the vegetal half develops into a small, somewhat misshapen, larva with an oversized gut. For fully normal development, factors from both the animal and vegetal halves of the embryo are necessary. By further experiments on eggs, Hörstadius showed that this unequal distribution of material between the animal and vegetal halves is already present in the unfertilized egg.

When a frog egg becomes fertilized, some components of the cytoplasm are again restricted to certain parts of the egg, resulting in—among other things—the formation of the gray crescent. In the early part of this century, Hans Spemann found that the gray crescent contains materials essential for normal embryonic development. In normal cleavage, the first cell division usually divides the gray crescent about equally between the daughter cells (Figure 17.17*a*). These two blastomeres, if separated, give rise to normal embryos. This process is regulative devel-

opment, as described earlier. Sometimes, however, the zygote divides in such a way that one of the blastomeres contains the entire gray crescent and the other none of this material (Figure 17.17*b*). If the two cells of an embryo that has divided this way are separated, the half with the gray crescent develops normally but the half lacking the gray crescent forms only an unorganized mass of ventral (belly) cells. These experiments by Hörstadius and by Spemann were among many that established that the unequal distribution of materials in the egg cytoplasm plays a role in directing embryonic development.

Cytoplasmic Factors in Polarity in *Drosophila*

In the eggs and larvae of the fruit fly *Drosophila melanogaster*, polarity is based on the distribution of more than a dozen mRNAs and proteins. These **cytoplasmic determinants** are products of specific genes in the mother and are distributed to the eggs, often in a nonuniform manner. They determine the dorsoventral (back–belly) and anteroposterior (head–tail) axes of the embryo. The discovery that this part of fruit fly development is determined by cytoplasmic segregation was due to the striking appearance of the mutant larvae produced when the determinants are distributed abnormally. For example, larvae formed by females homozygous for the *bicaudal D* mutation consist solely of two hind ends, joined at the middle and possessing no head (Figure 17.18).

We know that cytoplasmic determinants specify these axes from the results of experiments in which cytoplasm was transferred from one egg to another.

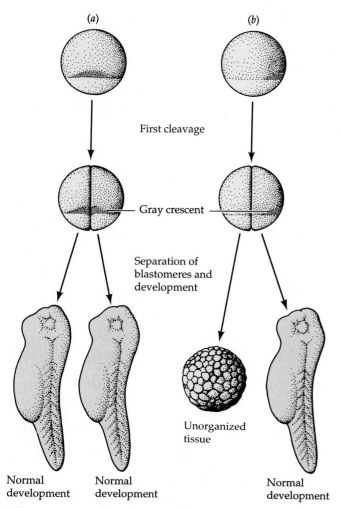

(a) *(b)*

First cleavage

Gray crescent

Separation of
blastomeres and
development

Unorganized
tissue

Normal
development   Normal
development   Normal
development

17.17   Developmental Importance of the Gray Crescent
The asymmetric distribution of materials in the egg establishes the egg's polarity and determines its developmental architecture. Without the material contained in the
gray crescent, a blastomere can form only unorganized
tissue from ventral cells, with no dorsal structures.

normal development. Let's consider one more example of determination by cytoplasmic segregation.

Germ-Line Granules in *Caenorhabditis*

Various cytoplasmic determinants play roles in the
development of the tiny nematode (roundworm) *Caenorhabditis elegans* (introduced in Figure 17.3). Tiny
particles called **germ-line granules** may be cytoplasmic determinants. Their positions in the zygote
and embryo are determined by the action of microfilaments. Before the zygote divides, the germ-line
granules collect at the posterior end of the cell, and
all the granules appear in only one of the first two
blastomeres (Figure 17.19). The germ-line granules
continue to be precisely distributed during the early
cell divisions, ending up in only those cells—the
germ cells—that will eventually give rise to eggs and
sperm. We do not know yet whether these granules
are cytoplasmic determinants themselves, or whether
they are simply distributed together with the "real"
cytoplasmic determinant.

We have seen several examples of determination
by cytoplasmic segregation. Now we will examine
another important general mechanism of determination, in which certain tissues induce the determination of other tissues.

Females homozygous for the *bicoid* mutation produce
larvae with no head and no thorax. However, if eggs
of homozygous-mutant *bicoid* females are inoculated
at the anterior end with cytoplasm from the anterior
region of a wild-type egg, the treated eggs develop
into normal larvae—with heads developing from the
part of the egg that receives the wild-type cytoplasm.
On the other hand, removal of 5 percent or more of
the cytoplasm from the anterior of a wild-type egg
results in an abnormal larva that looks like a *bicoid*
mutant larva.

Another gene, *nanos*, plays a comparable role in
the development of the posterior end of the larva.
Eggs from homozygous-mutant *nanos* females develop into larvae with missing abdominal segments;
injecting the mutant eggs with cytoplasm from the
posterior region of wild-type eggs, however, allows

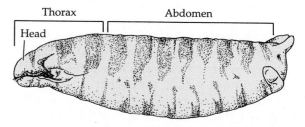

(a)   Wild-type

Thorax   Abdomen

Head

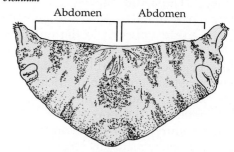

(b)   *bicaudal*

Abdomen   Abdomen

17.18   The *bicaudal* Mutation
The anteroposterior axis of *Drosophila* larvae arises from
the interaction of several cytoplasmic determinants. *(a)* A
larva produced by a wild-type female. *(b)* A larva produced by a female homozygous for the bicaudal D allele.
The larva has no head or middle—it consists of two hind
ends.

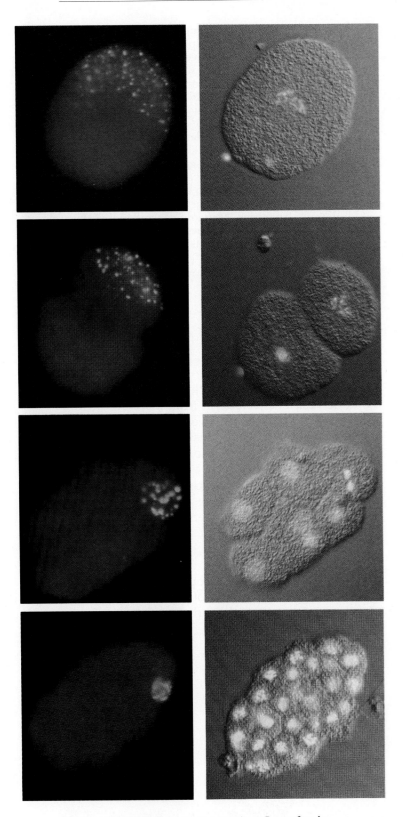

17.19 Distribution of Germ-Line Granules in *Caenorhabditis*

Micrographs show a developing embryo of *Caenorhabditis elegans*. In the left column, the germ-line granules (bright spots, stained with antibodies) move to the posterior end of the embryo. Eventually the granules are confined to the cell that gives rise to gametes. By contrast, the nuclei (right column; stained blue) of the same embryo are distributed evenly among its cells.

DETERMINATION BY EMBRYONIC INDUCTION

Experimental work has clearly established that the fates of particular tissues are determined by interactions with other specific tissues in the embryo. In the developing embryo there are many such instances of **induction**, in which one tissue causes an adjacent tissue to develop in a different manner.

Induction and the Organizer

In 1924 Hilde Mangold and Hans Spemann, working with newt embryos, performed a classic demonstration of induction. First Mangold transplanted a piece of the dorsal lip of the blastopore from an early gastrula of a lightly pigmented donor species to a particular place on the surface of an early gastrula of a heavily pigmented host species. She could follow the fate of the transplanted piece because it was lighter in color than the surrounding host tissue. As predicted from the known fate of the surrounding tissue—to become mesoderm—the graft itself developed into principally mesodermal products. But something unexpected also happened: In the region of the graft, an extra neural plate appeared, containing neural folds made from host tissue (Figure 17.20). The procedure was repeated many times, and in some instances the extra neural plate continued to develop into a secondary embryo attached to the main one! The grafted dorsal lip had induced the nearby host tissue to develop along completely different lines than it would have without the graft.

The dorsal lip of the blastopore not only has the fate of becoming part of the notochordal mesoderm, but in addition it induces any ectoderm that it contacts to organize into a neural tube. With the formation of the neural tube, the principal axes of the embryo—the anteroposterior, dorsoventral, and left–right axes—are formed. For this reason Spemann called the dorsal lip the **embryonic organizer**.

The development of the lens in the vertebrate eye is another classic example of induction. Lens formation is shown in Figure 17.21: The developing forebrain bulges out at both sides to form the **optic vesicles**, which expand until they come in contact with the ectoderm of the head. The head ectoderm in the region of contact with the optic vesicles thickens, forming a **lens placode** The lens placode invaginates, folds over on itself, and ultimately detaches from the surface to produce a structure that will develop into the lens. If the growing optic vesicle is cut away before it contacts the surface ectoderm, no lens forms in the head region from which the optic vesicle has been removed. An impermeable barrier placed between the optic vesicle and the ectoderm also prevents the lens from forming. These observations sug-

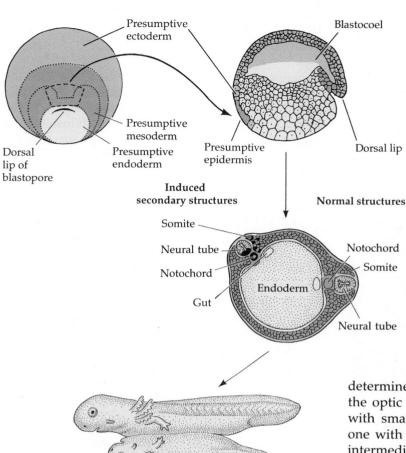

Induced secondary structures

Somite
Neural tube
Notochord
Gut

Endoderm

Notochord
Somite
Neural tube

Normal structures

17.20 Embryonic Induction

Mangold transplanted a donor's dorsal lip (presumptive mesodermal tissue) to an ectodermal region in the host gastrula. The donor and host organisms were differently pigmented, so she could distinguish graft tissue from host tissue. She expected the graft to become mesodermal tissue and it did. Unexpectedly, she saw that host ectoderm near the graft (that would normally have developed into epidermis) developed into a second neural plate, and in some instances into a second embryo. Mangold concluded that the dorsal lip induced nearby host ectoderm to alter its development.

gest that ectoderm begins to develop into a lens when it receives a signal—an **inducer**—from its contact with an optic vesicle.

The interaction of tissues in eye development is a two-way street: There is a "dialogue" between the developing optic vesicle and the ectoderm. The lens determines the size of the optic cup that forms from the optic vesicle. If ectoderm from a species of frog with small eyes is grafted over the optic vesicle of one with large eyes, both lens and optic cup are of intermediate size. The lens also induces the ectoderm over it to develop into a cornea. Thus an entire chain of inductive interactions participates in development of the parts required to make an organ such as the eye. Induction triggers a sequence of gene expression in the responding cells. Tissues do not induce themselves; rather, different tissues interact and induce each other.

One of the most resistant problems in developmental biology has been that of determining the specific chemical nature of the inducers. In some cases,

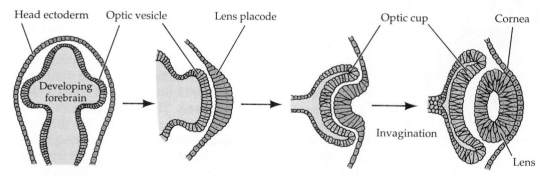

17.21 Inducers in the Vertebrate Eye

The vertebrate eye develops as inducers take their turns. Studies on frogs revealed that the optic vesicle induces overlying ectoderm to form placode tissue that, in turn, induces the formation of an optic cup. The optic cup then induces the placode to invaginate and form the eye's lens. Once the developing lens has separated from and become covered by surface ectoderm, it induces this ectoderm to form a cornea.

specific diffusible proteins may be involved; the inducer that acts earliest in frog gastrulas appears to be a growth factor. In other cases, however, insoluble extracellular materials such as collagen and other proteins may be involved in induction. Generally speaking, induction is a phenomenon confined to embryonic tissues; however, it continues in the production of certain white blood cells in the adult immune system.

Induction at the Cellular Level

The tiny worm *Caenorhabditis elegans* is an excellent organism for studying the mechanisms of induction because, as we saw in Figure 17.3, different parts of the worm's body form from different cell lineages in the developing organism. The hermaphroditic form (containing both male and female reproductive organs) of *C. elegans* lays eggs through a pore called the vulva on the ventral surface of the worm. A single cell, called the anchor cell, induces the vulva to form. If the anchor cell is destroyed by laser surgery, no vulva forms. The anchor cell controls the fates of six cells on the animal's ventral surface. Each of these cells has three possible fates. By the manipulation of

two genetic switches, a given cell becomes either a primary vulval precursor, a secondary vulval precursor, or simply part of the worm's surface, the epidermis.

The anchor cell produces an inducer. Cells that receive enough of the inducer become vulval precursors; those slightly farther from the anchor cell become epidermis. The first switch, controlled by the inducer from the anchor cell, determines whether a cell takes the "track" toward becoming part of the vulva or the track toward becoming epidermis. Now the cell closest to the anchor cell, having received the most inducer, becomes the primary vulval precursor and produces its own inducer, which acts on the two neighboring cells and directs them to become secondary vulval precursors. Thus the primary vulval precursor cell controls a second switch, determining whether a vulval precursor will take the primary track or the secondary track. The two inducers control the activation or inactivation of specific genes in the responding cells (Figure 17.22). These genes are explained in detail in Box 17.A.

There is an important lesson to draw from this example. Much of development is controlled by switches that allow a cell to proceed down either of

17.22 Two Switches Determine Vulva Formation in *C. elegans*
Two secreted proteins act as inducers. The primary inducer, produced by the anchor cell, activates a gene whose products determine that cells will develop as vulval precursors rather than epidermal cells; the second inducer activates another gene, thus determining that cells will develop as secondary precursors.

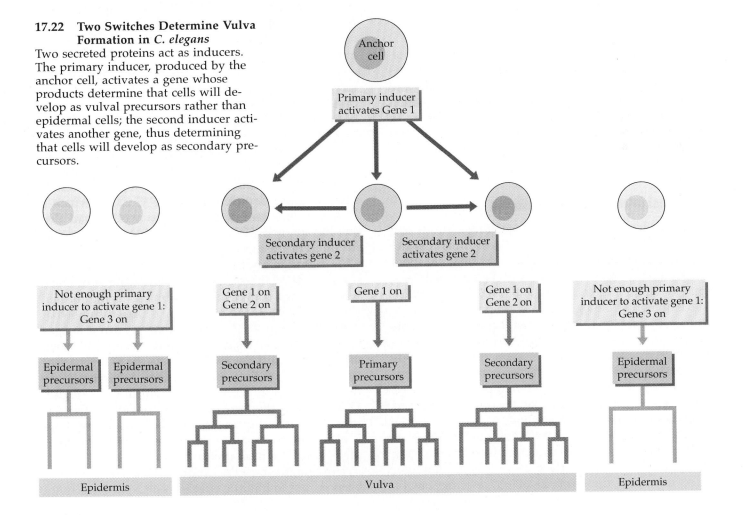

BOX 17.A

Genes Interact in Induction

Let us see how several genes interact to produce the vulva of *Caenorhabditis elegans*. In the anchor cell, a gene called *lin-3* is transcribed and translated to produce the first inducer, a protein similar to a growth factor (see Chapter 15). In the target cells, an-

other gene (called *let-23*) is responsible for producing a receptor for this inducer. The receptor is a protein that extends through the plasma membrane of the cell. The part of the receptor that lies outside the cell recognizes the inducer; when it binds the inducer, the part on the inside of the cell picks up a phosphate group from ATP. The phosphorylated receptor then binds an adapter protein produced by another gene, *sem-5*. Next, this complex interacts with another protein, a G-protein (itself encoded by another gene, *let-60*). Upon meeting the adaptor–receptor com-

plex, the G-protein acquires a phosphate group from GTP (hence the name G-protein). The G-protein thus gains the ability to direct another gene to produce the second inducer—which, in turn, activates yet another gene, *lin-12*, in the cells that become the secondary vulval precursors.

Many events at plasma membranes are controlled by receptor proteins that interact with ATP and with G-proteins. This type of membrane activity is described in more detail in our discussion of animal hormones (see Chapter 36).

two alternative tracks. One challenge for the developmentalist is to find these switches and determine how they work. Another challenge is to account for the larger-scale phenomenon of pattern formation.

PATTERN FORMATION

One area of current research in developmental biology is the study of **pattern formation**, the development of organs consisting of differentiated cells and tissues in ordered arrangements. The differentiation of cells is beginning to be understood in terms of molecular events, but what about the organization of multitudes of cells into specific body parts, such as a shoulder blade or a tear duct?

Positional Information in Developing Limbs

Many factors collaborate in regulating pattern formation. In addition to the genetic controls just described, **positional information**—information about where one group of cells lies in relation to others—plays a role. In the 1960s and 1970s, the English developmental biologist Lewis Wolpert developed a theory of positional information, based on gradients of **morphogens** in developing limb buds in chick embryos. A morphogen is a substance, produced in one place, that diffuses and produces a concentration gradient, with the result that different cells are exposed to different concentrations of the morphogen and thus develop along different lines.

It is a morphogen concentration gradient that results in the development of a chick wing from a wing bud, a bulge on the surface of a 3-day-old embryo. Like any three-dimensional object, a wing can be

described in terms of three perpendicular axes. We are interested here in the anteroposterior axis of the wing—the axis that corresponds to the axis of the body running from the head to the tail of the chick. (The proximodistal axis runs from the base of the limb to its tip, and the dorsoventral axis from back to belly.) Each axis has a corresponding type of positional information. Here we will consider just the anteroposterior axis.

Pattern formation along the anteroposterior axis can be modified experimentally in ways that suggest that it is controlled at least in part by a particular part of the wing bud, called the zone of polarizing activity, or **ZPA**, that lies on the posterior margin of the bud. Different parts of the limb appear to develop normally at *specific distances from the ZPA*. This hypothesis is supported by the results of grafting experiments such as those depicted in Figure 17.23, in which a ZPA from one bud is grafted onto another bud that still has its own ZPA. The donor ZPA can be placed in different positions on different hosts. If the extra ZPA is grafted on the anterior margin, opposite from the host ZPA, the distal part of the wing is duplicated, with two complete mirror images being formed (Figure 17.23b). If the extra ZPA is placed somewhat closer to the host ZPA, incomplete mirror images appear (Figure 17.23c): One digit is missing from each of the units, as if the missing units are of a type that can form only if they are more than some minimum distance from a ZPA. In the third case (Figure 17.23d), with the two ZPAs close together, there is room for some duplication between them, but there is also enough room on the anterior side for a complete, nearly normal unit to form.

How might the ZPA produce these effects? It has been proposed that the effects result from the activity

(a)

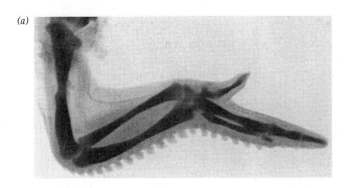

(b)

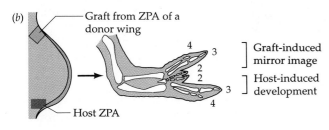

Graft from ZPA of a
donor wing

Host ZPA

4 3

2
2

3

4

Graft-induced
mirror image

Host-induced
development

(c)

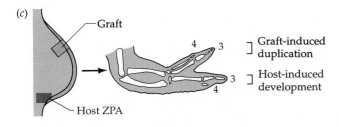

Graft

Host ZPA

4 3

3
4

Graft-induced
duplication

Host-induced
development

(d)

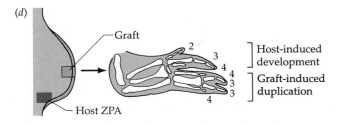

Graft

Host ZPA

2
3
4
4
3
4

3

Host-induced
development

Graft-induced
duplication

17.23 The ZPA Provides Positional Information
(a) By the age of 9.5 days, the embryonic chick wings
have developed to the stage shown in this preparation
that has been stained to reveal a full set of wing bones.
(b–d) A zone of polarizing activity (ZPA) is located on the
posterior margin of the chick wing bud—the area indicated as "Host ZPA." In each of the experiments diagrammed here, the ZPA from one bud was grafted onto
another wing bud which still had its own ZPA. *(b)* ZPA
grafted on the anterior margin causes mirror-image duplication of the distal part of the wing—digits 4, 3, and 2. *(c)*
A ZPA grafted closer to the host ZPA causes duplication
of digits 4 and 3, but no digit 2 develops. *(d)* A ZPA
grafted still closer to the host ZPA allows a nearly normal
set of digits 2, 3, and 4 to develop on the anterior side but
also results in partial duplication of digits 3 and 4 between the two ZPAs.

of an apparent morphogen, called retinoic acid, produced by the ZPA. Several pieces of experimental
evidence support this theory. Quite recently, however, contradictory evidence has convinced many
workers that retinoic acid plays a role other than that
of the ZPA morphogen. We still lack a satisfactory
explanation for the fascinating effects shown in Figure 17.23.

Establishing Body Segmentation

Developmental biologists have also studied pattern
formation using *Drosophila* fruit flies as experimental
subjects. Insects (and many other animals) develop
a highly modular body composed of different types
of modules, called segments. Complex interactions
of different sets of genes underlie the pattern for-

mation of segmented bodies. Unlike the body segments of segmented worms such as earthworms, the
segments of the *Drosophila* body are different from
one another. The *Drosophila* adult has a head (one
segment), three different thoracic segments, eight abdominal segments, and a genital segment at the posterior end. Thirteen segments in the *Drosophila* larva
correspond to these adult segments.

Key genes act sequentially to organize a developing *Drosophila* larva. An overall framework of anteroposterior and dorsoventral axes is laid down initially
by the activity of the genes that produce the cytoplasmic determinants referred to earlier (see Figure
17.18). As we saw, mutations in those genes result
in the duplication or deletion of body parts, such as
anterior and posterior halves of the embryo. The
number, boundaries, and polarity of the larval segments are determined by proteins encoded by several
more genes, the **segmentation genes**. Three classes
of segmentation genes participate, one after the
other, to regulate finer and finer details of the segmentation pattern. First, **gap genes** organize large
areas along the anteroposterior axis. Mutations in gap
genes result in the omission of several larval segments. Second, **pair-rule genes** divide the embryo
into two-segment-long units. Mutations in pair-rule
genes result in embryos missing every second segment. Third, **segment-polarity genes** determine the
boundaries and anteroposterior organization of the
segments themselves. Mutations in segment-polarity
genes result in segments in which some posterior
structures are replaced by reversed (mirror-image)
anterior structures.

Finally, after the basic pattern of segmentation has
been established by the segmentation genes, differences between the segments are mediated by the
activities of **homeotic genes**. These genes are expressed in different combinations along the length of
the body and tell each segment what to become.

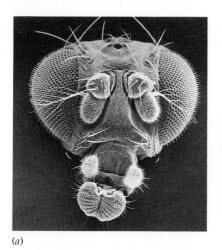

(a)

(b)

17.24   A Homeotic Mutation
Scanning electron micrographs of the heads of two *Drosophila melanogaster.* (a) A wild-type fly, with normal antennae. (b) An *antennapedia* mutant, with roughly normal legs in the positions usually occupied by antennae. This abnormality results from a homeotic mutation—a drastic mutation causing one structure to develop in the place of another.

Homeotic Mutations

Our present understanding of the genetics of pattern formation began with the discovery of dramatic mutations, called homeotic mutations, in *Drosophila.* Instead of a normal body part, the insect with a homeotic mutation has another part characteristic of another body segment. Two bizarre examples are the *antennapedia* mutant, in which legs grow in the place of antennae (Figure 17.24), and the *ophthalmoptera* mutant, in which wings grow in the place of eyes. Homeotic genes fall into a few tight clusters. Of these, the best characterized is the **bithorax complex**. The eight or more genes of the bithorax complex control development of the abdomen and posterior thorax of the fly. Development of the head and anterior thorax is controlled by another homeotic gene cluster, the **antennapedia complex**. The functions of the two complexes interact substantially.

Because many mutations in the bithorax complex are so severe they prevent development past the early larval stages, they can be studied only in larvae right after they hatch out of the egg. Figure 17.25 shows that if mutations have deleted all or part of the bithorax complex, the larvae have thirteen segments but the segments are abnormal. Other mutations in the bithorax complex cause the third thoracic segment of the adult to develop exactly like the second, resulting in a fly that has two pairs of normal wings and no halteres (Figure 17.26).

As we just saw, if the entire bithorax complex is deleted, the larva produced, although highly abnormal, still has the normal number of segments. Thus the bithorax complex clearly does not determine the number of segments. It is the segmentation genes that determine the number and polarity of segments. In the course of normal development, the homeotic genes give each segment its distinctive character.

The Homeobox

In the early 1980s Walter Gehring and his associates William McGinnis and Michael Levine, working in Switzerland, and Thomas Kaufman, at Indiana University, undertook a study of the antennapedia complex, using the techniques of recombinant DNA technology. They set out to isolate and clone the *antennapedia (Antp)* gene, a member of the antennapedia complex, from *Drosophila.* As part of this study, they prepared a DNA that could hybridize to this *Antp* gene. Surprisingly, the *Antp* DNA hybridized with both the *Antp* gene and a nearby segmentation gene (the *fushi tarazu* gene, *ftz*, from the Japanese for

17.25   *bithorax* Mutations in *Drosophila* Larvae
Genotypes at loci of the bithorax complex govern how segments develop in *Drosophila* larvae. The wild-type larva shown on the left has a head (H), three thoracic segments (T1–T3), eight abdominal segments (A1–A8), and one genital segment (G). The larva in the center lacks the entire bithorax complex. It has a normal head and first thoracic segment, but the other segments have developed like normal second thoracic segments. The larva on the right lacks part of the bithorax complex; all its segments are like normal eighth abdominal segments.

Head — H
Thorax — T1, T2, T3
Abdomen — A1, A2, A3, A4, A5, A6, A7, A8
Genital segment — G

Center larva: H, T1, T2, T2, T2, T2, T2, T2, T2, T2, T2, T2, T2, G

Right larva: A8, A8, A8, A8, A8, A8, A8, A8, A8, A8, A8, A8, A8

17.26 *bithorax* **Mutations in Adult** *Drosophila*
Because of two mutations in its bithorax complex, this fly's third thoracic segment developed as if it were a second thoracic segment. In wild-type *Drosophila* the second thoracic segment gives rise to wings and legs in the adult, while the third thoracic segment produces a pair of legs and a pair of small, winglike structures called halteres.

"too few segments"). The *Antp* and *ftz* genes must have DNA sequences of close similarity, because part of each gene hybridizes with the same DNA. Further hybridization studies demonstrated that the same shared stretch of DNA is also found in the *bicoid* gene of the bithorax complex, in some other parts of the *Drosophila* genome, and in genes in other insect species. In fact, this important sequence of 180 base pairs of DNA, called the **homeobox**, has now been shown to be part of a few genes of many animals and plants.

What does the homeobox, which is present in almost all eukaryotes, do? The homeobox sequence codes for a region of 60 amino acids—the homeodomain—that is part of some proteins. These proteins return to the nucleus and bind to DNA, regulating the transcription of other genes. A computerized search of published sequences of DNA from numerous species revealed a similarity between the homeobox and parts of certain regulatory genes in yeast—genes that produce proteins that also bind to specific DNA sequences. Some genes with homeoboxes are expressed only at certain times and in certain tissues as development proceeds, as would be expected if these proteins regulate development.

What are we to make of the presence of the homeobox in such diverse species as humans, fruit flies, frogs, *C. elegans*, and tomatoes—and of its presence in several genes in the same organism? The ubiquitous presence of the homeobox suggests that both the antennapedia and bithorax complexes may have arisen from a single ancestral gene. Further, it implies that a single gene in some ancient organism may have been the evolutionary progenitor of what is now a widespread controlling system for development.

YOU'VE SEEN THE PIECES; NOW LET'S BUILD A FLY

One of the most striking and important observations about development in *Drosophila*—and in other animals—is that it results from a *sequence* of changes, each change triggering the next. The sequence is a transcriptionally controlled cascade. To see this more clearly, let's "build" a *Drosophila* larva step by step, beginning with the unfertilized egg.

Before the egg is fertilized, mRNA for the *bicoid* protein is localized at the end destined to become the anterior end of the animal. The egg is fertilized and laid, and cleavage begins. At the same time, the *bicoid* mRNA is translated, forming *bicoid* protein that diffuses away from the anterior end, establishing a gradient of the protein (Figure 17.27a). Another morphogen, the *nanos* protein, diffuses from the posterior end, forming a gradient in the other direction. Thus, each nucleus in the developing blastula is exposed to a different concentration of *bicoid* protein and to a different ratio of *bicoid* protein to *nanos* protein (Figure 17.27b). What do these morphogens do?

The two morphogens regulate the expression of the gap genes. Both morphogens are transcription factors (see Chapter 13)—they enhance or repress the expression of the gap genes. High concentrations of *bicoid* protein turn on the gap gene called *hunchback*, but the *bicoid* protein also turns off another gap gene called *Krüppel*. The pattern of gap gene activity resulting from morphogen actions is shown in Figure 17.28. What is the function of the gap genes?

The proteins encoded by the gap genes are another set of transcription factors that control the expression of the pair-rule genes. Many of these in turn encode transcription factors that control the expression of the segment-polarity genes, giving rise to a complex, striped pattern that foreshadows the segmented body plan of *Drosophila*. By this point, each nucleus of the blastula is exposed to a distinctive set of transcription factors. The body pattern of the larva is established even before gastrulation begins, although no segmentation is visible yet. When the segments do appear, why aren't they all alike?

The homeotic genes give the different segments their different properties. Each homeotic gene is expressed over a characteristic portion of the embryo. Developmental biologists recently discovered that the six homeotic genes are arranged on the same chromosome—in the same order as the order of their function from the anterior to the posterior end of the larva! Each homeotic gene encodes a transcription factor, and each of these transcription factors has a homeobox as part of its primary structure.

And that is how genes and their products specify the structure of a *Drosophila* larva by a transcriptionally controlled cascade. Does this have any relevance

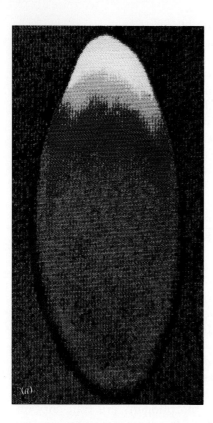

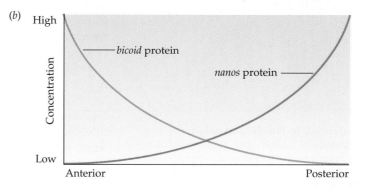

17.27 *bicoid* Protein Forms a Gradient
Translation of mRNAs at the ends of the *Drosophila* larva leads to gradients of the morphogen products. (*a*) The concentration of *bicoid* protein is highest at the embryo's anterior end—bright yellow in this photograph. The colorized gradient moves from orange to red as *bicoid* concentration decreases into the dark posterior end. (*b*) Morphogen gradients originating at the two ends of the embryo expose each cell to a unique environment.

17.28 Gap Genes in Action
Interactions of proteins coded by gap genes define domains of the larval body in *Drosophila*. In this larva, *hunchback* (orange) and *Krüppel* (green) proteins overlap, forming a boundary (yellow) between two domains.

to the development of other organisms—notably humans? In fact, it *is* relevant. Recent experiments have shown that mice and humans (the two best-studied mammals) have clusters of homeobox genes. Thirty-eight genes are divided into four clusters, each located on a different chromosome. As in *Drosophila*, these homeobox genes are arranged in the same order on each chromosome as the order of their expression from anterior to posterior of the developing animal. These genes are expressed in particular segments of the animal, just as in the "simple" fruit fly model. Thus there *are* some rules of development that apply to much of the animal kingdom. What might this and other findings of developmental biology imply for our understanding of the mechanisms of evolution?

DEVELOPMENTAL BIOLOGY AND EVOLUTION

As you are about to discover in Part Three, our current understanding of evolution is heavily dependent on contributions from the field of population genetics. More recently, developmental genetics has also begun to make key contributions to the study of evolution. In particular, homeotic genes are believed to contribute to "macroevolution"—the larger jumps involved in the appearance of new species and, long ago, of major groups of animals and other organisms. Figure 17.29 offers a speculative view of how the progressive addition of homeotic gene functions could account for some aspects of insect evolution. This model is meant simply to illustrate how homeotic genes may contribute to evolution and is not to be taken literally as a description of what actually happened.

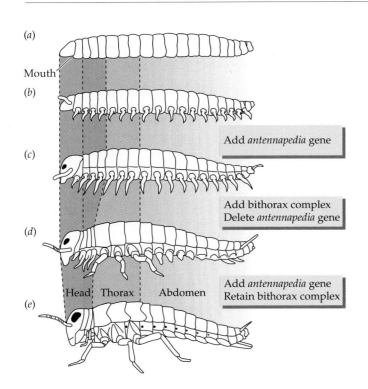

(a)

Mouth

(b)

Add *antennapedia* gene

(c)

Add bithorax complex
Delete *antennapedia* gene

(d)

Head Thorax Abdomen

Add *antennapedia* gene
Retain bithorax complex

(e)

17.29 How Homeotic Genes May Have Contributed to Insect Evolution
This hypothetical model illustrates how the addition of homeotic gene functions may have led progressively from (a) an earthwormlike animal to (e) the winged insects of today. The intermediate stages correspond roughly to animal groups alive today—(b) onychophorans, (c) centipedes, and (d) wingless insects (see Chapter 26). The same set of homeotic genes that establishes the anterior–posterior axis in insects appears to establish this major axis in all animals, including mammals.

Some developmental biologists are concerned with evolutionary questions, ranging from how the first embryos came into being to how macroevolution occurs. Let us, too, turn next to the study of evolution: its mechanisms, its consequences, and its history.

SUMMARY of Main Ideas About Animal Development

An animal develops by progressive changes from the zygote, through the embryonic and subsequent stages of its life cycle, until it dies.

The zygote divides mitotically, giving rise to thousands of cells by cleavage. The amount and distribution of yolk varies among animal groups, leading to different patterns of cleavage.
 Review Figure 17.2

Cleavage results in the formation of a hollow blastula.
 Review Figure 17.4

Redistribution of cells transforms the blastula into a gastrula, a two- or three-layered embryo with a gut.

The pattern of gastrulation varies among animal groups.
 Review Figures 17.5, 17.6, and 17.7

A gastrula has an outer layer (ectoderm) and an inner layer (endoderm); in most animals there is an intervening layer (mesoderm).

The ectoderm, mesoderm, and endoderm each give rise to particular parts of the adult animal.
 Review Table 17.1

The nervous system is one of the first embryonic organ systems to develop.

In neurulation, part of the surface ectoderm folds and becomes a neural tube within the embryo.
 Review Figure 17.9

Early products of the mesoderm are the notochord (a supporting rod); later products are the somites (which develop into muscles and the backbone).

Many animals grow throughout their lives; some develop continuously, and others develop discontinuously because they have external skeletons.
 Review Figure 17.11

Some animals, such as butterflies and frogs, undergo metamorphosis, in which a juvenile form reorients radically while developing into the adult form.

As the embryo develops, cells become determined—committed to developing into particular parts of the embryo and of the adult animal.

When a cell is determined, its fate is specified; its developmental flexibility is restricted.

At early stages a cell may have a particular fate, but the fate may change if the cell is transplanted to another part of an embryo.

Determination precedes differentiation.

Differentiation is the process in which cells change in structure, chemical makeup, and function. One source of differentiation is the segregation of some components of the egg cytoplasm into separate cells of the developing embryo.
 Review Figure 17.19

Chemical inducers are another source of differentiation.

Review Figures 17.21 and 17.22

Pattern formation is the process by which body parts become delineated and organized in predictable arrangements.

Pattern formation may result from the application of information about where different groups of cells lie in relation to one another.

Review Figure 17.23

Body segmentation in insects results from the successive expression of several types of genes; cytoplasmic segregation also plays a role.

Mutations in homeotic genes cause drastic changes in pattern formation.

SELF-QUIZ

1. Which statement is *not* true of cleavage?
   a. The blastomeres are produced by mitosis.
   b. Cleavage patterns depend in part on the distribution of yolk.
   c. The first few divisions often occur simultaneously.
   d. The first cleavage results in a cell toward the animal pole and one toward the vegetal pole.
   e. In eggs with a lot of yolk, daughter cells are not completely separated.

2. A blastula
   a. is a solid ball of cells.
   b. is surrounded by a sheet of cells called the blastocoel.
   c. develops over a few hours of cleavage.
   d. has a lower ratio of nuclear to cytoplasmic volume than did the zygote.
   e. is much larger than the fertilized egg.

3. Gastrulation
   a. is identical in all animal species.
   b. always results from ingression of cells at the vegetal pole.
   c. always produces a primitive streak.
   d. always results in a roughly spherical gastrula.
   e. is a process in which a two- or three-layered embryo is formed.

4. Which statement is *not* true of the dorsal lip of the amphibian blastopore?
   a. It is the first site of invagination as the blastula is forming.

   b. Cells turn in here and flow into the interior of the embryo.
   c. It serves as the embryonic organizer.
   d. It spreads, eventually forming a complete circle.
   e. Cells involuting here give rise to both endoderm and mesoderm.

5. Which statement is *not* true of cytoplasmic determinants in *Drosophila*?
   a. They specify the dorsoventral and anteroposterior axes of the embryo.
   b. Their positions in the embryo are determined by microfilament action.
   c. They are products of specific genes in the mother insect.
   d. They often produce striking effects in larvae.
   e. They have been studied by the transfer of cytoplasm from egg to egg.

6. The organizer of an amphibian embryo
   a. is a homeotic mutation.
   b. is the homeobox.
   c. induces adjacent ectoderm to organize into a neural tube.
   d. is the ventral lip of the blastopore.
   e. is a product of imaginal discs.

7. Which statement is *not* true of embryonic induction?
   a. One tissue induces an adjacent tissue to develop in a certain way.
   b. It triggers a sequence of gene expression in target cells.

   c. It may be either instructive or permissive.
   d. A tissue may induce itself.
   e. The chemical identification of specific inducers has been difficult.

8. In establishing body segmentation in *Drosophila* larvae,
   a. the first steps are specified by homeotic genes.
   b. mutations in pair-rule genes result in embryos missing every other segment.
   c. mutations in gap genes result in the insertion of extra segments.
   d. segment-polarity genes determine the dorsoventral axes of segments.
   e. segmentation is the same as in earthworms.

9. Homeotic mutations
   a. are often so severe that they can be studied only in larvae.
   b. cause subtle changes in the forms of larvae or adults.
   c. occur only in prokaryotes.
   d. do not affect the animal's DNA.
   e. are confined to the apical ectodermal ridge.

10. Which statement is *not* true of the homeobox?
    a. It is transcribed and translated.
    b. It is found only in animals.
    c. Proteins containing the homeodomain bind to DNA.
    d. It is a stretch of DNA shared by many genes.
    e. Its activities often relate to body segmentation.

FOR STUDY

1. Discuss the differences—and the reasons for the differences—in the gastrulas of a sea urchin, a frog, and a chicken.

2. Consider a cell in the lens of a frog eye. Trace its developmental history back to a particular part of the zygote. Describe all the ways in which other cells or tissues have interacted with it.

3. During development, the developmental potential of a tissue becomes ever more limited until, in the normal course of events, the developmental potential is the same as the original fate. On the basis of what you have learned in this chapter and in Chapter 13, discuss possible mechanisms for the progressive limitation of the developmental potential.

4. How was it possible for biologists to obtain such a complete accounting of all the cells in the roundworm *Caenorhabditis elegans*? Why can't we reason directly from studies of *C. elegans* to comparable problems in our own species?

READINGS

De Robertis, E. M., G. Oliver and C. V. E. Wright. 1990. "Homeobox Genes and the Vertebrate Body Plan." *Scientific American*, July. Extends material in this chapter to amphibian development; good discussion of evolution of anatomy.

Gehring, W. J. 1985. "The Molecular Basis of Development." *Scientific American*, October. A lucid account of homeotic mutations and the homeobox.

Gilbert, S. F. 1994. *Developmental Biology*, 4th Edition. Sinauer Associates, Sunderland, MA. An exceptionally well-balanced treatment of developmental biology, covering both molecular/cellular concepts and embryology. Gives a feeling for the history of the discipline.

Goodman, C. S. and M. J. Bastiani. 1984. "How Embryonic Nerve Cells Recognize One Another." *Scientific American*, December. How do brains develop their specific "wiring"? This article describes experimental methods for dealing with this question.

Holliday, R. 1989. "A Different Kind of Inheritance." *Scientific American*, June. Genes can be methylated in one cell, and the pattern of methylation is inherited by the products of cell division. Thus a pattern of gene activity is transmitted from one cell generation to the next.

Hynes, R. O. 1986. "Fibronectins." *Scientific American*, June. Proteins that guide migrating cells during development, and their possible role in cancer.

Stent, G. S. and D. A. Weisblat. 1982. "The Development of a Simple Nervous System." *Scientific American*, January. Details of a specific pattern of development in the leech.

GLOSSARY

Abdomen (ab´ duh mun) [L.: belly]   In arthropods, the posterior portion of the body; in mammals, the part of the body containing the intestines and most other internal organs, posterior to the thorax.

Abomasum (ab´ oh may´ sum)   The true stomach of ruminants (animals such as cattle, sheep, and goats).

Abscisic acid (ab sighs´ ik) [L. *abscissio*: breaking off]   A plant growth substance having growth-inhibiting action. Causes stomata to close.

Abscission (ab sizh´ un) [L. *abscissio*: breaking off]   The process by which leaves, petals, and fruits separate from a plant.

Absolute temperature scale   A temperature scale in which the degree is the same size as in the Celsius (centigrade) scale, and zero is the state of no molecular motion. Absolute zero is –273° on the Celsius scale.

Absorption   (1) Of light: complete retention, without reflection or transmission. (2) Of liquids: soaking up (taking in through pores or cracks).

Absorption spectrum   A graph of light absorption versus wavelength of light; shows how much light is absorbed at each wavelength.

Abyssal zone (uh biss´ ul) [Gr. *abyssos*: bottomless]   That portion of the deep ocean where no light penetrates.

Abzyme   An immunoglobulin (antibody) with catalytic activity.

Accessory fruit   A fruit derived from parts in addition to the ovary and seeds. (Contrast with simple fruit, aggregate fruit, multiple fruit.)

Accessory pigments   Pigments that absorb light and transfer energy to chlorophylls for photosynthesis.

Acclimatization   Changes in an organism that improve its ability to tolerate seasonal changes in its environment.

Acellular   Not composed of cells.

Acetylcholine   A neurotransmitter substance that carries information across vertebrate neuromuscular junctions and some other synapses. **Acetylcholinesterase** is an enzyme that breaks down acetylcholine.

Acetyl CoA (acetyl coenzyme A)   Compound that reacts with oxaloacetate to produce citrate at the beginning of the citric acid cycle; a key metabolic intermediate in the formation of many compounds.

Acid [L. *acidus*: sharp, sour]   A substance that can release a proton. (Contrast with base.)

Acid precipitation   Precipitation that has a lower pH than normal as a result of acid-forming precursors introduced into the atmosphere by human activities.

Acidic   Having a pH of less than 7.0 (a hydrogen ion concentration greater than 10^{-7} molar).

Acoelomate   Lacking a coelom.

Acquired Immune Deficiency Syndrome   See AIDS.

Acrosome (a´ krow soam) [Gr. *akros*: highest or outermost + *soma*: body]   The structure at the forward tip of an animal sperm which is the first to fuse with the egg membrane and enter the egg cell.

ACTH (adrenocorticotropin)   A pituitary hormone that stimulates the adrenal cortex.

Actin [Gr. *aktis*: a ray]   One of the two major proteins of muscle; it makes up the thin filaments. Forms the microfilaments found in most eukaryotic cells.

Action potential   An impulse in a neuron taking the form of a wave of depolarization or hyperpolarization imposed on a polarized cell surface.

Action spectrum   A graph of biological activity versus wavelength of light. It compares the effectiveness of light of different wavelengths.

Activation energy   The energy barrier that blocks the tendency for a set of chemical substances to react. A reaction is speeded up if this energy barrier is surmounted by adding heat energy, or if the barrier is lowered by providing a different reaction pathway with the aid of a catalyst. Designated by the symbol E_a.

Active site   The region on the surface of an enzyme where the substrate binds, and where catalysis occurs.

Active transport   The transport of a substance across a biological membrane against a concentration gradient—that is, from a region of low concentration (of that substance) to a region of high concentration. Active transport requires the expenditure of energy and is a saturable process. (Contrast with facilitated diffusion, free diffusion; see primary active transport, secondary active transport.)

Adaptation (a dap tay´ shun)   In evolutionary biology, a particular structure, physiological process, or behavior that makes an organism better able to survive and reproduce. Also, the evolutionary process that leads to the development or persistance of such a trait.

Adenosine triphosphate   See ATP.

Adenylate cyclase   Enzyme catalyzing the formation of cyclic AMP from ATP.

Adhesion molecules   See cell adhesion molecules.

Adrenal (a dree´ nal) [L. *ad-*: toward + *renes*: kidneys]   An endocrine gland located near the kidneys of vertebrates, consisting of two glandular parts, the cortex and medulla.

Adrenaline   See epinephrine.

Adrenocorticotropin   See ACTH.

Adsorption   Binding of a gas or a solute to the surface of a solid.

Aerenchyma (air eng´ kyma) [Gr. *aer*: air + *enchyma*: infusion]   Modified parenchyma tissue, with many air spaces, found in shoots of some aquatic plants. (See parenchyma.)

Aerobic (air oh´ bic) [Gr. *aer*: air + *bios*: life] In the presence of oxygen, or requiring oxygen.

Afferent (af´ ur unt) [L. *ad*: to + *ferre*: to bear] To or toward, as in a neuron that carries impulses to the central nervous system, or a blood vessel that carries blood to a structure. (Contrast with efferents.)

Age distribution The proportion of individuals in a population belonging to each of the age categories into which the population has been divided. The number of divisions is arbitrary.

Aggregate fruit A fruit developing from several carpels of a single flower. (Contrast with simple fruit, accessory fruit, multiple fruit.)

AIDS (Acquired immune deficiency syndrome) Condition in which the body's helper T lymphocytes are destroyed, leaving the victim subject to opportunistic diseases. Caused by the HIV-I virus.

Air sacs Structures in the avian respiratory system that facilitate unidirectional flow of air through the lungs.

Alcohol An organic compound with one or more hydroxyl (–OH) groups.

Aldehyde (al´ duh hide) A compound with a –CHO functional group. Many sugars are aldehydes. (Contrast with ketone.)

Aldosterone (al dahs´ ter own) A steroid hormone produced in the adrenal cortex of mammals. Promotes secretion of potassium and reabsorption of sodium in the kidney.

Aleurone layer (al´ yur own) [Gr. *aleuron*: wheat flour] In grass seeds, a specialized cell layer just between the seed coat and the endosperm, synthesizing hydrolytic enzymes under the influence of gibberellin, and thus helping mobilize reserves for the developing embryo.

Alga (al´ gah) (plural: algae) [L.: seaweed] Any one of a wide diversity of protists belonging to the phyla Pyrrophyta, Chrysophyta, Phaeophyta, Rhodophyta, and Chlorophyta (and, formerly, Cyanophyta—"blue-green algae"). Most live in the water, where they are the dominant autotrophs; most are unicellular, but a minority are multicellular ("seaweeds" and similar protists).

Allele (a leel´) [Gr. *allos*: other] The alternate forms of a genetic character found at a given locus on a chromosome.

Allele frequency The relative proportion of a particular allele in a specific population.

Allergy [Ger. *allergie*: altered reaction] An overreaction to an antigen in amounts that do not affect most people; often involves IgE antibodies.

Allometric growth A pattern of growth in which some parts of the body of an organism grow faster than others, resulting in a change in body proportions as the organism grows.

Allopatric (al´ lo pat´ rick) [Gr. *allos*: other + *patria*: fatherland] Pertaining to populations that occur in different places.

Allopatric speciation See geographical speciation.

Allostery (al´ lo steer´ y) [Gr. *allos*: other + *stereos*: structure] Regulation of the activity of an enzyme by binding, at a site other than the catalytic active site, of an effector molecule that does not have the same structure as any of the enzyme's substrates.

Alpha helix Type of protein secondary structure; a right-handed spiral.

Alternation of generations The succession of haploid and diploid phases in a sexually reproducing organism. In most animals (male wasps and honey bees are notable exceptions), the haploid phase consists only of the gametes. In fungi, algae, and plants, however, the haploid phase may be the more prominent phase (as in fungi and mosses) or may be as prominent as the diploid phase (see the life cycle of *Ulva*, for example). In vascular plants, the diploid phase is more prominent.

Altruistic act A behavior whose performance harms the actor but benefits other individuals.

Alveolus (al ve´ o lus) (plural: alveoli) [L. *alveus*: cavity] A small, baglike cavity, especially the blind sacs of the lung.

Amensalism (a men´ sul ism) Interaction in which one animal is harmed and the other is unaffected. (Contrast with commensalism, mutualism.)

Amine An organic compound with an amino group (see Amino acid).

Amino acid An organic compound of the general formula H_2N–CHR–COOH, where R can be one of 20 or more different side groups. An amino acid is so named because it has both a basic amine group, –NH_2, and an acidic carboxyl group, –COOH. Proteins are polymers of amino acids.

Ammonotelic (am moan´ o teel´ ic) [Gr. *telos*: end] Describes an organism in which the final product of breakdown of nitrogen-containing compounds (primarily proteins) is ammonia. (Contrast with ureotelic, uricotelic.)

Amniocentesis A medical procedure in which cells from the fetus are obtained from the amniotic fluid. The genetic material of the cells is then examined. (Contrast with chorionic villus sampling.)

Amniotic egg The eggs of birds and reptiles, which can be incubated in air because the embryo is enclosed by a fluid-filled sac.

Amoeba (a mee´ bah) [Gr. *amoibe*: change] Any one of a large number of different kinds of unicellular protists belonging to the phylum Rhizopoda, characterized among other features by its ability to change shape frequently through the protrusion and retraction of cytoplasmic extensions called pseudopods.

Amoeboid (a mee´ boid) Like an amoeba; constantly changing shape by the protrusion and retraction of pseudopodia.

Amphi- [Gr.: both] Prefix used to denote a character or kind of organism that occupies two or more states. For example, amphibian (an animal that lives both on the land and in the water).

Amphibian (am fib´ ee an) A member of the vertebrate class Amphibia, such as a frog, toad, or salamander.

Amphipathic (am´ fi path´ ic) [Gr. *amphi*: both + *pathos*: emotion] Of a molecule, having both hydrophilic and hydrophobic regions.

amu (atomic mass unit, or dalton) The basic unit of mass on an atomic scale, defined as one-twelfth the mass of a carbon-12 atom. There are 6.023×10^{23} amu in one gram. This number is known as Avogadro's number.

Amylase (am´ ill ase) Any of a group of enzymes that digest starch.

Anabolism (an ab´ uh liz´ em) [Gr. *ana*: up, throughout + *ballein*: to throw] Synthetic reactions of metabolism, in which complex molecules are formed from simpler ones. (Contrast with catabolism.)

Anaerobic (an ur row´ bic) [Gr. *an*: not + *aer*: air + *bios*: life] Occurring without the use of molecular oxygen, O_2.

Anagenesis See vertical evolution.

Analogy (a nal´ o jee) [Gr. *analogia*: resembling] A resemblance in function, and often appearance as well, between two structures which is due to convergence in evolution rather than to common ancestry. (Contrast with homology.)

Anaphase (an´ a phase) [Gr. *ana*: indicating upward progress] The stage in nuclear division at which the first separation of sister chromatids (or, in the first meiotic division, of paired homologues) occurs. Anaphase lasts from the moment of first separation to the time at which the moving chromosomes converge at the poles of the spindle.

Anaphylactic shock A precipitous drop in blood pressure caused by loss of fluid from capillaries because of an increase in their permeability stimulated by an allergic reaction.

Ancestral trait Trait shared by a group of organisms as a result of descent from a common ancestor.

Androgens (an´ dro jens) The male sex steroids.

Aneuploid (an´ you ploy dee) A condition in which one or more chromosomes or pieces of chromosomes are either lacking or present in excess.

Angiosperm (an´ jee oh spurm) [Gr. *angion*: vessel + *sperma*: seed] One of the flowering plants; literally, one whose seed is carried in a "vessel," which is the fruit. (See fruit.)

Angiotensin (an´ jee oh ten´ sin) A peptide hormone that raises blood pressure by causing peripheral vessels to constrict; maintains glomerular filtration by constricting efferent glomerular vessels; stimulates thirst; and stimulates the release of aldosterone.

Animal [L. *animus*: breath, soul]   A member of the kingdom Animalia. In general, a multi-cellular eukaryote that obtains its food by ingestion.

Animal pole   In some eggs, zygotes, and embryos, the pole away from the bulk of the yolk (contrast with vegetal pole).

Anion (an´ eye one)   An ion with one or more negative charges. (Contrast with cation.)

Anisogamy (an´ eye sog´ a mee) [Gr. *aniso*: unequal + *gamos*: marriage]   The existence of two dissimilar gametes (egg and sperm).

Annelid (an´ el id)   A member of the phylum Annelida; one of the segmented worms, such as an earthworm or leech.

Annual   Referring to a plant whose life cycle is completed in one growing season. (Contrast with biennial, perennial.)

Anorexia nervosa (an or ex´ ee ah) [Gr. *an*: not + *orexis*: appetite]   Severe malnutrition and body wasting brought on by a psychological aversion to food.

Anterior   Toward the front.

Anterior pituitary   The portion of the vertebrate pituitary gland that derives from gut epithelium and produces tropic hormones.

Anther (an´ thur) [Gr. *anthos*: flower]   A pollen-bearing portion of the stamen of a flower.

Antheridium (an´ thur id´ ee um) (plural: antheridia) [Gr. *antheros*: blooming]   The multicellular structure that produces the sperm in bryophytes and ferns.

Antibody   One of millions of blood proteins, produced by the immune system, that specifically recognizes a foreign substance and initiates its removal from the body.

Anticodon   A "triplet" of three nucleotides in transfer RNA that is able to pair with a complementary triplet (a codon) in messenger RNA, thus aligning the transfer RNA on the proper place on the messenger. The codon (and, reciprocally, the anticodon) codes for a specific amino acid.

Antidiuretic hormone   A hormone that controls water reabsorption in the mammalian kidney. Also called vasopressin.

Antigen (an´ ti jun)   Any substance that stimulates the production of an antibody or antibodies upon introduction into the body of a vertebrate.

Antigenic determinant   A specific region of an antigen, which is recognized by and binds to a specific antibody.

Antiparallel   Parallel but running in opposite directions. The two strands of DNA are antiparallel.

Antipodals (an tip´ o dulls) [Gr. *anti*: against + *podus*: foot]   Cells (usually three) of the mature embryo sac of a flowering plant, located at the end opposite the egg (and micropyle).

Antiport   A membrane transport protein that carries one substance in one direction and another in the opposite direction. (Contrast with symport.)

Antisense nucleic acid   A single-stranded RNA or DNA complementary to and thus targeted against the mRNA transcribed from a harmful gene such as an oncogene.

Anus (a´ nus)   Opening through which digestive wastes are expelled, located at the posterior end of the gut.

Aorta (a or´ tuh) [Gr. *aorte*: aorta]   The main trunk of the arteries leading to the systemic (as opposed to the pulmonary) circulation.

Apex (a´ pecks)   The tip or highest point of a structure, as the apex of a growing stem or root.

Apical (a´ pi kul)   Pertaining to the apex, as the apical meristem, which is the actively growing tissue at the tip of a stem or root.

Apomixis (ap oh mix´ is) [Gr. *apo*: away from + *mixis*: sexual intercourse]   The asexual production of seeds.

Apoplast (ap´ oh plast)   in plants, the continuous meshwork of cell walls and extracellular spaces through which material can pass without crossing a plasma membrane. (Contrast with symplast.)

Appendix   A vestigial portion of the human gut at the junction of the ileum with the colon.

Apterous   Lacking wings. (Contrast with alate: having wings.)

Aquatic [L. *aqua*: water]   Living in or on water, or taking place in or on water.

Aqueous [L. *aqua*: water]   Containing water, or dissolved in water.

Archaebacteria (ark´ ee bacteria) [Gr. *archaios*: ancient]   One of the two kingdoms of prokaryotes; the archaebacteria possess distinctive lipids and lack peptidoglycan. Most live in extreme environments. (Contrast with eubacteria.)

Archegonium (ar´ ke go´ nee um) [Gr. *archegonos*: first of a kind]   The multicellular structure that produces eggs in bryophytes, ferns, and gymnosperms.

Archenteron (ark en´ ter on) [Gr. *archos*: beginning + *enteron*: bowel]   The earliest primordial animal digestive tract.

Arteriole   One of the branches of an artery.

Arteriosclerosis   See atherosclerosis.

Artery   A muscular blood vessel carrying oxygenated blood away from the heart to other parts of the body. (Contrast with vein.)

Artifact [L. *ars*, *artis*: art + *facere*: to make]   Something made by human effort or intervention. In biology, something that was not present in the living cell or organism, but was unintentionally produced by an experimental procedure.

Ascospore (ass´ ko spor)   A fungus spore produced within an ascus.

Ascus (ass´ cuss) [Gr. *askos*: bladder]   In fungi belonging to the class Ascomycetes (sac fungi), the club-shaped sporangium within which spores are produced by meiosis.

Asexual   Without sex.

Associative learning   "Pavlovian" learning, in which an animal comes to associate a previously neutral stimulus (such as the ringing of a bell) with a particular reward or punishment.

Assortative mating   A breeding system under which mates are selected on the basis of a particular trait or group of traits. Results in more pairs of individuals sharing traits than would be the case if mating were random.

Assortment (genetic)   The random separation during meiosis of nonhomologous chromosomes and of genes carried on nonhomologous chromosomes. For example, if genes *A* and *B* are borne on nonhomologous chromosomes, meiosis of diploid cells of genotype *AaBb* will produce haploid cells of the following types in equal numbers: *AB*, *Ab*, *aB*, and *ab*.

Asymmetric   The state of lacking any plane of symmetry.

Asymmetric carbon atom   In a molecule, a carbon atom to which four different atoms or groups are bound.

Atherosclerosis (ath´ er oh sklair oh´ sis)   A disease of the lining of the arteries characterized by fatty, cholesterol-rich deposits in the walls of the arteries. When fibroblasts infiltrate these deposits and calcium precipitates in them, the disease become arteriosclerosis, or "hardening of the arteries."

Atmosphere   The gaseous mass surrounding our planet. Also: a unit of pressure, equal to the normal pressure of air at sea level.

Atom [Gr. *atomos*: indivisible]   The smallest unit of a chemical element. Consists of a nucleus and one or more electrons.

Atomic mass unit   See amu.

Atomic number   The number of protons in the nucleus of an atom, also equal to the number of electrons around the neutral atom. Determines the chemical properties of the atom.

Atomic weight   The average weight of an atom of an element on the amu scale. (The average depends upon the relative amounts of different isotopes of an element on Earth.)

ATP (adenosine triphosphate)   A compound containing adenine, ribose, and three phosphate groups. When it is formed, useful energy is stored; when it is broken down (to ADP or AMP), energy is released to drive endergonic reactions. ATP is a universal energy storage compound.

Atrium (a´ tree um)   A body cavity, as in the hearts of vertebrates. The thin-walled chamber(s) entered by blood on its way to the ventricle(s). Also, the outer ear.

Autocatalysis   An enzymatic reaction in which the inactive form of an enzyme is converted into its active form by the enzyme itself.

Autoimmune disease   A disorder in which the immune system attacks the animal's own body.

Autonomic nervous system   The system (which in vertebrates comprises sympathetic and parasympathetic subsystems) that controls such involuntary functions as those of guts and glands.

Autoradiography   The detection of a radioactive substance in a cell or organism by putting it in contact with a photographic emulsion and allowing the material to "take its own picture." The emulsion is developed, and the location of the radioactivity in the cell is seen by the presence of silver grains in the emulsion.

Autoregulatory mechanism   A feedback mechanism that enables a structure to regulate its own function.

Autosome   Any chromosome (in a eukaryote) other than a sex chromosome.

Autotroph (au´ tow trow´ fik) [Gr. *autos*: self + *trophe*: food]   An organism that is capable of living exclusively on inorganic materials, water, and some energy source such as sunlight or chemically reduced matter. (Contrast with heterotroph.)

Auxin (awk´ sin) [Gr. *auxein*: increase]   In plants, a substance (indoleacetic acid) that regulates growth and various aspects of development.

Auxotroph (awks´ o trofe) [Gr. *auxanein*: to grow + *trophe*: food]   A mutant form of an organism that requires a nutrient or nutrients not required by the wild-type, or reference, form of the organism. (Contrast with prototroph.)

Avogadro's number   The conversion factor between atomic mass units and grams. More usefully, the number of atoms in that quantity of an element which, expressed in grams, is numerically equal to the atomic weight in amu; 6.023×10^{23} atoms. (See mole.)

Axon [Gr.: axle]   Fiber of a neuron which can carry action potentials. Carries impulses away from the cell body of the neuron; releases a neurotransmitter substance.

Axon hillock   The junction between an axon and its cell body; where action potentials are generated.

Axon terminals   The endings of an axon; they form synapses and release neurotransmitter.

Axoneme (ax´ oh neem)   The complex of microtubules and their crossbridges that forms the motile apparatus of a cilium.

Bacillus (buh sil´ us) [L.: little rod]   Any of various rod-shaped bacteria.

Bacteriophage (bak teer´ ee o fayj) [Gr. *bakterion*: little rod + *phagein*: to eat]   One of a group of viruses that infect bacteria and ultimately cause their disintegration.

Bacterium (bak teer´ ee um) (plural: bacteria) [Gr. *bakterion*: little rod]   A prokaryote. An organism with chromosomes not contained in nuclear envelopes.

Balanced polymorphism [Gr. *polymorphos*: having many forms]   The maintenance of more than one form, or the maintenance at a given locus of more than one allele, at frequencies of greater than one percent in a population. Often results when heterozygotes are superior to both homozygotes.

Baroreceptor [Gr. *baros*: weight]   A pressure-sensing cell or organ.

Barr body   In mammals, an inactivated X chromosome.

Basal body   Centriole found at the base of a eukaryotic flagellum or cilium.

Basal metabolic rate   The minimum rate of energy turnover in an awake (but resting) bird or mammal that is not expending energy for thermoregulation.

Base   A substance which can accept a proton (H^+). (Contrast with acid.) In nucleic acids, a nitrogen-containing base (purine or pyrimidine) is attached to each sugar in the backbone.

Base pairing   See complementary base pairing.

Basic   having a pH greater than 7.0 (having a hydrogen ion concentration lower than 10^{-7} molar).

Basidium (bass id´ ee yum)   In fungi of the class Basidiomycetes, the characteristic sporangium in which four spores are formed by meiosis and then borne externally before being shed.

Batesian mimicry   Mimicry by a relatively harmless kind of organism of a more dangerous one, by which the mimic enjoys protection from predators that mistake it for the dangerous model. (Contrast with Müllerian mimicry.)

B cell   A type of lymphocyte involved in the humoral immune response of vertebrates. Upon recognizing an antigenic determinant, a B cell develops into a plasma cell, which secretes an antibody. (Contrast with a T cell.)

Benefit   An improvement in survival and reproductive success resulting from a behavior. (Contrast with cost.)

Benthic zone [Gr. *benthos*: bottom of the sea]   The bottom of the ocean. (Contrast with pelagic zone.)

Beta-pleated sheet   Type of protein secondary structure; results from hydrogen bonding between polypeptide regions running antiparallel to each other.

Biennial   Referring to a plant whose life cycle includes vegetative growth in the first year and flowering and senescence in the second year. (Contrast with annual, perennial.)

Bilateral symmetry   The condition in which only the right and left sides of an organism, divided exactly down the back, are mirror images of each other. (Contrast with radial symmetry.)

Bile   A secretion of the liver delivered to the small intestine via the common bile duct. In the intestine, bile emulsifies fats.

Binocular cells   Neurons in the visual cortex that respond to input from both retinas; involved in depth perception.

Binomial (bye nome´ ee al)   Consisting of two names; for example, the binomial nomenclature of biology which gives the name of the genus followed by the name of the species.

Biodiversity crisis   The current high rate of loss of species, caused primarily by human activities.

Biogenesis [Gr. *bios*: life + *genesis*: source]   The origin of living things from other living things.

Biogeochemical cycles   Movement of elements through living organisms and the physical environment.

Biogeography   The scientific study of the geographic distribution of organisms. Ecological biogeography is concerned with the habitats in which organisms live, historical biogeography with the complete geographic ranges of organisms and the historical circumstances that determine the ranges.

Biological species concept   The view that a species is most usefully defined as a population or series of populations within which there is a significant amount of gene flow under natural conditions, but which is genetically isolated from other populations.

Biology [Gr. *bios*: life + *logos*: discourse]   The scientific study of life in all its forms.

Bioluminescence   The production of light by biochemical processes in an organism.

Biomass   The total weight of all the living organisms, or some designated group of living organisms, in a given area.

Biome (bye´ ome)   A major division of the ecological communities of Earth; characterized by distinctive vegetation.

Biota (bye oh´ tah)   All of the organisms, including animals, plants, fungi, and microorganisms, found in a given area.

Biotic (bye ah´ tik)   Pertaining to any aspect of life, especially to characteristics of entire populations or ecosystems.

Bipedal locomotion (by ped´ ul) [L. *bipes*: two-footed]   Walking on two feet.

Biradial symmetry   Radial symmetry modified so that only two planes can divide the animal into similar halves.

Blastocoel (blass´ toe seal) [Br. *blastos*: sprout + *koilos*: hollow]   The central, hollow cavity of a blastula.

Blastodisc (blass´ toe disk)   A disk of cells forming on the surface of a large yolk mass, comparable to a blastula, but occurring in forms in which the massive yolk restricts cleavage to one side of the egg only.

Blastomere   A cell produced by the division of a fertilized egg.

Blastopore   The opening from the archenteron to the exterior of a gastrula.

Blastula (blass´ chu luh) [Gr. *blastos*: sprout]   An early stage in animal embryology; in many species, a hollow sphere of cells surrounding a central cavity.

Blood–brain barrier A property of the blood vessels of the brain that prevents most chemicals from diffusing from the blood into the brain.

Bloom A sudden increase in the density of phytoplankton, especially in a freshwater lake.

Body plan An entire animal, its organ systems, and the integrated functioning of its parts.

Bohr effect (boar) The reduction in affinity of hemoglobin for oxygen caused by acidic conditions, usually as a result of increased CO_2.

Bolting In rosetted angiosperms, a dramatic elongation of the stem, usually followed by flowering.

Bottleneck A combination of environmental conditions that causes a serious reduction in the size of the population.

Bowman's capsule An elaboration of kidney tubule cells that surrounds a know of capillaries (the glomerulus). Blood is filtered across the walls of these capillaries and the filtrate is collected into Bowman's capsule.

Brain A structure of nervous systems that provides the highest level of integration, control, and regulation.

Brain stem The portion of the vertebrate brain between the spinal cord and the forebrain.

Bronchus (plural: bronchi) The major airway(s) branching off the trachea into the vertebrate lung.

Browser An animal that feeds on the tissues of woody plants.

Bryophyte (bri´ uh fite´) [Gr. *bruon*: moss + *phyton*: plant] Any nonvascular plant, including mosses, liverworts, and hornworts.

Bud primordium [L. *primordium*: the beginning] In plants, a small mass of potentially meristematic tissue found in the angle between the leaf stalk and the shoot apex. Will give rise to a lateral branch under appropriate conditions.

Budding Asexual reproduction in which a more or less complete new organism simply grows from the body of the parent organism and eventually detaches itself.

Buffering A process by which a system resists change—particularly in pH, in which case added acid or base is partially converted to another form.

Bulb In plants, an underground storage organ composed principally of enlarged and fleshy leaf bases.

Bundle sheath In C_4 plants, a layer of photosynthetic cells between the mesophyll and a vascular bundle of a leaf.

C_3 photosynthesis The form of photosynthesis in which 3-phosphoglycerate is the first stable product, and ribulose bisphosphate is the CO_2 receptor.

C_4 photosynthesis The form of photosynthesis in which oxaloacetate is the first stable product, and phosphoenolpyruvate is the CO_2 acceptor. C_4 plants also perform the reactions of C_3 photosynthesis.

Caecum (see´ cum) [L. *caecus*: blind] A blind branch off the large intestine. In many nonruminant mammals, the caecum contains a colony of microorganisms that contribute to the digestion of food.

Calcitonin A hormone produced by the thyroid gland; it lowers blood calcium and promotes bone formation. (Contrast with parathormone.)

Callus [L. *calleo*: thick-skinned] In plants, wound tissue, of relatively undifferentiated proliferating cell mass, frequently maintained in cell culture.

Calmodulin (cal mod´ joo lin) A calcium-binding protein found in all animal and plant cells; mediates many calcium-regulated processes.

Calorie [L. *calor*: heat] The amount of heat required to raise the temperature of one gram of water by one degree Celsius (1°C) from 14.5°C to 15.5°C. In nutrition studies, "Calorie" (spelled with a capital C) refers to the kilocalorie (1 kcal = 1,000 cal), the amount of heat required to raise the temperature of one kilogram of water by 1°C.

Calvin–Benson cycle The stage of photosynthesis in which CO_2 reacts with RuBP to form 3PG, 3PG is reduced to a sugar, and RuBP is regenerated, while other products are released to the rest of the plant.

Calyptra (kuh lip´ tra) [Gr. *kalyptra*: covering for the head] A hood or cap found partially covering the apex of the sporophyte capsule in many moss species, formed from the expanded wall and neck of the archegonium.

Calyx (kay´ licks) [Gr. *kalyx*: cup] All of the sepals of a flower, collectively.

CAM See crassulacean acid metabolism.

Cambium (kam´ bee um) [L. *cambiare*: to exchange] A meristem that gives rise to radial rows of cells in stem and root, increasing them in girth; commonly applied to the vascular cambium which produces wood and phloem, and the cork cambium, which produces bark.

cAMP (cyclic AMP) A compound, formed from ATP, that mediates the effects of numerous animal hormones. Also needed for the transcription of catabolite-repressible operons in bacteria. Used for communication by cellular slime molds.

Canopy The leaf-bearing part of a tree. Collectively the aggregate of the leaves and branches of the larger woody plants of an ecological community.

Capacitance vessels Refers to veins because of their variable capacity to hold blood.

Capillaries [L. *capillaris*: hair] Very small tubes, especially the smallest blood-carrying vessels of animals between the termination of the arteries and the beginnings of the veins.

Capping In eukaryote RNA processing, the addition of a modified G at the 5´ end of the molecule.

Capsid The protein coat of a virus.

Capsule In bryophytes, the spore case. In some bacteria, a gelatinous layer exterior to the cell wall.

Carbohydrates Organic compounds with the general formula $C_nH_{2m}O_m$. Common examples are sugars, starch, and cellulose.

Carbon budget The amount of atmospheric carbon (from carbon dioxide) incorporated into organic molecules by a plant.

Carboxylic acid (kar box sill´ ik) An organic acid containing the carboxyl group, –COOH, which dissociates to the carboxylate ion, –COO⁻.

Carcinogen (car sin´ oh jen) A substance that causes cancer.

Cardiac (kar´ dee ak) [Gr. *kardia*: heart] Pertaining to the heart and its functions.

Carnivore [L. *carn*: flesh + *vovare*: to devour] An organism that feeds on animal tissue. (Contrast with detritivore, herbivore, omnivore.)

Carotenoid (ka rah´ tuh noid) [L. *carota*: carrot] A yellow, orange, or red lipid pigment commonly found as an accessory pigment in photosynthesis; also found in fungi.

Carpel (kar´ pel) [Gr. *karpos*: fruit] The organ of the flower that contains one or more ovules.

Carrier In facilitated diffusion, a membrane protein that binds a specific molecule and transports it through the membrane. In genetics, a person heterozygous for a recessive trait. In respiratory and photosynthetic electron transport, a participating substance such as NAD that exists in both oxidized and reduced forms.

Carrying capacity In ecology, the largest number of organisms of a particular species that can be maintained indefinitely in a given part of the environment.

Cartilage In vertebrates, a tough connective tissue found in joints, the outer ear, and elsewhere. Forms the entire skeleton in some animal groups.

Casparian strip A band of cell wall containing suberin and lignin, found in the endodermis. Restricts the movement of water across the endodermis.

Catabolism [Ge. *kata*: down + *ballein*: to throw] Degradational reactions of metabolism, in which complex molecules are broken down. (Contrast with anabolism.)

Catabolite repression The decreased synthesis of many enzymes that tend to provide glucose for a cell; caused by the presence of excellent carbon sources, particularly glucose.

Catalyst (cat´ a list) [Gr. *kata-*, implying the breaking down of a compound] A chemical substance that accelerates a reaction without itself being consumed in the overall course of the reaction. Catalysts lower the activation energy of a reaction. Enzymes are biological catalysts.

Cation (cat´ eye on)   An ion with one or more positive charges. (Contrast with anion.)

Caudal [L. *cauda*: tail]   Pertaining to the tail, or to the posterior part of the body.

cDNA   See complementary DNA.

Cell adhesion molecules   Molecules on animal cell surfaces that affect the selective association of cells during development of the embryo.

Cell cycle   The stages through which a cell passes between one division and the next. Includes all stages of interphase and mitosis.

Cell theory   The theory, well established, that organisms consist of cells, and that all cells come from preexisting cells.

Cell wall   A relatively rigid structure that encloses cells of plants, fungi, many protists, and most bacteria. The cell wall gives these cells their shape and limits their expansion in hypotonic media.

Cellular immune system   That part of the immune system that is based on the activities of T cells. Directed against parasites, fungi, intracellular viruses, and foreign tissues (grafts). (Contrast with humoral immune system.)

Cellular respiration   See respiration.

Cellulose (sell´ you lowss)   A straight-chain polymer of glucose molecules, used by plants as a structural supporting material. **Cellulase** is an enzyme that hydrolyzes cellulose.

Central dogma of molecular biology   The statement that information flows from DNA to RNA to polypeptide (in retroviruses, there is also information flow from RNA to cDNA).

Central nervous system   That part of the nervous system which is condensed and centrally located, e.g., the brain and spinal cord of vertebrates; the chain of cerebral, thoracic and abdominal ganglia of arthropods.

Centrifuge [L. *fugere*: to flee]   A device in which a sample can be spun around a central axis at high speed, creating a centrifugal force that mimics a very strong gravitational force. Used to separate mixtures of suspended materials.

Centriole (sen´ tree ole)   A paired organelle that helps organize the microtubules in animal and protist cells during nuclear division.

Centromere (sen´ tro meer) [Gr. *centron*: center + *meros*: part]   The region where sister chromatids join.

Cephalization (sef´ uh luh zay´ shun) [Gr. *kephale*: head]   The evolutionary trend toward increasing concentration of brain and sensory organs at the anterior end of the animal.

Cephalopod (sef´ a low pod)   A member of the mollusk class Cephalopoda, such as a squid or an octopus.

Cerebellum (sair´ uh bell´ um) [L.: diminutive of *cerebrum*: brain]   The brain region that controls muscular coordination; located at the anterior end of the hindbrain.

Cerebral cortex   The thin layer of gray matter (neuronal cell bodies) that overlays the cerebrum.

Cerebrum (su ree´ brum) [L.: brain]   The dorsal anterior portion of the forebrain, making up the largest part of the brain of mammals. In mammals, the chief coordination center of the nervous system; consists of two **cerebral hemispheres**.

Cervix (sir´ vix) [L.: neck]   The opening of the uterus into the vagina.

cGMP (cyclic guanosine monophosphate)   An intracellular messenger that is part of signal transmission pathways involving G-proteins. (See G-protein.)

Channel   A membrane protein that forms an aqueous passageway though which specific solutes may pass by simple diffusion; some channels are gated: they open and close in response to binding of specific molecules.

Character   In taxonomy, any trait of an organism used in creating a classification system.

Chemical bond   An attractive force stably linking two atoms.

Chemiosmotic mechanism   According to this model, ATP formation in mitochondria and chloroplasts results from a pumping of protons across a membrane (against a gradient of electrical charge and of pH), followed by the return of the protons through a protein channel with ATPase activity.

Chemoautotroph   An organism that uses carbon dioxide as a carbon source and obtains energy by oxidizing inorganic substances from its environment. (Contrast with chemoheterotroph, photoautotroph, photoheterotroph.)

Chemoheterotroph   An organism that must obtain both carbon and energy from organic substances. (Contrast with chemoautotroph, photoautotroph, photoheterotroph.)

Chemosensor   A cell or tissue that senses specific substances in its environment.

Chemosynthesis   Synthesis of food substances, using the oxidation of reduced materials from the environment as a source of energy.

Chiasma (kie az´ muh) (plural: chiasmata) [Gr.: cross]   An "x"-shaped connection between paired homologous chromosomes in prophase I of meiosis. A chiasma is the visible manifestation of crossing-over between homologous chromosomes.

Chitin (kye´ tin) [Gr. *chiton*: tunic]   The characteristic tough but flexible organic component of the exoskeleton of arthropods, consisting of a complex, nitrogen-containing polysaccharide. Also found in cell walls of fungi.

Chlorophyll (klor´ o fill) [Gr. *chloros*: green + *phyllon*: leaf]   Any of a few green pigments associated with chloroplasts or with certain bacterial membranes; responsible for trapping light energy for photosynthesis.

Chloroplast [Gr. *chloros*: green + *plast*: a particle]   An organelle bounded by a double membrane containing the enzymes and pigments that perform photosynthesis. Chloroplasts occur only in eukaryotes.

Choanocyte (cho´ an oh cite)   The collared, flagellated feeding cells of sponges.

Cholecystokinin (ko´ lee sis to kai nin)   A hormone produced and released by the lining of the duodenum when it is stimulated by undigested fats and proteins. It stimulates the gallbladder to release bile and slows stomach activity.

Chorion (kor´ ee on) [Gr. *khorion*: afterbirth]   The outermost of the membranes protecting mammal, bird, and reptile embryos; in mammals it forms part of the placenta.

Chorionic villus sampling   A medical procedure that extracts a portion of the chorion from a pregnant woman to enable genetic and biochemical analysis of the embryo. (Contrast with amniocentesis.)

Chromatid (kro´ ma tid)   Each of a pair of new sister chromosomes from the time at which the molecular duplication occurs until the time at which the centromeres separate at the anaphase of nuclear division.

Chromatin   The nucleic acid–protein complex found in eukaryotic chromosomes.

Chromatography   Any one of several techniques for the separation of chemical substances, based on differing relative tendencies of the substances to associate with a mobile phase or a stationary phase.

Chromatophore (krow mat´ o for) [Gr. *chroma*: color + *phoreus*: carrier]   A pigment-bearing cell that expands or contracts to change the color of the organism.

Chromosomal aberration   Any large change in the structure of a chromosome, including duplication or loss of chromosomes or parts thereof, usually gross enough to be detected with the light microscope.

Chromosome (krome´ o sowm) [Gr. *chroma*: color = *soma*: body]   In bacteria and viruses, the DNA molecule that contains most or all of the genetic information of the cell or virus. In eukaryotes, a structure composed of DNA and proteins that bears part of the genetic information of the cell.

Chromosome walking   A technique based on recognition of overlapping fragments; used as a step in DNA sequencing.

Chylomicron (ky low my´ cron)   Particles of lipid coated with protein, produced in the gut from dietary fats and secreted into the extracellular fluids.

Chyme (kime) [Gr. *chymus*, juice]   Created in the stomach; a mixture of ingested food with the digestive juices secreted by the salivary glands and the stomach lining.

Ciliate (sil´ ee ate)   A member of the protist phylum Ciliophora, unicellular organisms that propel themselves by means of cilia.

Cilium (sil´ ee um) (plural: cilia) [L. *cilium*: eyelash]   Hairlike organelle used for locomotion by many unicellular organisms and for moving water and mucus by many multicellular organisms. Generally shorter than a flagellum.

Circadian rhythm (sir kade´ ee an) [L. *circa*: approximately + *dies*: day]   A rhythm in behavior, growth, or some other activity that recurs about every 24 hours under constant conditions.

Circannual rhythm (sir can´ you al) [L. *circa*: approximately + *annus*: year]   A rhythm of behavior, growth, or some other activity that recurs on a yearly basis.

Citric acid cycle   A set of chemical reactions in cellular respiration, in which acetyl CoA reacts with oxaloacetate to form citric acid, and oxaloacetate is regenerated. Acetyl CoA is oxidized to carbon dioxide, and hydrogen atoms are stored as NADH and $FADH_2$.

Clade (clayd) [Gr. *klados*: branch]   All of the organisms, both living and fossil, descended from a particular common ancestor.

Cladistic classification   A classification based entirely on the phylogenetic relationships among organisms.

Cladogenesis (clay doh jen´ e sis) [Gr. *klados*: branch + *genesis*: source]   The formation of a new species by the splitting of an evolutionary lineage.

Cladogram   Graphic representation of a cladistic relationship.

Class   In taxonomy, the category below the phylum and above the order; a group of related, similar orders.

Clathrin   A fibrous protein on the inner surfaces of animal cell membranes that strengthens coated vesicles and thus participates in receptor-mediated endocytosis.

Clay   A soil constituent comprising particles smaller than 2 micrometers in diameter.

Cleavages   First divisions of the fertilized egg of an animal.

Climax   In ecology, a community that terminates a succession and which tends to replace itself unless it is further disturbed or the physical environment changes.

Climograph (clime´ o graf)   Graph relating temperature and precipitation with time of year.

Cline   A gradual change in the traits of a species over a geographical gradient.

Clitoris (klit´ er us, klite´ er us)   A structure in the human female reproductive system that is homologous with the male penis and is involved in sexual stimulation.

Cloaca (klo ay´ kuh) [L. *cloaca*: sewer]   In some invertebrates, the posterior part of the gut; in many vertebrates, a cavity receiving material from the digestive, reproductive, and excretory systems.

Clonal deletion   In immunology, the inactivation or destruction of lymphocyte clones that would produce immune reactions against the animal's own body.

Clonal selection   The mechanism by which exposure to antigen results in the activation of selected T-cell or B-cell clones, resulting in an immune response.

Clone [Gr. *klon*: twig, shoot]   Genetically identical cells or organisms produced from a common ancestor by asexual means.

Clutch   The number of offspring produced in a given batch.

Coacervate (ko as´ er vate) [L. *coacervare*: to heap up]   An aggregate of colloidal particles in suspension.

Coacervate drop   Drops formed when a mixture of large proteins and polysaccharides is shaken in water. The interiors of these drops, which are often very stable, contain most of the proteins and polysaccharides.

Coated vesicle   Vesicle, sometimes formed from a coated pit, with characteristic "bristly" surface; its membrane contains distinctive proteins, including clathrin.

Coccus (kock´ us) [Gr. *kokkos*: berry, pit]   Any of various spherical or spheroidal bacteria.

Cochlea (kock´ lee uh) [Gr. *kokhlos*: a land snail]   A spiral tube in the inner ear of vertebrates; it contains the sensory cells involved in hearing.

Codominance   A condition in which two alleles at a locus produce different phenotypic effects and both effects appear in heterozygotes.

Codon   A "triplet" of three nucleotides in messenger RNA that directs the placement of a particular amino acid into a polypeptide chain. (Contrast with anticodon.)

Coefficient of relatedness   The probability that an allele in one individual is an identical copy, by descent, of an allele in another individual.

Coelom (see´ lum) [Gr. *koiloma*: cavity]   The body cavity of certain animals, which is lined with cells of mesodermal origin.

Coelomate   Having a coelom.

Coenocyte (seen´ a sight) [Gr.: common cell]   A "cell" bounded by a single plasma membrane, but containing many nuclei.

Coenzyme   A nonprotein molecule that plays a role in catalysis by an enzyme. The coenzyme may be part of the enzyme molecule or free in solution. Some coenzymes are oxidizing or reducing agents, others play different roles.

Coevolution   Concurrent evolution of two or more species that are mutually affecting each other's evolution.

Cohort (co´ hort) [L. *cohors*: company of soldiers]   A group of similar-age organisms, considered as it passes through time.

Coitus (koe´ i tus) [L. *coitus*: a coming together]   The act of sexual intercourse.

Coleoptile (koe´ lee op´ til) [Gr. *koleos*: sheath + *ptilon*: feather]   A pointed sheath covering the shoot of grass seedlings.

Collagen [Gr. *kolla*: glue]   A fibrous protein found extensively in bone and connective tissue.

Collecting duct   In vertebrates, a tubule that receives urine produced in the nephrons of the kidney and delivers that fluid to the ureter for excretion.

Collenchyma (cull eng´ kyma) [Gr. *kolla*: glue + *enchyma*: infusion]   A type of plant cell, living at functional maturity, which lends flexible support by virtue of primary cell walls thickened at the corners. (Contrast with parenchyma, sclerenchyma.)

Colon [Gr. *kolon*: large intestine]   The large intestine.

Colostrum (koh los´ trum)   Substance secreted by the mammary glands around the time of an infant's birth. It contains protein and lactose but little fat, and its rate of production is less than the rate of milk production two or three days after birth.

Commensalism   The form of symbiosis in which one species benefits from the association, while the other is neither harmed nor benefited.

Common bile duct   A single duct that delivers bile from the gallbladder and secretions from the pancreas into the small intestine.

Communication   Action on the part of one organism (or cell) that alters the pattern of behavior in another organism (or cell) in an adaptive fashion.

Community   Any ecologically integrated group of species of microorganisms, plants, and animals inhabiting a given area.

Companion cell   Specialized cell found adjacent to a sieve tube element in some flowering plants.

Comparative analysis   An approach to studying evolution in which hypotheses are tested by measuring the distribution of states among a large number of species.

Compensation point   The light intensity at which the rates of photosynthesis and of cellular respiration are equal.

Competitive inhibitor   A substance, similar in structure to an enzyme's substrate, that binds the active site and thus inhibits a reaction.

Competition   In ecology, use of the same resource by two or more species, when the resource is present in insufficient supply for the combined needs of the species.

Competitive exclusion   A result of competition between species for a limiting resource in which one species completely eliminates the other.

Competitive inhibitor   A substance, similar in structure to an enzyme's substrate, that binds the active site and inhibits a reaction.

Complement system   A group of eleven proteins that play a role in some reactions of the immune system. The complement proteins are not immunoglobulins.

Complementary base pairing   The A–T (or A–U), T–A (or U–A), C–G and G–C pairing of bases in double-stranded DNA, in transcription, and between tRNA and mRNA.

Complementary DNA (cDNA)   DNA formed by reverse transcriptase acting with an RNA template; essential intermediate in the reproduction of retroviruses; used as a tool in recombinant DNA technology; lacks introns.

Complete metamorphosis A change of state during the life cycle of an organism in which the body is almost completely rebuilt to produce an individual with a completely different body form. Characteristic of insects such as butterflies, moths, beetles, ants, wasps, and flies.

Compound (1) A substance made up of atoms of more than one element. (2) Made up of many units, as the compound eyes of arthropods (as opposed to the simple eyes of the same group of organisms).

Compression wood See reaction wood.

Condensation reaction A reaction in which two molecules become connected by a covalent bond, and a molecule of water is released. ($AH + BOH \rightarrow AB + H_2O$.)

Cones (1) In the vertebrate retina: photoreceptors responsible for color vision. (2) In gymnosperms: reproductive structures consisting of many sporophylls packed relatively tightly.

Conidium (ko nid´ ee um) [Gr. *konis*: dust] An asexual fungus spore borne singly or in chains either apically or laterally on a hypha.

Conifer (kahn´ e fer) [Gr. *konos*: cone + *phero*: carry] One of the cone-bearing gymnosperms, mostly trees, such as pines and firs.

Conjugation (kahn´ jew gay´ shun) [L. *conjugare*: yoke together] The close approximation of two cells during which they exchange genetic material, as in *Paramecium* and other ciliates, or during which DNA passes from one to the other through a tube, as in bacteria.

Connective tissue An animal tissue that connects or surrounds other tissues; its cells are embedded in a collagen-containing matrix.

Connexon In a gap junction, a protein channel linking adjacent animal cells.

Consensus sequences Short stretches of DNA that appear, with little variation, in many different genes.

Constitutive enzyme An enzyme that is present in approximately constant amounts in a system, whether its substrates are present or absent. (Contrast with inducible enzyme.)

Consumer An organism that eats the tissues of some other organism.

Continental climate A pattern, typical of the interiors of large continents at high latitudes, in which bitterly cold winters alternate with hot summers. (Contrast with maritime climate.)

Continental drift The gradual drifting apart of the world's continents that has occurred over a period of billions of years.

Contractile vacuole An organelle, often found in protists, which pumps excess water out of the cell and keeps it from being "flooded" in hypotonic environments.

Cooperative act Behavior in which two or more individuals interact to their mutual benefit. No conscious awareness by the actors of the effects of their behavior is implied.

Cooption The act of capturing something for a particular use. In ecology refers to the diversion of ecological production for human use. Such production is said to be coopted.

Copulation Reproductive behavior that results in a male depositing sperm in the reproductive tract of a female.

Corepressor A low molecular weight compound that unites with a protein (the repressor) to prevent transcription in a repressible operon.

Cork A waterproofing tissue in plants, with suberin-containing cell walls. Produced by a cork cambium.

Corm A conical, underground stem that gives rise to a new plant. (Contrast with bulb.)

Corolla (ko role´ lah) [L.: diminutive of *corona*: wreath, crown] All of the petals of a flower, collectively.

Coronary (kor´ oh nair ee) Referring to the blood vessels of the heart.

Corpus luteum (kor´ pus loo´ tee um) [L. *corpus*: body + *luteum*: yellow] A structure formed from a follicle after ovulation; it produces hormones important to the maintenance of pregnancy.

Cortex [L.: bark or rind] (1) In plants: the tissue between the epidermis and the vascular tissue of a stem or root. (2) In animals: the outer tissue of certain organs, such as the adrenal cortex and cerebral cortex.

Corticosteroids Steroid hormones produced and released by the cortex of the adrenal gland.

Cost See energetic cost, opportunity cost, risk cost.

Cotyledon (kot´ ul lee´ dun) [Gr. *kotyledon*: a hollow space] A "seed leaf." An embryonic organ which stores and digests reserve materials; may expand when seed germinates.

Covalent bond A chemical bond that arises from the sharing of electrons between two atoms. Usually a strong bond.

Crassulacean acid metabolism (CAM) A metabolic pathway enabling the plants that possess it to store carbon dioxide at night and then perform photosynthesis during the day with stomata closed.

Crista (plural: cristae) A small, shelflike projection of the inner membrane of a mitochondrion; the site of oxidative phosphorylation.

Critical night length In the photoperiodic flowering response of short-day plants, the length of night above which flowering occurs and below which the plant remains vegetative. (The reverse applies in the case of long-day plants.)

Critical period The age during which some particular type of learning must take place or during which it occurs much more easily than at other times. Typical of song learning among birds.

Cross-pollination The pollination of one plant by pollen from another plant. (Contrast with self-pollination.)

Cross (transverse) section A section taken perpendicular to the longest axis of a structure.

Crossing over The mechanism by which linked markers undergo recombination. In general, the term refers to the reciprocal exchange of corresponding segments between two homologous chromatids. However, the reciprocity of crossing-over is problematical in prokaryotes and viruses; and even in eukaryotes, very closely linked markers often recombine by a nonreciprocal mechanism.

CRP The cAMP receptor protein that interacts with the promoter to enhance transcription; a lowered cAMP concentration results in catabolite repression.

Crustacean (crus tay´ see an) A member of the phylum Crustacea, such as a crab, shrimp, or sowbug.

Cryptic appearance The resemblance of an animal to some part of its environment, which helps it to escape detection by predators.

Culture A laboratory association of organisms under controlled conditions. Also the collection of knowledge, tools, values, and rules that characterize a human society.

Cuticle A waxy layer on the outer surface of a plant or an insect, tending to retard water loss.

Cutin (cue´ tin) [L. *cutis*: skin] A mixture of long, straight-chain hydrocarbons and waxes secreted by the plant epidermis, providing a water-impermeable coating on aerial plant parts.

Cyanobacteria (sigh an´ o bacteria) [Gr. *kuanos*: the color blue] A division of photosynthetic bacteria, formerly referred to as blue-green algae; they lack sexual reproduction, and they use chlorophyll *a* in their photosynthesis.

Cyclic AMP See cAMP.

Cyclins Proteins that activate maturation-promoting factor, bringing about transitions in the cell cycle.

Cyst (sist) [Gr. *kystis*: pouch] (1) A resistant, thick-walled cell formed by some protists and other organisms. (2) An abnormal sac, containing a liquid or semisolid substance, produced in response to injury or illness.

Cytochromes (sy´ toe chromes) [Gr. *kytos*: container + *chroma*: color] Iron-containing red proteins, components of the electron-transfer chains in photophosphorylation and respiration.

Cytokinesis (sy´ toe kine ee´ sis) [Gr. *kytos*: container + *kinein*: to move] The division of the cytoplasm of a dividing cell. (Contrast with mitosis.)

Cytokinin (sy´ toe kine´ in) [Gr. *kytos*: container + *kinein*: to move] A member of a class of plant growth substances playing roles in senescence, cell division, and other phenomena.

Cytoplasm   The contents of the cell, excluding the nucleus.

Cytoplasmic determinants   In animal development, gene products whose spatial distribution may determine such things as embryonic axes.

Cytoskeleton   The network of microtubules and microfilaments that gives a eukaryotic cell its shape and its capacity to arrange its organelles and to move.

Cytosol   The fluid portion of the cytoplasm, excluding organelles and other solids.

Cytotoxic T cells   Cells of the cellular immune system that recognize and directly eliminate virus-infected cells. (Contrast with helper T cells, suppressor T cells.)

Dalton   See amu.

Deciduous (de sid´ you us) [L. *decidere*: fall off]   Referring to a plant that sheds its leaves at certain seasons. (Contrast with evergreen.)

Degeneracy   The situation in which a single amino acid may be represented by any of two or more different codons in messenger RNA. Most of the amino acids can be represented by more than one codon.

Degradative succession   Ecological succession occuring on the dead remains of the bodies of plants and animals, as when leaves or animal bodies rot.

Dehydration   See condensation reaction.

Deletion (genetic)   A mutation resulting from the loss of a continuous segment of a gene or chromosome. Such mutations never revert to wild-type. (Contrast with duplication, point mutation.)

Deme (deem) [Gr. *demos*: common people]   Any local population of individuals belonging to the same species and among which mating is random.

Demographic processes   The events—such as births, deaths, immigration, and emigration—that determine the number of individuals in a population.

Demographic stochasticity   Random variations in the factors influencing the size, density, and distribution of a population.

Demography   The study of dynamical changes in the sizes, densities, and distributions of populations.

Denaturation   Loss of activity of an enzyme or nucleic acid molecule as a result of structural changes induced by heat or other means.

Dendrite [Gr. *dendron*: a tree]   A fiber of a neuron which often cannot carry action potentials. Usually much branched and relatively short compared with the axon, and commonly carries information to the cell body of the neuron.

Denitrification   Metabolic activity by which inorganic nitrogen-containing ions are reduced to form nitrogen gas and other products; carried on by certain soil bacteria.

Density dependence   Change in the severity of action of agents affecting birth and death rates within populations. Such changes may be directly or inversely related to population density.

Density independence   The state where the severity of action of agents affecting birth and death rates within a population does not change with the density of the population.

Deoxyribonucleic acid   See DNA.

Depolarization   A change in the electric potential across a membrane from a condition in which the inside of the cell is more negative than the outside to a condition in which the inside is less negative, or even positive, with reference to the outside of the cell. (Contrast with hyperpolarization.)

Desmosome (dez´ mo sowm) [Gr. *desmos*: bond + *soma*: body]   An adhering junction between animal cells.

Derived trait   A trait found among members of a lineage that was not present in the ancestors of that lineage.

Dermal tissue system   The outer covering of a plant, consisting of epidermis in the young plant and periderm in a plant with extensive secondary growth. (Contrast with ground tissue system and vascular tissue system.)

Determinate cleavage   A pattern of early embryological development in which the potential of cells is determined very early such that separated cells develop only into partial embryos. (Contrast with indeterminate cleavage.)

Determination   Process whereby an embryonic cell or group of cells becomes fixed into a predictable developmental pathway.

Detritivore (di try´ ti vore) [L. *detritus*: worn away + *vorare*: to devour]   An organism that eats the dead remains of other organisms.

Deuterium   An isotope of hydrogen possessing one neutron in its nucleus. Deuterium oxide is called "heavy water."

Deuterostome   One of two major lines of evolution in animals, characterized by radial cleavage, enterocoelous development, and other traits.

Deuterium   An isotope of hydrogen, possessing one neutron in its nucleus; deuterium oxide is called "heavy water."

Development   Progressive change, as in structure or metabolism; in most kinds of organisms, development continues throughout the life of the organism.

Dialysis (dye ahl´ uh sis) [Gr. *dialyein*: separation]   The removal of ions or small molecules from a solution by their diffusion across a semipermeable membrane to a solvent where their concentration is lower.

Diaphragm (dye´ uh fram) [Gr. *diaphrassein*, to barricade]   (1) A sheet of muscle that separates the thoracic and abdominal cavities in mammals; responsible for the action of breathing. (2) A method of birth control in which a sheet of rubber is fitted over the woman's cervix, blocking the entry of sperm.

Diastole (dye ahs´ toll ee) [Gr.: dilation]   The portion of the cardiac cycle when the heart muscle relaxes. (Contrast with systole.)

Dicot (short for dicotyledon) [Gr. *dis*: two + *kotyledon*: a cup-shaped hollow]   Any member of the angiosperm class Dicotyledones, flowering plants in which the embryo produces two cotyledons prior to germination. Leaves of most dicots have major veins arranged in a branched or reticulate pattern.

Differentiation   Process whereby originally similar cells follow different developmental pathways. The actual expression of determination.

Diffuse coevolution   The situation in which the evolution of a lineage is influenced by its interactions with a number of species, most of which exert only a small influence on the evolution of the focal lineage.

Diffusion   Random movement of molecules or other particles, resulting in even distribution of the particles when no barriers are present.

Digestion   Enzyme-catalyzed process by which large, usually insoluble, molecules (foods) are hydrolyzed to form smaller molecules of soluble substances.

Dihybrid cross   A mating in which the parents differ with respect to the alleles of two loci of interest.

Dikaryon (di care´ ee ahn) [Gr. *dis*: two + *karyon*: kernel]   A cell or organism carrying two genetically distinguishable nuclei. Common in fungi.

Dioecious (die eesh´ us) [Gr.: two houses]   Organisms in which the two sexes are "housed" in two different individuals, so that eggs and sperm are not produced in the same individuals. Examples: humans, fruit flies, oak trees, date palms. (Contrast with monoecious.)

Diploblastic   Having two cell layers. (Contrast with triploblastic.)

Diploid (dip´ loid) [Gr. *diploos*: double]   Having a chromosome complement consisting of two copies (homologues) of each chromosome. A diploid individual (or cell) usually arises as a result of the fusion of two gametes, each with just one copy of each chromosome. Thus, the two homologues in each chromosome pair in a diploid cell are of separate origin, one derived from the female parent and one from the male parent.

Diplontic life cycle   A life cycle in which every cell except the gametes is diploid.

Directional selection   Selection in which phenotypes at one extreme of the population distribution are favored. (Contrast with disruptive selection; stabilizing selection.)

Disaccharide   A carbohydrate made up of two monosaccharides (simple sugars).

Dispersal stage   Stage in its life history at which an organism moves from its birthplace to where it will live as an adult.

Displacement activity Apparently irrelevant behavior performed by an animal under conflict situations, especially when tendencies to attack and escape are closely balanced.

Display A behavior that has evolved to influence the actions of other individuals.

Disruptive selection Selection in which phenotypes at both extremes of the population distribution are favored. (Contrast with directional selection; stabilizing selection.)

Distal Away from the point of attachment or other reference point. (Contrast with proximal.)

Disturbance A short-term event that disrupts populations, communities, or ecosystems by changing the environment.

Diverticulum (di ver tic´ u lum) [L. *divertere*: turn away] A small cavity or tube that connects to a major cavity or tube.

Division A term used by some microbiologists and formerly by botanists, corresponding to the term phylum.

DNA (deoxyribonucleic acid) The fundamental hereditary material of all living organisms. In eukaryotes, stored primarily in the cell nucleus. A nucleic acid using deoxyribose rather than ribose.

DNA hybridization A process by which DNAs from two species are mixed and heated so that interspecific double helixes are formed.

DNA ligase Enzyme that unites Okazaki fragments of the lagging strand during DNA replication; also mends breaks in DNA strands. It connects pieces of a DNA strand and is used in recombinant DNA technology.

DNA methylation Addition of methyl groups to DNA; plays role in regulation of gene expression; protects a bacterium's DNA against its restriction endonucleases.

DNA polymerase Any of a group of enzymes that catalyze the formation of DNA strands from a DNA template.

Dominance In genetic terminology, the ability of one allelic form of a gene to determine the phenotype of a heterozygous individual, in which the homologous chromosome carries both it and a different allele. For example, if *A* and *a* are two allelic forms of a gene, *A* is said to be dominant to *a* if *AA* diploids and *Aa* diploids are phenotypically identical and are distinguishable from *aa* diploids. The *a* allele is said to be recessive.

Dominance hierarchy The set of relationships within a group of animals, usually established and maintained by aggression, in which one individual has precedence over all others in eating, mating, and other activities; a second individual has precedence over all but the highest-ranking individual, and so on down the line.

Dormancy A condition in which normal activity is suspended, as in some seeds and buds.

Dorsal [L. *dorsum*: back] Pertaining to the back or upper surface. (Contrast with ventral.)

Double fertilization Process virtually unique to angiosperms in which one sperm nucleus combines with the egg to produce a zygote, and the other sperm nucleus combines with the two polar nuclei to produce the first cell of the triploid endosperm.

Double helix Of DNA: molecular structure in which two complementary polynucleotide strands, antiparallel to each other, form a right-handed spiral.

Duodenum (doo´ uh dee´ num) The beginning portion of the vertebrate small intestine. (Contrast with ileum, jejunum.)

Duplication (genetic) A mutation resulting from the introduction into the genome of an extra copy of a segment of a gene or chromosome. (Contrast with deletion, point mutation.)

Dynein [Gr. *dunamis*: power] A protein that undergoes conformational changes and thus plays a part in the movement of eukaryotic flagella and cilia.

Ear pinnae (pin´ ee) [L. wings] External ear structures that surround the auditory canals.

Ecdysone (eck die´ sone) [Gr. *ek*: out of + *dyo*: to clothe] In insects, a hormone that induces molting.

Echinoderm (e kine´ oh durm) A member of the phylum Echinodermata, such as a seastar or sea urchin.

Ecological biogeography The study of the distributions of organisms from an ecological perspective, usually concentrating on migration, dispersal, and species interactions.

Ecological community The species living together at a particular site.

Ecological niche (nitch) [L. *nidus*: nest] The functioning of a species in relation to other species and its physical environment.

Ecology [Gr. *oikos*: house + *logos*: discourse, study] The scientific study of the interaction of organisms with their environment, including both the physical environment and the other organisms that live in it.

Ecosystem (eek´ oh sis tum) The organisms of a particular habitat, such as a pond or forest, together with the physical environment in which they live.

Ecto- (eck´ toh) [Gr.: outer, outside] A prefix used to designate a structure on the outer surface of the body. For example, ectoderm. (Contrast with endo- and meso-.)

Ectoderm [Gr. *ektos*: outside + *derma*: skin] The outermost of the three embryonic tissue layers first delineated during gastrulation. Gives rise to the skin, sense organs, nervous system, etc.

Ectotherm [Gr. *ektos*: outside + *thermos*: heat] An animal unable to control its body temperature. (Contrast with endotherm.)

Edema (i dee´ mah) [Gr. *oidema*: swelling] Tissue swelling caused by the accumulation of fluid.

Edge effect The changes in ecological processes in a community caused by physical and biological factors originating in an adjacent community.

Effector Any organ, cell, or organelle that moves the organism through the environment or else alters the environment to the organism's advantage. Examples include muscle, bone, and a wide variety of exocrine glands.

Efferent [L. *ex*: out + *ferre*: to bear] Away from, as in neurons that conduct action potentials out from the central nervous system, or arterioles that conduct blood away from a structure. (Contrast with afferent.)

Egg In all sexually reproducing organisms, the female gamete; in birds, reptiles, and some other vertebrates, a structure witin which early embryonic development occurs.

Elasticity The property of returning quickly to a former state after a disturbance.

Electrocardiogram (EKG) A graphic recording of electrical potentials from the heart.

Electroencephalogram (EEG) A graphic recording of electrical potentials from the brain.

Electromyogram (EMG) A graphic recording of electrical potentials from muscle.

Electron (e lek´ tron) [L. *electrum*: amber (associated with static electricity), from Gr. *slektor*: bright sun (color of amber)] One of the three most important fundamental particles of matter, with mass approximately 0.00055 amu and charge −1.

Electron microscope An instrument that uses an electron beam to form images of minute structures; the transmission electron microscope is useful for thinly-sliced material, and the scanning electron microscope gives surface views of cells and organisms.

Electrophoresis (e lek´ tro fo ree´ sis) [L. *electrum*: amber + Gr. *phorein*: to bear] A separation technique in which substances are separated from one another on the basis of their electric charges and molecular weights.

Electrotonic potential In neurons, a hyperpolarization or small depolarization of the membrane potential induced by the application of a small electric current. (Contrast with action potential, resting potential.)

Elemental substance A substance composed of only one type of atom.

Embolus (em´ buh lus) [Gr. *embolos*: inserted object; stopper] A circulating blood clot. Blockage of a blood vessel by an embolus or by a bubble of gas is referred to as an **embolism**. (Contrast with thrombus.)

Embryo [Gr. *en-*: in + *bryein*: to grow] A young animal, or young plant sporophyte, while it is still contained within a protective structure such as a seed, egg, or uterus.

Embryo sac In angiosperms, the female gametophyte. Found within the ovule, it consists of eight or fewer cells, membrane bounded, but without cellulose walls between them.

Emergent property A property of a complex system that is not exhibited by its individual component parts.

Emigration The deliberate and usually oriented departure of an organism from the habitat in which it has been living.

Endemic (en dem´ ik) [Gr. *endemos*: dwelling in a place] Confined to a particular region, thus often having a comparatively restricted distribution.

Endergonic reaction One for which energy must be supplied. (Contrast with exergonic reaction.)

Endo- [Gr.: within, inside] A prefix used to designate an innermost structure. For example, endoderm, endocrine. (Contrast with ecto-, meso-.)

Endocrine gland (en´ doh krin) [Gr. *endon*: inside + *krinein*: to separate] Any gland, such as the adrenal or pituitary gland of vertebrates, that secretes certain substances, especially hormones, into the body through the blood.

Endocrinology The study of hormones and their actions.

Endocytosis A process by which liquids or solid particles are taken up by a cell through invagination of the plasma membrane. (Contrast with exocytosis.)

Endoderm [Gr. *endon*: within + *derma*: skin] The innermost of the three embryonic tissue layers first delineated during gastrulation. Gives rise to the digestive and respiratory tracts and structures associated with them.

Endodermis [Gr. *endon*: within + *derma*: skin] In plants, a specialized cell layer marking the inside of the cortex in roots and some stems. Frequently a barrier to free diffusion of solutes.

Endomembrane system Endoplasmic reticulum plus Golgi apparatus plus, when present, lysosomes; thus, a system of membranes that exchange material with one another.

Endometrium (en do mee´ tree um) [Gr. *endon*: within + *metrios*: womb] The epithelial cells lining the uterus of mammals.

Endoplasmic reticulum [Gr. *endon*: within + L. *plasma*: form; L. *reticulum*: little net] A system of membrane-bounded tubes and flattened sacs, often continuous with the nuclear envelope, found in the cytoplasm of eukaryotes. Exists as rough ER, studded with ribosomes, and smooth ER, lacking ribosomes.

Endorphins Naturally occurring, opiatelike substances in the mammalian brain.

Endoskeleton A skeleton covered by other, soft body tissues. (Contrast with exoskeleton.)

Endosperm [Gr. *endon*: within + *sperma*: seed] A specialized triploid seed tissue found only in angiosperms; contains stored food for the developing embryo.

Endosymbiosis [Gr. *endon*: within + *syn*: together + *bios*: life] The living together of two species, with one living inside the body (or even the cells) of the other.

Endosymbiotic theory Theory that the eukaryotic cell evolved from a prokaryote that contained other, endosymbiotic prokaryotes.

Endotherm [Gr. *endon*: within + *thermos*: hot] An animal that can control its body temperature by the expenditure of its own metabolic energy. (Contrast with ectotherm.)

Energetic cost The difference between the energy an animal would have expended had it rested, and that expended in performing a behavior.

Energy The capacity to do work.

Enhancer In eukaryotes, a DNA sequence, lying on either side of the gene it regulates, that stimulates a specific promoter.

Enkephalins [Gr. *en-*: in + *kephale*: head] Two of the endorphins. (See endorphin.)

Enterocoelous development A pattern of development in which the coelum is formed by an outpocketing of the embryonic gut (enteron).

Enterokinase (ent uh row kine´ ase) An enzyme secreted by the mucosa of the duodenum. It activates the zymogen trypsinogen to create the active digestive enzyme trypsin.

Entrainment With respect to circadian rhythms, the process whereby the period is adjusted to match the 24-hour environmental cycle.

Entropy (en´ tro pee) [Gr. *en*: in + *tropein*: to change] A measure of the degree of disorder in any system. A perfectly ordered system has zero entropy; increasing disorder is measured by positive entropy. Spontaneous reactions in a closed system are always accompanied by an increase in disorder and entropy.

Environment An organism's surroundings, both living and nonliving; includes temperature, light intensity, and all other species that influence the focal organism.

Enzyme (en´ zime) [Gr. *en*: in + *zyme*: yeast] A protein, on the surface of which are chemical groups so arranged as to make the enzyme a catalyst for a chemical reaction.

Eon The largest division of geological time.

Epi- [Gr.: upon, over] A prefix used to designate a structure located on top of another; for example: epidermis, epiphyte.

Epicotyl (epp´ i kot´ il) [Gr. *epi*: upon + *kotyle*: something hollow] That part of a plant embryo or seedling that is above the cotyledons.

Epidermis [Gr. *epi*: upon + *derma*: skin] In plants and animals, the outermost cell layers. (Only one cell layer thick in plants.)

Epididymis (epuh did´ uh mus) [Gr. *epi*: upon + *didymos*: testicle] Coiled tubules in the testes that store sperm and conduct sperm from the seiminiferous tubules to the vas deferens.

Epinephrine (ep i nef´ rin) [Gr. *epi*: upon + *nephros*: a kidney] The "fight or flight" hormone. Produced by the medulla of the adren-al gland, it also functions as a neurotransmitter. Also known as adrenaline.

Epiphyte (ep´ e fyte) [Gr. *epi*: upon + *phyton*: plant] A specialized plant that grows on the surface of other plants but does not parasitize them.

Episome A plasmid that may exist either free or integrated into a chromosome. (See plasmid.)

Epistasis An interaction between genes, in which the presence of a particular allele of one gene determines whether another gene will be expressed.

Epithelium In animals, a layer of cells covering or lining an external surface or a cavity.

Equilibrium (1) In biochemistry, a state in which forward and reverse reactions are proceeding at counterbalancing rates, so there is no observable change in the concentrations of reactants and products. (2) In evolutionary genetics, a condition in which allele and genotype frequencies in a population are constant from generation to generation.

Era The second largest division of geological time.

Error signal In physiology, the difference between a set-point and a feedback signal that results in a corrective response.

Erythrocyte (ur rith´ row sight) [Gr. *erythros*: red + *kytos*: hollow vessel] A red blood cell.

Esophagus (i soff´ i gus) [Gr. *oisophagos*: gullet] That part of the gut between the pharynx and the stomach.

Essential amino acid An amino acid an animal cannot synthesize for itself and must obtain from its diet.

Essential element An irreplaceable mineral element without which normal growth and reproduction cannot proceed.

Estivation (ess tuh vay´ shun) [L. *aestivalis*: summer] A state of dormancy and hypometabolism that occurs during the summer; usually a means of surviving drought and / or intense heat. Contrast with hibernation.

Estrogen Any of several steroid sex hormones, produced chiefly by the ovaries in mammals.

Estrous cycle The cyclical changes in reproductive physiology and behavior in female mammals (other than some primates), culminating in estrus.

Estrus (es´ truss) [L. *oestrus*: frenzy] The period of heat, or maximum sexual receptivity, in some female mammals. Ordinarily, the estrus is also the time of release of eggs in the female.

Ethology (ee thol´ o jee) [Gr. *ethos*: habit, custom + *logos*: discourse] The study of whole patterns of animal behavior in natural environments, stressing the analysis of adaptation and evolution of the patterns.

Ethylene One of the plant hormones, the gas $H_2C{=}CH_2$.

Etiolation Plant growth in the absence of light.

Eubacteria (yew bacteria) Kingdom including the great majority of bacteria, such as the gram negative bacteria, gram positive bacteria, mycoplasmas, etc. (Contrast with Archaebacteria.)

Euchromatin Chromatin that is diffuse and non-staining during interphase; may be transcribed. (Contrast with heterochromatin.)

Eukaryotes (yew car´ ry otes) [Gr. *eu*: true + *karyon*: kernel or nucleus] Organisms whose cells contain their genetic material inside a nucleus. Includes all life other than the viruses, Archaebacteria, and Eubacteria.

Eusocial Term applied to insects, such as termites, ants, and many bees and wasps, in which individuals cooperate in the care of offspring, there are sterile castes, and generations overlap.

Eutrophication (yoo trofe´ ik ay´ shun) [Gr. *eu*-: well + *trephein*: to flourish] The addition of nutrient materials to water. Especially in lakes, the subsequent flourishing of algae and microorganisms can result in oxygen depletion and the eventual stifling of life in the water.

Evergreen A plant that retains its leaves through all seasons. (Contrast with deciduous.)

Evolution Any gradual change. Organic evolution, often referred to as evolution, is any genetic and resulting phenotypic change in organisms from generation to generation.

Evolutionary agent Any factor that influences the direction and rate of evolutionary changes.

Evolutionary biology The collective branches of biology that study evolutionary process and their products—the diversity and history of living things.

Evolutionary conservative Traits of organisms that evolve very slowly.

Evolutionary radiation The proliferation of species within a single evolutionary lineage.

Excitatory postsynaptic potential (EPSP) A change in the resting potential of a postsynaptic membrane in a positive (depolarizing) direction. (Contrast with inhibitory postsynaptic potential.)

Excretion Release of metabolic wastes by an organism.

Exergonic reaction A reaction in which free energy is released. (Contrast with endergonic reaction.)

Exo- (eks´ oh) Same as ecto-.

Exocrine gland (eks´ oh krin) [Gr. *exo*: outside + *krinein*: to separate] Any gland, such as a salivary gland, that secretes to the outside of the body or into the gut.

Exocytosis A process by which a vesicle within a cell fuses with the plasma membrane and releases its contents to the outside. (Contrast with endocytosis.)

Exon A portion of a DNA molecule, in eukaryotes, that codes for part of a polypeptide. (Contrast with intron.)

Exoskeleton (eks´ oh skel´ e ton) A hard covering on the outside of the body; the exoskeleton of insects and other arthropods has many of the same functions as the bony internal skeleton of vertebrates. (Contrast with endoskeleton.)

Experiment A scientific method in which particular factors are manipulated while other factors are held constant so that the potential influences of the manipulated factors can be determined.

Exploitation competition Competition that occurs because resources are depleted. (Contrast with interference competition.)

Exponential growth Growth, especially in the number of organisms in a population, which is a simple function of the size of the growing entity: the larger the entity, the faster it grows. (Contrast with logistic growth.)

Expressivity The degree to which a genotype is expressed in the phenotype— may be affected by the environment.

Extensor A muscle the extends an appendage.

Extinction The termination of a lineage of organisms.

Extrinsic protein A membrane protein found only on the surface of the membrane. (Contrast with intrinsic protein.)

F_1 generation The immediate progeny of a mating; the first filial generation.

F_2 generation The immediate progeny of a mating between members of the F_1 generation.

F-duction Transfer of genes from one bacterium to another, using the F-factor as a vehicle.

F-factor In some bacteria, the fertility factor; a plasmid conferring "maleness" on the cell that contains it.

Facilitated diffusion Passive movement through a membrane involving a specific carrier protein; does not proceed against a concentration gradient. (Contrast with active transport, free diffusion.)

Facultative Capable of occurring or not occurring, as in facultative aerobes. (Contrast with obligate.)

Family In taxonomy, the category below the order and above the genus; a group of related, similar genera.

Fat A triglyceride that is solid at room temperature. (Contrast with oil.)

Fatty acid A molecule with a long hydrocarbon tail and a carboxyl group at the other end. Found in many lipids.

Fauna (faw´ nah) All of the animals found in a given area. (Contrast with flora.)

Feces [L. *faeces*: dregs] Waste excreted from the digestive system.

Feedback control Control of a particular step of a multistep process, induced by the presence or absence of a product of one of the later steps. A thermostat regulating the flow of heating oil to a furnace in a home is a negative feedback control device.

Fermentation (fur men tay´ shun) [L. *fermentum*: yeast] The degradation of a substance such as glucose to smaller molecules with the extraction of energy, without the use of oxygen (i.e., anaerobically). Involves the glycolytic pathway.

Fertilization Union of gametes. Also known as syngamy.

Fertilization membrane A membrane surrounding an animal egg which becomes rapidly raised above the egg surface within seconds after fertilization, serving to prevent entry of a second sperm.

Fetus The latter stages of an embryo that is still contained in an egg or uterus; in humans, the unborn young from the eighth week of pregnancy to the moment of birth.

Fiber An elongated and tapering cell of vascular plants, usually with a thick cell wall. Serves a support function.

Fibrin A protein that polymerizes to form long threads that provide structure to a blood clot.

Filter feeder An organism that feeds upon much smaller organisms, that are suspended in water or air, by means of a straining device.

Filtration In the excretory physiology of some animals, the process by which the initial urine is formed; water and most solutes are transferred into the excretory tract, while proteins are retained in the blood or hemolymph.

First law of thermodynamics Energy can be neither created nor destroyed.

Fission Reproduction of a prokaryote by division of a cell into two comparable progeny cells.

Fitness The contribution of a genotype or phenotype to the composition of subsequent generations, relative to the contribution of other genotypes or phenotypes. (See inclusive fitness.)

Fixed action pattern A behavior that is genetically programmed.

Flagellate (flaj´ el late) A member of the phylum Mastigophora, unicellular eukaryotes that propel themselves by flagella.

Flagellin (fla jell´ in) The protein from which prokaryotic (but not eukaryotic) flagella are constructed.

Flagellum (fla jell´ um) (plural: flagella) [L. *flagellum*: whip] Long, whiplike appendage that propels cells. Prokaryotic flagella differ sharply from those found in eukaryotes.

Flexor A muscle that flexes an appendage.

Flora (flore´ ah) All of the plants found in a given area. (Contrast with fauna.)

Florigen A plant hormone (not yet isolated) involved in the conversion of a vegetative shoot apex to a flower.

Flower   The total reproductive structure of an angiosperm; its basic parts include the calyx, corolla, stamens, and carpels.

Fluorescence   The emission of a photon of visible light by an excited atom or molecule.

Follicle [L. *folliculus*: little bag]   In female mammals, an immature egg surrounded by nutritive cells.

Follicle-stimulating hormone   A gonadotropic hormone produced by the anterior pituitary.

Food chain   A portion of a food web, most commonly a simple sequence of prey species and the predators that consume them.

Food web   The complete set of food links between species in a community; a diagram indicating which ones are the eaters and which are consumed.

Forb   Any broad-leaved (dicotyledonous), herbaceous plant. Especially applied to such plants growing in grasslands.

Fossil   Any recognizable structure originating from an organism, or any impression from such a structure, that has been preserved over geological time.

Founder effect   Random changes in allele frequencies resulting from establishment of a population by a very small number of individuals.

Fovea [L. *fovea*; a small pit]   The area, in the vertebrate retina, of most distinct vision.

Frame-shift mutation   A mutation resulting from the addition or deletion of a single base pair in the DNA sequence of a gene. As a result of this, mRNA transcribed from such a gene is translated normally until the ribosome reaches the point at which the mutation has occurred. From that point on, codons are read out of proper register and the amino acid sequence bears no resemblance to the normal sequence. (Contrast with missense mutation, nonsense mutation.)

Free diffusion   Diffusion directly across a membrane without the involvement of carrier molecules. Free diffusion is not saturable and cannot cause the net transport from a region of low concentration to a region of higher concentration. (Contrast with facilitated diffusion and active transport.)

Free energy   That energy which is available for doing useful work, after allowance has been made for the increase or decrease of disorder. Designated by the symbol G (for Gibbs free energy), and defined by: $G = H - TS$, where H = heat, S = entropy, and T = absolute (Kelvin) temperature.

Frequency-dependent selection   Selection that changes in intensity when the proportion of individuals under selection increases or decreases.

Fruit   In angiosperms, a ripened and mature ovary (or group of ovaries) containing the seeds. Sometimes applied to reproductive structures of other groups of plants, and includes any adjacent parts which may be fused with the reproductive structures.

Fruiting body   A structure that bears spores.

Fundamental niche   The range of condition under which an organism could survive if it were the only one in the environment. (Contrast with realized niche.)

Fungus (fung´ gus)   A member of the kingdom Fungi, a (usually) multicellular eukaryote with absorptive nutrition.

G₁ phase   In the cell cycle, the gap between the end of mitosis and the onset of the S phase.

G₂ phase   In the cell cycle, the gap between the S (synthesis) phase and the onset of mitosis.

G-protein   A membrane protein involved in signal transduction; characterized by binding guanyl nucleotides. The activation of certain receptors activates the G-protein, which in turn activates adenylate cyclase. G-protein activation involves binding a GTP molecule in place of a GDP molecule.

Gametangium (gam i tan´ gee um) [Gr. *gamos*: marriage + *angeion*: vessel or reservoir]   Any plant or fungal structure within which a gamete is formed.

Gamete (gam´ eet) [Gr. *gamete*: wife, *gametes*: husband]   The mature sexual reproductive cell: the egg or the sperm.

Gametocyte (ga meet´ oh site) [Gr. *gamete*: wife, *gametes*: husband + *kytos*: cell]   The cell that gives rise to sex cells, either the eggs or the sperm. (See oocyte and spermatocyte.)

Gametogenesis (ga meet´ oh jen´ e sis) [Gr. *gamete*: wife, *gametes*: husband + *genesis*: source]   The specialized series of cellular divisions that leads to the production of sex cells (gametes). (Contrast with oogenesis and spermatogenesis.)

Gametophyte (ga meet´ oh fyte)   In plants with alternation of generations, the haploid phase that produces the gametes. (Contrast with sporophyte.)

Ganglion (gang´ glee un) [Gr.: tumor]   A group or concentration of neuron cell bodies.

Gap junction   A 2.7-nanometer gap between plasma membranes of two animal cells, spanned by protein channels. Gap junctions allow chemical substances or electrical signals to pass from cell to cell.

Gas exchange   In animals, the process of taking up oxygen from the environment and releasing carbon dioxide to the environment.

Gastrovascular cavity   Serving for both digestion (gastro) and circulation (vascular); in particular, the central cavity of the body of jellyfish and other cnidarians.

Gastrula (gas´ true luh) [Gr. *gaster*: stomach]   An embryo forming the characteristic three cell layers (ectoderm, endoderm, and mesoderm) which will give rise to all of the major tissue systems of the adult animal.

Gastrulation   Development of a blastula into a gastrula.

Gated channel   A channel (membrane protein) that opens and closes in response to binding of specific molecules or to changes in membrane potential.

Gene [Gr. *gen*: to produce]   A unit of heredity. Used here as the unit of genetic function which carries the information for a single polypeptide.

Gene amplification   Creation of multiple copies of a particular gene, allowing the production of large amounts of the RNA transcript (as in rRNA synthesis in oocytes).

Gene cloning   Formation of a clone of bacteria or yeast cells containing a particular foreign gene.

Gene family   A set of identical, or once-identical, genes, derived from a single parent gene; need not be on the same chromosomes; classic example is the globin family in vertebrates.

Gene flow   The exchange of genes between different species (an extreme case referred to as hybridization) or between different populations of the same species caused by migration following breeding.

Gene pool   All of the genes in a population.

Gene therapy   Treatment of a genetic disease by providing patients with cells containing wild-type alleles for the genes that are nonfunctional in their bodies.

Generative nucleus   In a pollen tube, a haploid nucleus that undergoes mitosis to produce the two sperm nuclei that participate in double fertilization. (Contrast with tube nucleus.)

Generator potential   A stimulus-induced change in membrane resting potential in the direction of threshold for generating action potentials.

Genet   The genetic individual of a plant that is composed of a number of nearly identical but repeated units.

Genetic drift   Changes in gene frequencies from generation to generation in a small population as a result of random processes.

Genetic stochasticity   Variation in the frequencies of alleles and genotypes in a population over time.

Genetics   The study of heredity.

Genetic structure   The frequencies of alleles and genotypes in a population.

Genome (jee´ nome)   The genes in a complete haploid set of chromosomes.

Genome project   An effort to map and sequence the entire genome of a species.

Genotype (jean´ oh type) [Gr. *gen*: to produce + *typos*: impression]   An exact description of the genetic constitution of an individual, either with respect to a single trait or with respect to a larger set of traits. (Contrast with phenotype.)

Genus (jean´ us) (plural: genera) [Gr. *genos*: stock, kind]   A group of related, similar species.

Geographical (allopatric) speciation Formation of two species from one by the interposition of (or crossing of) a physical barrier. (Contrast with parapatric, sympatric speciation.)

Geotropism See gravitropism.

Germ cell A reproductive cell or gamete of a multicellular organism.

Germination The sprouting of a seed or spore.

Gestation (jes tay′ shun) [L. *gestare*: to bear] The period during which the embryo of a mammal develops within the uterus. Also known as **pregnancy**.

Gibberellin (jib er el′ lin) [L. *gibberella*: hunchback (refers to shape of a reproductive structure of a fungus that produces gibberellins)] One of a class of plant growth substances playing roles in stem elongation, seed germination, flowering of certain plants, etc. Named for the fungus *Gibberella*.

Gill An organ for gas exchange in aquatic organisms.

Gill arch A skeletal structure that supports gill filaments and the blood vessels that supply them.

Gizzard (giz′ erd) [L. *gigeria*: cooked chicken parts] A very muscular port of the stomach of birds that grinds up food, sometimes with the aid of fragments of stone.

Gland An organ or group of cells that produces and secretes one or more substances.

Glans penis Sexually sensitive tissue at the tip of the penis.

Glia (glee′ uh) [Gr.: glue] Cells, found only in the nervous system, which do not conduct action potentials.

Glomerulus (glo mare′ yew lus) [L. *glomus*: ball] Sites in the kidney where blood filtration takes place. Each glomerulus consists of a knot of capillaries served by afferent and efferent arterioles.

Glucocorticoids Steroid hormones produced by the adrenal medulla. Secreted in response to ACTH, they inhibit glucose uptake by many tissues in addition to mediating other stress responses.

Glucagon A hormone produced and released by cells in the islets of Langerhans of the pancreas. It stimulates the breakdown of glycogen in liver cells.

Gluconeogenesis The biochemical synthesis of glucose from other substances, such as amino acids, lactate, and glycerol.

Glucose (glue′ kose) [Gr. *gleukos*: sweet wine mash for fermentation] The most common sugar, one of several monosaccharides with the formula $C_6H_{12}O_6$.

Glycerol (gliss′ er ole) A three-carbon alcohol with three hydroxyl groups, the linking component of phospholipids and triglycerides.

Glycogen (gly′ ko jen) A branched-chain polymer of glucose, similar to starch (which is less branched and may be of lower molecular weight). Exists mostly in liver and muscle;

the principal storage carbohydrate of most animals and fungi.

Glycolysis (gly kol′ li sis) [from glucose + Gr. *lysis*: loosening] The enzymatic breakdown of glucose to pyruvic acid. One of the oldest energy-yielding machanisms in living organisms.

Glycosidic linkage The connection in an oligosaccharide or polysaccharide chain, formed by removal of water during the linking of monosaccharides.by root pressure.

Glyoxysome (gly ox′ ee soam) A type of microbody, found in plants, in which stored lipids are converted to carbohydrates.

Golgi apparatus (goal′ jee) A system of concentrically folded membranes found in the cytoplasm of eukaryotic cells. Plays a role in the production and release of secretory materials such as the digestive enzymes manufactured in the pancreas. First described by Camillo Golgi (1844–1926).

Gonad (go′ nad) [Gr. *gone*: seed, that which produces seed] An organ that produces sex cells in animals: either an ovary (female gonad) or testis (male gonad).

Gonadotropin A hormone that stimulates the gonads.

Grade The level of complexity found in an animal's body plan.

Gram stain A differential stain useful in characterizing bacteria.

Granum Within a chloroplast, a stack of thylakoids.

Gravitropism A directed plant growth response to gravity.

Grazer An animal that eats the vegetative tissues of herbaceous plants.

Green gland An excretory organ of crustaceans.

Gross morphology The sizes and shapes of the major body parts of a plant or animal.

Gross primary production The total energy captured by plants growing in a particular area.

Ground meristem That part of an apical meristem that gives rise to the ground tissue system of the primary plant body.

Ground tissue system Those parts of the plant body not included in the dermal or vascular tissue systems. Ground tissues function in storage, photosynthesis, and support.

Groundwater Water present deep in soils and rocks; may be stationary or flow slowly eventually to discharge into lakes, rivers, or oceans.

Group transfer The exchange of atoms between molecules.

Growth Irreversible increase in volume (probably the most accurate definition, but at best a dangerous oversimplification).

Growth factors A group of proteins that circulate in the blood and trigger the normal growth of cells. Each growth factor acts only on certain target cells.

Growth stage That stage in the life history of an organism in which it grows to its adult size.

Guard cells In plants, paired epidermal cells which surround and control the opening of a stoma (pore).

Gut An animal's digestive tract.

Guttation The extrusion of liquid water through openings in leaves, caused by root pressure.

Gymnosperm (jim′ no sperm) [Gr. *gymnos*: naked + *sperma*: seed] A plant, such as a pine or other conifer, whose seeds do not develop within an ovary (hence, the seeds are "naked").

Habit The form or pattern of growth characteristic of an organism.

Habitat The environment in which an organism lives.

Habituation (ha bich′ oo ay shun) The simplest form of learning, in which an animal presented with a stimulus without reward or punishment eventually ceases to respond.

Hair cell A type of mechanosensor in animals.

Half-life The time required for half of a sample of a radioactive isotope to decay to its stable, nonradioactive form.

Halophyte (hal′ oh fyte) [Gr. *halos*: salt + *phyton*: plant] A plant that grows in a saline (salty) environment.

Haploid (hap′ loid) [Gr. *haploeides*: single] Having a chromosome complement consisting of just one copy of each chromosome. This is the normal "ploidy" of gametes or of asexual spores produced by meiosis or of organisms (such as the gametophyte generation of plants) that grow from such spores without fertilization.

Haplontic life cycle A life cycle in which the zygote is the only diploid cell.

Hardy–Weinberg rule The rule that the basic processes of Mendelian heredity (meiosis and recombination) do not alter either the frequencies of genes or their diploid combinations. The Law also states how the percentages of diploid combinations can be predicted from a knowledge of the proportions of alleles in the population.

Haustorium (haw stor′ ee um) [L. *haustus*: draw up] A specialized hypha or other structure by which fungi and some parasitic plants draw food from a host plant.

Haversian systems Units of organization in compact bone that reflect the action of intercommunicating osteoblasts.

Helper T cells T cells that participate in the activation of B cells and of other T cells; targets of the HIV-I virus, the agent of AIDS. (Contrast with cytotoxic T cells, suppressor T cells.)

Hematocrit (heme at o krit) [Gr. *haima*: blood + *krites*: judge] The proportion of 100 cc of blood that consists of red blood cells.

Hemizygous(hem´ ee zie´ gus) [Gr. *hemi*: half + *zygotos*: joined]   In a diploid organism, having only one allele for a given trait, typically the case for X-linked genes in male mammals and Z-linked genes in female birds. (Contrast with homozygous, heterozygous.)

Hemoglobin (hee´ mo glow´ bin) [Gr. *haima*: blood + L. *globus*: globe]   The colored protein of vertebrate blood (and blood of some invertebrates) which transports oxygen.

Hepatic (heh pat´ ik) [Gr. *hepar*: liver]   Pertaining to the liver.

Hepatic duct   The duct that conveys bile from the liver to the gallbladder.

Herbicide (ur´ bis ide)   A chemical substance that kills plants.

Herbivore [L. *herba*: plant + *vorare*: to devour]   An animal which eats the tissues of plants. (Contrast with carnivore, detritivore, omnivore.)

Heritable   Able to be inherited; in biology usually refers to genetically determined traits.

Hermaphroditism (her maf´ row dite´ ism) [Gr. *hermaphroditos*: a person with both male and female traits]   The coexistence of both female and male sex organs in the same organism.

Hertz (abbreviated as Hz)   Cycles per second.

Hetero- [Gr.: other, different]   A prefix used in biology to mean that two or more different conditions are involved; for example, heterotroph, heterozygous.

Heterochromatin   Chromatin that retains its coiling during interphase; generally not transcribed. (Contrast with euchromatin.)

Heterocyst   A large, thick-walled cell in the filaments of certain cyanobacteria; performs nitrogen fixation.

Heterogeneous nuclear RNA (hnRNA)   The product of transcription of a eukaryotic gene, including transcripts of introns.

Heterokaryon (het´ er oh care´ ee ahn) [Gr. *heteros*: different + *karyon*: kernel]   A cell or organism carrying a mixture of genetically distinguishable nuclei. A heterokaryon is usually the result of the fusion of two cells without fusion of their nuclei.

Heteromorphic (het´ er oh more´ fik) [Gr. *heteros*: different + *morphe*: form]   having a different form or appearance, as two heteromorphic life stages of a plant. (Contrast with isomorphic.)

Heterosporous (het´ er os´ por us)   Producing two types of spores, one of which gives rise to a female megaspore and the other to a male microspore. Heterosporous plants produce distinct female and male gametophytes. (Contrast with homosporous.)

Heterotherm   An animal that regulates its body temperature at a constant level at some times but not others, such as a hibernator.

Heterotroph (het´ er oh trof) [Gr. *heteros*: different + *trophe*: food]   An organism that

requires preformed organic molecules as food. (Contrast with autotroph.)

Heterozygous (het´ er oh zie´ gus) [Gr. *heteros*: different + *zygotos*: joined]   Of a diploid organism having different alleles of a given gene on the pair of homologues carrying that gene. (Contrast with homozygous.)

Hexose   A six-carbon sugar, such as glucose or fructose.

Hfr (for "high frequency of recombination")   Donor bacterium in which the F-factor has been integrated into the chromosome. This produces a bacterium that transfers its chromosomal markers at a very high frequency to recipient (F⁻) cells.

Hibernation [L. *hibernus*: winter]   The state of inactivity of some animals during winter; marked by a drop in body temperature and metabolic rate.

Hippocampus   A part of the forebrain that takes part in long-term memory formation.

Histamine (hiss; tah meen)   A substance released within a damaged tissue by a type of white blood cell. Histamines are responsible for aspects of allergice reactions, including the increased vascular permeability that leads to edema (swelling).

Histology   The study of tissues.

Histone   Any one of a group of basic proteins forming the core of a nucleosome, the structural unit of a eukaryotic chromosome. (See nucleosome.)

Historical biogeography   The study of the distributions of organisms from a long-term, historical perspective.

hnRNA   See heterogeneous nuclear RNA.

Holdfast   In many large attached algae, specialized tissue attaching the plant to its substratum.

Homeobox   A segment of DNA, found in a few genes, perhaps regulating the expression of other genes and thus controlling large-scale developmental processes.

Homeostasis (home´ ee o sta´ sis) [Gr. *homos*: same + *stasis*: position]   The maintenance of a steady state, such as a constant temperature or a stable social structure, by means of physiological or behavioral feedback responses.

Homeotherm (home´ ee o therm) [Gr. *homos*: same + *therme*: heat]   An animal that maintains a constant body temperature by virtue of its own heating and cooling mechanisms. (Contrast with heterotherm, poikilotherm.)

Homeotic genes (home´ ee ott´ ic)   Genes that determine what entire segments of an animal become.

Homeotic mutation   A drastic mutation causing the transformation of body parts in *Drosophila* metamorphosis. Examples include the *Antennapedia* and *ophthalmoptera* mutants.

Homolog (home´ o log´) [Gr. *homos*: same + *logos*: word]   One of a pair, or larger set, of chromosomes having the same overall genetic composition and sequence. In diploid organisms, each chromosome inherited from one parent is matched by an identical (except

for mutational changes) chromosome—its homolog—from the other parent.

Homology (ho mol´ o jee) [Gr. *homologi(a)*: agreement]   A similarity between two structures that is due to inheritance from a common ancestor. The structures are said to be homologous. (Contrast with analogy.)

Homoplasy (home´ uh play zee) [Gr. *homos*: same + *plastikos*: to mold]   The presence in several species of a trait not present in their most common ancestor. Can result from convergent evolution, reverse evolution, or parallel evolution.

Homosporous   Producing a single type of spore that gives rise to a single type of gametophyte, bearing both female and male reproductive organs. (Contrast with heterosporous.)

Homozygous (home´ o zie´ gus) [Gr. *homos*: same + *zygotos*: joined]   Of a diploid organism having identical alleles of a given gene on both homologous chromosomes. An organism may be a "homozygote" with respect to one gene and, at the same time, a "heterozygote" with respect to another. (Contrast with heterozygous.)

Hormone (hore´ mone) [Gr. *hormon*: excite, stimulate]   A substance produced in one part of a multicellular organism and transported to another part where it exerts its specific effect on the physiology or biochemistry of the target cells.

Host   An organism that harbors a parasite and provides it with nourishment.

Host–parasite interaction   The dynamic interaction between populations of a host and the parasites that attack it.

Humoral immune system   The part of the immune system mediated by B cells; it is mediated by circulating antibodies and is active against extracellular bacterial and viral infections.

Humus (hew´ muss)   The partly decomposed remains of plants and animals on the surface of a soil. Its characteristics depend primarily upon climate and the species of plants growing on the site.

Hyaluronidase (hill yew ron´ uh dase)   An enzyme that digests proteoglycans. Found in sperm cells, it helps digest the coatings surrounding an egg so the sperm can penetrate the egg cell membrane.

Hybrid (high´ brid) [L. *hybrida*: mongrel]   The offspring of genetically dissimilar parents.

Hybridoma   A cell produced by the fusion of an antibody-producing cell with a myeloma cell; it produces monoclonal antibodies.

Hydrocarbon   A compound containing only carbon and hydrogen atoms.

Hydrogen bond   A chemical bond which arises from the attraction between the slight positive charge on a hydrogen atom and a slight negative charge on a nearby fluorine, oxygen, or nitrogen atom. Weak bonds, but found in great quantities in proteins, nucleic acids, and other biological macromolecules.

Hydrological cycle   The sum total of movement of water from the oceans to the atmosphere, to the soil, and back to the oceans. Some water is cycled many times within compartments of the system before completing one full circuit.

Hydrolyze (hi´ dro lize) [Gr. *hydro*: water + *lysis*: cleavage]   To break a chemical bond, as in a peptide linkage, with the insertion of the components of water, –H and –OH, at the cleaved ends of a chain. The digestion of proteins is a hydrolysis.

Hydrophilic [Gr. *hydro*: water + *philia*: love]   Having an affinity for water. (Contrast with hydrophobic.)

Hydrophobic [Gr. *hydro*: water + *phobia*: fear]   Molecules and amino acid side chains, which are mainly hydrocarbons (compounds of C and H with no charged groups or polar groups), have a lower energy when they are clustered together than when they are distributed through an aqueous solution. Because of their attraction for one another and their reluctance to mix with water they are called "hydrophobic." Oil is a hydrophobic substance; phenylalanine is a hydrophobic animo acid in a protein. (Contrast with hydrophilic.)

Hydrophobic interaction   A weak attraction between highly nonpolar molecules or parts of molecules suspended in water.

Hydrostatic skeleton   The incompressible internal liquids of some animals that transfer forces from one part of the body to another when acted upon by the surrounding muscles.

Hydroxyl group   The –OH group, characteristic of alcohols.

Hymenopteran (high´ man op´ ter an)   A member of the insect order Hymenoptera, such as a wasp, bee, or ant.

Hyperpolarization   A change in the resting potential of a membrane so the inside of a cell becomes more electronegative. (Contrast with depolarization.)

Hypertension   High blood pressure.

Hypertonic [Gr.: higher tension]   Having a more negative osmotic potential, as a result of having a higher concentration of osmotically active particles. Said of one solution as compared with another. (Contrast with hypotonic, isotonic.)

Hypha (high´ fuh) (plural: hyphae) [Gr. *hyphe*: web]   In the fungi, any single filament. May be multinucleate (zygomycetes, ascomycetes) or multicellular (basidiomycetes).

Hypocotyl   That part of the embryonic or seedling plant shoot that is below the cotyledons.

Hypothalamus   The part of the brain lying below the thalamus; it coordinates water balance, reproduction, temperature regulation, and metabolism.

Hypothetico-deductive method   A method of science in which hypotheses are erected, predictions are made from them, and experiments and observations are performed to test the predictions. The process may be repeated many times in the course of answering a question.

Hypotonic [Gr.: lower tension]   Having a less negative osmotic potential, as a result of having a lower concentration of osmotically active particles. Said of one solution as compared with another. (Contrast with hypertonic, isotonic.)

Imaginal disc   In insect larvae, groups of cells that develop into specific adult organs.

Imbibition [L. *imbibo*: to drink]   The binding of a solvent to another molecule. Dry starch and protein will imbibe water.

Immunoglobulins   A class of proteins, with a characteristic structure, active as receptors and effectors in the immune system.

Immunological tolerance   A mechanism by which an animal does not mount an immune response to the antigenic determinants of its own macromolecules.

Imprinting   A rapid form of learning, in which an animal comes to make a particular response, which is maintained for life, to some object or other organism.

Inclusive fitness   The sum of an individual's own fitness (the effect of producing its own offspring: the individual selection component) plus its influence on fitness in relatives other than direct descendants (the kin selection component).

Incomplete dominance   Condition in which the heterozygous phenotype is intermediate between the two homozygous phenotypes.

Incomplete metamorphosis   Insect development in which changes between instars are gradual.

Incus (in´ kus) [L. *incus*: anvil]   The middle of the three bones that conduct movements of the eardrum to the oval window of the inner ear. (See malleus, stapes.)

Indeterminate cleavage   A pattern of development in which individual cells retain the potential to develop into complete organisms if separated from one another well into development.

Individual fitness   That component of inclusive fitness that results from an organism producing its own offspring. (Contrast with kin selection component.)

Indoleacetic acid   See auxin.

Induced fit   A change in the tertiary structures of some enzymes, caused by binding of substrate to the active site.

Inducer   In enzyme systems, a small molecule which, when added to a growth medium, causes a large increase in the level of some enzyme. Generally it acts by binding to repressor and changing its conformation so that the repressor does not bind to the operator. In embryology, a substance that causes a group of target cells to differentiate in a particular way.

Inducible enzyme   An enzyme that is present in much larger amounts when a particular compound (the inducer) has been added to the system. (Contrast with constitutive enzyme.)

Inflammation   A nonspecific defense against pathogens; characterized by redness, swelling, pain, and increased temperature.

Inflorescence   A structure composed of several flowers.

Ingestion   Taking in of food by swallowing.

Inhibitor   A substance which binds to the surface of an enzyme and interferes with its action on its substrates.

Inhibitory postsynaptic potential   A change in the resting potential of a postsynaptic membrane in the hyperpolarizing (negative) direction.

Initiation complex   Combination of a ribosomal light subunit, an mRNA molecule, and the tRNA charged with the first amino acid coded for by the mRNA; formed at the onset of translation.

Inositol triphosphate (IP3)   An intracellular second messenger derived from membrane phospholipids.

Insertion sequence   A large piece of DNA that can give rise to copies at other loci; a type of transposable genetic element.

Instar (in´ star) [L.: image, form]   An immature stage of an insect between molts.

Instinct   Behavior that is relatively highly sterotyped and self-differentiating, that develops in individuals unable to observe other individuals performing the behavior or to practice the behavior in the presence of the objects toward which it is usually directed.

Insulin (in´ su lin) [L. *insula*: island]   A hormone, synthesized in islet cells of the pancreas, that promotes the conversion of glucose to the storage material, glycogen.

Integrase   An enzyme that integrates retroviral cDNA into the genome of the host cell.

Integrated pest management   A method of control of pests in which natural predators and parasites are used in conjunction with sparing use of chemical methods to achieve control of a pest without causing serious adverse environmental side effects.

Integument [L. *integumentum*: covering]   A protective surface structure. In gymnosperms and angiosperms, a layer of tissue around the ovule which will become the seed coat. Gymnosperm ovules have one integument, angiosperm ovules two.

Intention movement   The preparatory motions that animals go through prior to a complete behavior response; for example, the crouch before flying, the snarl before biting, etc.

Intercalary meristem   A meristematic region in plants which occurs not apically, but between two regions of mature tissue. Intercalary meristems occur in the nodes of grass stems, for example.

Intercostal muscles Muscles between the ribs that can augment breathing movements by elevating and suppressing the rib cage.

Interference competition Competition resulting from direct behavioral interactions between organisms. (Contrast with exploitation competition.)

Interferon A glycoprotein produced by virus-infected animal cells; increases the resistance of neighboring cells to the virus.

Interkinesis The phase between the first and second meiotic divisions.

Interleukins Regulatory proteins, produced by macrophages and lymphocytes, that act upon other lymphocytes and direct their development.

Intermediate filaments Fibrous proteins that stabilize cell structure and resist tension.

Internode Section between two nodes of a plant stem.

Interphase The period between successive nuclear divisions during which the chromosomes are diffuse and the nuclear envelope is intact. It is during this period that the cell is most active in transcribing and translating genetic information.

Interspecific competition Competition between members of two or more species.

Interstitial fluid In vertebrates, the fluid filling the spaces between cells.

Intertropical convergence zone The tropical region where the air rises most strongly; moves north and south with the passage of the sun overhead.

Intraspecific competition Competition among members of a single species.

Intrinsic protein A membrane protein that is embedded in the phospholipid bilayer of the membrane. (Contrast with extrinsic protein.)

Intrinsic rate of increase The rate at which a population can grow when its density is low and environmental conditions are highly favorable.

Intron A portion of a DNA molecule that, because of RNA splicing, is not involved in coding for part of a polypeptide molecule. (Contrast with exon.)

Invagination An infolding.

Invasiveness Ability of a bacterium to multiply within the body of a host.

Inversion (genetic) A rare mutational event that leads to the reversal of the order of genes within a segment of a chromosome, as if that segment had been removed from the chromosome, turned 180°, and then reattached.

Invertebrate Any animal that is not a vertebrate, that is, whose nerve cord is not enclosed in a backbone of bony segments.

In vitro [L.: in glass] In a test tube, rather than in a living organism. (Contrast with in vivo.)

In vivo [L.: in the living state] In a living organism. Many processes that occur in vivo can be reproduced in vitro with the right selection of cellular components. (Contrast with in vitro.)

Ion (eye´ on) [Gr.: wanderer] An atom or group of atoms with electrons added or removed, giving it a negative or positive electrical charge.

Ionic channel A membrane protein that can let ions pass across the membrane. The channel can be ion-selective, and it can be voltage-gated or ligand-gated.

Ionic bond A chemical bond which arises from the electrostatic attraction between positively and negatively charged ions. Usually a strong bond.

Iris (eye´ ris) [Gr. iris: rainbow] The round, pigmented membrane that surrounds the pupil of the eye and adjusts its aperture to regulate the amount of light entering the eye.

Irruption A rapid increase in the density of a population. Often followed by massive emigration.

Islets of Langerhans Clusters of hormone-producing cells in the pancreas.

Isogamy (eye sog´ ah mee) [Gr. isos: equal + gamos: marriage] A kind of sexual reproduction in which the gametes (or gametangia) are not distinguishable on the basis of size or morphology.

Isolating mechanism Geographical, physiological, ecological, or behavioral mechanisms that lead to a reduction in the frequency of hybrid matings.

Isomers Molecules consisting of the same numbers and kinds of atoms, but differing in the way in which the atoms are combined.

Isomorphic (eye´ so more´ fik) [Gr. isos: equal + morphe: form] having the same form or appearance, as two isomorphic life stages. (Contrast with heteromorphic.)

Isotonic [Gr.: same tension] Having the same osmotic potential. Said of two solutions. (Contrast with hypertonic, hypotonic.)

Isotope (eye´ so tope) [Gr. isos: equal + topos: place] Two isotopes of the same chemical element have the same number of protons in their nuclei, but differ in the number of neutrons.

Isozymes Chemically different enzymes that catalyze the same reaction.

Jejunum (jih jew´ num) The middle division of the small intestine, where most absorption of nutrients occurs. (See duodenum, ileum.)

Joule (jool, or jowl) A unit of energy, equal to 0.24 calories.

Juvenile hormone In insects, a hormone maintaining larval growth and preventing maturation or pupation.

Karyogamy (care´ ee og´ uh me) [Gr. karyon: kernel, nut + gamos: marriage] Fusion of gamete nuclei.

Karyotype The number, forms, and types of chromosomes in a cell.

Kelvin temperature scale See absolute temperature scale.

Keratin (ker´ a tin) [Gr. keras: horn] A protein which contains sulfur and is part of such hard tissues as horn, nail, and the outermost cells of the skin.

Ketone (key´ tone) A compound with a C=O group attached to two other groups, neither of which is an H atom. Many sugars are ketones. (Contrast with aldehyde.)

Keystone species A species that exerts a major influence on the composition and dynamics of the community in which it lives.

Kidneys A pair of excretory organs in vertebrates.

Kin selection component The component of inclusive fitness resulting from helping the survival of relatives containing the same alleles by descent from a common ancestor.

Kinase (kye´ nase) An enzyme that transfers a phosphate group from ATP to another molecule. Protein kinases transfer phosphate from ATP to specific proteins, playing important roles in cell regulation.

Kinesis (ki nee´ sis) [Gr.: movement] Orientation behavior in which the organism does not move in a particular direction with reference to a stimulus but instead simply moves at an increasing or decreasing rate until it ends up farther from the object or closer to it. (Contrast with taxis.)

Kinetochore (kin net´ oh core) [Gr. kinetos: moving + khorein: to move] Specialized structure on a centromere to which microtubules attach.

Kingdom The highest taxonomic category in the Linnaean system.

Knockout mouse A genetically engineered mouse in which one or more functioning alleles have been replaced by defective alleles.

Lactic acid The end product of fermentation in vertebrate muscle and some microorganisms.

Lagging strand In DNA replication, the daughter strand that is synthesized discontinuously.

Lamella Layer.

Larynx (lar´ inks) A structure between the pharynx and the trachea that includes the vocal cords.

Larva (plural: larvae) [L.: ghost, early stage] An immature stage of any invertebrate animal that differs dramatically in appearance from the adult.

Lateral Pertaining to the side.

Lateral inhibition In visual information processing in the arthropod eye, the mutual inhibition of optic nerve cells; results in enhanced detection of edges.

Laterization (lat´ ur iz ay shun) The formation of a nutrient-poor soil that is rich in insoluble iron and aluminum compounds.

Law of independent assortment Alleles of different, unlinked genes assort independently of one another during gamete formation, Mendel's second law.

Law of segregation Alleles segregate from one another during gamete formation, Mendel's first law.

Leader sequence A sequence of amino acids at the N-terminal end of a newly synthesized protein, determining where the protein will be placed in the cell.

Leading strand In DNA replication, the daughter strand that is synthesized continuously.

Leaf axil The upper angle between a leaf and the stem, site of lateral buds which under appropriate circumstances become activated to form lateral branches.

Leaf primordium [L.: the beginning] A small mound of cells on the flank of a shoot apical meristem that will give rise to a leaf.

Lek A traditional courtship display ground, where males display to females.

Lenticel Spongy region in a plant's periderm, allowing gas exchange.

Leucoplast A colorless plastid that stores starch or fat.

Leukocyte (loo´ ko sight) [Gr. *leukos*: clear + *kutos*: hollow vessel] A white blood cell.

Leuteinizing hormone A peptide hormone produced by pituitary cells that stimulates follicle maturation in females.

Lichen (lie´ kun) [Gr. *leikhen*: licker] An organism resulting from the symbiotic association of a true fungus and either a cyanobacterium or a unicellular alga.

Life cycle The entire span of the life of an organism from the moment of fertilization (or asexual generation) to the time it reproduces in turn.

Life history The stages an individual goes through during its life.

Life table A table showing, for a group of equal-aged individuals, the proportion still alive at different times in the future and the number of offspring they produce during each time interval.

Ligament A band of connective tissue linking two bones in a joint.

Ligand (lig´ and) A molecule that binds to a receptor site of another molecule.

Light compass reaction A reaction of many invertebrates in which the angle between the direction of movement and the direction of the sun is kept constant.

Lignin The principal noncarbohydrate component of wood, a polymer that binds together cellulose fibrils in some plant cell walls.

Limiting resource The required resource whose supply most strongly influences the size of a population.

Linkage In genetics, association between markers on the same chromosome such that they do not show random assortment.

Linked markers recombine with one another at frequencies less than 0.5; the closer the markers on the chromosome, the lower the frequency of recombination.

Lipase (lip´ ase; lye´ pase) An enzyme that digests fats.

Lipids (lip´ ids) [Gr. *lipos*: fat] Substances in a cell which are easily extracted by organic solvents; fats, oils, waxes, steroids, and other large organic molecules, including those which, with proteins, make up the cell membranes. (See phospholipids.)

Litter The partly decomposed remains of plants on the surface and in the upper layers of the soil.

Littoral zone The coastal zone from the upper limits of tidal action down to the depths where the water is thoroughly stirred by wave action.

Liver A large digestive gland. In vertebrates, it secretes bile and is involved in the formation of blood.

Lobes Regions of the human cerebral hemispheres; includes the temporal, frontal, parietal, and occipital lobes.

Locus In genetics, a specific location on a chromosome. May be considered to be synonymous with "gene."

Logistic growth Growth, especially in the size of an organism or in the number of organisms that constitute a population, which slows steadily as the entity approaches its maximum size. (Contrast with exponential growth.)

Loop of Henle (hen´ lee) Long, hairpin loop of the mammalian renal tubule that runs from the cortex down into the medulla, and back to the cortex. Creates a concentration gradient in the interstitial fluids in the medulla.

Lophophore A U-shaped fold of the body wall with hollow, ciliated tentacles that encircles the mouth of animals in several different phyla. Used for filtering prey from the surrounding water.

Lordosis (lor doe´ sis) [Gk. *lordosis*: curving forward] A posture assumed by females of some mammalian species (especially rodents) to signal sexual receptivity.

Lumen (loo´ men) [L.: light] The cavity inside any tubular part of an organ, such as a piece of gut or a kidney tubule.

Lungs A pair of saclike chambers within the bodies of some animals, functioning in gas exchange.

Luteinizing hormone A gonadotropin produced by the anterior pituitary. It stimulates the gonads to produce sex hormones.

Lymph [L. *lympha*: water] A clear, watery fluid that is formed as a filtrate of blood; it contains white blood cells; it collects in a series of special vessels and is returned to the bloodstream.

Lymphocyte A major class of white blood cells. Includes T cells, B cells, and other cell types important in the immune response.

Lysis (lie´ sis) [Gr.: a loosening] Bursting of a cell.

Lysogenic The condition of a bacterium that carries the genome of a virus in a relatively stable form. (Contrast with lytic.)

Lysosome (lie´ so soam) [Gr. *lysis*: a loosening + *soma*: body] A membrane-bounded inclusion found in eukaryotic cells (other than plants). Lysosomes contain a mixture of enzymes that can digest most of the macromolecules found in the rest of the cell.

Lysozyme (lie´ so zyme) An enzyme in saliva, tears, and nasal secretions that attacks bacterial cell walls, as one of the body's nonspecific defense mechanisms.

Lytic Condition in which a bacterium lyses shortly after infection by a virus; the viral genome does not become stabilized within the bacterial cell. (Contrast with lysogenic.)

Macro- (mack´ roh) [Gr. *makros*: large, long] A prefix commonly used to denote something large. (Contrast with micro-.)

Macroevolution Evolutionary changes occurring over long time spans and usually involving changes in many traits. (Contrast with microevolution.)

Macroevolutionary time The time required for macroveolutionary changes in a lineage.

Macromolecule A giant polymeric molecule. The macromolecules are proteins, polysaccharides, and nucleic acids.

Macronutrient A mineral element required by plant tissues in concentrations of at least 1 milligram per gram of their dry matter.

Major histocompatibility complex (MHC) A complex of linked genes, with multiple alleles, that control a number of immunological phenomena; it is important in graft rejection.

Malleus (mal´ ee us) [L. *malleus*: hammer] The first of the three bones that conduct movements of the eardrum to the oval window of the inner ear. (See incus, stapes.)

Malpighian tubule (mal pee´ gy un) A type of protonephridium found in insects.

Mammal [L. *mamma*: breast, teat] Any animal of the class Mammalia, characterized by the production of milk by the female mammary glands and the possession of hair for body covering.

Mantle A sheet of specialized tissues that covers most of the viscera of mollusks; provides protection to internal organs and secretes the shell.

Mapping In genetics, determining the order of genes on a chromosome and the distances between them.

Marine [L. *mare*: sea, ocean] Pertaining to or living in the ocean. (Contrast with aquatic, terrestrial.)

Maritime climate Weather pattern typical of coasts of continents, particularly those on the western sides at mid latitudes, in which the difference between summer and winter is relatively small. (Contrast with continental climate.)

Marsupial (mar soo´ pee al) A mammal belonging to the subclass Metatheria, such as opossums and kangaroos. Most have a pouch (marsupium) that contains the milk glands and serves as a receptacle for the young.

Mass extinctions Geological periods during which rates of extinction were much higher than during intervening times.

Mass number The sum of the number of protons and neutrons in an atom's nucleus.

Maternal inheritance (cytoplasmic inheritance) Inheritance in which the phenotype of the offspring depends on factors, such as mitochondria or chloroplasts, that are inherited from the female parent through the cytoplasm of the female gamete.

Mating type In some bacteria, fungi, and protists, sexual reproduction can occur only between partners of different mating type. "Mating type" is not the same as "sex," since some species have as many as 8 mating types; mating may also be between hermaphroditic partners of opposite mating type, with both partners acting as both "male" and "female" in terms of donating and receiving genetic information.

Maturation The automatic development of a pattern of behavior, which becomes increasingly complex or precise as the animal matures. Unlike learning, the development does not require experience to occur.

Mechanosensor A cell that is sensitive to physical movement and generates action potentials in response.

Medulla (meh dull´ luh) [L.: narrow] (1) The inner, core region of an organ, as in the adrenal medulla (adrenal gland) or the renal medulla (kidneys). (2) The portion of the brain stem that connects to the spinal cord.

Medusa (meh doo´ suh) The tentacle-bearing, jellyfish-like, free-swimming sexual stage in the life cycle of a cnidarian.

Mega- [Gr. *megas*: large, great] A prefix often used to denote something large. (Contrast with micro-.)

Megareserve A large park or reserve; usually has associated buffer areas in which human use of the environment is restricted to activities that do not destroy the functioning of the ecosystem.

Megasporangium The special structure (sporangium) that produces the megaspores.

Megaspore [Gr. *megas*: large + *spora*:seed] In plants, a haploid spore that produces a female gametophyte. In many cases the megaspore is larger than the male-producing microspore.

Meiosis (my oh´ sis) [Gr.: diminution] Division of a diploid nucleus to produce four haploid daughter cells. The process consists of two successive nuclear divisions with only one cycle of chromosome replication.

Membrane potential The difference in electrical charge between the inside and the outside of a cell, caused by a difference in the distribution of ions.

Mendelian population A local population of individuals belonging to the same species and exchanging genes with one another.

Menopause The time in a human female's life when the ovarian and menstrual cycles cease.

Menstrual cycle The monthly sloughing off of the uterine lining if fertilization does not occur in the female. Occurs between puberty and menopause.

Meristem [Gr. *meristos*: divided] Plant tissue made up of actively dividing cells.

Mesenchyme (mez´ en kyme) [Gr. *mesos*: middle + *enchyma*: infusion] Embryonic or unspecialized cells derived from the mesoderm.

Meso- (mez´ oh) [Gr.: middle] A prefix often used to designate a structure located in the middle, or a stage that appears at some intermediate time. For example, mesoderm, Mesozoic.

Mesoderm [Gr. *mesos*: middle + *derma*: skin] The middle of the three embryonic tissue layers first delineated during gastrulation. Gives rise to skeleton, circulatory system, muscles, excretory system, and most of the reproductive system.

Mesoglea The jelly-like middle layer that constitutes the bulk of the bodies of the medusae of many cnidarians; not a true cell layer.

Mesophyll (mez´ a fill) [Gr. *mesos*: middle + *phyllon*: leaf] Chloroplast-containing, photosynthetic cells in the interior of leaves.

Mesosome (mez´ o soam´) [Gr. *mesos*: middle + *soma*: body] A localized infolding of the plasma membrane of a bacterium.

Messenger RNA (mRNA) A transcript of one of the strands of DNA, it carries information (as a sequence of codons) for the synthesis of one or more proteins.

Meta- [Gr.: between, along with, beyond] A prefix used in biology to denote a change or a shift to a new form or level; for example, as used in metamorphosis.

Metabolic compensation Changes in biochemical properties of an organism that render it less sensitive to temperature changes.

Metabolic pathway A series of enzyme-catalyzed reactions so arranged that the product of one reaction is the substrate of the next.

Metabolism (meh tab´ a lizm) [Gr. *metabole*: to change] The sum total of the chemical reactions that occur in an organism, or some subset of that total (as in "respiratory metabolism").

Metamorphosis (met´ a mor´ fo sis) [Gr. *meta*: between + *morphe*: form, shape] A radical change occurring between one developmental stage and another, as for example from a tadpole to a frog or an insect larva to the adult.

Metaphase (met´ a phase) [Gr. *meta*: between] The stage in nuclear division at which the centromeres of the highly supercoiled chromosomes are all lying on a plane (the metaphase plane or plate) perpendicular to a line connecting the division poles.

Metastasis (meh tass´ tuh sis) The spread of cancer cells from their original site to other parts of the body.

Methanogen Any member of a group of Archaebacteria that release methane as a metabolic product. This group is considered to be an extremely ancient one.

MHC See major histocompatibility complex.

Micelles (my sells´) [L. *mica*: grain, crumb] The small particles of fat in the small intestine, resulting from the emulsification of dietary fat by bile.

Micro- (mike´ roh) [Gr. *mikros*: small] A prefix often used to denote something small. (Contrast with macro-, mega-.)

Microbiology [Gr. *mikros*: small + *bios*: life + *logos*: discourse] The scientific study of microscopic organisms, particularly bacteria, unicellular algae, protists, and viruses.

Microbody A small organelle, bounded by a single membrane and possessing a granular interior. Peroxisomes and glyoxysomes are types of microbodies.

Microevolution The small evolutionary changes typically occurring over short time spans; generally involving a small number of traits and minor genetic changes. (Contrast with macroevolution.)

Microevolutionary time The time required for microevolutionary changes within a lineage of organisms.

Microfilament Minute fibrous structure generally composed of actin found in the cytoplasm of eukaryotic cells. They play a role in the motion of cells.

Micromorphology The structure of the macromolecules of an organism.

Micronutrient A mineral element required by plant tissues in concentrations of less than 100 micrograms per gram of their dry matter.

Microorganism Any microscopic organism, such as a bacterium or one-celled alga.

Micropyle (mike´ roh pile) [Gr. *mikros*: small + *pyle*: gate] Opening in the integument(s) of a seed plant ovule through which pollen grows to reach the female gametophyte within.

Microsporangium The special structure (sporangium) that produces the microspores.

Microspores [Gr. *mikros*: small + *spora*: seed] In plants, a haploid spore that produces a male gametophyte. In many cases the microspore is smaller than the female-producing megaspore.

Microtubules Minute tubular structures found in centrioles, spindle apparatus, cilia, flagella, and other places in the cytoplasm of eukaryotic cells. These tubules play roles in the motion and maintenance of shape of eukaryotic cells.

Microvilli (singular: microvillus)   The projections of epithelial cells, such as the cells lining the small intestine, that increase their surface area.

Middle lamella   A layer of derivative polysaccharides that separates plant cells; a common middle lamella lies outside the primary walls of the two cells.

Migration   The regular, seasonal movements of animals between breeding and nonbreeding ranges.

Mimicry (mim´ ik ree)   The resemblance of one kind of organism to another, or to some inanimate object; serves the function of making the organism difficult to find, of discouraging potential enemies or of attracting potential prey. (See Batesian mimicry and Müllerian mimicry.)

Mineral   An inorganic substance other than water.

Mineralocorticoid   A hormone produced by the adrenal cortex that influences mineral ion balance; aldosterone.

Minimal medium   A medium for the growth of bacteria, fungi, or tissue cultures, containing only those nutrients absolutely required for the growth of wild-type cells.

Minimum viable population.   The smallest number of individuals required for a population to persist in a region.

Missense mutation   A mutation that changes a codon for one amino acid to a codon for a different amino acid. (Contrast with frame-shift mutation, nonsense mutation.)

Mitochondrial matrix   The fluid interior of the mitochondrion, enclosed by the inner mitochondrial membrane.

Mitochondrion (my´ toe kon´ dree un) (plural: mitochondria) [Gr. *mitos*: thread + *chondros*: cartilage, or grain]   An organelle that occurs in eukaryotic cells and contains the enzymes of the ctric acid cycle, the respiratory chain, and oxidative phosphorylation. A mitochondrion is bounded by a double membrane.

Mitosis (my toe´ sis) [Gr. *mitos*: thread]   Nuclear division in eukaryotes leading to the formation of two daughter nuclei each with a chromosome complement identical to that of the original nucleus.

Mitotic center   Cellular region that organizes the microtubules for mitosis. In animals a centrosome serves as the mitotic center.

Mobbing   Gathering of calling animals around a predator; their calls and the confusion they create reduce the probability that the predator can hunt successfully in the area.

Modular organism   An organism which grows by producing additional units of body construction that are very similar to the units of which it is already composed.

Mole   A quantity of a compound whose weight in grams is numerically equal to its molecular weight expressed in atomic mass units. Avogadro's number of molecules: 6.023×10^{23} molecules.

Molecular clock   See radiometric clock.

Molecular formula   A representation that shows how many atoms of each element are present in a molecule.

Molecular weight   The sum of the atomic weights of the atoms in a molecule.

Molecule   A particle made up of two or more atoms joined by covalent bonds or ionic attractions.

Mollusk (mol´ lusk)   A member of the phylum Mollusca, such as a snail, clam, or octopus.

Molting   The process of shedding part or all of an outer covering, as the shedding of feathers by birds or of the entire exoskeleton by arthropods.

Monecious (mo nee´ shus) [Gr.: one house]   Organisms in which both sexes are "housed" in a single individual, which produces both eggs and sperm. (In some plants, these are found in different flowers within the same plant.) Examples: corn, peas, earthworms, hydras. (Contrast with dioecious, perfect flower.)

Moneran (moh neer´ un)   A bacterium. This term was coined when both archaebacteria and eubacteria were considered to be members of a single kingdom, Monera.

Mono- [Gr. *monos*: one]   Prefix denoting a single entity. (Contrast with poly.)

Monoclonal antibody   Antibody produced in the laboratory from a clone of hybridoma cells, each of which produces the same specific antibody.

Monocot (short for monocotyledon) [Gr. *monos*: one + *kotyledon*: a cup-shaped hollow]   Any member of the angiosperm class Monocotyledones, plants in which the embryo produces but a single cotyledon (seed leaf). Leaves of most monocots have their major veins arranged parallel to each other.

Monohybrid cross   A mating in which the parents differ with respect to the alleles of only one locus of interest.

Monomer   A small molecule, two or more of which can be combined to form oligomers (consisting of a few monomers) or polymers (consisting of many monomers).

Monophyletic (mon´ oh fih leht´ ik) [Gk. *monos*: single + *phylon*: tribe]   Being descended from a single ancestral stock.

Monosaccharide   A simple sugar. Oligosaccharides and polysaccharides are made up of monosaccharides.

Monosynaptic reflex   A neural reflex that begins in a sensory neuron and makes a single synapse before activating a motor neuron.

Morphogens   Diffusible substances whose concentration gradients determine patterns of development in animals and plants.

Morphogenesis (more´ fo jen´ e sis) [Gr. *morphe*: form + *genesis*: origin]   The development of form. Morphogenesis is the overall consequence of determination, differentiation, and growth.

Morphology (more fol´ o jee) [Gr. *morphe*: form + *logos*: discourse]   The scientific study of organic form, including both its development and function.

Mosaic development   Pattern of animal embryonic development in which each blastomere contributes a specific part of the adult body. (Contrast with regulative development.)

Motor end plate   The modified area on a muscle cell membrane where a synapse is formed with a motor neuron.

Motor neuron   A neuron carrying information from the central nervous system to an effector such as a muscle fiber.

Motor unit   A motor neuron and the set of muscle fibers it controls.

mRNA   (See messenger RNA.)

Mucosa (mew koh´ sah)   An epithelial membrane containing cells that secrete mucus. The inner cell layers of the digestive and respiratory tracts.

Müllerian mimicry   The resemblance of two or more unpleasant or dangerous kinds of organisms to each other; the mimicry gives each added protection because potential enemies that learn to avoid members of one group tend to avoid members of the others even though they lack prior experience with them.

Multicellular [L. *multus*: much + *cella*: chamber]   Consisting of more than one cell, as for example a multicellular organism. (Contrast with unicellular.)

Multiple fruit   A fruit formed from an inflorescence. (Contrast with accessory fruit, aggregate fruit, simple fruit.)

Muscle fiber   A single muscle cell. In the case of striated muscle, a syncitial, multinucleate cell.

Muscle spindle   Modified muscle fibers encased in a connective sheat and functioning as stretch sensors.

Muscle tissue   Contractile tissue containing actin and myosin organized into polymeric chains called microfilaments. In vertebrates, the tissues are either cardiac muscle, smooth muscle, or striated (skeletal) muscle.

Mutagen (mute´ ah jen) [L. *mutare*: change + Gr. *genesis*: source]   An agent, especially a chemical, that increases the mutation rate.

Mutation   In the broad sense, any discontinuous change in the genetic constitution of an organism. In the narrow sense, the word usually refers to a "point mutation," a change along a very narrow portion of the nucleic acid sequence.

Mutation pressure   Evolution (change in gene proportions) by different mutation rates alone.

Mutualism   The type of symbiosis, such as that exhibited by fungi and algae or cyanobacteria in forming lichens, in which both species profit from the association.

Mycelium (my seel´ ee yum) [Gr. *mykes*: fungus] In the fungi, a mass of hyphae.

Mycorrhiza (my´ ka rye´ za) [Gr. *mykes*: fungus + *rhiza*: root] An association of the root of a plant with the mycelium of a fungus.

Myelin (my´ a lin) A material forming a sheath around some axons. It is formed by Schwann cells that wrap themselves about the axon. It serves to insulate the axon electrically and to increase the rate of transmission of a nervous impulse.

Myofibril (my´ oh fy´ bril) [Gr. *mys*: muscle + L. *fibrilla*: small fiber] A polymeric unit of actin or myosin in a muscle.

Myogenic (my oh jen´ ik) [Gr. *mys*: muscle + *genesis*: source] Originating in muscle.

Myoglobin (my´ oh globe´ in) [Gr. *mys*: muscle + L. *globus*: sphere] An oxygen-binding molecule found in muscle. Consists of a heme unit and a single globiin chain, and carries less·oxygen than hemoglobin.

Myosin [Gr. *mys*: muscle] One of the two major proteins of muscle, it makes up the thick filaments. (See actin.)

NAD (nicotinamide adenine dinucleotide) A compound found in all living cells, existing in two interconvertible forms: the oxidizing agent NAD$^+$ and the reducing agent NADH.

NADP (nicotinamide adenine dinucleotide phosphate) Like NAD, but possessing another phosphate group; plays similar roles but is used by different enzymes.

Natal group The group into which an individual was born.

Natural killer cell A small leukocyte that nonspecifically kills certain tumor cells and virus-infected cells in tissue cultures.

Natural selection The differential contribution of offspring to the next generation by various genetic types belonging to the same population. The mechanism of evolution proposed by Charles Darwin.

Nauplius (no´ plee us) [Gk. *nauplios*: shellfish] The typical larva of crustaceans. Has three pairs of appendages and a median compound eye.

Negative control The situation in which a regulatory macromolecule (generally a repressor) functions to turn off transcription. In the absence of a regulatory macromolecule, the structural genes are turned on.

Negative feedback A pattern of regulation in which a change in a sensed variable results in a correction that opposes the change.

Nekton [Gr. *nekhein*: to swim] Animals, such as fish, that can swim against currents of water. (Contrast with plankton.)

Nematocyst (ne mat´ o sist) [Gr. *nema*: thread + *kystis*: cell] An elaborate, threadlike structure produced by cells of jellyfish and other cnidarians, used chiefly to paralyze and capture prey.

Nephridium (nef rid´ ee um) [Gr. *nephros*: kidney] An organ which is involved in excretion, and often in water balance, involving a tube that opens to the exterior at one end.

Nephron (nef´ ron) [Gr. *nephros*: kidney] The basic component of the kidney, which is made up of numerous nephrons. Its form varies in detail, but it always has at one end a device for receiving a filtrate of blood, and then a tubule that absorbs selected parts of the filtrate back into the bloodstream.

Nephrostome (nef´ ro stome) [Gr. *nephros*: kidney + *stoma*: opening] An opening in a nephridium through which body fluids can enter.

Nerve A structure consisting of many neuronal axons and connective tissue.

Net primary production Total photosynthesis minus respiration by plants.

Neural plate A thickened strip of ectoderm along the dorsal side of the early vertebrate embryo; gives rise to the central nervous system.

Neurohormone A hormone produced and secreted by neurons.

Neuron (noor´ on) [Gr. *neuron*: nerve, sinew] A cell derived from embryonic ectoderm and characterized by a membrane potential that can change in response to stimuli, generating action potentials. Action potentials are generated along an extension of the cell (the axon), which makes junctions (synapses) with other neurons, muscle cells, or gland cells.

Neurotransmitter A substance, produced in and released by one neuron, that diffuses across a synapse and excites or inhibits the postsynaptic neuron.

Neurula (nure´ you la) [Gr. *neuron*: nerve] Embryonic stage during formation of the dorsal nerve cord by two ectodermal ridges.

Neutral alleles Alleles that differ so slightly that the proteins for which they code function identically.

Neutron (new´ tron) [E.: neutral] One of the three most fundamental particles of matter, with mass approximately 1 amu and no electrical charge.

Nicotinamide adenine dinucleotide (See NAD.)

Nicotinamide adenine dinucleotide phosphate (See NADP.)

Nitrification The oxidation of ammonia to nitrite and nitrate ions, performed by certain soil bacteria.

Nitrogenase In nitrogen-fixing organisms, an enzyme complex that mediates the stepwise reduction of atmospheric N_2 to ammonia.

Nitrogen fixation Conversion of nitrogen gas to ammonia, which makes nitrogen available to living things. Carried out by certain prokaryotes, some of them free-living and others living within plant roots.

Node [L. *nodus*: knob, knot] In plants, a (sometimes enlarged) point on a stem where a leaf or bud is or was attached.

Node of Ranvier A gap in the myelin sheath covering an axons, where the axonal membrane can fire action potentials.

Noncompetitive inhibitor An inhibitor that binds the enzyme at a site other than the active site. (Contrast with competitive inhibitor.)

Nondisjunction Failure of sister chromatids to separate in meiosis II or mitosis, or failure of homologous chromosomes to separate in meiosis I. Results in aneuploidy.

Nonpolar molecule A molecule whose electric charge is evenly balanced from one end of the molecule to the other.

Nonsense (chain-terminating) mutation Mutations that change a codon for an amino acid to one of the codons (UAG, UAA, or UGA) that signal termination of translation. The resulting gene product is a shortened polypeptide that begins normally at the amino-terminal end and ends at the position of the altered codon. (Contrast with frameshift mutation, missense mutation.)

Nonvascular plants Those plants lacking well-developed vascular tissue; the liverworts, hornworts, and mosses. (Contrast with vascular plants.)

Normal flora The bacteria and fungi that live on animal body surfaces without causing disease.

Norepinephrine A neurotransmitter found in the central nervous system and also at the postganglionic nerve endings of the sympathetic nervous system. Also called noradrenaline.

Notochord (no´ tow kord) [Gr. *notos*: back + *chorde*: string] A flexible rod of gelatinous material serving as a support in the embryos of all chordates and in the adults of tunicates and lancelets.

Nuclear envelope The surface, consisting of two layers of membrane, that encloses the nucleus of eukaryotic cells.

Nucleic acid (new klay´ ik) [E.: nucleus of a cell] A long-chain alternating polymer of deoxyribose or ribose and phosphate groups, with nitrogenous bases—adenine, thymine, uracil, guanine, or cytosine (A, T, U, G, or C)—as side chains. DNA and RNA are nucleic acids.

Nucleoid (new´ klee oid) The region that harbors the chromosomes of a prokaryotic cell. Unlike the eukaryotic nucleus, it is not bounded by a membrane.

Nucleolar organizer (new klee´ o lar) A region on a chromosome that is associated with the formation of a new nucleolus following nuclear division. The site of the genes that code for ribosomal RNA.

Nucleolus (new klee´ oh lus) [from L. diminutive of *nux*: little kernel or little nut] A small, generally spherical body found within the nucleus of eukaryotic cells. The site of synthesis of ribosomal RNA.

Nucleoplasm (new´ klee o plazm) The fluid material within the nuclear envelope of a cell, as opposed to the chromosomes, nucleoli, and other particulate constituents.

Nucleosome A portion of a eukaryotic chromosome, consisting of part of the DNA molecule wrapped around a group of histone molecules, and held together by another type of histone molecule. The chromosome is made up of many nucleosomes.

Nucleotide The basic chemical unit (monomer) in a nucleic acid. A nucleotide in RNA consists of one of four nitrogenous bases linked to ribose, which in turn is linked to phosphate. In DNA, deoxyribose is present instead of ribose.

Nucleus (new´ klee us) [from L. diminutive of *nux*: kernel or nut] (1) In chemistry, the dense central portion of an atom, made up of protons and neutrons, with a positive charge. Surrounded by a cloud of negatively charged electrons. (2) In cells, the centrally located chamber of eukaryotic cells that is bounded by a double membrane and contains the chromosomes. The information center of the cell.

Nutrient A food substance; or, in the case of mineral nutrients, an inorganic element required for completion of the life cycle of an organism.

Obligate (ob´ li gut) Necessary, as in obligate anaerobe. (Contrast with facultative.)

Obligate anaerobe An animal that can live only in oxygenated environments.

Observational analysis A scientific method in which data are gathered in unmanipulated situations to test hypotheses. Often employed in the field where experimental manipulations are difficult or impossible.

Oceanic zone The deeper ocean basins.

Oil A triglyceride that is liquid at room temperature. (Contrast with fat.)

Okazaki fragments Newly formed DNA strands making up the lagging strand in DNA replication. DNA ligase links the Okazaki fragments to give a continuous strand.

Olfactory Having to do with the sense of smell.

Oligomer A compound molecule of intermediate size, made up of two to a few monomers. (Contrast with monomer, polymer.)

Omasum (oh may´ sum) The third division of the ruminant stomach. Its function is mostly the absorption of wastes. (Contrast with abomasum, rumen.)

Ommatidium [Gr. *omma*: an eye] One of the units which, collected into groups of up to 20,000, make up the compound eye of arthropods.

Omnivore [L. *omnis*: all, everything + *vorare*: to devour] An organism that eats both animal and plant material. (Contrast with carnivore, detritivore, herbivore.)

Oncogenic (ong´ co jen´ ik) [Gr. *onkos*: mass, tumor + *genes*: born] Causing cancer.

Ontogeny (on toj´ e nee) [Gr. *onto*: from "to be" + *genesis*: source] The development of a single organism in the course of its life history.

Oocyte (oh´ eh site) [Gr. *oon*: egg + *kytos*: cell] The cell that gives rise to eggs in animals.

Oogenesis (oh´ eh jen e sis) [Gr. *oon*: egg + *genesis*: source] Female gametogenesis, leading to production of the egg.

Oogonium (oh´ eh go´ nee um) In some algae and fungi, a cell in which an egg is produced.

Operator The region of an operon that acts as the binding site for the repressor.

Operon A genetic unit of transcription, typically consisting of several structural genes that are transcribed together; the operon contains at least two control regions: the promoter and the operator.

Opportunity cost The sum of the benefits an animal forfeits by not being able to perform some other behavior during the time when it is performing a given behavior.

Opsin (op´ sin) [Gr. *opsis*: sight] The protein protion of the visual pigment rhodopsin. (See rhodopsin.)

Optic chiasm Stucture on the lower surface of the vertebrate brain where the two optic nerves come together.

Optical isomers Isomers that differ in the configuration of the four different groups attached to a single carbon atom; so named because solutions of the two isomers rotate the plane of polarized light in opposite directions. The two isomers are mirror images of one another.

Order In taxonomy, the category below the class and above the family; a group of related, similar families.

Organ A body part, such as the heart, liver, brain, root, or leaf, composed of different tissues integrated to perform a distinct function for the body as a whole.

Organ of Corti Structure in the inner ear that transforms mechanical forces produced from pressure waves ("sound waves") into action potentials that are sensed as sound.

Organelles (or´ gan els´) [L.: little organ] Organized structures that are found in or on cells. Examples: ribosomes, nuclei, mitochrondria, chloroplasts, cilia, and contractile vacuoles.

Organic Pertaining to any aspect of living matter, e.g., to its evolution, structure, or chemistry. The term is also applied to any chemical compound that contains carbon.

Organism Any living creature.

Organizer, embryonic A region of an embryo which directs the development of nearby regions. In amphibian early gastrulas, the dorsal lip of the blastopore.

Osmoregulation Regulation of the chemical composition of the body fluids of an organism.

Osmosensor A neuron that converts changes in the osmotic potential of interstial fluids into action potentials.

Osmosis (oz mo´ sis) [Gr. *osmos*: to push] The movement of water through a differentially permeable membrane from one region to another where the water potential is more negative. This is often a region in which the concentration of dissolved molecules or ions is higher, although the effect of dissolved substances may be offset by hydrostatic pressure in cells with semi-rigid walls.

Osmotic potential A property of any solution, resulting from its solute content; it may be zero or have a negative value. A negative osmotic potential tends to cause water to move into the solution; it may be offset by a positive pressure potential in the solution or by a more negative water potential in a neighboring solution. (Contrast with pressure potential.)

Ossicle (ah´ sick ul) [L. *os*: bone] The calcified construction unit of echinoderm skeletons.

Osteoblasts Cells that lay down the protein matrix of bone. (Contrast with osteoclasts.)

Osteoclasts Cells that dissolve bone. (Contrast with osteoblasts.)

Otolith (oh´ tuh lith) [Gk.*otikos*: ear + *lithos*: stone[Structures in the vertebrate vestibular apparatus that mechanically stimulate hair cells when the head moves or changes position.

Outgroup A taxon that separated from another taxon, whose lineage is to be inferred, before the latter underwent evolutionary radiation.

Oval window The flexible membrane which, when moved by the bones of the middle ear, produces pressure waves in the inner ear

Ovary (oh´ var ee) Any female organ, in plants or animals, that produces an egg.

Oviduct [L. *ovum*: egg + *ducere*: to lead] In mammals, the tube serving to transport eggs to the uterus or to outside of the body.

Oviparous (oh vip´ uh rus) Reproduction in which eggs are released by the female and development is external to the mother's body. (Contrast with viviparous.)

Ovulation The release of an egg from an ovary.

Ovule (oh´ vule) [L. *ovulum*: little egg] In plants, an organ that contains a gametophyte and, within the gametophyte, an egg; when it matures, an ovule becomes a seed.

Ovum (oh´ vum) [L.: egg] The egg, the female sex cell.

Oxidation (ox i day´ shun) Relative loss of electrons in a chemical reaction; either outright removal to form an ion, or the sharing of electrons with substances having a greater affinity for them, such as oxygen. Most oxidation, including biological ones, are associated with the liberation of energy. (Contrast with reduction.)

Oxidative phosphorylation ATP formation in the mitochondrion, associated with flow of electrons through the respiratory chain. (Contrast with substrate-level phosphorylation.)

Oxidizing agent A substance that can accept electrons from another. The oxidizing agent becomes reduced; its partner becomes oxidized.

P generation The individuals that mate in a genetic cross. Their immediate offspring are the F_1 generation.

Pacemaker That part of the heart which undergoes most rapid spontaneous contraction, thus setting the pace for the beat of the entire heart. In mammals, the sinoatrial (SA) node. Also, an artificial device, implanted in the heart, that initiates rhythmic contraction of the organ.

Pacinian corpuscle A sensory neuron surrounded by sheaths of connective tissue. Found in the deep layers of the skin, where it senses touch and vibration.

Paleobiology The study of fossil evidence, the comparative biochemistry of living organisms, and conditions on the early Earth to determine the stages in the evolution of life.

Paleobotany The scientific study of fossil plants and all aspects of extinct plant life.

Paleontology (pale´ ee on tol´ oh jee) [Gr. *palaios*: ancient, old + *logos*: discourse] The scientific study of fossils and all aspects of extinct life.

Palisade parenchyma In leaves, one or several layers of tightly packed, columnar photosynthetic cells, frequently found just below the upper epidermis.

Pancreas (pan´ cree us) A gland, located near the stomach of vertebrates, that secretes digestive enzymes into the small intestine and releases insulin into the bloodstream.

Pangaea (pan jee´ uh) [Gk. *pan*: all, every] The single land mass formed when all the continents came together in the Permian period.

Parabronchi Passages in the lungs of birds through which air flows.

Paradigm A general framework within which some scientific discipline (or even the whole Earth) is viewed and within which questions are asked and hypotheses are developed. Scientific revolutions usually involve major paradigm changes.

Parapatric speciation Development of reproductive isolation among members of a continuous population in the absence of a geographical barrier. (Contrast with geographic, sympatric speciation.)

Paraphyletic taxon A taxon that includes some, but not all, of the descendants of a single ancestor.

Parasite An organism that attacks and consumes parts of an organism much larger than itself. Parasites sometimes, but not always, kill the host.

Parasitoid A parasite that is so large relative to its host that only one individual or at most a few individuals can live within a single host.

Parasympathetic nervous system A portion of the autonomic (involuntary) nervous system. Activity in the parasympathetic nervous system produces effects such as decreased blood pressure and decelerated heart beat. The neurotransmitter for this system is acetylcholine. (Contrast with sympathetic nervous system.)

Parathormone Hormone secreted by the parathyroid glands. Stimulates osteoclast activity and raises blood calcium levels.

Parathyroids Four glands on the posterior surface of the thyroid that produce and release parathormone.

Parenchyma (pair eng´ kyma) [Gr. *para*: beside + *enchyma*: infusion] A plant tissue composed of relatively unspecialized cells without secondary walls.

Parental investment Investment in one offspring or group of offspring that reduces the ability of the parent to assist other offspring.

Parsimony The principle of preferring the simplest among a set of plausible explanations of a phenomenon. Commonly employed in evolutionary and biogeographic studies.

Parthenocarpy Formation of fruit from a flower without fertilization.

Parthenogenesis (par´ then oh jen´ e sis) [Gr. *parthenos*: virgin + *genesis*: source] The production of an organism from an unfertilized egg.

Partial pressure The portion of the barometric pressure of a mixture of gases that is due to one component of that mixture. For example, the partial pressure of oxygen at sea level is 20.9% of barometric pressure.

Parturition (part uh rish un) [L. *parturire*, to give birth] Childbirth.

Pasteur effect The sharp decrease in rate of glucose utilization when

Pastoralism A nomadic form of human culture based on the tending of herds of domestic animals.conditions become aerobic.

Patch clamping A technique for isolating a tiny patch of membrane to allow the study of ion movement through a particular channel.

Pathogen (path´ o jen) [Gr. *pathos*: suffering + *gignomai*: causing] An organism that causes disease.

Pattern formation In animal embryonic development, the organization of differentiated tissues into specific structures such as wings.

Pedigree The pattern of transmission of a genetic trait in a family.

Pelagic zone (puh ladj´ ik) [Gr. *pelagos*: sea] The open waters of the ocean.

Pellicle (pell´ ik el) [L. *pellis*: skin] A thin, filmy covering.

Penetrance Of a genotype, the proportion of individuals with that genotype who show the expected phenotype.

Penis (pee´ nis) [L.: tail, penis] The male organ inserted into the female during coitus (sexual intercourse).

Pentose (pen´ tose) [Gk. *penta*: five] A five-carbon sugar, such as ribose or deoxyribose.

PEP carboxylase The enzyme that combines carbon dioxide with PEP to form a 4-carbon dicarboxylic acid at the start of C_4 photosynthesis or of Crassulacean acid metabolism (CAM).

Pepsin [Gr. *pepsis*: digestion] An enzyme, in gastric juice, that digests protein.

Peptide linkage The connecting group in a protein chain, –CO–NH–, formed by removal of water during the linking of amino acids, –COOH to –NH₂. Also called an amide linkage.

Peptidoglycan The cell wall material of many prokaryotes, consisting of a single enormous molecule that surrounds the entire cell.

Perennial (per ren´ ee al) [L. *per*: through + *annus*: a year] Referring to a plant that lives from year to year. (Contrast with annual, biennial.)

Perfect flower A flower with both stamens and carpels, therefore hermaphroditic.

Pericycle [Gr. *peri*: around + *kyklos*: ring or circle] In plant roots, tissue just within the endodermis, but outside of the root vascular tissue. Meristematic activity of pericycle cells produces lateral root primordia.

Periderm The outer tissue of the secondary plant body, consisting primarily of cork.

Period (1) A minor category in the geological time scale. (2) The duration of a cyclical event, such as a circadian rhythm.

Peripheral nervous system Neurons that transmit information to and from the central nervous system and whose cell bodies reside outside the brain or spinal cord.

Peristalsis (pair´ i stall´ sis) [Gr. *peri*: around + *stellein*: place] Wavelike muscular contractions proceeding along a tubular organ, propelling the contents along the tube.

Peritoneum The mesodermal lining of the coelom among coelomate animals.

Permafrost Soil that remains frozen for many years.

Permease A protein in membranes that specifically transports a compound or family of compounds across the membrane.

Peroxisome A microbody that houses reactions in which toxic peroxides are formed. The peroxisome isolates these peroxides from the rest of the cell.

Petal In an angiosperm flower, a sterile modified leaf, nonphotosynthetic, frequently brightly colored, and often serving to attract pollinating insects.

Petiole (pet´ ee ole) [L. *petiolus*: small foot] The stalk of a leaf.

pH The negative logarithm of the hydrogen ion concentration; a measure of the acidity of a solution. A solution with pH = 7 is said to be neutral; pH values higher than 7 characterize basic solutions, while acidic solutions have pH values less than 7.

Phage (fayj)   Short for bacteriophage.

Phagocyte   A white blood cell that ingests microorganisms by endocytosis.

Phagocytosis [Gr.: *phagein* to eat; cell-eating] A form of endocytosis, the uptake of a solid particle by forming a pocket of plasma membrane around the particle and pinching off the pocket to form an intracellular particle bounded by membrane. (Contrast with pinocytosis.)

Pharyngeal slits   Slits in the pharynx that originally functioned in gas exchange but became modified for other purposes among vertebrates.

Pharynx [Gr.: throat]   The part of the gut between the mouth and the esophagus.

Phenogram   Graphic representation of phenetic similarities.

Phenotype (fee´ no type) [Gr. *phanein*: to show + *typos*: impression]   The observable properties of an individual as they have developed under the combined influences of the genetic constitution of the individual and the effects of environmental factors. (Contrast with genotype.)

Pheromone (feer´ o mone) [Gr. *phero*: carry + *hormon*: excite, arouse]   A chemical substance used in communication between organisms of the same species.

Phloem (flo´ um) [Gr. *phloos*: bark]   In vascular plants, the food-conducting tissue. It consists of sieve cells or sieve tubes, fibers, and other specialized cells.

Phosphate group   The functional group –OPO₃H₂; the transfer of energy from one compound to another is often accomplished by the transfer of a phosphate group.

Phosphodiester linkage   The connection in a nucleic acid strand, formed by linking two nucleotides.

3-Phosphoglycerate   The first product of photosynthesis, produced by the reaction of ribulose bisphosphate with carbon dioxide.

Phospholipids   Cellular materials that contain phosphorus and are soluble in organic solvents. An example is lecithin (phosphatidyl choline). Phospholipids are important constituents of cellular membranes. (See lipids.)

Phosphorylation   The addition of a phosphate group.

Photoautotroph   An organism that obtains energy from light and carbon from carbon dioxide. (Contrast with chemoautotroph, chemoheterotroph, photoheterotroph.)

Photoheterotroph   An organism that obtains energy from light but must obtain its carbon from organic compounds. (Contrast with chemoautotroph, chemoheterotroph, photoautotroph.)

Photon (foe´ tohn) [Gr. *photos*: light]   A quantum of visible radiation; a "packet" of light energy.

Photoperiod (foe´ tow peer´ ee ud)   The duration of a period of light, such as the length of time in a 24-hour cycle in which daylight is present. The regulation of processes such as flowering by the changing length of day (or of night) is known as **photoperiodism.**

Photophosphorylation   Photosynthetic reactions in which light energy trapped by chlorophyll is used to produce ATP and, in noncyclic photophosphorylation, is used to reduce NADP⁺ to NADPH.

Photorespiration   Light-driven uptake of oxygen and release of carbon dioxide, the carbon being derived from the early reactions of photosynthesis.

Photosensor   A cell that senses and responds to light energy.

Photosynthesis (foe tow sin´ the sis) [literally, "synthesis out of light"]   Metabolic processes, carried out by green plants, by which visible light is trapped and the energy used to synthesize compounds such as ATP and glucose.

Phototropism [Gr. *photos*: light + *trope*: a turning]   A directed plant growth response to light.

Phylogenetic tree   Graphic representation of lines of descent among organisms.

Phylogeny (fy loj´ e nee) [Gr. *phylon*: tribe, race + *genesis*: source]   The evolutionary history of a particular group of organisms; also, the diagram of the "family tree" that shows genetic linkages between ancestors and descendants.

Phylum [Gr. *phylon*: tribe, stock]   In taxonomy, a high-level category just beneath kingdom and above the class; a group of related, similar classes.

Physiological time   The time required for significant changes in the physiological processes or states within an organism.

Physiology (fiz´ ee ol´ o jee) [Gr. *physis*: natural form + *logos*: discourse, study]   The scientific study of the functions of living organisms and the individual organs, tissues, and cells of which they are composed.

Phytoalexins   Substances toxic to fungi, produced by plants in response to fungal infection.

Phytochrome (fy´ tow krome) [Gr. *phyton*: plant + *chroma*: color]   A plant pigment regulating a large number of developmental and other phenomena in plants; can exist in two different forms, one of which is active and the other is not. Different wavelengths of light can drive it from one form to the other.

Phytoplankton (fy´ tow plangk´ ton) [Gr. *phyton*: plant + *planktos*: wandering]   The autotrophic portion of the plankton, consisting mostly of algae.

Pigment   A substance that absorbs visible light.

Piloting   Finding one's way by means of landmarks.

Pilus (pill´ us) [Lat. *pilus*: hair]   A surface appendage by which some bacteria adhere to one another during conjugation.

Pinocytosis [Gr.: drinking cell]   A form of endocytosis; the uptake of liquids by engulfing a sample of the external medium into a pocket of the plasma membrane followed by pinching off the pocket to form an intracellular vesicle. (Contrast with phagocytosis and endocytosis.)

Pistil [L. *pistillum*: pestle]   The female structure of an angiosperm flower, within which the ovules are borne. May consist of a single carpel, or of several carpels fused into a single structure. Usually differentiated into ovary, style, and stigma.

Pith   In plants, relatively unspecialized tissue found within a cylinder of vascular tissue.

Pituitary   A small gland attached to the base of the brain in vertebrates. Its hormones control the activities of other glands. Also known as the hypophysis.

Placenta (pla sen´ ta) [Gr. *plax*: flat surface] The organ, found in most mammals, that provides for the nourishment of the fetus and elimination of the fetal waste products. It is formed by the union of membranes of the mother's uterine lining with the membranes from the fetus.

Placental (pla sen´ tal)   Pertaining to mammals of the subclass Eutheria, a group that is characterized by the presence of a placenta and that contains the majority of living species of mammals.

Plankton [Gr. *planktos*: wandering]   The free-floating organisms of the sea and fresh water that for the most part move passively with the water currents. Consisting mostly of microorganisms and small plants and animals. (Contrast with nekton.)

Plant   A member of the kingdom Plantae. Multicellular, gaining its nutrition by photosynthesis.

Planula (plan´ yew la) [L. *planum*: something flat]   The free-swimming, ciliated larva of the cnidarians.

Plaque (plack) [Fr.: a metal plate or coin]   (1) A circular clearing in a turbid layer (lawn) of bacteria growing on the surface of a nutrient agar gel. Produced by successive rounds of infection initiated by a single bacteriophage. (2) An accumulation of prokaryotic organisms on tooth enamel. Acids produced by the metabolism of these microorganisms can cause tooth decay.

Plasma (plaz´ muh) [Gr. *plassein*: to mold] The liquid portion of blood, in which blood cells and other particulates are suspended.

Plasma cell   An antibody-secreting cell that developed from a B cell. The effector cell of the humoral immune system.

Plasma membrane The membrane that surrounds the cell, regulating the entry and exit of molecules and ions. Every cell has a plasma membrane.

Plasmid A DNA molecule distinct from the chromosome(s); that is, an extrachromosomal element. May replicate independently of the chromosome.

Plasmodesma (plural: plasmodesmata) [Gr. *plasma*: formed or molded + *desmos*: band] A cytoplasmic strand connecting two adjacent plant cells.

Plasmodium In the noncellular slime molds, a multinucleate mass of protoplasm surrounded by a membrane; characteristic of the vegetative feeding stage.

Plasmolysis (plaz mol´ i sis) Shrinking of the cytoplasm and plasma membrane away from the cell wall, resulting from the osmotic outflow of water. Occurs only in cells with rigid cell walls.

Plastid Organelle in plants that serves for food manufacture (by photosynthesis) or food storage; bounded by a double membrane.

Platelet A membrane-bounded body without a nucleus, arising as a fragment of a cell in the bone marrow of mammals. Important to blood-clotting action.

Pleiotropy (plee´ a tro pee) [Gr. *pleion*: more] The determination of more than one character by a single gene.

Pleural membrane [Gk. *pleuras*: rib, side] The membrane lining the outside of the lungs and the walls of the thoracic cavity. Inflammation of these membranes is a condition known as **pleurisy.**

Podocytes Cells of Bowman's capsule of the nephron that cover the capillaries of the glomerulus, forming filtration slits.

Poikilotherm (poy´ kill o therm) [Gr. *poikilos*: varied + *therme*: heat] An animal whose body temperature tends to vary with the surrounding environment. (Contrast with homeotherm, heterotherm.)

Point mutation A mutation that results from a small, localized alteration in the chemical structure of a gene. Such mutations can give rise to wild-type revertants as a result of reverse mutation. In genetic crosses, a point mutation behaves as if it resided at a single point on the genetic map. (Contrast with deletion.)

Polar body A nonfunctional nucleus produced by meiosis, accompanied by very little cytoplasm. The meiosis which produces the mammalian egg produces in addition three polar bodies.

Polar molecule A molecule in which the electric charge is not distributed evenly in the covalent bonds.

Polar nucleus One of two nuclei derived from each end of the angiosperm embryo sac, both of which become centrally located. They fuse with a male nucleus to form the primary triploid nucleus that will prduce the endosperm tissue of the angiosperm seed.

Pollen [L.: fine powder, dust] The fertilizing element of seed plants, containing the male gametophyte and the gamete, at the stage in which it is shed.

Pollination Process of transferring pollen from the anther to the receptive surface (stigma) of the ovary in plants.

Poly- [Gr. *poly*: many] A prefix denoting multiple entities.

Polygamy [Gr. *poly*: many + *gamos*: marriage] A breeding system in which an individual acquires more than one mate. In polyandry, a female mates with more than one male, in polygyny, a male mates with more than one female.

Polygenes Multiple loci whose alleles increase or decrease a continuously variable phenotypic trait.

Polymer A large molecule made up of similar or identical subunits called monomers. (Contrast with monomer, oligomer.)

Polymerase chain reaction (PCR) A technique for the rapid production of millions of copies of a particular stretch of DNA.

Polymerization reactions Chemical reactions that generate polymers by means of condensation reactions.

Polymorphism (pol´ lee mor´ fiz um) [Gr. *poly*: many + *morphe*: form, shape] (1) In genetics, the coexistence in the same population of two distinct hereditary types based on different alleles. (2) In social organisms such as colonial cnidarians and social insects, the coexistence of two or more functionally different castes within the same colony.

Polyp The sessile, asexual stage in the life cycle of most cnidarians.

Polypeptide A large molecule made up of many amino acids joined by peptide linkages. Large polypeptides are called proteins.

Polyploid (pol´ lee ploid) A cell or an organism in which the number of complete sets of chromosomes is greater than two.

Polysaccharide A macromolecule composed of many monosaccharides (simple sugars). Common examples are cellulose and starch.

Polysome A complex consisting of a threadlike molecule of messenger RNA and several (or many) ribosomes. The ribosomes move along the mRNA, synthesizing polypeptide chains as they proceed.

Polytene (pol´ lee teen) [Gr. *poly*: many + *taenia*: ribbon] An adjective describing giant interphase chromosomes, such as those found in the salivary glands of fly larvae. The characteristic, reproducible pattern of bands and bulges seen on these chromosomes has provided a method for preparing detailed chromosome maps of several organisms.

Pons [L. *pons*: bridge] Region of the brain stem anterior to the medulla.

Population Any group of organisms coexisting at the same time and in the same place and capable of interbreeding with one another.

Population density The number of individuals (or modules) of a population in a unit of area or volume.

Population dynamics The sum of the activities of the members of a population.

Population structure The proportions of individuals in a population belonging to different age classes (age structure). Also, the distribution of the population in space and the amount of migration between subpopulations.

Population vulnerability analysis A determination of the risk of extinction of a population given its current size and distribution.

Portal vein A vein connecting two capillary beds, as in the hepatic portal system.

Positive control The situation in which a regulatory macromolecule is needed to turn transcription of structural genes on. In its absence, transcription will not occur.

Positive cooperativity Occurs when a molecule can bind several ligands and each one that binds alters the conformation of the molecule so that it can bind the next ligand more easily. The binding of four molecules of O_2 by hemoglobin is an example of positive cooperativity.

Positive feedback A regulatory system in which an error signal stimulates responses that increase the error.

Postabsorptive period When there is no food in the gut and no nutrients are being absorbed.

Posterior Toward or pertaining to the rear.

Postsynaptic cell The cell whose membranes receive the neurotransmitter released at a synapse.

Postzygotic isolating mechanism Any factor that reduces the viability of zygotes resulting from matings between individuals of different species.

Predator An organism that kills and eats other organisms. Predation is usually thought of as involving the consumption of animals by animals, but in the broad usage of ecology it can also mean the eating of plants.

Pressure potential The actual physical (hydrostatic) pressure within a cell. (Contrast with osmotic potential, water potential.)

Presynaptic excitation/inhibition Occurs when a neuron modifies activity at a synapse by releasing a neurotransmitter onto the presynaptic nerve terminal.

Prey [L. *praeda*: booty] An organism hunted or caught as an energy source.

Prezygotic isolating mechanism A mechanism that reduces the probability that individuals of different species will mate.

Primary active transport Form of active transport in which ATP is hydrolyzed, yielding the energy required to transport ions against their concentration gradients. (Contrast with secondary active transport.)

Primary growth In plants, growth produced by the apical meristems. (Contrast with secondary growth.)

Primary producer A photosynthetic or chemosynthetic organism that synthesizes complex organic molecules from simple inorganic ones.

Primary succession Succession that begins in an areas initially devoid of life, such as on recently exposed glacial till or lava flows.

Primary structure The specific sequence of amino acids in a protein.

Primary wall Cellulose-rich cell wall layers laid down by a growing plant cell.

Primate (pry´ mate) A member of the order Primates, such as a lemur, monkey, ape, or human.

Primer A short, single-stranded segment of DNA serving as the necessary starting material for the synthesis of a new DNA strand, which is synthesized from the 3´ end of the primer.

Primitive streak A line running axially along the blastodisc, the site of inward cell migration during formation of the three-layered embryo. Formed in the embryos of birds and fish.

Primordium [L. *primordium*: origin] The most rudimentary stage of an organ or other part.

Principle of superposition The generalization that younger rocks lie on top of older rocks unless Earth movements have altered their positions.

Pro- [L.: first, before, favoring] A prefix often used in biology to denote a developmental stage that comes first or an evolutionary form that appeared earlier than another. For example, prokaryote, prophase.

Probe A segment of single stranded nucleic acid used to identify DNA molecules containing the complementary sequence.

Procambium Primary meristem that produces the vascular tissue.

Progesterone [L. *pro*: favoring + *gestare*: to bear] A vertebrate female sex hormone that maintains pregnancy.

Prokaryotes (pro kar´ ry otes) [L. *pro*: before + Gk. *karyon*: kernel, nucleus] Organisms whose genetic material is not contained within a nucleus. The bacteria. Considered an earlier stage in the evolution of life than the eukaryotes.

Prometaphase The phase of nuclear division that begins with the disintegration of the nuclear envelope.

Promoter The region of an operon that acts as the initial binding site for RNA polymerase.

Prophage (pro´ fayj) The noninfectious units that are linked with the chromosomes of the host bacteria and multiply with them but do not cause dissolution of the cell. Prophage can later enter into the lytic phase to complete the virus life cycle.

Prophase (pro´ phase) The first stage of nuclear division, during which chromosomes condense from diffuse, threadlike material to discrete, compact bodies.

Proplastid [Gr. *pro*: before + *plastos*: molded] A plant cell organelle which under appropriate conditions will develop into a plastid, usually the photosynthetic chloroplast. If plants are kept in the dark, proplastids may become quite large and complex.

Prostaglandin Any one of a group of specialized lipids with hormone-like functions. It is not clear that they act at any considerable distance from the site of their production.

Prosthetic group Any nonprotein portion of an enzyme.

Protease (pro´ tee ase) See proteolytic enzyme.

Protein (pro´ teen) [Gr. *protos*: first] One of the most fundamental building substances of living organisms. A long-chain polymer of amino acids with twenty different common side chains. Occurs with its polymer chain extended in fibrous proteins, or coiled into a compact macromolecule in enzymes and other globular proteins.

Proteolytic enzyme An enzyme whose main catalytic function is the digestion of a protein or polypeptide chain. The digestive enzymes trypsin, pepsin, and carboxypeptidase are all proteolytic enzymes (proteases).

Protist A member of the kingdom Protista, which consists of those eukaryotes not included in the kingdoms Animalia, Fungi, or Plantae. Many protists are unicellular. The kingdom Protista includes protozoa, algae, and fungus-like protists.

Protoderm Primary meristem that gives rise to epidermis.

Proton (pro´ ton) [Gr. *protos*: first] One of the three most fundamental particles of matter, with mass approximately 1 amu and an electrical charge of +1.

Proton motive force The proton gradient and electric charge difference produced by chemiosmotic proton pumping. It drives protons back across the membrane, with the concomitant formation of ATP.

Protonema (pro´ tow nee´ mah) [Gr. *protos*: first + *nema*: thread] The hairlike growth form that constitutes an early stage in the development of a moss gametophyte.

Proto-oncogenes The normal alleles of genes possessing oncogenes (cancer-causing genes) as mutant alleles. Proto-oncogenes encode growth factors and receptor proteins.

Protoplast A cell that would normally have a cell wall, from which the wall has been removed by enzymatic digestion or by special growth conditions.

Protostome One of two major lines of animal evolution, characterized by spiral, determinate cleavage of the egg, and by schizocoelous development. (Contrast with deuterostome.)

Prototroph (pro´ tow trofe´) [Gr. *protos*: first + *trophein*: to nourish] The nutritional wild-type, or reference form, of an organism. Any deviant form that requires growth nutrients not required by the prototrophic form is said to be a nutritional mutant, or auxotroph.

Protozoa A group of single-celled organisms classified by some biologists as a single phylum; includes the flagellates, amoebas, and ciliates. This textbook follows most modern classifications in elevating the protozoans to a distinct kingdom (Protista) and each of their major subgroups to the rank of phylum.

Provincialized A biogeographic term referring to the separation, by environmental barriers, of the biota into units with distinct species compositions.

Provirus See prophage.

Proximal Near the point of attachment or other reference point. (Contrast with distal.)

Pseudocoelom A body cavity not surrounded by a peritoneum. Characteristic of nematodes and rotifers.

Pseudogene A DNA segment that is homologous to a functional gene but contains a nucleotide change that prevents its expression.

Pseudoplasmodium [Gr. *pseudes*: false + *plasma*: mold or form] In the cellular slime molds such as *Dictyostelium*, an aggregation of single amoeboid cells. Occurs prior to formation of a fruiting structure.

Pseudopod (soo´ do pod) [Gr. *pseudes*: false + *podos*: foot] A temporary, soft extension of the cell body that is used in location, attachment to surfaces, or engulfing particles.

Pulmonary Pertaining to the lungs.

Pupa (pew´ pa) [L.: doll, puppet] In certain insects (the Holometabola), the encased developmental stage that intervenes between the larva and the adult.

Pupil The opening in teh vertebrate eye through which light passes.

Purine (pure´ een) A type of nitrogenous base. The purines adenine and guanine are found in nucleic acids.

Purkinje fibers Specialized heart muscle cells that conduct excitation throughout the ventricular muscle.

Pyramid of biomass Graphical representation of the total masses at different trophic levels in an ecosystem.

Pyramid of energy Graphical representation of the total energy contents at different trophic levels in an ecosystem.

Pyrimidine (peer im´ a deen) A type of nitrogenous base. The pyrimidines cytosine, thymine, and uracil are found in nucleic acids.

Pyrogen A substance that causes fever.

Pyruvate A three-carbon acid; the end product of glycolysis and the raw material for the citric acid cycle.

Q_{10} A value that compares the rate of a biochemical process or reaction over a 10°C range of temperature. A process that is not temperature-sensitive has a Q_{10} of 1. Values of 2 or 3 mean the reaction speeds up as temperature increases.

Quantum (kwon´ tum) [L. *quantus*: how great] An indivisible unit of energy.

Quaternary structure Of aggregating proteins, the arrangement of polypeptide subunits.

R factor (resistance factor) A plasmid that contains one or more genes that encode resistance to antibiotics.

Radial symmetry The condition in which two halves of a body are mirror images of each other regardless of the angle of the cut, providing the cut is made along the center line. Thus, a cylinder cut lengthwise down its center displays this form of symmetry. (Contrast with bilateral symmetry.)

Radioisotope A radioactive isotope of an element. Examples are carbon-14 (^{14}C) and hydrogen-3, or tritium (^{3}H).

Radiometric clock The use of the regular, known rates of decay of radioisotopes of elements to determine dates of events in the distant past.

Radiotherapy Treatment, as of cancer, with X- or gamma rays.

Radula The toothed feeding organ of many mollusks. Used to scrape prey from hard substrates.

Rain shadow A region of low precipitation on the leeward side of a mountain range.

Ramet The repeated morphological units of sessile, modular organisms. (Contrast with genet.)

Random drift Evolution (change in gene proportions) by chance processes alone.

Rate constant Of a particular chemical reaction, a constant which, when multiplied by the concentration(s) of reactant(s), gives the rate of the reaction.

Reactant A chemical substance that enters into a chemical reaction with another substance.

Reaction, chemical A process in which atoms combine or change bonding partners.

Reaction wood Modified wood produced in branches in response to gravitational stimulation. Gymnosperms produce compression wood that tends to push the branch up; angiosperms produce tension wood that tends to pull the branch up.

Realized niche The actual niche occupied by an organism; it differs from the fundamental niche because of the presence of other species.

Receptacle [L. *receptaculum*: reservoir] In an angiosperm flower, the end of the stem to which all of the various flower parts are attached.

Receptive field Of a neuron, the area on the retina from which the activity of that neuron can be influenced.

Receptor-mediated endocytosis A form of endocytosis in which macromolecules in the environment bind specific receptor proteins in the plasma membrane and are brought into the cell interior in coated vesicles.

Receptor potential The change in the resting potential of a sensory cell when it is stimulated.

Recessive See dominance.

Reciprocal altruism The exchange of altruistic acts between two or more individuals. The acts may be separated considerably in time.

Reciprocal crosses A pair of crosses, in one of which a female of genotype A mates with a male of genotype B and in the other of which a female of genotype B mates with a male of genotype A.

Recombinant An individual, meiotic product, or single chromosome in which genetic materials originally present in two individuals end up in the same haploid complement of genes. The reshuffling of genes can be either by independent segragation, or by crossing over between homologous chromosomes. For example, a human may pass on genes from both parents in a single haploid gamete.

Recombinant DNA technology The application of genetic tools (restriction endonucleases, plasmids, and transformation) to the production of specific proteins by biological "factories" such as bacteria.

Rectum The terminal portion of the gut, ending at the anus.

Redirected activity The direction of some behavior, such as aggression, away from the primary target and toward another, less appropriate object.

Redox reaction A chemical reaction in which one reactant becomes oxidized and the other becomes reduced.

Reducing agent A substance that can donate electrons to another substance. The reducing agent becomes oxidized, and its partner becomes reduced.

Reduction (re duk´ shun) Gain of electrons; the reverse of oxidation. Most reductions lead to the storage of chemical energy, which can be released later by an oxidation reaction. Energy storage compounds such as sugars and fats are highly reduced compounds. (Contrast with oxidation.)

Reflex An automatic action, involving only a few neurons (in vertebrates, often in the spinal cord), in which a motor response swiftly follows a sensory stimulus.

Refractory period Of a neuron, the time interval after an action potential, during which another action potential cannot be elicited.

Region In biogeography, a major division of the world distinguished by its peculiar animals or plants. For example, Africa south of the Sahara is recognized as constituting the Ethiopian region.

Regulative development A pattern of animal embryonic development in which the fates of the first blastomeres are not absolutely fixed. (Contrast with mosaic development.)

Regulatory gene A gene that contains the information for making a regulatory macromolecule, often a repressor protein.

Releaser A sensory stimulus that triggers a fixed action pattern.

Releasing hormone One of several hypothalamic hormones that stimulates the secretion of anterior pituitary hormone.

REM sleep A sleep state characterized by dreaming, skeletal muscle relaxation, and rapid eye movements.

Renal [L. *renes*: kidneys] Relating to the kidneys.

Replica plating A technique used in the selection of colonies of cells with a desired genotype.

Replication fork A point at which a DNA molecule is replicating. The fork forms by the unwinding of the parent molecule.

Repressible enzyme An enzyme whose synthesis can be decreased or prevented by the presence of a particular compound.

Repressor A protein coded by the regulatory gene. The repressor can bind to a specific operator and prevent transcription of the operon.

Reproductive isolating mechanism Any trait that prevents individuals from two different populations from producing fertile hybrids.

Reproductive isolation The condition in which a population is not exchanging genes with other populations of the same species.

Reproductive value The expected contribution of an individual of a particular age to the future growth of the population to which it belongs.

Resolving power Of an optical device such as a microscope, the smallest distance between two lines that allows the lines to be seen as separate from one another.

Resource Something in the environment required by an organism for its maintenance and growth that is consumed in the process of being used.

Resource defense polygamy A breeding system in which individuals of one sex (usually males) defend resources that are attractive to individuals of the other sex (usually females); individuals holding better resources attract more mates.

Respiration (res pi ra´ shun) [L. *spirare*: to breathe] (1) Cellular respiration; the oxidation of the end products of glycolysis with the storage of much energy in ATP. The oxidant in the respiration of eukaryotes is oxygen gas. Some bacteria can use nitrate or sulfate instead of O_2. (2) Breathing.

Respiratory chain The terminal reactions of cellular respiration, in which electrons are passed from NAD or FAD, through a series of intermediate carriers, to molecular oxygen, with the concomitant production of ATP.

Respiratory uncoupler   A substance that allows protons to cross the inner mitochondrial membrane without the concomitant formation of ATP, thus uncoupling respiration from phosphorylation.

Resting potential   The membrane potential of a living cell at rest. In cells at rest, the interior is negative to the exterior. (Contrast with action potential, electrotonic potential.)

Restoration ecology   The science and practice of restoring damaged or degraded ecosystems.

Restriction endonuclease   Any one of several enzymes, produced by bacteria, that break foreign DNA molecules at very specific sites. Some produce "sticky ends." Extensively used in recombinant DNA technology.

Restriction map   A partial genetic map of a DNA molecule, showing the points at which particular restriction endonuclease recognition sites reside.

Retina (rett´ in uh) [L. *rete*: net]   The light-sensitive layer of cells in the vertebrate or cephalopod eye.

Retinal   The light-absorbing portion of visual pigment molecules. Derived from β-carotene.

Retrovirus   An RNA virus that contains reverse transcriptase. Its RNA serves as a template for cDNA production, and the cDNA is integrated into a chromosome of the mammalian host cell.

Reverse transcriptase   An enzyme that catalyzes the production of DNA (cDNA), using RNA as a template; essential to the reproduction of retroviruses.

Reversion (genetic)   A mutational event that restores wild-type phenotype to a mutant.

RFLP (Restriction fragment length polymorphism)   Coexistence of two or more patterns of restriction fragments (patterns produced by restriction enzymes), as revealed by a probe. The polymorphism reflects a difference in DNA sequence on homologous chromosomes.

Rhizoids (rye´ zoids) [Gr. *rhiza*: root]   Hairlike extensions of cells in mosses, liverworts, and a few vascular plants that serve the same function as roots and root hairs in vascular plants. The term is also applied to branched, rootlike extensions of some fungi and algae.

Rhizome (rye´ zome) [Gr. *rhizoma*: mass of roots]   A special underground stem (as opposed to root) that runs horizontally beneath the ground.

Rhodopsin   A photopigment used in the visual process of transducing photons of light into changes in the membrane potential of photosensory cells.

Ribonucleic acid   See RNA.

Ribose (rye´ bose)   A sugar of chemical formula $C_5H_{10}O_5$, one of the building blocks of ribonucleic acids.

Ribosomal RNA (rRNA)   Several species of RNA that are incorporated into the ribosome.

Ribosome   A small organelle that is the site of protein synthesis.

Ribozyme   An RNA molecule with catalytic activity.

Ribulose 1,5-bisphosphate (RuBP)   The compound in chloroplasts which reacts with carbon dioxide in the first reaction of the Calvin-Benson cycle.

Risk cost   The increased chance of being injured or killed as a result of performing a behavior, compared to resting.

RNA (ribonucleic acid)   A nucleic acid using ribose. Various classes of RNA are involved in the transcription and translation of genetic information. RNA serves as the genetic storage material in some viruses.

RNA polymerase   An enzyme that catalyzes the formation of RNA from a DNA template.

RNA splicing   The last stage of RNA processing in eukaryotes, in which the transcripts of introns are excised through the action of small nuclear ribonucleoprotein particles (snRNP).

Rods   Light-sensitive cells (photosensors) in the retina. (Contrast with cones.)

Root cap   A thimble-shaped mass of cells, produced by the root apical meristem, that protects the meristem and that is the organ that perceives the gravitational stimulus in root gravitropism.

Root hair   A specialized epidermal cell with a long, thin process that absorbs water and minerals from the soil solution.

Round dance   The dance performed on the vertical surface of a honeycomb by a returning honeybee forager when she has discovered a food source less than 100 meters from the hive.

Round window   A flexible membrane between the middle and inner ear that distributes pressure waves in the fluid of the inner ear.

rRNA   See ribosomal RNA.

Rubisco (RuBP carboxylase)   Enzyme that combines carbon dioxide with ribulose bisphosphate to produce 3-phosphoglycerate, the first product of C_3 photosynthesis. The most abundant protein on Earth.

Rumen (rew´ mun)   The first division of the ruminant stomach. It stores and initiates bacterial fermentation of food. Food is regurgitated from the rumen for further chewing. (Contrast with abomasum, omasum.)

Ruminant   An herbivorous, cud-chewing mammal such as a cow, sheep, or deer, having a stomach consisting of four compartments.

S phase   In the cell cycle, the stage of interphase during which DNA is replicated. (Contrast with G_1 phase, G_2 phase.)

Sap   An aqueous solution of nutrients, minerals, and other substances that passes through the xylem of plants.

Saprobe [Gr. *sapros*:rotten + *bios*: life]   An organism (usually a bacterium or fungus) that obtains its carbon and energy directly from dead organic matter.

Sarcomere (sark´ o meer) [Gr. *sark*: flesh + *meros*: a part]   The contractile unit of a skeletal muscle.

Saturated hydrocarbon   A compound consisting only of carbon and hydrogen, with the hydrogen atoms connected by single bonds.

Schizocoelous development   Formation of a coelom during embryological development by a splitting of mesodermal masses.

Schwann cell   A glial cell that wraps around part of the axon of a peripheral neuron, creating a myelin sheath.

Sclereid   A type of sclerenchyma cell, commonly found in nutshells, that is not elongated.

Sclerenchyma (skler eng´ kyma)   A plant tissue composed of cells with heavily thickened cell walls, dead at functional maturity. The principal types of sclerenchyma cells are fibers and sclereids.

Scrapie-associated fibril   A type of protein fibril found in nervous tissues of mammals infected with certain diseases, notably scrapie, kuru, and Creutzfeld-Jacob disease. Little is known about these fibrils, including whether they are the causal agents of the diseases.

Scrotum (skrote´ um)   A sac of skin that contains the testicles in most species of mammals.

Secondary active transport   Form of active transport in which ions or molecules are transported against their concentration gradient using energy obtained by relaxation of a gradient of sodium ion concentration rather than directly from ATP. (Contrast with primary active transport.)

Secondary compound   A compound synthesized by a plant that is not needed for basic cellular metabolism. Typically has an antiherbivore or antiparasite function.

Secondary growth   In plants, growth produced by vascular and cork cambia, contributing to an increase in girth. (Contrast with primary growth.)

Secondary structure   Of a protein, localized regularities of structure, such as the α-helix and the β-pleated sheet.

Secondary wall   Wall layers laid down by a plant cell that has ceased growing; often impregnated with lignin or suberin.

Second law of thermodynamics   States that in any real (irreversible) process, there is a decrease in free energy and an increase in entropy.

Second messenger   A compound, such as cyclic AMP, that is released within a target cell after a hormone or other "first messenger" has bound to a surface receptor on a cell; the second messenger triggers further reactions within the cell.

Secretin (si kreet´ in) A peptide hormone secreted by the upper region of the small intestine when acidic chyme is present. Stimulates the pancreatic duct to secrete bicarbonate ions.

Section A thin slice, usually for microscopy, as a tangential section or a transverse section.

Seed A fertilized, ripened ovule of a gymnosperm or angiosperm. Consists of the embryo, nutritive tissue, and a seed coat.

Seed crop The number of seeds produced by a plant during a particular bout of reproduction.

Seedling A young plant that has grown from a seed (rather than by grafting or by other means.)

Segmentation genes In insect larvae, genes that determine the number and polarity of larval segments.

Segregation (genetic) The separation of alleles, or of homologous chromosomes, from one another during meiosis so that each of the haploid daughter nuclei produced by meiosis contains one or the other member of the pair found in the diploid mother cell, but never both.

Selective permeability A characteristic of a membrane, allowing certain substances to pass through while other substances are excluded.

Self-differentiating Behavior that develops without experience with the normal objects toward which it is usually directed and without any practice. (See also instinct.)

Selfish act A behavioral act that benefits its performer but harms the recipients.

Self-pollination The fertilization of a plant by its own pollen. (Contrast with cross-pollination.)

Semelparous organism An organism that reproduces only once in its lifetime. (Contrast with iteroparous.)

Semen (see´ men) [L.: seed] The thick, whitish liquid produced by the male reproductive organ in mammals, containing the sperm.

Semicircular canals Part of the vestibular system of mammals.

Semiconservative replication The common way in which DNA is synthesized. Each of the two partner strands in a double helix acts as a template for a new partner strand. Hence, after replication, each double helix consists of one old and one new strand.

Seminiferous tubules The tubules within the testes within which sperm production occurs.

Senescence [L. senescere: to grow old] Aging; deteriorative changes with aging.

Sensor A sensory cell; a cell transduces a physical or chemical stimulus into a membrane potential change.

Sensory neuron A neuron leading from a sensory cell to the central nervous system. (Contrast with motor neuron.)

Sepal (see´ pul) One of the outermost structures of the flower, usually protective in function and enclosing the rest of the flower in the bud stage.

Septum [L.: partition] A membrane or wall between two cavities.

Sertoli cells Cells in the seminiferous tubules that nuture the developing sperm.

Serum That part of the blood plasma that remains after clots have formed and been removed.

Sessile (sess´ ul) [L. sedere: to sit] Permanently attached; not moving.

Sertoli cells Cells in the seminiferous tubules that nuture the developing sperm.

Set point In a regulatory system, the threshold sensitivity to the feedback stimulus.

Sex chromosome In organisms with a chromosomal mechanism of sex determination, one of the chromosomes involved in sex determination. One sex chromosome, the X chromosome, is present in two copies in one sex and only one copy in the other sex. The autosomes, as opposed to the sex chromosomes, are present in two copies in both sexes. In many organisms, there is a second sex chromosome, the Y chromosome, that is found in only one sex—the sex having only one copy of the X.

Sexduction See F-duction.

Sex linkage The pattern of inheritance characteristic of genes located on the sex chromosomes of organisms having a chromosomal mechanism for sex determination. The sex that is diploid with respect to sex chromosomes can assume three genotypes: homozygous wild-type, homozygous mutant, or heterozygous carrier. The other sex, haploid for sex chromosomes, is either hemizygous wild-type or hemizygous mutant.

Sexuality The ability, by any of a multitude of mechanisms, to bring together in one individual genes that were originally carried by two different individuals. The capacity for genetic recombination.

Sexual selection Selection by one sex of characteristics in individuals of the opposite sex. Also, the favoring of characteristics in one sex as a result of competition among individuals of that sex for mates.

Shoot The aerial part of a vascular plant, consisting of the leaves, stem(s), and flowers.

Sibling A brother or sister.

Sieve plate In sieve tubes, the highly specialized end walls in which are concentrated the clusters of pores through which the protoplasts of adjacent sieve tube elements are interconnected.

Sieve tube A column of specialized cells found in the phloem, specialized to conduct organic matter from sources (such as photosynthesizing leaves) to sinks (such as roots). Found principally in flowering plants.

Sieve tube element A single cell of a sieve tube, containing cytoplasm but relatively few organelles, with highly specialized perforated end walls leading to elements above and below.

Sign stimulus The single stimulus, or one out of a very few stimuli, by which an animal distinguishes key objects, such as an enemy, or a mate, or a place to nest, etc.

Signal A component of a behavior that transmits information to another individual that influences the future behavior of the receiver.

Signal sequence N-terminal sequence of a protein that directs the protein through a particular cellular membrane.

Simple development In insects and other arthorpods, development in which eggs hatch into juveniles that are similar in form to adults.

Simple fruit A fruit that develops from a single ovary. (Contrast with accessory fruit, aggregate fruit, multiple fruit.)

Sinoatrial node (sigh´ no ay´ tree al) The pacemaker of the mammalian heart.

Sinus (sigh´ nus) [L. sinus: a bend, hollow] A cavity in a bone, a tissue space, or an enlargement in a blood vessel.

Skeletal muscle See striated muscle.

Sliding filament theory A proposed mechanism of muscle contraction based on formation and breaking of crossbridges between actin and myosin filaments, causing them to slide together.

Small intestine The portion of the gut between the stomach and the colon, consisting of the duodenum, the jejunum, and the ileum.

Small nuclear ribonucleoprotein particle (snRNP) A complex of an enzyme and a small nuclear RNA molecule, functioning in RNA splicing.

Smooth muscle One of three types of muscle tissue. Usually consists of sheets of mononucleated cells innervated by the autonomic nervous system.

Social insect One of the kinds of insect that form colonies with reproductive castes and worker castes; in particular, the termites, ants, social bees, and social wasps.

Society A group of individuals belonging to the same species and organized in a cooperative manner; in the broadest sense, includes parents and their offspring.

Sodium cotransport Carrier-mediated transport of molecules across membranes driven by sodium ions binding to the same carrier and moving down their concentration gradient.

Sodium–potassium pump The complex protein in plasma membranes that is responsible for primary active transport; it pumps sodium ions out of the cell and potassium ions into the cell, both against their concentration gradients.

Solute   A substance that is dissolved in a liquid (solvent).

Solution   A liquid (solvent) and its dissolved solutes.

Solvent   A liquid that has dissolved or can dissolve one or more solutes.

Somatic [Gr. *soma*: body]   Pertaining to the body, or body cells (rather than to germ cells).

Somite (so´ might)   One of the segments into which an embryo becomes divided longitudinally, leading to the eventual segmentation of the animal as illustrated by the spinal column, ribs, and associated muscles.

Southern blotting   Transfer of DNA fragments from an electrophoretic gel to a sheet of paper or other absorbent material for analysis with a probe.

Spatial summation   In the production or inhibition of action potentials in a postsynaptic neuron, the interaction of depolarizations and hyperpolarizations produced by several terminal boutons.

Spawning   The direct release of sex cells into the water.

Speciation (spee´ shee ay´ shun)   The process of splitting one population into two populations that are reproductively isolated from one another.

Species (spee´ shees) [L.: kind]   The basic lower unit of classification, consisting of a population or series of populations of closely related and similar organisms. The more narrowly defined "biological species" consists of individuals capable of interbreeding freely with each other but not with members of other species.

Species diversity   A weighted representation of the species of organisms living in a region; large and common species are given greater weight than are small and rare ones. (Contrast with species richness.)

Species pool   All the species potentially available to colonize a particular habitat.

Species richness   The number of species of organisms living in a region. (Contrast with species diversity.)

Specific heat   The amount of energy that must be absorbed by a gram of a substance to raise its temperature by one degree centigrade. By convention, water is assigned a specific heat of one.

Sperm [Gr. *sperma*: seed]   A male reproductive cell.

Spermatocyte (spur mat´ oh site) [Gr. *sperma*: seed + *kytos*: cell]   The cell that gives rise to the sperm in animals.

Spermatogenesis (spur mat´ oh jen´ e sis) [Gr. *sperma*: seed + *genesis*: source]   Male gametogenesis, leading to the production of sperm.

Spermatogonia   Undifferentiated germ cells that give rise to primary spermatocytes and hence to sperm.

Spermatophore   A package of sperm deposited in the environment by an invertebrate male, and then either inserted by him into the reproductive tract of the female or taken up by the female herself.

Sphincter (sfingk´ ter) [Gr. *sphinkter*: that which binds tight]   A ring of muscle that can close an orifice, for example at the anus.

Spindle apparatus   An array of microtubules stretching from pole to pole of a dividing nucleus and playing a role in the movement of chromosomes at nuclear division. Named for its shape.

Spiracle (spy´ rih kel) [L. *spirare*: to breathe]   An opening of the treacheal respiratory system of terrestrial arthorpods.

Spiteful act   A behavioral act that harms both the actor and the recipient of the act.

Spongy parenchyma   In leaves, a layer of loosely packed photosynthetic cells with extensive intercellular spaces for gas diffusion. Frequently found between the palisade parenchyma and the lower epidermis.

Spontaneous generation   The idea that life is generated continually from nonliving matter. Usually distinguished from the current idea that life evolved from nonliving matter under primordial conditions at an early stage in the history of earth.

Spontaneous reaction   A chemical reaction which will proceed on its own, without any outside influence. A spontaneous reaction need not be rapid.

Sporangiophore [Gr. *phore*: to bear]   Any branch bearing one or more sporangia.

Sporangium (spor an´ gee um) [Gr. *spora*: seed + *angeion*: vessel or reservoir]   In plants and fungi, any specialized stucture within which one or more spores are formed.

Spore [Gr. *spora*: seed]   Any asexual reproductive cell capable of developing into an adult plant without gametic fusion. Haploid spores develop into gametophytes, diploid spores into sporophytes. In prokaryotes, a resistant cell capable of surviving unfavorable periods.

Sporophyll (spor´ o fill) [Gr. *spora*: seed + *phyllon*: leaf]   Any leaf or leaflike structure that bears sporangia; refers to carpels and stamens of angiosperms and to sporangium-bearing leaves on ferns, for example.

Sporophyte (spor´ o fyte) [Gr. *spora*: seed + *phyton*: plant]   In plants with alternation of generations, the diploid phase that produces the spores. (Contrast with gametophyte.)

Stabilizing selection   Selection against the extreme phenotypes in a population, so that the intermediate types are favored. (Contrast with disruptive selection.)

Stamen (stay´ men) [L.: thread]   A male (pollen-producing) unit of a flower, usually composed of an anther, which bears the pollen, and a filament, which is a stalk supporting the anther.

Starch [O.E. *stearc*: stiff]   An α-linked polymer of glucose; used by plants as a means of storing energy and carbon atoms.

Stasis   Period during which little or no evolutionary change takes place within a lineage or groups of lineages.

Statocyst (stat´ oh sist) [Gk. *statos*: stationary + *kystos*: pouch]   An organ of equilibrium in some invertebrates.

Statolith(stat´ oh lith) [Gk. *statos*: stationary + *lithos*: stone]   A solid object that responds to gravity or movement and stimulates the mechanosensors of a statocyst.

Stele (steel) [Gr. *stele*: pillar]   The central cylinder of vascular tissue in a plant stem.

Stem cell   A cell capable of extensive proliferation, generating more stem cells and a large clone of differentiated progeny cells, as in the formation of red blood cells.

Step cline   A sudden change in one or more traits of a species along a geographical gradient.

Steroid   Any of numerous lipids based on a 17-carbon atom ring system.

Sticky ends   On a piece of two-stranded DNA, short, complementary, one-stranded regions produced by the action of a restriction endonuclease. Sticky ends allow the joining of segments of DNA from different sources.

Stigma [L.: mark, brand]   The part of the pistil at the apex of the style, which is receptive to pollen, and on which pollen germinates.

Stimulus   Something causing a response; something in the environment detected by a receptor.

Stolon   A horizontal stem that forms roots at intervals.

Stoma (plural: stomata) [Gr. *stoma*: mouth, opening]   Small opening in the plant epidermis that permits gas exchange; bounded by a pair of guard cells whose osmotic status regulates the size of the opening.

Stratosphere   The part of the atmosphere above the troposphere; extends upward to approximately 50 kilometers above the surface of the earth; contains very little water.

Stratum (plural strata)   A layer or sedimentary rock laid down at a particular time in a past.

Striated muscle   Contractile tissue characterized by multinucleated cells containing highly ordered arrangements of actin and myosin microfilaments. Also known as **skeletal muscle**.

Strobilus (strobe´ a lus) [Gr. *strobilos*: a cone]   The cone, or characteristic multiple fruit, of the pine and other gymnosperms. Also, a cone-shaped mass of sprophylls found in club mosses.

Stroma   The fluid contents of an organelle, such as a chloroplast.

Stromatolite   A composite, flat-to-domed structure composed of successive mineral layers. Some are known to be produced by the action of bacteria in salt or fresh water, and some ancient ones are considered to be evidence for early life on the earth.

Structural formula   A representation of the positions of atoms and bonds in a molecule.

Structural gene A gene that encodes the primary structure of a protein.

Style [Gr. *stylos*: pillar or column] In flowering plants, a column of tissue extending from the tip of the ovary, and bearing the stigma or receptive surface for pollen at its apex.

Sub- [L.: under] A prefix often used to designate a structure that lies beneath another or is less than another. For example, subcutaneous, subspecies.

Suberin A waxy material serving as a waterproofing agent in cork and in the Casparian strips of the endodermis in plants.

Submucosa (sub mew koe´ sah) The tissue layer just under the epithelial lining of the lumen of the digestive tract. (Contrast with mucosa.)

Substrate (sub´ strayte) The molecule or molecules on which an enzyme exerts catalytic action.

Substrate level phosphorylation ATP formation resulting from direct transfer of a phosphate group to ADP from an intermediate in glycolysis. (Contrast with oxidative phosphorylation.)

Succession In ecology, the gradual, sequential series of changes in species composition of a community following a disturbance.

Sulfhydryl group The –SH group.

Summation The ability of a neuron to fire action potentials in response to numerous subthreshold postsynaptic potentials arriving simultaneously at differentiated places on the cell, or arriving at the same site in rapid succession.

Supercoiling Coiling on coiling, as in DNA during prophase.

Supernormal stimulus Any stimulus, or any intensity of a variable stimulus, that is preferred by animals over the natural sign stimulus.

Suppressor T cells T cells that inhibit the res-ponses of B cells and other T cells to antigens. (Contrast with cytotoxic T cells, helper T cells.)

Surface tension A measure of the cohesiveness of the surface of a liquid. As a result of hydrogen bonding, water has a very high surface tension, allowing some insects to walk on the water surface.

Surface-to-volume ratio For any cell, organism, or geometrical solid, the ratio of surface area to volume; this is an important factor in setting an upper limit on the size a cell or organism can attain.

Surfactant A substance that decreases the surface tension of a liquid. Lung surfactant, secreted by cells of the alveoli, is mostly phospholipid and decreases the amount of work necessary to inflate the lungs.

Survivorship curve A plot of the logarithm of the fraction of individuals still alive, as a function of time.

Suspensor In plants, a cell or group of cells derived from the zygote, but not actually part of the embryo proper, which in some seed plants pushes the young embryo deeper into nutritive gametophyte tissue or endosperm by its growth.

Swim bladder An internal gas-filled organ that helps fishes maintain their position in the water column; later evolved into an organ for gas exchange in some lineages.

Symbiosis (sim´ bee oh´ sis) [Gr.: to live together] The living together of two or more species in a prolonged and intimate ecological relationship. (See parasitism, commensalism, mutualism.)

Symmetry In biology, the property that two halves of an object are mirror images of each other. (See bilateral symmetry and radial symmetry.)

Sympathetic nervous system A division of the autonomic (involuntary) nervous system. Its activities include increasing blood pressure and acceleration of the heartbeat. The neurotransmitter at the sympathetic terminals is epinephrine or norepinephrine. (Contrast with parasympathetic nervous system.)

Sympatric (sim pat´ rik) [Gr. *syn*: together + *patria*: homeland] Referring to populations whose geographic regions overlap at least in part.

Sympatric speciation Formation of new species even though members of the daughter species overlap in their distribution during the speciation process. (Contrast with geographic, parapatric speciation.)

Symplast The continuous meshwork of the interiors of living cells in the plant body, resulting from the presence of plasmodesmata. (Contrast with apoplast.)

Symport A membrane transport protein that carries two substances in the same direction across the membrane. (Contrast with antiport.)

Synapse (sin´ aps) [Gr. *syn*: together + *haptein*: to fasten] The narrow gap between the terminal bouton of one neutron and the dendrite or cell body of another.

Synapsis (sin ap´ sis) The highly specific parallel alignment (pairing) of homologous chromosomes during the first division of meiosis.

Synaptic vesicle A membrane-bounded vesicle, containing neurotransmitter, which is produced in and discharged by the presynaptic neuron.

Synergids (sin nur´ jids) Two cells found close to the egg cell in the angiosperm embryo sac; they disappear shortly after fertilization.

Syngamy (sing´ guh mee) [Gr. *sun-*: together + *gamos*: marriage] Union of gametes. Also known as fertilization.

Syrinx (sear´ inks) [Gr.: pipe, cavity] A specialized structure at the junction of the trachea and the primary bronchi leading to the lungs. The vocal organ of birds.

Systematics The scientific study of the diversity of organisms.

Systemic circulation The part of the circulatory system serving those parts of the body other than the lungs or gills.

Systole (sis´ tuh lee) [Gr.: contraction] Contraction of a chamber of the heart, driving blood forward in the circulatory system.

T cell A type of lymphocyte, involved in the cellular immune response. The final stages of its development occur in the thymus gland. (Contrast with B cell; see also cytotoxic T cell, helper T cell, suppressor T cell.)

T cell receptor A protein on the surface of a T cell that recognizes the antigenic determinant for which the cell is specific.

Target cell A cell which has the appropriate receptors to bind and respond to a particular hormone or other chemical mediator.

Taste bud A structure in the epithelium of the tongue that includes a cluster of chemosensors innervated by sensory neurons.

TATA box An eight-base-pair sequence, found about 25 base pairs before the starting point for transcription in many eukaryotic promoters, that binds a transcription factor and thus helps initiate transcription.

Taxis (tak´ sis) [Gr. *taxis*: arrange, put in order] The movement of an organism in a particular direction with reference to a stimulus. A taxis usually involves the employment of one sense and a movement directly toward or away from the stimulus, or else the maintenance of a constant angle to it. Thus a positive phototaxis is movement toward a light source, negative geotaxis is movement upward (away from gravity), and so on.

Taxon A unit in a taxonomic system.

Taxonomy (taks on´ oh me) [Gr. *taxis*: arrange, classify] The science of classification of organisms.

Telophase (tee´ lo phase) [Gr. *telos*: end] The final phase of mitosis or meiosis during which chromosomes became diffuse, nuclear envelopes reform, and nucleoli begin to reappear in the daughter nuclei.

Template In biochemistry, a molecule or surface upon which another molecule is synthesized in complementary fashion, as in the replication of DNA. In the brain, a pattern that responds to a normal input but not to incorrect inputs.

Template strand In a stretch of double-stranded DNA, the strand that is transcribed.

Temporal summation In the production or inhibition of action potentials in a postsynaptic neuron, the interaction of depolarizations or hyperpolarizations produced by rapidly repeated stimulation of a single point.

Tendon A collagen-containing band of tissue that connects a muscle with a bone.

Tension wood See reaction wood.

Tepal In an angiosperm flower, a sterile modified leaf. This term is used to refer to such flower parts when one is unable to distinguish between petals and sepals.

Terrestrial (ter res´ tree al) [L. *terra*: earth] Pertaining to the land. (Contrast with aquatic, marine.)

Territory A fixed area from which an animal or group of animals excludes other members of the same species by aggressive behavior or display.

Tertiary structure In reference to a protein, the relative locations in three-dimensional space of all the atoms in the molecule. The overall shape of a protein. (Contrast with primary, secondary, and quaternary structures.)

Test cross A cross of a dominant-phenotype individual (which may be either heterozygous or homozygous) with a homozygous-recessive individual.

Testis (tes´ tis) (plural: testes) [L.: witness] The male gonad; that is, the organ that produces the male sex cells.

Testosterone (tes toss´ tuhr own) A male sex steroid hormone.

Tetanus [Gr. *tetanos*: stretched] (1) In physiology, a state of sustained, maximal muscular contraction caused by rapidly repeated stimulation. (2) In medicine, an often-fatal disease ("lockjaw") caused by the bacterium *Clostridium tetani*.

Thalamus A region of the vertebrate forebrain; involved in integration of sensory input.

Thallus (thal´ us) [Gr.: sprout] Any algal body which is not differentiated into root, stem, and leaf.

Thermocline In a body of water, the zone where the temperatures change abruptly to about 4°C.

Thermoneutral zone The range of temperatures over that an endotherm does not have to expend extra energy to thermoregulate.

Thermosensor A cell or structure that responds to changes in temperature.

Thoracic cavity The portion of the mammalian body cavity bounded by the ribs, shoulders, and diaphragm. Contains the heart and the lungs.

Thorax In an insect, the middle region of the body, between the head and abdomen. In mammals, the part of the body between the neck and the diaphragm.

Thrombin An enzyme that converts fibrinogen to fibrin, thus triggering the formation of blood clots.

Thrombus (throm´ bus) [Gk. *thrombos*: clot] A blood clot that forms within a blood vessel and remains attached to the wall of the vessel. (Contrast with embolus.)

Thylakoid A flattened sac within a chloroplast. The membranes of the numerous thylakoids contain all of the chlorophyll in a plant, in addition to the electron carriers of photophosphorylation. Thylakoids stack to form grana.

Thymus A ductless, glandular portion of the lymphoid system, involved in development of the immune system of vertebrates.

Thyroid [Gr. *thyreos*: door-shaped] A two-lobed gland in vertebrates. Produces the hormone thyroxin.

Thyrotropic hormone A hormone that is produced in the pituitary gland of amphibia such as frogs and transported in the bloodstream to the thyroid gland, inducing the thyroid gland to produce the thyroid hormone that regulates metamorphosis from tadpole to adult frog.

Tight junction A junction between epithelial cells, in which there is no gap whatever between the adjacent cells. Materials may get through a tight junction only by entering the epithelial cells themselves.

Tissue A group of similar cells organized into a functional unit and usually integrated with other tissues to form part of an organ such as a heart or leaf.

Tonus A low level of muscular tension that is maintained even when the body is at rest.

Tornaria (tor nare´ e ah) [L. *tornus*: lathe] The free-swimming ciliated larva of certain echinoderms and hemichordates; its existence indicates the evolutionary relationship of these two groups.

Totipotency In a cell, the condition of possessing all the genetic information and other capacities necessary to form an entire individual.

Toxigenicity The ability of a bacterium to produce chemical substances injurious to the tissues of the host organism.

Trachea (tray´ kee ah) [Gr. *trakhoia*: a small rough artery] A tube that carries air to the bronchi of the lungs of vertebrates, or to the cells of arthropods.

Tracheid (tray´ kee id) A distinctive conducting and supporting cell found in the xylem of nearly all vascular plants, characterized by tapering ends and walls that are pitted but not perforated.

Trade winds The winds that blow toward the intertropical convergence zone from the northeast and southeast.

Transcription The synthesis of RNA, using one strand of DNA as the template.

Transcription factors Proteins that assemble on a eukaryotic chromosome, allowing RNA polymerase II to perform transcription.

Transduction (1) Transfer of genes from one bacterium to another, with a bacterial virus acting as the carrier of the genes. (2) In sensory cells, the transformation of a stimulus (e.g., light energy, sound pressure waves, chemical or electrical stimulants) into action potentials.

Transfection Uptake, incorporation, and expression of recombinant DNA.

Transfer cells A modified parenchyma cell that transports mineral ions from its cytoplasm into its cell wall, thus moving the ions from the symplast into the apoplast.

Transfer RNA (tRNA) A category of relatively small RNA molecules (about 75 nucleotides). Each kind of transfer RNA is able to accept a particular activated amino acid from its specific activating enzyme, after which the amino acid is added to a growing polypeptide chain.

Transformation Mechanism for transfer of genetic information in bacteria in which pure DNA extracted from bacteria of one genotype is taken in through the cell surface of bacteria of a different genotype and incorporated into the chromosome of the recipient cell. By extension, the term has come to be applied to phenomena in other organisms in which specific genetic alterations have been produced by treatment with purified DNA from genetically marked donors.

Transgenic Containing recombinant DNA incorporated into its genetic material.

Translation The synthesis of a protein (polypeptide). This occurs on ribosomes, using the information encoded in messenger RNA.

Translocation (1) In genetics, a rare mutational event that moves a portion of a chromosome to a new location, generally on a nonhomologous chromosome. (2) In vascular plants, movement of solutes in the phloem.

Transpiration [L. *spirare*: to breathe] The evaporation of water from plant leaves and stem, driven by heat from the sun, and providing the motive force to raise water (plus ions) from the roots.

Transposable element A segment of DNA that can move to, or give rise to copies at, another locus on the same or a different chromosome. May be a single insertion sequence or a more complex structure (transposon) consisting of two insertion sequences and one or more intervening genes.

Trichocyst (trick´ o sist) [Gr. *trichos*: hair + *kystis*: cell] A threadlike organelle ejected from the surface of ciliates, used both as a weapon and as an anchoring device.

Triglyceride A simple lipid in which three fatty acids are combined with one molecule of glycerol.

Triplet See codon.

Triplet repeat Occurrence of repeated triplet of bases in a gene, often leading to genetic disease, as does excessive repetition of CGG in the gene responsible for fragile-X syndrome.

Triploblastic Having three cell layers. (Contrast with diploblastic.)

Trisomic Containing three, rather than two members of a chromosome pair.

tRNA See transfer RNA.

Trochophore (troke´ o fore) [Gr. *trochos*: wheel + *phoreus*: bearer] The free-swimming larva of some annelids and mollusks, distinguished by a wheel-like band of cilia around the middle, and indicating an evolutionary relationship between these two groups.

Trophic level A group of organisms united by obtaining their energy from the same part of the food web of a biological community.

Tropic hormones Hormones of the anterior pituitary that control the secretion of hormones by other endocrine glands.

Tropism [Gr. *tropos*: to turn]   In plants, growth toward or away from a stimulus such as light (phototropism) or gravity (gravitropism).

Tropomyosin (troe poe my´ oh sin)   A protein that, along with actin, constitutes the thin filaments of myofibrils. It controls the interactions of actin and myosin necessary for muscle contraction.

Troposphere   The atmospheric zone reaching upward approximately 17 km in the tropics and subtropics but only to about 10 km at higher latitudes. The zone in which virtually all the water vapor in the atmosphere is located.

Trypsin   A protein-digesting enzyme. Secreted by the pancreas in its inactive form (trypsinogen), it becomes active in the duodenum of the small intestine.

T-tubules   A set of transverse tubes that penetrates skeletal muscle fibers and terminates in the sarcoplasmic reticulum. The T-system transmits impulses to the sacs, which then release CA^{2+} to initiate muscle contraction.

Tube foot   In echinoderms, a part of the water vascular system. It grasps the substratum, prey, or other solid objects.

Tube nucleus   In a pollen tube, the haploid nucleus that does not participate in double fertilization. (Contrast with generative nucleus.)

Tuber [L.: swelling]   A short, fleshy underground stem, usually much enlarged, and serving a storage function, as in the case of the potato.

Tubulin   A protein that polymerizes to form microtubules.

Tumor   A disorganized mass of cells, often growing out of control. Malignant tumors spread to other parts of the body.

Turgor   See pressure potential.

Twitch   A single unit of muscle contraction.

Tympanic membrane [Gr. *tympanum*: drum]   The eardrum.

Umbilical cord   Tissue made up of embryonic membranes and blood vessels that connects the embryo to the placenta in eutherian mammals.

Uncoupler   See respiratory uncoupler.

Understory   The aggregate of smaller plants growing beneath the canopy of dominant plants in a forest.

Unicellular (yoon´ e sell´ yer ler) [L. *unus*: one + *cella*: chamber]   Consisting of a single cell, as for example a unicellular organism. (Contrast with multicellular.)

Uniport   A membrane transport protein that carries a single substance. (Contrast with antiport, symport.)

Unitary organism   An organism that consists of only one module.

Unsaturated hydrocarbon   A compound containing only carbon and hydrogen atoms. One or more pairs of carbon atoms are connected by double bonds.

Upwelling   The upward movement of nutrient-rich, cooler water from deeper layers of the ocean.

Urea   A compound serving as the main excreted form of nitrogen by many animals, including mammals.

Ureotelic   Describes an organism in which the final product of the breakdown of nitrogen-containing compounds (primarily proteins) is urea. (Contrast with ammonotelic, uricotelic.)

Ureter (your´ uh tur) [Gr. *ouron*: urine]   A long duct leading from the vertebrate kidney to the urinary bladder or the cloaca.

Urethra (you ree´ thra) [Gr. *ouron*: urine]   In most mammals, the canal through which urine is discharged from the bladder and which serves as the genital duct in males.

Uric acid   A compound that serves as the main excreted form of nitrogen in some animals, particularly those which must conserve water, such as birds, insects, and reptiles.

Uricotelic   Describes an organism in which the final product of the breakdown of nitrogen-containing compounds (primarily proteins) is uric acid. (Contrast with ammonotelic, ureotelic.)

Urinary bladder   A structure structure that receives urine from the kidneys via the ureter, stores it, and expels it periodically through the urethra.

Urine (you´ rin) [Gk. *ouron*: urine]   In vertebrates, the fluid waste product containing the toxic nitrogenous by-products of protein and amino acid metabolism.

Uterus (yoo´ ter us) [L.: womb]   The uterus or womb is a specialized portion of the female reproductive tract in certain mammals. It receives the fertilized egg and nurtures the embryo in its early development.

Vaccination   Injection of virus or bacteria or their proteins into the body, to induce immunization. The injected material is usually attenuated (weakened) before injection.

Vacuole (vac´ yew ole) [Fr.: small vacuum]   A liquid-filled cavity in a cell, enclosed within a single membrane. Vacuoles play a wide variety of roles in cellular metabolism, some being digestive chambers, some storage chambers, some waste bins, and so forth.

Vagina (vuh jine´ uh) [L.: sheath]   In female mammals, the passage leading from the external genital orifice to the uterus; receives the copulatory organ of the male in mating.

Van der Waals interaction   A weak attraction between atoms resulting from the interaction of the electrons of one atom with the nucleus of the other atom. This attraction is about one-fourth as strong as a hydrogen bond.

Vascular (vas´ kew lar)   Pertaining to organs and tissues that conduct fluid, such as blood vessels in animals and phloem and xylem in plants.

Vascular bundle   In vascular plants, a strand of vascular tissue, including conducting cells of xylem and phloem as well as thick-walled fibers.

Vascular plants Those plants with xylem and phloem, including psilophytes, club mosses, horsetails, ferns, gymnosperms, and angiosperms. (Contrast with nonvascular plants.)

Vascular ray   In vascular plants, radially oriented sheets of cells produced by the vascular cambium, carrying materials laterally between the wood and the phloem.

Vascular tissue system   The conductive system of the plant, consisting primarily of xylem and phloem. (Contrast with dermal tissue system, ground tissue system.)

Vasopressin   See antidiuretic hormone.

Vector   (1) An agent, such as an insect, that carries a pathogen affecting another species. (2) A plasmid or virus that carries an inserted piece of DNA into a bacterium for cloning purposes in recombinant DNA technology.

Vegetal pole   In some eggs, zygotes, and embryos, the pole near the bulk of the yolk. (Contrast with animal pole.)

Vegetative   Nonreproductive, or nonflowering, or asexual.

Vein [L. *vena*: channel]   A blood vessel that returns blood to the heart. (Contrast with artery.)

Vena cava [L.: hollow vein]   One of a pair of large veins that carry blood from the systemic circulatory system into the heart.

Ventral [L. *venter*: belly, womb]   Toward or pertaining to the belly or lower side. (Contrast with dorsal.)

Ventricle   A muscular heart chamber that pumps blood through the body.

Vernalization [L. *vernalis*: belonging to spring]   Events occurring during a required chilling period, leading eventually to flowering. Vernalization may require many weeks of below-freezing temperatures.

Vertebral column   The jointed, dorsal column that is the primary support structure of vertebrates.

Vertebrate   An animal whose nerve cord is enclosed in a backbone of bony segments, called vertebrae. The principal groups of vertebrate animals are the fishes, amphibians, reptiles, birds, and mammals.

Vertical evolution   Evolutionary change in a single lineage over time. Also called anagenesis.

Vessel [L. *vasculum*: a small vessel]   In botany, a tube-shaped portion of the xylem consisting of hollow cells (vessel elements) placed end to end and connected by perforations. Together with tracheids, vessel elements conduct water and minerals in the plant.

Vestibular apparatus (ves tib´ yew lar) [L. *vestibulum*: an enclosed passage]   Structures associated with the vertebrate ear; these structures sense changes in position or momentum of the head, affecing balance and motor skills.

Vestigial (ves tij´ ee al) [L. *vestigium*: footprint, track]   The remains of body structures that are no longer of adaptive value to the organism and therefore are not maintained by selection.

Vicariance (vye care´ ee unce) [L. *vicus*: change]   The splitting of the range of a taxon by the imposition of some barrier to dispersal of its members. May lead to cladogenesis.

Vicariant distribution   A distribution resulting from the disruption of a formerly continuous range by a vicariant event.

Villus (vil´ lus) (plural: villi) [L.: shaggy hair]   A hairlike projection from a membrane; for example, from many gut walls.

Virion (veer´ e on)   The virus particle, the minimum unit capable of infecting a cell.

Viroid (vye´ roid)   An infectious agent consisting of a single-stranded RNA molecule with no protein coat; produces diseases in plants.

Virus [L.: poison, slimy liquid]   Any of a group of ultramicroscopic infectious particles constructed of nucleic acid and protein (and, sometimes, lipid) that can reproduce only in living cells.

Visceral mass   The major internal organs of a mollusk.

Vitamin [L. *vita*: life]   Any one of several structurally unrelated organic compounds that an organism cannot synthesize itself, but

nevertheless requires in small quantity for normal growth and metabolism.

Viviparous (vye vip´ uh rus) [L. *vivus*: alive]   Reproduction in which fertilization of the egg and development of the embryo occur inside the mother's body. (Contrast with oviparous.)

Waggle dance   The running movement of a working honey bee on the hive, during which the worker traces out a repeated figure eight. The dance contains elements that transmit to other bees the location of the food.

Water potential   In osmosis, the tendency for a system (a cell or solution) to take up water from pure water, through a differentially permeable membrane. Water flows toward the system with a more negative water potential. (Contrast with osmotic potential, pressure potential.)

Water vascular system   The array of canals and tubelike appendages that serves as the circulatory system, locomotory system, and food capturing system of many echinoderms; is in direct connection with the surrounding sea water.

Wavelength   The distance between successive peaks of a wave train, such as electromagnetic radiation.

Wild-type   Geneticists' term for standard or reference type. Deviants from this standard, even if the deviants are found in the wild, are said to be mutant.

Xanthophyll (zan´ tho fill) [Gr. *xanthos*: yellowish-brown + *phyllon*: leaf]   A yellow or orange pigment commonly found as an accessory pigment in photosynthesis, but found elsewhere as well. An oxygen-containing carotenoid.

X chromosome   See sex chromosome.

Xerophyte (zee´ row fyte) [Gr. *xerox*: dry + *phyton*: plant]   A plant adapted to an environment with a limited water supply.

Xylem (zy´ lum) [Gr. *xylon*: wood]   In vascular plants, the woody tissue that conducts water and minerals; xylem consists, in various plants, of tracheids, vessel elements, fibers, and other highly specialized cells.

Y chromosome   See sex chromosome.

Yolk   The stored food material in animal eggs, usually rich in protein and lipid.

Z-DNA   A form of DNA in which the molecule spirals to the left rather than to the right.

Zooplankton (zoe´ o plang ton) [Gr. *zoon*: animal + *planktos*: wandering]   The animal portion of the plankton.

Zoospore (zoe´ o spore) [Gr. *zoon*: animal + *spora*: seed]   In algae and fungi, any swimming spore. May be diploid or haploid.

Zygospore   A highly resistant type of fungal spore produced by the zygomycetes (conjugating fungi).

Zygote (zye´ gote) [Gr. *zygotos*: yoked]   The cell created by the union of two gametes, in which the gamete nuclei are also fused. The earliest stage of the diploid generation.

Zymogen   An inactive precursor of a digestive enzyme secreted into the lumen of the gut, where a protease cleaves it to form the active enzyme. Zymogens make it unnecessary for some digestive enzymes to be formed inside cells, which active ezymes might digest.

ANSWERS TO SELF-QUIZZES

Chapter 2
1. b	6. a
2. e	7. c
3. c	8. b
4. c	9. e
5. d	10. d

Chapter 3
1. e	6. a
2. d	7. c
3. c	8. e
4. d	9. a
5. b	10. d

Chapter 4
1. a	6. e
2. e	7. a
3. c	8. d
4. e	9. b
5. c	10. d

Chapter 5
1. e	6. b
2. d	7. c
3. a	8. b
4. d	9. e
5. c	10. c

Chapter 6
1. c	6. a
2. e	7. e
3. b	8. b
4. c	9. d
5. c	10. e

Chapter 7
1. a	6. d
2. d	7. a
3. c	8. e
4. e	9. c
5. c	10. e

Chapter 8
1. c	6. d
2. b	7. c
3. d	8. d
4. b	9. b
5. e	10. b

Chapter 9
1. e	6. a
2. c	7. e
3. b	8. d
4. d	9. b
5. c	10. a

Chapter 10*
1. d	6. d
2. a	7. b
3. e	8. a
4. d	9. b
5. d	10. c

Chapter 11
1. c	6. d
2. a	7. b
3. c	8. d
4. b	9. d
5. e	10. a

Chapter 12
1. c	6. d
2. d	7. c
3. b	8. a
4. b	9. b
5. e	10. d

Chapter 13
1. d	6. b
2. c	7. c
3. a	8. d
4. b	9. e
5. e	10. c

Chapter 14
1. c	6. c
2. b	7. a
3. e	8. b
4. d	9. c
5. a	10. d

Chapter 15
1. d	6. c
2. a	7. e
3. d	8. d
4. e	9. b
5. c	10. c

Chapter 16
1. d	6. a
2. b	7. d
3. e	8. d
4. e	9. a
5. c	10. e

Chapter 17
1. d	6. c
2. c	7. d
3. e	8. b
4. a	9. a
5. b	10. b

Chapter 18
1. e	6. c
2. d	7. e
3. d	8. c
4. e	9. a
5. e	10. a

Chapter 19
1. d	6. e
2. c	7. b
3. d	8. e
4. b	9. d
5. d	10. d

Chapter 20
1. c	6. a
2. a	7. b
3. e	8. a
4. d	9. c
5. c	10. a

Chapter 21
1. e	6. d
2. c	7. a
3. a	8. d
4. c	9. b
5. a	10. d

Chapter 22
1. e	6. a
2. e	7. c
3. b	8. c
4. d	9. b
5. b	10. d

Chapter 23
1. a	6. d
2. e	7. c
3. c	8. b
4. d	9. b
5. a	10. d

Chapter 24
1. b	6. a
2. d	7. e
3. e	8. a
4. c	9. c
5. d	10. c

Chapter 25
1. d	6. b
2. c	7. b
3. e	8. c
4. b	9. d
5. c	10. c

Chapter 26
1. c	7. d
2. d	8. c
3. b	9. d
4. e	10. e
5. e	11. a
6. b	

Chapter 27
1. b	7. d
2. c	8. e
3. a	9. a
4. c	10. e
5. c	11. c
6. b	12. c

*Answers to the Genetics "For Study" questions in Chapter 10 appear at the end of this section

Chapter 28

1. d	6. b
2. e	7. c
3. d	8. e
4. a	9. e
5. a	10. b

Chapter 29

1. d	6. b
2. b	7. b
3. e	8. c
4. e	9. a
5. a	10. d

Chapter 30

1. c	6. d
2. d	7. e
3. b	8. a
4. e	9. d
5. b	10. d

Chapter 31

1. d	6. a
2. a	7. c
3. b	8. c
4. c	9. e
5. d	10. b

Chapter 32

1. d	6. e
2. c	7. a
3. a	8. b
4. e	9. d
5. c	10. e

Chapter 33

1. a	6. c
2. e	7. e
3. c	8. c
4. d	9. a
5. b	10. b

Chapter 34

1. c	6. b
2. d	7. d
3. d	8. e
4. b	9. a
5. e	10. b

Chapter 35

1. c	6. e
2. a	7. a
3. d	8. e
4. b	9. d
5. b	10. c

Chapter 36

1. c	6. b
2. b	7. a
3. d	8. e
4. b	9. c
5. a	10. e

Chapter 37

1. (i) b	4. a
(ii) a	5. d
(iii) c	6. c
(iv) a,b,c	7. e
(v) a,b,c	8. d
2. c,e	9. b
3. e	10. d

Chapter 38

1. d	6. e
2. a	7. e
3. e	8. c
4. d	9. d
5. c	10. b

Chapter 39

1. d	6. e
2. d	7. b
3. a	8. c
4. b	9. c
5. e	10. c

Chapter 40

1. d	6. d
2. e	7. b
3. a	8. a
4. b	9. a
5. b	10. e

Chapter 41

1. e	6. b
2. d	7. c
3. a	8. c
4. b	9. a
5. c	10. d

Chapter 42

1. d	6. d
2. a	7. b
3. c	8. d
4. d	9. c
5. c	10. e

Chapter 43

1. b	6. d
2. e	7. a
3. c	8. b
4. a	9. d
5. b	10. d

Chapter 44

1. b	6. b
2. a	7. e
3. d	8. a
4. c	9. c
5. d	10. e

Chapter 45

1. a	6. c
2. e	7. d
3. b	8. c
4. b	9. d
5. a	10. a

Chapter 46

1. e	6. e
2. c	7. c
3. d	8. e
4. c	9. d
5. c	10. a

Chapter 47

1. b	6. c
2. c	7. a
3. d	8. c
4. a	9. d
5. d	10. a

Chapter 48

1. b	7. a
2. e	8. c
3. c	9. b
4. e	10. e
5. e	11. c
6. c	

Chapter 49

1. b	7. a
2. d	8. d
3. d	9. a
4. b	10. b
5. c	11. e
6. c	

Chapter 50

1. a	6. a
2. b	7. d
3. b	8. e
4. c	9. d
5. d	10. a

Chapter 51

1. b	7. d
2. c	8. e
3. d	9. e
4. e	10. a
5. e	11. a
6. a	

Answers to Genetics "For Study" Questions, Chapter 10

1. Each of the eight boxes should contain *Tt*.

2. See Figure 10.3.

3. The trait is autosomal. Mother *dp dp*, father *Dp dp*. If the trait were sex-linked, all daughters would be wild-type and sons would be *dumpy*.

4. All females wild-type; all males spotted.

5. F_1 all wild-type, *PpSwsw*; F_2 9:3:3:1 in phenotypes. See Figure 10.7 for analogous genotypes.

6a. Ratio of phenotypes in F_2 is 3:1 (double dominant to double recessive). See Figure 10.13.
6b. The F_1 are *Pby pBy*; they produce just two kinds of gametes (*Pby* and *pBy*). Combine them carefully and see the 1:2:1 phenotypic ratio fall out in the F_2.
6c. Pink-blistery.
6d. See Figures 9.16 and 9.17. Crossing over took place in the F_1 generation.

7. The genotypes are *PpSwsw*, *Ppswsw*, and *ppswsw* in a ratio of 1:1:1:1.

8a. 1 black:2 blue:1 splashed white.
8b. Always cross black with splashed white.

9a. $w^+ > w^e > w$
9b. Parents $w^e w$ and $w^+ Y$. Progeny $w^+ w^e$, $w^+ w$, $w^e Y$, and wY.

10. All will have normal vision because they inherit Dad's wild-type X chromosome, but half of them will be carriers.

11. Agouti parent: *AaBb*. Albino offspring *aaBb* and *aabb*; black offspring *Aabb*; agouti offspring *AaBb*.

12. The purple parent must be *AaB–*. If it were *AAB–*, there would be no white progeny—all the progeny contain *B* from the white parent. To be purple, that parent must contain at least one *B*; from the data given we cannot tell whether it is *AaBB* or *AaBb*.

13. Yellow parent = $s^Y s^b$; offspring 3 yellow (s^Y–): 1 black ($s^b s^b$). Black parent = $s^b s^b$; offspring all black ($s^b s^b$). Orange parent = $s^O s^b$; offspring 3 orange (s^O–): 1 black ($s^b s^b$). Both s^O and s^Y are dominant to s^b.

ILLUSTRATION CREDITS

Texas. 4.10: R. Rodewald, Univ. of Virginia/BPS. 4.11: Jim Solliday/BPS. 4.14: Hilton Mollenhauer, U.S.D.A. Research Unit, College Station, TX. 4.15: Runk/Schoenberger from Grant Heilman Photography, Inc. 4.16: E. H. Newcomb & W. P. Wergin, Univ. of Wisconsin/BPS. 4.17: D. J. Wrobel, Monterey Bay Aquarium/BPS. 4.19: B. F. King, Univ. of California, Davis, School of Medicine/BPS. 4.20a: G. T. Cole, Univ. of Texas, Austin/BPS. 4.21c: H. S. Pankratz, Michigan State Univ./BPS. 4.22a: R. Rodewald, Univ. of Virginia/BPS. 4.23: E. H. Newcomb & S. E. Frederick, Univ. of Wisconsin/BPS. 4.24: M. C. Ledbetter, Brookhaven National Laboratory. 4.25 left: Gopal Murti, Science Photo Library/Photo Researchers, Inc. 4.25 center: R. Alexley/Peter Arnold, Inc. 4.25 right: Gopal Murti, Science Photo Library/Photo Researchers, Inc. 4.26a,b: W. L. Dentler, Univ. of Kansas/BPS. 4.27a: B. F. King, Univ. of California, Davis, School of Medicine/BPS. 4.28a: J. R. Waaland, Univ. of Washington/BPS. 4.28b: E. H. Newcomb, Univ. of Wisconsin/BPS. Box 4.A: After N. Campbell, 1990, Biology, 2nd Ed., Benjamin Cummings Publishing Co.

Chapter 5 *Opener*: J. David Robertson, Duke Univ. Medical Center. 5.2: After L. Stryer, 1981, Biochemistry, 2nd Ed., W. H. Freeman. 5.4a: L. A. Staehelin, Univ. of Colorado/BPS. 5.4b: J. D. Robertson, Duke Univ. 5.7c: G. T. Cole, Univ. of Texas, Austin/BPS. 5.8 top: D. S. Friend, Univ. of California, San Francisco. 5.8 center: Darcy E. Kelly, Univ. of Washington. 5.8 bottom: Courtesy of C. Peracchia. 5.19a–d: M. M. Perry, J. Cell Sci. 39, p. 266, 1979.

Chapter 6 *Opener*: Jim Merli. 6.1: Nuridsany et Perennou/Photo Researchers, Inc. 6.9, 6.16b, 6.17b: Richard Alexander, Univ. of Pennsylvania

Chapter 7 *Opener*: Runk/Schoenberger from Grant Heilman Photography, Inc. 7.10a: Hilton Mollenhauer, U.S.D.A. Research Unit, College Station, Texas. 7.10b: E. Racker, Cornell Univ. Box 7.Aa: Ken Lucas/BPS. Box 7.Ab: Michael Fogden DRK PHOTO. Box 7.Ac: G. M. Thomas & G. Poinar, Univ. of California, Berkeley. Box 7.Ad: K. V. Wood, courtesy of M. DeLuca, Univ. of California, San Diego.

Chapter 8 *Opener*: Art Wolfe. 8.1, bottom: J. H. Troughton and L. A. Donaldson. 8.1, top: Runk/Schoenberger from Grant Heilman Photography, Inc. 8.19: J. R. Waaland, Univ. of Washington/BPS. 8.20a: J. A. Bassham, Lawrence Berkeley Lab., Univ. of California. 8.20b, 8.27b: E. H. Newcomb & S. E. Frederick, Univ. of Wisconsin/BPS.

Chapter 9 *Opener*: Scott Spiker/Adventure Photo. 9.1a: Phil Gates, Univ. of Durham/

BPS. 9.1b: R. Rodewald, Univ. of Virginia/BPS. 9.2a: G.F. Bahr, Armed Forces Inst. of Pathology. 9.3b: A. L. Olins, Univ. of Tennessee, Oak Ridge Grad. School of Biomedical Science. 9.5 insert: David Ward, Yale Univ. School of Medicine. 9.8: Andrew S. Bajer, Univ. of Oregon. 9.9: J. B. Rattner & S. G. Phillips, J. Cell Biol. 57, p. 359, 1973. 9.10b, c: C. L. Rieder, New York State Dept. of Health/BPS. 9.11a: T. E. Schroeder, Univ. of Washington/BPS. 9.11b: B. A. Palevitz & E. H. Newcomb, Univ. of Wisconsin/BPS. 9.12: G. T. Cole, Univ. of Texas, Austin/BPS. 9.14a,b: David Ward, Yale Univ. School of Medicine. 9.15, 9.17: C. A. Hasenkampf, Univ. of Toronto/BPS. 9.19: B. Schuh, Monmouth Medical Center. 9.20: © Ruth Kavenoff, Designergenes Ltd., P.O. Box 100, Del Mar, CA 90214. 9.21: J.J. Cardamone, Jr., Univ. of Pittsburgh/BPS.

Chapter 10 *Opener*: Runk/Schoenberger from Grant Heilman Photography, Inc. 10.12: Carl W. May/BPS. 10.17: Namboori B. Raju, Stanford Univ., Eur. J. Cell Biol. 23, p. 208, 1980. 10.20: Peter J.Bryant/BPS. 10.23: After N. Campbell, 1990, Biology, 2nd Ed., Benjamin Cummings Publishing Co. 10.24: © Walter Chandoha, 1991.

Chapter 11 *Opener*: Dan Richardson. 11.3 right, 11.5: Dan Richardson. 11.15: G. W. Willis, Ochsner Medical Instution/BPS. 11.19a: Dan Richardson. 11.22b: Courtesy of J. E. Edstrom and EMBO J.

Chapter 12 *Opener*: A. B. Dowsett, Science Photo Library/Photo Researchers, Inc. 12.1: Richard Humbert/BPS. 12.4: L. Caro & R. Curtiss. 12.17: Brian Matthews, Univ. of Oregon.

Chapter 13 *Opener*: Ken Edward, Science Source/Photo Researchers Inc. 13.7: Karen Dyer, Vivigen. 13.8: Joseph Gall, Carnegie Institution of Washington. 13.9: O. L. Miller, Jr., & B. R. Beatty. 13.12: After W. T. Keeton and J. L. Gould, Biological Science, 5th Edition, W. W. Norton & Co.

Chapter 14 *Opener*: Hank Morgan/Photo Researchers, Inc. 14.5a: N.Y. State Agricultural Experiment Station, Cornell Univ. 14.5b: J. S. Yun & T.E. Wagner, Ohio Univ. 14.11b: Mike Tincher, courtesy of Agracetus, Inc. (a subsidiary of W.R. Grace & Co.) 14.13: M. L. Pardue & J. G. Gall, Chromosomes Today 3, p. 47, 1972. 14.17a: Phil Gates, Univ. of Durham/BPS. 14.18: Paul F. Umbeck, courtesy of Agracetus, Inc. (a subsidiary of W.R. Grace & Co.) 14.19: N.Y. State Agricultural Experiment Station, Cornell Univ. 14.20: Advanced Genetic Sciences. 14.21: Larry Lefever from Grant Heilman Photography, Inc.

Chapter 15 *Opener*: Chip Mitchell. 15.1: From C. Harrison et al., J. Med. Genet. 20, p. 280, 1983. 15.9: Elaine Rebman/Photo Researchers, Inc. 15.10: P. P. H. DeBruyn, Univ. of Chicago.

Chapter 16 *Opener*: Painting by Keith Haring. 16.1: Z. Skobe, Forsyth Dental Center/BPS. 16.2: Courtesy of Lennart Nilsson. © Boehringer Ingelheim GmbH. 16.4: G. W. Willis, Ochsner Medical Institution/BPS. 16.9: R. Rodewald, Univ. of Virginia/BPS. 16.10b: Arthur J. Olson, Scripps Research Institute. 16.14: L. Winograd, Stanford Univ. 16.20: A. Liepins, Sloan-Kettering Research Inst. 16.25: A. Calin, Stanford Univ. School of Medicine. 16.26: After R. C. Gallo, The AIDS Virus, © 1987: by Scientific American, Inc.

Chapter 17 *Opener*: Norbert Wu. 17.1 sea urchin embryo: George Watchmaker. 17.1 sea urchin: D. J. Wrobel, Monterey Bay Aquarium/BPS. 17.1 tadpole, frog, chick embryo: © E. R. Degginger. 17.1 rooster: John Colwell from Grant Heilman Photography, Inc. 17.3a: After J. E. Sulston and H. R. Horvitz, Dev. Biol. 56, p. 110, 1977. 17.10: George M. Malacinski and A. W. Neff. 17.12: Peter J. Bryant/BPS. 17.19: Susan Strome. 17.28: C. Rushlow and M. Levine. 17.23a: From B. Alberts et al., 1983, Molecular Biology of the Cell, Garland Publishing Co. 17.24: F. R. Turner, Indiana Univ. 17.26: E. B. Lewis.

Chapter 18 *Opener*: Larry Ulrich/DRK PHOTO. 18.3: Stanley M. Awramik, U. of California/BPS.

Chapter 19 *Opener*: Frans Lanting/Minden Pictures. 19.6: Frank S. Balthis, Nature's Design. 19.7: Harold W. Pratt/BPS. 19.11a,b: Gary J.James/BPS. 19.12: After D. Futuyma, Evolutionary Biology, 2nd Ed., Sinauer Associates, Inc., 1987. 19.16a,b: Richard Alexander, Univ. of Pennsylvania.

Chapter 20 *Opener*: Raymond A. Mendez. 20.1: Des and Jen Bartlett, Bruce Coleman, Inc. 20.2a: Edward Ely/BPS. 20.2b: © John Shaw/NHPA. 20.4: Anthony D. Bradshaw, Univ. of Liverpool. 20.6: © E. R. Degginger. 20.7a–c: R.W. VanDevender. 20.8: Paul A. Johnsgard, Univ. of Nebraska. 20.9a: Peter J. Bryant/BPS. 20.9b: Kenneth Y. Kaneshiro, Univ. of Hawaii at Manoa. 20.9c: Peter J. Bryant/BPS. 20.11b: Heather Angel, BIOFOTOS. 20.11a: Virginia P. Weinland/Photo Researchers, Inc. 20.12a: Gary J. James/BPS. 20.12b,c: © Jim Denny.

Chapter 21 *Opener*: Joe McDonald. 21.1a: Helen E. Carr/BPS. 21.1b: Barbara J. Miller/BPS. 21.1c: © L. Campbell/NHPA. 21.5a: Jon Stewart/BPS. 21.5b: Barbara J. Miller/BPS. 21.9a,b: Peter J. Bryant/BPS. 21.10: Illustration by Marianne Collins.

21.12*a,b*: Paul A. Johnsgard, Univ. of Nebraska. 21.14*a,b*: Art Wolfe.

Chapter 22   *Opener*: Tony Brain, Science Photo Library/Photo Researchers, Inc. 22.1: Alfred Pasieka/Science Photo Library Photo Researchers, Inc. 22.2: H. W. Jannasch, Woods Hole Oceanographic Institution. 22.3, 22.4, 22.5: T. J. Beveridge, Univ. of Guelph/BPS. 22.6*a*: Leonard Lessin , Peter Arnold Inc. 22.7*b*: C. Forsberg & T. J. Beveridge, Univ. of Guelph/BPS. 22.7*a*: Paul W. Johnson/BPS. 22.7*b*: K. Stephens, Stanford Univ./BPS. 22.8*a*: J.A. Breznak & H. S. Pankratz, Michigan State Univ./BPS. 22.8*b*: G. W. Willis, Ochsner Medical Institution/BPS. 22.9: T. J. Beveridge, Univ. of Guelph/BPS. 22.10: G. W. Willis, Ochsner Medical Institution/BPS. 22.11: D. A. Glawe, Univ. of Illinois/BPS. 22.12: G. W. Willis, Ochsner Medical Institution/BPS. 22.13*a*: W. Burgdorfer, Rocky Mountain Lab. 22.13*b*: Nat. Animal Disease Center, Ames, IA. 22.14: S. C. Holt, Univ. of Texas Health Science Center, San Antonio/BPS. 22.15*a*: Paul W. Johnson/BPS. 22.15*b*: H. S. Pankratz, Michigan State Univ./BPS. 22.15*c*: © E. R. Degginger. 22.16*a,b*: Leon J. Le Beau/BPS. 22.17*a*: G. W. Willis, Ochsner Medical Institution/BPS. 22.17*b*: G. W. Willis, Ochsner Medical Institution/BPS. 22.18: Centers for Disease Control, Atlanta. 22.19: M. G. Gabridge, cytoGraphics, Inc./BPS. 22.20: Arthur J. Olson, Scripps Research Institute. 22.21: D. T. Brown et al., *J. Virol.* 10, p. 524, 1972. 22.22*a*: D. L. D. Caspar, Brandeis Univ. 22.22*b*: D. S. Goodsell & A. J. Olson, Scripps Research Institute. 22.22*c*: S. C. Holt, Univ. of Texas Health Science Center, San Antonio/BPS. 22.22*d*: F. A. Murphy, Centers for Disease Control, Atlanta. Box 22.A1 *upper left*: Centers for Disease Control, Atlanta. Box 22.A *upper center*: S. C. Holt, Univ. of Texas Health Science Center, San Antonio/BPS. Box 22.A *lower left*: Leon J. Le Beau/BPS. Box 22.A *lower center*: A J. J. Cardamone, Jr., Univ. of Pittsburgh/BPS.

Chapter 23   *Opener*: Jan Hinsch, Science Photo Library/Photo Researchers, Inc. 23.3*a*: Animals Animals/Oxford Scientific Films. 23.3*b*: © E. R. Degginger. 23.3*c*: James Solliday/BPS. 23.4: Eric V. Gravé, Science Source/Photo Researchers, Inc. 23.6: G. W. Willis, M.D./BPS. 23.26*a*: Dennis D. Kunkel/BPS. 23.7: Paul W.Johnson/BPS. 23.9*a*: Jim Solliday/BPS.23.9*b*: Eric V. Gravé/Photo Researchers, Inc. 23.11*a*: Jim Solliday/BPS. 23.11*b–d*: Paul W. Johnson/BPS. 23.12*b*: M. A. Jakus, NIH.23.14: Eric V. Gravé. 23.16*a*: Barbara J. Miller/BPS. 23.16*b*: Henry Aldrich, Institute of Food and Agricultural Sciences, Univ. of Florida. 23.17*a*: D. W. Francis, Univ. of Delaware. 23.17*b*: © David Scharf. 23.18: J. R. Waaland, Univ. of Washington/BPS. 23.19: Dwight R. Kuhn. 23.20*a*: Paul W. Johnson/BPS. 23.20*b*: Gary J.James/BPS. 23.21*a*: Charles Gellis/Photo Researchers, Inc.

23.21*b*: V. Cassie. 23.24*a*: J.N. A. Lott, McMaster Univ./BPS. 23.24*b*: J. R. Waaland, Univ. of Washington/BPS. 23.25*a*: Maria Schefter/BPS. 23.25*b*: J. N. A. Lott, McMaster Univ./BPS. 23.26*a*: Dennis D. Kunkel/BPS. 23.26*b*: Harold W. Pratt/BPS. 23.26*c*: J. R. Waaland, Univ. of Washington/BPS. Box 23.A*a*: Gerald Corsi, Tom Stack and Associates. Box 23.A*b*: J. N. A. Lott, McMaster Univ./BPS.

Chapter 24   *Opener*: Stephen J. Kraseman/DRK PHOTO. 24.1*a*: D. A. Glawe, Univ. of Illinois/BPS. 24.1*b*: L. E. Gilbert, Univ. of Texas, Austin/BPS. 24.1*c*: G. L. Barron, Univ. of Guelph/BPS. 24.2*b*: G. T.Cole, Univ. of Texas, Austin/BPS. 24.3: D. A. Glawe, Univ. of Illinois/BPS. 24.4: G. L. Barron, Univ. of Guelph/BPS. 24.5: Barbara J. Miller/BPS. 24.7 *upper & lower left*: W. F.Schadel, Small World Enterprises/BPS. 24.8: J. R. Waaland, Univ. of Washington/BPS. 24.9*b*: D. A. Glawe, Univ. of Illinois/BPS. 24.10: W. F. Schadel, Small World Enterprises/BPS. 24.11*a*: Jim Solliday/BPS. 24.11*b*: D. A. Glawe, Univ. of Illinois/BPS. 24.11*c*: Richard Humbert/BPS. 24.12: Centers for Disease Control, Atlanta. 24.13*a*: Michael Fogden DRK PHOTO. 24.13*b*: Photography by Rannels, Grant Heilman Photography, Inc. 24.13*c*: M. Graybill and J. Hodder/BPS. 24.14 *inset*: © Biophoto Associates. 24.15*a*: E. I. Friedmann, Florida State Univ. 24.15*b*: Barbara J. O'Donnell/BPS. 24.16*a*: Grant Heilman, Grant Heilman Photography, Inc. 24.16*b*: J. N. A. Lott, McMaster Univ./BPS. 24.16*c*: Barbara J. Miller/BPS. 24.17*a*: J. N. A. Lott, McMaster Univ./BPS. Box 24.A: R. L. Peterson, Univ. of Guelph/BPS.

Chapter 25   *Opener*: Art Wolfe. 25.1: J. Robert Stottlemeyer/BPS. 25.3: Gary J. James/BPS. 25.4*a,b*: J. R. Waaland/BPS. 25.5*a*: © E. R. Degginger. 25.5*b*: Runk/Schoenberger from Grant Heilman Photography, Inc. 25.5*c*: J. N. A Lott, McMaster Univ./BPS. 25.7: J. H. Troughton. 25.8: © E. R. Degginger. 25.9: Fig. information provided by Prof. Hermann Pfefferkorn, Dept. of Geology, Univ. of Pennsylvania. Original oil painting by John Woolsey. 25.14*a*: Runk/Schoenberger from Grant Heilman Photography, Inc. 25.14*b*: J. N. A. Lott, McMaster Univ./BPS. 25.15*a*: Carl W. May/BPS. 25.15*b*, 25.16: J. N. A. Lott, McMaster Univ./BPS. 25.17*a*: Barbara J. Miller/BPS. 25.17*b*: J. N.A. Lott, McMaster Univ./BPS. 25.17*c*: Art Wolfe. 25.18: J. N. A. Lott, McMaster Univ./BPS. 25.21: Phil Gates, Univ. of Durham/BPS. 25.22*a*: John Cancalosi/DRK PHOTO. 25.22*b*: Ken Lucas/BPS. 25.22*c*: Gary J. James/BPS. 25.22*d*: Joel Simon. 25.25*a*: B. Miller/BPS. 25.25*b*: Roger de la Harpe/BPS. 25.25*c*: Grant Heilman Photography 25.26*a*: Barbara J. Miller/BPS. 25.26*b*: Barbara J. Miller/BPS. 25.28*a*: J. N.A. Lott, McMaster Univ./BPS. 25.28*b*: J. N.A. Lott, McMaster Univ./BPS. 25.28*c*: Catherine M. Pringle/BPS. 25.28*d*:

C.S. Lobban/BPS. 25.30*a*: Jon Mark Stewart/BPS. 25.30*b*: Lara Hartley, TER-RAPHOTOGRAPHICS/BPS. 25.30*c*: © E. R. Degginger. 25.31*a*: Jon Mark Stewart/BPS. 25.31*b*: Lefever/Grushow from Grant Heilman Photography, Inc. 25.31*c*: Jon Mark Stewart/BPS. 25.32: Barbara J. O'Donnell/BPS.

Chapter 26   *Opener*: Norbert Wu. 26.8*a*: Ken Lucas/BPS. 26.8*b*: Robert Brons/BPS. 26.8*c*: Joel Simon. 26.10, 26.11, 26.12, 26.13: Adapted from F. M. Bayerand, H. B. Owre, 1968, *The Free-Living Lower Invertebrates*, Macmillan Pubishing Co. 26.14: After G. and R. Brusca, 1990, *Invertebrates*, Sinauer Associates, Inc. 25.14*a*: Andrew J. Martinez/Photo Researchers,Inc. 25.14*b*: Douglas Faulkner/Photo Researchers, Inc. 26.16*a*: Robert Brons/BPS. 26.16*,c*: After G. and R. Brusca, 1990, *Invertebrates*, Sinauer Associates, Inc. 26.17*b*, 26.18*a*: D. J. Wrobel, Monterey Bay Aquarium/BPS. 26.18*c*, 26.21*b*: Robert Brons/BPS. 26.21*c*: Jim Solliday/BPS. 26.22*a*: After G. and R. Brusca, 1990, *Invertebrates*, Sinauer Associates, Inc. 26.22*b*: Jim Solliday/BPS. 26.23: R. R. Hessler, Scripps Institute of Oceanography. 26.25*a*: Andrew J. Martinez/Photo Researchers, Inc. 26.25*c*: Roger K. Burnard/BPS. 26.25*d*: © Robert & Linda Mitchell. 26.27*a*: M. P. L. Fogden/Bruce Coleman, Inc. 26.27*b*: From Kristensen and Hallas, 1980. 26.29*a*: Joel Simon. 26.28: J. N. A. Lott, McMaster Univ./BPS. 26.29*b*: Barbara J. Miller/BPS. 26.30*a*: Ken Lucas/BPS. 26.30*b*: Peter J. Bryant/BPS. 26.30*c*: L. E. Gilbert, Univ. of Texas, Austin/BPS. 26.30*d*: Robert Brons/BPS. 26.32*a*: Gregory Ochocki/Photo Researchers, Inc. 26.32*b*: Peter J.Bryant/BPS. 26.32*c*: D. J. Wrobel, Monterey Bay Aquarium/BPS. 26.32*d*: C. R. Wyttenbach, Univ. of Kansas/BPS. 26.33*a*: Ken Lucas/BPS. 26.33*b*: Roger K. Burnard/BPS. 26.34*a*: Richard Humbert/BPS. 26.34*b*: Peter J. Bryant/BPS. 26.34*c– g*: Peter J. Bryant/BPS. 26.34*h*: © E. R. Degginger. 26.36*a*: Ken Lucas/BPS. 26.36*b*: Harold W. Pratt/BPS. 26.36*c*: D. J. Wrobel, Monterey Bay Aquarium/BPS. 26.36*d,e,g*: Ken Lucas/BPS. 26.36*f*: J. W. Porter, Univ. of Georgia/BPS. 26.37: D. J. Wrobel, Monterey Bay Aquarium/BPS.

Chapter 27   *Opener*: Heather Angel, BIOFO-TOS. 27.2*a*: D. J. Wrobel, Monterey Bay Aquarium/BPS. 27.2*b*: Robert Brons/BPS. 27.4: D. J. Wrobel, Monterey Bay Aquarium/BPS. 27.5*b*: C.R. Wyttenbach, Univ. of Kansas/BPS. 27.8*a*: Doug Perrine/DRK PHOTO. 27.8*b*: Joel Simon 27.8*c–e*: D. J. Wrobel, Monterey Bay Aquarium/BPS. 27.9: Robert Brons/BPS. 27.10*b*: © Heather Angel, BIOFOTOS. 27.12: Tom Stack/Tom Stack and Associates. 27.14*a*: Tom McHugh/Photo Researchers, Inc. 27.14*b*: D. J. Wrobel, Monterey Bay Aquarium/BPS. 27.15*a*: Ken Lucas/BPS. 27.16*a*: Peter Scoones, Planet Earth Pictures. 27.15*b–d*: Ken Lucas/BPS.

27.15e: Animals Animals/Breck P. Kent. 27.16a: Peter Scoones/Planet Earth Pictures. 27.17a: Ken Lucas/BPS. 27.17b: Art Wolfe. 27.17c: E.D. Brodie, Jr., Univ. of Texas, Arlington/BPS. 27.20a: Doug Perrine DRK PHOTO. 27.20b: Carl Gans, Univ. of Michigan/BPS. 27.20c: Joe McDonald, Bruce Coleman Inc. 27.20d: Michael P. Fogden, Bruce Coleman Inc. 27.21a: © E. R. Degginger. 27.21b: Wayne Lankinen/DRK PHOTO. 27.22: Courtesy of Carnegie Museum of Natural History, Pittsburgh. 27.24a: Johnny Johnson/DRK PHOTO. 27.24b: D. Cavagnaro/DRK PHOTO. 27.24c,d: Stephen J. Kraseman/DRK PHOTO. 27.26a: John Cancalosi/Tom Stack and Associates. 27.26b: Joel Simon. 27.26c: M. P. L. Fogden, Bruce Coleman, Inc. 27.27a: J. N. A. Lott, McMaster Univ./BPS. 27.27b: Merlin D. Tuttle, Bat Conservation International. 27.27c: Stephen J. Kraseman/DRK PHOTO. 27.27d: Robert Stottlemyer/BPS. 27.29a: Art Wolfe. 27.29b: Stanley Breeden/DRK PHOTO. 27.29c: Frans Lanting/Minden Pictures. 27.30a,b: Steve Kaufman/DRK PHOTO. 27.31a: © E. R. Degginger. 27.31b: © Peter Drowne, E. R. Degginger. 27.31c: Kennan Ward/DRK PHOTO. 27.31d: © Peter Drowne, E. R. Degginger. 27.34a: Edward S. Ross. 27.34b,c: Robert E. Ford, TERRAPHO-TOGRAPHICS/BPS. Box 27A: Anne Marie Weber/Adventure Photo.

Chapter 28 *Opener*: Frans Lanting/Minden Pictures. 28.3: Peter Ward, Univ. of Washington 28.4: Mark R. Meyer. 28.7a: Ken Lucas/BPS. 28.7b: S. M. Awramik, Univ. of California/BPS. 28.8a: S. Conway Morris. 23.8b: From S. J.Gould, 1989, Wonderful Life, © W. W. Norton. 28.9: Courtesy of the Natural History Museum of London. 28.10: Courtesy of the Smithsonian Institution. 28.11: Ken Lucas/BPS. 28.12: Painting by Rudolph Zallinger; courtesy of the Peabody Museum of Natural History, Yale Univ. 28.14: Painting by Chip Clark; courtesy of the Smithsonian Institution.

Chapter 29 *Opener*: Ed Reschke, Peter Arnold, Inc. 29.1: Joel Simon. 29.2: Michael P. Gadomski/Bruce Coleman, Inc. 29.3: Gary J. James/BPS. 29.4: Thomas Hovland from Grant Heilman Photography, Inc. 29.7a: Runk/Schoenberger from Grant Heilman Photography, Inc. 29.7b: © E. R. Degginger. 29.8: Grant Heilman Photography. 29.9a: J. R. Waaland, Univ. of Washington/BPS. 29.9b: © E. R. Degginger. 29.9c: Carl W. May/BPS. 29.13a,b: Phil Gates, Univ. of Durham/BPS. 29.13c: Runk/Schhoenberger from Grant Heilman Photography, Inc. 29.13d: Phil Gates, Univ. of Durham/BPS. 29.13e: © E. R. Degginger. 29.13f, 29.15b, 29.17 top & bottom: J. R. Waaland, Univ. of Washington/BPS. 29.18a: L. Elkin, Hayward State Univ./BPS. 29.18b,c: Jim Solliday/BPS. 29.18d: Phil Gates, Univ. of Durham/BPS. 29.20 top: Dwight R. Kuhn. 29.20 bottom: © E. R. Degginger. 29.20: J. R. Waaland, Univ. of

Washington/BPS. 29.21a,b: Phil Gates, Univ. of Durham/BPS. 29.23: J. N.A. Lott, McMaster Univ./BPS. 29.24: Jim Solliday/BPS. 29.25: J. N. A. Lott, McMaster Univ., BPS. 29.26a,b: Phil Gates, Univ. of Durham/BPS. 29.27b: W. F. Schadel, Small World Enterprises/BPS. 29.27c: E. J. Cable/Tom Stack and Associates.

Chapter 30 *Opener*: Gary Gray/DRK PHOTO. 30.2a,b: Runk/Schoenberger from Grant Heilman Photography, Inc. 30.7: Phil Gates, Univ. of Durham/BPS. 30.10: J. H. Troughton & L. A. Donaldson. 30.12: Jon Stewart/BPS. 30.13: Thomas Eisner, Cornell Univ. 30.17: Larry Lefever from Grant Heilman Photography, Inc.

Chapter 31 *Opener*: Thomas Eisner, Cornell Univ. 31.1: Jon Mark Stewart/BPS. 31.2: J. N. A. Lott, McMaster Univ./BPS. 31.3a: J. N. A. Lott, McMaster Univ./BPS. 31.3b: J. N. A. Lott, McMaster Univ./BPS. 31.4: Carl W. May/BPS. 31.5: Gary J. James/BPS. 31.6: Joel Simon. 31.7: J. N. A. Lott, McMaster Univ./BPS. 31.8: © Robert & Linda Mitchell. 31.9: J. Antonovics, Duke Univ. 31.10: Jane Grushow from Grant Heilman Photography, Inc. 31.13: Richard Alexander, Univ. of Pennsylvania.

Chapter 32 *Opener*: Runk/Schoenberger from Grant Heilman Photography, Inc. 32.1: Lou Jacobs, Jr. from Grant Heilman Photography, Inc. 32.4: © William E. Ferguson. 32.7: Runk/Schoenberger from Grant Heilman Photography, Inc. 32.8: Barbara J. O'Donnell/BPS. 32.10: E. H. Newcomb & S. R. Tandon, Univ. of Wisconsin/BPS. 32.12, 32.13: Runk/Schoenberger from Grant Heilman Photography, Inc. Box 32.A: Aladar A. Szalay, Univ. of Alberta.

Chapter 33 *Opener*: Andrew Taylor Photography. 33.4: Art Wolfe. 33.5: Barbara J. Miller/BPS. 33.6: Barry L. Runk from Grant Heilman Photography, Inc. 33.10: Runk/Schoenberger from Grant Heilman Photography, Inc. 33.18 inset: Biophoto Associates/Photo Researchers, Inc. 33.20: Grant Heilman Photography. 33.22: J. N. A. Lott, McMaster Univ./BPS. 33.21: Grant Heilman from Grant Heilman Photography, Inc. 33.23: J. N. A. Lott, McMaster Univ./BPS. 33.28: R. Last, Cornell Univ. Courtesy of the Society for Plant Physiology.

Chapter 34 *Opener*: Art Wolfe. 34.1 bottom: J. R. Waaland, Univ. of Washington/BPS. 34.1 top: Jim Solliday/BPS. 34.2: Dennis D. Kunkel/BPS. 34.3: J. N. A. Lott, McMaster Univ./BPS. 34.7a: Gary J. James/BPS. 34.7b: John Colwell from Grant Heilman Photography, Inc. 34.16a: J. N. A. Lott, McMaster Univ./BPS. 34.16b: Phil Gates, Univ. of Durham/BPS. 34.18: Plant Genetics, Inc.

Chapter 35 *Opener*: Animals Animals/Gerald L. Kooyman. 35.14a: © Cherry Alexander/NHPA. 35.14b: Belinda Wright/DRK PHOTO. 35.19b,c: G. W. Willis, Ochsner Medical Institution/BPS. 35.19d Fran Thomas, Stanford Univ. 35.20a: Stephen J. Kraseman/DRK PHOTO. 35.20b: Art Wolfe.

Chapter 36 *Opener*: R. D. Fernald, Stanford Univ. 36.7a: AP Wide World Photos. 36.7b: The Bettmann Archive Inc. 36.9: S. H. Ingbar, Harvard Medical School. Box 36.B: James Sugar, Black Star.

Chapter 37 *Opener*: Belinda Wright/DRK Photo. 37.1a: © M. Walker/NHPA. 37.1b: J. Greenfield, Planet Earth Pictures. 37.1c: Geoff du Feu, Planet Earth Pictures. 37.2: David M. Phillips/Photo Researchers, Inc. 37.6, *center insert*: P. Motta, Univ. La Sapienza, Rome, Science Photo Library/Photo Researchers, Inc. 37.6, *bottom insert*: David M. Phillips/Photo Researchers, Inc. 37.7: P. Bagavandoss. 37.12a: Animals Animals/Fritz Prenzel. 37.12b (embryo): John Cancalosi/DRK PHOTO. 37.12b (adult): Animals Animals/Mickey Gibson. 37.16a,b: From *A Child Is Born*. Photos © Lennart Nilsson, Bonnier Fakta.

Chapter 38 *Opener*: Scott Camazine/Photo Researchers, Inc. 38.3: From R.G. Kessel and R. H. Kardon, *Tissues and Organs*. © 1979, W. H. Freeman and Co. 38.21: Dan McCoy, Rainbow. Boxes 38.A and 38.B: William F. Gilly, Hopkins Marine Station.

Chapter 39 *Opener*: Shelly Katz/Time Magazine. 39.4a: R. A. Steinbrecht. 39.4b: Animals Animals/G. I. Bernard, Oxford Scientific Films. 39.5 center: Peter J. Bryant/BPS. 39.8 top: P. Motta, Univ. La Sapienza, Rome, Science Photo Library/Photo Researchers, Inc. 39.17 right: S. Fisher, Univ. of California, Santa Barbara. 39.21a: Animals Animals/G. I. Bernard, Oxford Scientific Films. 39.24: © E. R. Lewis, Y. Y. Zeevi & F. S. Werblin, Univ. of California, Berkeley/BPS. 39.30: Michael Fogden/DRK PHOTO. 39.31: © Stephen Dalton/NHPA.

Chapter 40 *Opener*: Mik Dakin/Bruce Coleman, Inc. 40.1: D. J. Wrobel, Monterey Bay Aquarium/BPS. 40.2: CNRI Science Photo Library Photo Researchers. 40.5a: Secchi-Lecaque-Roussel, UCLAF/CNRI, Science Photo Library/Photo Researchers, Inc. 40.5b: P. Motta, Univ. La Sapienza, Rome, Science Photo Library/Photo Researchers, Inc. 40.7a: M. I. Walker, Science Source/Photo Researchers, Inc. 40.7b: Michael Abbey, Science Source/Photo Researchers, Inc. 40.7c: CNRI, Science Photo Library/Photo Researchers, Inc. 40.8b: F. A. Pepe, Univ. of Pennsylvania School of Medicine/BPS. 40.11: G.W. Willis, Ochsner Medical Institution/BPS. 40.14: John Dudak/Phototake. 40.19: G. Mili. 40.20: Robert Brons/BPS. 40.24a: David J. Wrobel/BPS.

Chapter 41 *Opener*: Courtesy of NASA. 41.1*a*: © Sea Studios, Inc. 41.1*b*: Robert Brons/BPS. 41.1*c*: © Robert & Linda Mitchell. 41.3: © 1991, Eric Reynolds/Adventure Photo. 41.5*b*: Peter J. Bryant/BPS. 41.5*c*: Thomas Eisner, Cornell Univ. 41.9: Walt Tyler, Univ. of California, Davis. 41.13 *top insert*: Science Photo Library, Science Source/Photo Researchers, Inc. 41.13 *bottom insert*: P. Motta, Univ. La Sapienza, Rome, Science Photo Library/Photo Researchers, Inc. 41.16: George Holton/Photo Researchers, Inc.

Chapter 42 *Opener*: Larry Mulvehill, Science Source/Photo Researchers, Inc. 42.11: © Ed Reschke. 42.13: UNICEF, Maggie Murray-Lee. 42.15: After N. Campbell, 1990, Biology, 2nd Ed., Benjamin Cummings Publishing Co. 42.16*a*: From R. G. Kessel & R. H. Kardon, *Tissues and Organs*, © 1979, W. H. Freeman and Co. 42.17*b*: Secchi-Lecaque-Roussel, UCLAF/CNRI, Science Photo Library/Photo Researchers, Inc. Box 42.A: Jon Feingersh/Tom Stack and Associates.

Chapter 43 *Opener*: Animals Animals/Carl Roessler. 43.1*a*: Timothy O'Keefe/Tom Stack and Associates. 43.1*b*: Animals Animals/A. G.(Bert) Wells, Oxford Scientific Films. 43.1*c*: Animals Animals/Bruce Watkins. 43.1*d*: Jack Stein Grove/Tom Stack and Associates. 43.6: D. J. Wrobel, Monterey Bay Aquarium/BPS. 43.7: © Robert & Linda Mitchell. 43.8: © James D. Watt. 43.11: © Ed Reschke. 43.14 *insert*: E. S. Strauss.

Chapter 44 *Opener*: Art Wolfe. 44.1*a*: Helen E. Carr/BPS. 44.1*b*: © E. R. Degginger. 44.3*b*: Marc Chappell, Univ. of California, Riverside. 44.10 *a–d*: From R. G. Kessel & R. H. Kardon, *Tissues and Organs*, © 1979, W.H. Freeman and Co. 44.12: Gregory G. Dimijian, M.D./Photo Researchers, Inc. 44.17: Lise Bankir, I.N.S.E.R.M. Unit, Hopital Necker, Paris. Box 44.A: © Custom Medical Stock Photos. Box 44.B: © Robert & Linda Mitchell.

Chapter 45 *Opener*: John Gerlach/DRK PHOTO. 45.2: Marc Chappell, Univ. of California, Riverside. 45.8: W. C. Dilger. 45.11*b*: Anup & Manos Shah/Planet Earth Pictures. 45.12: Anthony Mercieca/Photo Researchers, Inc. 45.15: H. Craig Heller. 46.17: Animals Animals/Michael Dick. 45.19: H. Craig Heller. 45.22, 45.24: © Jonathan Blair, Woodfin Camp & Associates.

Chapter 46 *Opener*: © E. R. Degginger. 46.1: Kennan Ward/DRK PHOTO. 46.4: Aileen N. C. Morse, Marine Science Institute, Univ of California at Santa Barbara. 46.5: T. G. Whitham, Northern Arizona Univ. 46.7*a*: Art Wolfe. 46.7*b*: Robert and Jean Pollock/BPS. 46.7*c*: Drawing by Sally Landry. 46.8: Erwin and Peggy Bauer. 46.10: John Alcock, Arizona State Univ. 46.12: John Alcock, Arizona State Univ. 46.11: Natalie J. Demong. 46.13: Michael Fogden/DRK PHOTO. 46.15*a*: Elizabeth N. Orians. 46.16: Clem Haagner/Bruce Coleman, Inc. 46.19: S. G. Hoffman. 46.20: J. Erckmann. 46.21: Georgette Douwma/Planet Earth Pictures. 46.23: Patricia Moehlman. 46.24*a*: Jeremy Burgess, Science Photo Library/Photo Researchers, Inc. 46.24*b*: D. Houston/Bruce Coleman, Inc. 46.25: Stanley Breeden/DRK PHOTO. 46.26: Art Wolfe. 46.27: Jonathan Scott, Planet Earth Pictures. 46.28: K. and K. Ammann/Bruce Coleman, Inc.

Chapter 47 *Opener*: © E. R. Degginger. 47.2: Frans Lanting/Minden Pictures. 47.5: © Brian Hawkes/NHPA. 47.9: Dennis Johns. 47.14: Richard Root/Cornell Univ. 47.16: Elizabeth N. Orians. 47.17: Douglas Sprugel, College of Forest Resources, Univ. of Washington. 47.18: Michio Hoshino/Minden Pictures. 47.19: G.Tortoli, Food and Agriculture Organization of the United Nations. 47.20: Animals Animals Earth Scenes/George H. H. Huey.

Chapter 48 *Opener*: Michael Fogden/DRK PHOTO. 48.2*a*: S. K. Webster, Monterey Bay Aquarium/BPS. 48.2*b*: E. S. Ross. 48.3: Mark Mattock/Planet Earth Pictures. 48.7: John R. Hosking, NSW Agriculture, Australia. 48.8*a*: Thomas Eisner, Cornell Univ. 48.8*b*: Ken Lucas/BPS. 48.9: © James Carmichael/NHPA. 48.10: Thomas Eisner, Cornell Univ. 48.11: M.Freeman/Bruce Coleman, Inc. 48.12: Kim Heacox Photography/DRK PHOTO. 48.15: Charlie Ott/Photo Researchers, Inc. 48.16: G. T. Bernard, Oxford Scientific Films/Animals Animals Earth Scenes 48.17: Jonathan Scott/Planet Earth Pictures. 48.18: A. Kerstich/Planet Earth Pictures. 48.19*a*: Larry E. Gilbert, Univ. of Texas, Austin. 48.19*b,c*: Daniel Janzen, Univ. of Pennsylvania. 48.20*a*: S. Nielsen/DRK PHOTO. 48.20*b*: © Raymond A. Mendez. 48.21: John Alcock, Arizona State Univ. 48.22*a*: © Larry Ulrich. 48.22*b*: E. S. Ross. 48.23*a*: Thomas Eisner, Cornell Univ. 48.24: Joel Simon. 48.26: After M. Begon, J. Harper, and C. Townsend, 1986, *Ecology*, Blackwell Scientific Publications. 48.28*a–c*: Robert and Jean Pollock/BPS.

Chapter 49 *Opener*: Courtesy of NASA. 49.6: Animals Animals/Len Rue, Jr. 49.17*a*: Brian A. Whitton. 49.17*b*: J. N. A. Lott, McMaster Univ./BPS.

Chapter 50 *Opener*: Norman Owen Tomalin/Bruce Coleman, Inc. 50.10: E. O. Wilson, Harvard Univ. 50.14*a,b*: Art Wolfe. 50.15*a*: J. N. A. Lott, McMaster Univ./BPS. 50.15*b*: M. Sutton/Tom Stack and Associates. 50.16*a,b*: Paul W. Johnson/BPS. 50.17: D. W. Kaufman, Kansas State Univ. 50.18: Edward Ely/BPS. 50.19: Kim Heacox Photography/DRK PHOTO. 50.21*a*: Elizabeth N. Orians. 50.21*b*: Frans Lanting/Minden Pictures. 50.22: Elizabeth N. Orians. 50.23: Art Wolfe. 50.24*a*: © E. R. Degginger. 50.24*b*: Jett Britnell/DRK PHOTO. 50.26*a*: Norbert Wu. 50.26*b*: M. C. Chamberlain/DRK PHOTO. 50.28: Gary Gray/DRK PHOTO. 50.29: Art Wolfe.

Chapter 51 *Opener*: Kevin Schafer, Tom Stack and Associates. 51.1: Paintings by H. Douglas Pratt. Courtesy of the Bernice P. Bishop Museum, Honolulu, Hawaii. 51.2: David S. Boynton. 51.4: Joe McDonald/Tom Stack and Associates. 51.5: Danny R. Billings, Missouri Dept. of Conservation. 51.6: Animals Animals/Fred Whitehead. 51.8: Kenneth W. Fink/Bruce Coleman, Inc. 51.9: Edward S. Ross. 51.10*a*: John Gerlach/DRK PHOTO. 51.11: Richard P. Smith, Tom Stack and Associates. 51.12*a*: © Walt Anderson. 51.12*b*: Frans Lanting/Minden Pictures. 51.12*c*: Animals Animals/Paul Freed. 51.14*a,b*: Art Wolfe. 51.15, 51. 16: Courtesy of the Smithsonian Institution, Office of Environmental Awareness/Richard Bierregaard, photographer. 51.18: Animals Animals/Peter Weimann. 51.19*a,b*: © Bill Gabriel, BIOGRAPHICS. Box 51.A *top*: Steven L. Hilty/Bruce Coleman, Inc. Box 51.A *bottom*: N. H. Cheatham/DRK PHOTO.

INDEX

Numbers in *italic* refer to information in an illustration, caption, table, or box.

Abalones, 604, *1059–1060*
Abandonment, of nests, *1059*, 1071
Abdomen, *599*
ABO blood group system, 224–*226*
Abomasum, *1000*–1001
Abortion, 344, 842
Abscisic acid, 733, *734–735*, *739*, 750–751
   and dormancy, 736, *751*
   and seed germination, 736
   and stomatal cycles, 697
Abscission, 743–*744*, 749–750
Absolute dating, 413–414
Absolute temperature, 118
Absorption, of nutrients, 997–999, 1002–1005
Absorption spectrum
   of chlorophyll, 167–168
   of cone cells, *900*
   of phytochromes, 751–752
Absorptive nutrition, 514
   in fungi, 530–531
Absorptive period, 1002–1005
Abyssal zone, *1171*
Abzymes, 133
Acacias, *1076*, *1115*
Accessory fruits, *567*, 569
Accessory pigments, 168–169
Accessory sex organs, 826, 828
Acclimatization, *785*
   role of thyroid, 810–*811*
   See also Seasonal environments
Acellular organisms, 497
Acellular slime molds, 513–515
Acer rubrum, 666–667
Acetabularia, 69, 72–73, 523
Acetaldehyde, *35*, *156*, *418*
Acetate, 149
Acetic acid, *35*, 116, *418*
   in citric acid cycle, *148*–149
Acetone, *35*
Acetyl coenzyme A
   allosteric controls, 158–160
   in citric acid cycle, *148–149*, 157
   functions, 157, 158
Acetyl group, *983*, 984
Acetylcholine
   and blood flow, 974
   and control of heartbeat, *965*
   as neurotransmitter, 866–868, 869, 872
   receptors, *108*, 869–870, 929
Acetylcholinesterase, 127, 870
Acetylsalicylic acid, *714*, 794, *795*
Achenes, *746*
Achondroplastic dwarfism, 336
Acid precipitation, 1138, *1142*
Acids
   defined, 29–30

names of, 147
Acinetobacter, 493
Acmaea, *427*
Acoelomates, 576–*578*
Acorn worms, *613–614*
Acquired immune deficiency syndrome. *See* AIDS
Acrasiomycota, *514*, *515*–516
Acromegaly, 311, 808
Acrosomal reaction, 836
Acrosome, 830–*831*, 836
Actin, 83–*84*, 307, 913–914
   in cytokinesis, 201
   in muscle cells, *915–918*
Actinomyces israelii, *496*
Actinomycetes, *496*, 536, 724
Actinopoda. *See* Radiolarians
Action potentials, 849, 851, 858–866
   in cardiac muscle, 920, 963–*965*
   in muscle contraction, 917–*918*
   in sensory cells, 881–*882*, *883–884*
   synaptic transmission, 866–869
Action spectrum, *168–169*
Activating enzymes, 261–262
Activation, of sperm and eggs, 836–837
Activation energy, *119*–121, 122–*123*
Activation gates, 861
Active form, of enzymes, 129–*130*, 993–999
Active reabsorption, 1009, 1017, *1020–1021*
Active secretion, 1009, 1017, *1020–1021*
Active site, of enzymes, *121*, 127–128
Active transport, 102–103, 104–*106*
   across gut, 998
   in osmoregulation, 1009, 1012, 1013, 1021–1022
   in plants, 690–692, *699–701*, 719–720
Acute renal failure, *1023*
Acute toxins, 1109
ADA, 342
Adam's apple. *See* Larynx
Adaptation, 3, 12–14, 435–437. *See also* Evolution; Natural selection
Adaptation, to sensations, 884, 888–889
Adaptor molecules. *See* tRNA
Addressing, of proteins, 78, 80, 264–265, 306–*307*
Adélie penguins, *1011*
Adenine
   in ATP, 137
   in genetic code, 266–267
   in nucleic acids, *57–58*, 245–*249*, 247
   in organic compounds, 417, *419*
Adenomas, *347*

Adenosine 3´,5´-cyclic monophosphate (cAMP). *See* cAMP
Adenosine deaminase (ADA), 342
Adenosine diphosphate. *See* ADP
Adenosine monophosphate. *See* AMP
Adenosine triphosphate. *See* ATP
Adenoviruses, 499
Adenylate cyclase, 818–820, *886–887*
ADH. *See* Vasopressin
Adipose tissue, 778. *See also* Fat
Adolescence. *See* Puberty
ADP
   chemical structure, *137*
   in glycolysis, *145*, 146
   in respiratory control, 155, 158–*159*
Adrenal glands, 802, 807, 813–815
Adrenaline. *See* Epinephrine
Adrenocorticotropin, *806*, 808, 809, *810*, 813, 814, *815*
Adrenocorticotropin-releasing hormone, *795*, *810*, 814, 815
Adventitious roots, 671
Adzuki bean weevil, *1105*
Aeciospores, *535–536*
Aecium, *535–536*
Aequipecten irradians, 604
Aerenchyma, *707–708*
Aerobic bacteria, 152, 484, *491*
Aerobic exercise, *919*–920
Aerobic respiration, 136, *627*
   origin of, *415*
Afferent blood vessels
   in fishes, *938–939*
   in kidneys, *1015–1016*, *1019–1020*, 1024
Afferent pathways, in nervous systems, 853
Agapornis, *1037*
Agar, 523
Agaricus campestris, 539
Agassiz, Louis, 600
Agave, *1082*
Age distributions, 1083, *1085–1086*
Aggregate fruits, *567*–568
Aggregation, in slime molds, 515, 1040
Aggressive behavior, *800*, *1031*. *See also* Territoriality; Threat displays
Aging
   and cancer, 345–346
   and deafness, 894
   and death rates, 1083–*1084*
   effects of stress, *815*
   and growth hormone, 809
   and vision, 899
Aging, of soils, 723–724
Agnatha, 619–620
Agouti pattern, 234–235

Agricultural ecosystems, *1144*–1145, 1178–*1179*
Agriculture
  environmental effects, *1144*–1145, 1178–*1179*
  fertilizers in, 722–723, 726, 1138–*1139*, 1144
  and human population growth, 639, 1097
  recombinant DNA technology in, 329–331, 726, *728*, 745
  soils in, 721–722, 1167, 1169
  vegetative reproduction in, 757–758, 771–772
  See also Crop plants
Agrobacterium tumefaciens, 329–330, *492*
Agrostis, 449, 451, 452, 710
AIDS, 259, 350, 375–378, *839*
  transmission of, 376, 378
AIDS-related complex, 376
Air, compared with water, 934–935, 1169
Air bubbles, breathing, 938
Air capillaries, *938*, 940, *942*
Air circulation. *See* Atmospheric circulation
Air pollution, and cancer, 345, 349
Air pressure, 1145
Air sacs, 935, 939–*942*
Alanine, *26*, *50*, *267*, *418*
  isomers of, *36*
Albatrosses, 1048–*1049*
Albersheim, Peter, 746
Albinos, 234–*235*
Albizzia, 767–768
Albugo, 517
Albumin, 973
Alcohol
  absorption of, 996
  chemical structure, 33–35
  effects on humans, 852, 1025
Alcoholic fermentation, *156*, 707
Aldehydes, 33, *35*
Alders, 724, 725, *1119*
Aldolase, *144*
Aldosterone, *807*, 813–*814*, 1024
Aleuria aurantia, *538*
Aleurone layer, 738–*739*
Algae, 504, 517–526
  absorption spectrum, *484*
  compared with plants, 504, 517, 523
  endosymbiotic, *76*, 505–506, 517
  in lichens, 531, 540–542, 1114
  parasitic, 523
  in studies of photosynthesis, *175*–180
  as substrate, *1060*
  in sulfur cycle, 1140
Algal blooms, 1140
Algal mats, 580
Alginic acid, 521

Alkaloids, as plant defenses, 712
Alkaptonuria, 255
All-*trans*-retinal, *894*–895
All-*trans*-retinoic acid, *44*
Allantois, *625*, 841–*842*
Allard, H. A., 764–765
Alleles, 217–*218*, 224–225
  frequencies, 428–*431*
  multiple, 224–*226*
Allergic reactions, 366–367, 374–375, 801, 814, 968–969
Alligators, *625*, *626*
Allomyces, *516*
Allopatric speciation, 448–449, 452
Allosteric activators, 158
Allosteric control
  of cellular respiration, 157–160
  of enzymes, *129*–132
Allosteric inhibitors, 158
Alluaudia, *1187*
Alpha carbons, in amino acids, 34, *36*
Alpha helix, 52, *53*, *56*, 96
Alpine tundra biome, *1163*.
  See also Tundra
Alternation of generations
  in algae, *506*, 521, 523–525
  in flowering plants, *568*–569, 758–759
  in funguslike protists, *506*, *516*
  in plants, 546
  in seedless plants, *556*
Altitude
  adaptations to, 935–936, *949*–950
  and temperature, 1145
Altricial nestlings, 628
Altruism, 1064–1066, 1072–1075. *See also* Eusociality
Aluminum ions, 25
Alvarez, Luis, 647–648
Alveoli, 944–*945*, 946
  in *Paramecium*, 511
Alzheimer's disease, 500
Amacrine cells, *901*
Amanita, 870
Amber, fossils in, *649*
Amboebophilus, *529*
Amensalisms, *1101*, 1113
Ames, Bruce, 349
Ames test, 349–*350*
Amines, 33, *35*
Amino acid substitutions, systematics and, 440–443
Amino acids
  assembly into proteins, 261–265
  chemical structure, 34–36, 157
  evolution of, 417–*419*
  genetic codes for, 266–268
  nutritional requirements for, 983–*984*
  in proteins, 47–55
Amino groups, 33–*35*, 49–51
Amino sugars, 47, *49*

Aminoacyl-tRNA synthases, 261–262
α-Aminoisobutyric acid, *418*
Aminomaleonitrile, *419*
3-Aminotriazole (3AT), 745
Ammonia (NH_3)
  in early atmosphere, 414, *415*, *417*, *418*, 420
  excretion of, 1010–1012
  in nitrogen cycle, 724, *725*, 1138–*1139*
  in plants, 728–729
Ammonites, 605, 654
Ammonium (NH_4^+), 25
  in nitrogen cycle, 728–729
  in plant nutrition, *718*
Ammonotelic animals, 1010–1011
Amniocentesis, 344, 841–842, 843
Amnion, *625*, 841–*842*, 844
Amniotic egg, *625*, 841
Amoebas, *505*, 507, 508–509, 913–914
  parasites of, *529*
Amoeboid movement, 913–914
AMP (Adenosine monophosphate), *137*, 139, 158–*159*, 261–262. *See also* cAMP
Amphibians, 622–625
  circulatory systems, *959*, 960
  gene amplification, 302–303
  lungs, 943
  osmotic balance, 1011, 1012, 1017, 1018
Amphipods, 600
Amphisbaenas, *625*, *626*
Amplification
  of genes, 302–303, 326–*328*, 346
  of hormonal effects, 819
  of sensory signals, *883*–884
  Amplitude, of circadian rhythms, 767–768
Ampullae, 890–*891*
amu (atomic mass unit), 19
Amylase, 121–122, *132*
  in coacervate drops, *421*
  in seeds, 738
  in starch digestion, 995, 997
Amylose, 46
Anabaena, *495*
Anabolic steroids, 815, *817*
Anabolism, defined, 114
Anacharis, 168
Anaerobic bacteria, 422, 484, 489–490
  as nitrogen fixers, 725–726
  respiration, 155, 156
Anaerobic respiration, 136, 156, *627*
Anagenesis, 448
Anal pore, 587–588
Anaphase, *198*, 200, 205–208
Anaphylactic shock, 801
Ancestral traits, 466–472
Anchor cells, 399–400

Androgens, *807*, 814–817
Androstenedione, *814*
Anemia
  hemolytic, 109
  pernicious, 988
  sickle-cell, 256–257, 268, 335, 337, *340*–341
Anesthetics, 852
Aneuploidy, 209–210, 270, 334
Angelfishes, 522
Angiosperms. *See* Flowering plants
Angiotensin, 974–975, 1024
Angular gyrus, 877
Animal models, 342
Animal pole, 383, 394–395
Animals, 15
  cells, 71
  effects on communities, 1120–1121
  evolution of, 503–504, 578–580
  taxonomy, 574–580, 606, 633
  use in research, 9, 342, 350
  See also Protostomes; Deuterostomes
Anion channels, *104*
Anions, 25
Anisogamy, 525, 526
Ankyrin, 109–*110*
Annelids, 577, 578, 579, 594–595
  circulatory systems, *958*
  digestive systems, *992*–993
  excretory systems, *1013*
  locomotion, 594–595, 921
  regeneration, 825
  reproduction, 594–595, 828
Annual cycles. *See* Circannual rhythms; Seasonal environments
Annual plants, 677, 757–758
  in desert biomes, *705*, 1110, 1166, 1167
  parental investment, *1082*
Annual rings, 666, 682–683
Anopheles, 510–511
Anopleura, 602
Anorexia nervosa, 982
Antbirds, 1189–*1190*
Anteaters, 456–457
Antennae, 600
  chemosensitive, 884–885
  in communication, 1042
Antennapedia complex, 402
antennapedia mutation, 402
Anterior, defined, 388
Anterior pituitary, 805–806, 808–809, 810
Anteroposterior axis, 395, *396*, 400–*401*
Antheridium, 549, 556
Anthers, 565–567, 758–759
Anthocerophyta, 546, 547, 550–551
Anthoceros, 550
Anthocerotae, 546–547, 550–551
Anthocyanins, 83
Anthophyta. *See* Flowering plants

Anthozoa, 585–587
Anthrax, 494
Anthropoids, 634–640
Antibiotics
    bacterial resistance to, 274, 284
    effect on intestinal microorganisms, 1000
    function, 487
    in recombinant DNA technology, 316–318, 317
    sources, 496, 540
Antibodies, 357–358, 372–373
    and blood type, 225
    chemical structure, 361–366
    classes of, 365–366, 369–370
    genetic diversity, 367–370
    monoclonal, 364–365
    production of, 374
Anticodons, 261, 262–263
Antidiuretic hormone. See Vasopressin
Antigen, processed, 371–372
Antigen–antibody complex, 362, 365
Antigen-presenting cells, 371–373
Antigenic determinants, 357–358
Antigens, 357–358
Antihistamines, 968–969
Antimicrobial proteins, 358. See also Complement system; Interferons
Antiparallel flow. See Countercurrent flow
Antiparallel strands, 245–246, 251, 254
Antipodal cells, 758–759, 760
Antiports, 103, 104
Antiself lymphocytes, 361
Antisense nucleic acids, 348–349
Ants, 602
    in desert ecosystems, 1110–1111, 1167
    leaf-cutting, 531–533, 989
    mutualisms, 699, 1114, 1115
    in rainforest ecosystems, 1189
    as seed dispersers, 1116
    social systems, 1074–1075
Anura, 622–624. See also Frogs; Toads
Anus, 991–992, 995
    and animal classification, 387, 549–550, 589, 610
Aorta, 958–960, 961–962
    chemosensors in, 952, 975
    in hot fish, 790
    stretch receptors in, 974–975, 1024
Aortic body, 952, 975
Aortic valves, 961, 962
Apatite, 1138–1139
Apes, 632, 635–640
Aphasias, 877–878
Aphids, 602, 699

habitat selection, 1060–1061
    mutualisms with ants, 1114
    parthenogenesis, 825
    photoperiodism, 766
Apical buds, 672
Apical dominance, 743, 745
Apical hook, 750–751, 753–754
Apical meristems, 677–678, 761. See also Root apical meristem; Shoot apical meristem
Apicomplexa, 507, 509–511
Aplysia, 875–876, 1039, 1046
Apomixis, 770–771
Apoplast, 692–693, 700
Appearance. See Morphology; Phenotype
Appendages
    of apes, 636
    of arthropods, 595–596
    of crustaceans, 599–600
    development, 400–401
    evolution, 12–13, 466–467, 621, 623
    shape, 10, 792
    of vertebrates, 619, 620, 627, 629–630
    See also Fins; Wings
Appendicular skeleton, 922
Appendicularia, 618
Appendix, 997, 1000
Apterygotes, 600–602
Aquatic animals
    gas exchange, 934–935, 938–939, 940
    osmoregulation, 1009, 1010–1011
Aquatic ecosystems, 1169–1173
    oxygen depletion, 1102, 1169
    productivity, 1128, 1131–1132
Aquatic plants, 706–707, 708
Arabidopsis thaliana, 754
Araceae, 714
Arachnids, 598. See also Spiders
Arboreal lifestyle, 632–635
Arboviruses, 499
Archaebacteria, 15, 62, 483, 485–486, 487, 489–490
Archaeopteris, 561–562
Archaeopteryx, 627–628
Archean eon, 416, 483, 646
Archegonium, 549, 556
Archenteron, 386–390
Archosauria, 625
Arctic fox, 792, 1104
Arctic hare, 792
Arctic tundra biome, 1163. See also Tundra
Area cladograms, 1154
Argia, 829
Arginase, 132
Arginine, 50, 267
    and canavanine, 712–713
Argon, 1136
    isotopes, 648
Argopecten irradians, 239
Argyroxiphium, 457–458

Arms, of echinoderms, 615–617
Army ants, 18, 1189
Arrow worms, 578, 579, 611, 614, 633
Artemia, 1009, 1010
Arteries, 958, 966–967
    chemosensors in, 952, 975
Arterioles, 958, 966–968
    in kidneys, 1015–1016, 1019–1020
    regulation of blood flow, 973–975
Arteriosclerosis, 43, 967, 999, 1003
Arthrobotrys, 531
Arthropods, 595–602, 640–641
    chemosensation, 884–885
    circulatory systems, 957, 958
    compound eyes, 897–898
    digestive systems, 992
    as disease vectors, 493, 497, 499, 510–511, 600, 1109
    excretory systems, 1013–1014
    exoskeleton, 595–596, 922
    gas exchange, 596, 600, 937–938
    See also Crustaceans; Insects
Artificial chromosomes, 297
Artificial mRNAs, 266
Artificial seeds, 772
Artificial selection, 14, 1038
    of crop plants, 329, 428, 429
    experiments, 337–338
Artiodactyla, 631
Asci, 229–230, 533–534, 537
Ascidiacea, 618
Asclepias syriaca, 704
Ascomycetes, 502, 529, 537–539
Ascorbic acid. See Vitamin C
Ascospores, 537
Asexual reproduction
    advantages and disadvantages, 202, 435, 770, 824
    in animals, 824–826
    in bacteria, 488
    in flowering plants. See Vegetative reproduction
    in protists, 506
Ashkenazic Jews, and Tay-Sachs disease, 335
Asparagine, 50, 267
Aspartic acid, 50, 267, 418
    in crassulacean acid metabolism, 698
    as neurotransmitter, 869
Aspens, 770, 1089, 1120
Aspergillus, 540
Aspirin, 714, 794, 795
Association cortex, 857, 858, 876, 894
Associations. See Coloniality
Associative learning, 875–876
Assortative mating, 434
Asteraceae, 475

Asterias rubens, 825
Asteroidea. See Sea stars
Asters (flowers), 566
Asters (in mitosis), 197–199, 200–201
Astrocytes, 852
Asymmetry, 395, 575
Atherosclerosis, 43, 967, 999, 1003
Athletes, muscle types, 919–920
Athlete's foot, 540
Atmosphere
    of early Earth, 414–415
    as ecosystem component, 1135–1136
Atmospheric circulation, 1135, 1145–1146
Atomic mass unit (amu), 19
Atomic number, 20–21
Atomic weight, 21
Atoms, 19–23
ATP, 137–139
    in active transport, 104–106
    as allosteric inhibitor, 159
    in Calvin–Benson cycle, 178–182
    in cell movements, 911–912
    in citric acid cycle, 147–149, 150
    in fermentation, 144, 155–156
    formation in chloroplasts, 172–173
    formation in mitochondria, 74–75, 152–155
    in glycolysis, 142, 144–147, 150
    in induction, 400
    in muscle contraction, 915–918, 920
    in photophosphorylation, 164–165, 169–175
    in respiratory chain, 143–144, 151, 152
    in translation of RNA, 261–262
    yields, 136–139, 150, 156–157
ATP synthetase, 152–153, 154, 172–173
Atria, of heart, 958–960, 961–965
Atrial natriuretic hormone, 807, 817
Atrioventricular node, 965
Atrioventricular valves, 961, 962
Atriplex halimus, 708
Atropine, 870
Atta cephalotes, 533
Auditory canal, 890, 892
Auditory cortex, 877–878, 894
Auditory nerve, 892–893
Auditory signals, 929, 1042
Auditory systems, 890–894
Australia, species in, 1152, 1156, 1177, 1178
Australian vampire bat, 1025
Australopithecines, 637–638
Autocatalysis, 996, 997

Autoimmune diseases, 362,
    374–375
Autonomic nervous system,
    853, 870–872, 914
    and blood flow, 974–975
    in digestion, 870–872
    and heartbeat, 920, 965
    and ventilation, 951–953
Autonomic reflexes
    as communication, 1044
    conditioning of, 875–876
Autoradiography, 21, 22,
    175–178
    in recombinant DNA tech-
        nology, 323
Autoregulatory mechanisms,
    of capillary beds,
    973–975
Autosomal traits, 334–336, 339
Autosomes, 227
    and genetic disorders,
        334–335, 338–340
Autotomy, 462, 467, 469
Autotrophs, 1, 2, 162, 718, 979
Autumn leaves, 749–750
Auxin, 733, 734–735, 739,
    742–748, 771
Auxotrophic mutants
    of E. coli, 274–278
    of Neurospora, 255–257
Avena, 1112
Avery, Oswald T., 242, 495
Aves. See Birds
Avogadro's number, 27
Axes, in development,
    395–396
Axial filaments, 490–491
Axial skeleton, 922
Axis, of Earth, 1145–1146
Axon hillocks, 868–869
Axon terminals, 851–852,
    866–869
Axonemes, 911–912
Axons, 851–853
    and action potentials,
        861–866

B cells, 356, 357–361
    activation, 361, 371–373
    disorders of, 375
    genetic diversity, 359–361,
        367–370
    and major histocompatibil-
        ity complex, 371–373
B vitamins, 667, 985–986, 987,
    988
    in carrier compounds, 141
Baboons, 1077
Baboons (Black howler mon-
    keys), 1191
Baby boom, 1085–1086
Bacilli, 66, 487, 491–492, 494,
    495
    as denitrifiers, 728
Bacillus anthracis, 488–489, 494
Bacillus megaterium, 282
Bacillus subtilis, 494, 495
Bacillus thuringiensis, 494
Backbones, of DNA, 243, 244,
    251
Bacteria, 15, 62–68, 483–496
    aerobic, 152, 484, 491

anaerobic. See Anaerobic
    bacteria
and bacteriophages, 274,
    281–284
bioluminescence, 138, 728
cell structure, 63–64
chemosynthetic, 484–485,
    718, 729–730
conjugation, 274–276,
    278–281
fossils, 414, 483
genes, 211, 273–289
intestinal, 77, 992, 1000
metabolic diversity,
    483–485
nitrogen-fixing. See
    Nitrogen–fixing bacte-
    ria
oxygen requirements, 579
as pathogens, 488–489,
    713–714
photosynthetic, 64-65,
    173–174, 422, 484,
    493–494
population growth, 1089
reproduction, 211, 486, 488,
    1081–1082
taxonomy, 485–486
transforming principle,
    242–243, 275–276
See also Prokaryotes;
    Archaebacteria;
    Eubacteria
Bacteriochlorophyll, 422, 484
Bacteriophages
    genes, 265, 273–274,
        281–284
    life cycles, 281–283, 497
    in recombinant DNA tech-
        nology, 312
    transduction, 282–284
    transforming principle,
        243, 244
    See also Viruses
Bacteriorhodopsin, 174
Bacteroids, 726–727
Bakanae disease, 739–740
Baker's yeast, 538
Balaenoptera musculus, 630,
    989, 1095–1096
Balance, sensation of, 889–891
Balanus, 1101–1102
Baldwin, Thomas, 728
Baleen plates, 989–990
Baleen whales, 641, 989–990,
    1103
Ball-and-hill analogy, 119–120
Ball-and-socket joint, 926–927
Balsa, 686
Bamboo, 770
BamHI, 317
Band patterns, chromosomal,
    204, 233–234, 323
Banting, Frederick, 813
Barberry, 535–536
Bark. See Cork
Bark beetles, 1103–1104
Barley, 739
Barn swallows, 1069–1070
Barnacles, 599–600, 990, 1172
    competition among,
        1101–1102

Barometric pressure, 935
Barr, Murray Llewellyn, 301
Barr body, 301–302
Barrel cactus, 468
Barriers
    between biogeographic
        regions, 1156, 1158
    and speciation, 448–449
    and species distributions,
        1155–1156
Barron, G. L., 531
Basal body, of flagellum, 85,
    86, 491
Basal body temperature, and
    contraception, 839
Basal metabolic rate, 790–791,
    981
Base pairing, 57–58, 245–248
    in nucleic acid hybridiza-
        tion, 294–295
    in recombinant DNA tech-
        nology, 312–314
    rules, 57–58
Base sequences
    determining. See DNA,
        sequencing
    as genetic code, 266–267
Base substitution mutations,
    268–269
Bases (alkaline), 29–30
Bases (nitrogenous), 55–58
    analogs, 269
    in DNA, 245–249
    in RNA, 246–247
Basidia, 539–540
Basidiomycetes, 502, 529,
    539–540
Basidiospores, 535–536,
    539–540
Basilar membrane, 892–893
Basking, 786–788
Bassham, James, 175
Batesian mimicry, 1107–1108
Bateson, William, 234
Bats, 631, 632
    echolocation, 881, 904–905
    osmotic balance, 1025
    as pollinators, 1117
    teeth, 991
Bayliss, Sir William Maddock,
    1001
Bdellovibrio, 491
Beach communities, 522, 1172
Beach strawberry, 671
Beadle, George W., 255–257
Beaks
    of birds. See Bills
    of cephalopods, 605
Bean weevils, 1105
Beans, 753
Beard worms. See
    Pogonophores
Bears, 980, 1189
Beavers, 991, 1120–1121, 1188
Beck, Charles, 562
Bee-eater, white-fronted, 1065
Bee hummingbird, 629
Beeches, 1083, 1185, 1186
Beer, 538
Bees, 601, 602
    male–male competition,
        1064

as pollinators, 1117, 1118,
    1184–1185
social systems, 1074–1076
See also Honeybees
Beetles, 600, 601, 602
    defenses, 929, 1107
    development, 577
    mating behavior, 788–789,
        826
    plant defenses against,
        704, 1103
    species diversity, 446
Beggiatoa, 490
Behavior, 610, 1029–1077
    and classification, 473–474
    and evolution, 610, 628
    genetics of, 1029–1030,
        1036–1039, 1052
    hormones in, 1034–1037,
        1039
    sexual. See Sexual behavior
    social. See Social behavior
    thermoregulatory, 786–788,
        1044, 1045
Behavioral ecology, 1057
Behavioral isolation, 452–453
Belladonna, 870
Belly button, 845
Benefits. See Cost–benefit
    analysis
Benign tumors, 345–347
Benson, Andrew, 175
Bent grass, 710
Benthic zone, 1171
Benzodiazepines, 870
Bergey's Manual, 485
Beriberi, 987
Bermuda grass, 709
"Berries," of gymnosperms,
    563
Best, Charles, 813
Beta-pleated sheet, 52, 53, 56
Bicarbonate ions (HCO_3^-), 30
    in carbon cycle, 1137–1138
    in digestion, 997
    transport by circulatory
        system, 104, 950–951
bicaudal mutation, 395–396
bicoid gene, 395–396
bicoid protein, 403–404
Bicoordinate navigation,
    1051–1052
Biennials, 677, 741
Big Bang, 414
Big brown bat, 631
Bighorn sheep, 631
Bignonia capreolata, 469
Bilateral symmetry, 575, 588,
    614
    in flowers, 565, 566
Bile, 43, 354, 996–999
Bile duct, 996–997
Bills, of birds, 436–437, 449,
    450, 628
Binal viruses, 498–499
Binocular vision, 903–904
Binomial nomenclature, 464,
    466–467
Biochemical pathways,
    136–137, 142–144

Biodiversity crisis, 1176. *See also* Extinctions; Species diversity
Biogeochemical cycles, 1136–1143
Biogeographic regions, 1156, *1158*
Biogeography, 1152–1173
Biological clock. *See* Circadian rhythms; Circannual rhythms
Biological rhythms. *See* Circadian rhythms; Circannual rhythms
Biological species concept, 446–447
Bioluminescence, *136, 138,* 517–518, *728, 1171*
    in communication, 929, 1041–1042
Biomass, 1084
    effects of climate, *1148–1149*
    of plants, 723, *1149*
    pyramids of, 1131–*1132*
Biomes, 4, 1160–1173
Biosphere, 4, *5*
Biosphere reserves, 1190–1191
Biota, 648
Biotechnology. *See* Recombinant DNA technology
Biotin, *126, 986,* 1000
Bipedalism, *637–638*
Bipolar cells, 900–*903*
Biradial symmetry, 575
Birches, 1109–1110
Birds, 627–629, *1040–1041*
    bills, 436–437, 449, *450,* 628
    circadian rhythms, 1046–1047
    coloniality, *1069, 1076*
    colonization of islands, *1159–1160*
    eggs, 628, *841,* 1066–1067
    embryonic development. *See* Chickens, embryonic development
    extinctions, *1177, 1180*
    helping behavior, *1065, 1072–*1073
    lungs, 939–*942*
    migration, 1047–1052, 1094
    osmoregulation, 1010, *1011,* 1012, 1019
    parental care, 628, *1031–1032, 1058–1059,* 1071
    phylogeny, 465, *468, 627–629*
    as pollinators, 566, 568
    population densities, 1093
    as predators, 7, 1104
    as seed dispersers, 629, *1116,* 1117
    sex determination, 227, 228
    song learning, 1033–*1034, 1036–1037*
    survivorship, 1087–*1088*
    wings, 3, 400–*401,* 627–628
    See also Songbirds
Bird's beak (plant), 1193

Birth, 807, 844–*845*
Birth control, 835, 837, *838–839*
Birth defects, 843. *See also* Genetic disorders
Birth rates, *1085–1088, 1091–1092,* 1181
Bison, 1177
1,3-Bisphosphoglycerate, *145,* 146
Bithorax complex, 402–*403*
Bivalves, *603, 604–605*
    predators of, 616–617
Black bear, 1189
Black-bellied seedcracker, 436–437
Black-browed albatross, *1049*
Black-eyed susan, *1118*
Black howler monkey, *1191*
Black lemur, *1187*
Black stem rust, 534–536, *535,* 539
Black-tailed deer, *2*
Black-throated warblers, 1156–*1157*
Black-winged damselfly, 1067–*1068*
Blackbird, red-winged, 447–448, 1044, *1068–1069*
Blackbirds, 1072–*1073*
Blackburnian warbler, 464, 466–*467*
"Blackwater" rivers, 1109–*1110*
Bladders, urinary, 1018, 1019–*1021*
Blades, of leaves, 672–673
Blastocoel, *385,* 841–*842*
    and animal classification, 576–577, 591
Blastocyst, 387, *389,* 833, 835, 841–*842,* 843
Blastoderm, *385*
Blastomeres, 383–385, *394–395*
Blastopore, *386–390,* 577, 610
Blastula, *385,* 393
Bleached cells, 520
Bleaching, of retinal, 895
Blending theory, of inheritance, 217, 222
Blind spot, 899
Blocks to polyspermy, 836–*837*
Blood, 956, *971–973*
    cells. *See* Red blood cells; White blood cells
    filtration by kidneys, 1021–*1022*
    immune system components, 355, 356–*357,* 371–372
    osmolarity, 1025–1026
    pH, 30
    transport of respiratory gases, 938–*939,* 946–953, 969, 974–975
    See also Circulatory systems
Blood–brain barrier, 852, 969
Blood pressure, 962–*963*
    elevated, 814–*815,* 967

and glomerular filtration, *1015–1017,* 1024–*1025*
    regulation, 27, 966–968, 973–976, *1024*
Blood transfusions, 224–226
Blood types, 224–226, 477
Blood vessels, 956, 958, *966–971. See also* Arteries; Capillaries; Veins
Blood volume, *1024–1025*
Blooms, in eutrophic lakes, 1140
Blowdowns, 1093–*1094*
Blubber, 630
Blue-backed grassquit, *450*
Blue goose, *454*
Blue-green bacteria. *See* Cyanobacteria
Blue jay, 7
Blue light, *697*
Blue mussels, *434*
Blue pike, 1141
Blue whale, 630, 989, 1095–*1096*
Bluebells, *463*
Bluebonnets, *763*
Bluefin tuna, 789–*790*
Bluegill sunfish, 1061–*1063*
Body cavities
    and animal classification, 576–578, 589–591
    and locomotion, 589–591, 592–593, 640, *921*
Body plans, 575–578
Body shapes, 10, *792*
Body temperature. *See* Thermoregulation
Bohemian waxwing, *1116*
Bohr effect, 949
Bolting, *741*
Bolus, *995*
Bombardier beetle, 929, *1107*
Bombykol, *884–885*
Bombyx, 884–885
Bonds
    chemical, 23–33
    high-energy, 139
Bone marrow, 356, 924–*925,* 972
Bones, 780, 922–928
    and animal size, 10
    of birds, 628
    calcium storage in, 811–*812*
    evolution of, *12–13*
    in human skeleton, *923*
    of mammals, 629–630
Bonner, James, 766
Bony fishes, 621, *622*
Boreal forest biome, 562, 1163–*1164*
Bormann, F. H., 1193
Boron, *719*
Bottlenecks, genetic, 432–*433*
Botulism, 494
Bowerbirds, 1032, *1067,* 1069
Bowman's capsule, 1015–1017, *1020*
Brachinus, 1107
Brachiopods, 578, *579,* 612–613, 633, 652, 654
Bracket fungi, *539*

Brackish environments, *1124*
Bracts, *468,* 764
Brain, *849, 850,* 855–858, *874–878*
    and behavior, 1045–*1047*
    in birds, 628, 1036–*1037*
    blood-brain barrier, 852, 969
    development, 388–390, 853–*854*
    evolution, 638, 858
    oxygen requirements, 155, 933
    and vision, 902–904
    See also Nervous systems
Brain hormone, 802–*804*
Brain stem, *854,* 951–953
Branch roots, 679–680
Branched pathways, 130–132
Branches
    damage by, *1113,* 1118
    support of, 684–*686*
    See also Stems
Branching patterns, 554–*555,* 672
    regulation of, 743, 745, 749
Branchiostoma lanceolatum, 618
Brassica oleracea, 428, 429
Brassica rapa, 741
Braxton–Hicks contractions, 844
Brazil, conservation issues, 1188, 1191–1192
Bread molds, 533–534, 536, *537,* 538
    in genetics studies. *See* *Neurospora crassa*
Bread palm, 562
Breakage, of chromosomes, *269–270*
Breasts, 305, 630, 845–*846*
    cancer of, 345
Breathing. *See* Gas exchange; Ventilation
Breeding. *See* Artificial selection; Sexual behavior
Breeding seasons, and reproductive isolation, 449, 452–*453*
Brenner, Sydney, 258
Brewer's yeast, *62, 538*
Briggs, Robert, 393–394
Brightness, of light, 166
Brine shrimp, 1009–*1010*
Bristles
    of annelids, *594–595,* 921
    of *Drosophila,* 436
Brittle stars, 616–617
Broad fish tapeworm, *590*
Broad-tailed hummingbird, *1121*
Broca's area, 877–878
Broccoli, 428, *429*
5-Bromouracil, 269
Bromoxynil, *745*
Bronchi, 940, 944–*945*
Bronchioles, 944–*945*
Brood parasites, 1058–*1059,* 1180
Brown algae, *518, 520–521*
Brown fat, 154, *791–792*

Brown-headed cowbird, *1059*, 1180

Brown volcano sponge, *582*

Browsing, *1076*, 1109, 1133
　and speciation, 456–*457*

Brush-footed butterfly, 470–471

Brussels sprouts, 428, *429*

Bryophyta, *546*, *547–548*, *549*, *551–552*

Bryopsis, 523

Bryozoa, *578*, *579*, 611–*612*, 633

Buccal cavity, 992

Buckeye butterfly, *601*

Bud scales, *751*

Budding
　in animals, 618, 824–*825*
　in viruses, 497, *498*
　in yeasts, 538

Budding (grafting technique), 771

Buds, 671–*672*
　dormancy, *751*

Buffalo, *1114*, 1177

Buffer areas, *1190–1191*

Buffers, 30–*31*

Buffon, George Louis Leclerc de, 12–13

Bufo periglenes, 624

Bugs, 602

Bulbs, 770–*771*

Bullseye electric ray, *620*

Bumblebees, *601*

Bundle of His, *965*

Bundle sheath, 684, *685*
　in C₄ plants, 183–*185*

Bungarotoxin, 929

Buoyancy control, 640–*641*
　in cephalopods, 605–606
　in fishes, 621

Burdock, 762–*763*

Burgess Shale, *652*

Burnet, MacFarlane, 360, 361

Burrs, *762*

Butterflies, *1*, *7*, *601*, 602
　classification, 470–471
　mimicry, *1108*
　See also Caterpillars;
　　Metamorphosis

C₃ plants, *183–184*

C₄ plants, 183–*185*, 684

C segments, *369*

C-terminus, 49, 263

Cabbage, 428, *429*, *741*

Cacao, 1179

Cacti, *468*, *570*, 705–706, 1106–*1107*, *1166*

Cactoblastis cactorum, 1106–*1107*

Caddisflies, 602

Caecilians, 622–*623*

Caecum, *992*, 1000, *1001*

Caenorhabditis elegans, 384, 387, 396–*397*, 399–*400*

Caesarian section, 845

Caffeine, 1025

Caimans, 625

Calcarea, *582*

Calciferol, 43, *44*, 986, *988*

Calcitonin, *806*, 811

Calcium

　as animal nutrient, 984–*985*

　blood level regulation, 811–*812*, 923

　as plant nutrient, 710–711, *719*, *720*, 722–723

　in shells, 602–*605*, 612–*613*, 922, *1137–1138*

Calcium carbonate (CaCO₃)
　in carbon cycle, 1143
　in coral reefs, 523, 585–*586*
　in shells, 509, 922
　in soils, 723–*724*

Calcium channels
　in nervous system function, 859, 866–*867*
　in sarcoplasmic reticulum, 917–*918*

Calcium ions (Ca²⁺)
　in blocks to polyspermy, *837*
　effect on microtubules, 911–*912*
　in muscle function, 917–*918*
　in neuron function, 858, 866–*867*
　in pollination, 760
　in responses to hormones, *820*

Calcium phosphate, 923, 984

Calcium pumps, 917, 918

Calendar, geological, 415, *416*, *646*

California condor, 1183–*1184*

California spiny lobsters, *599*

Callixylon, 561–562

Calmodulin, 821

Calories, 27, 981–*982*

Calostoma cinnabarina, *539*

Calvin, Melvin, 175

Calvin–Benson cycle, 164, *175–181*

Calyptra, 552

Calyx, *565*, 758

CAM. *See* Crassulacean acid metabolism

Cambium
　cork, 677, *678*, 681, *683*, 684
　and grafting, *771*
　vascular, 677, *678*, 681–*683*

Cambrian period, *646–647*, 652
　animals, 589, 605, *659–660*
　mass extinction, 658

Camembert cheese, 540

Camera, analogy with eye, *897–898*

Camouflage, *7*, *928–929*

cAMP
　in gene transcription, *289*
　as pheromone, 515, 1040
　as second messenger, 818–821, *886*

cAMP receptor protein, *289*

Campanula, 463

Campylobacter fetus, 491, *492*

Canadian lynx, *1104–1105*

Canary grass, 742

Canavanine, 712–*713*, 715

Cancellous bone, 924–*925*

Cancer, 196, 344–350

　and aging, *1083–1084*

　defenses against, 355, 970

　medical treatment, 44, 348–350, *365*

　plasma membranes in, 97, 100

Canine teeth, 990–*991*

Cannel coal, 557

Cannibalism, sexual, 1030

Canopy, *8*, 446, 1118, 1123

Capacitance vessels, 970

Capacitation, of sperm, 836

Capillaries, 958, *966*, 967–969, 973–975
　and inflammation, 355
　in nephrons, *1015–1017*, 1019–1021

Capillary action, 31–32, 693

Capping, of RNA, 299–*300*, 305

Capra ibex, 1181

Capsids, 497

Capsules
　bacterial, 64, 242
　of nonvascular plants, 549

Captive propagation, *1182–1184*

Carapace, *599*

Carbohydrases, 993, 996

Carbohydrates, 45–47
　digestion of, 995–996
　in membranes, *93*, 97, 108
　metabolism of, *1010–1011*
　as nutrients, 981

Carbon
　alpha (α carbon), 34, *36*
　in animal nutrition, *983*
　covalent bonds, 24–25
　isotopes, *21*, *175–178*, 645
　numbering system, 45
　in organic compounds, 33–36
　in plant nutrition, 717–719

Carbon cycle, 1137–1138, 1143

Carbon dioxide
　atmospheric, 1136, *1138*, *1143*
　and C₄ plants, 183–185
　in Calvin–Benson cycle, *175–182*
　in carbon cycle, *1137–1138*
　in cellular respiration, 142
　chemical structure, 25
　in citric acid cycle, 146–148
　in early atmosphere, 414, *415*, *418*, 422
　exchange in animals, 933, 935–936, 969
　in fermentation, *156*
　and fruit ripening, 749
　in photorespiration, *181–183*
　in photosynthesis, *163–164*, 164
　regulation of blood content, *951–953*, 974–975
　transport by circulatory system, *104*, *950–951*
　uptake by plants, 684, 696–697, 717, *718*

Carbon-14 (¹⁴C), *175–178*, 645

Carbon monoxide
　in algal bladders, 521
　in early atmosphere, 414
　as neurotransmitter, 869

Carbon reduction cycle. *See* Calvin–Benson cycle

Carbon skeletons, *983*

Carbonate ions, 1137

Carbonic acid (H₂CO₃), 30
　in soils, 722
　transport by circulatory system, *950–951*

Carbonic anhydrase, *950–951*

Carboniferous period, *646*, 653
　plants, 545, *553*, 557

Carbonyl groups, *35*

Carboxyhemoglobin, 950

Carboxyl groups, 34–*35*
　in amino acids, 49–51
　in Calvin–Benson cycle, 175
　and pH, 29

Carboxylase, in C₄ plants, *184*

Carboxylic acids, *36*, 147

Carboxypeptidase, *124–125*

Carcinogens, 345, 349–*350*

Carcinomas, 345, *347*

Cardiac cycle, *962–965*

Cardiac muscle, 920, *963–965*, 971

Cardiovascular disease, 43, *967*, 999, *1003*

Cardiovascular systems. *See* Circulatory systems

Caribou, *38*, *1094*

Carnivores, 580, 631, 980, 1103–1104
　food acquisition, 988
　teeth, *991*
　as trophic level, *1128–1129*, *1132*

Carnivorous plants, *717*, 730–*731*

β-Carotene, *43*, *167*, 523

Carotenoids, *43*, 76
　in algae, 518–519, 520–521, 523
　in bacteria, 422, 490
　in photosynthesis, 168–169, *174*
　in plants, 545

Carotid arteries, stretch receptors, 974–975, *1024*

Carotid body, 952, 975

Carpels, 563, *565–567*, 569, *758–759*, 760

Carrier molecules, 123, *140–141*
　in membranes, 101, *102*, 104
　in respiratory chain, 143, 144, 150–152
　in transport across gut, 998
　and transport in plants, 690–691
　and transport of RNA, 299
　See also Electron carriers

Carriers, of genetic disorders, 228–230, 334

Carrying capacity, 39, *1090*, *1093*, 1096–1097

Cartilage, 47, 621, 922–925
Cartilage bone, 924–925
Cartilaginous fishes, 620–621, 923, 1011, 1018
Cascades of activation, 196–197
Casein, 305
Casparian strips, 692–693
Castration, 1035–1036
Catabolism, defined, 114
Catabolite repression, 289
Catalysis
    in coacervate drops, 421
    by enzymes, 121–123
    evolution of, 419–421
    by RNA, 299, 419–421
Catalyst, defined, 121
Catalytic subunit, 129–130
Catastrophic events, 644–645
Caterpillars, 602
    bioluminescence, 138
    coloration, 7
    as energy-gathering stage, 1081
    metamorphosis, 391–392
    as predators, 115, 1081, 1106–1107, 1109–1110
    See also Butterflies; Moths; Metamorphosis
Catfishes, 622, 905, 929
Cations, 25, 105
Cats, 631
    Siamese, 224, 235–236
    threat display, 1044
    vision, 900
Cattle egrets, 1113, 1114
Caucasians
    blood types, 225
    genetic disorders, 334–335
Caudata. See Salamanders
Cauliflower, 428, 429
Cave paintings, 639
CD4, in AIDS, 375–377
cdc2, 196–197, 198
cDNA, 318–320, 319
cDNA library, 320
Ceanothus, 725
Cecropia, 1082
Cell adhesion molecules, 109
Cell body, of neurons, 851–852
Cell cycle, 194–196
Cell division
    in cancer, 345–346
    in early development, 382–385
    regulation in plants, 748–749
    See also Mitosis; Meiosis; Cytokinesis
Cell expansion, regulation in plants, 734, 746–748
Cell fractionation, 87–89
Cell-free extract, 242
Cell fusion, 95–96
Cell layers. See Germ layers
Cell plate, 201
Cell theory, 12, 61
Cell walls
    of bacteria, 64, 485, 486–487
    of fungi, 528, 536, 539
    microtubules in, 84

    of plants, 201–202, 546, 673–674, 690–693, 746–748
    of protists, 505, 519
    removing, 315
    structure, 70, 86–87
Cellobiose, 46–47
Cells, 61–89
    characteristics, 61–63
    eukaryotic, 62–63, 68–87
    evolution of, 421
    number in body, 191
    as organizational level, 4, 5
    of plants, 673–676
    prokaryotic, 62–66, 87
Cellular immune response, 357–358, 370–374. See also T cells
Cellular respiration, 136–137, 138, 142–143, 149–55
    in bacteria, 484–485
    compared with photosynthesis, 185
    energy yield, 141–142, 143, 154–155
    in plants, 185
Cellular slime molds, 513–516
Cellulases, 1000
Cellulose, 46–48
    digestion of, 1000, 1114
    in plants, 546, 547, 673, 746–748
    in wood, 1103
Cenozoic era, 416, 646, 655
Center, of receptive field, 901–902
Centipedes, 600
Central dogma, of molecular biology, 258, 420
Central nervous system, 779–780, 849, 850, 852–853. See also Nervous systems
Central sulcus, 856–857
Central vacuole, 674
Centrifugation, 87–88, 153
    of blood, 971
Centrioles, 86, 196–200, 201–202
Centris pallida, 1064
Centromeres, 192–193, 198–200, 204
    in meiosis, 205–208
    in recombinant DNA technology, 314
    repetitive DNA in, 296–297
Centrosomes, 196–198
Century plant, 1082
Cephalization, 575
Cephalopods, 603
    camouflage, 928–929
    eyes, 897–898
    jet propulsion, 921
Cereals, germination, 739
Cerebellum, 853–854
    of birds, 628
Cerebral cortex, 856–858, 857, 876–878, 894, 902–904
Cerebral hemispheres, 854, 856–858, 857, 877–878
Cerebrospinal fluid, 1025

Cerebrum. See Cerebral hemispheres
Cervical cap, 839
Cervix, 832–833, 844–845
Cesium chloride (CsCl), 250
Cestoda, 588
CFCs, 1127, 1143
CGG triplet, 336
cGMP
    in photoreceptors, 896
    as second messenger, 820
Chaetognatha. See Arrow worms
Chaffinches, 1034
Chain terminators, 267, 268
Chambered hearts, 958, 959
Chance. See Probability
Channels. See Ion channels
Chaparral, 1160–1161, 1167
Chargaff, Erwin, 245
Charged tRNA, 261–262, 263
Chase, Martha, 243, 244
Cheek teeth, 991
Cheeses, and molds, 540
Cheetahs, 988, 1040, 1041, 1181
Chelicerates, 578, 579, 597–598, 606, 633
Chelonia, 625, 626
Chelonia mydas, 626
Chemical basis, of biology, 11
Chemical bonds, 23–33
Chemical communication. See Hormones; Pheromones
Chemical compounds, defined, 26
Chemical defenses
    of animals, 929, 1107
    of plants, 83, 356, 437–438, 704, 712–715, 1109–1110
Chemical elements, 20–21
    abundances, 20
    in biogeochemical cycles, 1133, 1136–1140
    covalent bonding capabilities, 25
Chemical equilibrium, 115–117
Chemical isolation, 453
Chemical lift, 605–606
Chemical reactions, defined, 22, 27
Chemical symbols, 20–21
Chemical synapses, 866–870
Chemical weathering, 723
Chemically gated channels, 859, 866
Chemiosmotic mechanism, 143–144, 152–155, 172–174
Chemoautotrophs, 484–485
Chemoheterotrophs, 484–485
Chemosensation, 881–882, 884–888
    of blood gas content, 952–953, 975
    in communication, 1040–1041
Chemosynthetic bacteria, 484–485, 718, 729–730
Chemotherapy, 348

Chenopodium rubrum, 768
Chestnut blight, 538, 1178
Chestnut cowry, 604
Chiasmata, 204, 207. See also Crossing over
Chickens
    embryonic development, 382, 383–384, 385, 387–388, 841
    Rous sarcoma, 259
    wing development, 400–401
Childbirth, 807, 844–845
Chimaeras, 620–621
Chimpanzees, 475, 635, 636
    phylogeny, 470–471, 635–636
Chipmunks, 1112
Chironex, 583
Chironomidae, 1156
Chiroptera. See Bats
Chitin
    chemical structure, 47, 49
    in exoskeletons, 595, 922
    in fungi, 528
    in funguslike protists, 515, 516
    as gut lining, 993
Chitons, 603–605
Chlamydias, 493
Chloranthaceae, 570
Chlorella, 175, 523, 524
Chloride channels
    in cystic fibrosis, 335
    in nervous system function, 859, 868
Chloride ions (Cl⁻), 27
    in nervous system function, 858
    transport by red blood cells, 104
Chlorine (Cl)
    as animal nutrient, 984–985
    as plant nutrient, 719, 720
Chlorine cycle, 1143
Chlorobium, 422
Chlorofluorocarbons (CFCs), 1127, 1143
Chlorophyll, 167, 168–169
    in algae, 521, 523
    in bacteria, 422, 484
    in chloroplasts, 75–76
    in coacervate drops, 421
    in photophosphorylation, 164–165, 169–175
    in plants, 546
    synthesis of, 157
Chlorophyll a, 168–169, 484, 493–494
Chlorophyll b, 168–169
Chlorophyta, 517, 518, 523–525
    as ancestors of plants, 523, 547
Chloroplasts, 62, 70, 75–76, 162
    ATP formation in, 172–174
    and classification, 474–475
    compared with mitochondria, 172, 173
    DNA in, 237, 297–298, 475

evolution of, 76–77, 297–298, 415
of hornworts, 551
membranes, 75–76, 108
nitrate reduction in, 729
Chlorosis, 717
Choanocytes, 581
Choanoflagellida, 508, 581
Cholecystokinin, 807, 1001–1002
Cholera, 492, 818, 1109
Cholesterol
in bile, 999
and cardiovascular disease, 967
chemical structure, 43, 44
in diet, 1003
in membranes, 94–95
in steroids, 814
Chondrichthyes. See Cartilaginous fishes
Chondrus crispus, 523
Chordates, 578, 579, 611, 617–618, 633
Choreocolax, 523
Chorion, 625, 841–842
Chorionic villus sampling, 344, 842
Chorus frogs, 1105–1106
Christmas tree worm, 595
Chromatic adaptation, 523
Chromatids, 192–193
crossing over, 205, 231–233
in meiosis, 205–209
in mitosis, 197–200
Chromatin, 69, 72, 192–195
in mitosis, 197–199
transcription of, 300–302
Chromatography, 474
Chromatophores, 928–929
Chromium, 985
Chromomeres, 232–233
Chromoplasts, 76
Chromosomal mutations, 268–270
Chromosome jumping, 340
Chromosome maps. See Gene mapping
Chromosome-mediated gene transfer, 339–340
Chromosome walking, 323–325, 340
Chromosomes, 69, 72, 192–195
artificial, 297
of bacteria, 211, 279–281
of bacteriophages, 281–283
band patterns, 204, 233–234, 323
breakage, 269–270
crossing over, 205, 231–233, 298
of eukaryotes, 293, 296–297
gene linkage, 225–226, 228–230
gene location on, 323, 337–338, 403–404
and genetic disorders, 334, 335–336
of humans, 203–204, 227–230, 636
inactivation, 300–302

independent assortment, 219–223
and Mendelian inheritance, 218
in mitosis, 197–201
number, 203–204, 208, 209–210
polytene, 232–234, 301–302, 303
of prokaryotes, 211
structure, 192–195
studies with recombinant DNA, 320–327
of viruses, 265
See also DNA; Genes
Chrysaora melanaster, 587
Chrysolaminarin, 519
Chrysophyta, 518–520
Chthamalus, 1101–1102
Chylomicrons, 999, 1003
Chyme, 996, 997
Chymotrypsin, 124–125
Chytrids, 516
Cicadas, 602
Cichlids, 455, 458–459, 800, 1071
Cilia, 84–86, 910–912
of Paramecium, 512
Ciliary muscles, 899
Ciliates, 507, 511–513
Ciliophora, 507, 511–513
Ciona intestinalis, 618
Circadian rhythms
of animals, 1001, 1045–1048
of plants, 734–736, 767–768
Circannual rhythms, 796–797, 1047–1048
Circuits
in circulatory systems, 958–960
in nervous systems, 852–853, 870–874
Circular chromosomes, 211, 279–281
Circular muscle, 994
Circulating hormones, 801–802
Circulatory systems, 781, 956–976
in fishes, 790, 938–939
and thermoregulation, 787–788, 792
transport of respiratory gases, 946–951, 969
Circumcision, 831
Cirri, 513
Cirripedia, 599–600
11-cis-retinal, 894–895
Cities, and energy flow, 1144
Citrate, 148–149, 157, 983
Citric acid cycle, 142–143, 147–149
allosteric control, 158–160
energy yield, 150, 156–157
Clades, 465
Cladistic systematics, 465–477
in biogeography, 1153–1154, 1156–1157
Cladogenesis. See Speciation
Cladograms, 465, 467–477
Cladonia subtenuis, 540

Cladophora, 523
Clams, 604, 911, 980
Clapper rail, 1193
Claret cup cactus, 570
Class switching, 369–370
Class (taxonomic level), 464, 466–467
Classification. See Taxonomy
Clathrin, 108, 109
Clay, in soils, 721, 722, 723
Cleaning mutualisms, 1114
Clear-cutting, 1122, 1189
Cleavage, 382–385, 842
and animal classification, 576, 610
Cleaving, of DNA, 312–313
Cleft grafting, 771
Clematis afoliata, 469
Climates, 1145–1149, 1160–1162
Climatic changes, 647, 1138, 1140, 1143
and extinctions, 647, 657, 1185–1187
and speciation, 456–457
Climbing structures, convergent evolution, 469, 471
Clines, 438–439
Clitoris, 816, 832, 835
Cloaca, 830
Clocks, biological. See Circadian rhythms; Circannual rhythms
Clonal deletion theory, 361
Clonal selection theory, 360–361
Clones
defined, 202, 274
of lymphocytes, 360–361
of plants, 770, 772
Cloning. See Recombinant DNA technology
Cloning vectors, 314–315
Closed circulatory systems, 957–960
Closed systems, 114–115
Clostridium, 155, 488, 494
Clothing, and thermoregulation, 787, 788, 792
Clotting, of blood, 972–973
Clotting factor, in hemophilia, 336
Cloud forests, 1169
Clouds, 28, 1140
Clovers, 224, 437–438
Club fungi, 502, 529, 539–540
Club mosses, 546, 553, 555, 556–557
Clumped distributions, 1084–1085
Clumped resources, and social behavior, 1076
Clutches, 1082
Cnidarians, 575, 576, 578, 579, 583–587, 606, 633, 651–652
budding, 825
exchanges with environment, 777, 957, 991, 993
nematocysts, 583, 927–928
nervous systems, 850
Cnidocytes, 583

Coacervate drops, 421
Coal, 557, 653
Coated pits, 108, 109
Coated vesicles, 108, 109
Cobalamin, 986
Cobalt, 984–985
Cocci, 66, 487, 492–493, 494–496
Cochlea, 891, 892–894
Cockleburs, 738, 766–767, 768–769
Cockroaches, 992
Coconut milk, 748
Coconut palms, 666, 762–763
Cocoons, 391–392
Cocos nucifera, 666
Codfish, 1089
Codominance, 224
Codons, 260, 263, 266–268, 267
Coefficient of relatedness, 1066, 1074–1075
Coelom, 576–577, 610, 613, 614, 619
Coelomates, 576–578, 593
Coelomic fluid, 1013–1014, 1017
Coenocytes
in algae, 517, 523
in funguslike protists, 514, 516
Coenzymes, 125–126, 984–985
Coevolution, 1116–1118
of mutualists, 1178
of plants and pollinators, 566–568, 1116–1118
in rare species, 1184–1185
Cofactors. See Coenzymes
Cohesion
of species, 447
of water, 695
Cohorts, 1087–1088
Coin toss, and probability, 219, 432
Coir, 666
Coitus interruptus, 838–839
Cold, adaptations to, 792–793, 795–797, 810–811. See also Hibernation; Seasonal environments; Winter dormancy
Cold-blooded animals, 785
Cold desert biome, 1166
Colds, 885, 888, 891
Coleman, Annette, 523
Coleoptera, 601, 602
Coleoptile, 742–744, 747–748, 753
Coleus blumei, 748
Collagen, 52, 53, 985–986
in blood clotting, 972–973
in extracellular matrix, 98
in induction, 399
in skeletal system, 922–923
Collapsed lung, 946
Collar, in hemichordates, 613
Collared lizard, 626, 1182
Collecting ducts, 1019–1022, 1024–1025
Collenchyma, 674–675, 680–681, 684
Colon, 997–1000
cancer of, 345, 347

Coloniality
  in algae, 523–524
  and animal evolution, 578, 580
  in bacteria, 487–488, 494
  in birds, 1068–1069, 1076
  costs and benefits, 1064
  evolution of, 1074–1076
  in invertebrates, 584–585, 611–612, 613, 618
  in sessile organisms, 606–607
  See also Eusociality
Colonization, 1080
  of islands, 1157–1160
  by plants, 571, 725, 737, 770, 1119
Color blindness, 336, 338
Color change, 809, 928–929
Color vision, 882, 900
Coloration
  cryptic, 7, 928–929
  warning, 7, 623, 1107–1108
Colors, visible, 165–166
Colostrum, 845
Columnar cells, in blastula, 386
Columnar epithelial cells, 789
Comb jellies. See Ctenophores
Comet-tailed moth, 392
Commensalisms, 1101, 1113, 1114
Committed steps, 131–132
Common ancestor, 464, 465
Common bile duct, 996–997
Common names, 463
Common species, 1090–1091, 1184–1185
Communication, 929, 1040–1045
  evolution of, 638–640, 1044–1045
Communities, 4–5, 1100–1124, 1187
Community Baboon Sanctuary, 1191
Community ecology, 1100–1124
Compact bone, 924–925
Companion cells, 676, 700
Compartments, in biogeochemical cycles, 1133–1136
Compass sense, 1050–1051
Competition, 1101, 1110–1113
  among males, 1057, 1058, 1064, 1067–1069, 1072–1073
  among plants, 672, 711, 718, 1110, 1111, 1112
  among sessile organisms, 606–607, 1101–1102
  interspecific, 1102, 1111–1112
  intraspecific, 1059, 1085, 1110, 1111
  with introduced species, 1177–1178
  and rare species, 1184
  See also Territoriality
Competitive exclusion, 1111–1112, 1123

Competitive inhibition, 127–128
Complement system, 355, 358, 366–367, 371
Complementarity, among genes, 234–235
Complementary base pairing. See Base pairing
Complementary DNA (cDNA), 318–320, 319
Complementary foods, 983–984
Complete cleavage, 383
Complete metamorphosis, 391–392, 600, 602
Complex cells, of visual cortex, 902–903
Complex leaves, 555–556
Complexity, of plant communities, 1118–1119, 1123
Compound eyes, 897–898
Compound leaves, 672–673
Compound tissues, 672, 677
Compounds, chemical, 26
Compression wood, 684–686
Computers, in genome projects, 332
Concentration gradient, 100–101
Concentricycloidea, 612
Conception. See Fertilization
Concerted feedback inhibition, 131–132
Condensation reaction, 45, 417–419
Conditioned reflexes, 875–876, 1001
Condoms, 839
Condors, 1183–1184
Conduction deafness, 894
Cone cells, 899–901
Cones, of gymnosperms, 561, 563–564
Conidia, 532, 538
Coniferophyta. See Conifers
Coniferous forests, 1163–1164
Conifers, 546, 547, 561–564
  humus, 723, 1120, 1136
Conjugating fungi. See Zygomycetes
Conjugation
  in bacteria, 274–276, 278–281
  in fungi, 528, 536–537
  in Paramecium, 512–513
Connective tissue, 778–779, 923–925
Connexons, 99–100, 868
Conotoxin, 929
Conscious nervous system functions, 853
Consensus sequences, in RNA splicing, 299–301
Copulatory plugs, 1068
Conservation biology, 1176
Conservative replication theory, 248–249, 251
Conspicuous coloration, 7, 623, 1107–1108
Constant region
  of antibodies, 361–363, 365, 368–370
  of T-cell receptors, 370

Constellations, navigation by, 1051–1052
Constipation, 999–1000
Constitutive enzymes, 285
Constricting ring, 531
Constrictors (snakes), 988
Consumers, 1101, 1129
Contagious diseases. See Infectious diseases
Continental climate, 1147–1148
Continental drift, 646–647, 654
  and forest evolution, 553
  and species distributions, 1152–1153
Continental shelf, 1173
Continuous variation, 235–236
Contour feathers, 628
Contraception, 835, 837, 838–839
Contractile vacuoles, 83, 504, 512
Contraction, muscular, 889, 916–919
Contragestational drugs, 839
Control (homeostatic), 782–784
  systems for, 778, 779–780
Controlled systems, 782, 793
Convergent evolution, 469, 471, 536, 1160–1161
Convergent extension, 387, 388–389
Convolutions, of brain, 856–857
Cooling mechanisms, 791, 792, 793
Cooperative acts, 1064–1065
Cooperative breeding, 1065, 1072–1075
Cooperative hunting, 638, 1062–1063
Cooption, by humans, 1179–1180, 1188
Copepods, 590, 598, 599–600
Copper
  as animal nutrient, 984–985
  ions, 25
  as plant nutrient, 719, 723
  in soils, 709–710
Copra, 666
Coprophagy, 1000, 1001
Copulation, 829–830
  in earthworms, 595
  in humans, 835
  in insects, 829–830, 1067–1068
  in parthenogenetic species, 826
  in rats, 1035–1036
  See also Sexual behavior
Copulatory plugs, 1068
Coral reef communities, 5, 622, 1173
  fish cleaning mutualisms, 1114
Coralline algae, 1060
Corals, 574, 583, 585–587, 1173
  chemosensation, 884
  mutualisms, 517, 523, 586, 587, 1114

Cordycepin triphosphate, 259, 326–327
Core, of Earth, 414
Corepressors, 288–289
Cork, 673, 674, 713
  cambia, 650, 677
  growth of, 677, 683, 684
Cormorants, 1085
Corms, 770
Corn
  in genetics studies, 226–227, 230
  gibberellins in, 740
  seedling development, 753
Corn smut, 529
Cornea, 897–898
Corolla, 565, 758
Corona, 592–593
Coronary arteries, 967
Coronary infarction, 956, 967
Corpora allata, 803–804
Corpora cardiaca, 802–804
Corpus callosum, 877
Corpus luteum, 832, 833–834, 843
Correns, Karl, 215
Cortex
  of adrenal gland, 807, 813–815
  cerebral, 856–858, 857, 876–878, 894, 902–904
  of kidneys, 1019–1021
  of roots, 678–680, 692–693, 726–727
  of stems, 680–681
Corti, organ of, 892–894, 913
Cortical vesicles, 837
Corticosteroids, 813–815
Corticosterone, 814
Corticotropin-releasing hormone, See Adrenocorticotropin-releasing hormone
Cortisol, 807, 813–815, 1004–1005
Cortisone, 44
Corynebacterium diphtheriae, 488
Corynebacterium parvum, 486
Cosgrove, David, 747–748
Cost–benefit analysis, of social behavior, 1058–1059, 1064
Costa Rica, national parks, 1190–1192
Cottonwoods, 1060
Cotyledons, 569, 668–669, 753, 761–762
Countercurrent exchangers, 790, 941
Countercurrent flow, in fish gills, 938–939, 941
Countercurrent multiplier system, 1021–1022
Counting individuals, 1084, 1088
Coupled reactions, 123–124, 137
Coupled transport systems, 103, 104, 105

Courtship displays, 1032–1033, 1044, *1045*, *1067–1069*
  species-specific, 452, 473–474, *1037–1038*
  tactile, 1030, 1042
  territories for, *1069*
  See also Mate choice; Sexual behavior
Courtship feeding, *1067*
Courtship pheromones, 452, *884–885*, *1040–1041*
Covalent bonds, 23–25, *24*, 30–31, *33*
Cowbirds, *1059*, 1180
Cowries, *604*
cpDNA, 474–475
Crab-eating frog, 1018
Crabs, 600
  soft-shelled, 596, 922
Cranial nerves, 852, *853*
Crassulaceae, 184, *698*
Crassulacean acid metabolism, 184, 697–698
Crayfishes, 600, *883–884*
  excretory systems, *1014*
Creation science, 15
Cretaceous period, 553, *646*, 647, 655
  mass extinction, *646*, 648, 655, 658, 659
Cretinism, 810
Creutzfeld–Jakob disease, 500
Cri-du-chat syndrome, 210
Crick, Francis, 245–246, *248*, 258
Crickets, 602
Crimson rosella, *629*
Crinoidea, *615–616*
Cristae, *74*, *143*, 144
Cristatella mucedo, *612*
Critical day length, *765–767*
Critical period, 1034
Critical temperature, *791*, *793*
Cro-Magnons, *638–639*
Crocodiles, 625, *626*
Crocodilians, 625, *626*
  circulatory systems, 960
  cleaning mutualisms, *1114*
  phylogeny, 465, *468*
Crop (digestive), *992*
Crop milk, 805
Crop plants, 667–668
  artificial selection, 329, 428, *429*
  defenses, 713
  domestication, 639
  high-yielding, 329, 1144
  pathogens of, 516–517, 534–536, 538, 539
  recombinant DNA technology, *329–331*, 713, 726, *728*, *745*
  senescence, 750
  See also Agriculture
Cross-breeding
  of crop plants, 329
  of endangered species, 1181–1182
  in Mendel's experiments, 214–222

Cross walls, in fungi, 529, 536, 539
Crosscurrent flow, 942
Crossing over, *205*, 208–209, 226, 231–233
  in bacteria, 280, *281*
  and gene duplication, 298
Crossopterygians, 621–623
Crotaphytus collaris, *626*
Crown gall, 329–330, *492*
CRP (cAMP receptor protein), 289
Crust, of Earth, 414
Crustaceans, *578*, *579*, 598–600, *606*, 633
  excretory systems, *1014*
Crustose lichens, *541*
Cryptic coloration, *7*, 928–929
Crystallization, of bird song, 1036
Crystallized viruses, 497
Crystals, formation from ions, 25–26
Ctenes, *587–588*
Ctenophores, 575, 576, *578*, *579*, 582, 587–588, *606*, 633
Cucumaria, *616*
Cud, 1000
Cumulus, *836*
Cup fungi, *538*
Cupric ions, 25
Cuprous ions, 25
Cupulae, *889–891*
Curare, 870
Currents, oceanic, *1147*
Cuticle
  of invertebrates, *591–592*, 596, *922*
  of leaves, *683–684*, 696, 705, 709
  of plants, 677
Cutin, 713
Cuttings, of plants, 671, *733*, 743, 771
Cyanide, as plant defense, 437–438, 715
Cyanoacetylene, *418*
Cyanobacteria, *62*, 64, *415*, 484, 493–494
  evolution of, 422
  in hornworts, 551
  in lake ecosystems, 1140
  in lichens, 531, 540–542, *1114*
  in nitrogen cycle, 724–725, 1138
  as oxygen producers, 422, 578
  photosynthetic pigments, 169, 174–175
Cyanogenic plants, 437–438
Cycadeoids, 570
Cycadophyta. *See* Cycads
Cycads, *546*, 560, 561, *562*
Cyclic adenosine 3´,5´ monophosphate. *See* cAMP
Cyclic guanosine monophosphate. *See* cGMP
Cyclic photophosphorylation, 164, 171–172

Cyclins, 196–*197*, 198
Cyclosporin, 374
Cypraea spadicea, *604*
Cypresses, 737
Cysteine, 49, *50–51*, 267, 729
Cystic fibrosis, 332, 334–335, *339*, 340, 342, 841
Cysts
  as resting structures, 490
  of *Trichinella*, 591–592
Cytochrome *c*
  evolution of, 440–*443*, 474, 656
  in respiratory chain, 150–152
Cytochrome complex, 171–172, *173*
Cytochrome oxidase, 150–*151*, 152
Cytochrome reductase, 150–*151*, 152, *154*
Cytogenetics, 232–233
Cytokinesis, *201*–202, 913
Cytokinins, 733, *734–735*, *739*, 748–749
Cytoplasm, *64*, 69
Cytoplasmic determinants, 395–397
Cytoplasmic plaques, *99*–100
Cytoplasmic segregation, 394–397
Cytoplasmic streaming, in slime molds, 514–515
Cytosine, *57–58*, 245–249, 269
  in genetic code, 266–267
Cytoskeleton, 83–86, 912
Cytosol, 83–86
  in eukaryotic cells, 69
  in prokaryotic cells, 64
Cytotoxic T (T$_C$) cells, 370, 372–373

D segments, *368–369*
Dabbling ducks, *1037–1038*
Dactyella, 531
Dactylaria, 531
Daddy longlegs, 598
DAG (diacylglycerol), as second messenger, *820*
Daily cycles. *See* Circadian rhythms
Daily torpor, 795–796
Dalamites limulurus, 597
Dall sheep, *1087–1088*
Dalton, John, 19
Daltons, 19
Dams, beaver, 1120–*1121*
Damselfishes, *1071*
Damselflies, 602, *829*, *1067–1068*
Danaidae, 470–471
Dances, of honeybees, 1042–*1043*
Dandelions, 670, 762
DAP, *145*, 146
Daphnia, 427
Dark, and circadian rhythms, 1045–1047
Dark reactions. *See* Calvin–Benson cycle

Darwin, Charles, *13–14*, 426, 448, 742, 1044, 1065, 1111
Darwin, Francis, 742
Darwins, 657–658
Darwin's finches, 449, *450*
Dating, of rocks, 413–414, 509
Daughter chromosomes, *198*–200
Davis, Bernard D., 275–276
Day gecko, *1187*
Day length
  and circannual rhythms, 1047–1048
  and flowering, 765–767
Day-neutral plants, 766
DDT, and extinctions, 1183
ddTNPs, *326–327*
Dead space, in lungs, 944
"Dead spots," as plant defense, 714
Deafness, 894
Death, causes in humans, 350, 1083–*1084*
Death rates, 1083–*1084*, *1085*–1088, 1091–1092
  and evolution, *14*
  and risk of extinction, 1181
Decanal, *728*
Decapod crustaceans, *599*–600
Decentralization, in plants, 669
Deciduous forests, 1164–*1165*, 1167, *1168*
  and global warming, 1185–*1186*
Deciduous teeth, 990
Decomposers, 516–*517*, 1120, 1121
  fungi as, 528, 530, *1120*, 1121
deCoursey, Patricia, 1046
Deep-sea organisms, *1171*
Deer
  antlers, 1047
  black-tailed, *2*
  death rates, 1083, *1088*
Defecation, 992
Defenses
  chemical. *See* Chemical defenses
  against pathogens, 353–361, 713–714, 794, 795
  of plants. *See* Plants, defenses
  of prey. *See* Prey defenses
  of rare species, 1184
  of territories. *See* Territoriality
Deficiency diseases, *987*, 988
Deficiency symptoms, in plants, 719–720
Deforestation, and carbon cycle, 1143
Degeneracy, of genetic code, 267
Degradative succession, 1119–*1120*
Dehydration reaction. *See* Condensation reaction

Delbrück, Max, 273
Deletions, of genetic material, 233, *269–270*
Delivery (birth), 845
Demes. *See* Subpopulations
Demographic events, 1086–1087
Demographic stochasticity, 1181
Demospongiae, *582*
Denaturation
    of DNA, 294–*295*, 325–*326*
    of enzymes, *132*
    of proteins, 783
Dendrites, *851–852*
Dendroica fusca, 466–467
Dendrotoxin, 929
Denitrification, 728, *729*
Denitrifying bacteria, 484, 494
Density. *See* Population density
Density dependence, 1091–*1092*, 1095
Density gradient centrifugation, 88, *250*
Density independence, 1091–*1092*
Dentine, *990*
Deoxyribonucleic acid. *See* DNA
Deoxyribonucleoside triphosphates. *See* Nucleoside triphosphates
Deoxyribose, *46*, 55, *57*
Depolarization, of membranes, 859–862
Deprivation experiments, 1030
Depth vision, 903–*904*
Derivative carbohydrates, 47, *49*
Derived traits, 466–472
Dermal tissue system, 673, 677
Dermaptera, 602
Desert biomes, *1166–1167*
    adaptations of animals, *792*, 1021, *1026*
    adaptations of plants, *705–707*
    annual plants, *705*, 1110, 1166
    climate, 1145–*1146*
    competition experiments, *1110–1111*
Desertion, of nests, *1059*, 1071
Desmids, 523, *524*
Desmodus, 1025
Desmosomes, *99–100*
Desulfovibrio, 422
Determinate cleavage, 576
Determinate growth, 677, 764
Determination, developmental, 391–393
Detritivores, 979–*980*, *1128–1129*, *1132*, 1133
Deuterium, *21*
Deuteromycetes, *529*, 540
Deuterostomes, 610–*611*
    evolution, 640–*641*
    taxonomy, 574–580, *611*, *633*
Development, 381–406

cell adhesion molecules in, 109
and classification of animals, 575–*578*
    determination, 391–*393*
    differentiation, 392–*400*
    gap junctions in, 100
    gene expression in, 300, *302*, 393–396, *399–400*, 401–*405*
    in insects, 600–602, 802–805. *See also* Metamorphosis
    in mammals, 387–389, *840–844*
    of nervous system, 388–390, *853–854*, 912
    in plants, 733–754, *760–762*, 767
    research tools, 22
    sex determination, 814–*816*, 1034–1036, *1035*
Devonian period, *646*, 653
    animals, 605, 619–620, 622
    mass extinction, *646*, *658*
    plants, 553–555, 558–559, 561
de Vries, Hugo, 215
Diabetes insipidus, 1025
Diabetes mellitus, 311, 812–813, 1025
    nitric oxide in, 27
Diacylglycerol (DAG), as second messenger, *820*
Diadophus punctatus, 626
Diageotropica (*dgt*) mutation, 748
Dialysis, *1023*
Diaminomaleonitrile, 417, *419*
Diaphragm (contraceptive), *839*
Diaphragm (ventilatory), *945–947*
Diarrhea, 1000
Diastole, *962–963*
Diatomaceous earth, 520
Diatoms, *503*, 518–520
Dichotomous branching, *554–555*
Dicots, 569–*570*
    compared with monocots, 668–669
    roots, 679–680
    seedling development, 750–751, 753–754, *761*
    and selective herbicides, 744, *745*
    stems, 680–683
    trees, 667, 681–683
Dictyostelium, 515
Die rolls, 219
Diencephalon, *854*
Diener, Theodore, 499
Diet
    and evolution, 437, 449, *450*
    requirements. *See* Nutrition
    and social systems, *1076–1077*
    and speciation, 456–457

and teeth, *991*
    weight loss, 982
Differential centrifugation, 88
Differentiation
    in algae, 521
    in animal development, 392–400
    in animal evolution, 578
    in plants, 678–679, 734, 748
    zone of, 678–679
Difflugia, 505
Diffuse coevolution, 1117
Diffusion, 100–*101*
    across membranes, 101–107
    facilitated, 101–*103*, 104, 690–691
    Fick's law, 936, 937
Diffusion path length, *934–935*, 937
Digestion, 157, 980–981
    in protists, 504–*505*
Digestive enzymes, 990, 992–993, 996–*999*
    chemical action, *120–122*, *124–125*, *132*
    in lysosomes, 82–83
    and seed germination, 738–739
Digestive hormones, 802, *807*, 813, 817, 1001–*1002*
Digestive systems, *781*, 981, 990–*1002*
    defenses against pathogens, 354
    regulation of, *1002–1005*
Digests, 340–341
Dihybrid cross, 221–222
Dihydroxyacetone phosphate (DAP), 145, 146
Diisopropylphosphofluoridate (DPF), 127
Dikaryosis, 533–534, 537
Dimethyl sulfide, *1140*
Dinitrogen (N_2), 717, 724, 725–726
Dinoflagellates, 510, 517–*518*, 583
    mutualisms with corals, *586*, 587
Dinosaurs, 625–627, 629, *654–655*
Dioecious species, 227, 565, 828
Dionaea, 730–731
Dipeptidase, 997, *998*
Diphosphoglyceric acid, 949–950
Diphtheria, 488
Diphyllobothrium latum, 590
Diplobatis ommata, 620
Diploblastic animals, 576, 582
Diplococcus pneumoniae, 495
Diplodinium dentatum, 513
Diploid generation. *See* Sporophytes
Diploidy, 202–*203*
Diplontic life cycle, 525
Diptera. *See* Flies
Directional selection, 435–436
Dirona albolineata, 604
Disaccharides, 46, 47

Disc flowers, *566*
Discontinuous growth, *390–391*
Discontinuous variation, 235
Discs, in rod cells, 895
Disease-resistance genes, in plants, 714
Diseases
    as cost of social behavior, 1064
    evolution of, 1109
    genetic. *See* Genetic disorders
    in humans, 350, 1083–*1084*, 1097
    introductions of, 1178
    See also Pathogens
Disjunct distribution, *1155*
Disorder. *See* Entropy
Dispersal
    and geographic distribution, 1152–1155
    and habitat selection, *1059–1061*, 1093–1094
    as life history stage, *1081–1082*
    rule of parsimony, *1154–1155*
    of seeds. *See* Seed dispersal
Dispersal distribution, 1155
Dispersive replication theory, 248–*249*, 251
Displacement behavior, 1044
Display sites, *1069*
Displays, 1040
    and classification, 473–474
    courtship. *See* Courtship displays
    evolution of, *1044–1045*
    threat. *See* Threat displays
Disruptive selection, 435–436
Distal, defined, 394
Distal convoluted tubule, 1019–1022
Distance and direction navigation, 1050–*1052*
Distribution patterns, 1058, 1084–*1085*, 1152–1156
    and historical biogeography, 1152–1156
    and phylogeny, 477
    and resource availability, 1076–1077
Disturbances
    colonization after, 1118–1119, *1159–1160*
    and population dynamics, 1092–1094
    and species richness, 1123
Disulfide bonds, in immunoglobulins, 361–363
Disulfide bridge, 45, 51
Diversity. *See* Genetic diversity; Species diversity
Diving beetle, 601
Diving mammals, 948, 975–976
Diving reflex, 975–976
Division of labor. *See* Differentiation

Divisions (taxonomic level), 464, *466–467*, 485 DNA, 241–270
  chemical structure, 55–58, 243–*248*
  in chromosomes, 192–*195*
  and classification, 474–476, 477
  cleaving and splicing, 312–314
  complementary, 318–320, *319*
  discovery of, 241–243
  in eukaryotes, 69, 293–308
  in fossils, 474–475
  genetic code, 266–268
  hybridization, 294–*295*, 476
  libraries, 318–320
  methylation, 302, *312*
  mitochondrial, 237, 297–298, *324*, 640, *1156–1157*
  nonnuclear, 231
  origin of, *415*, 421
  in prokaryotes, *211*, 279–*281*, 284–289, 485–486
  repair of damage to, 254
  repetitive, 296
  replication, 248–254
  sequencing, 323–327, 332
  synthetic, 320
  technology. *See* Recombinant DNA technology
  transposable elements, 284–285
  in viruses, 498–*499*
  See also Genes; Chromosomes
DNA-binding proteins, *253*
DNA ligase, 252, *253*, 313–314
DNA polymerase
  in recombinant DNA technology, 320, 326–*328*
  and replication, 248, 251–254
  and transposable elements, 285
dNTPs, *326–327*
Dodder, *531*, 730
Dogs
  breeds, 1038
  senses, 881, 885
Dogs, wild, *1062–1063*
Domestic animals, 639
Dominance (genetic), 217–218
  incomplete, 222–223
Dominance (social)
  in females, 1077
  hierarchies, 1077
  in males, 1057, *1058*, 1064, 1077
Dominant competitors, 1112–*1113*
Dopamine, 869
Dormancy, of seeds, *734*, 735–737, 762
Dorsal, defined, 575
Dorsal aorta, *790*
Dorsal horn, 854–*855*

Dorsal lip, 387, *388–389*, *397–398*
Dorsoventral axis, 395, 400
Double bonds, *24–25*
Double fertilization, 563, *568–569*, 760–761
Double helix, *58*, 245–246, 247–248
Double membrane, *74–75*
  evolution of, 75–77
Doublets, 83–*84*
Doubling time, 273, 274
Douching, *839*
Doves, 805
Down feathers, 628
Down syndrome, 209–*210*, 334, *344*, 841
Downy mildews, 516
DPF, *127*
Dragonflies, 602, 1105–*1106*
Dreams, 874
Drosera, 730
Drosophila
  artificial selection, *436*
  behavioral studies, 1038
  chemosensation, *885*
  developmental studies, *22*
  gene amplification, 302
  in genetics studies, 230
  growth and metamorphosis, *392*
  inheritance, 224–225, 227–228
  mutations, 395–*396*, 401–*403*
  polytene chromosomes, 232–234
  segmentation, 401–*405*
  speciation, *455*
  Z-DNA, 246
Drought, adaptations to, *705–707*
Drowning, 795
Drugs, and biodiversity, *1176*
Dry matter, 720
Dry weight, 720
Dryas drummondii, 725
Drying, and seed germination, 736
Dubautia, *458*
Duchenne muscular dystrophy, 335
Duck-billed platypus, 630, *839–840*, 881, 887
Ducks, courtship displays, *473–474*, *1037–1038*
Ducts, of glands, 802
Dugesia, 589
Dulse, *521*
Dunes, 770
Duodenum, *996–997*
Duplications, of chromosomes, 269–270
Dussourd, David, 704
Dutch elm disease, 538, 1178
Dwarfism
  in humans, 336, *808*
  in plants, 740–741
Dyenin, 84–86, 200, 911–*912*
Dynamics, of populations, 1086–1097
Dysentery, 284, 492, 1109

Ear pinnae, 890, *892*
Eardrum, 890, *892*
Ears, 883, 890–894, 905
Earth
  age of, 412–*413*
  prebiotic, 414–419
Earthworms, 577, 594–595, 1000
  circulatory system, *958*
  digestive system, *992–993*
  excretory system, *1013*
  hermaphroditism, *595*, 828
  locomotion, *921*
  nervous system, *850*
  See also Annelids
Earwigs, 602
Eating. *See* Feeding
Ebola virus, *499*
Ecdysone, 303, 803–805
Echidnas, 630, *631*, 839
Echinoderms, 575, *578*, *579*, 611, 614–617, 633
  regeneration, *825*
Echinoidea. *See* Sea urchins
Echolocation, 881, 904–*905*
Ecological biogeography, 1154, 1157–*1160*
Ecological communities. *See* Communities
Ecological niches, 988, 1101–*1102*
Ecological succession. *See* Succession
Ecology
  defined, 1057–1058
  and evolution, 456–*457*
Economics
  of ecosystem services, 1193
  of food selection, 1061–*1062*
  of species reintroductions, 1183–1184
*Eco*RI, 312–314
Ecosystem services, 1176, 1193
Ecosystems, 1127–1149
  compartments, 1133–1136
  economic value, 1193
Ectocarpus, *520*, 521
Ectoderm, 385–*390*, 576–577
Ectoprocta. *See* Moss animals
Ectotherms, 785–790, 935
Ectothiorhodospira, 422
Eczema, 366, *367*
Edelman, Gerald M., 361
Edema, 968–*969*, 970
Edge effects, *1188–1190*
Edges, and vision, 903
Ediacaran fauna, *651*
Eels
  electric, *905*, 929
  migration, 621
Effector cells, 360–361, 370
Effector molecules, 129–130
Effectors, 778, 849, 909–929
Efferent blood vessels
  in fishes, 938–*939*
  in kidneys, *1015–1016*, 1019–*1020*, 1024
Efferent pathways, in nervous system, 853, 872
Efts, *1080*

Egg cocoons, of spiders, 1030
Egg-laying hormone, 1039
Egg trading, *1070*
Eggs, 202–*203*, 824, *826*
  amniotic, *625*, 841
  of amphibians, 623–624
  of birds, 628, *841*, 1018, 1066–1067
  development, 826–828, 830, *832–833*
  enucleation, 393–*394*
  fertilization, 836–*837*. *See also* Zygotes
  of fishes, 621
  of flowering plants, 758–759, 760–761
  gene amplification in, 302
  of oviparous animals, 837–*841*
  and parental investment, 1066–1067, 1071
  polarity, 394–*396*
  of reptiles, *625*, 1018
  translational control in, 305
Egrets, 1113, *1114*
Eigen, Manfred, 420
Eigenmannia, 1043–1044
Eijkman, Christian, *987*
Eisner, Thomas, 704
Ejaculation, 830, 835
Elaiosomes, *1116*
Elaphe obsoleta, 438–439
Elasmobranchs. *See* Cartilaginous fishes
Elastase, *124–125*
Elastic fibers, in blood vessels, *966–967*
Elasticity, in plant cell walls, 747–748
Elaters, 549–550
Electric charges
  across membranes, 107–108, 858–866
  of atoms, 19–20
  in nervous system function, 858–866
Electric fishes, *905–906*, 929, 1043–1044
Electric organs, 929
Electrocardiogram, 963, *964*
Electrocommunication, 1043–1044
Electrodes, 861–*862*
Electroencephalogram, *874*
Electromagnetic spectrum, *165–166*
Electromyogram, 874
Electron carriers, 108
  in bacteria, 484
  in chemiosmotic mechanism, 172–174
  in noncyclic photophosphorylation, 170–171
  in respiratory chain, 143, 144, 150–152
Electron microprobe, 696
Electron microscope, 67–68, 97–98
Electrons, 19–20, 22–23
  and light energy, *166–168*
  in redox reactions, 139–*141*

Electrophoresis, 320–322, 340–341, 430, 433
Electroporation, 316
Electroreception, 905–906
Electrosensors, 882
Elemental substances, 26
Elements. See Chemical elements
Elephant seal, 433, 1057, 1058
Elephants, 10, 788, 793, 988, 1177
Eleutherozoans, 615–617
Elevation, and temperature, 1145
Ellipsoid joints, 927
Elongation
    of plant cells, 671, 753–754
    of polypeptides, 262–265
    of shoots, 741
    zone of, 678–679
Embolism, 967
Embryo sac, 758–759, 760–761, 762
Embryonic development. See Development
Embryonic induction, 397–400
Embryonic organizer, 397–400
Embryophyta, 545–546
Embryos
    of animals, 841–844
    defined, 381
    of flowering plants, 569, 738–739, 760–762
    of plants, 545, 548
    polarity, 394–395
    of seed plants, 561
Emergent properties, 4
Emigration, 432, 1086–1087
Emission, of sperm, 830
Emlen, Stephen, 1051
Emotions, and brain, 856, 1052
Emperor angelfish, 622
Emperor penguin, 777
Enamel, 990
Encephalitis, 499
End-chain codons. See Chain terminators
End sac, 1014
Endangered species, 1180–1185. See also Rare species; Extinctions
Endemism, 457–458, 1155–1156, 1187–1188
Endergonic reactions, 117–118
Endocrine systems, 779–780, 800–802. See also Hormones
Endocrinology, defined, 800
Endocuticle, 922
Endocytosis, 80–81, 83
    by phagocytes, 355
    receptor-mediated, 108
Endoderm, 385–390, 576–577
Endodermis, 679, 692–693
Endomembrane system, 78–82
Endometrium, 833–835, 841–842
Endonucleases. See Restriction endonucleases
Endoplasmic reticulum, 70–71, 78–79

and membrane synthesis, 109
    and protein synthesis, 264–265
    See also Rough endoplasmic reticulum
Endorphins, 806, 808, 809
Endoscopy, 838
Endoskeleton, 922–928
Endosperm, 565, 568–569, 738–739
    of coconuts, 666
    development of, 760–761
Endospores, 488, 494, 495
Endosymbiosis, 77, 505–506
Endosymbiotic theory, 76–77, 297–298, 415, 1113
Endotherms, 785–786, 790–797
Endymion, 463
Energetic costs, 1058
    of reproduction, 1081–1083
Energy, 11, 114–133
    of chemical bonds, 34, 139
    in chemical reactions, 27–28, 115–121
    conversion, 115–116
    flow through ecosystems, 1127–1133, 1144–1149
    light, 164–168
    and nitrogen fixation, 726
    for origin of life, 415, 417
    pyramids of, 1131–1132
    and reaction rates, 119–121
    storage, in plants, 546, 547, 571, 674, 701
    storage compounds, 47–48, 139, 981–982, 1002–1005
    sublevels, 167–168
    thermodynamics, 114–119
    yields, 150, 156–157
    See also Fossil fuels; Free energy; Solar energy
Energy barrier, 119–120
Energy efficiency, 1143
Energy-gathering stage, 1081–1082
Energy maximizers, 1061–1062
Enhancers, 304
Enkephalins, 806, 808, 809
Enterogastrone, 807
Enterokinase, 997, 998
Enteropneusta, 613–614
Entrainment, 767–768, 1045–1047
Entropy, 115, 118
Enucleated eggs, 393–394
Environment
    defined, 1057
    versus heredity, 235–236, 427–428
    and natural selection, 434–437
    and social systems, 1076–1077
Environmental cues, in plant growth, 733, 751–754, 764–767
Environmental damage, 1058
    by agriculture, 1144–1145, 1178–1179

to fish spawning grounds, 1173
    and human population growth, 1097
    and recombinant DNA technology, 330–331
    See also Pollution
Enzyme kinetics, 123
Enzyme–substrate complex, 121
Enzymes, 121–133
    allosteric, 129–132
    catalysis by, 120–123
    chemical structure, 124–126
    digestive. See Digestive enzymes
    diversity, 58
    environmental sensitivity, 132–133, 784–785
    genetic coding for, 222–223
    inducible, 285, 287
    in lysosomes, 81–82
    regulation of, 126–133
Eons, 646
Ephedra, 563, 570
Ephemeroptera, 601, 602
Ephialtes, 1103
Epicotyl, 753
Epicuticle, 922
Epidemiology, 350
    of AIDS, 376, 378
    of genetic disorders, 337
Epidermis, 713
    of leaves, 683–685, 696
    of plants, 677
    of roots, 679
    of stems, 681, 683
Epididymis, 830–831
Epilepsy, 876, 877
Epinephrine, 801, 807, 813–814
    and blood flow, 974
    and glucose metabolism, 1004–1005
    and heartbeat, 965
    second messengers, 818–820
Epiphanes senta, 593
Epiphytes, 1169
Episomes, 283–284
Epistasis, 234–235
Epithelial cells, junctions, 98–100, 99
Epithelial tissue, 777–779
Epithelium, 386
EPSPs, 868
Eptesicus fuscus, 631
Equator, 1145–1147
Equatorial migrants, 1047–1048
Equatorial plate, 198–200, 205–208
Equilibrium
    chemical. See Chemical equilibrium
    Hardy–Weinberg, 430–431
    sensation of, 889–891
Equilibrium centrifugation, 88
Equilibrium constant, 116–117, 118–119
Equilibrium model, of species richness, 1157–1160

Equisetum, 557
Eras, 646
Erectile tissue, 830–831, 832
Erection, of penis, 27, 830, 835
Erosion, 671, 770
Error rate, in DNA replication, 253, 254
Error signal, 782–783
Erysiphe grammis, 530
Erythrocytes. See Red blood cells
Erythropoietin, 972
Escape behavior, 1033
Escape movements, 873–874
Escherichia coli, 65, 67, 492
    and bacteriophages, 274, 281–284
    chromosomes, 211, 279–281
    conjugation, 276, 278–281
    DNA replication, 252, 253–254
    in DNA studies, 250–251, 273
    in eutrophic lakes, 1141
    gene map, 279–281
    in human intestines, 77, 1000
    lac operon, 287–288, 289
    nutritional mutants, 274–278
    in recombinant DNA technology, 311, 312
    transposable elements, 284–285
Esophagus, 995, 997
Essential amino acids, 983–984
Essential elements
    for animals, 984–985
    for plants, 719–721
Essential fatty acids, 984
Esters, 139
Estivation, 1018
Estradiol, 814
Estrogens, 807, 814–817
    in contraceptives, 838
    in menstrual cycle, 833–835
    in pregnancy, 835, 843, 844, 845
Estrous cycle, 833
Estrus, 833, 1035
Estuaries, in fish life cycles, 621, 1173
Ethane, 33–34
Ethanol, 33, 35
    as product of fermentation, 143, 156, 538
    as research tool, 175
Ethene. See Ethylene
Ethics
    medical, 344, 809
    in scientific method, 6, 9
Ethology, 1031
Ethyl alcohol. See Ethanol
Ethylene, 733, 734–735, 739, 749–750
    chemical structure, 33–34
Etiolated seedlings, 753–754
Euascomycetes, 537–539
Eubacteria, 62, 422, 483, 485–486, 487

Eucalyptus, 693, *706*, 988, *1093*
Eucheda, 599
Euchromatin, 300–302
Euglena, *62*, *520*
Euglenophyta, *518*, *520*
Eukaryotes, 503
  cell division, 192–211
  cells, 62–63, 68–87
  compared with prokaryotes, *15*, *87*, 485–486
  evolution of, 76–77, 293–294, 297–298
  genes, 69, 293–308
  origin of, *415*
Eulaema, *1184*
Eumycota, 529–542
Euphorbia, *1187*
Euplectella aspergillum, *582*
Euplotes, *511*
European robin, *1031*
European starling, *1050*
Eurosta solidaginis, *1092*
Eusociality, *1074–1076*
  and communication, *1042–1043*
Eustachian tubes, 891–892
Eutamias, *1112*
Eutherian mammals, 630–632, 840
Eutrophication, *495*, 1140–1142, *1141*
Evaporation
  in hydrological cycle, *1137*
  from plants, *See* Transpiration
  and thermoregulation, 793
Evaporation–cohesion–tension mechanism, *694–696*
"Eve," 640
Evergreen forest, tropical, 1167, *1169*
Evolution, 2–3, 12–14, 426–443
  agents of, 432–437
  calendar, 415, *416*, 1153–1154
  and classification. *See* Phylogeny; Systematics
  convergent, 469, *471*, 536, 1160–1161
  and developmental processes, 403, *404–405*
  of humans, *416*, 632–640
  mutations in, 270, 274, 432
  parallel, 467
  patterns of, 644–660
  rates of, 656–658
  theory of, 12–14
  transposable elements in, 297–298
  See also Coevolution; Natural selection; Speciation
Evolutionary conservatism, 575–576
Evolutionary radiations, 456–459
Evolutionary trees, 465. *See also* Cladograms
Excision repair, 254
Excitatory postsynaptic potential (EPSP), 868

Excitatory synapses, 867–868
Excited state, *166–168*, 169, *170*
Excitement phase (sexual), *835*
Excretory systems, 781–782, 1008–1009, 1012–1026
Exergonic reactions, *117–118*
Exhalation, *946–947*, 951
Exocrine glands, 802
Exocytosis, *80–81*
Exons, 295–296
  splicing, 299–*301*
Exoskeletons, 595–596, 640–641, 921–922
  fossilization, 649
  See also Molting; Shells
Expanded-tip tactile receptors, 889
Experimentation, in scientific method, 8–9
Expiratory reserve volume, *943–944*
Exploitation competition, 1111
Explosions, evolutionary, *659–660*
Exponential growth, *1089–1090*
Expressivity, of genotype, 235
Extensors, 873, *922*, 926
External armor, in early fishes, 618–619
External fertilization, 606, 828–829
Extinctions
  of amphibians, 624
  causes, 1111, *1176–1180*
  human-caused, 629, 630, 639, 655, 1097, 1176–1180
  mass, *646*, 647–648, 654, 655, *658–659*
  probability of, *1180–1181*
  rates of, 658–659
  and species richness, 1158–1160
Extracellular digestion, 981, 990, 992
Extracellular fluids, 777, 781, 956, 973
  filtration, 1009, 1012, *1015–1017*
  in lymphatic systems, 969–970
  in open circulatory systems, 957
  regulation of, 1008–1009, 1023–1026
Extracellular matrix, 98, 778, 923–925
Extractive agriculture, 1144
Extraembryonic membranes, *625*, *841–842*
Extrinsic proteins, *94–96*
Eye color, in *Drosophila*, *224–228*, 229
Eye cups, *897*
Eyelash viper, *905*
Eyes, 882–883, 897–904
  of cephalopods, 605
  development, *397–398*
  of potatoes, *671–672*
  of primates, 632

of scallops, *604*

F factor, in bacteria, *278–281*, 285
F-pili, *276*, 278
Face recognition, 856–*857*
Facial expressions, 1052
Facilitated diffusion, 101–*103*, 104
  and transport in plants, *690–691*
Facultative anaerobes, 484
Facultative parasites, 531
FAD, 123–124, 140, 141, 985
  in citric acid cycle, *148–149*
  in respiratory chain, 149–155
FADH$_2$, 123–124, 140, 141
Fainting, 965, 970
Fallopian tubes, 830, *832*, 833, *841–842*
Family (taxonomic level), 464, *466–467*
Family trees, 337–339
Far-red light, 751–752, *767*
Faroe Islands, 434
Fasciola hepatica, *589*
Fast block to polyspermy, 837
Fast-twitch fibers, *919–920*, 948
Fat, 778
  brown, 154, *791–792*
  as insulation, *40*, 630, 792
Fat-soluble vitamins, 986–988
Fates, of germ layers, *386*, 391–*392*
Fats, 40–42
  breakdown and synthesis, 157, 1002–1005, *1003*
  in diet, 43, *967*, 981
  digestion of, 996, *999*
  products of metabolism, 1010–*1011*
  as storage compounds, 139, 981–982
Fatty acids, 40–41
  breakdown and synthesis, 157, 158
  in membranes, *93–94*
  nutritional requirements, 984
Faunas
  defined, 651
  evolutionary, *659–660*
Feather stars, 615–616
Feather tracts, *628*
Feathers, 627–628
  as insulation, 792
Feces, 991, 992, 999–1000
Feedback, *782–783*
  in metabolic pathways, *131–132*, 157–160
  in thermoregulation, 793–794
Feeder cells, *762*
Feedforward information, *782–783*, 794
Feeding. *See* Diet; Foraging; Ingestion
Feeding stage, *1081–1082*
Feet, chemosensors on, 884–*885*
Feldspar, in soils, 723

Females
  development, 814–*816*
  mate choice, 1069–1070
  reproductive systems, 830, *832–833*
Fermentation, 136–*137*, *143*, 144, *155–156*
  allosteric inhibition of, 158
  in plant roots, 707
  in sulfur cycle, *1140*
Ferns, *546*, *547*, *553*, *558–560*
Ferredoxin, 170–172
Ferric ions, 25
  as prosthetic groups, *126*
  in respiratory chain, 151
Ferrocactus acanthodes, *468*
Ferrous ions, 25, *126*
  in respiratory chain, 151
Fertility factor, in bacteria, *278–281*, 285
Fertilization, 202–203
  in animals, 606, *826*, 828–829, 836–837. *See also* Copulation
  dependence upon water, 533, 549, *551*, 552, 558, 560, 563, 758
  in vitro, 762
  in plants, 560–*561*, 563–564, *568*, 758–761. *See also* Pollination
  in seed plants. *See* Double fertilization
  and translational control, 305
Fertilization membrane, *837*
Fertilizers
  in agriculture, 722–723, 726, 1144
  and lake eutrophication, *1141–1142*
  in nitrogen cycle, 1138–*1139*
Fetal hemoglobin, 298, 948, *949*
Fetus, 800, 814, 843–*844*
Fever, *714*, 794, *795*
Fiber, dietary, 999
Fiber cells, 563, 674–*675*
Fibrin threads, 972–973
Fibrinogen, 972–973
Fibronectin, 387
Fibrous root systems, *670–671*
Fick's law of diffusion, 936, 937
Fiddleheads, *558*, 559
Field studies, *8*, 786–787
Fight-or-flight response, 801, 813–814, 870, 974
Figs
  as keystone species, 1188
  mutualisms with wasps, *1100–1101*, 1116, 1117
Filaments
  algal, 517
  bacterial, 488, 494, *495*, *496*
  of stamen, 758–*759*
Filicinae. *See* Ferns
Filopodia, 386
Filter feeders
  cilia and flagella, *910–911*

evolution, 617, 640–641, 652
food acquisition, 979, 980, 989–990, 1103, 1104
Filtration
in capillary beds, 968–969
of extracellular fluids, 1009, 1012, 1015–1017
of sensory signals, 882–883
Fin whale, 1096
Finches
bills, 436–437
Darwin's, 449, 450
Fingernails, 632
Fins
of arrow worms, 614
of fishes, 619–621, 623
Firebrat, 601
Fireflies, 138, 929, 1041–1042
Fires
as disturbances, 1093, 1165
and forest restoration, 1192
and seed germination, 563, 736, 737, 1093, 1185
Fireweed, 737
First filial generation (F₁), 215
First law of thermodynamics, 114–115
Fischer, Emil, 124, 125
Fischer's lovebird, 1037
Fishes, 619–621
acclimatization, 133, 784–785
aggression, 800
bony, 621, 622
cartilaginous. See Cartilaginous fishes
circulatory systems, 790, 958–959
cleaning mutualisms, 1114
communication, 1042, 1043–1044
effects of acid precipitation, 1142
in eutrophic lakes, 1141
evolution, 619–623
in freshwater ecosystems, 1170
gas exchange, 621–622, 935, 938–939, 940
genetic variation, 439–440
gills, 938–939, 940
hermaphroditism, 828, 1070
life cycles, 621, 1173
in marine ecosystems, 1171–1173
osmoregulation, 1017, 1018
parental investment, 1071
population dynamics, 1095
sensation, 881, 887, 889–890, 905–906
thermoregulation, 789–790
Fishing, environmental effects, 1095
Fission
in bacteria, 211, 486, 488
in yeasts, 538
Fitness, 437, 1058–1059
and social behavior, 1065–1066
See also Natural selection

5′ end
of DNA, 245, 247, 251–254, 296–297
of mRNA, 259–260
Fixed action patterns, 1030–1036, 1039
Flagella
of eukaryotes, 84–86, 910–912
of prokaryotes, 64–66, 65, 86, 486
of protists, 504, 517, 520
of sperm, 830–831
Flagellates, 507–508
Flagellin, 65, 486
Flame cells, 1012
Flamingos, 989
Flashlight fish, 138
Flatulence, 489–490, 1000
Flatworms, 574, 577, 578, 579, 588–590, 606, 633
excretory systems, 1012
gas exchange, 934
gastrovascular cavity, 957, 991
nervous systems, 850
as parasites, 588–590
Flavin adenine dinucleotide. See FAD
Flavin mononucleotide (FMN), 985
Flavonoids, 712
Flavors, 887–888
Fleas, 602, 909
Flexors, 873, 922, 926
Flies, 600, 601, 602
chemosensation, 884–885
Flight, in birds, 3, 627–628
Flight muscles, and thermoregulation, 788–789
Floating plants, 707–708
Flocking, as defense against predators, 1059, 1063–1064
Flora, defined, 651
Floral meristems, 764
Florida scrub jay, 1072–1073
Floridean starch, 523
Florigen, 735, 768–769
Flowering, 714, 734–736, 765–769
Flowering plants, 546, 547, 553, 563–571
evolution, 553, 570
taxonomy, 668–669
Flowers, 563–567, 757, 758–759
coevolution with pollinators, 565–568, 1117–1118
evolution of, 565–567
monocot vs. dicot, 669
Fluid feeders, 980
Flukes, 588–589
Fluorescence
defined, 169, 170
in recombinant DNA technology, 320–321, 326–327
Fluoride, 985
FMN, 985
Focusing, of eyes, 898–899
Foliar fertilizers, 723

Folic acid, 986
Foliose lichens, 541
Follicle-stimulating hormone, 806, 808, 809–810
in menstrual cycle, 833–834
in puberty, 815–816, 833
Follicles, 832, 833–834
Fontanel, 924
Food chain, 1129
Food intake, regulation of, 783
Food poisoning, 492, 494
Food solicitation behavior, 1031–1032
Food supply, and territoriality, 1068–1069
Food vacuoles, 83, 504–505, 993
Food web, 1129–1131
Foods
caloric content of, 982
defined, 1–2
selection, 1061–1062
See also Diet; Nutrition; Foraging
Foolish seedling disease, 739–740
Foot, molluscan, 603, 605
Foraging, 1033, 1060–1062, 1103–1107
adaptations for, 979–980, 988–990
cooperative, 1062–1063
Foraging theory, 1060–1062
Foraminifera, 505, 507, 509
Forbs, 1165
Forebrain, 853–854
Foreskin, 830
Forest analogs, 1179
Forest reserves, 1191–1192
Forestry, and habitat loss, 1121–1122, 1176, 1179–1180, 1189–1190
Forests
boreal, 562, 1163–1164
Carboniferous, 545, 553
complexity, 1123
deciduous, 1164–1165, 1167, 1168, 1185–1186
effects of acid precipitation, 1142
effects of global warming, 1185–1186
effects on microclimate, 1118
old-growth, 1189
productivity, 1133, 1144–1145
succession, 1118–1120
tropical, 8, 1144–1145, 1167–1169, 1189–1190, 1192
Formaldehyde, 417–418
Formamide, 417
Formic acid, 418
Formica, 1116
Forward reaction, 115
Fossil fuels, 1127
and acid precipitation, 1142
in agriculture, 1144
in carbon cycle, 1138, 1143

and human population growth, 1097
Fossilization, 648–649
Fossils, 645–646, 648–649, 656
DNA in, 474–475
in historical biogeography, 1153–1154
and relative dating, 413–414, 416
in systematics, 472
teeth, 990
Founder effect, 433–434
Four-minute mile, 627
Four-o'-clock plant, 236–237
Fovea, 899–900
Fox, arctic, 792, 1104
Fractionation, of cells, 87–89
Fraenkel-Conrat, Heinz, 258–259
Fragile-X syndrome, 335–336
Fragmentation
of geographic ranges, 448
of habitat, 1179–1180, 1187, 1188–1190
Frame-shift mutations, 268–269
Frank–Starling law, 971
Franklin, Rosalind, 243
Free energy, 115, 117–119, 123–124
from ATP, 138–139
in citric acid cycle, 149–150
in glycolysis, 150
in redox reactions, 140
Free ribosomes, 70–71, 78, 263, 264
Freeze-etching, 97–98
Freeze-fracture technique, 97–98
Freezing, 28–29
of cyanogenic plants, 437–438
damage to cells, 784
and recombinant DNA technology, 330, 331
and seed germination, 736, 737
Frequencies
of electric fishes, 1043–1044
of light waves, 165
Fresh water
communities, 1124, 1170
as ecosystem compartment, 1134–1135
osmoregulation in, 1009, 1012, 1014, 1017–1018
Friedman, William, 563
Frigate bird, 1044, 1045
Frogs, 6, 622–624
embryonic development, 382, 383, 385, 387–389, 390, 391–393, 394, 395–396, 398
flying, 610
metamorphosis, 391
oxygen requirements, 933
reproductive rates, 14
water conservation, 1018
See also Tadpoles
Fronds, of ferns, 558–559
Front ratio, 176–177

Frontal lobes, 856–857
Frost. See Freezing
Fructose, 46, 144, 146
Fructose 1,6-bisphosphate
  (FBP), 49, 144, 158–159
Fructose 6-phosphate (F6P),
  144, 158–159
Frugivores, 736, 737, 762–764,
  1188
Fruit flies. See Drosophila
Fruiting, 735–736
Fruiting bodies, of bacteria,
  490
Fruiting structures
  of fungi, 530, 533–535, 537,
    538, 539–540
  of slime molds, 514–515
Fruits, 567–569
  agricultural production,
    772
  development, 735–736,
    741, 746, 762–764
  dispersal, 762–764, 1116,
    1117, 1188
  ripening, 735, 749
Fruticose lichens, 541, 542
Fucoxanthin, 520–521
Fucus, 521
Fuel metabolism, 1002–1005
Fulcrum, of joints, 926, 928
Fumarate, 123–124, 127–128
  in citric acid cycle, 147, 151
Fumaroles, 1139
Functional groups, 33–36
Fundamental niches,
  1101–1102
Fungi, 15, 528–542
  bioluminescent, 138
  characteristics, 528–536
  compared with funguslike
    protists, 504, 513, 516,
    528
  as decomposers, 528, 530,
    1120, 1121
  evolution of, 503–504
  in lichens, 531, 540–542,
    1114
  mutualisms, 528, 531–533,
    559, 989
  in mycorrhizae, 531–532,
    679
  as parasites, 529, 531
  as pathogens, 529,
    534–536, 538, 539, 540,
    713–714
  as predators, 531
Funguslike protists, 504,
  513–517
  compared with fungi, 504,
    513, 516, 528
Funk, Casimir, 987
Fur, 792
Fusarium moniliforme, 540
fushi tarazu gene, 402

G cap, 299–300, 305
G-proteins
  in chemosensation,
    886–887
  in induction, 400
  in photoreceptors, 896–897

and second messengers,
    818–820
GABA, as neurotransmitter,
    869
Galactosamine, 47, 49
Galactose, 46, 997–998
α-galactosidase, 285, 316–319
Galapagos finches, 449, 450
Galapagos Islands, 449, 450,
    1178
Galerina, 539
Gall wasps, 1123
Gallbladder, 996–997
Galls, 329–330, 492,
    1060–1061, 1092
Gama, Vasco da, 987
Gametangia
  in fungi, 537
  in green algae, 525
  in Protomycota, 516
Gametes, 202–203
  in animals, 824, 826
  and parental investment,
    1066–1067, 1071
  See also Eggs; Sperm
Gametocytes, 510
Gametogenesis, 826–828
Gametophytes
  of algae, 524–525
  evolution of, 537
  of ferns, 559–560
  of flowering plants,
    568–569, 758–761
  of nonvascular plants, 549
  of plants, 546
  of seed plants, 559–560
  of seedless plants, 552, 556
Gamma-aminobutyric acid
    (GABA), 869
Gamma rays, 165
Ganglia, 850, 872
Ganglion cells, 900–903
Gap genes, 401, 403–404
Gap junctions, 99–100
  in muscles, 914, 920, 963
  in nervous system, 852,
    868–869
Gap phases, in cell cycle, 194,
    196–197
Garden spider, 1029
Garner, W. W., 764–765
Garrod, Archibald, 255
Gas exchange, 781, 933–953
  in amphibians, 622, 624
  in arthropods, 596, 600,
    937–938
  in birds, 939–942
  by direct diffusion, 934,
    956–957
  and evolution, 578–580,
    617
  in fishes, 621–622, 935, 937,
    938–939, 940
  See also Respiratory sys-
    tems; Ventilation
Gases, properties of, 1145
Gasoline, 33
Gasterosteus aculeatus, 656–657
Gastric pits, 996
Gastrin, 807, 1002

Gastrocolic reflex, 1001
Gastrointestinal tract.
    See Digestive systems
Gastropods, 603, 604–605
Gastrovascular cavity,
    583–584, 956–957,
    990–991
Gastrulation, 385–389
Gated channels, 859
Gause, G. F., 1111–1112
Gavials, 625
Geckos, 114, 1187
Geese, hybrid, 454
Gehring, Walter, 402
Gel electrophoresis. See
    Electrophoresis
Gemmae, 550
Gemmules, 824
Gene families, 298–299, 342,
    368
Gene flow, 432, 447
Gene libraries, 318–320, 319,
    323–325
Gene mapping, 230, 231–234,
    232, 323–324, 332
  of bacteria, 278–281
  of bacteriophages, 281–284
  of humans, 337–342
Gene pool, 428–429
Gene substitution tree, 477
Gene swapping, 308
Gene therapy, 342–344
Gene transfer, chromosome-
    mediated, 339–340
General homologous traits,
    466–467, 469
General transduction, 282–284
Generation time, and specia-
    tion, 456
Generative nucleus, 760–761
Generator potential, 883–884
Genes, 217–218
  activation by steroids, 303,
    821
  amplification, 302–303,
    326–328, 346
  and behavior, 1029–1030,
    1036–1039, 1052
  chemical structure,
    241–249
  for disease resistance, 714
  in eukaryotes, 293–308
  expression during devel-
    opment, 300, 302,
    393–396, 399–400,
    401–405
  inactivation, 300–302
  interaction among,
    233–236
  interaction with environ-
    ment, 235–236, 427–428
  linkage, 225–226, 229–230,
    335–339
  locating. See Gene map-
    ping
  number in humans, 226
  overlapping, 265
  and phenotypes, 255–256
  of prokaryotes, 211,
    273–289
  regulatory, 287–288
  synthetic, 320

transcription of, 258,
    259–260
  of white blood cells,
    367–370, 371
  See also DNA;
    Chromosomes;
    Recombinant DNA
    technology
Genetic code, 11–12, 266–268
  age of, 413–414
  and DNA sequencing,
    323–327
Genetic disorders, 334–344
  diagnosis, 324, 344,
    841–842, 843
  and meiotic errors,
    209–210
  sex-linked, 227–230,
    335–339
  treatment, 342–344
Genetic drift, 432–434, 433,
    1182
Genetic engineering. See
    Recombinant DNA
    technology
Genetic markers, 229, 278–281,
    282–284
Genetic recombination. See
    Genetic variation;
    Recombination
Genetic relatedness, and
    social behavior,
    1065–1066, 1072–1074,
    1077
Genetic screening, 344
Genetic stochasticity, 1181
Genetic structure, of popula-
    tions, 428–431
Genetic variation, 2–3,
    428–437
  among species, 455
  in animal taxa, 430
  calculating, 428–431
  in crop plants, 329
  and evolution, 437–442
  geographic, 437–438
  and natural selection, 435
  and risk of extinction,
    1104–1105, 1181–1182
  and sexual reproduction,
    202–203, 435, 758, 826
Genetics, 214–237
Genitalia
  lock-and-key, 829
  See also Penis; Vagina
Genome, 296, 329
Genome projects, 332, 337
Genotype, defined, 217, 427
Genotype frequencies, calcu-
    lating, 430–431
Genus (taxonomic level), 464,
    466–467
Geographic isolation, and spe-
    ciation, 448–449,
    452–453
Geographic populations, 428
Geographic ranges, 1080,
    1090–1091
  and extinction rates,
    658–659
  shifts in, 1185–1187

Geographic speciation, 448–449, 452
Geographic variation, 437–438
Geological calendar, 415, 416, 646
Georges Bank, 1095
Gerbils, 1026
Germ cells, 203, 396–397, 824, 826
    gene amplification in, 302
    modifying, 344
Germ layers, 385–390, 576–577
    fates, 386, 391–392
Germ line, 824, 826
Germ-line granules, 396–397
Germ plasm, 329
Germination, of seeds, 734, 735–739, 751–752, 1093
Gestation, 630, 841–844
Giant panda, 475, 476
Giant sequoia, 562
Giant squid axons, 860, 864
Giardia lamblia, 508
Giardiasis, 508
Gibberella fujikuroi, 540, 740
Gibberellins, 733, 734–735, 738–742, 748
Gibbons, 635
Gibbs, Josiah Willard, 117
Gibbs free energy. See Free energy
Gigantism, 808
Gilbert, Walter, 324
Gilia, 711–712
Gill arches, 938–939
Gill filaments, 938–939
Gill slits, 617
Gilliam, T. Conrad, 334
Gills
    of arthropods, 596, 937
    blood flow in, 938–939, 958–959
    external, 910, 934, 936–937
    of fishes, 938–939, 940
    internal, 621, 936–937, 938–939, 940
    of mollusks, 602–603, 910
    of mushrooms, 539–540
Gingkos, 546, 561, 562
Ginkgophyta. See Ginkgos
Giraffes, 989
Gizzards, 992
Glaciers, 1156–1157
    and evolution, 647, 655
    and primary succession, 1119
Glades, 1182
Glands, 779, 802
    as effectors, 929
Glans penis, 816, 830–831, 835
Glass knife fish, 1043–1044
Glass sponge, 582
Glia, 851, 852, 865–866
Gliding, in frogs, 610
Gliding bacteria, 490, 494
Global warming, 1138, 1140, 1143
    and range shifts, 1185–1187
βglobin, 340–341
Globin family, 298–299
Globins
    genes for, 302, 340–341

and hemoglobin synthesis, 305–306
Glomerular filtration rate, 1016–1017, 1021, 1024
Glomeruli, 1015–1017, 1019–1022
Glucagon, 806, 813, 1002–1005
Glucocorticoids, 807, 813–815
Gluconeogenesis, 1002–1005
Glucosamine, 47, 49
Glucose, 45–48
    absorption of, 997–998
    active transport of, 105, 106
    blood level regulation, 812–813, 814, 818–820, 1002–1005
    chemical structure, 26, 45–46
    energy yield, 156–157
    in fermentation, 155–156
    in glycolysis, 136–137, 142, 144–147
    from photosynthesis, 181
Glucose 1-phosphate, 116–117
Glucose 1-phosphorylase, 421
Glucose 6-phosphate, 115–117, 144
Glucose 6-phosphate dehydrogenase, 268
Glue lines, 925
Glutamic acid, 50, 267, 418
    as neurotransmitter, 869
Glutamine, 50, 267
Glutarate, 128
Glyceraldehyde, 46
Glyceraldehyde 3-phosphate (G3P)
    in Calvin–Benson cycle, 177–182
    in glycolysis, 145, 146, 157
Glyceraldehyde 3-phosphate dehydrogenase, 126
Glycerol, 40–42, 157
Glycine, 49, 50, 267, 418
    in cytochrome c, 440–442
    as neurotransmitter, 869
Glycine max, 668
Glycogen
    breakdown and synthesis, 813, 818–820, 1002–1005
    chemical structure, 47–48, 52
    as energy-storage compound, 139, 981
Glycogen phosphorylase, 818–820
Glycogen synthetase, 818–820
Glycolic acid, 418
Glycolipids, 94, 97
Glycolysis, 136–137, 142–143, 144–147, 627, 933
    allosteric control of, 158–160
    in CAM plants, 184
    energy yield, 150, 156–157
    evolution of, 142, 150, 152
    and fermentation, 155–156
Glycolytic pathway. See Glycolysis
Glycoproteins, 94, 97

Glycosidic linkages, 46
    Glyoxysomes, 82
Gnetophyta, 546, 561, 562, 570
Goats, offspring recognition, 1034
Gobies, 620
Goff, Lynda, 523
Goiter, 810–811, 988
Golden feather star, 616
Golden lion tamarin, 634
Golden-mantled ground squirrel, 631
Golden snake plant, 1116
Golden toad, 623, 624
Goldenrods, 1092
Goldman equation, 860
Golgi, Camillo, 78
Golgi apparatus, 70–71, 78–81
    and membrane synthesis, 109–111
    in plant cytokinesis, 201
    in plasma cells, 363
Golgi tendon organ, 889
Gonadotropin-releasing hormone, 809–810, 815
    in menstrual cycle, 833–834
    in puberty, 815–816, 833
Gonadotropins. See Follicle-stimulating hormone; Luteinizing hormone
Gonads, 780, 826
    hormone production, 814–817
    See also Ovaries; Testes
Gondwanaland, 647, 1156
Gonium, 523
Gonorrhea, 492, 839
Gonothyraea loveni, 587
Gonyaulax, 517–518
Gooseneck barnacles, 599
Gorillas, 635–636
Goshawks, 1063–1064
Gout, 1012
gp120, 375
GP3D, 126
Gradual metamorphosis, 391–392
Grafts, 771–772
    immune reactions to, 361, 362
    and induction of flowering, 768–769
Grains, essential amino acids in, 983–984
Gram, Hans Christian, 487
Gram-negative bacteria, 485–487, 490–494
Gram-positive bacteria, 485–487, 494–496
Gram stain, 487
Grana, 75–76, 547
    in C4 plants, 183–185
Grapes, 741, 746, 772
Graptolites, 652–653
Grass pike, 463
Grasses, 1165
    effects of grazing, 711
    flowers, 565, 566
    germination, 739
    growth, 681

heavy metal tolerance, 449, 451, 452, 709–710
    population studies, 1087–1088
    roots, 670
Grasshoppers, 227, 602
Grassland biomes, 1123, 1131–1133, 1165
Gravitropism, 743–744, 748
Gray crescent, 383, 387–389, 395–396
Gray matter, 854–855, 856
Gray whale, 1048
Grazing, 1076, 1109, 1131–1133
    effect on algae, 522
    effect on plants, 632, 711–712, 1123
    in habitat restoration, 1192
    and speciation, 456–457
Great white shark, 979
Green algae. See Chlorophyta
Green glands, 1014
Green Revolution, 329, 1144
Green sea turtle, 626
Green sulfur bacteria, 422
Greenhouse effect, 1136, 1138. See also Global warming
Griffith, Frederick, 242
Gristle. See Cartilage
Grooves, in DNA, 245, 246
Gross morphology, 473–474
Gross primary production, 1128
Ground meristem
    of roots, 677–679
    of shoots, 680
Ground squirrels, 631
    hibernation, 791, 796–797
    thermoregulation, 794
Ground state, 166–168
Ground tissue system, 673, 677, 680–681
Groundwater, 1134
Group living. See Social behavior
Growth
    in animals, 389–391
    defined, 1
    as life history stage, 1081–1083
    in plants, 669, 677–683, 734–735, 742–748, 764
    regulated, 2
Growth factors
    in cancer, 345–346
    in induction, 399, 400
Growth hormone, 806, 808–809, 810
Growth hormone release-inhibiting hormone. See Somatostatin
Growth hormone-releasing hormone, 810
GTP
    in citric acid cycle, 148–149
    in G cap, 299–300
    in induction, 400
    and second messengers, 818–820, 886
Guanacaste National Park, 1192

Guanine, *57–58*, 245–249, 417
  in genetic code, 266–267
Guanosine triphosphate. *See*
  GTP
Guard cells, 684–685, 696–697
Guarding
  of eggs and young, *1071*
  of mates, 1067–1068, 1070
Guayale, *1094–1095*
Guillemin, Roger, 809
Gulls
  food solicitation,
    1031–1032
  salt glands, 1010, *1011*
Gurdon, John B., 394
Gustation, *887–888*
Gut, 745, 980, 991–1001
  lining of, 100
  loss of, *593–594*
  tissue layers, *779*, 993–995
Guttation, *694*
Gymnophiona, 622–623
Gymnophis mexicana, *623*
Gymnosperms, 546, 547, 553,
    559, *561–564*, 676
  reaction wood, 684–686
Gyplure, *1040–1041*
Gypsy moths, *1040–1041*
Gyri, 856

H1 histone, *193*
Haber process, 724, 726
Habitat, 1059–1063
  fragmentation of,
    1179–1180, 1187,
    1188–1190
  islands, 1084–1085, 1159,
    1179, *1182*
  loss of, 1176, 1178–*1180*,
    1189–1190
  and social systems,
    1076–1077
  and species preservation,
    1096–1097
  and species richness, *1124*
  See also Environment
Habitat selection, 1059–*1061*,
    1091, 1093–*1094*
Habituation, 875–876
Hadean eon, *416*, 646
Hagfishes, *470–471*, 619
Hair cells, 889–*891*, 892–894,
    913
Hair follicles, *888–889*
Hairs, *52*, 630
  chemosensitive, 884–*885*
  on leaves, *705*
Half-lives, 22, 645
Hallucinogena, *652*
Halobacteria, *174*, 490
Halophiles, *485–486*
Halophytes, *708–709*
Halteres, *402–403*
Hamilton, W. D., 1074–1075
Hamlet fish, *1070*
Hamner, Karl, 766
Handling time, 1061, 1103
Hands, evolution of, 637, *638*
Hanging flies, 1067, *1069*
Hangovers, 1025
Haplo-diploidy, *227*, 826,
    1074–1075

Haploid generation. *See*
  Gametophytes
Haploidy, *202–203*
Haplontic life cycle, *525*
Hardy, G. H., 430
Hardy–Weinberg equilibrium,
    430–431
Hare, arctic, *792*
Harems, 828, 1077
Harlequin bugs, *601*
Harvestmen, *598*
Haustoria, 529–*530*
Haversian systems, *925*
Hawaiian fruit flies, *455*
Hawaiian Islands, *455*,
    457–458
  extinctions, *1177*, 1178
Hawks
  prey defenses against,
    1059, 1063–*1064*
  vision, 899
  wings, *3*
Hayes, William, 278
Head
  in arthropods, *569*
  in crustaceans, *599*
Headaches, 1025
Heads, flower, 565–566
Hearing, 890–894
Heart, 956, 958–965
  endocrine functions, 807,
    817
  evolution of, 617, 619
  of invertebrates, *957–958*
  muscles, 920, 963–965, 971
  of reptiles, 625
Heart attacks, 956, 967
Heart disease, 350, 1083–*1084*
Heart murmurs, 962
Heartbeat, 920, 962–965
  regulation of, 872–874, 971,
    974–976
  and thermoregulation, 788
Heartwood, 683
Heat
  as energy. *See*
    Thermodynamics
  in energy conversions, 115,
    118
  flow through ecosystems,
    1129, *1132*, 1145
  specific, 1147
  See also Infrared radiation
Heat exchange, and ther-
    moregulation, 787–788,
    789–*790*, 792
Heat loss mechanisms, 791,
    792, 793
Heat production, metabolic,
    785–786, 788–789,
    791–792
Heat (sexual cycle). *See* Estrus
Heavy chains, 361–*366*,
    368–370
Heavy metals, tolerance in
    plants, 449, *451*, 452,
    709–710
Heavy nitrogen, *250–251*
Heavy oxygen, *163*
Heavy subunit, of ribosomes,
    261–264
Heavy water, 22

Helical viruses, 498–499
Helicase, *253*
Helicona rostrata, *468*
Heliozoans, *509*
Helium, 1136
Helix
  alpha (α helix), 52, *53*, 56,
    96
  double, *58*, 245–246,
    247–248
Hellriegel, H., 724
Helper T cells (T$_H$), 361, 372
  in AIDS, 375–377
Helping, by siblings, 1065,
    1072–1075
Hematocrit, *971*
Heme groups, *54*, 55
  in cytochrome *c*, *440–442*
  in hemoglobin, 947–948
  in respiratory chain, 151
  synthesis, 305–306
Hemiascomycetes, 537–538
Hemichordates, 578, 579, 611,
    633
Hemiptera, 601, 602
Hemizygotes, 228–229
Hemocoel, *596*
Hemocyanin, 984
Hemoglobin, *55*, 972, 984
  of apes, *636*
  genes for, *298*
  genetic defects in, 256–257
  in nitrogen fixation, 726
  in sickle-cell anemia, 335
  synthesis of, 305–306
  transport of respiratory
    gases, 947–951, 952
Hemolymph, *957*, 1014
Hemolytic anemia, 109
Hemophilia, 229–230, 336,
    339, 972
Henle, loop of, 1019–*1022*,
    1026
Hennig, Willi, 469
Hepatic duct, 996, *998*
Hepaticophyta, *546*, 547,
    549–550
Hepatitis, 1109
Herbicides, 744, 745, 1144
Herbivores, *980*, 1103
  digestion, *1000–1001*
  effects on communities,
    1120, 1167
  evolution of, 641
  food acquisition, 988–989
  as keystone species, 1188
  salt balance, 1010
  social behavior, *1076–1077*
  as trophic level, 1128–1129,
    1132
Herbs, 571, 668
Herding, *1076–1077*
Heredity, 191. *See also*
  Inheritance
Hering–Breuer reflex, 951
Heritable traits, 215–216, 427
Hermaphroditism, *828*
  in annelids, *594–595*, 828
  in *C. elegans*, 384, *399*
  mate choice, *1070*
Hermaphroditism (hormonal
    disorder), 815

Herring gulls, *1031–1032*
Hershey, Alfred D., 243, *244*,
    281
Hershey-Chase experiment,
    243, *244*
Hertz, 165
Heterochromatin, 300–302
Heterocysts, 494, *495*
Heterogenous nuclear RNA
    (hnRNA), 299–301
Heterokaryosis, in fungi,
    533–534
Heteromorphic alternation of
    generations, 506
  in algae, 521, *525*
  in plants, 546
Heteromorphy, 522
Heteropods, 604
Heterospory, 556, 559
Heterotherms, 785
Heterotrophs, 1, 2, 162, 718,
    979–981
  plants, 717, 730–731
Heterozygotes, 217
Hexactinella, *582*
Hexagenia, *1141*
Hexokinase, 122, 144, 146
Hexosaminidase A, 335
Hexoses, 46
Heyn, A. J. N., 747
Hfr mutants, 278–281, 285
Hibernation, 796–797, 982,
    1047–1048, 1093
Hicks, Glenn, 748
Hierarchical organization, 3–6
Hierarchy, of classification,
    464, *466–467*
High-energy bonds, in ATP,
    139
Highly repetitive DNA, 296
Hindbrain, 853–*854*
Hindgut, *992*
Hinduism, 412, 465
Hinge joint, 926–927
Hippocampus, *815*, 856, 876
Hirudinea, 595, 992
Hirundo, 992
Histamine, 366, 801, 968–969
Histidine, *50*, 266, *267*
Histones
  in chromosomes, 193–195
  genes for, 299, 302
Historical biogeography,
    1154–*1157*
HIV. *See* Human immunodefi-
    ciency virus
hnRNA, 299–301
Hodgkin, A. L., *864*
Holdfasts, 521, *522*, 523
Holley, Robert, 261
Holothruoidea, *616–617*
Holotrichs, *511*
Home ranges, 1188–1189. *See
    also* Territoriality
Homeobox, 402–403
Homeodomain, 403
Homeostasis, 2, 778, 782–784
Homeotherms, 785
Homeotic genes, 401–405
  in plants, 754, 764–765
Homing, 906, 1048–*1049*
Hominids, 637–638

Homo erectus, 638
Homo habilis, 638, *1062*
Homo sapiens. See Humans
Homologous chromosomes, 204, *205*
Homologous traits, 466–468
Homoplasy, 467–470
Homoptera, 602
Homospory, *556,* 559
Homozygotes, 217
Honey guides. *See* Nectar guides
Honeybees
    communication, 639, *1042–1043*
    hygienic behavior, 1038–*1039*
    sex determination, 227
    social systems, *1074–1075*
    thermoregulation, 789
Honeycreepers, *1178*
Honeyeaters, *980*
Hood, Leroy, 326
Hooke, Robert, *12*
Hooved mammals, *631*
    social systems, 1076–*1077*
    speciation, 456–*457,* 655
Hooves, evolution of, 442, 466, 472
Horizons, in soil, 721–722
Horizontal cells, *901*
Hormones, 800–821, *806–807*
    and behavior, 1034–*1037,* 1039
    chemical action, 817–821
    chemical structure, 43, *44*
    in digestion, 802, *807,* 813, 817, 1001–*1002*
    in growth and metamorphosis, 802–*804*
    in menstrual cycle, 833–835
    and nervous systems, 853
    in plants, 733–735, 738–751
    in pregnancy, 835, 843
    in regulation of blood flow, 974–*975*
    in stress responses, 751, *795,* 801, 814–*815*
    See also Endocrine systems
Hormosira, 520
Hornworts, *546–547,* 550–551
Horses
    evolution, 442, *472,* 649, *650*
    vision, 899
Horseshoe crabs, *597–598,* 656
Horsetails, *546,* 553, 555, *556–557*
Hörstadius, Sven, 394–395
Hosts
    interactions with parasites, 451–452, *1101, 1103,* 1108–1109
    multiple, 534–536, *535,* 539, 588, *590*
    recognition by viruses, 498
Hot desert biome, *1166–1167*
Hot fish, 789–*790, 941*
Hot springs, organisms in, *489*
Hot water vents. *See* Hydrothermal vents

House sparrows, 657–658
"How" questions, 4, 6, 1029
Howler monkeys, *1191*
HTLV-III. *See* Human immunodeficiency virus
Hubel, David, 902
Human chorionic gonadotropin, 835, 843
Human Genome Project, 332, 337
Human immunodeficiency virus, 259, 375–377. *See also* AIDS
Humans
    behavior, 1052, *1062–1063*
    as cause of extinctions, 629, 630, 639, 655, 1097, 1176–*1180*
    classification, 464
    compared with chimpanzees, 475, 635–636
    development, 381, 387, 841–844
    effects on biogeochemical cycles, *1138,* 1140–1145
    effects on environment, *1127*
    effects on other species, 1058, 1094–1097, *1179–1180,* 1188
    evolution, *416,* 632–640
    genetic variation, 434, *438*
    growth, 390–*391*
    migration patterns, *477*
    pigmentation, 235–*236*
    population dynamics, 639, 1083–*1084,* 1085–*1086,* 1089, 1097
    sex-linked inheritance, 227–230, 335–339
Hummingbirds, *3,* 629, 988, *1121*
Humoral immune response, 357–*358,* 361–366, 370, 371–*372. See also* B cells; T cells
Humpback whale, 1042
Humus, 723–724, *1120*
hunchback protein, 403–404
Hunger, regulation of, 783
Hunting
    cooperative, 638, *1062–1063*
    effect on populations, *433,* 629, 630, 639, *1177*
Hunting dogs, *1062–1063*
Huntington's disease, 336, 337, *339*
Huso huso, 622
Huxley, A. F., *864*
Huxley, Andrew, 915
Huxley, Hugh, 915
Hyalophora cecropia, 803–*804*
Hyaluronidase, 836
Hybrid zones, 453–454
Hybridization
    of nucleic acids. *See* Nucleic acid hybridization
    somatic-cell, 338–340, *339*
Hybridomas, 364–365
Hybrids, 214, 447, 452–454

behavior, *1037–1038*
Hydra, 824–*825, 957*
Hydration, *31*
Hydrocarbons, 33–34
Hydrochloric acid (HCl)
    ionization, 25, 29–30
    in stomach, 354, 996, 1001
Hydrogen
    atmospheric, 1136
    in cellular respiration, 142, 143
    covalent bonding, 23–25
    in early atmosphere, 414, *415, 417, 418,* 422
    in fermentation, 144
    ions, 25, 29–30
    isotopes, *21*
    and nitrogen fixation, 725–726
    and pH, 29–30
    and plant nutrition, 717–719, 722
    in redox reactions, 139–140
    in respiratory chain, 140, 143–144, 152
Hydrogen bonds, 31–*32, 33*
    in DNA, *247, 313*
    in ribozymes, 420
Hydrogen chloride. *See* Hydrochloric acid
Hydrogen cyanide, 417–*419*
Hydrogen peroxide, 82
Hydrogen sulfide (H_2S)
    in bacterial metabolism, 422, 484, *593–594,* 718, 729–730, 1000, 1148–1149
    in sulfur cycle, 1139–*1140*
Hydrogenated fatty acids, 40–42, *41*
Hydroids, 552, 583
Hydrological cycle, *1137*
Hydrolysis
    of ATP, 139
    of peptides, *125*
    in soil weathering, 723
Hydrolytic enzymes, 992–993
Hydrophilic compounds
    defined, 42–43
    in digestion, 996
Hydrophobic compounds
    defined, 42–43
    in digestion, 996
Hydrophobic interactions, 33
Hydrophobic regions, in cytochrome *c, 440–442*
Hydrostatic skeletons, 589–591, 606, 640, *921, 922*
Hydrothermal vents, 484–*485,* 490, *593–594,* 1148–1149
Hydroxide ions, and pH, 29–30
Hydroxyl groups, 31, 33–35
    in DNA, 245, 251–252
Hydrozoans, *584–585, 587*
Hygienic behavior, 1038–*1039*
Hylobates. See Gibbons
Hymen, 832–833
Hymenoptera, *601,* 602
    communication, *1042–1043*
    sex determination, *227,* 826

social systems, *1074–1076*
    See also Ants; Bees; Wasps
Hyperpolarization, of membranes, 859
Hypersensitive reaction, in plants, 714
Hypertension, 814–*815, 967*
Hyperthyroidism, 810–811
Hypertonic, defined, 106
Hypertonic osmoregulators, *1009–1010*
Hyphae
    in fungal reproduction, 533–534
    of fungi, 529–*530,* 536–*537,* 539
    of Protomycota, 516
Hypochytrids, 516
Hypocotyl, *753, 761*
Hypoplectrus, 1070
Hypothalamic releasing neurohormones, 809–*810,* 813
Hypothalamus, 805, 808, 809–*810, 854*
    and osmoregulation, 1024–1026
    and thermoregulation, 793–794, *795*
Hypothermia, 795–*796*
Hypotheses, 6–9
Hypothetico-deductive method, 6–9
Hypothyroidism, 810–811
Hypotonic, defined, 106, 690
Hypotonic osmoregulators, *1010*
Hyracotherium, 442

i gene, 287–288
Ibexes, 1181
Ice, *28–29, 31. See also* Freezing
Ice caps, 1143
Ice plants, 184, 705
Icosahedral viruses, 498, *499*
Identical twins, 384
    and organ transplants, 374
    in studies of inheritance, 236
IgA, 365, *366,* 369–370
IgD, 365–*366,* 369–370
IgE, 365–*366,* 369–370
IgG, 365–*366, 367,* 369–370
IgM, 365–*366,* 368–370, *373*
Iguanas, 787–788, 792
Iiwi, *1178*
Ileum, 996–*997*
Illicit receivers, 1042
Imaginal discs, 391–*393,* 394
Imbibition, 738
Iminoaceticpropionic acid, *418*
Immediate memory, 876
Immigration, 1086–1087
    and gene flow, 432, 447
    and species richness, 1158–*1160*
Immune systems, 353, 356–378
    and cancer, 348
    disorders of, 342, 374–*378*
    effect of stress, 814–*815*
    evolution of, 370, 371

genetic diversity in, 367–370
nitric oxide in, 27
See also B cells; T cells
Immunization, 359
Immunoglobulins (Igs). *See* Antibodies
Immunological memory, 359–361
Immunological tolerance, 361, 362
Immunosuppressants, 374
Impala, 10
Imperfect flowers, 564
Imperfect fungi, 529, 540
Implantation, 841–842
Imprinting, 1034
In situ hybridization, 323
In vitro fertilization, 762
Inactivation, of genes, 300–302
Inactivation gates, 861
Inactive forms, of enzymes, 129–130, 993–999
Inbreeding, 1104–1105, 1181
Incisors, 990–991
Inclusive fitness, 437, 1065–1066
Incomplete dominance, 222–223
Incomplete metamorphosis, 602
Incubation, of eggs, 628
Incus, 891–892
Independent assortment, of chromosomes, 219–223
Indeterminate cleavage, 576, 610
Indeterminate growth, 677, 764
Indian pipe, 545–546, 730
Indian python, 789
Indigenous people, in national parks, 1190
Individual fitness, 437, 1065–1066
Indoleacetic acid. *See* Auxin
Induced fit, 121–122
Inducers
  embryonic, 398–399
  of protein synthesis, 285, 287–288, 318
Inducible enzymes, 285, 287
Inducible systems, 286–288
Induction, embryonic, 397–400
Inert elements, 23
Infection thread, 726–727
Infectious diseases, 350, 354, 1064, 1083–1084, 1109
  defenses against, 353–361
Inferior vena cava, 961
Inflammation, 355, 358, 366, 801, 814, 968–969
Inflorescence meristem, 764
Inflorescences, 565–566, 568
Information, and maintenance of homeostasis, 782–783
Information reduction, in vision, 901

Information systems, 778, 779–780. *See also* Endocrine systems; Nervous systems
Information transfer. *See* Communication
Infrared radiation, 165–166
  perception of, 904–905
  trapping by atmosphere, 1136
Ingestion, 718, 979–980, 988–990
Ingression, in gastrulation, 386
Inhalation, 946–947, 951
Inheritance
  and meiosis, 222–237
  Mendelian, 214–222
  non-Mendelian, 236–237
  sex-linked, 227–230, 335–339
  See also Genes
Inhibitors, of enzymes, 126–128
Inhibitory postsynaptic potential (IPSP), 868
Inhibitory synapses, 868
Initial bodies, in chlamydias, 493
Initiation complexes, 262–263
Initiation sites, 259–260
Inner ear, 889–894
Inorganic fertilizers, 722–723
Inositol triphosphate (IP3), 820
Insectivora, 632
Insectivorous plants, 717, 730–731
Insects, 596, 600–602
  development, 391–392, 802–805. *See also* Metamorphosis
  excretory systems, 1013–1014
  fossilization, 649
  gas exchange, 596, 600, 937–938
  hormones, 802–805
  plant defenses against, 704, 712–713, 1103, 1109–1110
  as pollinators, 566, 568, 714, 1117–1118
  population densities, 1093
  sexual behavior, 829–830, 1067–1068
  social systems, 1074–1076
  speciation, 451–452
  species richness, 1123
  thermoregulation, 788–789
  See also Arthropods
Insomnia, 874
Inspiratory reserve volume, 943–944
Instars, 391–392, 602, 802–803
Instinctive behavior, 856, 1030–1036, 1039
Insulation, 628, 630, 792–793
Insulin, 806, 810–811, 1002–1005
  activation, 306
  chemical structure, 26

Insurance, 1086
Integrase, 375
Integrated pest management, 1144
Integuments
  of flowering plants, 758, 760
  of gymnosperms, 563
Intelligence, evolution of, 858
Intensity
  of light, 166
  of sensations, 887
Intention movements, 1037, 1044
Interbreeding, 446–447
Intercalated disks, 920
Intercostal muscles, 946
Interference competition, 1111
Interferons, 355–356, 358
Interkinesis, 205
Interleukins, 374, 795
Intermediate filaments, 83–84
Internal environment, 777–778
  regulation of, 952–953, 956, 1003–1010, 1023–1026
Internal fertilization, 630, 828–830, 840
Internal skeleton
  in chordates, 612, 641
  in echinoderms, 613–617
Interneurons, 854–855, 873–874
Internodes, 669–670, 681, 764
Interphase, 192, 194, 196–197, 198, 201
Interrupted mating, in bacteria, 279–281
Interruptions, of night length, 766–767
Interspecific competition, 1102, 1111–1112
Interstitial fluids. *See* Extracellular fluids
Intertidal zone, 522, 1101–1102, 1172
Intertropical convergence zone, 1146, 1166, 1167
Intestinal microorganisms, 77, 513, 992, 1000, 1114
Intestines, 992–993, 996–1000
  in osmotic balance, 1009
Intracellular digestion, 991
Intraspecific competition, 1059, 1085, 1110, 1111
Intrauterine device, 838–839
Intrinsic nervous system, 1001
Intrinsic proteins, 94–97
Intrinsic rate of increase, 1089–1090
Introduced species, 1141–1142, 1177–1178
Introns, 294–296
  removal of, 299–301
  in ribozymes, 419–420
Inuit people, 986, 987
inv gene, 492
Invagination, in development, 385–389
Invasiveness, of bacteria, 487
Inversions, of chromosomes, 233, 269–270
Invertebrates

excretory systems, 1012–1014
nervous systems, 850, 864
visual systems, 897–899
Inverted pyramid of biomass, 1131–1132
Involution, 387, 388–389
Iodine
  in Gram stain, 487
  nutritional requirements, 811, 985, 988
Ion channels, 103–104
  in cystic fibrosis, 335
  in muscle function, 917–918, 920
  in nervous system function, 859–866
  in photoreceptors, 895–897, 896
  in sensors, 881–882, 883–884, 886–887
  in synaptic transmission, 866–870
  See also specific types
Ion exchange, in plant nutrition, 722
Ion pumps, 858–859
Ionic bonds, 25–26, 33
Ionic conformers, 1009
Ionic regulators, 1009
Ions, 25–26
  in blood plasma, 973
  and pH, 29–30, 147
  in plant nutrition, 718–719, 722
  as prosthetic groups, 125–126
  regulation in animals. *See* Salt balance
  in solution, 25
  uptake by plants, 690–692
IPSPs, 868
IP3, 820
Iridium layer, 648
Iris, 897–898
Iron
  as animal nutrient, 984–985
  competition with pathogens for, 354, 358
  in hemoglobin, 972
  ions, 25
  as plant nutrient, 718–719, 720, 723
  in soils, 723
Iron oxides, 414
Irreversible inhibition, 126–127
Irruptions, 1094–1095
Isaacs, Alick, 355
Islands
  colonization of, 1157–1160
  endemism, 457–458, 1187–1188
  extinctions, 1177, 1178
  speciation, 449, 450, 456–458
  species richness, 1123
Isle Royale, 1104–1105, 1120, 1121
Islets of Langerhans, 813
Isocitrate, 148–149, 158–159

Isogamy, 525, 526
Isolating mechanisms, 452–453
Isoleucine, 50, 267
    nutritional requirements, 983–984
Isomerase, 144
Isomers, 34, 36
Isometric contractions, and thermoregulation, 788–789
Isomorphic alternation of generations, 506
    in *Allomyces*, 516
    in *Ulva*, 523–525, 524
Isopentenyl adenine, 749
Isopods, 599–600, 1103
Isoptera, 602
Isotonic solutions, 106, 690
Isotopes, 21–22
    and dating, 413, 645, 648
    in DNA studies, 243, 244, 250–251
    in recombinant DNA technology, 321–323
    as tracers, 22, 163, 175–178, 657
Isozymes, 132–133, 785
Isthmus, in desmids, 524
IUD, 838–839
Ivanovsky, Dmitri, 497
Ixodes ricinus, 598

J segments, 368–369
Jack pine, 1185–1186
Jackals, 1074
Jackrabbits, 792
Jacob, François, 258, 279, 285
Jaegers, 1104
Jagendorf, Andre, 172
Janzen, Daniel, 1192
Japanese macaques, 634
Jawless fishes, 619–620
Jaws, 924
    asymmetrical, 439
    evolution of, 619–620, 629–630
    joints, 928
Jays, 7, 1072–1073
Jejunum, 996–997
Jellyfishes, 583, 585, 587
    development, 385
Jerne, Niels K., 359–360
Jet lag, 1045
Jet propulsion, 921, 922
Jewel flower, 710–711
Jews, and Tay-Sachs disease, 335
Joints
    of arthropods, 595–596, 922
    evolution of, 621–623
    function of, 926–928
    stretch receptors in, 889
Joshua tree, 1117
Joules, 981
Jowett, D., 709–710
Jumping genes, 285, 297–298
Jumping (locomotion), 909
Juncos, 1059

Junctions, between cells. *See* Desmosomes; Gap junctions; Tight junctions
Jurassic Park, 474
Jurassic period, 646, 654–655
Juvenile hormone, 803–804

K (carrying capacity), 1090
Kalanchoe, 767, 770–771
Kale, 428, 429
Kandel, Eric, 875
Kangaroo rats, 1021, 1033
Kangaroos, 630, 631, 840, 909
Karyotype, 203–204
Katydids, 601, 922
Kaufman, Thomas, 402
Kelps, 521, 522, 689
Kelvin, Lord, 413
Kelvin units, 118
Kennedy's disease, 337
Keratin, 51, 99–100, 625
α-Ketoglutarate
    in amino acids, 157
    in citric acid cycle, 148–149, 158–159
Ketones, 33, 35
Keystone species, 1121, 1188
Khorana, Har Gobind, 266
Kidneys, 1015–1017, 1019–1026
    artificial, 1023
    erythropoietin production, 972
    regulation of, 807, 1023–1026
Kidston, Robert, 554
Kilocalories, 27, 981–982
Kin selection, 437, 1065–1066
Kinases, 146
Kinesin, 85
Kinetin, 739, 749
Kinetochores, 193–200
    in meiosis, 205–206
    microtubules, 199–200
King, Thomas J., 393–394
King cormorant, 1085
King penguin, 1069
King snake, 1033
Kingdoms, 14–15, 464, 466–467, 485, 503–504
Kirtland's warbler, 1185–1186
Klebsiella ozaenae, 745
Klinefelter syndrome, 227, 228
Kluger, Matthew, 795
Knee-jerk reflex, 872–873, 926
Knees, 926, 928
Knife fish, 929
Knockout mice, 342
Knots, in wood, 683
"Knuckle walk," 635
Koala, 988, 1152
Kobs, 1069
Koch, Robert, 487
Koch's postulates, 487
Köhler, Georges, 364–365
Kohlrabi, 428, 429
Kölreuter, Joseph G., 214–215
Konishi, 739–740
Kornberg, Arthur, 248, 251
Krakatau, 644, 1159–1160
Krause's end bulb, 888

Krebs cycle. *See* Citric acid cycle
Krill, 989
Krüppel protein, 403–404
Krypton, 1136
Kuna Indians, 438
Kurosawa, Eiichi, 740
Kuru, 500
Kwashiorkor, 969, 988

Labia, 816, 832
Labidomera clivicollis, 704
Labor (birth), 844–845
Laboratory studies, 8, 786–787
Labroides dimidiatus, 828
Labyrinth, 1014
lac operon, 287–288, 289
    in recombinant DNA technology, 316, 318
Lacewings, 602
Lactase, 997–998
Lactate. *See* Lactic acid
Lactation, 630, 845–846
    role of oxytocin, 807
    role of prolactin, 305, 805
Lactic acid, 144, 158, 418
    fermentation, 144, 155–156
Lactobacillus, 494, 495
Lactose, 46–47
    digestion of, 998
    intolerance for, 978
    and *lac* operon, 285, 287–288
Lagging strand, 251–254, 297
Lake Erie, 1141–1142
Lake Washington, 1142
Lakes, 1170
    and acid precipitation, 1142
    eutrophication, 1140–1142, 1141
    turnovers, 1134–1135, 1141
Lamarck, Jean Baptiste de, 13
Lamellae, in fish gills, 938–939
Lamina, of nucleus, 196
    Laminaria, 521
Lamp shells. *See* Brachiopods
Lampreys, 619
Lancelets, 618
Landmarks, in navigation, 1048
Landsteiner, Karl, 225
Lang, William H., 554
Language, 856–857, 877–878, 1043, 1052
    evolution of, 477, 639
Laqueus, 613
Large intestine, 997, 999–1000
Larvaceans, 618
Larvae, 391, 392, 606, 640
    of amphibians. *See* Tadpoles
    in aquatic ecosystems, 1170, 1172
    and classification, 473
    of insects. *See* Caterpillars; Metamorphosis
    nauplius, 599–600
    of sea stars, 615
    settling, 434, 1059–1060
    trochophore, 577, 594–595
    of tunicates, 618

Larynx, 923, 944–945
    cancer of, 348
Lasers, for separating chromosomes, 320–321
Lateral bud meristem, 678
Lateral buds, 671, 711–712
    inhibition of, 742, 743, 745
Lateral line, 889–890, 905–906, 1042
Lateral lip, 387, 388–389
Lateralization, of brain, 877
Laterization, of soil, 723
Latex, as chemical defense, 704
Laticifers, 704, 715
Latimeria, 621–623
Latitude
    and climate, 1145, 1148–1149, 1167
    in navigation, 1052
Laurasia, 647
Lawn, bacterial, 274
Leaching
    of seeds, 736, 737
    of soils, 721–722, 723
Lead, effects on plants, 449, 451, 452
Leader sequences, 306–307
Leading strand, 251–254, 297
Leaf-cutting ants, 531–533, 989
Leaf primordia, 672, 678, 680–681
Leafhoppers, 602
Learning, 875–876, 1030, 1033–1034
    of bird song, 1033–1034
    in humans, 1052 Least chipmunk, 1112
Least tern, 1193
Leaves
    abscission, 743–744, 749–750
    adaptations of, 467–468, 705–706, 709
    chemical defenses, 704, 1109–1110
    environmentally induced variation, 427, 428
    evolution of, 555–556
    as insect habitats, 1060–1061
    monocot vs. dicot, 669
    shapes, 427, 428, 672–673
    stomatal cycles, 696–698
    structure and function, 672–673, 683–685
    transpiration, 695–696
    and vegetative reproduction, 770–771
Lederberg, Esther, 273, 275, 277
Lederberg, Joshua, 273, 275, 277, 282
Leeches, 595, 992
Left brain, 877–878
Left-handed double helix, 246, 248
Leghemoglobin, 726
Legs. *See* Appendages
Legumes
    essential amino acids in, 983–984

nitrogen fixation, *724–725*
Lekking, *1069*
Lemmings, 1104
Lemurs, *632, 634, 1187*
Lens, *897–899*
   induction of, *397–398*
Lens placode, *397–398*
Lenticels, *683–684*
Leontopithecus rosalia, 634
Lepas anatifera, 599
Lepidodendron, 557
Lepidoptera, 228, *601*, 602. *See also* Butterflies; Moths
Lepomis, 657
Lesser panda, 475, *476*
Leucine, *50, 267*
   nutritional requirements, *983–984*
Leucine zippers, *305*
Leucoplasts, 76, 674
Leucosolenia, 582
Leukemia, *44,* 345, *1176*
Leukocytes. *See* White blood cells
Levels of organization, 3–6
Levers, skeletal systems as, 926, *928*
Levine, Michael, 402
Lice, 602, 772
Lichens, 531, *540–542,* 1114
Life
   defined, 1
   origins, *397,* 411–423
Life cycles, 202–203
   of amphibians, *623–624*
   of broad fish tapeworm, *588, 590*
   of cnidarians, *583–585*
   of ferns, *559–560*
   of fig wasps, *1100*
   of flowering plants, *568–569, 758–761*
   of fungi, 533–536
   of green algae, *523–525, 524*
   of gymnosperms, *563–564*
   of insects, 600, 602. *See also* Metamorphosis
   of nonvascular plants, *548–549, 551–552*
   of *Plasmodium, 510*
   of red-spotted newt, *1080*
Life histories, 1080–1084
Life insurance, 1086
Life tables, *1087–1088*
Ligaments, 889, *926*
Ligands, 108
Light
   absorption by algae, 523
   chemical generation. *See* Bioluminescence
   and circadian rhythms, *767–768, 1045–1047*
   and flowering, *765–767, 768–769*
   in photosynthesis, *162–163*
   physics of, *164–168*
   and plant growth, *718, 742–744, 751–754*
   and seed germination, 737
   and vision, *894–904*
Light chains, 361–366, 368

Light-dark contrasts, and vision, 900–903
Light-footed clapper rail, 1193
Light microscope, 66–67
Light reactions. *See* Photophosphorylation
Light subunit, *261–264*
Lightning, and origin of life, *415,* 417
Ligia occidentalis, 599
Lignier, E. A. O., 554
Lignins, 673, 1103
Lilacs, *671*
Limbic system, *856*
Limbs. *See* Appendages
Limbs, phantom, 882
Limestone, 509
Liming, of soils, 722–723
Limiting resources, 1102
Limpets, *427,* 605
Lind, James, 987
Lindermann, Jean, 355
Linkage, 225–226, *229–230, 335–339*
   and gene mapping, 233
Linnaean system, 464–467
Linnaeus, Carolus, 446, 464
Linoleic acid, *40–41,* 984
Lionfish, *1107*
Lions, 2, *631*
Lipases, 993, 996, *998, 999, 1003*
Lipid-soluble hormones, 817, 821
Lipids, 39–45
   in coacervate drops, *421*
   in membranes, 92–94
   in seeds, 738
   storage in plants, 674
Lipophilic compounds, 996
Lipophobic compounds, 996
Lipoproteins, *1003*
Liquid hybridization, 295–296
Liquid scintillation counting, *21–22*
Listeria, 494
Litter (offspring), 1082
Littoral zone, *1172*
Liver
   cancer of, 345
   in digestion, *996–998,* 999
   processing of carcinogens, 349–350
   production of clotting factors, 972
   regulation of metabolism, 1002–1005
Liverworts, 546, 547, *549–550*
"Living fossils," 656
Lizards, 625, *626*
   phylogeny, 465, *468, 470–471*
   tail autotomy, *462*
   thermoregulation, *785–787,* 795
Llamas, 949, 950
Load arm, 926, *928*
Loading, of phloem, *700–701*
Lobe-finned fishes, 621–623
Lobelia, 1178
Lobsters, *599,* 600

equilibrium sensation, *889–890*
excretory systems, 1014
   growth, *390–391*
Local hormones, *801*
Local species richness, *1122–1124*
Lock-and-key model, of enzymes, 124
Locomotion, 606, 909–929
   in amphibians, 623, *627*
   in annelids, *594–595, 921*
   and body cavities, *589–591, 592–593, 640, 921*
   in cephalopods, 605
   ciliary, *512*
   in echinoderms, 615–617
   in fishes, 619–621
   in humans, 637
   in leeches, 595
   in reptiles, 627
   and symmetry, 575, 614
   in terrestrial vertebrates, 627
   See also Muscles; Effectors
Locus, defined, 218
Locusts, *1094–1095*
Lodgepole pine, 563
Logging, 1121–1122, 1176, *1179, 1189–1190*
Logistic growth, *1090*
Lollipop flask, 175, *178*
Lomax, Terri, 748
Long-day plants, *765–767*
Long-tailed widow bird, 1072
Long-term memory, 856, 876
Longitude, in navigation, 1052
Longitudinal muscle, *994–995*
Loop of Henle, 1019–1022, *1026*
Loops
   in nucleic acid hybridization, 294–296
   in RNA splicing, *299–301*
Lophophorate animals, *610–613*
Lophophores, *611–613,* 617
Lordosis, *1035–1036*
Lorenz, Konrad, 473–474, *1031, 1037–1038*
Loricifera, 578, *579,* 591
Loris tardigradis, 634
Lorises, *632, 634*
Lotus, *569*
Lovebirds, *1037*
Low-frequency sound, and navigation, 1051
Luciferase, *728,* 929
"Lucy" (fossil), 637
Lumbricus. See Earthworms
Lumen, of gut, *993–994*
Lung cancer, 345
Lungfishes, 621–623, *943*
   circulatory systems, *959*
   geographic distribution, *1154*
Lungs, 937
   of amphibians, *943*
   of birds, *939–942*
   evolution of, *942–943*
   of fishes, 621, *943, 959*

of mammals, *942–946*
   of reptiles, 625, 627
   and running speed, *627,* 628
Lupinus texensis, 763
Luria, Salvador, 273
Luteinizing hormone, *806, 808, 809–810*
   in menstrual cycle, *833–834*
   in puberty, 815–816, 833
Lwoff, Andre, 282
Lycophyta. *See* Club mosses
Lycopodium obscurum, 557
Lycopods, *553,* 555
Lymph, *356–357, 969–970*
Lymph nodes, 345, 970
Lymphatic systems, *356–357, 781, 969–970*
   and cancer, *345–346*
Lymphocytes, *356–361,* 970.
   See also B cells; T cells
Lymphomas, 345
Lynxes, *1104–1105*
Lyon, Mary, 301
Lyperobius huttoni, 1154–1155
Lysine, *50, 266, 267*
   nutritional requirements, *983–984*
Lysogenic cycle, *281–283,* 498
Lysosomes, *81–82,* 355
Lysozyme, 52, *54, 121–122,* 124, 354
Lytic complex, 367
Lytic cycle, *281–283,* 498

M-cyclin, *197*
M phase. *See* Mitosis
Macaca fuscata, 634
Macaques, 634
Macaroni penguins, *629*
MacArthur, Robert, 1157
MacLeod, Colin, 242, 495
Macrocystis, 521
Macroevolution, 404–*405,* 442
Macromolecules, 44–45
Macronucleus, in Ciliophora, *511–513*
Macronutrients
   for animals, *984–985*
   for plants, 719, 720
Macrophages, 27, 365, *366, 371–372, 374,* 795
Macropus rufus, 631
Madagascar, 634, *1187*
Madagascar rosy periwinkle, *1176*
Maggots, and spontaneous generation theory, 411–412
Magnesium
   as animal nutrient, *984–985*
   ions, 25
   as plant nutrient, 718, *719, 720,* 723
   tolerance in plants, 710
Magnetic sense, 906, 1051
Magnolias, 566, *567,* 570
   Mahogany, 686
Maidenhair tree, 561, *562*
Maize. *See* Corn

Major histocompatibility complex (MHC), *371–373*, *1181*
Malaria, *510–511*, 1109
  allergy to drugs for, 268
  and sickle-cell anemia, 337
Malate, *148*, 149
Malathion, 870
Males
  competition among, 1057, *1058, 1064, 1067–1069,* 1072–1073
  development of, 228, *814–816*
  mating tactics, *1064,* 1067–1069
  reproductive system, 830–831
Malic acid, 698
Malignant tumors, 345–347
  See also Cancer
Mallards, *1037–1038*
Malleus, *891–892*
Malloch, D. W., *532*
Mallophaga, 602
Malnutrition, 983, *986, 987*
Malonate, *127–128*
Malpighian tubules, *1013–1014*
Maltase, *997–998*
Maltose, *46–47, 421*
Mammals, 629–640, 655
  colonization by, 1159–1160
  embryonic development, 387–389, 840–844
  extinctions, *657,* 1177
  phylogeny, 465, *468*
  as seed dispersers, 1117
Mammary glands, 305, 630, 845–846
Manatees, *980*
Mandibles, *992*
Manganese
  as animal nutrient, 984–985
  as plant nutrient, *719, 720,* 723
Mangold, Hilde, *397–398*
Mangrove islands, *1160*
Mangroves, *707, 708,* 1173
Mannose, *46*
Mantids, 602
Mantispid fly, *1108*
Mantle, of Earth, *414*
Mantle, molluscan, *603,* 605
Map distance, *232, 340*
Map sense, 1052
Maples, 666–667, *701*
Marchantia, 550
Margulis, Lynn, 76
Marine animals, osmoregulation, 1009, 1012, 1018
Marine ecosystems, 1170–1173
  production, 1148–1149
  species richness, *1124*
Marine iguana, 787–788, 792
Maritime climate, 1148
Mark and recapture studies, 1084
Markers, genetic, 229, 278–281, 282–284
Marmosa, 631

Marmots, *1081–1082*
Marsilea, 558
Marsileales, 559
Marsupials, 630–631, *840*
Marsupium, *840*
Maryland Mammoth tobacco, 765
Masai, *438*
Mass extinctions, *646, 647–648, 654, 655, 658–659*
Mass number, 21
Mass spectroscopy, 22, 474
Mast cells, 366, 801
Mastax, 592–593
Mastectomy, 345
Mastigamoeba aspera, 508
Mate choice, 434, *1066–1072*
  by females, *1069–1070*
  in plants, 566
  and reproductive isolation, 452–453, 455–456
  See also Sexual selection
Maternal effects, in flowering plants, 763
Maternal energy, 1081–1082
Maternal inheritance, 237, 640
Mating, nonrandom, *433,* 456.
  See also Copulation;
  Sexual behavior
Mating systems
  and sexual dimorphism, 1057, *1072–1073*
  and social organization, 1076–1077
Mating types
  in fungi, 533
  in green algae, 525
  in slime molds, 516
  in yeasts, *308*
Matrix
  extracellular, 98, 778, *923–925*
  mitochondrial, *74, 152–155*
Matter, as energy, 114
Matthaei, J. H., 266
Maturation-promoting factor (MPF), 197, 198
Mawsonites, 651
Maxam, Allan, 324
Mayflies, *601,* 602, *1141*
Mayr, Ernst, 446–447
McCarty, MacLyn, 242, 495
McClintock, Barbara, 297
McGinnis, William, 402
Measurement, units of, *63*
Mechanical isolation, 452–453
Mechanical weathering, 723
Mechanosensors, *881–882, 888–889*
Medawar, Peter B., 361, *362*
Mediterranean biome, *See* Chaparral
Medulla
  of adrenal gland, *807,* 813
  of brain, 853–854, 951–953, *974–975*
  of kidneys, *1019–1022*
Medusae, 583–585
Megagametophytes, *556, 758–759*

Megapodes, 628
Megareserves, *1190–1191*
Megasporangia, *556,* 560, *758–759,* 760–761
Megaspores, *556, 758–759*
Megasporocytes, 759
Meiosis, 191–192, 202–209
  compared with mitosis, *208–209*
  in gametogenesis, 826–828
  and genetic disorders, 209–210
  and inheritance, 218, *220,* 222–237
  in *Neurospora,* 230–233
Meiosis I, *204–209*
Meiosis II, *207–208*
Meissner's corpuscles, *888–889*
Melanin
  and color change in animals, 809
  and human skin color, 235–236
Melanocyte-stimulating hormone, *806, 808,* 809, *810*
Melanocyte-stimulating hormone release-inhibiting hormone,*810*
Melanocytes, 809
Melatonin, *807,* 809, 1046–1047
Membrane potential
  in muscle cells, *917–918*
  in nervous system function, 858–866, *860*
  in sensors, *881–882*
Membranes, 93–111
  in bacteria, *486, 487*
  in chemical reactions, 107–108
  chemical structure, 39–43, 92–97
  of chloroplasts, 75–76, 108
  and diffusion, 101–107
  electric charges across, 107–108, 858–866
  of endoplasmic reticulum, *78–82*
  in eukaryotic cells, 68–69
  evolution of, *421*
  formation and breakdown, *109–111*
  functions, 101–111
  microscopy, *97–98*
  mitochondrial, *74–75, 94–95,* 108, *152–155*
  movement across, 100–107
  nuclear, *69–70*
  of organelles, 94–95
  thylakoids, *95, 172–174*
  transport proteins in, *103–106*
  vesicle formation, *79–81*
  See also Plasma membranes
Membranous bone, 924
Memory, 27, 856, 875–876
Memory cells, of immune system, 360–361
Menadione, *986–987,* 1000

Mendel, Johann Gregor, 214–222, 241, 426
Mendelian genetics, 214–222, 230
  and behavior, *1038–1039*
Mendelian populations. *See* Subpopulations
Menopause, 833
Menstrual cycle, 833–835
Menstruation, 833–835
Mental retardation, 209–210, *335, 336, 344*
Mercaptan, 929
Mercaptoethanol, *35*
Meristems, 669, 677–683
  in flowering, *764*
  in nodules, 726–727
Merostomata, *597–598*
Merozoites, of *Plasmodium,* 510
Mertensia, 463
Meselson, Matthew, *250–251*
Meselson–Stahl experiment, 250–251
Mesenchyme, 386, *576–577*
Mesenteries, *585–586*
Mesocoel, 610, 615
Mesoderm, 385–390, 610
  and classification, *576–577*
Mesoglea, *584,* 586, 587
Mesophyll, *683–685*
  in C$_4$ plants, 183–185
  in phloem loading, *700*
  and transpiration, *694–695*
Mesosomes, 64, 65, *486,* 610, 613
Mesozoic era, *416,* 629, *646, 647,* 654, 1153
Mesquite, *706–707*
Messenger hypothesis, 258–259
Messenger RNA. *See* mRNA
Metabolic compensation, *784–785*
Metabolic heat production, 785–786, 788–789, *791–792*
Metabolic pathways, *131–132, 136–137,* 142–144, 157–160
Metabolic rates, 981
  Q$_{10}$ concept, *784–785*
  and surface-to-volume ratio, *9–10*
  of terrestrial vertebrates, *627,* 628
  and thermoregulation, 627–628, *784–786,* 790–791, 793–797
Metabolism, 1–2, 114
  evolution of, 421
  of nutrients, 1002–1005
  regulation of, 130–132, 810, 1002–1005
  waste products of, *1010–1011*
Metabolism chamber, *785–786*
Metacoel, 610
Metals
  as prosthetic groups, *125–126*
  recovery of, 331

tolerance in plants, 449, *451, 452, 709–710*
Metamorphosis, 391–392, 1081
in frogs, 391
hormonal control of, 803–804
in insects, 391–392, 600, 602
Metanephridia, *1013*, 1017
Metaphase
in meiosis, 205–*209*
in mitosis, *198*, 200, 203
Metaphase plate, *198–200*, 205–208
Metasome, 610, 613
Metastasis, 345–*347*
Meteorites, and mass extinctions, 644–645, 647–*648*, 655
Methane (CH_4)
atmospheric, 1136
chemical structure, *24*, 33–34
in early atmosphere, 414, *415, 417, 418*, 422
production by bacteria, *489–490*, 1000
Methanogens, 485–*486*, *489–490*
Methanopyrus, 490
Methionine, *50*, 267, 983–*984*
Methylamine, *35*
Methylases, *312*
Methylation, of DNA, 302, *312*
Mexico, human population, *1085*
MHC, *371–373*, *1181*
Mice
coat color inheritance, 234–235
colonization by, 1159–1160
in genetic research, 241–*242*, 338–*339*, 342
in immunology research, 361, *362*, 368–369
phylogeny, 470–471
in thermoregulation experiments, 785–786
Micelles, 746–747, 996, *999*
Micrasterias, 524
Microbodies, *82*, 181–183
Microclimate, 1118
Microevolution, 442, 657–658
Microfibrils, 746–747
Microfilaments, 83–*84*, 910, 912–914, *913*
in chromatophores, *928–929*
in cytokinesis, 913
in development, 396
in slime molds, 515
Microgametophytes, *556*
Micronuclei, in Ciliophora, *512–513*
Micronutrients
for animals, 984–*985*
for plants, *719–720*
Micropipettes, *316*
Microprojectiles, 315–*316*
Micropyle
in flowering plants, *758–759*

in gymnosperms, 563–*564*
Microscopy, *63*, 66–68
invention of, *12*
and membranes, 97–98
Microsporangia, *556*, 560, 758–759
Microspores, *556*, 560, *759*
Microsporocytes, 758–759
Microtubules, 83–86, 910–912
in cytokinesis, *201*
in development, 912
in meiosis, 205–207
in mitosis, 197–200, 912
Microvilli
in gut, *913*, 992, *994*, 997
in renal tubules, *1020–1021*
in taste buds, *887*
Microwaves, 165
Midbrain, 853–854
Middle ear, 890–892
Middle lamella, 673–674
Midges, *1156*
Midgut, *992*
Miescher, Friedrich, 241
Migration
of cells in gastrulation, 385–389
of cellular proteins, 95–*96*
and evolution, 432, 434
by modular organisms, 1088–1089
seasonal, *38*, 777, 1047–*1052, 1094*
and spawning, 621, 829
and species cohesion, 447
tracing, 477, 640, 649, 656
vertical, 614
Migratory restlessness, *1050*
Mildews, 516–517, 531, 538
Milk
digestion of, 998
production of. *See* Lactation
Milkweeds, *704*
Miller, Carlos, 749
Miller, Stanley, *417, 418*
Millipedes, *600*
Milstein, Cesar, 364–365
Mimicry, 1107–*1108, 1116*, 1123
Mine tailings, plants on, 449, *451, 452, 709–710*
Mineralocorticoids, *807*, 813–*814*
Minerals
requirements in animals, 981, *984–985*
requirements in plants, *718–720*
uptake by plants, 690–692, 695
Minimum viable population (MVP), 1181
Missense mutations, 268
Mistletoe, 730
Mites, 598, 829
as parasites, 1069–1070
Mitochondria, 70–71, 74–75
and cellular respiration, 142, *143, 152–155*
chemical reactions in, 123–*124*

and classification, 474–475
compared with chloroplasts, 172, *173*
DNA in, 237, 268, 297–298, *324*, 640, 1156–*1157*
evolution of, 76–77, 297–298, *415*
membranes, 74–75, 94–*95*, 108, *152–155*
in photorespiration, 181–*183*
transport of proteins to, 307
Mitochondrial matrix, *74*, *152–155*
Mitosis, 191–192, 196–201
compared with meiosis, 208–209
in early development, 385, 386
effectors, 197–200, 912, 913
Mitotic center, 197, 202
Mixotricha paradoxa, 77
Mnium, 549
Moas, 1177
Models, mimics and, 1107–*1108*
Moderately repetitive DNA, 296, 302
Modern fauna, 659–660
Modular organisms, 1088–*1089*
Modules, 1088–*1089*
Molar solution, 27
Molars (teeth), 990–991
Mole rats, 1075–1076
Molecular clocks, 1154
Molecular formula, 26
Molecular traits, and classification, 474–476, 1154
Molecular weight, 26–27
Molecules, defined, 23, 26–27
Moles (mammal), 632
Moles (unit of measure), 26–27
Mollusks, *578, 579*, 602–606, *633*
circulatory systems, 957
digestive systems, *992*
filter feeding, 602–603, 910
rates of evolution, 658
shells, 602–605, 922
Molting, 596, 602, 640–641, 922
as growth pattern, *391*
hormonal control of, 802–804
Molybdenum, *719*, 985
Monadenia infumata, 604
Monarch butterfly, *1, 8*, 470–471
Monera. *See* Prokaryotes; Bacteria; Archaebacteria; Eubacteria
Monitor lizards, 625, 1129, *1131*
Monkeys, *632*, 634–635
endangered, 1189
Monoamines, as neurotransmitters, 869

Monoclonal antibodies, *364–365*
Monocots, *569*, 668–669
roots, *679–680*
seedling development, *753*
and selective herbicides, 744, *745*
stems, *680–681*
trees, 666, 682
Monocytes, *357*
Monod, Jacques, 130, 285, 288
Monoecious species, 227, 565, *828*
Monogamy, *1073, 1076*
Monogenea, 588
Monoglycerides, *999*
Monohybrid crosses, 216–220
Monomers, 44
Monoplacophorans, 605
Monosaccharides, 45–*46*
Monosomies, 210
Monosynaptic reflex loop, 872–873
Monotremes, 630–*631*, 839
Monsoons, 1148
Montezuma oropendola, 1072
Moose, 1104–*1105*, 1120, 1121
Mor humus, 723, 1136
Moraines, and plant succession, *1119*
Moraxella, 493
Morchella esculenta, 537, 538
Morels, 538
Morgan, Thomas Hunt, 228, 229, 230
Morphogens, 400–401
Morphology, 446–447, 465, 473–474, 610
Mortality, 350, 1083–*1084*
Mortality rates. *See* Death rates
Mosaic development, *384–385*
Mosquitoes, as vectors, *510*–511, 1109
Moss animals, *578, 579*, 611–612, 633
Moss rose, 466–467
Mosses, 546, 547–*548*, 549, *551–552*
Moths, 602, 1070
flight muscles, 788–789
growth and metamorphosis, 391–392
life histories, 1081–*1082*
pheromones, 884–*885*, 1040–*1041*
as pollinators, *1117*
Motivational state, 1032–1033
and evolution of displays, *1044*
Motor cortex, 857, 878
Motor end plate, *866–867*
Motor neurons, 866–868, *917–918*
Motor oils, 33
Motor scores, 1029–1030, 1034, 1039
Motor units, 917–919
Mountain lilac, 725
Mountains
biomes, *1163–1164*, 1169

effects on climate, 1145, *1146*
endemism, 1188
Mouse opossum, *631*
"Mouthedness," *439*
Mouths, *991–992*
and animal classification, 387, *576–578*
normal flora, *354, 496*
Movement. *See* Locomotion
MPF, 197, 198
mRNA, 258–265
artificial, *266*
in cancer, *348–349*
hybridization, *294*
polycistronic, *285–286*
in recombinant DNA technology, 318, *321*
splicing, *299–301*
stability, 299–300, 305
synthesis of. *See* Transcription
translation of. *See* Translation
*Mst*II, *340–341*
Mucosal cells, *993–994, 999*
Mucous membranes, as defense mechanism, 354, *358*
Mucus
of gut lining, *993–994, 996*
nasal, 885
for trapping prey, 613, 989
Mucus escalator, 354, 910, 946
Mules, *453*
Mull humus, 723, 1136
Müllerian mimicry, *1108*
Multicellularity, 9, 294, 578
Multiple alleles, 224–225
Multiple fruits, *567–568*
Multiple sclerosis, 375
Muscarine, 870
Muscarinic receptors, 869–870
Muscle building, and steroids, 815, *817*
Muscle fibers, 915–920
plasma membranes, 96, *916–918*
synapses with nerves, *866–868, 917–918*
Muscle spindles, 873, 889
Muscle tissue, *778–779, 914*
Muscles, *780, 914–920*
cardiac, *920, 963–965*, 971
fermentation in, 144, 155
of gut, 993–995
heat production, 788–790, 791
and joints, *926–928*
myoglobin in, 948, *949*
reflexes, 872–874
Muscular dystrophy, 335
Mushrooms, *539–540*
bioluminescent, *138*
as predators, 531
Musk oxen, *1063*
Mussels, 604
in communities, 1141–1142, 1172
competition, *434*
predators of, *1112–1113*
Mustard

artificial selection, *429*
in genetic research, *741*
Mutagens, 269, 345, 349–*350*
as plant chemical defenses, *712–713*
Mutants
conditional, 268
isolation of, 274, 276–278
nutritional, 255–257, 274–278
Mutations, 224–225, 268–270
in AIDS virus, 375, *376–378*
in *Drosophila melanogaster*, *395–396*, 401–*403*
and evolution, 270, 274, 432
in lymphocytes, 368–369
in plants, 754, *764–765*
and transposable elements, *297–298*
See also Genetic disorders
Mutualisms, *1101*, 1113–*1116*
endangered, *1178*
of keystone species, 1188
lichens, 531, *540–542, 1114*
mycorrhizae, *531–532, 539*, 679
and nitrogen fixation, 726–727, 1114
See also Coevolution; Endosymbiosis
Mycelium
of fungi, *529–530*
of funguslike protists, 516–517
Mycobacterium tuberculosis, 496
Mycoplasmas, *485, 487, 496*, 497
Mycorrhizae, 531–532, 539, 679, 1114, 1121
Myelin, 375, *865–866*
Myelomas, 364–365
Myofibrils, *914–918*
Myogenesis, of heartbeat, 920, *963–965*
Myoglobin, 920, 948, *949*
chemical structure, 52, *54*, 55
genes for, *298*
Myonemes, 511
Myosin, 83, *913–914*, 915–*918*
in cytokinesis, 201
Myotonic dystrophy, 336–337
Myriapods, 600
Myrica, 725
Mytilus californianus, 1112–1113
Mytilus edulis, 434
Myxamoebas, 515–516
Myxomycetes, *514–515*
Myxomyosin, 515
Myxotricha paradoxa, 505

N-Methylalanine, *418*
N-terminus, 49, 263
NAD, *126, 985*
in citric acid cycle, 142, 144, 147–149, *150*
as energy carrier, 140–*141*
in fermentation, *155–156*

in glycolysis, 142, *144–147*, 150
in respiratory chain, 143–144, 149–155
NADH-Q reductase, 150–*151*, 152, 158
NADP, *985*
in Calvin–Benson cycle, *178–182*
in photophosphorylation, *164–165*, 169–175
NADP reductase, *173*
Naegleria, 508
Naiads, 391
Naked mole rats, 1075–1076
Names, of organisms, 464, *466–467*
Nanometer, 165
nanos mutation, 396
nanos protein, 403–404
Nasal cavity, 885–886
Nasal salt glands, 1010, *1011*
Natal group, 1074, 1077
National Forest Management Act, 1181 National parks, 1190–1192
Native Americans, agriculture, 726
Natural history, 1058, 1077
Natural killer (NK) cells, 355, *358*, 374
Natural selection, 12–14, *434–437*, 1058–1059. *See also* Evolution
Nauplius larvae, *599–600*
Nautilus, 604, 605, 656
Navel oranges, 757–758
Navigation, *1048–1052*
in bees, *1042–1043*
Neanderthals, 638–639
Necrotic lesions, 714
Nectar, 1115–1116
Nectar guides, 568, 904, 1117–*1118*
Needles, decomposition, *1120*
Negative feedback, 782–783
in metabolic pathways, 130–132, 157–160
in thermoregulation, 793–794
Negative-pressure pump, 939, *940*
Neher, Erwin, *860*
Neidium iridus, 519
Neisseria gonorrhoeae, 492–493
Neisseria meningitidis, 493
Nematocysts, 583, 927–928
Nematodes, 577, 578, 579, 591–592, 606, 633
bioluminescence, *138*
digestive systems, *992*
as parasites, 591–592, 1109
predators of, *531*
in studies of development. *See Caenorhabditis elegans*
Nemertea. *See* Ribbon worms
Neon, 1136
Neopilina, 605
Nephridial canal, *1014*
Nephridiopores, *1013*

Nephrons, *1015–1017*, 1019–*1022*
Nephrostomes, *1013, 1017*
Nernst equation, *860*
Nerve cells. *See* Neurons
Nerve deafness, 894
Nerve endings, in skin, *888–889*
Nerve gases, 127, 870
Nerve impulses. *See* Action potentials
Nerve nets, *850*
Nerves, 852–853
Nervous systems, *779–780*, *849–850*
and behavior, 1034–*1037*
of chordates, 617
circuits, 852–853, 870–874
development, 388–390, 853–854, 912
energy requirements, 1002
intrinsic, 1001
oxygen requirements, 155, 933
See also Brain; Central nervous system
Nervous tissue, 778–779
Nest cleaning, 1038–1039
Nest holes, 1121
Nest parasites. *See* Brood parasites
Nest sites, and mate choice, *1068–1069*
Nesting
behavior, *1037, 1059*
colonial, 1068–*1069*
Nestlings, 628
Net primary production, *1128*
Neural axis, 854
Neural folds, 388–390
Neural groove, 388–390
Neural plate, 388–390
Neural reflexes. *See* Reflexes
Neural tube, 388–390, *853–854*
Neurobiology, 850
Neurofibrils, 513
Neurohormones, 802, 806, 853
Neuromuscular junctions, *866–868, 917–918*
Neurons, 778, 851–852
development, 912
function, 858–870
motor, 866–868, 872–873, *917–918*
oxygen requirements, 155
sensory, 872–873, 881, *886*, 895–896
Neuropharmacology, 870, 929
Neuroptera, 602
Neurospora crassa, 533–534, 538
crossing over in, 230–233
nutritional mutants, 255–257
Neurotransmitters, 852, 866–870
in autonomic nervous system, 872
and habituation, 875
in sensation, *883–884*
Neurulation, 388–390, *397–398*
Neutral pH, 29–30
Neutrons, 19–21

Neutrophils, *357*
New World monkeys, *632, 634–635*
New Zealand, species distributions, 1154–*1155*
Newts
external gills, *934*
population studies, *1080*
in studies of development, 397–398
Niacin, 141, 985, *986,* 988
Niches, 988, 1101–*1102*
Nickel, 721
Nicotinamide, in carrier compounds, 141
Nicotinamide adenine dinucleotide. *See* NAD
Nicotinamide adenine dinucleotide phosphate. *See* NADP
Nicotine, 869–870, 1109
Nicotinic acid. *See* Niacin
Nicotinic receptors, 869–870
Night length, and flowering, 734–736, 765–767
Night navigation, 1051–*1052*
Night vision, 900
Nirenberg, Marshall W., 266–267
Nitrate (NO₃⁻)
in anaerobic respiration, 156
in early oceans, 422
in nitrogen cycle, 728–729, 1138–*1139*
and plant nutrition, *718*
reduction of, 729
Nitric acid, in acid precipitation, *1142*
Nitric oxides, 27
as neurotransmitter, 869
in nitrogen cycle, 1138–*1139*
Nitrification, 728–729
Nitrifiers (bacteria), 484, 492
Nitrite (NO₂⁻)
in anaerobic respiration, 156
in nitrogen cycle, 729
Nitrobacter, 491, 729
Nitrocellulose filters, 321–322
Nitrogen
atmospheric, *24,* 1136
in early atmosphere, 414, *415, 418,* 422
fixation, 484–485, 717, 724–729, 1138–*1139*
heavy, 250–251
as metabolic waste product, 1010–*1012*
oxides of, *27*
as plant nutrient, 717, *719, 720,* 722, 729
Nitrogen cycle, 724–729, 1138–*1139*
Nitrogen-fixing bacteria, 484–485, 491–492, 493, 726–729, 1114
in nitrogen cycle, 717, 728–729, 1138–*1139*

and recombinant DNA technology, 329–330, *728*
and succession, 724–725, *1119,* 1121–*1122*
Nitrogenase, 494, 725–726, *728*
Nitrosococcus, 729
Nitrosomonas, 729
Nitsch, Jean-Pierre, 746
Nocturnal animals, vision, 900
Nodes, of plants, 669–670
Nodes of Ranvier, 865–866
Nodules, in nitrogen-fixing plants, 724–725, 726–727
Noise, and deafness, 894
Noncompetitive inhibitors, 127–*128*
Noncyclic photophosphorylation, 164, 169–*171*
Nondisjunction, 209
Nongated channels, 859
Non-Mendelian inheritance, 236–237
Non-nuclear genes, 237
Nonpolar molecules, 30–*31,* 33
Nonrandom mating, *433,* 456
Nonself, recognition of, 359–360, *361*
Nonsense mutations, 268
Nonshivering heat production, 791–792
Nonspecific defense mechanisms, 353–356, *358,* 366–367
Nonvascular plants, *546, 547*–552
Nonvirulent strains, 242
Noradrenaline. *See* Norepinephrine
Norepinephrine, *807,* 813
in autonomic nervous system, 872
and control of heartbeat, *965*
as neurotransmitter, 869
and regulation of blood flow, 974
Normal flora, as nonspecific defense, 354, *358*
North Star, 1051
Northern blotting, 321
Northern spotted owl, *1189*
Nose, 885–*886*
Nothofagus, 1163–*1164*
Notochord, 387, *388*–390, 617
in sea squirts, *473*
Notorhyncus cepedianus, 620
Notropis, 657
Nuclear envelope, *68–69, 70, 71*
in cell division, *192,* 196, *198*–201
in meiosis, 205–207
Nuclear pores, *68–69, 70, 71*
Nucleases, 993
Nucleic acid hybridization, 294–295
in recombinant DNA technology, 321–322
Nucleic acids

chemical structure, 55–58, *243–248*
and classification, 474–476
origins, 417–419
products of metabolism, 1010–*1011*
See also DNA; RNA
Nucleoids, 63–64
Nucleolus, *70*–72, *192, 196, 198,* 201
in meiosis, 205
in rRNA synthesis, 302–303
Nucleoplasm, 69, 71
Nucleoside triphosphates, 248–249, 251–254
Nucleosides, 55, *57*
Nucleosomes, *193*–195
Nucleotides, 55–58, *57,* 245–249
Nucleus (atomic), 19
Nucleus (cellular), *68–69, 70, 71*–73
in cell division, *192*–195
in cleavage, 382–383
developmental potency, 393–394
in mitosis, 196–201
of muscle cells, *914, 915,* 920
in plant reproduction, *759, 760*–761
Nudibranchs, *604*
Null hypothesis, 7–8
Nursing. *See* Lactation
Nutrients
defined, 717
in lake eutrophication, 1140–1142
Nutrition
in animals, 981–988
in fungi, 530–531
in germinating seedlings, *738*–739
in plants, 717–731
Nutritional mutants
of *E. coli,* 274–278
of *Neurospora,* 255–257
Nutritive body, of seeds, *1117,* 1118
Nymphalidae, 470–*471*
Nymphs, 391, 1105–*1106*
Nyssa aquatica, 737

Oaks
galls, *492,* 1123
variation in leaves, 427, *428*
Oats, 742–743, 1112
Obelia, 584
Obesity, 982
Obligate aerobes, 484
Obligate anaerobes, 421–423, 484, 489–490
Obligate parasites, 509–510, 517, 531
intracellular, 493, 496, 497
Occipital lobes, 856–*857*
Ocean sunfish, 621
Oceans
biomes, 1170–*1173*
circulation, *1147*–1148

as ecosystem compartment, 1133–*1134*
pyramid of biomass, 1131–*1132*
Ocotillo, 705–706
Octopuses, 603, *604*
camouflage, *928*–929
Odonata, 602
Odorant molecules, 885–887
Oedogonium, 523, 526
Off-center receptive field, 901–*903*
Offspring
helping behavior in, *1065, 1072*–1075
recognition of, 1034
–OH radical, 1136
Oil production, 331
Oil spills, 330–331
Oils, 40–42
Okazaki, Reiji, 252
Okazaki fragments, 252–254, 297
Old age. *See* Aging
Old-growth forests, *1189*
Old World monkeys, *632, 634*–635
Olduvai Gorge, *637,* 638
Oleic acid, 40–*41*
Olfaction, 885–887
and communication, 1040–*1041*
Olfactory bulb, 885–*886*
Olfactory hairs, 885–*886*
Oligo-dT primer, 318–320
Oligochaetes, 594–595
in eutrophic lakes, *1141*
Oligomers, 44
Oligonucleotides, as primers, 327
Oligosaccharides, 45
Olives, 1167
Omasum, *1000*–1001
Ommatidia, 897–*898*
Omnivores, 980
speciation, 456–457
teeth, *991*
as trophic level, 1128–*1129*
Omphalolotus illudens, 138
On-center receptive field, 901–*903*
Onchorhynchus, 1082
Oncogenes, 196, 346–*347*
One-gene, one-polypeptide theory, 255–256
Onychophora, 578, 579, 596–597, *606,* 633
Oocytes, *827*–828, *832, 833*
gene amplification in, 302
translational control in, 305
Oogamy, 526, 547
Oogonia, 826–828
Oomycota, 514, 516–517
Ootids, *827*–828
Oparin, Alexander I., 421
Open channels. *See* Nongated channels
Open circulatory systems, 603, *957*
Operators, *287*–289
Opercular flaps, *938–939, 940*

Operons, 286–289
   absence in eukaryotes, 303
   in recombinant DNA technology, 316, 318
Ophiuroidea, 616–617
ophthalmoptera mutation, 402
Oporina autumnata, 1109–1110
Opossum shrimp, 1171
Opossums, 630, 631
Opportunistic infections, 376
Opportunity cost, 1058
Opposable thumbs, 632
Opsin, 894–895, 900
Optic chiasm, 903–904
Optic cup, 397–398
Optic nerves, 865, 900, 902–904
Optic vesicle, 397–398
Optical isomers, 34, 36
Optimal temperature, for enzymes, 132
Opuntia, 1106–1107
Oral contraceptives, 835, 838–839
Orangutans, 635
Orb webs, 1029
Orbitals, 22–23
Orchids, 566
   mutualisms with fungi, 531, 532
   pollination, 1116, 1184–1185
Orders (taxonomic level), 464, 466–467
Ordovician period, 646, 652
Organ formation, in plants, 748
Organ of Corti, 892–894, 913
Organ systems, 778–782
Organ transplants, 371, 373–374
Organelles, 5, 68–69, 94–95
   analogs in prokaryotes, 486
   in cytokinesis, 201
   microscopy, 97–98
   separating, 87–89
   of unicellular protists, 512–513
Organic compounds, 33–36
Organic fertilizers, 722
Organisms, defined, 4–5
Organizers, in developmental induction, 397–398
Organizing concepts, of biology, 11–14
Organogenesis, 843–844
Organs
   defined, 4, 5
   development, in plants, 748
   development of, 386, 400–403, 843–844
Orgasm, 830, 835
Orientation, in navigation, 1042–1043, 1048, 1052
Origin of replication, 252, 254
Oropendolas, 1072–1073
Orthasterias koehleri, 616
Orthodox systematics, 464–467
Orthoptera, 601, 602

Oryza sativa, 667–668
Oscillations
   of population size, 1080
   of predators and prey, 1104–1105, 1107, 1164
Oscilloscope, 862
Osmoconformers, 1009–1010
Osmolarity, 1009
Osmoregulators, 1009–1010
Osmosensors, 1026
Osmosis, 105–107
   and transport in plants, 689–690, 694–695, 699, 700
Osmotic balance
   in animals, 1003–1010, 1011, 1024–1026. See also Excretory systems
   in fungi, 530
   in mussels, 434
   in plants, 708–709
   in protists, 505
Osmotic potential, 106–107
   in capillaries, 968–969
   in plants, 689–690
Osmunda cinnamomea, 558
Ossicles, 891–892, 894
Ossification, 924–925
Osteichthyes, 621, 622
Osteoblasts, 811–812, 923–925
Osteoclasts, 811–812, 923–924
Osteocytes, 923–925
Ostia, 957
Ostracoderms, 619
Ostriches, 629
Otoliths, 890–891
Outbreaks. See Irruptions
Outgroups, 469–471
Ova. See Eggs
Oval window, 891–894
Ovalbumin, and RNA splicing, 300
Ovarian cycle, 832, 833
Ovaries
   of animals, 826–828, 830, 832–835
   of flowering plants, 565, 759
   hormone production, 802, 807
Overexploitation, 1177
Overhunting, 1177
   effect on populations, 433, 629, 630, 639, 1177
   of whales, 1095–1096
Overnourishment, 982
Overturning, in lakes, 1134–1135, 1141, 1142
Oviducts, 830, 832, 833, 841–842
Oviparity, 837–841
Ovipositor, 1103
Ovoviviparity, 840
Ovulation, 830, 832, 833
Ovules
   of flowering plants, 565, 569, 758–759
   of gymnosperms, 563–564
Owen, Ray D., 361
Owls, 1104
Oxalate, 128
Oxaloacetate, 128

allosteric controls, 158–160
   in amino acids, 157
   in C_4 plants, 183–184
   in citric acid cycle, 148–149, 157
Oxidation, defined, 139–140
Oxidation-reduction reactions. See Redox reactions
Oxidative phosphorylation, 143, 144, 152–153
   uncoupling, 154, 792
Oxidizing agents, 140
Oxygen
   animal requirements for, 144, 155, 933–934, 1102
   atmospheric, 1136
   in chemical reactions, 27–28, 140
   chemical structure, 24, 25
   and evolution of life, 174, 414–415, 417, 422, 578–580
   exchange, 933–935, 936–946, 969
   isotope ratios, 657
   nervous system requirements, 155, 933
   and nitrogen fixation, 725–726
   in photorespiration, 181–183
   in photosynthesis, 162–163, 169–170, 174
   plant requirements for, 706–707, 708, 717, 719
   production by plants, 163
   regulation of blood content, 951–953, 974–975
   in respiratory chain, 143–144, 150–152
   in soils, 706–707, 721
   transport by circulatory system, 947–951, 952
   in water, 934–935, 1102, 1134, 1135, 1141, 1169
Oxygen-dissociation curve, of hemoglobin, 947–948, 949
Oxytocin, 806–807, 844–845, 846, 1034
Oxyura, 474
Oystercatchers, 1031
Oysters, 604, 828
Ozone hole, 1127, 1135–1136, 1143
Ozone (O^3), 1127, 1135–1136
   and evolution of life, 415

p53 gene, 346
Pacemaker cells, 874, 920, 963–965
Pacinian corpuscles, 888–889
Packed cell volume, of blood, 971
Paige, Ken, 711–712
Pain sensation, 856, 866, 888
Pair-rule genes, 401, 403
Paleozoic era, 416, 545, 646–647
Paleozoic fauna, 659–660
Palisade mesophyll, 683–685

   in C_3 plants, 184
Palmitic acid, 40–41, 983
Palms, 569, 666, 682
Pancreas
   in digestion, 996–998, 1001
   endocrine functions, 802, 806–807, 812–813
   regulation of metabolism, 1002–1005
Pancreatic duct, 997–998
Pancreatitis, and somatostatin, 311
Pandas, 475, 476
Pandorina, 523
Pangaea, 553, 646–647, 654, 1153
Panting, 791, 793
Pantothenic acid, 986
Panulirus interruptus, 599
Paper partition chromatography, 175–178
Papilionidae, 470–471
Papillae, 887
Parabronchi, 940, 942
Parachuting, in frogs, 610
Paracineta, 511
Paradigms, 14
Paramecium, 62, 504–505, 511–513
   interspecific competition experiments, 1111–1112
   reproduction, 202
Parapatric speciation, 449, 451, 452
Parapodia, 594
Parasites, 531, 607, 1101, 1103
   absorptive nutrition, 991
   coevolution with hosts, 451–452, 1108–1109
   as cost of sociality, 1064
   life cycles, 510–511, 534–536, 535, 539
   obligate, 509–511, 517, 531
   obligate intracellular, 493, 496, 497
   and reproductive success, 1067, 1069–1070
   speciation, 451–452
Parasitodiplogaster, 1109
Parasitoids, 1103, 1105
Parasympathetic nervous system, 870–872
   and heartbeat, 920, 965
   and regulation of blood flow, 974–975
Parathormone, 806, 811–812
Parathyroid glands, 802, 806, 811–812
Parenchyma, 674–675, 677
Parental care
   by birds, 628, 1031–1032, 1058–1059, 1071
   courtship feeding as, 1067
   and evolution of sociality, 1072–1074
   in oviparous animals, 625, 839
   in primates, 632
   releasers of, 1031–1032
   sex roles, 1070–1071

by siblings, *1065,*
1072–1075
Parental generation (P), 215
Parental investment,
1070–1071, 1081–1084
in plants, *763*
Parietal lobes, *856–857*
Parks, 1190–1192
Parrots, *629, 1037*
Parsimony, 472–473, 476–477,
1154–1155
Parthenium argentatum,
1094–1095
Parthenocarpy, 746
Parthenogenesis, *825–826*
Partial pressure of carbon
dioxide, 951–953
Partial pressure of oxygen,
935–936, 947–950,
951–953
Particulate theory, of
Mendelian inheritance,
217
Parturition, 807, 844–845
Parvovirus, 1104–1105
Passenger pigeon, 1177
Passeriformes. *See* Songbirds
Passiflora rubra, 469
Passive immunization, 365
Passive transport, *103*
Pasteur, Louis, 12, *412*
Pasteur effect, 157
Pastoralism, 639
Patch clamping, 691–692,
696–697, 860
Patchiness, 1059–1060, 1112
and predator–prey interac-
tions, 1105–1107
and risk of extinction,
1180–1181, *1188–1190*
Pathogenesis-related (PR) pro-
teins, *714*
Pathogens
animal defenses against,
353–361, 366, 794, *795*
bacterial, 488–489, 493,
713–714
of crop plants, 516–517,
534–536, 538, 539
flatworms, 588–590
fungi, *529,* 534–536, *535,*
538, 539, 540, 713–714
plant defenses against,
356, 713–714
protozoa, 508, 509–511
See also Viruses
Pattern formation, 401–*405*
Patterns, in vision, 901–*903*
Pauling, Linus, 245, 256
Pavlov, Ivan, 875–876, 1001
Peach-faced lovebird, *1037*
Peach palms, 569
Peafowl, *1041*
Peanut butter, 42
Peas
in Mendel's experiments,
215–222
in plant growth experi-
ments, 743–*745*
seedling development, *753*
Peat bogs, 552

Pecking, as releaser,
1031–1032
Pedigree analysis, 337–*339*
Pelagic zone, *1171–1172*
Pelicans, *1062–1063*
Pellagra, 988
Pellicle, *511*
Pelmetazoans, 615–616
Pelvic girdle, *630*
bipedalism and, 637
Penetrance, of genotype, 235
Penguins, *629*
colonial nesting, 1068–*1069*
migration, 777
salt glands, 1010, *1011*
Penicillin, 276–277
Penicillium, 530, 540
Penis, *829–830,* 1067–1068
development of, *816*
in humans, 830–*831,* 835
Pentoses, *46*
in nucleic acids, 55, *57,* 419
PEP
in C₄ plants, 183–*185*
in CAM plants, 184
in glycolysis, *144, 145*
Peppered moth, *7*
Pepsin, *132,* 996, *998*
Pepsinogen, 996
Peptidases, 993
Peptide linkage, 49, *51,* 261
Peptides, origin of, 419
Peptidoglycan, *64,* 486–*487*
Perches, 470–*471,* 622
Perching birds. *See* Songbirds
Perdeck, A. C., 1031
Peregrine falcon, *1182–1183*
Perennials, 677, 735
in tundra biome, 1163
Perfect flowers, *565*
Perfusion, 937, 946–951
Pericycle, 679–680
Periderm, 677, 683–684
Perilla, 679
Period
geological, *646*
of photoperiodic fluctua-
tions, *733*
Periods, of circadian rhythms,
767–768, 1045–1047
Peripatus, 597
Peripheral nervous system,
849, 852–*853*
Peripheral populations, 449
Perisodus microlepis, 439
Peristalsis, *995,* 999
Perithecium, 537
Peritoneum, 576–*577,* 994–995
Peritrichs, *511*
Peritubular capillaries,
1015–1017, 1019–1021
Periwinkles (plants), *1176*
Permafrost, 1163
Permian period, 646–647, 653
mass extinction, *646,* 654,
658
Pernicious anemia, 988
Peroxisomes, 82
Pesticides
in agriculture, 1144
breakdown by fungi, 528
and extinctions, 1182–1183

hormones as, 803
and lake eutrophication,
1141–1142
resistance to, 329, *331*
Pests
control of, 1094–1096, 1144
introduced, 1178, 1180
Petals, *565, 566,* 757, *758–759*
Petioles, 672
abscission, 743–744
Petromyzon marinus, 619
PGA. *See* 3PG
pH, 29–31
in chemiosmotic mecha-
nism, 172–173
and enzyme activity, *132*
influence on hemoglobin,
949
ion forms, 147
of soils, 722, 723, 728
Phaeophyta, *518, 520–521*
Phage coats, 244, 282–283
Phages. *See* Bacteriophages
Phagocytes, *355, 356, 358,* 913,
970
and complement system,
367
and IgG, 365, *366*
interaction with lympho-
cytes, 371–374
Phagocytosis, 80, 81, 913
Phalaropes, *1071*
Phanerozoic eon, *646*
Phantom limb sensations, 882
Pharyngeal basket, *628*
Pharyngeal slits, 617
Pharynx, 944–*945*
of acorn worms, 613–614
evolution of, 619
of flatworms, *991*
of humans, *945,* 995
of nematodes, 591–*592*
Phase-shifting, of circadian
rhythms, 1045–1046,
1050–1051
Phelloderm, 683 *Phelsuma*
madagascariensis, 1187
Phenolics, as plant defenses,
712
Phenotype, 217, 427
Phenylalanine, *50,* 266, *267,*
335
nutritional requirements,
983–984
Phenylketonuria, 335, 344
Pheophytin-I, 170–*171*
Pheromones, 884–*885,* 929,
1040–*1041*
in cooperative foraging,
1103
in fungi, 537
in funguslike protists, 515,
516
Phillips, David, 124
Philodina roseola, 593
Phinney, Bernard O., 740
Phloem, 548, 698–701
cell types, *676*
growth of, 681–*683*
of leaves, 683–*685*
of roots, 679–680

Phoronids, *578, 579,* 611–*612,*
633
Phoronis, 612
Phosopholipase C, *820*
Phosphate esters, 139
Phosphate groups, 34–35, 984
in ATP, *137,* 139
in DNA, 245–249, 251–*252*
in nucleotides, 55, *57*
Phosphate ions, 25
Phosphates, in early oceans,
422
Phosphates, in phosphorus
cycle, 1138
Phosphatidylinositol, *820*
Phosphodiester bonds, in
DNA, 245–*247*
Phosphodiester linkages, 55
Phosphodiesterases, 819,
896–897
Phosphoenolpyruvate car-
boxylase (PEP). *See* PEP
Phosphoenolpyruvic acid, 698
Phosphofructokinase, *144,*
146, 158–*159*
Phosphoglucomutase, 144
Phosphoglycerate kinase, 145
2-Phosphoglycerate (2PG),
145
3-Phosphoglycerate (3PG), *35,*
145
in C₃ plants, 183
in Calvin–Benson cycle,
175–*182*
Phosphoglyceromutase, *145*
Phospholipid bilayer, 42–*43,*
92
microscopy, 97–*98*
Phospholipids, *42–43,* 109
Phosphoprotein phos-
phatases, 819–820
Phosphorus
as animal nutrient,
984–*985*
chemical structure, 25
in DNA, 243, 244
in fertilizers, 722
and lake eutrophication,
1140–1142
in membranes, 42–*43,* 92
as plant nutrient, 718, *719,*
720
Phosphorus-32 (³²P), 243, *244,*
645
Phosphorus cycle, 1138–*1139*
Phosphorylase, in coacervate
drops, 421
Phosphorylase kinase,
818–*820*
Phosphorylation, 419
oxidative, *143,* 144,
152–153, 792
substrate-level, *141,* 146
Photoautotrophs, *484*
Photochemical reactions, 1136
Photoheterotrophs, 484
Photomicrographs, 66
Photons, 165–168
Photoperiodic behavior,
1047–1048
Photoperiodic flowering,
734–*736,* 765–767

Photophosphorylation, 164–165, 169–175
Photoreceptors
  in animals, 894–904
  in plants, 742–744
Photorespiration, 82, 181–183
  avoidance by C4 plants, 183–185
Photosensors, 881–882, 894–904
Photosynthesis, 136–137, 162–185
  in algae, 517, 520, 521–523
  in bacteria, 64–65, 173–174, 422, 484, 493–494
  in C4 plants, 183–185
  Calvin–Benson cycle, 175–181
  compared with cellular respiration, 185
  discovery of, 163–164, 172–183
  energy yield, 180–181
  and primary production, 1128
Photosynthesizers, as trophic level, 1128–1129
Photosynthetic autotrophs, 718
Photosynthetic carbon reduction cycle. See Calvin–Benson cycle
Photosynthetic lamellae, 494
Photosynthetic pigments, 167–169, 484
  in algae, 169, 517, 518–519, 520–524
  in cyanobacteria, 169, 174–175
  in plants, 546, 547
Photosystems, 164, 170–172, 173–174
Phototropism, 742–744
Phycobilins, 169
Phycocyanin, 167, 169, 521
Phycoerythrin, 167, 169, 521, 523
Phycomyces, 536
Phyllomedusa, 1018
Phylloxera, 772
Phylogeny, 462–477
  in biogeography, 1153–1154, 1156–1157
Phylum, 464, 466–467
Physalia, 583
Physarum, 514
Physical basis, of biology, 11
Physical factors, in environments, 1057
Physiology, 777–778
Phytoalexins, 714
Phytochrome, 733, 752–754, 767
Phytophthora infestans, 517
Phytoplankton, 580
Pickerel, 463
Pieridae, 470–471
Pigeons
  crop milk, 805
  defenses against predators, 1063–1064

homing, 906, 1048
phylogeny, 470–471
Pigmentation
  and camouflage, 928–929
  in human skin, 235–236, 986
  inheritance, 222–223, 234–236
Pigments
  chemical properties, 167–168
  genetic coding for, 222–223
  photosynthetic. See Photosynthetic pigments
Pileus, of mushrooms, 540
Pili, 65, 66, 276, 278
Pill, birth control, 835, 838–839
Piloting, 1048, 1052
Pineal gland, 802, 807, 1046–1047
Pines
  defenses, 1103–1104
  life cycle, 563–564
  needles, 1120
Pink molds, 538
Pinnae, 890, 892
Pinocytosis, 81, 913
Pinus sylvestris, 1120
Pioneering. See Colonization
Piper pseudobumbratum, 1088–1089
Pirozynzki, K. A., 532
Pisaster ochraceus, 1112–1113, 1188
Pistils, 565–566, 758–759
Pit organs, 904–905
Pit pairs, 674
Pit vipers, 904–905
Pitch, 893, 1103
Pitcher plants, 717, 730
Pith
  differentiation in, 748
  in monocot roots, 679–680
  in stems, 680–681
Pith rays, 681
Pits, in plant cell walls, 674
Pituitary gland, 802, 805–809, 810
Pivotal joint, 926–927
Placenta, 387, 625, 833, 840, 841–842, 843–845
Placental mammals, 630–632, 840
Placoderms, 619–620
Placozoa, 578, 582, 588, 606
Plague, 350, 492
Planarians
  excretory systems, 1012
  eye cups, 897
  gastrovascular cavity, 957, 991
Planck, Max, 165
Planck's constant, 165
Plane joints, 927
Planetarium, 1051–1052
Planets, formation of, 414, 415
Plankton, 580, 1103, 1170, 1172
Plant communities, and species richness, 1123
Plants, 15

in biogeochemical cycles, 1137–1138
cell division, 201–202
cell structure, 70, 673–676
cell walls, 201–202, 546, 673–674, 690–693, 746–748
and classification of biomes, 1161–1162
common vs. rare, 1184–1185
compared with algae, 504, 517, 523
competition, 672, 711, 718, 1110, 1111, 1112
defenses against herbivores, 83, 437–438, 704, 712–713, 1103, 1109–1110
defenses against pathogens, 356, 713–714
DNA, 474–475, 754
effects on communities, 1118–1120, 1123
evolution of, 503–504, 523, 547
fossils, 554, 561–562
heavy metal tolerance, 449, 451, 452, 709–710
parasitic, 545, 546, 730
taxonomy, 546–548
turgor, 83, 107, 684, 690–691
Planula larvae, 583–585
Plaque
  in blood vessels, 967
  and tooth decay, 354
Plaques, on bacterial lawn, 274
Plasma (blood), 356, 781, 971, 973
  osmotic potential, 105–106
  transport of respiratory gases, 947, 950
Plasma cells, 372–373
  antibodies, 357, 360
  development, 361, 363
  diversity, 359–361
Plasma membranes
  of cardiac muscle, 920, 963–965
  of eggs, 836–837
  electroporation, 316
  junctions between, 98–100, 99
  of muscle fibers, 96, 916–918
  of neurons, 851, 858–866
  of prokaryotes, 63–64
  and second messengers, 818–820
  of sensory cells, 881–882, 883–884, 894–897
  at synapses, 866–870
  transport across, 80–81
  and uptake in plants, 690–692
  and viruses, 375–377, 497, 498
  See also Membranes
Plasmagel, 913–914

Plasmasol, 913–914
Plasmids, 283–284
  in crown gall, 492
  in recombinant DNA technology, 311, 313–315, 329–330
Plasmodesmata, 86–87, 674, 692
Plasmodium, 510–511, 1109
Plasmodium, of slime molds, 514–515
Plasmolysis, 106
Plastic, 33
Plasticity, in plant cell walls, 747–748
Plastids, 75–76
Plastocyanin, 171–172
Plastoquinone (PQ), 171–172
Plateau phase, 835
Platelets, 967, 972–973
Platinum black, 121
Platyhelminthes. See Flatworms
Platypus, 630, 839–840, 881, 887
Pleasure centers, 856
β-pleated sheet, 52, 53, 56
Plecoptera, 602
Pleiotropy, 224
Pleistocene epoch, 655, 658, 659
Plethodon, 453–454
Pleural cavities, 945–946
Pleural membranes, 945–946
Ploidy levels, 202–203, 210–211
Plugs, copulatory, 1067–1068
Pneumatophores, 707
Pneumococcus, 242–243
Pneumonia, 242–243
Poa annua, 1087–1088
Podocytes, 1015–1016
Pogonophores, 578, 579, 593–594, 606, 633
Poikilotherms, 785
Point mutations, 268–269, 340, 346
Poisons
  as defenses. See Chemical defenses
  production by animals, 929
Polar bear, 980, 987
Polar bodies, 827–828
Polar microtubules, in mitosis, 197–200
Polar molecules, 30–31
Polar nuclei, 758–759
Polaris, 1051
Polarity, 30–33
  of auxin transport, 742
  of eggs, 383, 394–396
  of membranes, 859
Polarized light, and navigation, 1051
Poles, and climate, 1145
Pollen grains, 560–561
  of flowering plants, 568–569, 758–761
  of gymnosperms, 563–564
Pollen tubes, 560–561

of flowering plants, 568–569, 735, 758–761, 762
of gymnosperms, 563–564
Pollination, 565–568
by animals, 566–568, 760, 1100–1101, 1115–1117, 1178
and apomixis, 770–771
by bees, 1117, 1118, 1184–1185
in cross-breeding experiments, 214–215
by insects, 566, 568, 714, 1117–1118
of rare species, 1184–1185
by wind, 561, 563, 758, 760
See also Fertilization; Self-pollination
Pollinia, 1184
Pollution
breakdown by fungi, 528
and carcinogens, 345, 349
effect on lichens, 540
and eutrophication, 1141
by herbicides, 744
by pesticides, 1144, 1183
and recombinant DNA technology, 330–331
Poly A tail, 299–300, 318–320
Polyadenylation, 299–300
Polychaetes, 594–595
Polycistronic messengers, 285–286
Polydactyly, 336
Polyethylene glycol, 338
Polygamy
and display territories, 1069
and social organization, 1076–1077
Polygenic inheritance, 235–236 Polygyny
and resource defense, 1069
and sexual dimorphism, 1057, 1072, 1073
Polymerase chain reaction, 326–328
and classification, 474
medical applications, 344
Polymerization reactions, 45
and origin of life, 417–419
Polymers, 44–45
Polymorphic populations, 430, 432
Polymorphism, 439–440
Polynucleotides, in DNA, 245–249
Polypeptides, 49
genetic coding for, 256–258
synthesis of, 261–265
synthetic genes for, 320
Polyplacophora, 603–605
Polyploidy, 449–452, 453
Polyporus sulphureus, 539
Polyps, 583–586
Polyribosomes, 263–264
Polysaccharides, 45–49, 48
in coacervate drops, 421
and origin of life, 417–419
Polysiphonia, 521, 523
Polysomes, 263–264

Polyspermy, blocks to, 836–837
Polysynaptic reflexes, 873–874
Polytene chromosomes, 232–234, 301–302, 303
Polyunsaturated fatty acids, 40–42, 41
Pomocanthus imperator, 522
Pongo, 635
Pons, 853–854
Poorwill, 796
Poplar aphids, 1060–1061
Population density, 1084–1085, 1091–1092
and geographic range, 1090–1091
and habitat selection, 1091
oscillations, 1080, 1104–1105, 1107, 1164
Population ecology, 1080–1097
Population growth, 1089–1092
human, 639, 1097
Population vulnerability analysis (PVA), 1181
Populations, 4–5
defined, 428–430
dynamics, 1086–1097
genetics, 428–431, 432–443, 658
regulation of, 1091–1097
risk of extinction, 658, 1181
speciation, 448–449
structure, 429–431, 1084–1086
Pores, in nuclear envelope, 192, 194
Porifera. See Sponges
Portal blood vessels, 809–810
Porter, Rodney M., 361
Portuguese man-of-war, 583, 928
Positional information, 889–891
in development, 400–401
Positive cooperativity, 947–948
Positive feedback, 158–160, 782
Positive-pressure pump, 939, 940
Postabsorptive period, 1002–1005
Postelsia palmaeformis, 521, 522
Posterior, defined, 388
Posterior pituitary, 805–807, 809, 810
Postganglionic neurons, 872
Postsynaptic cells, 866–869
Postsynaptic membranes, 852, 866–870
Posttranslational control, 306–307
Postzygotic isolating mechanisms, 452–453
Potassium
as animal nutrient, 984–985
isotopes, 645, 648
as plant nutrient, 719–720, 722
Potassium channels
in cardiac muscles, 920

in neuron function, 859–865
in postsynaptic membranes, 868
Potassium equilibrium potential, 860
Potassium ions (K+)
active transport of, 104–105, 123
in nervous system function, 858–866
and stomatal cycles, 696–697
Potato beetle, 577
Potato famine, 517
Potatoes, 671–672, 770
Potency, nuclear, 393–394
Pottos, 634
Pouch, in marsupials, 630, 631, 840
Powdery mildews, 538
Power arm, 926, 928
Power stroke, of cilium, 911
PR proteins, 714
Prairie dog colonies, 1133–1134
Prawns, 1114
Precambrian organisms, 586, 649–652
Precapillary sphincters, 974–975
Precipitation
acid, 1138, 1142
and biological productivity, 1128, 1148–1149
and classification of biomes, 1160–1162, 1167
effect of mountain ranges, 1145–1146
in hydrological cycle, 1137
Precocial nestlings, 628, 1071
Predator–prey interactions, 1060–1063, 1101, 1103–1110, 1188
and evolution, 433, 439–440, 580, 602, 606, 641, 652, 655, 656–657
in food webs, 1129–1132
plants and herbivores, 711–715
Predators, 981–982, 1101, 1103–1110
defenses against. See Predator–prey interactions; Prey defenses
food acquisition, 988–989, 1033
foraging ecology, 1060–1063
introductions of, 1177–1178
as keystone species, 1188
prey location, 1042
Preflight warmup, 788–789
Preganglionic neurons, 872
Pregnancy, 841–844
Pregnancy tests, 843
Prehensile tails, 635
Premature babies, 946
Premolars, 990–991
Premutations, 336

Preparatory reactions, in metabolic pathways, 146
Pressure, sensation of, 856–857, 888–889
Pressure bomb, 696
Pressure flow model, 699–701
Pressure potential, 107
in plants. See Turgor
Pressure pump hypothesis, 693, 694
Presynaptic cells, 866–869
Presynaptic excitation, 868
Presynaptic inhibition, 868
Presynaptic membranes, 852
Prey defenses, 462, 1033, 1103, 1107–1110
social behavior as, 1059, 1063–1064, 1077
See also Predators; Predator–prey interactions; Plants, defenses
Prey selection, 1061–1063
Prezygotic isolating mechanisms, 452–453
Primaquine, allergy to, 268
Primary active transport, 104–105
Primary auditory cortex, 877–878, 894
Primary carnivores, 1128–1129, 1132
Primary growth, 677
Primary immune response, 359
Primary meristems, 677–679
Primary mesenchyme, 386
Primary motor cortex, 857, 878
Primary oocytes, 827–828, 832, 833
Primary producers, 1129
Primary production, 1128
Primary products, 712
Primary somatosensory cortex, 856–857
Primary spermatocytes, 827–828, 831
Primary structure, of proteins, 51–52, 56
Primary visual cortex, 877–878
Primary walls, of plant cells, 673–674
Primases, in DNA replication, 253
Primates, 632–640
social behavior, 1077
Primers, 253, 297, 318–320, 326–328
Primitive streak, 387–388
Primosomes, in DNA replication, 253
Prions, 499–500
Privacy, in medicine, 344
Pro-opiomelanocortin, 809
Probability, in Mendelian genetics, 218–219, 221, 230, 432–434
Probes, 321, 323–327, 340–341
Proboscis
in hemichordates, 613

in ribbon worms, 589, *591*
Procambium
    of roots, 677–679
    of shoots, 680
Processed antigen, 371–372
Processed pseudogenes, 298
Processes, of neurons, *851–852*
Prochloron, 494
Production, *1128*, 1148–1149
    cooption by humans, *1179–1180*, 1188
    effects of climate, *1148–1149*
    efficiency, 1129–1131
Progesterone, *807*, 814–816
    in contraceptives, *838*
    in menstrual cycle, *834–835*
    in pregnancy, 835, *839*, 843, 844, 845
Programmed cell death, 391
Programs, behavioral. *See* Motor scores
Prokaryotes, 15, 87, 483, 485–486
    cell division, *211*
    cell structure, 62–66, 87, 485–486
    compared with eukaryotes, 15, 485–486
    flagella, 64–66, *65*, 86, *486*
    genetic material, *211*, 273–289
    See also Bacteria
Prolactin, 805, *806*, 808, 809, *810*, 845–846
    translational control, 305
Prolactin release-inhibiting hormone, *810*
Prolactin-releasing hormone, *810*
Proline, 49, *50*
    genetic code for, 266, *267*
    in plants, 706, 709
Prometaphase
    in meiosis, 205
    in mitosis, *198–200*
Promoters
    and cancer, 346
    in eukaryotes, 303–304
    in prokaryotes, 286–289
    in recombinant DNA technology, *745*
Pronghorn, *991*
Proofreading, of DNA, 252–254
Propane, *27–28*
Prophages, 282–284
Prophase
    in meiosis, 204–209, *231–233*
    in mitosis, 196–199
Prophylactics, *839*
Propionaldehyde, *418*
Propionic acid, *418*
Propithecus verreauxi, *634*
Proplastids, 76
Prosimians, *632*, 634
Prosome, 610, 613
Prosopis, 706–707
Prostaglandins, *807*
    in contraception, *839*

and fever, *795*
    in seminal fluid, 830, 836
Prostate gland, 830–*831*
Prostheceraeus bellostriatus, *589*
Prosthetic groups, 124, 125–*126*
Proteases, 307, 738, 993
Protein complementarity, 983–*984*
Protein kinases, 817–821
Protein starvation, *969*
Proteins, 47–55, *51–56*
    activation, 306–307
    and central dogma, 258
    and classification, 474
    in coacervate drops, *421*
    digestion of, *124*–125
    genetic coding for, 222–223, 324
    inactivation, 306–307
    in membranes *93*, *94–97*, *103–106*
    as nutrients, 981, 983–*984*, 988
    and origin of life, 417–419
    products of metabolism, 1010–*1011*
    synthesis of. *See* Translation
    targeting, 78, 80, 264–265, 306–307
    transport, 103–106
Proterozoic eon, *416*, 646
Prothonotary warbler, *1059*
Prothoracic gland, 803–804
    Prothrombin, 972–973
Protists, 15, 503–526
    characteristics, 503
    evolution, 503–504, 506, 579
    mutualisms, 77, 505–*506*, 989, 1114
    size scales, *497*
Proto-oncogenes, 196, 346–347
Protocoel, 610
Protoderm, 677–679, 680
Protomycota, 514, 516
Proton-motive force, 141–142, 154–155
Proton pumps, 141–142, 152–155
    in photophosphorylation, 172–173
    in plants, 690–692, 700–*701*
Protonemata, *551*
Protonephridia, *1012*
Protons, 19–21
Protoplast, 315
Protopteridium, *559*
Protopterus aethiopicus, *622*
Protostomes, 387, 574–*580*, *606*, 633
    evolution, 606–607
Prototheria, 630, *631*
Prototrophs, 255, *275–278*
Protozoa, *504–505*, *507*–513
    as pathogens, 508, 509–511
Provincialization, 654
Proviruses, 500
Proximal, defined, 394
Proximal convoluted tubule, 1019–*1022*

Proximodistal axis, 400
Pruning, 742
Prusiner, Stanley, 500
Pseudacris triseriata, 1105–*1106*
Pseudocoel, 576–577, 591–592
Pseudocoelomates, 576–578, 591–*593*
Pseudogenes, 298–299
Pseudohermaphrodites, 815–*816*
Pseudomonas, as denitrifiers, 728
Pseudomonas aeruginosa, 64, 211
Pseudomonas syringae, 331
Pseudomyrmex, 1115
Pseudoplasmodium, 515–*516*
Pseudopods, 509, 913–914
Psilophyta, 557–558
Psilotum, 557–558
Pterobranchs, *613*
Pterois volutans, 1107
Pterophyta. *See* Ferns
Pterygotes, 602
Puberty, 815–816, 833
Puccinia graminis, 534–536, *535*, 539
Puffballs, *539*
Puffs, polytene chromosome, *302*, 303
Pulmonary arteries, 961–962
Pulmonary circuit, 958–960, 961
Pulmonary valves, *961*
Pulmonary veins, *961*
Pulp, of teeth, *990*
Pulse, 962. *See also* Heartbeat
Punctuated equilibrium, 656–658
Punnett, Reginald C., 218, 225, 234
Punnett square, *218*
Pupae, 391–392, 602, 803–*804*, 1081
Pupils, 897–*898*
Purines
    as mutagens, 269
    in nucleic acids, 58, 245–249
    origins, 417
Purkinje fibers, 965
Purple nonsulfur bacteria, 422, 484
Purple sea urchin, *616*
Purple sulfur bacteria, 422, *484*
Pycniospores, 535–536
Pycnogonids, 597–598
Pygmy monitor lizard, 1129, *1131*
Pyloric sphincter, 995, 996
Pyramids
    of biomass, 1131–*1132*
    of energy, 1131–1132
Pyrenestes ostrinus, 436–437
Pyridoxine, *986*
Pyrimidines
    as mutagens, 269
    in nucleic acids, 58, 245–249
    origins, 417
Pyrogens, 794

Pyrrophyta, 517–*518*
Pyruvate
    allosteric controls, 158–160
    in fermentation, *143*, 144, 147, *155–156*
    in glycolysis, 142, *144–147*
Pyruvate dehydrogenase complex, 149
Pyruvate kinase, *144*
Pythons, 789

Q. *See* Ubiquinone
Q_{10} effects, *784–785*, 795
Q_{10} effects, and gas exchange, 935
Quaking aspens, *1089*
Quanta, of light, 165–168
Quarantines, 1064
Quaternary period, 646, 655
Quaternary structure, of proteins, 55–56
Queen Anne's lace, 566
Queens, in insect societies, 1075
Quills (feathers), *628*
Quinones, as plant defenses, 712

R factors, 284–285
R groups, 49, *50*
Rabbits
    digestion, 992, 1001
    effects on plants, 1123
    thermoregulation, 792
Radial cleavage, 576
Radial symmetry, *575*, 614
    in flowers, 565, *566*
Radiation spectrum, 165–*166*, 904
Radiations, evolutionary, 456–459
Radicals, 1136
Radicle, *738–739*, 753
Radio telemetry, 786
Radio waves, 165
Radiocarbon dating, 645
Radioisotopes, 21–22
    use in dating, 11, *413*, 645
Radiolarians, 505–*506*, *507*, 509, 575
Radiometric clocks, 413
Radiotherapy, for cancer, 348
Radula, 602–*603*, 992
Raffini's corpuscle, *888*
Rails (birds), 1193
Rain. *See* Precipitation
Rain shadow, 1145, *1146*, 1166
Rainbow sea star, 616
Rainbow trout, 133, *622*
Rainforests. *See* Tropical forests
Ramapithecus, 634
Ranges. *See* Geographic ranges; Home ranges
Ranvier, nodes of, *865–866*
Rapid-eye-movement (REM) sleep, 874–875
Rare alleles, 334
Rare species, 1181, 1184–1185
    population regulation, 1091, 1097
Rat snake, 438–439

Rate constants, 121–122
Ratios, in Mendelian genetics, 230
Rats
    as pests, 1096
    sexual behavior, 1034–1036, *1035*
    tails, 52
Rattlesnakes, *1033*
Ray flowers, *566*
Rayle, David, 748 Rays (fishes), 620–621, 923
    electric, 929
Reabsorption, active, 1009, 1017, *1020–1021*
Reactants, 45
Reaction center, 169, 170
Reaction rates, 119–121
    effects of enzymes, 122–123, *129–132*
    temperature sensitivity, *784–785*
Reaction wood, 684–*686*
Reactivity, of elements, *23*
Reading, *877–878*
Realized niches, 1101–*1102*
Reannealing, of nucleic acids, 294–296
Rearrangement, of genes in lymphocytes, *368–370*
Receptacle
    of flowering plants, 758–*759*
    of strawberries, *746*
Receptive field, 901–*903*
Receptor carbohydrates, 108
Receptor-mediated endocytosis, 108
Receptor potential, *883–884*
    of rod cells, 895–*896*
Receptor proteins, 107–*108*
    in cancer, 346
    in chemosensory cells, 881–882, *883–884*, *886–887*
    in endoplasmic reticulum, 264–*265*
    for hormones, 800–*801*
    for inducers, *400*
    in nervous system function, 859
    in plants, 748
    in postsynaptic membranes, 866–870
    and second messengers, 817–821
    See also Ion channels
Recessive traits, 217–218, 334
Reciprocal altruism, 1066
Reciprocal crosses, 214–222
Reciprocal translocations, *269–270*
Reclassification, 465, *468*
Recognition, of self and nonself, 359–360, *361*, 371
Recognition sites, of restriction endonucleases, *312–314*, 315
Recombinant DNA technology, 311–332

agricultural applications, 329–*331*, 713, 714, 726, *728*, *745*
applications, 328–332
and bioluminescence, *138*, *728*
medical applications, 311, 324, 328–329, 332, 336, 342–344, 772, 809
and nitrogen fixation, 329–330, 726, *728*
safety of, 328–329, 331
Recombinant phenotypes, 222
Recombination, 231–233
    in bacteria, 274–276, *278–281*
    in bacteriophages, 281–284
    in sexual reproduction, 202–203, 435, 758, 770
Recombination frequencies, 231, 232
Recovery stroke, of cilium, *911*
Rectal glands, 1018
Rectum, 992
Red abalone, 1059–*1060*
Red algae, 517, *518*, 521–523
    photosynthetic pigments, 169
Red blood cells, 356, *357*, *971–972*
    development, 393
    disorders of, 109, 256–*257*, 268, 335
    hemoglobin synthesis, 305–*306*
    and malaria, *510–511*
    membranes, 97, *104*, 109–*110*
    osmotic balance, 105–*106*
    replacement, *293*
    transport of respiratory gases, *104*, 947–951
Red deer, 1083, *1088*
Red file clam, *980*
Red-green color blindness, 336
Red kangaroo, 630, *631*
Red light, and plant development, 751–752, *767*
Red maple, 666–*667*
Red muscle, *919–920*, 948
Red-naped sapsucker, *1121*
Red phalarope, *1071*
Red-spotted newt, *1080*
Red-tailed hawk, *3*
Red tide, 517–*518*
Red-winged blackbird, 447–448, 1044, *1068–1069*
Redi, Francesco, *412*
Redox reactions, 139–*141*
    in photophosphorylation, 169–173
Reducing agents, 139–*140*
Reducing environment, of prebiotic Earth, 414–415, 417
Reduction, defined, 139–*140*
Redwoods, 561, *689*, 693
Reefs. *See* Coral reef communities
Reflexes

autonomic, 875–876, 1044
    in digestion, 1001
    spinal, 872–874, 1019
Refractory period
    of neurons, 861–864
    sexual, *835*
Regeneration, 393, *825*
Regional species richness, *1123–1124*
Regulation, defined, 782–784
Regulative development, 384–385
Regulatory enzymes, 130–132
Regulatory genes, 287–*288*
Regulatory subunits, *129–130*
Regulatory systems, 793
Reindeer, *38*, *1094*
Reindeer moss, 541
Reinertsen, R. 796
Reintroductions, of endangered species, 1181–*1184*
Reissner's membrane, *892–893*
Rejection, of transplants, 358, 371
Relatedness, genetic. *See* Genetic relatedness
Releasers, 1030–1033
Releasing hormones, 809–*810*, 813
Reliable signals, 1069–*1070*
Religion
    on age of Earth, 412–413
    evolution of, 634
    and science, 15, 462
REM sleep, *874–875*
Renal artery, 1019–*1020*
Renal corpuscles, *1015–1017*
Renal pyramids, 1019–*1020*, 1022, *1026*
Renal tubules, *1015–1017*, 1019–*1022*
Renal vein, 1019–*1020*
Renin, *975*, 1024
Reorganization stage, 1081–*1082*
Repair, of DNA, 254
Repetitive DNA, 296
Replica plating, 277–278
Replication
    of DNA, 248–254
    evolution of, 419–421
Replication fork, 251–254
Replicons, 284, 314
Repressible enzymes, 288–*289*
Repressible systems, 288–*289*
Repressors, *287–289*
Reproduction
    defined, 2–3
    energetic costs, 1081–1083
    as life history stage, 1081–*1082*
    See also Asexual reproduction; Life cycles; Sexual reproduction
Reproductive isolation, 449, 452–453, 829–830
Reproductive rates, *14*, 628
Reproductive success, 435

and social behavior, 1058–1059, 1064–*1066*, 1067, 1069–1070, 1072–1077
Reproductive value, 1083
Reptiles, 625–627
    circulatory systems, 625, *959*, 960
    phylogeny, 465, *468*
    water conservation, 625, 1018–1019
Reserves, wildlife, 1190–1192
Residual volume, *943–944*
Residue groups, 49, *50*
Resistance factors, in bacteria, 284–285
Resistance vessels, *966–967*
Resistant stages. *See* Resting stages
Resolution, of microscopes, 66–68
Resolution phase (sexual), *835*
Resolving power, 66–68
Resource defense polygyny, 1069
Resources
    competition for, *1110*, 1111
    defined, 1101–*1102*
    and ecological niches, 1101–*1102*
    and population regulation, 1089–1092, 1096–1097
    and social behavior, *1076–1077*
    and species richness, 1122–1123
Respiration, defined, 933. *See also* Cellular respiration; Gas exchange; Ventilation
Respiratory chain, *143–144*, 149–155
    allosteric controls, 157–160
    evolution of, 422
Respiratory distress syndrome, 946
Respiratory systems, *781*, *933–953*
Respiratory uncouplers, 154
Resting potential, 858–863, *859*
Resting stages, 1081–*1082*
    in bacteria, 488, 494
    seeds as, 561, *705*
    in slime molds, 515
    in tardigrades, 596
Restoration ecology, *1192–1193*
Restricted transduction, 282–284
Restriction digests, *340–341*
Restriction endonucleases, *312–314*, 323–324
    in gene mapping, 340–342
Restriction fragment length polymorphisms (RFLPs), 340–342, 344
Restriction mapping, 323–324
Reticular system (brain), *855–856*, 893
Reticulum (digestive), *1000*
Retina, 895, 897–903

Retinal, *894–895*, 900
  in halobacteria, *174*
Retinoblastoma, 346
Retinoic acid, *44*, *401*
Retinol. *See* Vitamin A
Retinula cells, 897
Retroviruses, 259, 298
  in AIDS, *375–377*
  and cancer, *346–348*
Reverse reactions, 115
Reverse transcriptase, 259, 298
  in AIDS, *375–377*
  and cancer, *346–348*
  in recombinant DNA technology, *318–320*
Reversible reactions, 29
RFLPs, 340–342, 344
Rh factor, 225
Rhabdom, *897–898*
Rhabdopleura, *613*
Rhagoletis, *451–452*
Rheumatoid arthritis, 374–375
Rhincodon typhus, 621
Rhinoceroses, 1177
Rhizidiomyces apophysatus, 516
Rhizobium, 491, 1114
  nitrogen fixation by, *724–725, 726–727*
Rhizoids
  of *Acetabularia*, *72–73*
  of bryophytes, 551
  of fungi, 529
  of Protomycota, 516
  of psilopsids, *554*
Rhizomes
  of psilopsids, *554*
  as storage organs, 571
  in vegetative reproduction, 770
Rhizopoda, 505, *507*, *508–509*
Rhizopus stolonifer, *536, 537*
Rhodinus, *802–804*
Rhodophyta, 517, *518, 521–523*
Rhodopsin, *894–897*
Rhynchocoel, 576, 589, *591*
Rhynia, 554
Rhyniophyta, *554–555, 557–558*
Rhythm method, *839*
Rhythms, biological. *See* Circadian rhythms; Circannual rhythms
Ribbon worms, 576, *578, 579, 589, 591, 606, 633*
Riboflavin
  in carrier compounds, 141
  nutritional requirements, *985–986*
Ribonuclease P, 420
Ribonucleases, in seedlings, 738
Ribonucleic acid. *See* RNA
Ribonucleotides, 58
Ribose, *46*
  in RNA, 55, *57, 58, 246–247, 419*
Ribosomal RNA. *See* rRNA
Ribosomes
  in endoplasmic reticulum, *78–79*

  in eukaryotes, 69, *70–71, 73–74*
  in oocytes, 302
  in plasma cells, 361
  in prokaryotes, *64*
  in translation, 260–265, *286*
Ribozymes, 133, *415, 419–421, 420*
Ribulose bisphosphate, 180–183
Ribulose monophosphate, 180–183
Rice, *667–668*, 987
  bakanae disease, 739–740
Rice rats, 1178
Richness, of species. *See* Species richness
Rickets, 43, 988
Rickettsias, *493*
Right brain, 877
Rigor mortis, 915
Ring canals, *615–617*
Ring form, of glucose, *45*
Ring-necked snake, *626*
Ring structure, of steroids, 44
Ringworm, 540
Risk cost, 1058
Rivers
  in biogeochemical cycles, 1134, 1137
  "blackwater," *1109–1110*
  ecosystems, *1171*
RNA
  antisense, *348–349*
  catalysis by, 419–421
  chemical structure, 55–58, 246
  and classification, 474–476
  evolution of, 419–421
  hybridization, 294–*295*
  and origin of life, *415, 419–421*
  splicing, 299–301
  synthesis of. *See* Transcription
  translation of. *See* Translation
  in viruses, 258–259, *498–499*
  See also mRNA; rRNA; tRNA
RNA polymerase, 259–260, 420
  and promoters, 286–287, 289, *303–304*
  in transcription, *303*
RNA primers, *253*, 297
RNA viruses. *See* Retroviruses
Roaches, 602
Robber flies, *601*
Robins, *629*, 1031
Rock dudleya, *698*
Rocks
  dating, *413–414*, 509
  sedimentary. *See* Sedimentary rocks
Rocky Mountain spotted fever, 493
Rod cells, *895–897, 899–901*
Rod-shaped bacteria. *See* Bacilli Rodents, *631, 632*

  in ecosystems, 1167
  interspecific competition, *1110–1111*
  teeth, *991*
Rods, resistance to forces, 924
Root apical meristem, *671, 677–679*
Root cap, *671, 677–679*
Root hairs, *678–679*
Root pressure, *693–696*
Root suckers, 770
Root systems, *669–670*
Roots, *669–671, 677–680*
  adaptations to dry environments, *706–707*, 709
  embryonic, *761–762*
  evolution of, *554–555*
  of nitrogen-fixing plants, *724–725, 726–727*
  secondary growth in, *681–683*
  uptake of water, *692–693, 695*
Rosa gallica, *466–467*
Roses, 570
  classification, 464, *466–467*
Rosy periwinkle, *1176*
Rotifers, *578, 579, 592–593, 606, 633*
Rotman, Raquel, 281
Rough endoplasmic reticulum, *70–71, 78–79*
  and membrane synthesis, *109–111*
  in plasma cells, 361, *363*
  and protein synthesis, *264–265*
Round dance, *1042–1043*
Round window, *893*
Roundworms. *See* Nematodes
Rous sarcoma, 259
rRNA, 260, 261
  in archaebacteria, 489
  catalytic, 299
  and classification, 476
  gene family, 299
  transcription, *302–303*
Ru486, *839*
Rubber tapping, *1191–1192*
Rubisco
  in C_4 plants, *183–185*
  in Calvin–Benson cycle, *180–183*
  in photorespiration, *181–183*
RuBP, *180–183*
RuBP carboxylase. *See* Rubisco
Ruddy duck, *474*
Rumen, *1000–1001*
RuMP, *180–183*
Runners, of plants, *671–672*
Running speed, *627*
  Rupturing, of cells, 87
Russell, Liane, 301
Rusts, 531, *534–536, 535*, 539

S-cyclin, *197*
S phase, 194, *196–197*
S-shaped growth curve, *390–391, 1090*
Saber-tooth cat, *991*

Sac fungi, *529, 537–539*
Saccharomyces cerevisiae, 314, 538
Saccoglossus kowaleski, *613*
Saddle joint, *927*
Safranine, in Gram stain, *487*
Sake, 540
Sakmann, Bert, 860
Salamanders, 622–624
  fertilization, 829
  hybrids, 453–454
  in old-growth forests, 1189
  phylogeny, *470–471*
  population structure, *1080*
  tail autotomy, 462, 467, *469*
Saline environments, *1124*
  adaptations in animals, *1009–1011*
  adaptations in plants, *707–709*
  bacteria in, 489, 490
Salinity, *434*
Salinization, of cropland, 708
Saliva
  lysozyme in, 354
  and starch digestion, *120–122*, 995–996
Salivary glands, 802, 995
Salivation reflex, 1001
Salix, *714*
Salmo gairdneri, 622
Salmon, 621, 829, 1082, *1170*
Salmonella, *492*
Salmonids, *622*
Salsifies, *451–452*
Salt balance, 1003–1010, *1011*, 1018
Salt concentration, in plants, 708
Salt crystals, *25–26*
Salt glands
  of animals, 1010, *1011*
  of plants, *708*
Salt licks, 984, 1010
Salt marshes, adaptations by plants, 709
Salt water. *See* Marine ecosystems; Saline environments
Saltatory conduction, *865–866*
Salviniales, 559
Salycilic acid, *714*
Samara, *667*, 762
Sanctuaries, *1190–1192*
Sand, in soils, *721*
Sand fleas, 600
Sandpipers, *1071*
Sanger, Frederick, *265*, 324
Sap
  in phloem, *698–700*
  in xylem, *693–696, 701*
Sapolsky, Robert, *815*
Saprobes, *516–517, 528, 529, 530*
Saprolegnia, *516–517*
Saprophytes, 979
Sapsuckers, *1121*
Sapwood, 683
Sarcomas, 345
Sarcomeres, *915–916*
Sarcoplasm, *917–918*

Sarcoplasmic reticulum, 917–918
Sarcosine, 418
Sargassum, 521
Sarracenia, 717, 730
Satin bowerbird, 1067, 1069
Saturated fatty acids, 40–42, 41
Saturated hydrocarbons, 33–34
Saturation, 122–123
Savannas, 456–457, 1076, 1167, 1168
Sawflies, 602
　larvae, 115
Scale insects, 532
Scales, of fishes, 621
Scallops, 225, 239, 604, 605, 922
Scanning, for predators, 1059, 1065–1066
Scanning electron microscopy, 68
Scarab beetle, 788–789
Scarification, 736
Scarlet gilia, 711–712
Scarlet maple, 666–667
Schaal, Barbara A., 763
Schally, Andrew, 809
Schistosomiasis, 366
Scholander, Per, 696
Schools, of fishes, 621
Schwann, Theodor, 12
Schwann cells, 865
Scientific method, 6–9
Scions, 771
Scissor-tailed flycatcher, 1090–1091
Sclera, 897–898
Sclereids, 674–675
Sclerenchyma, 674–675, 684
Scleria, 436–437 Sclerotium, 515
Scolopendra heros, 600
Scorpions, 598
Scots pine, 1120
Scouring rushes. See Horsetails
Scrapie-associated fibrils, 499–500
Scrotum, 816, 830–831, 835
Scrub jays, 1072–1073
Scurvy, 985, 987
Scyphozoans. See Jellyfishes
Sea anemones, 575, 583, 585–587
　hydrostatic skeleton, 921
　mutualisms, 76
Sea butterflies, 604
Sea cucumbers, 616–617
Sea daisies, 617
Sea gooseberries. See Ctenophores
Sea gulls. See Gulls
Sea hare, 875–876, 1039, 1046
Sea lettuce, 523–525, 524
Sea level changes, 646–647, 652, 654
Sea lilies, 615
Sea nettle jellyfish, 587
Sea otters, 991
Sea palms, 521, 522

Sea slugs, 875–876, 1039, 1046
Sea spiders, 597–598
Sea squirts, 473, 618, 1114
Sea stars, 615–617
　as keystone species, 1188
　nervous systems, 850
　as predators, 988–989, 1112–1113, 1188
　regeneration, 825
Sea turtles, 625, 626
Sea urchins, 616–617
　embryonic development, 201, 382, 383, 385, 386–387, 394–395, 577
　fertilization, 836–837
　genes, 302, 305
　parasites of, 589
　as predators, 522
Sea walnuts. See Ctenophores
Sea wasp, 583
Seals, 975–976
　male-male competition, 1057, 1058
　population bottlenecks, 433
Searching behavior, 1033
Seasonal dormancy. See Winter dormancy
Seasonal environments
　adaptations in animals, 38, 785, 796–797, 810–811, 1047–1052, 1093
　adaptations in plants, 750, 751, 766, 1093
　death rates in, 1091, 1093
　deciduous forests in, 1164–1165
　See also Thermoregulation
Seasons, and climate, 1146
Seaweeds, 520–526
Second filial generation (F₂), 215
Second-hand smoke, 349
Second law of thermodynamics, 115, 118
Second messengers
　in chemosensation, 886–887
　of hormones, 817–821
　at slow synapses, 868
Secondary active transport, 105, 106, 123, 700–701
Secondary carnivores, 1128–1129, 1132
Secondary compounds, 712–713, 1109
Secondary growth, 681–683
　in gymnosperms, 561–562
Secondary immune response, 359, 365
Secondary mesenchyme, 386
Secondary oocytes, 827–828, 832, 833
Secondary products, 712–713, 1109
Secondary sexual characteristics, 814, 815–816
Secondary spermatocytes, 827–828, 831
Secondary structure, of proteins, 52–54, 56
Secondary tissues, 681–683

Secondary walls, of plant cells, 673–674
Secretin, 807, 1001–1002
Secretion, active, 1009, 1017, 1020–1021
Sedges, 436–437
Sedimentary rocks
　in carbon cycle, 1137–1138
　dating, 413, 509
　fossils in, 645, 648
Sedimentation, in centrifuge, 87–88
Sediments, in biogeochemical cycles, 1134, 1136, 1137–1138
Seed coat, 561, 736, 760–762
Seed crop, 1082
Seed dispersal, 737, 762–764, 1115–1117
　by animals, 563, 1101, 1115–1117, 1188
　by birds, 629, 1116, 1117
　and habitat restoration, 1192
Seed ferns, 562
Seed leaves. See Cotyledons
Seed plants, 547, 559–571
Seed predators, 1100, 1116
Seedless grapes, 741, 746
Seedless plants, 552–559
Seedlings
　development of, 735–739, 750–754, 760–762
　phototropism, 742–744
Seeds, 561, 568–569
　artificial, 772
　in desert ecosystems, 1110–1111, 1166, 1167
　development, 761–762, 763
　dormancy, 734, 735–737, 762
　and fruit development, 741
　germination, 734, 735–739, 751–752, 760–762, 1093
　as resting structures, 561, 705
Segment-polarity genes, 401, 403
Segmentation, 401–405, 593
　abnormal, 395–396, 401–405
　in annelids, 594–595
　in crustaceans, 598
　and locomotion, 921
　in unirames, 600
Segmentation genes, 401–405
Segmented worms. See Annelids
Segregation, Mendel's law of, 217–218, 220
Sei whale, 1096
Selaginella, 557
Selective herbicides, 744, 745
Selective permeability, of membranes, 92
Selective transcription, 303–305
Selective uptake, in plants, 718
Selenium, 985
Self, recognition of, 359–360, 361, 371

Self-antigens, 361, 371
Self-fertilization, 447
Self-pollination, 215, 434, 757, 758
　in Mendel's experiments, 218–221
　and speciation, 451–452
Self-tolerance, 373–374
Selfish acts, 1064–1065
Semelparous organisms, 1082
Semen, 830
Semicircular canals, 890–891
Semiconservative replication, 248–250
Seminal fluid, 830
Seminal vesicles, 830–831
Seminiferous tubules, 827–828, 830–831
Senescence, in plants, 329, 735–736, 749–750
Sensation, 882–884
　and brain, 856–857
Sensitization, 875–876
Sensors, 881–882
Sensory circuits, 882
Sensory organs, 849, 882–883
Sepals, 565, 758–759
Separation, of DNA strands. See Denaturation
Septa, in annelids, 921
Septate hyphae, 539
Septic shock, 27
Septobasidium, 532
Sequencing, of DNA, 323–327, 332
Sequential hermaphroditism, 828
Sequoia sempervirens, 693
Sequoiadendron giganteum, 562
Serine, 50, 127, 267
Serosa, 994–995
Serotonin, 869
Serpentine soils, 710–711, 1124
Sertoli cells, 827–828, 831
Sessile organisms, 580, 606–607
　external fertilization, 606, 829
　nutrition, 718, 989, 1103, 1104
Set point, 782–783
　thermoregulatory, 793–794, 795, 796
Setae. See Bristles
Settling, 434, 1059–1060. See also Colonization
Seven-gill shark, 620
Sewage, and lake eutrophication, 1140–1142
Sex, defined, 202
Sex cells. See Gametes
Sex chromosomes, 204, 226–230, 301–302
　and genetic disorders, 335–339
　interaction with hormones, 815
Sex determination, 226–230
　in Hymenoptera, 227, 826, 1074–1075
　role of hormones, 814–816

Sex differences, in death rates, 1088
Sex-linked inheritance, 227–230
   and genetic disorders, 335–339
Sex pheromones. See Courtship pheromones
Sex roles
   and hormones, 1035–1036
   in parental care, 1071
   reversal, 1071
Sex steroids, 807, 813, 814–817
   and sexual behavior, 1034–1036, 1035
   See also specific hormones
Sex types, 828–830. See also Mating types
Sexduction, in bacteria, 280–281
Sexual behavior, 828–830
   female receptivity, 833
   positive feedback in, 782
   role of hormones, 1034–1036, 1035
   See also Copulation; Courtship displays
Sexual dimorphism, 1057, 1072–1073
Sexual reproduction, 294
   in flowering plants, 757–764
   in fungi, 528, 532–536
   and genetic variation, 202–203, 435, 758, 826
   organ systems in animals, 780, 826, 830–833
   and ploidy levels, 202–203
   and social behavior, 1057, 1058–1059
   See also Conjugation; Recombination
Sexual responses, in humans, 833, 835
Sexual selection, 1071–1073
Sexually transmitted diseases, 354, 376–378, 839. See also AIDS
Shade leaves, 428
Shading, of competing plants, 672, 718
Shadowcasting technique, 97–98
Shallow torpor, 795–796
Shape. See Body shape
Sharks, 620–621, 923, 979
   digestive system, 992–993
Sharp-pain sensation, 866
Sheep, 631, 1087–1088
Sheep liver fluke, 589
Shellfish. See Mollusks
Shells
   of brachiopods, 612–613
   in carbon cycle, 1137–1138
   of eggs, 625, 839
   of mollusks, 602–605, 922
Shells (electron), 22–23
Shigella, 284, 492
Shipworms, 1000
Shivering, 789, 791
Shoemaker, Vaughan, 1018

Shoot apical meristem, 672, 678, 680–681
   in flowering, 764
Shoot systems, 669
Shoots, 669, 671–672, 680–681
   development, 753–754
   elongation, 741
   embryonic, 761–762
Short-day plants, 765–767
Short-term memory, 856, 876
Shotgunning, 318–319
Shrews, 630, 632
   colonization by, 1159–1160
Shrimps, 600, 1114, 1171
Siamese cats, 224, 235–236
Siblings, helping behavior, 1065, 1072–1075
Sickle-cell anemia, 256–257, 268, 335, 337, 340–341
Side chains
   of amino acids, 49
   in cytochrome c, 440–442
   and protein structure, 52
Sieve plates, 676, 700
Sieve tube elements, 676
Sieve tubes, 698–701
Sifakas, 634
Sign stimuli, 1030–1033
Signals, 1040
Signal sequences, 78, 80, 264–265
Silicates, in Earth's mantle, 414
Silicon, in diatoms, 519
Silk
   chemical structure, 52
   of spiders, 598
Silkworm moth, 803–804
   gas exchange, 938
   pheromones, 884–885
Silurian period, 554, 646, 653
Silverfish, 1000
Silverswords, 457–458
Simple cells, of visual cortex, 902–903
Simple development, in insects, 600
Simple diffusion, 101–103
Simple fruits, 567–568
Simple leaves, 555, 672–673
Simple lipids. See Triglycerides
Simple sugars. See Monosaccharides
Simple tissues, 672, 677
Simultaneous hermaphroditism, 828
Sindbis virus, 498
Single bonds, 24–25
Single-copy sequence, 296
Single-gene disorders, 334, 337
Sinks, and sources, 677, 698–701
Sinoatrial node, 963–965
Siphon, of cephalopods, 603, 605
Siphonaptera, 602
Siphonophores, 585, 594
Sister species, 1156
Size, of organisms
   evolutionary trends, 641

   and reproductive success, 1072–1073
   and social systems, 1076–1077
   See also Surface-to-volume ratio
Size effects, 1188
Size scales, 9–10, 61, 63, 497
   of birds, 629
   of mammals, 630
   of microorganisms, 497
Skates, 620–621
Skeletal muscles, 914, 915–920
   in venous return, 970–971
Skeletal systems, 780, 920–928
   of birds, 628
   of mammals, 629–630
   See also Bones Skin, 780
   of amphibians, 622, 624, 1018
   as defense against pathogens, 354, 358
   pigmentation in humans, 235–236, 986
   of reptiles, 625, 1018
   sensors in, 882, 888–889
   and thermoregulation, 787–788, 792, 793–794
   vitamin D synthesis in, 986
Skin cancer, 345
Skin grafts, 361, 362
Skoog, Folke, 749
Skulls, 618–619, 629, 924
Skunks, 929
Sleep, 855–856, 874–875
Sleep movements, in plants, 767
Sleeping sickness, 507–508
Slender loris, 634
Sliding-filament theory, 915–918
Slime, in bacterial locomotion, 490
Slime molds, 514–516, 1040
Slow block to polyspermy, 837
Slow synapses, 868
Slow-twitch fibers, 919–920, 948
Slow-wave sleep, 874–875
Slugs, 604
Small intestine, 994, 996–999
Small populations
   genetic drift, 432–433
   risk of extinction, 1181
Smallpox, 359
Smell. See Olfaction
Smokestack gases, and acid precipitation, 1142
Smoking, 345, 349, 870, 946, 967, 1003
Smooth endoplasmic reticulum, 71, 78
Smooth muscle, 914–915
   in blood vessels, 966–967, 971, 974
   in gut, 993–995
Smut fungi, 529, 539
Snails, 603, 604
   digestive systems, 992
Snakes, 625, 626
   phylogeny, 465, 468
   as predators, 988, 1033

   sensation, 888, 904–905
   step clines, 438–439
Snapdragons, 222–223, 570
Sneezing, 354
Snow. See Precipitation
Snow goose, 454
Snowshoe hare, 1104–1105
Snowy owl, 1104 snRNA, 299–301
snRNPs, 299–301
Social behavior, 1058–1059, 1062–1066
   costs and benefits, 1058–1059, 1064
   evolution, 1072–1076
   evolution in humans, 638
   and resource availability, 1076–1077
   types, 1064–1066
Social insects, 1014–1016
Sockeye salmon, 1170
Sodium, nutritional requirements, 984–985, 1010
Sodium channels
   in muscle fibers, 917
   in neuron function, 859–865
   in photoreceptors, 895–897, 896
   in postsynaptic membranes, 866–867
   in sensory cells, 881–882, 883–884, 886–887
Sodium chloride, 25–26
Sodium cotransport, 998
Sodium ions (Na+), 25–26
   active transport of, 104–105
   in blocks to polyspermy, 837
   in nervous system function, 858–866
Sodium-potassium pump, 104–105, 123
   in nervous system function, 858–866, 859
Soft-shelled crabs, 596, 922
Soil microorganisms, 1121–1122
Soil profile, 721–722
Soils, 721–724, 1136
   effects of plants on, 723–724, 1119
   formation of, 542, 723–724, 1136
   heavy metals in, 709
   and parapatric speciation, 449, 451, 452
   serpentine, 710–711, 1124
   and species richness, 1124
   tropical, 724, 1144–1145, 1167, 1169
Solar energy, 162, 1127–1128, 1145–1149
Solar system, formation of, 414, 415
Soldierfish, 1103
Solicitation, of food, 1031–1032
Solidago, 1092
Solvents, 28, 31
Somatic-cell genetics, 338–340

Somatomedins, 808
Somatosensory cortex, 856–857
Somatostatin, *807, 810, 813*
  and recombinant DNA technology, 311, *318,* 320
Somites, *389–390*
Song sparrows, 1033, *1068*
Song-spread, 1044, *1068*
Songbirds, 628, *629*
  endangered, *1180*
  migration, *1047–1052*
  song learning, 1033–1034, *1036–1037*
  species richness, *1123,* 1156–1157
  territoriality, *1068–1069*
Songs, of whales, 1042
Sonoran Desert, *1110–1111*
Soredia, *542*
Sori, *559*
Sound, 890–894
  as communication, 929, 1042
  in echolocation, 881, 905
  navigation by, 1051
Sources, and sinks, 677, *698–701*
Southern, E. M., 321
Southern beech, *1163–1164*
Southern blotting, 321–322, *340–341*
Sowbugs, *599–600*
Soy sauce, 540
Soybeans, *668, 728*
Space scales, in biogeography, 1153–1154
Spacers, on DNA, 303
Spacing, among organisms, *1084–1085*
Spallanzani, Lazzaro, *412*
Sparrows
  microevolution, 657–658
  song learning, *1033–1034*
  territories, *1068*
Spatial patchiness, 1106
Spatial summation, 868–869, *917–919*
Spawning, 621, 1173
Spaying, *1035–1036*
Special homologous traits, 466–467, 469
Speciation, 446–459, *1156–1157*
  defined, 448
  and historical biogeography, *1154–1157*
  mechanisms of, 448–452
  rates of, 656–658
Species
  defined, *4–5,* 446–447
  lifespans of, 649
  names of, *463,* 464
  numbers of, 648–649
Species diversity, 1122, 1176. *See also* Extinctions
Species pool, 1158
Species richness, 1122–1124, *1156–1160*
  in agricultural ecosystems, 1178–1179

over evolutionary time, 648–649, *651*
  and rarity, 1184
  in tropical ecosystems, *446,* 1187
Species-specific behaviors, 1030, *1037–1038*
Species-specific coevolution, 1116–1117
Specific defense mechanisms. *See* Immune system
Specific heat, 1147
Spectrin, *109–110*
Spectrum, electromagnetic, *165–166*
Speech, brain function in, *877–878*
Spemann, Hans, 395, 397–398, 786–787
Sperm, *202–203,* 824, *826, 830–831*
  fertilization, 836–837
  flagella, 911
  and parental investment, 1066–1067, 1071
  production of. *See* Spermatogenesis
Sperm competition, 829–830, 1067–1068, 1070
Sperm nuclei, 758–761
Sperm whale, *1096*
Spermatids, 827–828, *830–831*
Spermatocytes, 827–828, 831
Spermatogenesis, 826–828, *830–831*
Spermatogonia, 826–828, *830–831*
Spermatophores, 829, 1070
Spermicides, *839*
Sperry, Roger, 877–878
Sphagnum, 552
Sphenodon punctatus, 626
Sphenodontida, 625, *626*
Sphenopsids. *See* Horsetails
Spherical symmetry, 575
Sphincters
  in digestive system, 995, 996
  urinary, 1019
Sphinx moth, *789*
Sphygmomanometer, 962–963
Sphyrapicus nuchalis, 1121
Spicules, *581, 582,* 602
Spiders, *598,* 909
  fertilization, 829
  prey capture, 988, 1103
  tactile communication, 1030, 1042
  web building, *1029–1030*
Spike, of action potential, 861
Spikes, (flowers), 565–566
Spinal circuits, 854–855, *872–874*
Spinal cord, *619,* 850, 854–855, *872–873*
  development, 853–854
Spinal muscular atrophy, *334*
Spinal nerves, 852, 853, *854–855*
Spinal reflexes, 872–874, 1019
Spindle diagram, *651*
Spindles

in meiosis, 205–207
  in mitosis, 197–201, 202, 912
  muscle, 873, 889 Spines of cacti, 706
  as defenses, 656–657, *1107*
  evolution, in plants, 467, *468*
  of sea urchins, 617
Spiny anteaters, 630, *631,* 839
Spiny-cheeked honeyeater, *980*
Spiny lobster, 599
Spiracles, 937–938
Spiral bacteria, 487, 491, *492*
Spiral cleavage, *578*
Spirilla. *See* Spirochetes
Spirillum, 486, 491, *492*
Spirobranchus grandis, 595
Spirochetes, 66, 490–491
Spiteful acts, *1064–1065*
Spleen, 364–365
Spliceosomes, 299–301
Splicing
  of DNA, *313–314*
  in lymphocytes, *369*
  by ribozymes, 419–420
  of RNA, 299–301
Split-brain patients, 877–878
Sponges, 575, 576, 578, 579, 580–582, 606, 633
  budding, 824
  evolution, 508
  exchange with environment, 777, *934*
Sponges (contraceptive), *839*
Spongin, 581
Spongy mesophyll, 683–685
  in C3 plants, *184*
Spontaneous generation, 411–412
Spontaneous reactions, 115–118
Sporangia
  of bryophytes, 549
  of ferns, *559*
  of fungi, 532
  of plants, 546
  of Protomycota, 516
  of psilopsids, *554*
  of slime molds, *514,* 515
  of zygomycetes, *536*
Sporangiophores
  of slime molds, *514,* 515
  of zygomycetes, 536–537
Spores, *202,* 556
  of fungi, 532–536, *537,* 539–540
  of *Neurospora,* 230–233
  as resting structures, *488,* 490, 494
  of seedless plants, 554, *556*
Sporocytes, 506, 523–525
  Sporophytes
  of algae, 523–525
  evolution of, 563
  of ferns, 559–560
  of nonvascular plants, 549
  of plants, 546
  of seed plants, 559–560
  of seedless plants, 552, 556

Sporozoites, of *Plasmodium, 510*
Sports, and muscle types, *919–920*
Spotted owl, *1189*
Spreo supurbus, 629
Spriggina floundersi, 651
Spring wheat, 769
Spyrometer, *943*
Squamata, 625, *626*
Squid, 604
  eyes, *898*
  fertilization, 829
  nervous system, 850, 860, *864*
Squirrels, instinctive behavior, 1030
Stabilizing selection, *435–436*
Staghorn coral, *586*
Stahl, Franklin, 250–251
Staining, of bacteria, 487
Stamens, 565–567, 758–759
Standing biomass, 1131
Stanley, Wendell, 497
Stapes, 891–892
Staphylococci, 494, *496*
Starch, 46–48
  in algae, 523
  in C4 plants, *183–185*
  digestion of, 157, 995, *997*
  formation in coacervate drops, *421*
  in seeds, 738
  storage, in plants, 139, 546, 547, 674
Starfish. *See* Sea stars
Starling, Ernest H., 1001
Starlings, 629, *1050*
Stars, navigation by, *1051–1052*
Start codons, 267
Starvation, *969,* 981–982
Stasis, evolutionary, 456, 458, 654
Statistical methods, in Mendelian genetics, 230
Statocysts, 889–890
Statoliths, *889–890*
Steady state, 778, 782–784
  in citric acid cycle, 149
Stearic acid, 40–41
Stele, *678–680,* 692–693
Steller's sea cow, 1177 Stem cells, 342, *972*
Stems, 669–670, 671–672, *680–683*
  elongation, 741
  secondary growth, 681–683
  in vegetative reproduction, 770
  See also Shoots
Stenotaphrum secundatum, 530
Step clines, 438–439
Stereocilia, 888–891, 913
Stereotypic behaviors, 1030–1033, 1034, 1039
Sterile classes, *1074–1075*
Sterile hybrids, 453
Sterilization, as contraception, *838–839*

Sternum, of birds, 628
Steroids, 813–817
   abuse of, 817
   of adrenal glands, 813–815
   chemical structure, 43–44, 814, 983
   and gene transcription, 303, 821
   as plant defenses, 712
   See also Sex steroids
Sterol, 814
Sticklebacks, 439–440, 656–657, 1044
Sticky ends, 313–314
Stigmas, 565, 758–759
Stigmatella aurantiaca, 490
Stimuli, and behavior, 1030–1033
Stock (graft), 771
Stolons, 770
Stomach, 992, 995–998, 1002
   defenses against pathogens, 354
   eversion in sea stars, 617, 988, 989
   of ruminants, 1000–1001
   tissue layers, 779
Stomata, 163, 684–685, 694–695, 696–698
   in bryophytes, 552
   in C₄ plants, 183–185
   in CAM plants, 184
Stomatal crypts, 705, 709
Stomatal cycles, 696–698
Stone cells, 674
Stone flies, 602
Stoneworts, 547
Stop codons, 263, 267. See also Chain terminators
Straight-chain form, of glucose, 45
Strasburger, Eduard, 693, 694
Strata, 411, 413
Stratification, and seed germination, 736, 737
Stratosphere, 1135
Strawberries, 671
   fruit formation, 746
   vegetative reproduction, 758, 770
Streptanthus glandulosis, 710–711
Streptococci, 494–496
Streptococcus pneumoniae, 242
Streptomyces, 496
Stress, effects of, 814–815
Stress hormones
   in animals, 801, 813–815
   in plants, 751
Stress response. See Fight-or-flight response
Stretch receptors
   in blood vessels, 1024
   in muscles, 873, 889
Stretch sensors, 883–884, 888–889
   in blood vessels, 974–975
   in lungs, 951
   in muscles, 883–884, 888, 889
Striated muscle. See Skeletal muscles

Strict halophiles, 485–486
Strip-cutting, 1179
Strobili
   of seed plants, 560
   of seedless plants, 556–557
Strokes (cardiovascular), 967
Stroma, 75–76, 173, 180
Strongylocentrotus purpuratus, 616
Structural environment, plants as, 1118–1120, 1123
Structural formulas, of atoms, 26
Structural genes, 285–287
Structure, of populations, 429–431, 1084–1086
Stubby rose anemone, 587
Sturgeon, 622
Style, 565, 758–759
Suberin
   in Casparian strips, 692
   in cork, 673, 683, 713
   in roots, 679
Submucosa, 993–995, 998
Subpopulations, 428, 446, 1084
   and speciation, 448–449
Substrate, 121
Substrate-level phosphorylation, 141, 146
Succession, 571, 724–725, 1118–1120, 1122
Succinate, 123–124, 127–128, 418
   in citric acid cycle, 148–149
   in respiratory chain, 151–152
Succinate dehydrogenase, 123–124, 152
   inhibitors, 127–128
Succinate-Q reductase, 151–152
Succinyl coenzyme A
   in chlorophyll, 157
   in citric acid cycle, 148, 159
   Succulents, 672, 697–698, 709, 1166
Sucrase, 997–998
Sucrose, 26, 46–47
Suctorians, 511
Sugar maples, 701
Sugar phosphates, 47, 49
Sugars
   chemical structure, 33–35, 45–47
   and origin of life, 417–419
   solubility, 31
Sulci, 856–857
Sulfate ions (SO₄²⁻), 25
   in bacterial metabolism, 422
   in early oceans, 422
   as plant nutrient, 729–730
Sulfhydryl groups, 34–35
Sulfolobus, 422, 489
Sulfur
   as animal nutrient, 984–985
   chemical structure, 25
   as plant nutrient, 718, 719, 722, 723, 729–730
   in proteins, 243, 244

Sulfur bacteria, 422, 484, 729–730, 1148–1149
   mutualisms with pogonophores, 593–594
Sulfur butterfly, 470–471
Sulfur cycle, 729–730, 1139–1140
Sulfur dioxide, 1139–1140
Sulfuric acid, in acid precipitation, 1142
Summation, 868–869, 917–919
Summit metabolism, 791
Sun, navigation by, 1050–1052. See also Solar energy
Sunblocks, 349
Sundews, 730
Sunfishes, 1061–1063
Sunflowers, 475, 566
Sunlight
   and cancer, 345, 349
   and vitamin D synthesis, 986
Superb starling, 629
Superficial cleavage, 383
Superior vena cava, 961
Supernormal releasers, 1031–1032
Superposition, and dating, 413
Support
   in animals, 10, 780
   in plants, 547, 548, 552, 684–686, 689
Suppressor T cells (T_S), 361, 371
Suprachiasmatic nuclei, 1045–1047
Surface tension, 31–32
   effect on lungs, 944, 946
Surface-to-volume ratio, 9–10
   and absorptive nutrition, 530
   and gas exchange, 578–580
   and thermoregulation, 9–10, 792, 795–796
Surfactant, in lungs, 944, 946
Surgery, for cancer, 348
Surround, of receptive field, 901–902
Survivorship, 1087–1088
Suspension feeders. See Filter feeders
Suspensions, 69
Suspensors, 761
Suspensory ligaments, of eyes, 899
Sutherland, E. W., 818
Swallowing, 995
Swallows, 1069–1070
Swallowtails, 470–471
Swamps, adaptations by plants to, 706–707, 708
Swarm cells, 515
Sweat glands, 802
Sweating, 29, 791, 793
Sweet gale, 725
Sweet peas, 234–235
Sweetlips, 1114
Swim bladders, 621, 622, 640–641, 943
Switches, in induction, 399–400

Sycamores, 455–456
Symbiosis. See Commensalisms; Mutualisms
Symmetry, 575, 588, 614
Sympathetic nervous system, 870–872
   and heartbeat, 920, 965
   regulation of blood flow, 974–975
Sympatric speciation, 449–452, 455
Symplast, 692–693, 700
Symports, 103, 104, 691, 700–701
Synapses, 851, 852, 866–870
   in learning, 875–876
   neuromuscular, 866–868, 917–918
Synapsis, of homologous chromosomes, 204–205, 208–209
Synaptic cleft, 866–870
Synconia, 1100
Syndesmis, 589
Synergids, 758–759, 760–761, 762
Syngamy, 546
Syphilis, 66, 491, 839
Syrinx, 1037
Systematics, 462–477
   in biogeography, 1153–1154, 1156–1157
Systemic circuits, 958–960, 961
Systole, 962–963
Szalay, Aladar, 728
Szent-Györgyi, Albert, 987

T2 bacteriophage, 243, 244, 281, 499
T4 bacteriophage, 274
T7 bacteriophage, 312, 476–477
T cells, 356, 358–361, 370–374
   activation, 361, 374
   disorders of, 375–377
   diversity of, 359–361
   and major histocompatibility complex, 371–373
   receptors, 358, 370–371
T-tubules, 917–918
Tactile communication, 1030, 1042–1043
Tactile receptors, 888–889
Tadpoles
   metamorphosis, 82, 391, 624
   osmotic balance, 1011, 1012
   as prey, 1106
Tails
   autotomy, 462, 467, 469
   in chordates, 617, 619
   loss in metamorphosis, 82
   and mate choice in birds, 1069–1070, 1071
   prehensile, 635
Tails, of viruses, 497
Talus slopes, colonization of, 725
Tamarins, 634
Tamarisks, 708

Tambora, 644
Tandemly repetitive regions, 302–303
Tannins, 1109–1110
Tapetum, 900
Tapeworms, 588, 590, 828, 991
Taproots, 670
    in desert plants, 706–707
Tardigrades, 578, 579, 596–597, 606, 633
Target cells
    of hormones, 800–801
    receptors, 817–821
Targeting, of proteins, 78, 80, 264–265, 306–307
Tarsiers, 632, 634
Tarsius bancanus, 634
Tarweeds, 457–458
Taste, 887–888
Taste buds, 887–888
TATA box, 304
Tatum, Edward, 255–257, 273, 275
Taxon, defined, 463, 464
Taxonomy, 463–467 Tay-Sachs disease, 335, 841
T_C cells, 370, 372–373
Tealia coriacia, 587
Tears, lysozyme in, 354
Technology, economics of, 1193
Tectorial membrane, 892–893
Teeth, 630, 990–991, 992
    decay, 354, 494
    evolution, 466–467, 619
    of humans, 990–991
    loss of, 924
    of primates, 635, 636
    of sharks, 979
Tegeticula, 1117
Telencephalon, 854, 856, 858
Teliospores, 535–536
Telomeres, 296–297
Telophase
    in meiosis, 201, 205–207
    in mitosis, 199, 201
Temin, Howard, 259
Temperate deciduous forest biome, 1164–1165
Temperate regions
    adaptations in plants, 750, 751
    soils, 723–724
Temperature
    absolute, 118
    and biological productivity, 1128, 1148–1149
    and classification of biomes, 1160–1162, 1167
    effect of atmosphere, 1136
    effects of atmosphere. See also Global warming
    and enzyme activity, 132–133, 784–785
    limits for organisms, 784
    regulation in animals. See Thermoregulation
    regulation in plants, 695
    sensation of, 866, 888
Template strand, 259
Temporal isolation, 452–453
Temporal lobes, 856–857

Temporal patchiness, 1106–1107
Temporal summation, 868–869, 917–919
Tendons, 889, 926
Tendrils, 467, 469
Tension, in xylem sap, 694–696
Tension wood, 684–686
Tentacles
    of cephalopods, 603
    of cnidarians, 583–584
    of ctenophores, 587–588
    of lophophorates, 611–613
    of polychaetes, 594–595
Tepals, 565–566 Termination signals, 286
Termination sites, 259–260
Termites, 602
    digestion, 77, 1114
    mutualisms, 505–508, 989
    social systems, 1074–1075
Terns, 1193
Terpenes, 712
Terrestrial ecosystems
    adaptations by animals, 596, 602, 621–623, 625, 627, 640–641, 652–653, 777, 829, 841, 1018–1019
    adaptations by plants, 545, 547, 548, 552–553, 560–561, 684–686, 689
    classification of, 1160–1162, 1167
    energy flow, 1131–1132
    osmoregulation in, 1009–1010, 1011, 1018–1019
Territoriality, 1068–1069
    aggression in, 800, 1031
    marking, 1040, 1041
    and mate choice, 1068–1069
    and population dynamics, 1085
Tertiary period, 646, 655
Tertiary structure, of proteins, 52, 54–55, 56
Test crosses, 219–222
Testes, 830–831, 835
    hormone production, 802–807
    sperm production, 826–828
Testing, of hypotheses, 7–9
Testosterone, 814–817
    and bird song, 1036–1037
    chemical structure, 43, 44, 814
    and sexual behavior, 1035–1036
Tetanus, 488, 494, 918–919
Tetrahymena thermophila, 299, 419–420
Tetraploidy, 210–211, 451–452
Tetrapogon, 451–452
Tetrodoxin, 929
Texas bluebonnet, 763
T_H cells, 361, 372, 375–376
Thalamus, 854, 856
    and vision, 902–904
Thallus
    in brown algae, 520

in lichens, 542
Thecodonts, 625
Theories, 8–9
Therapsids, 629, 630
Theria, 630–632
Thermal gradients, 786, 795
Thermal insulation, 628, 630, 792–793
Thermal vents. See Hydrothermal vents
Thermoacidophiles, 485–486, 489
Thermobia domestica, 601
Thermocline, 1042, 1134–1135
Thermodynamics, laws of, 114–119
Thermogenin, 791–792
Thermolysin, 126
Thermoneutral zone, 790–791
Thermoregulation, 784–797
    behavioral, 786–788, 1044, 1045
    and size, 9–10
Thermosensors, 882
Thermostat, 782–783
    in animals, 793–797
Thermus aquaticus, 328
Thiamin, 986, 987
Thiobacillus, 422, 491–492
Thiocystis, 484, 486
Thiols, 35
Third eye. See Pineal gland
Thirst, 1024, 1026
Thoracic cavity, 945–946
Thoracic duct, 969
Thoracic girdle, 630
Thorax, in crustaceans, 599
Thorn, R. G., 531
Thorn forest biome, 1167, 1168
Threat displays, 1044, 1057, 1068
Three-dimensional vision, 903–904
Three-factor cross, 232
3′ end
    of DNA, 245, 247, 251–254, 296–297
    of mRNA, 259–260
Three-spined stickleback, 439–440, 656–657
Threonine, 50, 266, 267
    nutritional requirements, 983–984
Thrips, 602
Throat pouch display, 1044, 1045
Thrombin, 972–973
Thrombus, 967
Thumbs, evolution of, 636
Thylakoids, 75–76, 95, 172–174, 547
    in cyanobacteria, 422, 494
Thymine, 57–58, 245–249, 269
Thymus, 356, 373, 802, 806
Thyroid gland, 802, 806, 810–812
Thyrotropin, 806, 808, 809–811
Thyrotropin-releasing hormone, 809–811
Thyroxine, 805, 806, 810–811
Thysanoptera, 602
Thysanura, 601

Ti plasmid, 329–330
Ticks, 598
Tidal volume, 942–944
Tide pools, 1009
Tight junctions, 99–100, 387, 389
Timberline, 1163
Time lags, in population growth, 1090
Time scales, 10–11, 415, 416
    in biogeography, 1153–1154
Tinbergen, Niko, 1031
Tissue systems
    of animals, 778–779
    of plants, 672–673, 677–686
Tissue typing, 365
Tissues, 4, 5, 672
Tmesipteris, 557
Toads, 622–624
Toadstools, 539
Tobacco
    flowering, 765
    in plant growth experiments, 745
    See also Smoking
Tobacco hornworm, 305
Tobacco mosaic virus, 258–259, 497, 499, 714
Tocopherol, 986–987
Toes, vestigial, 13
Tolerance
    immunological, 361, 362, 371
    of self, 373–374
Tomatoes
    in plant growth experiments, 748
    on serpentine soils, 710–711
Tonegawa, Susumu, 368
Tongue, 887–888, 995
Tonicella lineata, 604
Tonus, 919
Tools, of early humans, 638–640
Toothless whales. See Baleen whales
Torpedo ray, 929
Torpedo stage, 761
Torpor, 795–796
Torsion, in gastropods, 605
Tortoises, 625, 1178
Total lung capacity, 943–944
Totipotency, 393, 825
Touch
    and communication, 1030, 1042–1043
    sensation of, 856–857, 888–889
Tourism, in national parks, 1191, 1193
Toxic shock syndrome, 494
Toxic waste. See Pollution
Toxigenicity, of bacteria, 488
Toxins. See Chemical defenses
Tracheae, 600, 937–938, 940, 944–945, 995
Tracheary elements, 674, 676
Tracheids, 393, 552, 561, 675–676
    in psilopsids, 554

See also Xylem
Tracheoles, *938*
Tracheophytes. *See* Vascular plants
Trachoma, 493
Tracking, of sun by leaves, 672
Trade-offs, life history, 1081–1084
Trade winds, *1146*
Traditional agriculture, 1179–1180
Trail marking, 1040
Transcription, *258*, 259–260
    control by steroids, 303, *821*
    control in eukaryotes, 300–305
    control in prokaryotes, 285–289
    reverse. *See* Reverse transcription
Transcription factors, *304–305*
    and development, 403–404
Transdetermination, 394
Transducin, *896–897*
Transduction
    of bacterial genes, 282–284
    of sensory stimuli, *883–884*, 893–894, 895–897
Transfection, 315–316
Transfer cells, 690, 700
Transfer RNA. *See* tRNA
Transferrin, 973
Transformation
    in bacteria, *242–243*, 275–276
    in recombinant DNA technology, 315, 317
Transforming principle, *242–243*, *275–276*
Transgenic cells, 314, 316–318, *317*
Transgenic organisms, 314, 329–330, 342
Transition-state species, 119–120
Translation, *258*, 261–265, 285–286
    control of, 305–306
    of polycistronic mRNA, 285–286
Translocation
    of chromosomes, 209–210, *269–270*
    in phloem, 698–*701*
Transmission electron microscopy, 68, 97–*98*
Transpiration, 695–696
    and energy flow, 1128
Transplants, and immune system, 358, 371, 373–374
Transport
    in animals. *See* Circulatory systems
    of auxin, 743
    in plants, 689–701
    of proteins, 264–*265*
Transport proteins, 103–106
Transposable elements, 284–285
    and cancer, 346–348

in eukaryotes, 296, 297–298
in pathogenic bacteria, 493
and viroids, 499
Transposase, 285
Transposons, 284–*285*
    in recombinant DNA technology, 329–330
Transverse tubules, 917–*918*
Tree falls, 1093–*1094*, 1123
Tree ferns, 553, 559
Tree frogs, *1018*
Trees
    defenses, 356, 714, 1103
    diversity, *666–667*
    effect on communities, 1118–*1120*, 1123
    effects of global warming, 1185–1187
    evolution, 553, 570
    growth, 681–683, 684–686
    life histories, *1082*, *1083*
    transport in, 693–696
Trematoda, 588
Treponema pallidum, 66, 491
Triangular web spider, 1030
Triassic period, *646*, 647, 654
    mass extinction, *646*, 658
Tricarboxylic acid cycle. *See* Citric acid cycle
Trichinella, 591–592
Trichinosis, 591–592
Trichocysts, in *Paramecium*, 511
Trichomonas vaginalis, 508
Trichoptera, 602
Trifolium repens, 437–*438*
Triglochin maritima, 709
Triglycerides, 40–42, *41*, 999
Trilobites, *597*, 652
Trimesters, of pregnancy, 843–844
Triose phosphate dehydrogenase, *145*, 146, 147
Tripartite body plan
    in arthropods, *599*
    in deuterostomes, 610–613, *614*
Triple bonds, 24-25
Triple helix, 52, *53*
Triplet repeats, *336–337*
Triploblastic animals, 576, 610
Triploid nuclei, in flowering plants, 563, 760–761
Triploidy, 210–211
Trisaccharides, 47
Trisomies, 209–*210*
Tristearin, *41*
Tritium (³H), 21, 22, 645
tRNA, 258, 259–264
    and dating, 413–414
    terminal, *286*
    transcription, 299, 302, 303
Trochophore larvae, 594–595
Troop size, in primates, *1077*
Trophic levels, 1128–*1132*
Trophoblast, 387, *389*
Tropic hormones, *806*, 808
Tropic of Cancer, 1146
Tropic of Capricorn, 1146
Tropical ecosystems, species richness, *446*, 1187

Tropical forests, *8*, 1167–*1169*, 1189–*1190*, 1192
    nutrient cycling, 1144–1145
Tropical soils, 724, 1144–1145, 1167, 1169
Tropocollagen, *53*
Tropomyosin, 915–*918*
Troponin, 821, 917–*918*
Troposphere, *1135*
True-breeding strains, 215
True bugs, 602
Truffles, 538
Trumpet cells, 521
Trunk, in deuterostomes, 613–614
Trunk, of elephants, 988
Trunks
    of dicot trees, *667*, 681–*683*
    of monocot trees, 666
Trypanosoma, 507, 508
Trypsin, 124–125, 997, *999*
    inhibitors, *127*
Trypsinogen, 997, *999*
Tryptophan, *50*, 267
    and bacterial operons, 288–289
    nutritional requirements, 983–984
T_S cells, 361, 371
Tsetse fly, 508
Tuataras, 625, *626*
Tubal ligation, *838–839*
Tube feet, in echinoderms, 615–617
Tube nucleus, *759*, 760–*761*
Tuberculosis, 496
Tubers, 571, 671–672, 770
Tubular gut, 991–992
Tubulin, 83–86, 307, 911–*912*
Tumor-suppressor genes, 346–347
Tumors, 345–347
    defenses against, 355
    and monoclonal antibodies, *364–365*
    and recombinant DNA technology, 329–330
Tuna, 789–790
Tundra, *1163*
    herbaceous plants, *571*
    seasonal migrations, *1094*
Tungsten, 316
Tunicates, *473*, 618, 1114
Turbellarians, 588–*589*
Turbulence, water, *427*
Turgor, 83, *107*, 684, 690–*691*
Turner syndrome, 227, 228
Turnovers, in lakes, 1134–*1135*, 1141, 1142
Turtles, 625, *626*
    phylogeny, 465, *468*
Twins, 361. *See also* Identical twins
Twitches, 917–920, *919*
2,4-D, 744, *745*
Two-dimensional chromatography, *176–177*
Two-point discrimination test, 889
Tylototriton verrucosus, *623*
Tympanic membrane, 890, *892*, 894

Typhlosole, *992–993*
Typhoid fever, *492*
Typhus, 350, *493*
Tyrosine, *50*, 255, *267*

U-shaped gut, 610, 613
U-tube experiment, 275–*276*
Ubiquinone (Q), in cellular respiration, 150–152
Uganda kob, *1069*
Ulcers, 993
Ulothrix, 525
Ultraviolet radiation, 165–*166*
    as mutagen, 276, 345, 349
    and origin of life, 415
    and ozone layer, *1135–1136*
    perception by insects, 568, 904, 1117–*1118*
    and prophages, 282
    and vitamin D synthesis, 986
Ulva, 484
Ulva lactuca, 523–525, *524*
Umbels, 565–566
Umbilical cord, 841–*842*, 845
Umbilicus, 845
Unbranched polymers, 52
Uncaria gambir, 469
Unconditioned reflexes, 1001
Undernourishment, 981–982
Understory, 1118
Uniports, *103*, 104
Uniramia, *578*, 579, 600–602, *606*, 633
Unitary organisms, 1088
United States, population distribution, *1085–1086*
Units of measurement, *63*
Unsaturated fatty acids, 40–42, *41*
Unsaturated hydrocarbons, 33–34
Upwelling, *1134*, 1139, 1171
Uracil, 57–58, 247
    in genetic code, 266–267
Uranium-238 (²³⁸U), 413, 645, 648
Urea, *418*
    excretion of, 1010–1012
Uredospores, 535–536
Ureotelic animals, 1011–1012
Ureter, 1019–*1021*
Urethra, 830–*831*, 832, 1019–*1021*
Uribe, Ernest, 172
Uric acid, excretion of, 1010–1012
Uricotelic animals, 1012
Urinary systems. *See* Excretory systems
Urination, 1019
Urine, 1009
    of annelids, *1013*
    of bats, *1025*
    concentration of, 807, 1021–*1022*, 1024–1025
    of humans, 1012
Urochordata, *473*, 618, 1114
Uroctonus mordax, 598
Urodela. *See* Salamanders
Ussher, James, 413
Ustilago maydis, 529

Uterus, 630, *832–835*, 840, 841–845
Utethesia ornatrix, 1070

V segments, *368–369*
Vaccination, 359
    against AIDS, 375, 376–378
    against cancer, 349
Vacuoles, 70, *82–83*, 993
    in plant cells, 674, 715, 746
    in protists, 504–*505*, 512
Vagal reaction, 965
Vagina, 829
    of humans, *832*, 835, 836
Vagus nerve, 951, 965
Valine, *50, 267*
    nutritional requirements, 983–984
Valves
    in heart, 958, *959, 961–965*
    in lymph vessels, 969
    in veins, *966, 970–971*
Vampire bats, *991, 1025*
van der Waals interactions, *33*
van Leeuwenhoek, Anton, 12
Variable regions
    of antibodies, *361–363*
    of lymphocytes, *368–369*
    of T-cell receptors, *370*
Variation, *426–428*
    environmentally induced, *427, 428*
    genetic. *See* Genetic varia-tion
Varicose veins, 970
Vas deferens, 830–*831*, 838
Vasa recta, 1019–1022
Vascular bundles, 680–681
    monocot vs. dicot, *667, 669*
Vascular cambium, 677, *678*, 681–683
Vascular cylinder, of roots. *See* Stele
Vascular plants, *546, 547*, 548, 552–556
    evolution of, *553*
Vascular rays, *682*
Vascular systems
    in animals, *966–971*
    in plants, 548
Vascular tissue, 548, 673, *676–677*. *See also* Phloem; Tracheids; Xylem
Vasectomy, *838–839*
Vasopressin, *806–807*, *974–975, 1024–1025*
Vectors
    cloning, *314–315*
    and parasite virulence, 1109
    of pathogens, *493, 497, 499, 510–511*, 600, 1109
Vegetal pole, *383, 394–395*
Vegetarian diets, *983–984*
Vegetative growth, 764
    role of auxin, 743–744
Vegetative organs, 770
Vegetative reproduction, 488, 672, 757–758, 770–772
Veins (blood vessels), 958, *961*, *966, 970–971*

Veins (of leaves), 683–*685*
    patterns, *669, 672*
Velocity, of light, *165*
Vena cavae, *961*
Venereal diseases. *See* Sexually transmitted diseases
Venom, in snakes, 625
Venous return, *970–971*
Ventilation, 937–946
    regulation of, 951–953
    and running speed, *627*
Ventral, defined, 575
Ventral horn, *854–855*
Ventral lip, 387, *388–389*
Ventricles, of heart, 958–960, *961–965*
Vents. *See* Hydrothermal vents
Venules, 958, *966, 968*
    in kidneys, 1019–*1020*
Venus's-flytrap, 730–731
Vermiform animals, 591
Vernalization, 769–770
Vertebrae, and swimming speed, 439–440
Vertebral column, 618
Vertebrates, 610, *618–621*
Vertical evolution, 448
Vesicles, *79–81, 85, 201*
    coated, 108, *109*
    cortical, *837*
    and neurotransmitters, *866–867*
Vessel elements, 563, 570, *675–676*
    in Gnetophyta, 561, 570
    See also Xylem
Vestibular apparatus, 890–*891*
Vestigial organs, 1000
Viability, of populations, 1181
Vibrations
    as communication, 1030, 1042
    sensation of, 889
Vibrio cholerae, 492
Vibrio harveyi, 728
Vicariant distributions, 1155
Vicuñas, 949
Villi, 992, *994*
Vinograd, Jerry, 250
Viperfish, *1171*
Virchow, Rudolf, 12
Virions, 497, *498*
Viroids, 499
Virulence, of parasites, 1108–1109
Virulent strains, 242
Viruses, 87, 483, *497–500*
    in AIDS, 375–377
    in cancer, 345–348
    as cloning vectors, 315
    in DNA studies, 243, *244*
    genetic material, 243, 258–259, 264, *265*, 273, 298, 498–*499*
    plant defenses against, 713–714
    response of interferon, 355–356
    response of T cells, 358, *370, 371, 372–373*

structure, 87, 375–377
    taxonomy, 476–477, 498–499, *500*
    See also Bacteriophages
Visceral mass, in mollusks, *603*
Visible light, 165–166
Vision, 856–*857*, 882, *894–904*
    and prey selection, *1062*
Visual cortex, *877–878*, 902–904
Visual fields, 903–*904*
Visual signals, 1040–1042
Vitamin A, *43, 44, 983, 986*, 987
Vitamin C, 985–*986*, 987
    Vitamin D, *43, 44, 986*, 988
Vitamin E, *986–987*
Vitamin K, *986–987*, 1000
Vitamins, 984–988
    in carrier compounds, 141
    as coenzymes, 125, *126*, 985
    requirements in animals, *43*, 981, 984–988, *986*
    requirements in fungi, 531
Vitelline envelope, *837*
Vitis vinifera, 772
Viviparity, 840
Vocalizations, and communi-cation, 1042
Volcanic vents. *See* Hydrothermal vents
Volcanoes
    and early atmosphere, *415*, 422
    and mass extinctions, *644–645*, 648
    recolonization, 1159–*1160*
    in sulfur cycle, 1139–*1140*
Voltage, defined, 858
Voltage clamping, *864*
Voltage-gated channels, 859, 861
    in muscles, 917
    in postsynaptic cells, 866–867, 868
    in sensors, 883
Volume, and surface area. *See* Surface-to-volume ratio
Voluntary nervous system functions, 853
Volvox, 523, 524, 578
Vomiting, 995
Von Frisch, Karl, *1042–1043*
von Tschermak, Erich, 215
Vorticella, 511, 513
Vulva, in *C. elegans*, 384, *399–400*

Waggle dance, 1042–*1043*
Wakefulness, 855–856, *874–875*, 893
Wald, George, 25
Walking sticks, 602
Wallace, Alfred Russel, 13
Wallowing, 793
Warblers
    classification, 464, *466–467*
    endangered, 1185–*1186*
    speciation, 1156–*1157*

Wardrop, A. B., 685
Warm-blooded animals, 785
Warning coloration, 7, 623, 1107–1108
Wasps, 602
    mimics of, *1108*
    parasites of, 1109
    parasitoid, *1103, 1105*
    as pollinators, *1100–1101*, 1116, 1117
    social systems, 1074–1075
Waste products, of plants, 718
Water
    chemical properties, *26*, 28–29, 30–32
    compared with air, 934–935, 1169
    conservation by animals, 625, *1018*–1019, 1021, *1026*
    conservation by plants, 684–685, 696–698, 705–706
    fertilization and, 533, 549, *551*, 552, 558, 560, 563, 758
    in early atmosphere, 414, *415*, 417, 418
    global hydrological cycle, 1137
    oxygen content, 934–935, 1102, 1134, *1135*, 1141, 1169
    as solvent, 28, *31*
    temperature, 29, 1147
    uptake by plants, 689–696
Water balance. *See* Osmotic balance
Water bears, 596–597
Water fleas, *427*, 1061–1063
Water holes, *1008*
Water hyacinths, 770
Water molds, 516–*517*
Water potential, 690, 738
Water-soluble hormones, 817–821
Water-soluble vitamins, 985–987
Water striders, *32*
Water vapor, atmospheric, 1136, *1137*
Water vascular system, 615–617
Watson, James D., *245–246*, 248
Wave action, and seaweeds, *522*
Wavelengths, of light, 165–168
Waxy coatings, and water conservation, *1019*
Weathering, of soils, 723–724, 1136
Webbed feet, *13*
Webs, of spiders, 598, *1029–1030*, 1103
Weedkillers, 744, *745*, 1144
Weeds, 737
Weevils, *1105*, 1154–*1155*
Wegener, Alfred, 1152–1153
Weight
    regulation of, 783, 982

and volume, *10*
Weinberg, W., 430
Welwitschia, 561, *562*
Went, Frits W., 742–743
Wernicke's area, *877–878*
West coasts, 1163, 1167
Westerly winds, *1146*
Western blotting, 321
Wetlands
    adaptations by plants to,
       706–707, *708, 709*
    economic value of, 1193
    restoration of, 1193
Whale shark, 621, 979
Whales
    communication, 1042
    filter feeding, 641, *989–990,*
       1103
    navigation, *1048*
    overhunting, *1095–1096*
"What" questions, 1029
Wheat, vernalization of,
    769–770
Wheat rust, 534–536, *535, 539*
Whelks, 604
Whip grafting, *771*
Whisk ferns, 557–558
White blood cells, *971, 972*
    genes, 367–370, *371*
    in immune responses, 342,
       355, *356–357*
    See also Lymphocytes;
       Natural killer (NK)
       cells; Phagocytes
White clover, *224*
White-crowned sparrows,
    1033–1034
White-fronted bee-eater, *1065*
White-handed gibbon, *635*
White-headed duck, *474*
White light, *167*
White matter, 854–*855,* 856,
    865
White muscle, *919–920,* 948
White pelican, *1062–1063*
White-plumed antbird,
    1189–1190
Whitefish, 1086
Whitham, Thomas, 711–712
Whorls, 764

"Why" questions, 4, 6, 1029
Widowbirds, *1072*
Wiesel, Torsten, 902
Wigglesworth, Sir Vincent,
    802–804
Wild dogs, *1062–1063*
Wild-type, defined, 224
Wildebeests, *1076–1077*
    migration, 1094
Wilfarth, 724
Wilkesia, 458
Wilkins, Maurice, 243
Willow tit, *796*
Willows, *714*
Wilson, E. O., 1157
Wilting, 684, 690–*691*
Wind
    global patterns, *1145–1146,*
       1147
    and lake turnovers,
       1134–1135
    and pollination, *561,* 563,
       758, *760*
Windpipe. *See* Trachea
Wine
    grapes, *772*
    and yeast, 538
Wingless insects, 600, *601, 602*
Wings
    of birds, *3,* 400–*401,*
       627–628
    of insects, 600, *601, 602*
Winter dormancy, 1093
    in animals. *See*
       Hibernation
    in plants, 735, *750, 751,* 766
Winter survival, 1091, 1093
Winter wheat, 769–770
Wisdom teeth, 990
Woese, Carl, 485, 489
Wolf spiders, *598*
Wollman, Elie, 279, 285
Wolpert, Lewis, 400
Wolves, 792, *1104–1105*
Womb. *See* Uterus
Wood
    biomass, 1163
    decomposition, 1121
    defenses, 356, 714, 1103,
       1133

    growth of, 673, *681–683*
    as supporting tissue,
       684–686
Wood pigeons, *1063–1064*
Wood rose, *570*
Wood ticks, *598*
World Health Organization,
    359
Worms, 591. *See also* Annelids;
       Earthworms;
       Flatworms; Nematodes

X chromosomes, *227–230*
    inactivation, *300–302*
    interaction with hormones,
       815
X-linked traits, 335–336, 338
X-ray crystallography, 124,
    243
X rays, 165, *166*
    as mutagens, 255–*256,* 276,
       345, 349
ψX174 virus, *265*
Xanthophylls, in green algae,
    523
Xenon, atmospheric, 1136
Xeroderma pigmentosum, 254
Xerophytes, *705–707*
Xgal, *317–318*
Xiphosura, *597–598*
Xylem, 548
    cell types, 674–676, *677*
    growth of, *681–683*
    of roots, *679–680*
    transport in, *693–696, 701*

Y chromosomes, *227–230,* 301
    interaction with hormones,
       815
Yeasts, 62, *538*
    fermentation by, 156, 538
    genetic material, *324*
    mating types, *308*
    in recombinant DNA tech-
       nology, 311, 314, 331
    regulatory mechanisms,
       157
Yellow-eyed junco, 1059
Yellow-pine chipmunk, 1112
Yersinia, 492

Yolk, *383–385*
    in amniotic eggs, *625,*
       838–841
    and polarity, 394
Yolk plug, 387, *388–390*
Yolk sac, *841*
Yucca, 1082, *1117*

Z-DNA, 246, *248*
z gene, 316–318
Zea mays, in genetics studies,
    226–227, 230
Zeatin, 739, 749
Zebra finches, *1037*
Zebra mussels, 1141–1142
Zebroids, *447*
Zeevaart, Jan A. D., 769
Zenith tragedy, 952
Zinc
    as animal nutrient,
       984–985
    in enzymes, *126*
    as plant nutrient, 719, *720*
Zinc fingers, *305*
Zona pellucida, *836*
Zone of polarizing activity
    (ZPA), *400–401*
Zones of upwelling. *See*
    Upwelling
Zooflagellates, *507–508*
Zoomastigophora, *507–508*
Zooplankton, 580
Zoos, 1182
Zoospores
    in funguslike protists, 516
    in green algae, 523–525
ZPA, *400–401*
Zygomycetes, *529, 536–537*
Zygospores, *536–537*
Zygotes, *202–203,* 826, 835,
    836
    in alternating generations,
       506, 760–762
    cleavage, 382–385
    polarity, 394–396
    See also Embryos; Gametes
Zymogens, *993–999*